✦ *Understanding Our Universe*

Understanding Our Universe

Stacy Palen
Weber State University

Laura Kay
Barnard College

Brad Smith
Santa Fe, New Mexico

George Blumenthal
University of California—Santa Cruz

W. W. Norton & Company, Inc.
New York • London

W. W. NORTON & COMPANY has been independent since its founding in 1923, when William Warder Norton and Mary D. Herter Norton first published lectures delivered at the People's Institute, the adult education division of New York City's Cooper Union. The firm soon expanded its program beyond the Institute, publishing books by celebrated academics from America and abroad. By mid-century, the two major pillars of Norton's publishing program—trade books and college texts—were firmly established. In the 1950s, the Norton family transferred control of the company to its employees, and today—with a staff of four hundred and a comparable number of trade, college, and professional titles published each year—W. W. Norton & Company stands as the largest and oldest publishing house owned wholly by its employees.

Editor: Erik Fahlgren
Project Editor: Christine D'Antonio
Editorial Assistant: Mina Shaghaghi
Developmental Editor: Andrew Sobel
Director of Production, College: Jane Searle
Marketing Manager: Stacy Loyal
Managing Editor, College: Marian Johnson
Book Designer: Brian Salisbury
Design Director: Rubina Yeh
Associate Design Director: Hope Miller Goodell
Photo Editor: Michael Fodera
Photo Researcher: Donna Ranieri
Associate Editor, Media: Matthew Freeman
Media Editor: Rob Bellinger
Composition and Layout: Carole Desnoes
Illustrations: Penumbra Design, Inc.
Manufacturing: Quad/Graphics, Versailles

Library of Congress Cataloging-in-Publication Data

Understanding our universe / Stacy Palen . . . [et al.]. — 1st ed.
    p. cm.
  Includes index.
  ISBN 978-0-393-91210-4 (pbk.)
  1. Astronomy—Textbooks. I. Palen, Stacy E. II. Smith, Brad, 1931– III. Kay, Laura.
  QB43.3.U565 2012
  520—dc22

                                                    2010051752

W. W. Norton & Company, Inc., 500 Fifth Avenue, New York, N.Y. 10110-0017
www.wwnorton.com
W. W. Norton & Company Ltd., Castle House, 75/76 Wells Street, London W1T 3QT
3 4 5 6 7 8 9 0

Stacy Palen thanks the wonderful colleagues in her Department and the crowd at Bellwether Farm, all of whom made room for this project in their lives, even though it wasn't their project.

Laura Kay thanks her partner, M.P.M.

Brad Smith dedicates this book to his patient and understanding wife, Diane McGregor.

George Blumenthal gratefully thanks his wife, Kelly Weisberg, and his children, Aaron and Sarah Blumenthal, for their support during this project. He also wants to thank Professor Robert Greenler for stimulating his interest in all things related to physics.

✦ BRIEF TABLE OF CONTENTS

✦ CONTENTS

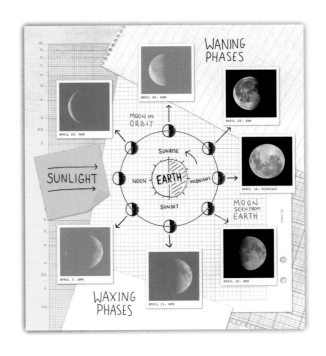

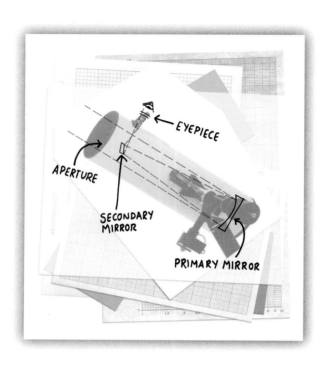

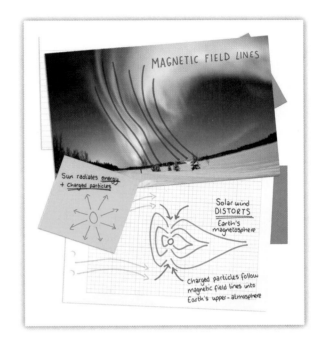

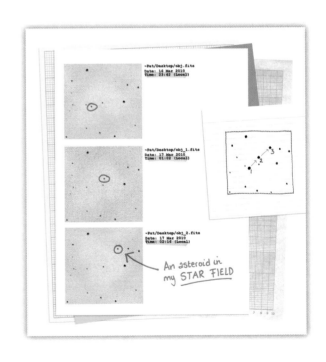

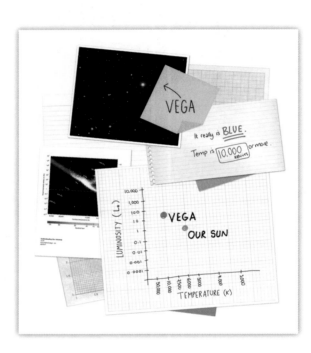

Part III Stars and Stellar Evolution

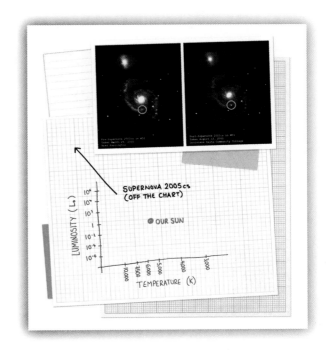

Part IV Galaxies, the Universe, and Cosmology

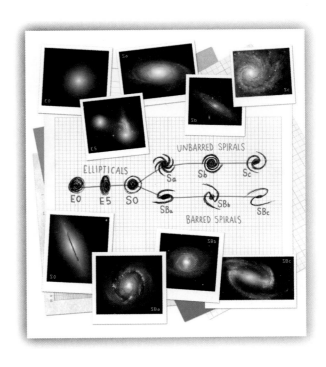

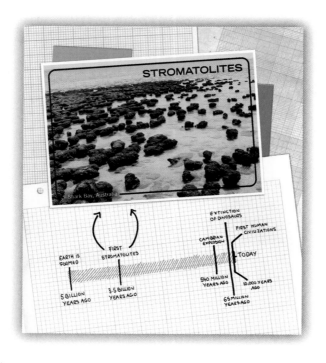

Dear Student,

You may reasonably wonder why it is a good idea to take a general education science course. Throughout your education, you have been exposed to different ways of thinking—different approaches to solving problems, different definitions of *understanding*, and different meanings of the verb *to know*. Astronomy offers one example of the scientific viewpoint. Astronomers, and all scientists, have a specific approach to problem solving (sometimes called the scientific method, although the common understanding of this term only skims the surface of the process). Astronomers "understand" something when they can make correct predictions about what will happen next. Astronomers "know" something when it has been tested dozens or even hundreds of times, and the idea stands the test of time. Your instructor likely has two basic goals in mind for you as you take this course. The first is to understand some basic physical concepts and be familiar with the night sky. The second is to think like a scientist and learn to use the scientific method not only to answer questions in this course but also to make decisions in your life. We have written *Understanding Our Universe* with these two goals in mind.

Throughout this book, we emphasize not only the content of astronomy (the masses of the planets, the compositions of stellar atmospheres) but also the process of arriving at that knowledge—how and why we know what we know. We believe that an understanding of the scientific method is a valuable tool that you can carry with you for the rest of your life.

Astronomy is one of the purest expressions of one of the more distinctive impulses of humanity—curiosity. Astronomy does not capture the public interest because it is profitable, will cure cancer, or build better bridges. People choose to learn about astronomy because they are curious about the universe in which we live. The most effective way to learn something is to "do" it. Whether it's playing an instrument, a sport, or becoming a good cook, reading "how" can take you only so far. The same is true of learning about astronomy. We have developed a number of tools in this book to facilitate "doing" as you learn. We start with the figures at the beginning of each chapter. These chapter-opening figures demonstrate the types of activities you might engage in to fulfill that curiosity. They are presented from the viewpoint of a student who is wondering about the universe, asking questions, and keeping a journal of experiments that investigate the answers. Your instructor may ask you to

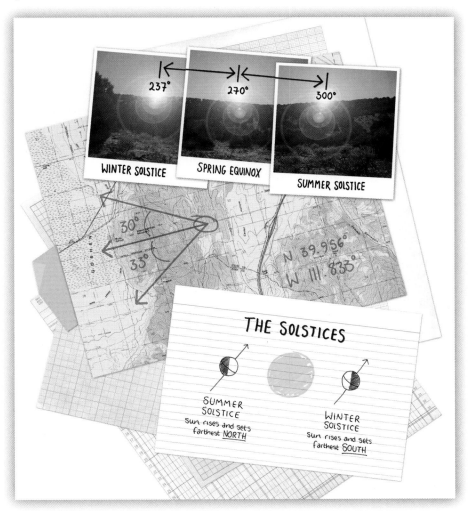

keep such a journal. Or you may choose to do it on your own, as a useful way of investigating the world around you.

As you learn any new subject, one of the stumbling blocks is often the language of the subject itself. This can be jargon—the specialized words unique to that study—*supernova*, say, or *Cepheid variable*. But it can also be ordinary words that are used in a special way. As an example, the common word *inflation* usually applies to balloons or tires in everyday life, but economists use it very differently, and astronomers use it differently still. Throughout the book, we have included **Vocabulary Alerts** that point out the astronomical uses of common words to help you recognize how that term is used by astronomers.

In learning science, there's another language issue. The language of science is mathematics, and it can be as challenging to learn as any other language. The choice to use mathematics as the language of science is not arbitrary; nature "speaks" math. To learn about nature, you will also need to speak this language. We don't want the language of math to obscure the concepts, so we have placed this book's mathematics in **Working It Out** boxes to make it clear when we are beginning and ending a mathematical argument.

Vocabulary Alert

Revolve: In common language, the words *revolve* and *rotate* are sometimes used synonymously to describe something that spins. Astronomers distinguish between the two terms, using *rotate* to mean that an object spins around an axis through its center, and *revolve* to mean that one object orbits another. Earth rotates about its axis (causing our day) and revolves around the Sun (causing our year).

Working It Out 2.2 | Kepler's Third Law

Just as *squaring* a number means that you multiply it by itself, as in $a^2 = a \times a$, *cubing* it means that you multiply it by itself again, as in $a^3 = a \times a \times a$. Kepler's third law states that the square of the period of a planet's orbit, P_{years}, measured in years, is equal to the cube of the semimajor axis of the planet's orbit, A_{AU}, measured in astronomical units. Translated into math, the law says

$$(P_{years})^2 = (A_{AU})^3$$

Here, astronomers use nonstandard units as a matter of convenience. Years are handy units for measuring the periods of orbits, and astronomical units are handy units for measuring the sizes of orbits. When we use years and astronomical units as our units, we get the relationship just shown (**Figure 2.29**). It is important to realize that *our choice of units in no way changes the physical relationship* we are studying. For example, if we instead chose seconds and meters as our units, this relationship would read

$$(3.2 \times 10^{-8} \text{ years/second} \times P_{seconds})^2$$
$$= (6.7 \times 10^{-12} \text{ meters/AU} \times A_{meters})^3$$

which simplifies to

$$(P_{seconds})^2 = 3.2 \times 10^{-19} \times (A_{meters})^3$$

Suppose that we want to know the average radius of Neptune's orbit, in AU. First, we need to find out how long Neptune's period is in Earth years, which can be determined by careful observation of Neptune's position relative to the fixed stars. Neptune's period is 165 years. Plugging this into Kepler's third law, we find that

$$(P_{years})^2 = (A_{AU})^3$$
$$(165)^2 = (A_{AU})^3$$

To solve this equation, we must first square the left side to get 27,225 and then take its cube root.

Calculator hint: A scientific calculator usually has a cube root function. It sometimes looks like $x^{1/y}$ and sometimes like $\sqrt[x]{y}$. You use it by typing the base number, hitting the button, and then typing the root you are interested in (2 for square root, 3 for cube root, and so on). Occasionally, a calculator will instead have a button that looks like x^y (or y^x). In this case, you need to enter the root as a decimal. For example, if you want to take the square root, you type 0.5 because the square root is denoted by ½. For

the cube root, you type 0.333333333 (repeating) because the cube root is denoted by ⅓.

To find the length of the semimajor axis of Neptune's orbit, we might type 27,225 $[x^{1/y}]$ 3. This gives

$$30.1 = A_{AU}$$

so the average distance between Neptune and the Sun is 30.1 AU.

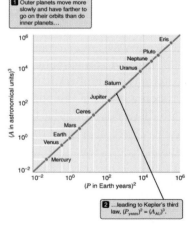

1 Outer planets move more slowly and have farther to go on their orbits than do inner planets…

2 …leading to Kepler's third law, $(P_{years})^2 = (A_{AU})^3$.

Figure 2.29 A plot of A^3 versus P^2 for the eight classical and three of the dwarf planets in our Solar System shows that they obey Kepler's third law. (Note that by plotting powers of 10 on each axis, we are able to fit both large and small values on the same plot. We will do this frequently.)

This way you can spend time with the concepts in the chapter text and then revisit the mathematics to study the formal language of the argument. Our suggestion is to read through a Working It Out box once; then cover the worked example with a piece of paper, and try to work through the example on your own. When you can do this, you will have learned a bit of the language of science;

and even if you are not fluent in it today, you can work with data and identify when someone else's data isn't quite right. We want you to be comfortable reading, hearing, and speaking the language, and we will provide you with tools to make it easier. Be patient.

As a citizen of the world, you make judgments about science, determining what is good science and what is pseudoscience. You use these judgments to make decisions in the grocery store, pharmacy, car dealership, and voting booth. You base these decisions on the presentation of information you receive through the media, which is very different from the presentation you get in class. You must learn to judge what is credible and what is not. To help you hone this skill, we have provided **Reading Astronomy News** boxes in every chapter. These boxes include a news article with questions to help you make sense of how science is presented to you. It is important that you learn to be critical of the information you receive. We hope that these boxes help you refine your critical thinking skills.

At the end of each chapter, we have provided several types of questions, problems, and activities for you to practice your skills. Summary Self-Test questions may be used to check your understanding. If you can answer these multiple-choice, true/false, and fill-in-the-blank questions correctly, you have a basic grasp of the information in the chapter. Next, a separate group of true/false and multiple-choice questions focuses on more detailed facts and concepts from the chapter. Conceptual questions ask you to synthesize information and explain

---

 READING ASTRONOMY News

You might think that information as fundamental as the motions of Earth and the Moon is fixed and unchanging, but given the right circumstances, these values can change in an instant. This article appeared in the March 1, 2010, online edition of Bloomberg Businessweek.

Chilean Quake Likely Shifted Earth's Axis, NASA Scientist Says

By **ALEX MORALES**, *Bloomberg Businessweek*

The earthquake that killed more than 700 people in Chile on Feb. 27 probably shifted the Earth's axis and shortened the day, a National Aeronautics and Space Administration scientist said.

Earthquakes can involve shifting hundreds of kilometers of rock by several meters, changing the distribution of mass on the planet. This affects the Earth's rotation, said Richard Gross, a geophysicist at NASA's Jet Propulsion Laboratory in Pasadena, California, who uses a computer model to calculate the effects.

"The length of the day should have gotten shorter by 1.26 microseconds (millionths of a second)," Gross, said today in an e-mailed reply to questions. "The axis about which the Earth's mass is balanced should have moved by 2.7 milliarcseconds (about 8 centimeters or 3 inches)."

The changes can be modeled, though they're difficult to physically detect given their small size, Gross said. Some changes may be more obvious, and islands may have shifted, according to Andreas Rietbrock, a professor of Earth Sciences at the U.K.'s Liverpool University who has studied the area impacted, though not since the latest temblor.

Santa Maria Island off the coast near Concepcion, Chile's second-largest city, may have been raised 2 meters (6 feet) as a result of the latest quake, Rietbrock said today in a telephone interview. He said the rocks there show evidence pointing to past earthquakes shifting the island upward in the past.

Ice Skater Effect

"It's what we call the ice-skater effect," David Kerridge, head of Earth hazards and systems at the British Geological Survey in Edinburgh, said today in a telephone interview. "As the ice skater puts her arms out when she's going around in a circle, and she pulls her arms in, she gets faster and faster. It's the same idea with the Earth going around if you change the distribution of mass, the rotation rate changes."

Rietbrock said he hasn't been able to get in touch with seismologists in Concepcion to discuss the quake, which registered 8.8 on the Richter scale.

"What definitely the earthquake has done is made the Earth ring like a bell," Rietbrock said.

The magnitude 9.1 Sumatran earthquake in 2004 that generated an Indian Ocean tsunami shortened the day by 6.8 microseconds and shifted the axis by about 2.3 milliarcseconds, Gross said.

The changes happen on the day and then carry on "forever," Benjamin Fong Chao, dean of Earth Sciences of the National Central University in Taiwan, said in an e-mail.

"This small contribution is buried in larger changes due to other causes, such as atmospheric mass moving around on Earth," Chao said.

Evaluating the News

1. What has happened to Earth's rotation rate? What caused this?

2. David Kerridge refers to something called the "ice-skater effect." What is that? Does this comparison to an everyday phenomenon make the argument that the rotation rate has changed easier or more difficult to understand? Does it make the claim more or less convincing?

3. Study the numbers in the article. Gross claims that the length of day should have gotten shorter by 1.26 microseconds. Use the list of prefixes in the inside cover to express this number in scientific notation. Is it a big or small number? Does this change meaningfully affect the amount of time you have to get your next homework assignment done?

4. Is this a unique event? Have similar incidents happened before?

5. Is the changing of Earth's rotation rate a scientific hypothesis? Recall the discussion of falsifiability in Chapter 1. Is this hypothesis falsifiable? If so, how might we test it?

6. Does this article have anything to do with business? Why might *Bloomberg Businessweek* report on this story?

7. Connect the ice-skater effect in this news story to the orbit of the Moon. The Moon is moving away from us by about 1 cm every century. Is its orbital period getting longer or shorter? Would you guess that its speed is increasing or decreasing?

Exploration | Kepler's Laws

wwnorton.com/studyspace

In this exploration, we will be exploring how Kepler's Laws apply to the orbit of Mercury. Visit StudySpace, and open the Planetary Orbit Simulator applet in Chapter 2. This simulator animates the orbits of the planets, allowing you to control the simulation speed as well as a number of other parameters. Here, we focus on exploring the orbit of Mercury, but you may wish to spend some time examining the orbits of other planets as well.

Kepler's First Law

To begin exploring the simulation, in the "Orbit Settings" panel, use the drop-down menu next to "set parameters for" to select "Mercury" and then click "OK." Click the "Kepler's 1st Law" tab at the bottom of the control panel. Use the radio buttons to select "show empty focus" and "show center."

1. **How would you describe the shape of Mercury's orbit?**

Deselect "show empty focus" and "show center," and select "show semiminor axis" and "show semimajor axis." Under Visualization Options, select "show grid."

2. **Use the grid markings to estimate the ratio of the semiminor axis to the semimajor axis.**

3. **Calculate the eccentricity of Mercury's orbit from this ratio using the equation $e = \sqrt{1-(\text{ratio})^2}$.**

Kepler's Second Law

Click the "reset" button near the top of the control panel, set parameters for Mercury, and then click on the "Kepler's 2nd Law" tab at the bottom of the control panel. Slide the "adjust size" slider to the right, until the fractional sweep size is ⅛.

Click "start sweeping." The planet moves around its orbit, and the simulation fills in area until ⅛ of the ellipse is filled. Click "start sweeping" again as the planet arrives at the right-most point in its orbit (that is, at the point in its orbit farthest from the Sun). You may need to slow the animation rate using the slider under "Animation Controls." Click "show grid" under the visualization options. (If the moving planet annoys you, you can pause the animation.) One easy way to estimate an area is to count the number of squares.

4. **Count the number of squares in the yellow area and in the red area. You will need to decide what to do with fractional squares. Are the areas the same? Should they be?**

Kepler's Third Law

Click the "reset" button near the top of the control panel, set parameters for Mercury, and then click on the "Kepler's 3rd Law" tab at the bottom of the control panel. Select "show solar system orbits" in the Visualization Options panel. Study the graph in the Kepler's 3rd Law tab. Use the eccentricity slider to change the eccentricity of the simulated planet. First, make the eccentricity smaller and then larger.

5. **Did anything in the graph change?**

6. **What does this tell you about the dependence of the period on the eccentricity?**

Set parameters back to those for Mercury. Now use the semimajor axis slider to change the semimajor axis of the simulated planet.

7. **What happens to the period when you make the semimajor axis smaller?**

8. **What happens when you make it larger?**

9. **What does this tell you about the dependence of the period on the semimajor axis?**

---

the "how" or "why" of a situation. Problems give you a chance to practice the quantitative skills you learned in the chapter and to work through a situation mathematically. Finally, the **Exploration** is an activity that asks you to use the concepts and skills you learned in an interactive way. About half of the book's Explorations ask you to use animations and simulations on StudySpace, while the others are hands-on, paper-and-pencil activities that use everyday objects such as ice cubes or balloons.

If you think of human knowledge as an island, each scientific experiment makes the island a little bigger by adding a pebble or a grain of sand to the shoreline. But each of those pebbles also increases our exposure to the ocean of the unknown: the bigger the island of knowledge, the longer the shoreline of ignorance. Throughout this book, we have tried to show clearly which pebbles of knowledge are on the shore, which we are just catching a glimpse of under the water, and which are only thought to be there because of the way the water smoothes out as it passes over them. Sometimes the most speculative ideas are the most interesting, because they show how astronomers approach unsolved problems and explore the unknown. As astronomers, we authors know that one of the greatest feelings in the world is to forge a pebble yourself, and place it on the shoreline.

Astronomy gives you a sense of perspective that no other field of study offers. The universe is vast, fascinating, and beautiful, filled with a wealth of objects that, surprisingly, can be understood using only a handful of principles. By the end of this book, you will have gained a sense of your place in the universe—both how incredibly small and insignificant you are, and how incredibly unique and important you are.

Dear Instructor,

We wrote this book with a few overarching goals: to inspire students, to make the material interactive, and to create a useful and flexible tool that can support multiple learning styles.

As scientists and as teachers, we are passionate about the work we do. We hope to share that passion with students and inspire them to engage in science on their own. As authors, one way we do this is through the "student notebook"–style sketches at the beginning of each chapter. These figures model student engagement, and the Learning Goals on the facing page challenge them to try something similar on their own.

Through our familiarity with education research and surveys of instructors, we have come to know a great deal about how students learn and what goals teachers have for their students. We have explicitly addressed many of these goals and learning styles in this book, sometimes in large, immediately visible ways like the inclusion of feature boxes but also through less obvious efforts like questions and problems that relate astronomical concepts to everyday situations or take fresh approaches to organizing material.

For example, many teachers state that they'd like their students to become "educated scientific consumers" and "critical thinkers," or that their students should "be able to read a newspaper story about science and understand its significance." We have specifically addressed these goals in our Reading Astronomy News feature, which presents a news article and a series of questions that guide a student's critical thinking about the article, the data presented, and the sources.

Many teachers want students to develop better spatial reasoning and visualization skills. We address this explicitly by teaching students to make and use spatial models. One example is in Chapter 2, when we ask students to use an orange and a lamp to understand the celestial sphere and the phases of the moon. In nearly every chapter, we have Visual Analogy figures that compare astronomy concepts to everyday events or objects. Through these analogies, we strive to make the material more interesting, relevant, and memorable.

Education research shows that for many students, the most effective way to learn is by doing. Exploration activities at the end of each chapter are hands-on, asking students to take the concepts they've learned in the chapter and apply them as they interact with animations and simulations on the StudySpace website or work through pencil-and-paper activities. Many of these Explorations incorporate everyday objects and can be used either in your classroom or as activities at home.

To learn astronomy, students must also learn the language of science—not just the jargon, but the everyday words we scientists use in special ways. *Theory* is a famous example of a word that students think they understand, but their definition is very different from an astronomer's. The first time we use an ordinary word in a special way, a Vocabulary Alert in the margin calls attention to it, helping to reduce student confusion. This is in addition to the back-of-the-book glossary, which includes all the text's boldface words in addition to other terms students may be unfamiliar with.

Students must also be fairly fluent in the more formal language of science—mathematics. We have placed the math in Working It Out boxes, so it does not interrupt the flow of the text or get in the way of students' understanding of conceptual material. But we've gone further by beginning with fundamental ideas in early math boxes and building in complexity through the book. We've also worked to remove some of the stumbling blocks that crush student confidence by providing calculator hints, references to earlier boxes, and detailed, fully worked examples.

In our overall organization, we have made several efforts to encourage students to engage with the material and build confidence in their scientific skills as they proceed through the book. We organize the physical principles with a "just-in-time" approach: for example, we cover the Stefan-Boltzmann law in Chapter 6, when it is used for the first time in an astronomical context. For both stars and galaxies, we have organized the material to cover the general case first and then delve into more details with specific examples. Thus you will find "stars" before the Sun, and "galaxies" before the Milky Way. This allows us to avoid frustrating students by making assumptions about what they know about stars or galaxies. Planets have been organized comparatively, to emphasize that science is a process of studying individual examples that lead to collective conclusions. All of these organizational choices were made with the student perspective in mind, and a clear sense of the logical hierarchy of the material.

Even our layout has been designed to maximize student engagement—one wide text column is interrupted as seldom as possible.

SmartWork, Norton's online tutorial and homework system, puts student assessment at your fingertips. SmartWork contains over 1,000 questions and problems that are tied directly to this text, including the Summary Self-Test questions and versions of the Reading Astronomy News and Exploration questions. Any of these could be used as a reading quiz to be completed before class or as homework. Every question in SmartWork has hints and answer-specific feedback so that students are coached to work toward the correct answer. Instructors can easily modify any of the provided questions, answers, and feedback, or you can create your own questions.

We approached this text by asking: What do teachers want students to learn, and how can we best help students learn those things? Where possible, we consulted the education research to help guide us, and that guidance has led us down some previously unexplored paths. Please let us know how these features and structures work for your students. We'll be interested to hear from the field.

Sincerely,
Stacy Palen
Laura Kay
Brad Smith
George Blumenthal

Ancillaries for Students

SmartWork Online Homework

Over 1,000 questions support *Understanding Our Universe*—all with answer-specific feedback, hints, and ebook links. Questions include Summary Self-Tests and versions of the Explorations (based on AstroTours and the University of Nebraska simulations) and Reading Astronomy News questions. Image-labeling questions based on NASA images allow students to apply course knowledge to images that are not contained in the text.

StudySpace

W. W. Norton's free and open student website features

- Study Plans and Outlines for each chapter
- 28 AstroTour animations. These animations, some of which are interactive, use art from the text to help students visualize important physical and astronomical concepts.
- University of Nebraska interactive simulations (sometimes called applets; or NAAPs, for Nebraska Astronomy Applet Program). Well known by introductory astronomy instructors, the simulations allow students to manipulate variables and see how physical systems work.
- Quiz+ diagnostic multiple-choice quizzes, which provide students with feedback on any incorrect answers, also include links to the ebook, AstroTours, and University of Nebraska simulations.
- Vocabulary flashcards
- Astronomy in the News feed

Starry Night Planetarium Software (College version) and Workbook

Steven Desch, Guilford Technical Community College, and Donald Terndrup, The Ohio State University

Starry Night is a realistic, user-friendly planetarium simulation program designed to allow students in urban areas to perform observational activities on a computer screen. Norton's unique accompanying workbook offers observation assignments that guide students' virtual explorations and help them apply what they've learned from the text reading assignments. The workbook is fully integrated with *Understanding Our Universe.*

FOR INSTRUCTORS

Instructor's Manual

Kristine Washburn, Everett Community College, Ben Sugerman, Goucher College, Gregory D. Mack, Ohio Wesleyen University, and Steven Desch, Guilford Technical Community College

This resource includes brief chapter overviews, suggested classroom demonstrations/activities with handouts, lecture outlines to accompany the PowerPoint lecture slides, notes on the AstroTour animations contained on the Norton Resource Disc and StudySpace, worked solutions to all end-of-chapter problems, instructor's notes for the *Starry Night Workbook* activities, and teaching suggestions for how to use the Reading Astronomy News and the Exploration activity elements found in the textbook.

Test Bank

Trina Van Ausdal and Jonathan Barnes, Salt Lake Community College

The Test Bank has been developed using the Norton Assessment Guidelines and provides a quality bank of over 1,300 items. Each chapter of the Test Bank consists of three question types classified according to Norton's taxonomy of knowledge types:

1. Factual questions test students' basic understanding of facts and concepts.
2. Applied questions require students to apply knowledge in the solution of a problem.

3. Conceptual questions require students to engage in qualitative reasoning and to explain why things are as they are.

Questions are further classified by section and difficulty, making it easy to construct tests and quizzes that are meaningful and diagnostic. The question types are short answer, multiple choice, and true/false.

Norton Instructor's Resource Site

This web resource contains the following resources to download:
- Test Bank, available in ExamView, Word RTF, and PDF formats
- Instructor's Manual in PDF format
- Lecture PowerPoint slides with lecture notes
- All art and tables in JPEG and PPT formats
- Starry Night College, W. W. Norton Edition, Instructor's Manual
- 28 AstroTour animations. These animations, some of which are interactive, use art from the text to help students visualize important physical and astronomical concepts.
- University of Nebraska simulations (sometimes called applets; or NAAPs, for Nebraska Astronomy Applet Program). Well known by introductory astronomy instructors, the simulations allow students to manipulate variables and see how physical systems work.
- Coursepacks, available in BlackBoard, Angel, Desire2Learn, and Moodle formats

Coursepacks

Norton's Coursepacks, available for use in various Learning Management Systems (LMS), feature all Quiz+ and Test Bank questions, links to the AstroTours and Applets, plus discussion questions from the Reading Astronomy News features. Coursepacks are available in BlackBoard, Angel, Desire2Learn, and Moodle formats.

Instructor's Resource Folder

This two-disc set contains the Instructor's Resource DVD—which contains the same files as the Instructor's Resource Website—and the Test Bank on CD-ROM in ExamView format.

Acknowledgments

The authors would like to acknowledge the extraordinary efforts of the staff at W. W. Norton: Mary Lynch and Mina Shaghaghi, who kept the mail flowing smoothly; Christine D'Antonio, who kept track of us in the final stages; the copy editor, Chris Thillen, who managed to decipher our meaning even when it wasn't clear. We would especially like to thank Andrew Sobel, whose editing improved every page and nearly every paragraph; and Erik Fahlgren, who has been a fount of ideas and encouragement and occasionally cracked the whip to keep us from wandering.

We'd also like to thank Ron Proctor, of the Ott Planetarium, whose early efforts on the chapter openers were invaluable; and Kiss Me I'm Polish, who turned his sketches into a finished product. Jane Searle managed the producton. Kelly Paralis at Penumbra Design turned our rough sketches into finished art. Hope Miller Goodell was the design director. Brian Salisbury created the lovely book design, and Carole Desnoes worked within that framework to lay out all the components. Rob Bellinger and Matthew Freeman edited the media and supplements, and we welcome Stacy Loyal, who will help get this book in the hands of people who can use it. We would also like to recognize the contributions made by Howard Voss, a well-known expert on science education.

And we would like to thank the reviewers, whose input at every stage improved the final product:

James S. Brooks, *Florida State University*
Edward Brown, *Michigan State University*
James Cooney, *University of Central Florida*
Kelle Cruz, *Hunter College*
Robert Friedfeld, *Stephen F. Austin State University*
Steven A. Hawley, *University of Kansas*
Dr. Eric R. Hedin, *Ball State University*
Scott Hildreth, *Chabot College*
Emily S. Howard, *Broward College*
Dain Kavars, *Ball State University*
Kevin Krisciunas, *Texas A&M University*
Duncan Lorimer, *West Virginia University*
Jane H. MacGibbon, *University of North Florida*
James McAteer, *New Mexico State University*
Ian McLean, *University of California–Los Angeles*
Jo Ann Merrell, *Saddleback College*
Michele Montgomery, *University of Central Florida*
Christopher Palma, *Pennsylvania State University*
Nicolas A. Pereyra, *University of Texas–Pan American*
Vahe Peroomian, *University of California–Los Angeles*
Dwight Russell, *Baylor University*
Ulysses J. Sofia, *American University*
Michael Solontoi, *University of Washington*
Trina Van Ausdal, *Salt Lake Community College*
Nilakshi Veerabathina, *University of Texas at Arlington*

✦ ASTROTOURS

Stacy Palen is an award-winning professor in the physics department and the director of the Ott Planetarium at Weber State University. She received her BS in physics from Rutgers University and her PhD in physics from the University of Iowa. As a lecturer and postdoc at the University of Washington, she taught Introductory Astronomy more than 20 times over 4 years. Since joining Weber State, she has been very active in science outreach activities ranging from star parties to running the state Science Olympiad. Stacy does research in formal and informal astronomy education and the death of Sun-like stars. She spends much of her time thinking, teaching, and writing about the applications of science in everyday life. She then puts that science to use on her small farm in Ogden, Utah.

Laura Kay is Ann Whitney Olin Professor and Chair of the Department of Physics and Astronomy at Barnard College, where she has taught since 1991. She received a BS degree in physics and an AB degree in feminist studies from Stanford University, and MS and PhD degrees in astronomy and astrophysics from the University of California–Santa Cruz. Laura spent 13 months at the Amundsen-Scott South Pole Station in Antarctica as a graduate student and has had fellowships in Chile and Brazil. She studies active galactic nuclei, using optical and X-ray telescopes. At Barnard she teaches courses on astronomy, astrobiology, women and science, and polar exploration.

Brad Smith is a retired professor of planetary science. He has served as an associate professor of astronomy at New Mexico State University, a professor of planetary sciences and astronomy at the University of Arizona, and as a research astronomer at the University of Hawaii. Through his interest in Solar System astronomy, he has participated as a team member or imaging team leader on several U.S. and international space missions, including *Mars Mariners 6, 7*, and *9*; *Viking*; *Voyagers 1* and *2*; and the Soviet *Vega* and *Phobos* missions. He later turned his interest to extrasolar planetary systems, investigating circumstellar debris disks as a member of the Hubble Space Telescope NICMOS experiment team. Brad has four times been awarded the NASA Medal for Exceptional Scientific Achievement. He is a member of the IAU Working Group for Planetary System Nomenclature and is Chair of the Task Group for Mars Nomenclature.

George Blumenthal is chancellor at the University of California–Santa Cruz, where he has been a professor of astronomy and astrophysics since 1972. He received his BS degree from the University of Wisconsin–Milwaukee and his PhD in physics from the University of California–San Diego. As a theoretical astrophysicist, George's research encompasses several broad areas, including the nature of the dark matter that constitutes most of the mass in the universe, the origin of galaxies and other large structures in the universe, the earliest moments in the universe, astrophysical radiation processes, and the structure of active galactic nuclei such as quasars. Besides teaching and conducting research, he has served as Chair of the UC–Santa Cruz Astronomy and Astrophysics Department, has chaired the Academic Senate for both the UC–Santa Cruz campus and the entire University of California system, and has served as the faculty representative to the UC Board of Regents.

✦ Understanding Our Universe

1 Our Place in the Universe

What are the Sun and Moon made of? How far away are they? How do stars shine? Do they have anything to do with me? How did the universe begin? How will it end? You may have asked yourself such questions while looking up at the sky. Astronomers seek the answers to these and many other compelling questions. Loosely translated, the word **astronomy** means "patterns among the stars." But modern astronomy—the astronomy we will talk about in this book—has become far more than looking at the sky and cataloging what is visible there. Now is the most fascinating time to be studying this most ancient of sciences. The answers we are finding are changing not only our view of the cosmos but also our view of ourselves.

◇ LEARNING GOALS

Astronomers seek knowledge using a very specific set of processes, sometimes collectively called the scientific method. Part of this process stems from recognizing patterns in nature. In the image at right, a student has been observing where the Sun sets over the course of six months, as viewed from his house. By the end of this chapter, you should understand how observations like this fit into the development of a scientific law and a scientific theory. You should also be able to:

- Relate our place in the universe to the rest of the cosmos

- Explain how the physical laws that shape our daily lives are related to those that govern the universe

- Describe where we come from, astronomically speaking

- Compare the scientific way of viewing the world with other ways of viewing the world

- Distinguish between understanding and accumulation of fact

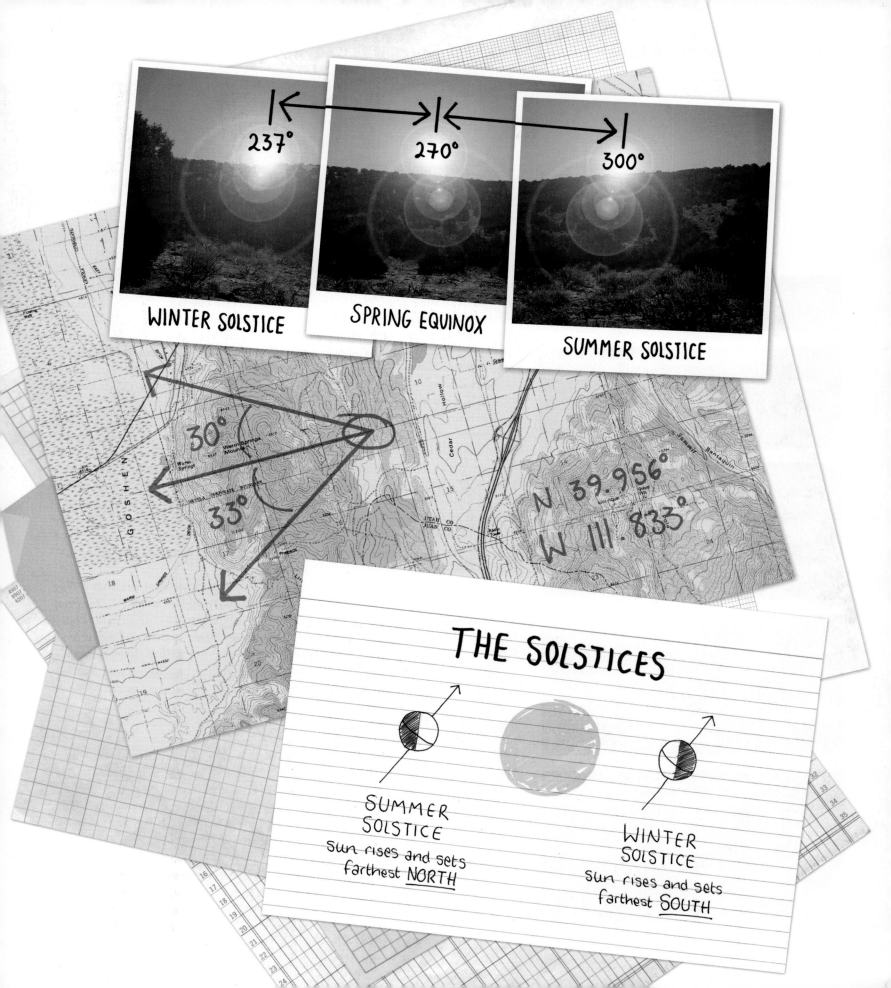

237°

270°

300°

WINTER SOLSTICE

SPRING EQUINOX

SUMMER SOLSTICE

30°

33°

GOSHEN

N 39.956°
W 111.833°

THE SOLSTICES

SUMMER
SOLSTICE
sun rises and sets
farthest <u>NORTH</u>

WINTER
SOLSTICE
sun rises and sets
farthest <u>SOUTH</u>

1.1 Getting a Feel for the Neighborhood

Glimpsing Our Place in the Universe

Most of us have an address where the mail carrier delivers our mail—street number, street, city, state, country. But let's expand our view to include the enormously vast universe we live in. What is our "cosmic address"? It might look something like this: planet, star, galaxy, galaxy group, galaxy cluster.

We reside on a planet called Earth, which is orbiting under the influence of gravity about a star called the Sun. The Sun is an ordinary, middle-aged star, more **massive** and luminous than some stars but less massive and luminous than others. The Sun is extraordinary only because of its importance to us within our own Solar System. Our Solar System consists of eight classical planets—Mercury, Venus, Earth, Mars, Jupiter, Saturn, Uranus, and Neptune. It also contains many smaller bodies, such as dwarf planets (for example, Pluto, Ceres, or Eris), asteroids (for example, Pallas or Vesta) and comets (for example, Halley or Hyakutake). We might define the size of our Solar System by its most distant known planetary body, the dwarf planet Eris, although others could rightfully include the realm of small, icy bodies that extends far beyond the orbit of Eris.

The Sun is located about halfway out from the center of the Milky Way Galaxy, a flattened collection of stars, gas, and dust. Our Sun is just one among several hundred billion stars scattered throughout our galaxy, and astronomers are finding that many of these stars are themselves surrounded by planets, suggesting that other planetary systems may be common.

The Milky Way, in turn, is a member of a small collection of a few dozen galaxies called the Local Group. The Milky Way Galaxy and the Andromeda Galaxy are true giants within the Local Group. Most others are "dwarf galaxies." The Local Group itself is part of a vastly larger collection of thousands of galaxies—a supercluster—called the Virgo Supercluster.

We can now define our "cosmic address"—Earth, Solar System, Milky Way Galaxy, Local Group, Virgo Supercluster—as illustrated in **Figure 1.1**. Yet even this address is not complete, because the vast structure we just described is only the *local universe*. The part of the universe that we can see extends far beyond—a distance that light takes 13.7 billion years to travel—and within this volume we estimate that there are *several hundred billion galaxies*, roughly as many galaxies as there are stars in the Milky Way.

The Scale of the Universe

One of the first conceptual hurdles we face as we begin to think about the universe is its sheer size. A hill is big, and a mountain is really big. If a mountain is really big, then Earth is enormous. But where do we go from there? We run out of words as the scale of the universe comes to dwarf our human experience. One technique that helps us develop a sense of scale is to use a little sleight of hand

Earth

Solar System

Milky Way Galaxy

Local Group

Virgo Supercluster

Figure 1.1 Our place in the universe—our cosmic address: Earth, Solar System, Milky Way Galaxy, Local Group, Virgo Supercluster. We live on Earth, a planet orbiting the Sun in our Solar System, which is a star in the Milky Way Galaxy. The Milky Way is a large galaxy within the Local Group of galaxies, which in turn is located in the Virgo Supercluster.

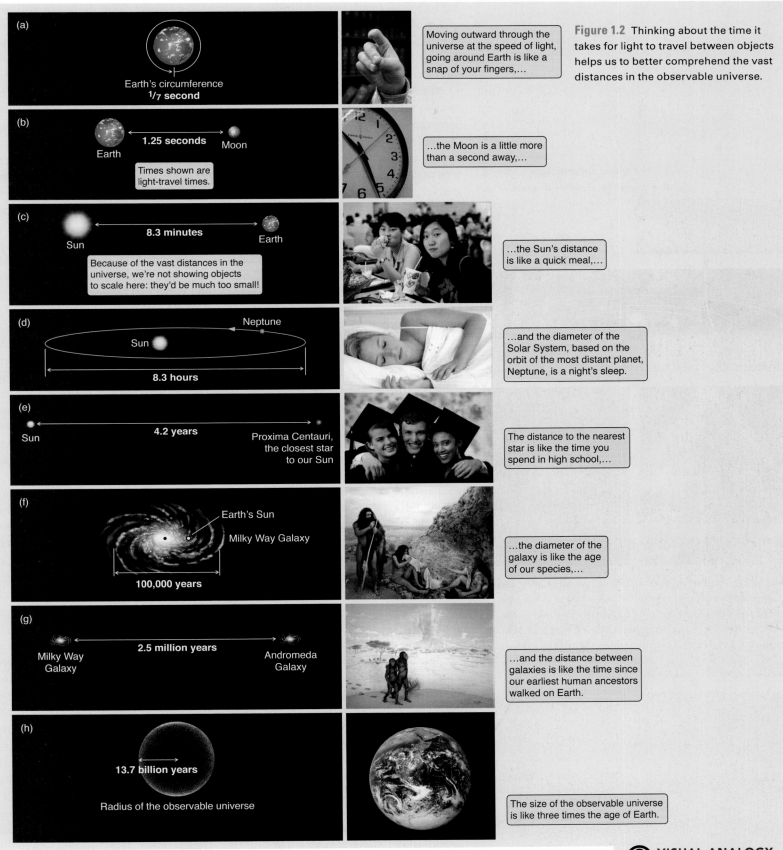

Figure 1.2 Thinking about the time it takes for light to travel between objects helps us to better comprehend the vast distances in the observable universe.

(a) Earth's circumference
1/7 second

Moving outward through the universe at the speed of light, going around Earth is like a snap of your fingers,…

(b) Earth — 1.25 seconds → Moon
Times shown are light-travel times.

…the Moon is a little more than a second away,…

(c) Sun ←— 8.3 minutes —→ Earth
Because of the vast distances in the universe, we're not showing objects to scale here: they'd be much too small!

…the Sun's distance is like a quick meal,…

(d) Neptune / Sun
8.3 hours

…and the diameter of the Solar System, based on the orbit of the most distant planet, Neptune, is a night's sleep.

(e) Sun ←— 4.2 years —→ Proxima Centauri, the closest star to our Sun

The distance to the nearest star is like the time you spend in high school,…

(f) Earth's Sun / Milky Way Galaxy
100,000 years

…the diameter of the galaxy is like the age of our species,…

(g) Milky Way Galaxy ←— 2.5 million years —→ Andromeda Galaxy

…and the distance between galaxies is like the time since our earliest human ancestors walked on Earth.

(h) 13.7 billion years
Radius of the observable universe

The size of the observable universe is like three times the age of Earth.

VISUAL ANALOGY

and change from discussing distance to talking about time. Suppose you are driving down the highway at 60 miles per hour (mph). At this **speed**, you travel a mile in 1 minute. Sixty miles is how far you go in an hour. Six hundred miles is how far you go in 10 hours. To get a feel for the difference between 600 miles and 1 mile, you can think about the difference between 10 hours and 1 minute.

We can play this same game in astronomy, but the speed of a car on the highway is far too slow to be useful. Instead we will use the greatest speed in the universe—the speed of light. Light travels at 300,000 kilometers per second (km/s). At that speed, light can circle Earth (a distance of 40,000 kilometers [km]) in just under one-seventh of a second—about the time it takes you to snap your fingers. Fix that comparison in your mind. The size of Earth is like a snap of your fingers. Follow along in **Figure 1.2** as we move outward to the entire universe. (This figure is the first of this book's Visual Analogies, illustrations that relate astronomical phenomena to everyday objects.)

Pause to let the distances described in the figure sink in. The size of Earth is to the universe as a snap of your fingers is to the age of the universe (about 13.7 billion years). Even relatively small distances in astronomy are so vast that they are measured in light-years: the distance light travels in one year.

The travel time of light helps us understand the scale of the universe.

The Origin and Evolution of the Universe

While seeking knowledge about the universe and how it works, modern astronomy and physics have repeatedly come face-to-face with ideas long thought to be solely within the domain of philosophers. Throughout most of history, philosophers saw the universe as remote and different from Earth—disconnected from our terrestrial existence. When modern astronomers look at the universe, they see that we are a part of a network of ongoing processes. Astronomy begins by looking out at the universe, but that outward gaze turns introspective as we find that our very existence is a consequence of those same processes.

The chemical evolution of the universe is such a case. Both theory and observation tell us that the universe began at the **Big Bang**, 13.7 billion years ago. The only chemical elements found in abundance in the early universe were hydrogen and helium, plus tiny amounts of lithium, beryllium, and boron. Yet we live on a planet with a core consisting mostly of iron and nickel, surrounded by a mantle made up of rocks containing large amounts of silicon and various other elements. Our bodies contain carbon, nitrogen, oxygen, sodium, phosphorus, and a host of other chemical elements. If the elements that make up Earth and our bodies were not present in the early universe, where did they come from?

Figure 1.3 You and everything around you, including beautiful waterfalls contain atoms that were forged in the interior of stars that lived and died before the Sun and Earth were formed. The supermassive star Eta Carinae, shown on the right, is currently ejecting a cloud of enriched material.

The answer to this question lies deep within stars where less massive atoms combine to form more massive atoms. When a star nears the end of its life, it loses much of its mass—including some of the new atoms formed in its interior—by blasting it back into interstellar space. Our Sun and Solar System formed from a cloud of gas and dust that had been enriched by earlier generations of stars. This chemical legacy supplies the building blocks for the interesting chemical processes, such as life, that go on around us. **Figure 1.3** symbolizes this intimate relationship between the world around us and our heritage in the stars. Look around you. The atoms that make up everything you see were formed in stars. Poets sometimes say that "we are stardust," but this is not just poetry. It is literal truth.

We are stardust.

From earliest history, humans have speculated about our beginnings: How was Earth created? Who or what is responsible for our existence? In the modern world, scientists seek answers to slightly different questions: How and when did the universe begin? What combination of events led to our existence on a small, rocky planet orbiting a typical middle-aged star? Was this a unique happening, or are there others like us scattered throughout the galaxy? In this book, we'll look into the origins of terrestrial life, but we'll also examine the possibilities of life elsewhere in our Solar System and beyond—a subject called **astrobiology**. For life to exist around other stars in the galaxy, there must be life-sustaining planets to support it. We will include the discovery of planets orbiting stars other than our Sun and compare them with the planets of our own Solar System.

Astronomers Use a Variety of Tools

As we view the universe through the eyes of astronomers, we also learn something of how science works. It is beyond the scope of this book for us to provide a detailed justification for all that we will say. However, we will try to offer some explanation of where an idea comes from and why we think it is valid. We will not present something as fact unless there is a compelling reason to believe it. We will be honest when we are uncertain, and we will admit it when the truth is that we really do not know. This book is not a compendium of revealed truth or a font of accepted wisdom. Rather, it is an introduction to a body of knowledge and understanding that was painstakingly built (and sometimes torn down and rebuilt) piece by piece.

It is almost impossible to overstate the importance of science in our civilization. One obvious manifestation of science is the technology that has enabled us to explore well beyond our planet. Since the 1957 launch of Sputnik, the first human-made **satellite**, we have lived in an age of space exploration, and humans have walked on the Moon (**Figure 1.4**). Our unmanned probes have visited all of the classical planets. Spacecraft have flown past asteroids, comets, and even the Sun. Our inventions have landed on Mars, Venus, Titan (Saturn's largest moon), and an asteroid, and have plunged into the atmosphere of Jupiter and the heart of a comet. Most of what we know of the Solar System has resulted from these past five decades of exploration.

Satellite observatories have given us a new perspective on the universe. The atmosphere that shields us from harmful solar radiation also blinds us to much of what is going on around us. Space astronomy shows vistas hidden from ground-based telescopes by our atmosphere. Satellites detect many types of radiation—from the highest-energy gamma rays and X-rays, through ultraviolet and infrared radiation, and to the lowest-energy radio waves—and have brought us discovery after surprising discovery. Each has forever altered our perception of the universe, further expanding the domain of the human mind. Since the closing years of the 20th century, we have witnessed a renewed vigor in astronomical observations from the surface of Earth. The view of the sky seen by radio telescopes, as shown in **Figure 1.5**, illustrates the new perspectives that have been made possible by our growing technological prowess.

When we think of astronomy, telescopes—both on the ground and in space—immediately come to mind. You may be surprised to learn that much cutting-edge astronomy is now carried out in large physics facilities like the one shown in **Figure 1.6**. Astronomers work with colleagues in related fields, such as physics, chemistry, geology, and planetary science, to sharpen our understanding of physical laws and to use this understanding to make sense of our observations of the cosmos. Astronomers have also benefited from the computer revolution. The 21st-century astronomer spends far more time staring at

Vocabulary Alert

Satellite: In common language, *satellite* refers to a human-made object. Astronomers use this word to describe any object, human-made or natural, that orbits another.

Figure 1.4 *Apollo 15* astronaut James B. Irwin stands by the lunar rover during an excursion to explore and collect samples from the Moon.

G X U V I **R**

Figure 1.5 In the 20th century, advances in telescope technology opened new windows on the universe. This is the sky as we would see it if our eyes were sensitive to radio waves, shown as a backdrop to the National Radio Astronomy Observatory site in Green Bank, West Virginia.

Figure 1.6 The Tevatron high-energy particle collider at Fermilab has provided clues about the physical environment during the birth of the universe. Laboratory astrophysics, in which astronomers model important physical processes under controlled conditions, has become an important part of astronomy.

a computer screen than peering through the eyepiece of a telescope. Today's astronomers use computers to collect and analyze data from telescopes, calculate physical models of astronomical objects, and prepare and disseminate the results of their work.

1.2 Science Is a Way of Viewing the World

Given the visible importance of technology in our lives, you might be tempted to say that science *is* technology. It is true that science forms the basis of technology through a mutually supportive relationship in which each enables advances in the other. Yet science is much more than technology. We should understand science as a fundamental way of viewing the world through the **scientific method**, a worldview that has survived constant upheavals and challenges.

The Scientific Method and Scientific Principles

If science is more than technology, you might instead suggest that science is the *scientific method*. Suppose a scientist has an idea that might explain a particular observation or phenomenon. She presents the idea to her colleagues as a **hypothesis**. Her colleagues then look for testable predictions capable of *disproving* her hypothesis. *This is an important property of the scientific method*, because any hypothesis that is not **falsifiable** must be based on faith alone. (Note that a *falsifiable* hypothesis—one capable of being shown false—may not be testable using current technology, but we must at least be able to outline an experiment or observation that could prove the idea wrong.) As continuing tests fail to disprove a hypothesis, scientists will come to accept it as a **theory**, but *never* as an undisputed fact. Scientific theories are accepted only as long as their predictions are correct. A classic example is Einstein's theory of relativity, which has withstood more than a century of scientific efforts to disprove its predictions.

Science is sometimes misunderstood because of the special ways that scientists use everyday words. An example is the word **theory**. In everyday language, "theory" may mean something that is little more than a conjecture or a guess: "Do you have a theory about who might have done it?" "My theory is that a third party could win the next election." In everyday language, a theory is something worthy of little serious regard. "After all," we say, "it is only a theory."

In stark contrast, a *scientific* theory is a carefully constructed proposition that takes into account all the relevant data and all our understanding of how the world works. It makes testable predictions about the outcome of future observations and experiments. A well-corroborated theory has survived many such tests. Scientific theories represent and summarize bodies of knowledge and understanding that provide fundamental insights into the world around us. Rather than simple speculation, a successful theory is the pinnacle of human knowledge about the world.

Theories are near the top of the loosely defined hierarchy of scientific knowledge. In science, the word *idea* has its everyday use: an idea is just a notion about how something might be. In science, a *hypothesis* is an idea that leads to testable predictions. A hypothesis may be the forerunner of a scientific theory, or it may be based on an existing theory, or both. In science, a *theory* is an idea that has been examined carefully, is consistent with all existing theoretical and experimental knowledge, and makes testable predictions. Ultimately, the success of the

> The scientific method involves trying to *falsify* ideas.

Vocabulary Alert

Falsify: In common language, we are likely to think of "falsified" evidence as having been *manipulated* to misrepresent the truth. Astronomers (and scientists in general) use *falsify* in the sense of "proving a hypothesis false," as we will throughout this book.

Theory: In common language, a *theory* is weak—just an idea or a guess. Astronomers use this word to label the most well-known, well-tested, and well-supported principles in science.

predictions is the deciding factor between competing theories. Some theories become so well tested and are of such fundamental importance that we come to refer to them as **physical laws**.

A scientific **principle** is a general idea or sense about how the universe *is* that guides our construction of new theories. **Occam's razor**, for example, is a guiding principle in science stating that when we are faced with two hypotheses that explain a particular phenomenon equally well, we should adopt the simpler of the two. At the heart of modern astronomy is another principle: the **cosmological principle**. The cosmological principle states that on a large scale, the universe is the same everywhere. In other words, there is nothing special about our particular location. By extension, matter and energy obey the same physical laws throughout space and time as they do today on Earth. The same physical laws that we learn about in laboratories can be used to understand what goes on in the centers of stars or in the hearts of distant galaxies. Each new success that comes from applying the cosmological principle to observations of the universe around us—each new theory that succeeds in explaining or predicting patterns and relationships among celestial objects—adds to our confidence in the validity of this cornerstone of our worldview. We will discuss the cosmological principle in more detail in Chapter 14.

> There is nothing special about our place in the universe.

Science as a Way of Knowing

The path to scientific knowledge is solidly based on the scientific method. This concept is so important to your understanding of how science works that we should emphasize it once again. The scientific method consists of an observation or idea, followed by a hypothesis, followed by a prediction, followed by further observation or experiments to test the prediction, and ending as a tested theory, as illustrated in **Figure 1.7**. Look back at the chapter-opening figure. Where does the activity represented in that figure fit onto this simplified flowchart? For all practical purposes, Figure 1.7 defines what we mean when we use the verb *to know*. It is sometimes said that the scientific method is how scientists prove things to be true, but actually it is a way of proving things to be *false*. Before scientists accept something as true, they work hard to show that it is false. Only after repeated attempts to disprove an idea have failed do scientists begin to accept its likely validity. For a theory to be given serious scientific consideration, it *must be falsifiable*. Scientific theories are accepted only as long as they are able to be tested and are not shown to be false. In this sense, *all scientific knowledge is provisional*.

Despite being tied so closely to the scientific method, science can no more be said to *be* the scientific method than music can be said to *be* the rules for writing down a musical score. The scientific method provides the rules for asking nature whether an idea is false, but it offers no insight into where the idea came from in the first place or how an experiment was designed. If you were to listen to a group of scientists discussing their work, you might be surprised to hear them using words such as *insight*, *intuition*, and *creativity*. Scientists speak of a beautiful theory in the same way that an artist speaks of a beautiful painting or a musician speaks of a beautiful performance. Yet science is not the same as art or music in one important respect: whereas art and music are judged by a human jury alone, in science it is nature (through application of the scientific method) that decides which theories can be kept and which theories must be discarded. Nature is completely unconcerned about what we *want* to be true. In the history

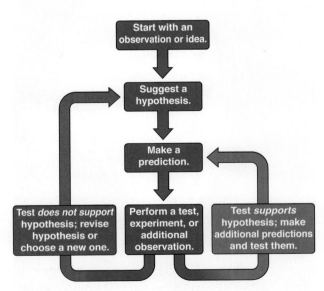

Figure 1.7 The scientific method. This is the path by which an idea or observation leads to a falsifiable hypothesis that is either accepted as a tested theory or rejected, on the basis of observational or experimental tests of its predictions. Note that the green loop goes on indefinitely as scientists continue to test the hypothesis.

Nature is the arbiter of science.

of science, many a beautiful and beloved theory has been abandoned. At the same time, however, science has an aesthetic that is as human and as profound as any found in the arts.

Scientific Revolutions

As we have seen, science cannot be defined as either technology or the scientific method, although it is related to both. It is also incorrect to say that science is simply a collection of facts. If we limited our definition of science to the existing facts, we would fail to convey the dynamic nature of scientific inquiry. Scientists do not have all the answers and must constantly refine their ideas in response to new data and new insights. This vulnerability of knowledge may seem like a weakness at first. "Gee, you really don't know anything," the cynical person might say. But this vulnerability is actually science's great strength. It is what keeps us honest. In science, even our most cherished ideas about nature remain fair game, subject to challenge by evidence according to the rules of the scientific method. Many of history's best scientists earned their status by falsifying a well-loved idea.

Scientists spend most of their time working within an established framework of understanding, extending and refining that framework and testing its boundaries. Occasionally, however, major shifts occur in the framework of an entire scientific field. Many books have been written about how science progresses, perhaps the most influential being *The Structure of Scientific Revolutions*, by Thomas Kuhn. In this work, Kuhn emphasizes the constant tension between the scientist's human need to construct a system of beliefs within which to interpret the world and the occasional (and likely painful) need to drastically overhaul that system of beliefs.

For example, the classical physics developed by Sir Isaac Newton in the 17th century withstood the scrutiny of scientists for over 200 years. It seemed that little remained but cleanup work—filling in the details. Yet during the late 19th and early 20th centuries, physics was rocked by a series of scientific revolutions that shook the very foundations of our understanding of the nature of reality. If one face represents these scientific revolutions, it is that of Albert Einstein (**Figure 1.8**). Einstein's special and general theories of relativity replaced Newton's mechanics. Einstein did not prove Newton wrong, but instead showed that Newton's theories were a special case of a far more general and powerful set of physical laws. Einstein's new ideas unified our concepts of mass and energy and destroyed our conventional notion of space and time as separate things.

We will encounter many other discoveries and successful ideas that forced scientists to abandon their treasured notions or be left behind. Einstein himself was unable to accept the implications of a second revolution he helped start— quantum mechanics—and he went to his grave unwilling to embrace the view of the world it offered. The point is this: In physical science, we are not just paying lip service to a hollow ideal when we say that the rigorous standards of scientific knowledge respect no authorities. No theory, no matter how central or how strongly held, is immune from the rules.

The scientist's axiom is that there is an order in the universe and that the human mind is capable of grasping the essence of the rules underlying that order. The scientist's creed is that nature, through observation and experiment, is the final arbiter of *objective truth*. Science is an exquisite blend of aesthetics

Figure 1.8 Albert Einstein, perhaps the most famous scientist of the 20th century and *Time* magazine's selection as Person of the Century. Einstein helped usher in two different scientific revolutions, one of which he was never able to accept.

and practicality. In the final analysis, science has found such a central place in our civilization because *science works*.

Challenges to Science

Science has not achieved its prominence without criticism and controversy. Philosophers and theologians of the 16th century did not easily give up the long-held belief that Earth is the center of the universe. For many, it came as a rude awakening that our Milky Way does not, in fact, constitute the entire universe—that it is only a tiny part of a vastly larger universe.

You might feel that science is sometimes arbitrary. For example, the reclassification of Pluto as a dwarf planet may seem subjective. This controversial decision was really more a matter of semantics than of science. Pluto is still the same scientifically important Solar System body that it always was; it just has a new label. As you will see in discussions throughout this book, the path to knowledge embodied in science is no more arbitrary than the logic it is built on.

Critics of science have pointed out that science is influenced by culture. Political and cultural considerations strongly influence which scientific research projects are funded. This choice of funding channels the directions in which scientific knowledge advances and can lead to serious ethical issues. For example, progress in stem cell research and solutions to human-caused climate change have been hindered by widely distributed misinformation on the web and by constant political bickering.

Some critics even carry this view a step further, arguing that scientific knowledge itself is an arbitrary cultural construct. Yet successful scientific theories are *never* arbitrary. Scientific theories must be consistent with all that we know of how nature works, and turning a clever idea into a real theory with testable predictions is a matter of careful thought and effort. One of the most remarkable aspects of scientific knowledge is its *independence* from culture. Scientists are people, and politics and culture enter into the day-to-day practice of science. But in the end, *scientific theories are judged not by cultural norms but by whether their predictions are borne out by observation and experiment*. As long as the results of experiments are repeatable and do not depend on the culture of the experimenter—that is, as long as there is such a thing as objective physical reality—scientific knowledge cannot be called a cultural construct. The foundations of the scientific worldview have withstood centuries of fine minds trying to prove them false.

It is significant that no philosopher critical of science has ever offered a viable alternative for obtaining reliable knowledge of the workings of nature. Furthermore, no other category of human knowledge is subject to standards as rigorous and unforgiving as those of science. For this reason, scientific knowledge is reliable in a way that no other form of knowledge can claim. Whether you want to design a building that will not fall over, consider the most recent medical treatment for a disease, or calculate the orbit of a spacecraft on its way to the Moon, you had better consult a scientist rather than a psychic—regardless of your cultural heritage.

Finally, we must admit that some disreputable scientists purposefully try to influence results by inventing or ignoring data, often when claiming a "major breakthrough" or challenging a well-established scientific principle. Fortunately, attempts by others to repeat the experiment inevitably expose such scientific misconduct.

1.3 Patterns Make Our Lives and Science Possible

We experience patterns in our everyday life, and we find comfort in them. But imagine what life would be like if sometimes when you let go of an object, it fell up instead of down. What if, unpredictably, one day the Sun rose at noon and set at 1:00 P.M., the next day it rose at 6:00 A.M. and set at 10:00 P.M., and the next day the Sun did not rise at all? In fact, objects do fall toward the ground. The Sun rises, sets, and then rises again at predictable times, and in predictable locations, as shown in the chapter-opening figure. Spring turns into summer, summer turns into autumn, autumn turns into winter, and winter turns into spring. The rhythms of nature produce patterns in our lives, and we count on these patterns for our very survival. If nature did not behave according to regular patterns, then our lives—indeed, life itself—would not be possible.

The patterns that make our lives possible also make science possible. The goal of science is to identify and characterize these patterns and to use them to understand the world around us. Some of the most regular and easily identified patterns in nature are the patterns that we see in the sky. What in the sky will look different or the same a week from now? A month from now? A year from now? Most of us probably lead an indoor and in-town existence, removed from an everyday awareness of the patterns in the sky. Away from the smog and glare of our cities, however, the patterns and rhythms of the sky are as easy to see today as they were in ancient times. Patterns in the sky mark the changing of the seasons (**Figure 1.9**) and determine the planting and harvesting of crops. It is no surprise that astronomy, which is the expression of our human need to understand these patterns, is the oldest of all sciences.

One important tool that astronomers use to analyze these patterns is mathematics. There are many branches of mathematics, most of them dealing with more than just numbers. Arithmetic is about counting things. Algebra is about manipulating symbols and the relationships between things. Geometry is about shapes. Calculus is about change. Other types of mathematics include topology (the properties of surfaces) and statistics (groups of objects and their relationships). What do all these branches have in common? Why do we consider all of them to be part of a single discipline called mathematics? All share one thing: they deal with patterns. The best working definition of mathematics is "the language of patterns."

If patterns are the heart of science, and mathematics is the language of patterns, it should come as no surprise that *mathematics is the language of science*. Trying to study science while avoiding mathematics is the practical equivalent of trying to study Shakespeare while avoiding the written or spoken word. It quite simply cannot be done, or at least cannot be done meaningfully. Distaste for mathematics is one of the most common obstacles standing between a nonscientist and an appreciation of the world as seen through the eyes of a scientist.

Part of the responsibility for moving beyond this obstacle lies with us, the authors. It is our job to take on the role of translators, using words to express concepts as much as possible, even when these concepts are more concisely and accurately expressed mathematically. When we do use mathematics, we will explain in everyday language what the equations mean and show you how equations express concepts that you can connect to the world. We will also limit the mathematics to a few tools that we would like you to know.

Changing patterns in the sky echo changing patterns on Earth.

Figure 1.9 Since ancient times, our ancestors have recognized that patterns in the sky change with the seasons. These and other patterns shape our lives.

These mathematical tools, a few of which are described in **Working It Out 1.1**, enable scientists to convey complex information. *Scientific notation* enables astronomers to express the vast range of sizes of astronomical objects. *Units* distinguish among time, distance, mass, and energy. *Geometry* illuminates the sizes, shapes, and volumes of astronomical objects and the distances between them. *Algebra* expresses the patterns that relate one physical quantity to another. We use the most accessible tools to make our journey of discovery as comfortable and informative as possible.

Your responsibility is to accept the challenge and make an honest effort to think through the mathematical concepts that we use. The mathematics in this book are on a par with what it takes to balance a checkbook, build a bookshelf that stands up straight, check your gas mileage, estimate how long it will take to drive to another city, figure your taxes, or buy enough food to feed an extra guest or two at dinner.

Working It Out 1.1 | Units and Scientific Notation

Mathematics is the language of patterns, and the universe "speaks" math. If you want to ask questions of the universe and understand the answers, you have to speak math too. Throughout the text, we present the math as needed, with worked examples to help you understand how to work it out on your own.

Units

Astronomers use the metric system of units. We prefer this system because many conversion factors are powers of 10, and simply involve moving the decimal place. For example, to convert from meters to centimeters, move the decimal place two spaces to the right. Compare this with converting from yards to inches by multiplying by an awkward factor of 36 (3 to convert from yards to feet, then 12 to convert feet to inches). The United States is one of very few countries in the world still using the English system of units, but gradual acceptance of the metric system can be seen in everyday life. We buy milk in quart-sized containers, but our soft drinks come in liter-sized bottles. Bank signs often display temperature in both Fahrenheit (°F) and Celsius (°C), also known as centigrade. Some road signs show distances in both miles and kilometers. Many of the units we use will be unfamiliar or new to you; but if you remember that a meter is about 3 feet, a centimeter is a bit under half an inch, and a kilometer is about two-thirds of a mile, you will start to develop an intuition about these units.

- **Approximately how many feet are in 2.5 meters?** Since a meter is about 3 feet, 2.5 meters is *about* 7.5 feet. This is about the diameter of the Hubble Space Telescope's primary mirror.
- **Approximately how many centimeters are in a foot?** We can estimate this value in two ways. Since a foot is 12 inches, and an inch is about 2.5 centimeters, there are about 30 centimeters in a foot. We know this is a rough approximation, because we can calculate an estimate another way. A meter is 100 centimeters, and a foot is about $^1/_3$ meter, so there are roughly 33 cm in a foot. If we take a ruler and measure precisely, we find that one foot is 30.48 centimeters, in between our two estimates. Often, these quick estimates are all that is needed to understand the fundamental point. In this case, a foot is about an "order of magnitude" (a power of 10) larger than a centimeter—in other words, there are a few tens (but not hundreds) of centimeters in a foot. The telescope used to discover Pluto was just over a foot in diameter.

Scientific Notation

Scientists deal with numbers of vastly different sizes using scientific notation. Writing out 7,540,000,000,000,000,000,000 in standard notation is very inefficient, especially since most of the digits are unimportant. Scientific notation uses the first few digits (the "significant" ones) and counts the number of decimal places to create the condensed form 7.54×10^{21}. Similarly, rather than writing out 0.000000000005, we write 5×10^{-12}. The exponent on the 10 tells you where to move the decimal place. If it is positive, move the decimal place that many places to the right. If it's negative, move the decimal place that many places to the left. For example, the average distance to the Sun is 149,598,000 km, but astronomers usually express it as 1.49598×10^8 km.

- **Write the number 13.7×10^9 in standard notation.** Since the exponent of the 10 is nine, and it is positive, we must move the decimal point nine places to the right: 13,700,000,000. This number is the age of the universe in years. Usually, we would say "thirteen point seven billion years," where the word *billion* stands in for "times 10 to the nine."
- **Write the number 2×10^{-10} in standard notation.** Since the exponent of the 10 is 10, and it is negative, we must move the decimal point 10 places to the left: 0.0000000002. This is about the size of an atom in meters.
- **Write the number 50,000 in scientific notation.** To put this number in scientific notation, we must move the decimal place four places to the left: 5×10^4. This is about the radius of the Milky Way Galaxy, in light-years. To convert back to standard notation, we would move the decimal place to the right, so the exponent is positive.
- **Write the number 0.000000570 in scientific notation.** To put this in scientific notation, we must move the decimal place seven places to the right: 5.7×10^{-7}. We could also write this as 570×10^{-9}. Why would we do that? Because we have special names for units every three decimal places. The special name for 10^{-9} meters is the nanometer (abbreviated nm). So 570×10^{-9} is 570 nanometers, or 570 nm. This is the wavelength of yellow light. The *nano–* part of this special name is called a prefix. Other useful prefixes can be found inside the cover of the book.

Calculator hint: How do you put numbers in scientific notation into your calculator? Most scientific calculators have a button that says EXP or EE. These mean "times 10 to the." So for 4×10^{12}, you would type [4][EXP][1][2] or [4][EE][1][2] into your calculator. Usually, this number shows up in the window on your calculator either just as you see it written in this book, or as a 4 with a smaller 12 all the way over on the right of the window.

1.4 Thinking Like an Astronomer

Almost everyone harbors a spark of interest in astronomy. Because you are reading this book, you probably share this spark as well. The prominence of the Sun, Moon, and stars in cave paintings and rock drawings (such as those in **Figure 1.10**) dating back thousands of years tells us that these objects have long occupied the human imagination. Your initial interest may have grown over the years as you saw or read news reports about spectacular discoveries made in your lifetime—some so amazing that they seemed to blur the line between science fact and science fiction.

If you are like many people, your conception of astronomy may not yet have gone much beyond learning about the constellations and the names of the stars in them. We hope this book will take you to places you never imagined going and will lead you to insights and understandings you never imagined having.

Reading the book is only part of the journey. This is only a guidebook. We can lead you, but *you* have to walk the path! If you become an active participant in this adventure, then what you gain from the journey will remain a part of you long after the final exam is forgotten. And if along the way you find yourself applying your understanding to new situations and new information, and if you learn to combine your understandings and arrive at new insights that are greater than the sum of their parts, then you will have learned something far more than just astronomy.

We may ask you to think in ways that are different from the ways you are accustomed to thinking, and to learn to view the world from new and unfamiliar perspectives. Remember that building understanding is always an *active* process. Here are a few practical suggestions for using this text:

Figure 1.10 Ancient petroglyphs often include depictions of the Sun, Moon, and stars.

- **Read the text actively.** Pause to think about each section after you have read it. What major concepts were discussed? How are they related to what you have read so far? Have you run into similar concepts elsewhere? *In your class notes, briefly summarize the section and your thoughts about it.* At the end of the chapter, check to see if you can perform the learning goals.
- **Visualize the concept.** Visualizing physical and mathematical concepts is often the key to understanding them. Review the photos, diagrams, and charts in each chapter to study key concepts. *Draw a picture of the concept.* If you understand a concept well enough to draw a picture that expresses it, then you probably understand the concept fairly well. Trying to sketch a picture will also help you better identify what you understand and what you do not.
- **Ask yourself "What if?"** when trying to understand a concept. What if Earth were more massive? How would that affect Earth's gravity? What if the Sun were hotter? How would that affect the color of the light from the Sun or the amount of energy that the Sun radiates? If you cannot "what if" a concept, then you probably do not really understand it yet.
- **Try to teach it.** You often do not really understand something until you try to explain it to someone else. At the end of a reading assignment, talk about the material with a friend or family member. Find a partner or a group in your class with whom you can meet, and *take turns teaching each other the material.* Ask each other questions—the harder the better! Make a game of "stump the instructor." Even explaining a concept out loud to yourself can help.

- **Share your ideas and insights** with your discussion group. If a concept really "clicks" for you, or if you really think it is cool, share both your understanding and your enthusiasm.
- **Focus your discussion on concepts, relationships, and connections.** You have to know the facts, but the facts are the starting point, not the end. Use the key concepts, study questions, and other study aids to identify and concentrate your effort on the most important ideas.
- **Be honest with yourself** about what you understand and what you do not. Do not avoid concepts that you find difficult. Facing and mastering these challenging concepts are important steps in this journey. You will find that personal growth comes from this real accomplishment.
- **Do not let discomfort with math keep you from succeeding.** Mathematical formulas are expressions of logical ideas. We will always present a verbal discussion of an idea. Begin by focusing on this discussion. After you grasp the idea, then look at the math and any equations presented in "Working It Out" boxes. Study how the relationships between the quantities in the equation embody the concept and follow along with the worked examples. If you need help with math skills, visit StudySpace (wwnorton.com/studyspace) for other aids available to you. Your instructor can also help.

Each chapter includes tools to help guide you on your journey. "Learning Goals" appear at the beginning of each chapter and are a checklist you can use to test your understanding after you have read the chapter. "Working It Out" boxes provide math information with worked examples. Throughout each chapter, you will find "Trailmarkers," one-sentence summaries of fundamental concepts. "Vocabulary Alerts" identify ordinary words that scientists use in a special way. Chapter summaries can be used as reminders when you are studying. You can also compare your own section summaries to these summaries to see if you've picked out the main concepts.

Astronomy commonly appears in the news. Evaluating science news is similar to evaluating other kinds of news: you should consider the relevance of the article, the evidence presented, the qualifications and credentials of the researchers, the author, the publisher, the timeliness and comprehensiveness. This is a lot to think about all at once, but the skills you learn will help you become a more informed citizen who can make intelligent decisions about science. To help you learn these skills, we've included newspaper articles of astronomical interest throughout the book, along with follow-up questions so that you can practice reading critically.

In the 20th century, our knowledge of the universe expanded at an ever-accelerating pace, due to advances in technology and in our understanding of matter, energy, space, and time. Many fundamental questions about the origin and fate of the universe—and about the threads that tie our existence to the cosmos—moved from the realm of philosophical speculation into the realm of rigorous scientific inquiry. The insights of this age of exploration and discovery are far more profound and startling than what was dreamed of even a few decades ago.

What has yet to be determined is whether future historians will remember us for reaching out to touch the universe and embracing what we found, or whether we will instead be remembered for a loss of spirit—for stepping back and turning away from the frontier of exploration and discovery. The direction that we take from here is a decision in which you—and your willingness to embrace science—will play a part.

READING ASTRONOMY News

This article from the New York Times describes the astronomical controversy that developed around the change of Pluto's classification from "planet" to "dwarf planet." You may remember this controversy, and may even have had strong feelings about it at the time.

Pluto Is Demoted to "Dwarf Planet"

By **DENNIS OVERBYE**, *New York Times*

Pluto got its walking papers today.

Throw away the placemats. Grab a magic marker for the classroom charts. Take a pair of scissors to the solar system mobile.

After years of wrangling and a week of bitter debate, astronomers voted on a sweeping reclassification of the solar system. In what many of them described as a triumph of science over sentiment, Pluto was demoted to the status of a "dwarf planet."

In the new solar system, there are eight planets, at least three dwarf planets and tens of thousands of so-called "smaller solar system bodies," like comets and asteroids.

For now, the dwarf planets include, besides Pluto, Ceres, the largest asteroid, and an object known as UB 313, nicknamed "Xena," that is larger than Pluto and, like it, orbits out beyond Neptune in a zone of icy debris known as the Kuiper Belt. But there are dozens more potential dwarf planets known in that zone, planetary scientists say, and the number in that category could quickly swell.

In a nod to Pluto's fans, the astronomers declared Pluto to be the prototype for a new category of such "trans-Neptunian" objects but failed in a close vote to approve the name Plutonians for them.

"The new definition makes perfect sense in terms of the science we know," said Alan Boss, a planetary theorist at the Carnegie Institution of Washington, adding that it doesn't go too far in cultural terms. "We have a duty to satisfy the whole world."

The vote completed a stunning turnaround from only a week ago when the assembled astronomers had been presented with a proposal that would include 12 planets, including Pluto, Ceres, Xena and even Pluto's moon

Charon. Dr. Boss said today's decision spoke to the integrity of the planet-defining process. "The officers were willing to change their resolution and find something that would stand up under the highest scientific scrutiny and be approved," he said.

Jay Pasachoff, a Williams College astronomer who favored somehow keeping Pluto a planet, said, "The spirit of the meeting was of future discovery and activity in science rather than any respect for the past."

Mike Brown of the California Institute of Technology, who as the discoverer of Xena had the most to lose personally from Pluto's and Xena's downgrading, said he was relieved. "Through this whole crazy circus-like procedure, somehow the right answer was stumbled on," he said. "It's been a long time coming. Science is self-correcting eventually, even when strong emotions are involved."

It has long been clear that Pluto, discovered in 1930, stood apart from the previously discovered planets. Not only was it much smaller than them, only about 1,600 miles in diameter, smaller than the Moon, but its elongated orbit is tilted with respect to the other planets and it goes inside the orbit of Neptune part of its 248-year journey around the Sun.

Pluto makes a better match with the other ice balls that have since been discovered in the dark realms beyond Neptune, they have argued. In 2000, when the new Rose Center for Earth and Space opened at the American Museum of Natural History, Pluto was denoted in a display as a Kuiper Belt Object and not a planet.

Two years ago, the International Astronomical Union appointed a working group of astronomers to come up with a definition that would resolve this tension. The group, led by Iwan Williams of Queen Mary University in

London, deadlocked. This year a new group with broader roots, led by Owen Gingerich of Harvard, took up the problem.

According to the new rules a planet [must] meet three criteria: it must orbit the Sun, it must be big enough for gravity to squash it into a round ball, and it must have cleared other things out of the way in its orbital neighborhood. The latter measure knocks out Pluto and Xena, which orbit among the icy wrecks of the Kuiper Belt, and Ceres, which is in the asteroid belt.

Dwarf planets only have to be round.

"I think this is something we can all get used to as we find more Pluto-like objects in outer solar system," Dr. Pasachoff said.

The final voting came from about 400 to 500 of the 2,400 astronomers who were registered at the meeting of the International Astronomical [Union] in Prague. Many of the astronomers, Dr. Pasachoff explained, had already left, thinking there would be nothing but dry resolutions to decide in the union's final assembly.

It was hardly the first time that astronomers have rethought a planet. The asteroid Ceres was hailed as the eighth planet when it was first discovered in 1801 by Giovanni Piazzi floating in the space between Mars and Jupiter. It remained a "planet" for about half a century until the discovery of more and more things like it in the same part of space led astronomers to dub them asteroids.

In the aftermath, some astronomers pointed out that the new definition only applies to our own solar system and that there was so far no such thing as an extra-solar planet.

The decision was bound to have both a cultural and economic impact on the industry of astronomical artifacts and toys, publishing and education. The *World Book Encyclopedia*,

SEE **PLUTO**

for example, had been holding the presses for its new 2007 edition until Pluto's status could be clarified.

Neil deGrasse Tyson, director of the Hayden Planetarium in New York, said children are flexible when asked about the cultural impact of today's redefinition. He said that he had not bothered to watch the International Astronomical Union's vote in the Internet, as many astronomers did. "Counting planets is not an interesting exercise to me," he said. "I'm happy however they choose to define it. It doesn't really make any difference to me."

Dr. Tyson said the continuing preoccupation with what the public and schoolchildren would think about this was a concern and a troubling precedent. "I don't know any other science that says about its frontier, 'I wonder what the public thinks,' " he said. "The frontier should move in whatever way it needs to move."

Evaluating the News

1. Why was Pluto reclassified? Is Pluto any different now than it was before?

2. Mike Brown makes the statement in this article that "Science is self-correcting eventually, even when strong emotions are involved." What does he mean by this? How does it relate to what you learned about the nature of science in Section 1.2?

3. Is the reclassification of Pluto a unique event? Have similar "self-corrections" happened before? Are these self-corrections a weakness or a strength of science? (Consider Section 1.2.)

4. One way to evaluate scientific claims is to consider a what-if question. What if Pluto had not been reclassified? How would we classify Ceres, Charon, and the thousands of other Pluto-like objects? How many planets would there be in the Solar System if we included all the ones that are Pluto-like? Would we consider all these objects to be similar to planets like Jupiter or Earth for the purposes of studying them further?

5. How does this article fit into the scheme of the scientific method that you learned about in Section 1.2? Explain how the reclassification of Pluto is an example of the scientific method at work.

SUMMARY

1.1 Our Solar System is but a tiny speck in a vast universe. We are stardust: our particles were made at the Big Bang and were processed inside stars to become the elements that form our bodies today.

1.2 The scientific method is a way of trying to *falsify*, not prove, ideas; thus all scientific knowledge is provisional. Once we have an accumulation of facts, we can begin to form and test the ideas that lead to understanding. We know we understand when we find the underlying connections that allow us to predict the results of new observations. There is nothing special about our particular place in the universe; the entire universe is governed by the same physical laws.

1.3 Mathematics is the language of patterns, and thus it is the language of science.

◆ SUMMARY SELF-TEST

1. Our Solar System is to the universe as _____ is to three times the age of Earth.

2. The following astronomical events led to the formation of you. Place them in order of their occurrence over astronomical time.
   a. Stars die and distribute heavy elements into the space between the stars.
   b. Hydrogen and helium are made in the Big Bang.
   c. Enriched dust and gas gather into clouds in interstellar space.
   d. Stars are born, and process light elements into heavier ones.
   e. The Sun and planets form from a cloud of interstellar dust and gas.

3. The scientific method is a way of trying to _____, not prove, ideas; thus all scientific knowledge is provisional.

4. *Understanding* in science means that
   a. we have accumulated lots of facts.
   b. we are able to connect facts through an underlying idea.
   c. we are able to predict events based on accumulated facts.
   d. we are able to predict events based on an underlying idea.

5. If we compare our place in the universe with a very distant place, we can say that
   a. the laws of physics are different in each place.
   b. some laws of physics are different in each place.
   c. all of the laws of physics are the same in each place.
   d. some laws of physics are the same in each place, but we don't know about others.

6. _____ is the language of patterns, and thus it is the language of science.

QUESTIONS AND PROBLEMS

True/False and Multiple-Choice Questions

7. T/F: A scientific theory can be tested by observations and proven to be true.

8. T/F: A pattern in nature can reveal an underlying physical law.

9. T/F: A theory, in science, is a guess about what might be true.

10. T/F: Once a theory is proven in science, scientists stop testing it.

11. T/F: Astronomers use instruments from many branches of science to investigate the universe.

12. The Solar System contains
  a. classical planets, dwarf planets, asteroids, and galaxies.
  b. classical planets, dwarf planets, comets, and billions of stars.
  c. classical planets, dwarf planets, asteroids, comets, and one star.
  d. classical planets, dwarf planets, one star, and many galaxies.

13. The Sun is part of
  a. the Solar System.   b. the Milky Way Galaxy.
  c. the universe.   d. all of the above.

14. A light-year is a measure of
  a. distance.   b. time.
  c. speed.   d. mass.

15. Which of the following was *not* made in the Big Bang?
  a. hydrogen   b. lithium
  c. beryllium   d. carbon

16. Occam's razor states that
  a. the universe is expanding in all directions.
  b. the laws of nature are the same everywhere in the universe.
  c. if two hypotheses fit the facts equally well, choose the simpler one.
  d. patterns in nature are really manifestations of random occurrences.

Conceptual Questions

17. Suppose you lived on imaginary planet Zorg that orbits Alpha Centauri, a nearby star in our galaxy. How would you write your cosmic address?

18. Draw a diagram representing your cosmic address. How can you best show the difference in size scales represented in the pieces of this address? How do these differences in scale compare to the differences in scale in your postal address?

19. Figure 1.2 gives examples of time scales that are analogous to the distance scales in the universe. Pick one of the distances in the figure, and think of a different analogy. That is, think of a different activity that takes the same amount of time.

20. Examine Figure 1.2. When an event occurs on the Sun, how soon after do we know about it?

21. When a star explodes in the Andromeda Galaxy, how long until we see it here on Earth? (Hint: study Figure 1.2.)

22. Scientists say that we are "made of stardust." Explain what they mean.

23. Author Erich von Däniken claims that Earth was visited by extraterrestrials (that is, aliens) in the remote past. Can you think of any tests that could support or refute his hypothesis? Is it falsifiable? Would you regard this as science or pseudoscience?

24. What does the word *falsifiable* mean? Give an example of an idea that is not falsifiable. Give an example of an idea that is falsifiable.

25. Explain how the word *theory* is used differently by a scientist than in common everyday language.

26. What is the difference between *hypothesis* and *theory*?

27. Astronomers commonly work with scientists in other fields. Use what you have learned about the scientific method to predict what would happen if a discrepancy were found between fields.

28. Suppose the tabloid newspaper at your local supermarket claims that, compared to average children, children born under a full Moon become better students.
  a. Is this theory falsifiable?
  b. If so, how could it be tested?

29. A textbook published in 1945 stated that it takes 800,000 years for light to reach us from the Andromeda Galaxy. In this book we say that it takes 2,500,000 years. What does this tell you about a scientific "fact" and how our knowledge changes with time?

30. Astrology makes testable predictions. For example, it predicts that the horoscope for your star sign on any day should fit you better than horoscopes for other star signs. Without indicating which sign is which, read the daily horoscopes to a friend, and ask how many of them might fit the day that your friend had yesterday. Repeat the experiment every day for a week and keep records. Did one particular horoscope sign consistently describe your friend's experiences?

31. A scientist on television states that it is a known fact that life does not exist beyond Earth. Would you consider this scientist reputable? Explain your answer.

32. Some astrologers use elaborate mathematical formulas and procedures to predict the future. Does this mean that astrology is a science? Why or why not?

33. Imagine yourself living on a planet orbiting a star in a very distant galaxy. What does the cosmological principle tell you about the way you would perceive the universe from this distant location?

34. Describe a pattern in your own life that repeats regularly. How does this pattern affect you? Is this pattern of your own making, set by others, or determined by nature?

35. Should Pluto have been reclassified as a dwarf planet? List two arguments for and two arguments against. Evaluate these arguments on their scientific merit. Then evaluate them on some other basis (cultural, historical, emotional). How do these two evaluations compare?

36. You run across an old newspaper with the headline "EINSTEIN PROVES NEWTON WRONG!" Did the newspaper get this story right? Explain your answer.

37. Find an astronomy article in your local newspaper or online. Summarize the article and analyze it. What has happened and why? Is the source of the article reputable? Can you think of a what-if question? Are the claims in the article scientific (that is, falsifiable)?

38. Draw a cartoon that expresses your reaction to the size of the universe, as discussed in this chapter.

39. Make a concept map from the information presented in Section 1.2. Use this map to test your understanding of the scientific use of the word *theory*.

40. Some people claim that a tray of hot water will freeze more quickly than a tray of cold water when both are placed in a freezer.
    a. Does this claim make sense to you?
    b. Is the claim falsifiable?
    c. Do the experiment yourself. Note the results. Was your intuition borne out?

Problems

41. Estimate the number of stars in the Local Group. (Assume that the number of stars in all the dwarf galaxies in our Local Group is negligible compared to the number of stars in our Milky Way. Assume also that the Andromeda Galaxy and the Milky Way are similar.)

42. Astronomers often express distances in terms of the amount of time it takes light to travel that distance. We call such distances light-minutes, light-hours, or light-years. Others do this as well. Express the distance to a nearby town in terms of both the distance in miles and the time it takes to get there by car traveling at 60 mph (car-hours). Express the distance to a nearby friend's house in terms of distance and in terms of time. Invent your own units for the travel time measurement to your friend's house.

43. New York is 2,444 miles from Los Angeles. What is that distance in car-hours? In car-days? (Assume a reasonable number for the speed limit.) What if you had to walk it? How far is New York from Los Angeles in foot-days, foot-months, or foot-years?

44. (a) It takes about 8 minutes for light to travel from the Sun to Earth. Pluto is 40 times as far from us as the Sun. How long does it take light to reach Earth from Pluto? (b) Radio waves travel at the speed of light. What problems would you have if you tried to conduct a two-way conversation between Earth and a spacecraft orbiting Pluto?

45. The times that it takes light to travel across the diameter of the Milky Way Galaxy, from the Milky Way Galaxy to the Andromeda Galaxy, and halfway across the universe are all given in Figure 1.2. Express each of these travel times of light in standard notation and scientific notation.

46. Astronomers use the speed of light and the time it takes light to travel a given distance to express astronomical distances. From the travel time of light from the Milky Way to the Andromeda Galaxy, find the distance to the Andromeda Galaxy in kilometers and then in miles. Express both numbers first in standard notation and then in scientific notation.

47. Find the diameter of the Milky Way Galaxy in miles. You will need to reference Figure 1.2 to find the travel time of light across the galaxy. Do you have a "feel" for this number?

48. One way to get a feel for very large numbers is to "scale them" to everyday objects, much like making a map of a large area on a small piece of paper. Imagine the Sun is the size of a grain of sand and Earth a speck of dust 0.083 meters (83 mm) away. (Each light-minute of distance is represented by 10 mm.) On this scale, how many millimeters (mm) represent a light-second? How many represent a light-hour? On this scale, what is the distance from Earth to the Moon? From the Sun to Neptune?

49. Using the same scale as problem 48 (1 light-minute equals 10 mm), how many millimeters represent a light year? On this scale, what is the distance from Earth to the nearest star? To the Andromeda Galaxy? (Note that 1 meter = 10^3 mm and that 1 km = 10^3 meters.)

50. The average distance from Earth to the Moon is 384,000 km. How many hours would it take, traveling at 800 kilometers per hour (km/h)—the typical speed of jet aircraft—to reach the Moon? How many days is this? How many months?

51. The average distance from Earth to the Moon is 384,000 km. In the late 1960s, astronauts reached the Moon in about three days. How fast must they have been traveling (in km/hr) to cover this distance in this time? Compare this speed to the speed of a jet aircraft (800 km/hr).

52. Estimate your height in meters.

53. How many nanometers are there in a micrometer?

54. How many centimeters are there in a mile?

55. Write 86,400 (the number of seconds in a day) and 0.0123 (the Moon's mass compared to Earth's) in scientific notation.

56. Write 1.60934×10^3 (the number of meters in a mile) and 9.154×10^{-3} (Earth's diameter compared to the Sun's) in standard notation.

57. The information from an Internet web host may sometimes reach your computer by traveling as light to a satellite approximately 3.6×10^7 meters above Earth's surface and back again. What is the minimum time, in seconds, that the information takes to reach your computer?

SmartWork, Norton's online homework system, includes algorithmically generated versions of these questions, plus additional conceptual exercises. If your instructor assigns questions in SmartWork, log in at **smartwork.wwnorton.com**.

StudySpace is a free and open website that provides a Study Plan for each chapter of **Understanding Our Universe**. Study Plans include animations, reading outlines, vocabulary flashcards, and multiple-choice quizzes, plus links to premium content in SmartWork and the ebook. Visit **wwnorton.com/ studyspace**.

Exploration | Logical Fallacies

Logic is fundamental to the study of science and to scientific thinking. A logical fallacy is an error in reasoning, which good scientific thinking avoids. For example, "because Einstein said so" is not an adequate argument. No matter how famous the scientist is (even if he is Einstein), he must still supply a logical argument and evidence to support his claim. When someone claims that something must be true because Einstein said it, they have committed the logical fallacy known as an appeal to authority. There are many types of logical fallacies, but a few of them crop up often enough in discussions about science that you should be aware of them.

Ad hominem. In an *ad hominem* fallacy, you attack the person who is making the argument, instead of the argument itself. Here is an extreme example of an ad hominem argument: "A famous politician says Earth is warming. But I think this politician is an idiot. So Earth can't be warming."

Appeal to belief. This fallacy has the general pattern "Most people believe X is true, therefore X is true." So, for example, "Most people believe Earth orbits the Sun. Therefore Earth orbits the Sun." Note that even if the conclusion is correct, you may still have committed a logical fallacy in your argument.

Begging the question. In this fallacy (also known as circular reasoning), you assume the claim is true and then use this assumption as your evidence to prove the claim is true. For example, "I am trustworthy, therefore I must be telling the truth." No real evidence is presented for the conclusion.

Biased sample. If a sample drawn from a larger pool has a bias, then conclusions about the sample cannot be applied to the larger pool. For example, imagine you poll students at your university and find that 30 percent of them visit the library one or more times per week. Then you conclude that 30 percent of Americans visit the library one or more times per week. You have committed the biased sample fallacy, because university students are not a representative sample of the American public.

Post hoc ergo propter hoc. *Post hoc ergo propter hoc* is Latin for "After this, therefore because of this." Just because one thing follows another doesn't mean that one caused the other. For example, "There was an eclipse and then the king died. Therefore, the eclipse killed the king." This fallacy is often connected to the inverse reasoning: "If we can prevent an eclipse, the king won't die."

Slippery slope. In this fallacy, you claim that a chain reaction of events will take place, inevitably leading to a conclusion that no one could want. For example, "If I fail astronomy, I will not ever be able to get a college degree, and then I won't ever be able to get a good job, and then I will be living in a van down by the river until I'm old and toothless and starve to death." None of these steps actually follows inevitably from the one before.

Following are some examples of logical fallacies. Identify the type of fallacy represented. Each of the fallacies is represented once.

1. **You get a chain email, threatening terrible consequences if you break the chain. You move it to your spam box. On the way home, you get in a car accident. The following morning, you retrieve the chain email and send it along.**

2. **If I get question one on the assignment wrong, then I'll get question two wrong as well, and before you know it, I will fail the assignment.**

3. **All my friends love the band Degenerate Electrons. Therefore, all people my age love this band.**

4. **Eighty percent of Americans believe in the Tooth Fairy. Therefore, the Tooth Fairy exists.**

5. **My professor says that the universe is expanding. But my professor is a nerd, and I don't like nerds. So the universe can't be expanding.**

6. **When applying for a job, you use a friend as a reference. When your prospective employer asks you how they know your friend is trustworthy, you say, "I can vouch for her."**

2

Patterns in the Sky— Motions of Earth

The stars first found a special place in legend and mythology as the realm of gods and goddesses, holding sway over the lives of humankind. From these legends and myths, the constellations were born—patterns in the sky that we told stories about to help us keep track of passing time. The coming of night and day, the changing seasons, and the rising and falling tides all mirror other patterns in the sky. Through careful observation, our distant ancestors found that they could use these changing patterns to predict when the seasons would change and the rains would come. Knowledge of the sky offered knowledge of the world, and knowledge of the world is power.

The patterns that captured the attention and imagination of our ancestors still serve as beacons on dark, cloudless nights. Yet unlike our ancestors, we see them with the perspective of centuries of hard-won knowledge, and we can explain how patterns of change in the sky are the unavoidable consequences of the motions of Earth and the Moon. This discovery is an example of science at its best, showing us the way outward into a universe far more vast and awe-inspiring than our ancestors could ever have imagined.

✦ LEARNING GOALS

In this chapter, we look at patterns both in the sky and on Earth. Then we look beyond appearances to the underlying motions that cause patterns. A student is working this out for herself in the figure at right. Photos of the Moon taken from Earth at different times during its cycle show that its appearance changes. The sketches show how this appearance is related to the positions of the Earth, Moon, and Sun. By the end of this chapter, you should be able to reproduce such a figure, draw similar diagrams for other photos you might see, and understand why the appearance of the Moon changes as it does. You should also be able to:

- Explain how the stars appear to move through the sky as Earth rotates on its axis, and how those motions differ when seen from different latitudes on Earth

- Visualize how Earth's motion around the Sun and the tilt of Earth's axis relative to the plane of its orbit combine to determine which stars we see at night and which seasons we experience through the year

- Connect the motion of the Moon in its orbit about Earth to the phases we observe and to the spectacle of eclipses

- Understand the fundamental concept of a frame of reference, and how Earth's rotating frame of reference affects our perception of celestial motions

- Describe the elliptical orbits of planets around the Sun

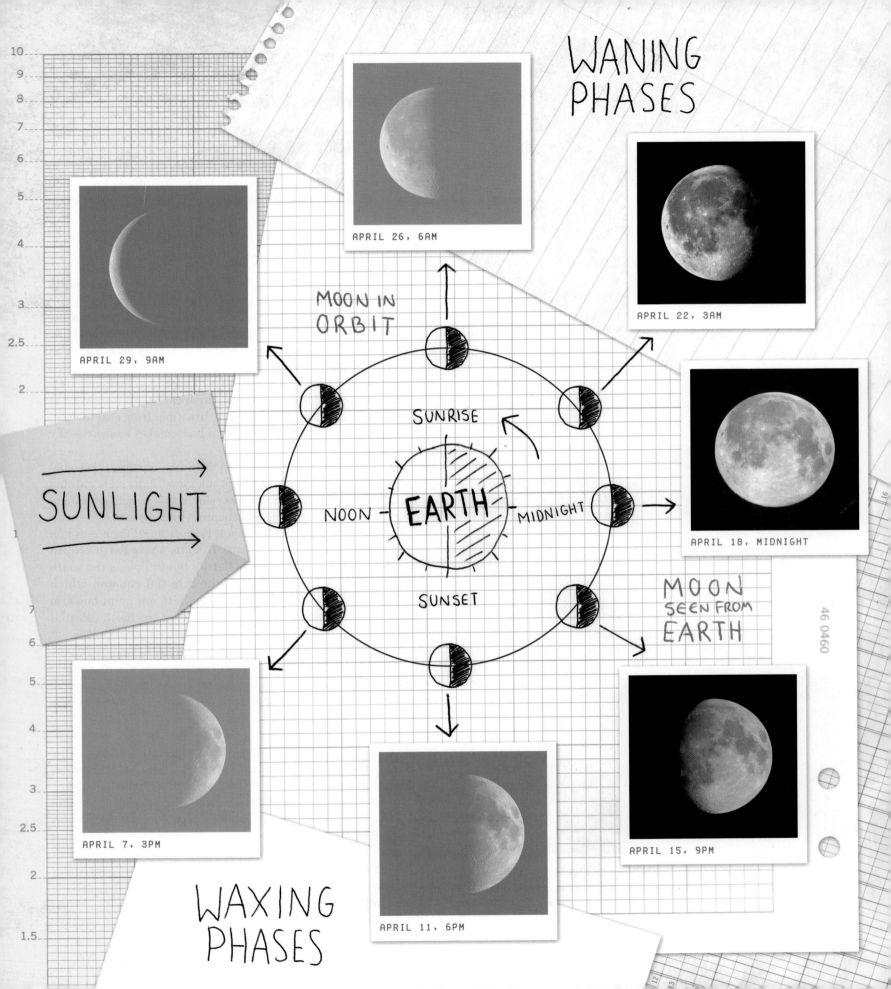

WANING PHASES

WAXING PHASES

SUNLIGHT

MOON IN ORBIT

MOON SEEN FROM EARTH

SUNRISE

NOON — EARTH — MIDNIGHT

SUNSET

APRIL 26, 6AM

APRIL 22, 3AM

APRIL 18, MIDNIGHT

APRIL 15, 9PM

APRIL 11, 6PM

APRIL 7, 3PM

APRIL 29, 9AM

46 0460

2.1 Earth Spins on Its Axis

Long before Christopher Columbus journeyed to the New World, Aristotle and other Greek philosophers knew that Earth is a sphere. The theory that motions of Earth cause daily and monthly changes in the sky was far more difficult to accept because Earth seems stationary. However, we have come to know that Earth's rotation on its axis sets the very rhythm of life on Earth—the passage of day and night.

The Celestial Sphere Is a Useful Fiction

The ancients could not perceive the spinning motion of Earth, and so they did not believe it rotated. In fact, as Earth rotates its surface is moving quite fast—about 1,670 kilometers per hour (km/h) at the equator. We do not "feel" that motion any more than we would feel the speed of a car with a perfectly smooth ride cruising down a straight highway. Nor do we feel the *direction* of Earth's spin, although it is revealed by the hourly motion of the Sun, Moon, and stars across the sky. As viewed from above Earth's **North Pole**, Earth rotates counterclockwise (**Figure 2.1**) once each 24-hour period. As the rotating Earth carries us from west to east, objects in the sky *appear* to move in the other direction, from east to west. As seen from Earth's surface, the path each celestial body makes across the sky is called its apparent daily motion.

To help visualize the apparent daily motions of the Sun and stars, it is sometimes useful to think of the sky as if it were a huge sphere with the stars painted on its surface and Earth at its center. Astronomers refer to this imaginary sphere as the **celestial sphere (Figure 2.2)**. The celestial sphere is useful because it is easy to draw and visualize, but always remember that it is imaginary! Each point on the celestial sphere indicates a *direction* in space. Directly above Earth's North Pole is the **north celestial pole**. Directly above Earth's South Pole is the **south celestial pole**. Directly above Earth's equator is the **celestial equator**, which divides the sky into a northern half and a southern half. If you point one arm toward a point on the celestial equator and one arm toward the north celestial pole, your arms will always form a right angle, so the north celestial pole is 90° away from the celestial equator. The angle between the celestial equator and the south celestial pole is also 90°. Take a moment to think about this. You might want to draw an equator and north and south pole on an orange, and visualize those markings projected onto the walls of your room (**Figure 2.3**). (We will use this orange again throughout this chapter, so don't eat it!)

Wherever you are, you can also divide the sky into an east half and a west half with an imaginary north–south line called the **meridian (Figure 2.4)**. This line runs from due north through a point directly overhead, called the **zenith**, to a point due south. It then continues around the far side of the celestial sphere, through the **nadir** (the point directly below you), and back to the starting point due north. Take a moment to visualize this. You may want to draw a little person on your orange, and visualize that person's meridian on the walls of your room. Where is the zenith for that person? Where is the nadir? How is the meridian oriented relative to the celestial equator? Now, think about the meridian for you, where you are today, and point at all the locations you've just learned about: north celestial pole, zenith, celestial equator, south celestial pole, nadir. When you can do all this, you have oriented yourself to the sky.

Earth rotates on its axis,
causing day and night.

▶⏸ AstroTour: **The Earth Spins and Revolves**

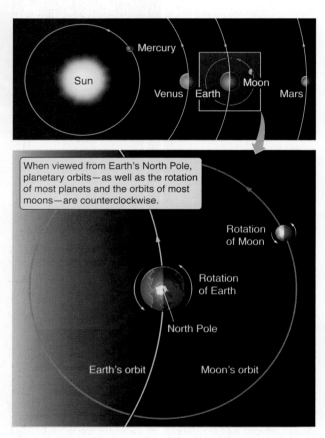

When viewed from Earth's North Pole, planetary orbits—as well as the rotation of most planets and the orbits of most moons—are counterclockwise.

Figure 2.1 The rotation of Earth and the Moon, the revolution of Earth and the planets about the Sun, and the orbit of the Moon about Earth are counterclockwise as viewed from above Earth's North Pole. (Not drawn to scale.)

To see how to use the celestial sphere, let's consider the Sun at noon and at midnight. Astronomers say that "local noon" occurs when the Sun crosses the meridian at their location. This is the highest point above the horizon that the Sun will reach on any given day. (This highest point is almost never the zenith. You have to be in a special place on a special day for the Sun to be directly over your head at noon; for example, at 23.5° north latitude on June 21.) "Local midnight" occurs when the Sun again crosses the meridian on the other side of Earth. From our perspective on Earth, the celestial sphere appears to rotate, carrying the Sun across the sky to its highest point at noon and around through the meridian again at midnight. What is really happening? The Sun remains in the same place in space throughout the entire 24-hour period. Earth rotates, so that our spot faces a different direction at every moment. At noon, our spot on Earth has rotated to face most directly toward the Sun. Half a day later, at midnight, our spot on Earth has rotated to face most directly away from the Sun.

The View from the Poles

The apparent daily motions of the stars and the Sun depend on where you live. The apparent daily motions of celestial objects in northern Europe, for example, are quite different from the apparent daily motions seen from an island in the tropics. To understand why your location matters, let's examine the special case of the North Pole. This is known as a limiting case, meaning we are looking at a limit where we've gone as far as we can go in some way. In this case, we've gone as far north as possible, which puts us in a special place with the north celestial

Vocabulary Alert

Horizon: In common language, the horizon is the line where the sky meets the ground. Astronomers mean the line you would see if you held your eyes perfectly level and turned all the way around. A line to the horizon always makes a right angle with a line to the zenith.

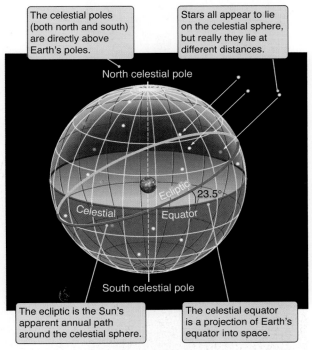

The celestial poles (both north and south) are directly above Earth's poles.

Stars all appear to lie on the celestial sphere, but really they lie at different distances.

North celestial pole

Ecliptic

23.5°

Celestial Equator

South celestial pole

The ecliptic is the Sun's apparent annual path around the celestial sphere.

The celestial equator is a projection of Earth's equator into space.

Figure 2.2 The celestial sphere is a useful fiction for thinking about the appearance and apparent motion of the stars in the sky.

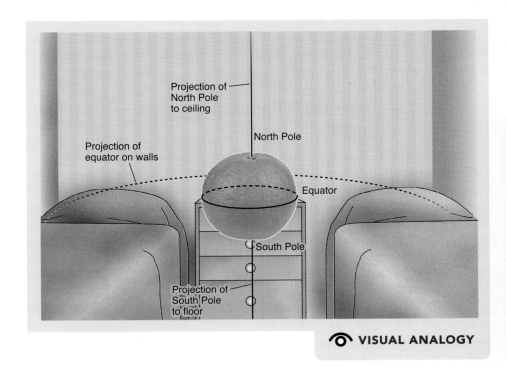

Projection of North Pole to ceiling

North Pole

Projection of equator on walls

Equator

South Pole

Projection of South Pole to floor

👁 **VISUAL ANALOGY**

Figure 2.3 You can draw poles and an equator on an orange, and imagine these points projected onto the walls of your room. Similarly, we imagine the poles and equator of Earth projected onto the celestial sphere. (Not drawn to scale.)

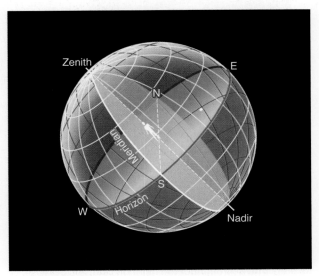

Zenith

E

N

Meridian

S

W   Horizon

Nadir

Figure 2.4 Principal features of an observer's coordinate system projected onto the celestial sphere.

(a)

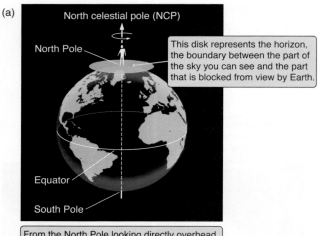

North celestial pole (NCP)

North Pole

This disk represents the horizon, the boundary between the part of the sky you can see and the part that is blocked from view by Earth.

Equator

South Pole

From the North Pole looking directly overhead, the **north celestial pole** is at the zenith.

(b)

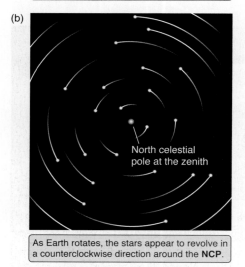

North celestial pole at the zenith

As Earth rotates, the stars appear to revolve in a counterclockwise direction around the **NCP**.

(c)

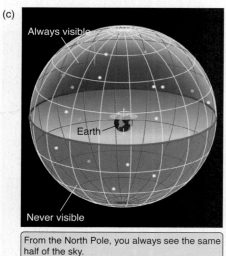

Always visible

Earth

Never visible

From the North Pole, you always see the same half of the sky.

Figure 2.5 (a) An observer standing at the North Pole sees (b) stars move throughout the night on counterclockwise, circular paths about the zenith. (c) The same half of the sky is always visible from the North Pole.

pole directly overhead at the zenith. One way scientists test their theories is by considering such limiting cases.

Imagine that you are standing on the North Pole watching the sky as shown in **Figure 2.5a**. (Ignore the Sun for the moment and pretend that you can always see stars in the sky.) You are standing where Earth's axis of rotation intersects its surface, which is much the same as standing at the center of a rotating merry-go-round. As Earth rotates, the spot directly above you remains fixed over your head while everything else in the sky appears to revolve in a counterclockwise direction around this spot (**Figure 2.5b**). (If you are having trouble visualizing this, use your orange. As you spin it, imagine standing at the pole and looking at the objects in your room.) Notice that objects close to the pole appear to follow small circles, while the largest circles are followed by objects nearest to the horizon.

Everywhere on Earth, all the time, half of the sky is "below the horizon"; your view is blocked by Earth itself. The view from the North Pole is special because nothing rises or sets each day as Earth turns; from there we always see the *same* half of the celestial sphere (**Figure 2.5c**). Everywhere but the poles, the visible half changes as Earth rotates. In contrast, Earth's North Pole points in the *same* direction, hour after hour and day after day. For this reason, the objects visible there follow circular paths that always have the same **altitude**, or angle above the horizon.

The view from Earth's **South Pole** is much the same, but with two major differences. First, the South Pole is on the opposite side of Earth from the North Pole, so the visible half of the sky at the South Pole is precisely the half that is hidden from the North Pole. The second difference is that instead of appearing to move counterclockwise around the sky, stars appear to move *clockwise* around the south celestial pole. Think about this for a moment to completely visualize it. It might help to sit in a swivel chair and spin it around from right to left. As you look at the ceiling, things appear to move in a counterclockwise direction; but as you look at the floor, they appear to be moving clockwise.

Away from the Poles, the Part of the Sky We See Is Constantly Changing

Now let's imagine that we leave the North Pole and travel south to lower latitudes. **Latitude** is a measure of how far north or south we are on the face of Earth. Imagine a line from the center of Earth to your location on the surface of the planet. Now imagine a second line from the center of Earth to the point on the equator closest to you. (Refer to **Figure 2.6** for help imagining these lines.) The angle between these two lines is your latitude. At the North Pole, these two imaginary lines form a 90° angle. At the equator, they form a 0° angle. So the latitude of the North Pole is 90° north, and the latitude of the equator is 0°. The South Pole is at latitude 90° south.

Our latitude determines the part of the sky that we can see throughout the year. At the North Pole, our horizon makes a 90° angle with the north celestial pole, at our zenith. As we move south, our horizon tilts and our zenith moves away from the north celestial pole. At a latitude of 60° north (as shown in Figure 2.6b), our horizon is tilted 60° from the north celestial pole. In Figure 2.6d, we have reached Earth's equator, at a latitude of 0°. The horizon is tilted 0° from the north celestial pole. This relationship between these angles remains the same everywhere in between as well. Also at the equator, we get our first look at the south celestial pole, on the southern horizon. Continuing into the Southern Hemisphere, the south celestial pole is now visible above the southern horizon, while the north celestial pole is hidden from view by the northern horizon. At a latitude of 45° south (Figure 2.6e), the south celestial pole has an altitude of 45°. At the South Pole (latitude 90° south—Figure 2.6f), the south celestial pole is at the zenith, 90° above the horizon.

Probably the best way to cement your understanding of the view of the sky at different latitudes is to draw pictures like those in Figure 2.6. If you can draw a picture like this for any latitude—filling in the values for each of the angles in the drawing and imagining what the sky looks like from that location—then you will be well on your way to developing a working knowledge of the appearance of the sky. That knowledge will prove useful later when we discuss a variety of phenomena, such as the changing of the seasons. When practicing

Vocabulary Alert

Altitude: In common language, altitude is the height of an object, such as an airplane, above the ground. Astronomers use the word to refer to the angle formed between an imaginary line from an observer to an object and a second line from the observer to the point on the horizon directly below the object.

▶❚❚ **AstroTour: The View from the Poles**

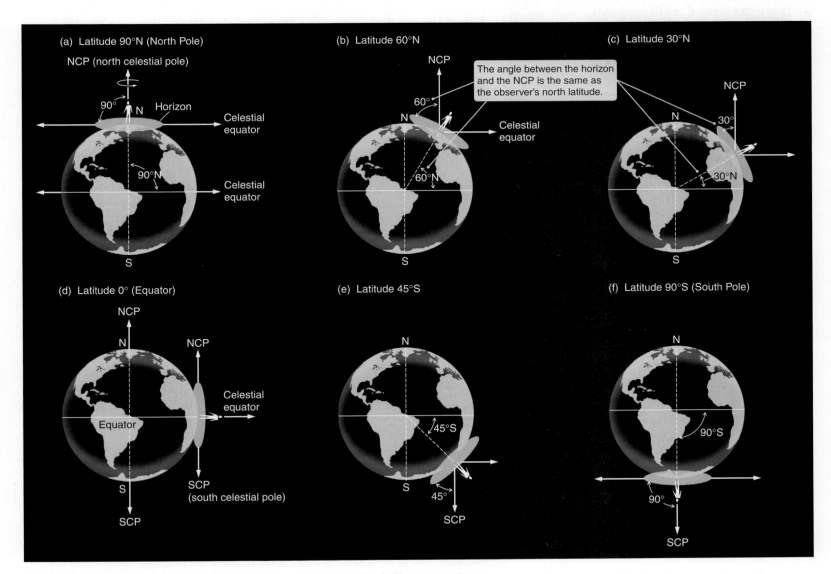

Figure 2.6 Our perspective on the sky depends on our location on Earth. Here we see how the locations of the celestial poles and celestial equator depend on an observer's latitude.

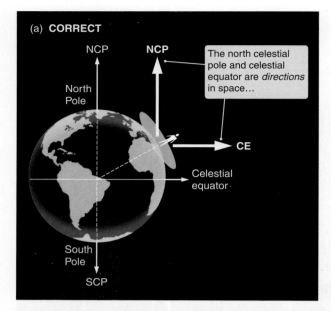

(a) **CORRECT**

The north celestial pole and celestial equator are *directions* in space...

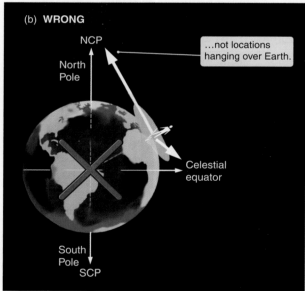

(b) **WRONG**

...not locations hanging over Earth.

Figure 2.7 The celestial poles and the celestial equator are directions in space, not fixed locations hanging above Earth. (The red *X* means "avoid this misconception.")

▶❙❙ AstroTour: **The Celestial Sphere and the Ecliptic**

Circumpolar stars are always above the horizon.

your sketches, however, take care not to make the common mistake illustrated in **Figure 2.7**. *The north celestial pole is not a location* in space, hovering over Earth's North Pole. Instead, *it is a direction* in space—the direction parallel to Earth's axis of rotation.

Now that we have shown how the horizon is oriented at different latitudes, let's see how the apparent motions of the stars about the celestial poles differ from latitude to latitude. **Figure 2.8a** shows our view if we are at latitude 30° north. As Earth rotates, the visible part of the sky constantly changes. From this perspective the horizon appears fixed, and the stars appear to move. If we focus our attention on the north celestial pole, we see much the same thing we saw from Earth's North Pole. The north celestial pole remains fixed in the sky, and all of the stars appear to move throughout the night in counterclockwise, circular paths around that point.

Stars located close enough to the north celestial pole never dip below the horizon. How close is close enough? Remember that the latitude angle is equal to the altitude of the north celestial pole. Stars within this angle from the north celestial pole never dip below the horizon (even if we can't see them while the Sun is up) as they complete their apparent paths around the pole (see Figure 2.8a and **Figure 2.9**). These stars are called **circumpolar** ("around the pole") stars. Another group of stars, close to the south celestial pole, never rise above the horizon and can *never* be seen from this latitude. And stars between this region and the circumpolar region can be seen for *only part* of each day. Stars in this intermediate region appear to rise above and set below Earth's shifting horizon as Earth turns. The only place on Earth where you can see the entire sky over the course of 24 hours is the equator. From the equator (**Figure 2.8b**), the north and south celestial

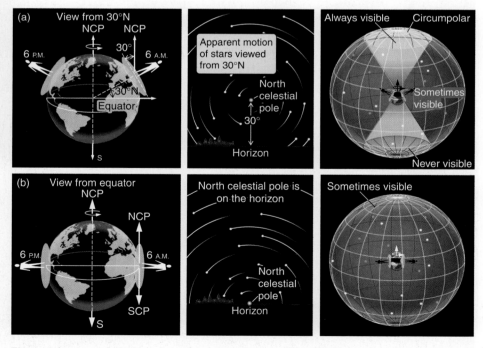

Figure 2.8 (a) As viewed from latitude 30° north, the north celestial pole is 30° above the northern horizon. Stars appear to move on counterclockwise paths around this point. At this latitude some parts of the sky are always visible, while others are never visible. (b) From the equator, the north and south celestial poles are seen on the horizon, and the entire sky is visible over a period of 24 hours.

poles sit on the northern and southern horizons, respectively, and the whole of the heavens passes through the sky each day.

The celestial equator intersects the horizon due east and due west (**Figure 2.10**). (The exception is at the poles, where the celestial equator lies along the horizon and so doesn't intersect the horizon at all.) A star on the celestial equator rises due east and sets due west. Stars located north of the celestial equator on the celestial sphere rise north of east and set north of west. Stars located south of the celestial equator rise south of east and set south of west.

Regardless of where you are on Earth (again with the exception of the poles), half of the celestial equator is always visible above the horizon. You can therefore see any object that lies along the celestial equator half of the time. An object that is in the direction of the celestial equator rises due east, is above the horizon for 12 hours, and sets due west. This is not true for objects that are not on the celestial equator. Figure 2.10b shows that from the Northern Hemisphere, you can see more than half of the apparent circular path of any star that is north of the celestial equator. If you can see more than half of a star's path, then the star is above the horizon for more than 12 hours each day.

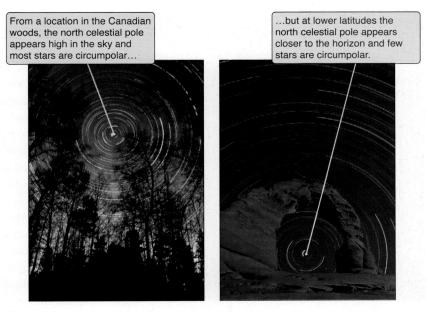

From a location in the Canadian woods, the north celestial pole appears high in the sky and most stars are circumpolar…

…but at lower latitudes the north celestial pole appears closer to the horizon and few stars are circumpolar.

Figure 2.9 Time exposures of the sky showing the apparent motions of stars through the night. Note the difference in the circumpolar portion of the sky as seen from the two different latitudes.

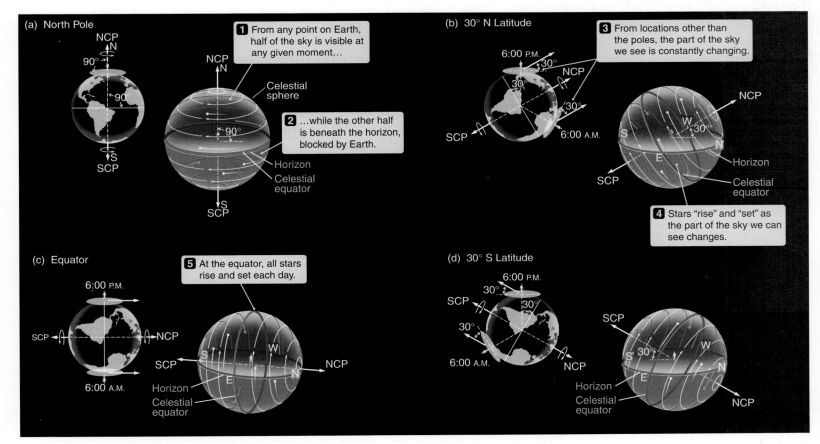

Figure 2.10 The celestial sphere is shown here as viewed by observers at four different latitudes. At all locations other than the poles, stars rise and set as the part of the celestial sphere that we see changes during the day.

As seen from the Northern Hemisphere, stars north of the celestial equator remain above the horizon for more than 12 hours each day. The farther north the star is, the longer it stays up. Circumpolar stars are the extreme example of this phenomenon; they are up 24 hours a day. In contrast, stars south of the celestial equator are above the horizon for less than 12 hours a day. The farther south a star is, the less time it stays up. Stars located close to the south celestial pole never rise above the horizon in the Northern Hemisphere.

If you are in the Southern Hemisphere (Figure 2.10d), the same phenomenon holds with the directions reversed. Stars on the celestial equator are still up for 12 hours a day; but stars south of the celestial equator stay up for more than 12 hours, and stars north of the celestial equator stay up for less than 12 hours.

Since ancient times, travelers including sailors at sea have used the stars for navigation. We can find the north or south celestial poles by recognizing the stars that surround them. In the Northern Hemisphere, a moderately bright star happens to be located close to the north celestial pole. This star is called Polaris, the "North Star." If you can find Polaris in the sky and measure its altitude, then you know your latitude. If you are in Phoenix, Arizona, for example (latitude 33.5° north), the north celestial pole has an altitude of 33.5°. In Fairbanks, Alaska (latitude 64.6° north), the altitude of the north celestial pole is 64.6°. Once a navigator locates the North Star, she not only knows what direction north is (and therefore also south, east, and west), but she also knows her latitude. She can determine whether she needs to head north or south to reach her destination. Figuring out your *longitude* (east–west location) astronomically is much more complicated because of Earth's eastward rotation.

▶❙❙ **AstroTour: The Earth Spins and Revolves**

Earth orbits the Sun, causing annual changes.

2.2 Revolution About the Sun Leads to Changes during the Year

Earth's average distance from the Sun is 1.50×10^8 km. This distance is called an **astronomical unit** (AU) and is used for measuring distances in the Solar System. Earth orbits the Sun in a nearly circular orbit in the same direction that Earth spins about its axis—counterclockwise as viewed from above Earth's North Pole. Earth orbits once around the Sun in one **year**, by definition. This motion is responsible for many of the patterns we see in the sky and on Earth, such as the night-to-night changes in the stars we see overhead. As Earth **revolves** around the Sun, our view of the night sky changes. Six months from now, Earth will be on the other side of the Sun. The stars overhead at midnight will be those in the opposite direction from the stars overhead at midnight tonight. The stars overhead at midnight in 6 months will be the same stars that are overhead today at noon, but of course we cannot see them today because of the glare of the Sun. Take a moment to visualize this motion. You can again use your orange as Earth—and perhaps your desk lamp, with the shade removed, to represent the Sun. Move the orange around the lamp, and notice which parts of your room walls are visible from the "nighttime side" of the orange at different points in the orbit.

Also take a moment to notice the location of your desk lamp relative to the walls of your room for the "observer" on your orange. If we correspondingly note the position of the Sun relative to the stars each day for a year, we find that it traces out a path against the background of the stars called the **ecliptic** (**Figure 2.11**). On September 1, the Sun is in the direction of the constellation

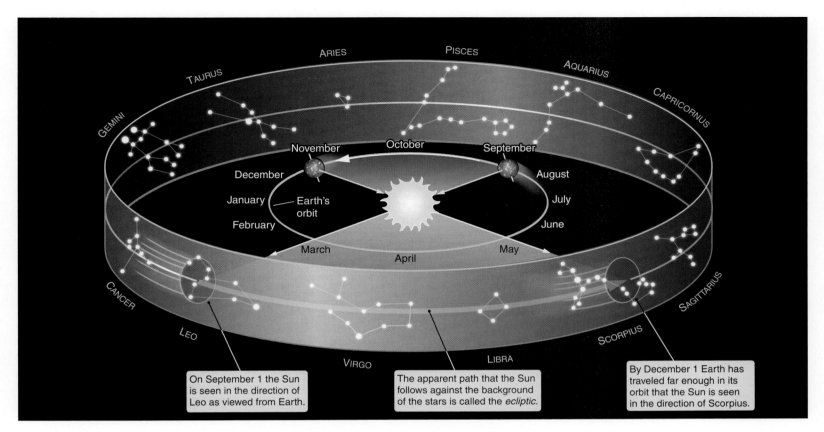

On September 1 the Sun is seen in the direction of Leo as viewed from Earth.

The apparent path that the Sun follows against the background of the stars is called the *ecliptic*.

By December 1 Earth has traveled far enough in its orbit that the Sun is seen in the direction of Scorpius.

Figure 2.11 As Earth orbits the Sun, the Sun's apparent position against the background of stars changes. The imaginary circle traced by the annual path of the Sun is called the ecliptic. Constellations along the ecliptic form the zodiac.

of Leo. Six months later, on March 1, Earth is on the other side of the Sun, and the Sun is in the direction of the constellation of Aquarius. The constellations that lie along the ecliptic are called the constellations of the **zodiac**. Ancient astrologers assigned special mystical significance to these stars because they lie along the path of the Sun. Actually, the constellations of the zodiac are nothing more than random patterns of distant stars that happen by chance to be located near the ecliptic.

The ecliptic is the Sun's apparent yearly path against the background of stars.

Seasons Are Due to the Tilt of Earth's Axis

So far we have discussed the rotation of Earth on its axis and the revolution of Earth about the Sun. To understand the changing of the seasons, we need to consider the combined effects of these two motions. Many people believe that Earth is closer to the Sun in the summer and farther away in the winter, and that this is the cause for the seasons. What if this idea were a hypothesis? Can it be falsified? Yes, we can make a prediction. If the distance from Earth to the Sun causes the seasons, all of Earth should experience summer at the same time of year. But the United States experiences summer in June, while Chile experiences summer in December. We have just falsified this hypothesis, and we need to go look for another one that explains *all* of the available facts. We will find it by investigating Earth's axial tilt.

▶‖ **AstroTour:** The Celestial Sphere and the Ecliptic

Working It Out 2.1 | Manipulating Equations

So far, we have discussed scientific notation and units. Now we want to relate quantities to each other, such as distance and time. In Section 2.1, we mention the speed of the surface of Earth as it rotates. How do we know that speed? In Section 2.2, we talk about finding the size of Earth's orbit from the speed and the period. How do we do that? Both of these calculations use the same relationship. Once you have the relationship in hand, you can apply it to many different situations without having to think about it from scratch every time.

How are distance, time, and speed related? If you travel a distance of 60 miles, and it takes you a time of 1 hour, you have traveled at a speed of 60 miles per hour. That's how we say it in English. How do we translate this sentence into math? We write the equation:

$$\text{speed} = \frac{\text{distance}}{\text{time}}$$

But that takes up a lot of space, so we abbreviate it by using the letter s to represent speed, the letter d to represent distance, and the letter t to represent time:

$$s = \frac{d}{t}$$

To find the speed of the surface of Earth as it rotates, we imagine a spot on Earth's equator. The time it takes to go around once is 1 day, and the distance traveled is Earth's circumference. So the speed is

$$s = \frac{\text{circumference}}{1 \text{ day}}$$

The circumference is given by $2 \times \pi \times r$, where r is the radius. Earth's radius is 6,378 km. To find the circumference, multiply that by 2 and then by π to get 40,074 km. (Does it matter if you multiply first by π and then by 2? If you can't remember the rule, try it and see.)

Now we have the circumference, and we can find the speed:

$$s = \frac{\text{circumference}}{1 \text{ day}} = \frac{40,074 \text{ km}}{1 \text{ day}} = 40,074 \ \frac{\text{km}}{\text{day}}$$

But that's not the way we usually write a speed. We usually use km/hr. One day = 24 hours, so (1 day)/(24 hours) is equal to one. We can always divide or multiply by one, so let's multiply our speed by (1 day)/(24 hours):

$$s = 40,074 \ \frac{\text{km}}{\text{day}} \times \frac{1 \text{ day}}{24 \text{ hours}}$$

The unit of "day" divides (or cancels) out, and we have

$$s = \frac{40,074 \text{ km}}{24 \text{ hr}} = 1,670 \text{ km/hr}$$

This agrees with the value we learned earlier.

How can we manipulate this equation to solve for distance? We start with the original equation:

$$s = \frac{d}{t}$$

The d, which is what we are looking for, is "buried" on the right side of the equation. We want to get it all by itself on the left. That's what we mean when we say "solve for." The first thing to do is to flip the equation around:

$$\frac{d}{t} = s$$

We can do this because of the equals sign. It's like saying, "2 quarters is equal to 50 cents" versus "50 cents is equal to 2 quarters." It has to be true on both sides. We still have not solved for d. To do this, we need to multiply by t. This cancels the t on the bottom of the left side of the equation:

$$\frac{d}{\cancel{t}} \times \cancel{t} = s$$

But if you do something only to one side of an equation, the two sides are not equal anymore. So whatever we do on the left, we also have to do on the right:

$$\frac{d}{\cancel{t}} \times \cancel{t} = s \times t$$

The t on the left cancels, to give us our final answer:

$$d = s \times t$$

We often write this as

$$d = st$$

and so we have left out the ×. When two terms are written side by side, with no symbols between them, it means you should multiply them together.

Follow along with the discussion in Section 2.2, and see if you can find the circumference of Earth's orbit for yourself.

To understand how the combination of Earth's axial tilt and its path around the Sun creates seasons, let's consider a limiting case. If Earth's spin axis were exactly perpendicular to the plane of Earth's orbit (the **ecliptic plane**), then the Sun would always be on the celestial equator. Because the position of the celestial equator in our sky is determined by our latitude, the Sun would follow the same path through the sky every day, rising due east each morning and setting due west each evening. The Sun would be above the horizon exactly half the time, and days and nights would always be exactly 12 hours long. In short, if Earth's axis were exactly perpendicular to the plane of Earth's orbit, there would be no seasons.

Since we do have seasons, we can conclude that Earth's axis of rotation is *not* exactly perpendicular to the plane of the ecliptic. In fact, it is tilted 23.5° from the perpendicular. As Earth moves around the Sun, its axis points in the same direction throughout the year. Sometimes the Sun is in the direction of Earth's North Pole tilt, and at other times the Sun is opposite to that direction. When Earth's North Pole is tilted toward the Sun, the Sun is *north* of the celestial equator. Six months later, when Earth's North Pole is tilted away from the Sun, the Sun is *south* of the celestial equator. If we look at the circle of the ecliptic, we see that it is tilted by 23.5° with respect to the celestial equator. Take a moment to visualize this with your orange and your desk lamp. Tilt the orange slightly, so that its north pole no longer points at the ceiling but at some distant point in the sky through a wall or window. Keep the North Pole dot on your orange pointing in that direction, and "orbit" it around your desk lamp. At one point, the lamp will be in the direction of the tilt. Halfway around the orbit, the lamp will be opposite the direction of the tilt.

Figure 2.12a shows the orientation of Earth relative to the Sun on June 21, the day that the Sun is most nearly in the direction of Earth's North Pole tilt. The Sun is north of the celestial equator. We found earlier in the chapter that, in the Northern Hemisphere, a star north of the celestial equator can be seen above the horizon for more than half the time. This is also true for the Sun, so the Northern Hemisphere's night is shorter than 12 hours. These are the long **days** of the northern summer. Six months later, on December 22 (**Figure 2.12b**), the Sun is most nearly opposite to the direction of Earth's North Pole tilt, so the Sun appears in the sky south of the celestial equator. Night will be longer than 12 hours; it is winter in the north.

In the preceding paragraph we were careful to specify the *Northern* Hemisphere because seasons are opposite in the Southern Hemisphere. Look again at Figure 2.12. On June 21, while the Northern Hemisphere is enjoying the long days and short nights of summer, the Sun is opposite the direction of Earth's South Pole tilt, so it is winter in the Southern Hemisphere. Similarly, on December 22, the Sun is in the direction of Earth's South Pole tilt, and the southern summer nights are short.

The differing length of the night through the year is part of the explanation for seasonal temperature changes, but it is not the whole story. There is another important effect: the Sun appears higher in the sky during the summer than it does during the winter, so sunlight strikes the ground *more directly* during the summer than during the winter. To see why this is important, hold your hand near your desk lamp and look at the size of its shadow on the wall. If your hand is held so that the palm faces the lamp, then your hand's shadow is large. As you tilt your hand, the shadow shrinks. The size of your hand's shadow tells you that it catches less energy when it is tilted than it does when your palm faces the

▶❚❚ **AstroTour: The Earth Spins and Revolves**

The tilt of Earth's axis causes the seasons.

Vocabulary Alert

Day: In common language, this word means both the time during which the Sun is up in the sky and the length of time it takes Earth to rotate once (from midnight to midnight). The context of the sentence tells us which meaning is intended. Unfortunately, astronomers use this word in both senses as well. You will have to consider the context of the sentence to know which meaning is being used in each instance.

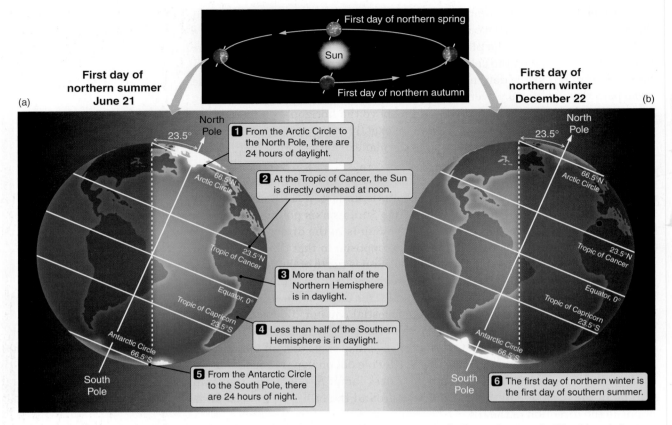

First day of
northern summer
June 21

First day of
northern winter
December 22

Figure 2.12 (a) On the first day of the northern summer (June 21, the summer solstice), the northern end of Earth's axis is tilted most nearly toward the Sun, while the Southern Hemisphere is tipped away. (b) Six months later, on the first day of the northern winter (December 22, the winter solstice), the situation is reversed. Seasons are opposite in the Northern and Southern Hemispheres.

> Sunlight strikes the ground more directly in summer, so there is more heating per unit area than in winter.

lamp. If you hold your hand close enough, you may notice that it is hotter when your palm faces the lamp and less hot when tilted. This is exactly what happens with the changing seasons. During the summer, Earth's surface at your location is more nearly face-on to the incoming sunlight, so more energy falls on each square meter of ground each second. During the winter, the surface of Earth where you are is more tilted with respect to the sunlight, so less energy falls on each square meter of the ground each second. This is the main reason why it is hotter in the summer and colder in the winter.

Together, these two effects—the directness of sunlight and the differing length of the night—mean that during the summer there is more heating from the Sun, and during the winter there is less.

Four Special Days Mark the Passage of the Seasons

As Earth orbits the Sun, the Sun moves along the ecliptic, which is tilted 23.5° with respect to the celestial equator. The day when the Sun is most directly in the direction of the North Pole's tilt is called the **summer solstice**. This occurs each year around June 21, the first day of summer in the Northern Hemisphere.

Six months later, the Sun is most directly opposite to the direction of the North

Pole's tilt. This day is the **winter solstice**. This occurs each year around December 22, the shortest day of the year and the first day of winter in the Northern Hemisphere. Almost all cultural traditions in the Northern Hemisphere include a major celebration of some sort in late December. These winter festivals all share one thing: they celebrate the return of the source of Earth's light and warmth. The days have stopped growing shorter and are beginning to get longer. Spring will come again.

Between these two special days, there are days when the Sun lies directly above Earth's equator, so that the entire Earth experiences 12 hours of daylight and 12 hours of darkness. These are called the equinoxes (*equinox* means "equal night"). The **autumnal equinox** occurs in the fall, around September 23, halfway between summer solstice and winter solstice. The **vernal equinox** occurs in the spring, around March 21, between winter solstice and summer solstice.

Figure 2.13 shows these four special points from two perspectives. The first has a stationary Sun, which is what's actually happening, and the second shows a moving Sun along the celestial sphere, which is how things appear to us as observers on Earth. In both cases, we are looking at the plane of Earth's orbit from an angle, so that it is shown in perspective and looks quite flattened. We have also tilted the picture so that the North Pole of Earth points straight up. Use your orange and your desk lamp to reproduce these pictures and the motions implied by the arrows. Practice shifting between these two perspectives. Once you can look at one of the positions in Figure 2.13a and predict the corresponding positions of the Sun and Earth in Figure 2.13b (and vice versa), you will know that you really understand these differing perspectives.

Just as it takes time for a pot of water on a stove to heat up when the burner is turned up and time for the pot to cool off when the burner is turned down, it takes time for Earth to respond to changes in heating from the Sun. The hottest months of northern summer are usually July and August, which come *after* the summer solstice, when the days are growing shorter. Similarly, the coldest months of northern winter are usually January and February, which occur *after* the winter solstice, when the days are growing longer. The climatic seasons on Earth lag behind changes in the amount of heating we receive from the Sun.

Today's calendar is the **Gregorian calendar**, which is based on the **tropical year**. The tropical year measures the time from one vernal equinox to the next—from the start of Northern Hemisphere spring to the start of the next Northern Hemisphere spring—and is 365.242189 days long. Notice that the tropical year is not a whole number of days long, but has an "extra" almost 0.25 day. **Leap years**—years in which a 29th day is added to the month of February—are used in our calendar to make up for most of the extra fraction of a day. This prevents the seasons from becoming out of sync with the months and giving us a northern winter in August. The Gregorian calendar also takes care of the remaining 0.007811 days, but those add up so slowly that we make adjustments on much longer time scales than a few years.

Some Places on Earth Experience Seasons Differently

Our picture of the seasons must be modified somewhat near Earth's poles. At latitudes north of 66.5° north (the **Arctic Circle**) and south of 66.5° south (the **Antarctic Circle**), the Sun is circumpolar for part of the year. When the Sun is circumpolar, it is above the horizon 24 hours a day, earning the polar regions

Motion of Earth around the Sun

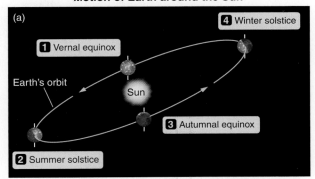

Apparent motion of the Sun seen from Earth

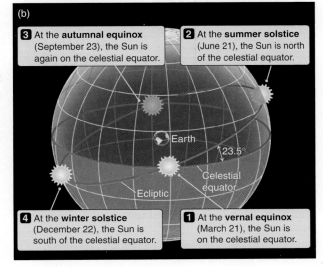

Figure 2.13 The motion of Earth about the Sun as seen from the frame of reference of (a) the Sun and (b) Earth.

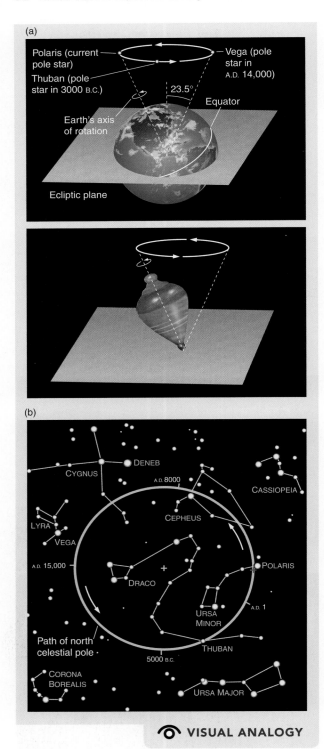

the nickname "land of the midnight Sun." The Arctic and Antarctic regions pay for these long days with an equally long period when the Sun never rises and the nights are 24 hours long. The Sun never rises very high in the Arctic or Antarctic sky. This means that sunlight is never very direct; so even with the long days at the height of summer, the Arctic and Antarctic regions remain relatively cool.

The seasons are also different near the equator. On the equator, days and nights are 12 hours long throughout the year. The Sun passes directly overhead on the first day of spring and the first day of autumn because these are the days when the Sun is on the celestial equator. Sunlight is most direct at the equator on these days. On the summer solstice, the Sun is at its northernmost point along the ecliptic. It is on this day, and on the winter solstice, that the Sun is *farthest* from the zenith at noon, and therefore sunlight is *least* direct. The Sun is up for 12 hours a day year-round, and the Sun is always close to overhead at noon.

If you live between the latitudes of 23.5° south and 23.5° north—in Rio de Janeiro or Honolulu, for example—the Sun will be directly overhead at noon twice during the year. The band between these two latitudes is called the **Tropics**. The northern limit of this region is called the Tropic of Cancer; the southern limit is called the Tropic of Capricorn (see Figure 2.12).

Earth's Axis Wobbles

When the ancient Egyptian astronomer Ptolemy (Claudius Ptolemaeus) and his associates were formalizing their knowledge of the positions and motions of objects in the sky 2,000 years ago, the Sun appeared in the constellation of Cancer on the first day of northern summer and in the constellation of Capricorn on the first day of northern winter—hence, the names of the Tropics. Today, the Sun is in Taurus on the first day of northern summer, and in Sagittarius on the first day of winter. Which leads us to this question: Why has this change occurred? There are *two* motions associated with Earth and its axis. Earth spins on its axis, but its axis also wobbles like the axis of a spinning top (**Figure 2.14a**). The wobble is very slow, taking about 26,000 years to complete one cycle. During this time, the north celestial pole makes one trip around a large circle. In Section 2.1 we saw that the North Star, Polaris, currently lies very near the north celestial pole. However, if you could travel several thousand years into the past or future, you would find that the point about which the northern sky appears to rotate is no longer near Polaris (**Figure 2.14b**).

Recall that the celestial equator is the set of directions in the sky that are perpendicular to Earth's axis. As Earth's axis wobbles, then, the celestial equator must tilt with it. And as the celestial equator wobbles, the locations where it crosses the ecliptic—the equinoxes—change as well. During each 26,000-year wobble of Earth's axis, the locations of the equinoxes make one complete circuit around the celestial equator. Together, these shifts in position are called the **precession of the equinoxes**.

Figure 2.14 (a) Earth's axis of rotation changes orientation in the same way that the axis of a spinning top changes orientation. (b) This precession causes the projection of Earth's rotation axis to move in a circle 47° in diameter, centered on the north ecliptic pole (orange cross), with a period of 25,800 years. The red cross shows the projection of Earth's axis on the sky in the early 21st century.

2.3 The Motions and Phases of the Moon

The Moon is the most prominent object in the sky after the Sun. Earth and the Moon orbit around each other, and together they orbit the Sun. The Moon takes just under 30 days to move through one full set of phases, from full Moon to full Moon. As it passes through the phases, the Moon's appearance constantly changes; but we begin our discussion of the motion of the Moon by talking about an aspect of the Moon's appearance that does *not* change.

We Always See the Same Face of the Moon

The lighted shape and position of the Moon in the sky are constantly changing, but one thing that does *not* change is the face of the Moon that we see. If you were to go outside next week or next month, or 20 years from now, or 20,000 *centuries* from now, you would still see the "Man in the Moon," or whatever shape you choose to see in the pattern of light and dark on the near side of the Moon. This observation is responsible for the common misconception that the Moon does not rotate. In fact, the Moon *does* rotate on its axis—exactly once for each revolution it makes about Earth.

Once again, use your orange to help you visualize this idea. This time, the orange represents the Moon, not Earth. Use your chair or some other object to represent Earth. Face the person you drew on the orange toward the chair. First, make the orange "orbit" around the chair without rotating on its axis. This means that the person is always facing the same direction relative to the walls of your room. When you do it this way, the side of the orange with the person is not always facing the chair. Now, make the orange orbit the chair, with the person always facing the chair. You will have to turn the orange to make this happen, which means that the orange is rotating around its axis. By the time the orange completes one orbit, it will have rotated exactly once. The Moon does exactly the same thing, rotating on its axis once per revolution around Earth, always keeping the same face toward Earth, as shown in **Figure 2.15**. This phenomenon is called **synchronous rotation** because the revolution and the rotation are synchronized (or in sync) with each other. The Moon's synchronous rotation is not an accident. The Moon is elongated, which causes its *near side* always to fall toward Earth.

The Moon's *far side*, facing away from Earth, is often improperly called the dark side of the Moon. In fact, the far side spends just as much time in sunlight as the near side. The far side is not dark as in "unlit," but until the middle of the 20th century it was dark as in "unknown." Until spacecraft orbited the Moon, we had no knowledge of what it was like.

The Changing Phases of the Moon

The constantly changing aspect of the Moon is fascinating. Unlike the Sun, the Moon has no light source of its own; like the planets, including Earth, it shines by reflected sunlight. Like Earth, half of the Moon is always in bright daylight, and half is always in darkness. Our view of the illuminated portion of the Moon is constantly changing, causing the phases of the Moon. Sometimes (during a new Moon) the side facing away from us is illuminated, and sometimes (during a full Moon) the side facing toward us is illuminated. The rest of the time, only a part of the illuminated portion can be seen from Earth. Sometimes the Moon appears as a circular disk in the sky. At other times, it is nothing more than a sliver.

To help you visualize the changing phases of the Moon, use your orange, your desk lamp (with the shade removed), and your head. Your head is Earth, the orange is the Moon, and the lamp is the Sun. Turn off all the other lights in

▶‖ **AstroTour:** The View from the Poles

The Moon rotates on its axis once for each orbit around Earth.

▶‖ **AstroTour:** The Moon's Orbit: Eclipses and Phases

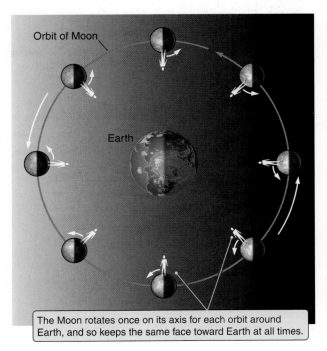

The Moon rotates once on its axis for each orbit around Earth, and so keeps the same face toward Earth at all times.

Figure 2.15 The Moon rotates once on its axis for each orbit around Earth—an effect called *synchronous rotation.*

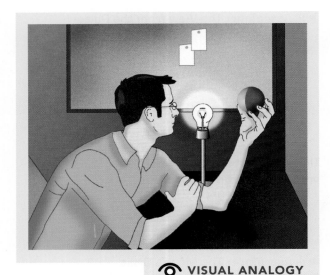

👁 **VISUAL ANALOGY**

Figure 2.16 You can experiment with illumination effects by using an orange as the Moon, a lamp with no shade as the Sun, and your own head as Earth. As you move the orange around your head, viewing it in different relative locations, you will see that the illuminated part of the orange mimics the phases of the Moon.

▶❙❙ **AstroTour: The Moon's Orbit: Eclipses and Phases**

The phase of the Moon is determined by how much of its bright side we can see.

---

the room, and push your chair back as far from the lamp as you can. Hold up the orange slightly above your head so that it is illuminated from one side by the lamp. Move the orange clockwise (this would be counterclockwise if viewed from the ceiling), and watch how the appearance of the orange changes. When you are between the orange and the lamp, the face of the orange that is toward you is fully illuminated. The orange appears to be a bright, circular disk. As the orange moves around its circle, you will see a progression of lighted shapes, depending on how much of the bright side and how much of the dark side of the orange you can see. This progression of shapes exactly mimics the changing phases of the Moon (**Figure 2.16**).

Figure 2.17 shows the changing phases of the Moon. When the Moon is between Earth and the Sun, the far side of the Moon is illuminated, the near side is in darkness, and we cannot see it. This is called a **new Moon**. It is up in the daytime, since it is in the direction of the Sun. A new Moon is never visible in the night-time sky. It appears close to the Sun in the sky, so it rises in the east at sunrise, crosses the meridian near noon, and sets in the west near sunset.

A few days later, as the Moon orbits Earth, a small curve of its illuminated half becomes visible. This shape is called a **crescent**. Because the Moon appears to be "filling out" from night to night at this time, this phase of the Moon is called a **waxing** crescent Moon. (*Waxing* here means "growing in size and brilliance." It may help you to remember this if you think about a puddle of wax getting larger at the base of a burning candle—waxing is when the Moon appears to be growing.) During the week that the Moon is in this phase, the Moon is visible east of the Sun. It is most noticeable just after sunset, near the western horizon. The Moon is illuminated on the bottom right (as viewed from the Northern Hemisphere), so the "horns" of the crescent always point away from the Sun.

As the Moon moves farther along in its orbit and the angle between the Sun and the Moon grows, more and more of the near side becomes illuminated, until half of the near side of the Moon is in brightness and half is in darkness. This phase is called a **first quarter Moon** because the Moon is one quarter of the way through its cycle of phases. The first quarter Moon rises at noon, crosses the meridian at sunset, and sets at midnight.

As the Moon moves beyond first quarter, more than half of the near side is illuminated. This phase is called a waxing **gibbous** Moon, from the Latin *gibbus*, meaning "hump." The gibbous Moon waxes until finally we see the entire bright side of the Moon—a **full Moon**. The Sun and the Moon are now opposite each other in the sky. The full Moon rises as the Sun sets, crosses the meridian at midnight, and sets in the morning as the Sun rises.

The second half of the Moon's cycle of phases is the reverse of the first half. The Moon appears gibbous, but now the near side is becoming less illuminated. This phase is called a **waning** gibbous Moon (*waning* means "becoming smaller"). When the Moon is waning, the left side (as viewed from the Northern Hemisphere) appears bright. A **third quarter Moon** occurs when half the near side is in brightness and half is in darkness. A third quarter Moon rises at midnight, crosses the meridian near sunrise, and sets at noon. The cycle continues with a waning crescent Moon in the morning sky, west of the Sun, until the new Moon once more rises and sets with the Sun and the cycle begins again.

Do not try to memorize all the possible combinations of where the Moon is in the sky at each phase and at every time of day. You do not have to. Instead, work on *understanding* the motion and phases of the Moon. Use your orange and your desk lamp, or draw a picture like the chapter-opening figure and fol-

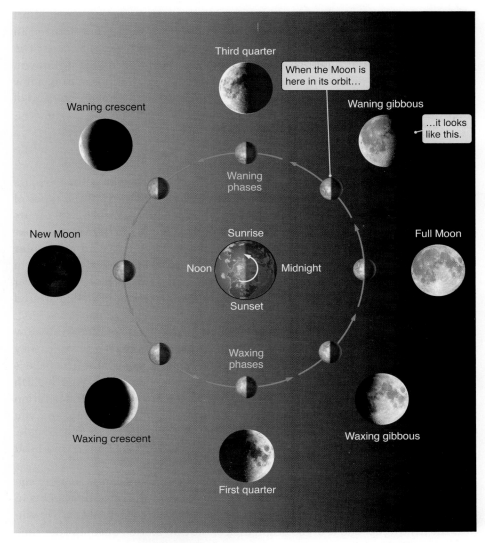

Figure 2.17 The inner circle of images (connected by blue arrows) shows the Moon as it orbits Earth, as seen by an observer far above Earth's North Pole. The outer ring of images shows the corresponding phases of the Moon as seen from Earth.

low the Moon around its orbit. From your drawing, figure out what phase you would see and where it would appear in the sky at a given time of day. Now return to this chapter's opening figure and compare it to Figure 2.17, which is more compact but contains the same information. Does the earlier illustration make more sense to you now? As an extra test of your understanding, think about the phases of Earth an astronaut on the Moon would see when looking back at our planet.

2.4 Eclipses: Passing through a Shadow

Can you imagine any celestial event that would strike more terror in our ancestors' hearts than to look up and see the Sun, giver of light and warmth, being eaten away as if by a giant dragon, or the full Moon turning the ominous color

Figure 2.18 Stonehenge is an ancient artifact in the English countryside, used 4,000 years ago to keep track of celestial events.

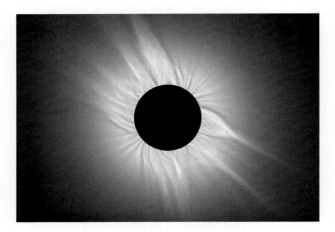

Figure 2.19 The full spectacle of a total eclipse of the Sun.

Figure 2.20 An annular solar eclipse, in which the Moon does not quite cover the Sun. Note that Earth's atmosphere distorts the shape of the Sun when close to the horizon.

▶‖ **AstroTour: The Moon's Orbit: Eclipses and Phases**

of blood? Archaeological evidence suggests that our ancestors put great effort into studying the pattern of **eclipses**. For example, Stonehenge, pictured in **Figure 2.18**, may have been used to predict when eclipses would occur. The motivations of the builders of Stonehenge, 4,000 years dead, are unknown. Yet certainly they desired to overcome the terror of eclipses by learning a few of their secrets—assuring themselves that an eclipse did not mean the end of the world.

Varieties of Eclipses

Eclipses are caused by shadows projected into space. A **solar eclipse** occurs when Earth moves through the shadow of the Moon. There are three types of solar eclipse: *total*, *partial*, and *annular*. A **total solar eclipse** (**Figure 2.19**) occurs when the Moon completely blocks the disk of the Sun. A **partial solar eclipse** occurs when the Moon partially covers the disk of the Sun. An **annular solar eclipse** (**Figure 2.20**) occurs when the Moon is slightly farther away from Earth in its noncircular orbit, so it appears slightly smaller in the sky. It is centered over the disk of the Sun but does not block the entire disk. A ring is visible around the blocked portion.

Figure 2.21a shows the geometry of a solar eclipse, when the Moon's shadow falls on the surface of Earth. Figures like this are seldom drawn to scale; they show Earth and the Moon much closer together than they really are. The page is too small to draw them correctly and still see the critical details. **Figure 2.21b** shows the geometry of a solar eclipse with Earth, the Moon, and the separation between them drawn to scale. Compare this drawing to Figure 2.21a, and you will understand why drawings of Earth and the Moon are rarely drawn to scale. If the Sun were drawn to scale in Figure 2.21b, it would be bigger than your head and located almost 64 meters off the left side of the page.

A total solar eclipse never lasts longer than 7½ minutes and is usually significantly shorter. Even so, it is one of the most amazing and awesome sights in nature. People all over the world flock to the most remote corners of Earth to witness the fleeting spectacle of the bright disk of the Sun blotted out of the daytime sky, leaving behind the eerie glow of the Sun's outer atmosphere.

Lunar eclipses are very different in character from solar eclipses. The geometry of a lunar eclipse is shown in **Figure 2.21c** (and is drawn to scale in **Figure 2.21d**). Because Earth is much larger than the Moon, Earth's shadow at the distance of the Moon is over 2 times the diameter of the Moon. A **total lunar eclipse** lasts as long as 1 hour and 40 minutes. A total lunar eclipse is often called a blood-red Moon in literature and poetry (**Figure 2.22a**); the Moon appears red because it is being illuminated by red light from the Sun that is bent as it travels through Earth's atmosphere and hits the Moon.

Many more people have experienced a total lunar eclipse than have experienced a total solar eclipse. To see a total solar eclipse, you must be located within the very narrow band of the Moon's shadow as it moves across Earth's surface. On the other hand, when the Moon is immersed in Earth's shadow, anyone located in the hemisphere of Earth that is facing the Moon can see it.

If Earth's shadow incompletely covers the Moon, some of the disk of the Moon remains bright and part of it is in shadow. This is called a **partial lunar eclipse**. **Figure 2.22b** shows a composite of images taken at different times during a partial lunar eclipse. In the center image, the Moon is nearly completely eclipsed by Earth's shadow.

(a) Solar eclipse geometry (not to scale)

Total or annular eclipse   Partial eclipse

Sun

Moon

Earth

(b) Solar eclipse to scale

Moon   Earth

(c) Lunar eclipse geometry (not to scale)

Sun

Earth   Moon

(d) Lunar eclipse to scale

1/2°

Moon

Earth   1/2°

Figure 2.21 (a, b) A solar eclipse occurs when the shadow of the Moon falls on the surface of Earth. (c, d) A lunar eclipse occurs when the Moon passes through Earth's shadow. Note that (b) and (d) are drawn to proper scale.

(a)

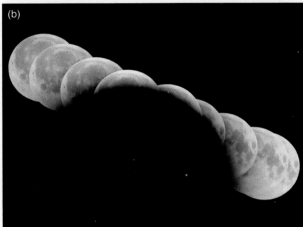

(b)

Figure 2.22 (a) A total lunar eclipse. (b) The progress of a partial lunar eclipse. Note the size of Earth's shadow compared to the size of the Moon.

We don't see a lunar eclipse every time the Moon is full, nor do we observe a solar eclipse every time the Moon is new. If the Moon's orbit were in exactly the same plane as the orbit of Earth (imagine Earth, the Moon, and the Sun all sitting on the same flat tabletop), then the Moon would pass directly between Earth and the Sun at every new Moon. The Moon's shadow would pass across the face of Earth, and we would see a solar eclipse. Similarly, each full Moon would be marked by a lunar eclipse.

Solar and lunar eclipses do *not* happen every month, because the Moon's orbit does not lie in exactly the same plane as the orbit of Earth. The plane of the Moon's orbit about Earth is inclined by about 5° with respect to the plane of Earth's orbit about the Sun (**Figure 2.23**). Most of the time, the Moon is "above" or "below" the line between Earth and the Sun. About twice per year, the orbital planes line up and eclipses can occur.

2.5 The Motions of the Planets in the Sky

Astronomers and philosophers in ancient times hypothesized that the Sun might be the center of the Solar System, but they did not have the tools to test the

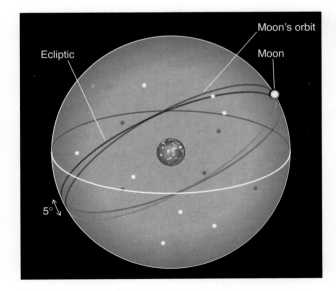

Figure 2.23 The orbit of the Moon is tilted with respect to the ecliptic, so we do not see eclipses every month.

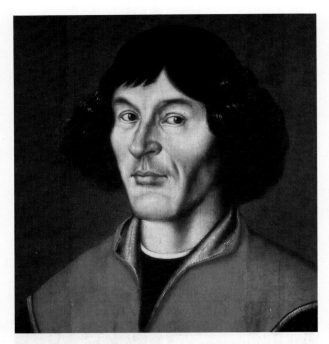

Figure 2.24 Nicolaus Copernicus rejected the ancient belief in an Earth-centered universe and replaced it with one centered on the Sun.

hypothesis or the mathematical insight to formulate a more complete and testable model. Instead, in large part because we can't feel Earth's motion through space, a **geocentric** (Earth-centered) model of the Solar System prevailed. For nearly 1,500 years, most educated people believed that the Sun, the Moon, and the known planets (Mercury, Venus, Mars, Jupiter, and Saturn) all moved in circles around a stationary Earth.

Nicolaus Copernicus (1473–1543; **Figure 2.24**) not only revived the idea that the Sun rather than Earth lies at the center of the Solar System, but he also developed a mathematical model that made predictions that later astronomers would be able to test. This work was the beginning of what was later called the Copernican Revolution. Through the work of scientists such as Tycho Brahe (1546–1601), Galileo Galilei (1564–1642), Johannes Kepler (1571–1630), and Sir Isaac Newton (1642–1727), the **heliocentric** (Sun-centered) theory of the Solar System has become one of the most well-corroborated theories in all of science.

Ancient peoples were aware that planets move in a generally eastward direction among the "fixed stars." Ancient astronomers also knew that these planets would occasionally exhibit **apparent retrograde motion**—that is, they would seem to turn around, move westward for a while, and then return to their normal eastward travel. This odd behavior of the five known planets created a puzzling problem for the geocentric model.

In 1543, Copernicus proposed a heliocentric model that explained retrograde motion much more simply than the geocentric model did. In the Copernican model, the outer planets Mars, Jupiter, and Saturn move in apparent retrograde motion when Earth overtakes them in their orbits. Likewise, the inner planets Mercury and Venus move in apparent retrograde motion when overtaking Earth. Except for the Sun, all Solar System objects exhibit apparent retrograde motion. The magnitude of the effect diminishes with increasing distance from Earth. **Figure 2.25** shows a time-lapse sequence of Mars going through its retrograde "loop."

Retrograde motion is only apparent, not real. If we are in a car or train and we pass a slower-moving car or train, it can seem to us that the other vehicle is moving backward. Without an external **frame of reference**—a coordinate system within which an observer measures positions and motions—it can be hard to tell which vehicle is moving and in what way. Copernicus provided this frame of reference for the Sun and its planets.

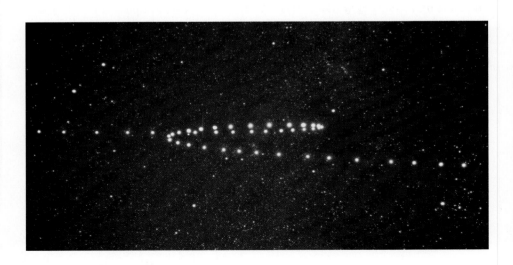

Figure 2.25 Mars moves in apparent retrograde motion.

Combining geometry with observations of the positions of the planets in the sky (their altitudes, and the times they rise and set), Copernicus estimated the planet–Sun distances in terms of the Earth–Sun distance. These relative distances are remarkably close to those obtained by modern methods. From these observations, he also found when each planet, Earth, and the Sun were in alignment. He used this information plus geometry to figure out how long it took each planet to orbit the Sun. This model made testable predictions of the location of each planet on a given night. The heliocentric model was so much simpler than the geocentric model that Occam's razor made it the favored model.

Tycho (conventionally referred to by his first name) was the last great observational astronomer before the invention of the telescope. He carefully measured the precise positions of planets in the sky, developing the most comprehensive set of planetary data available at that time. His assistant, Kepler, received these records when Tycho died. Kepler used the data to deduce three rules that elegantly and accurately describe the motions of the planets. These three rules are now generally referred to as **Kepler's laws**.

Kepler's First Law

When Kepler compared Tycho's observations with predictions from Copernicus's model, he expected the data to confirm circular orbits. Instead, he found disturbing disagreements between his predictions and the observations. He was not the first to notice such discrepancies. Rather than simply discarding the model, Kepler adjusted Copernicus's idea until it matched the observations.

Kepler discovered that if he replaced circular orbits with *elliptical* orbits, the predictions fit the observations almost perfectly. This is **Kepler's first law** of planetary motion: the orbit of a planet is an **ellipse** with the Sun at one **focus**. An ellipse (**Figure 2.26a**) is a specific kind of oval whose shape is determined by two foci (the plural of *focus*). It is symmetric from right to left and from top to bottom. As the two foci approach each other, the figure becomes a circle (**Figure 2.26b**). Correspondingly, as the foci move farther apart, the ellipse becomes more elongated.

Figure 2.26a illustrates the vocabulary of ellipses. The dashed lines represent the two main axes of the ellipse. Half of the length of the long axis is called the **semimajor axis**, often denoted by the letter A. The semimajor axis of an orbit is a handy way to describe a planet's orbit because it is the same as the average distance between that planet and the Sun.

The shape of an ellipse is determined by its **eccentricity**, e. A circle has an eccentricity of 0. The more elongated the ellipse becomes, the closer its eccentricity gets to 1. Most planets have nearly circular orbits with eccentricities close to 0. Earth's orbit is very nearly a circle centered on the Sun, with an eccentricity of 0.017, as shown in **Figure 2.27a**. By contrast, Pluto's orbit has an eccentricity of 0.249. The orbit is noticeably oblong (**Figure 2.27b**), with the Sun offset from center. This is one of the characteristics that distinguish the dwarf planet Pluto from its classical planet cousins.

Kepler's Second Law

From Tycho's observations of planetary motions through the sky, Kepler found that a planet moves fastest when it is closest to the Sun and slowest when it is farthest from the Sun. For example, the average speed of Earth in its orbit about

Vocabulary Alert

Focus: In common language, this word is used in several ways, to indicate directed attention, or the place where light is concentrated by a lens. In this mathematical context, it refers to a special point within an ellipse. An ellipse has two of these special points, and the sum of the distance from these points to any point on the ellipse is constant.

(a)

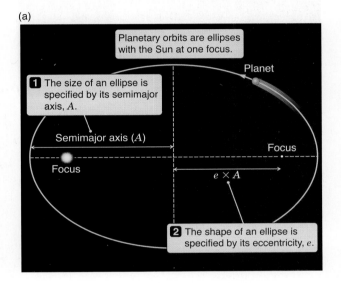

Planetary orbits are ellipses with the Sun at one focus.

1 The size of an ellipse is specified by its semimajor axis, A.

Semimajor axis (A)

Planet

Focus

Focus

$e \times A$

2 The shape of an ellipse is specified by its eccentricity, e.

(b)

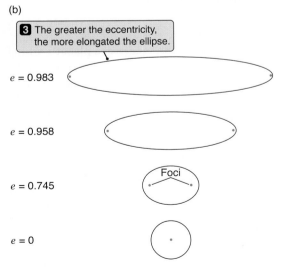

3 The greater the eccentricity, the more elongated the ellipse.

$e = 0.983$

$e = 0.958$

$e = 0.745$

Foci

$e = 0$

Figure 2.26 (a) Kepler's first law: Planets move in elliptical orbits with the Sun at one focus. (b) Ellipses range from circles to elongated eccentric shapes.

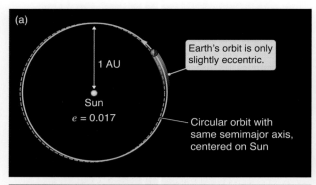

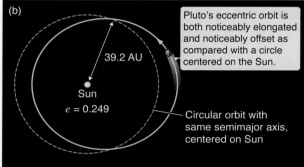

Figure 2.27 The shapes of the orbits of (a) Earth and (b) Pluto compared with circles centered on the Sun.

▶❚❚ AstroTour: **Kepler's Laws**

Planets move fastest when they are closest to the Sun.

The square of a planet's orbital period equals the cube of the orbit's semimajor axis.

Vocabulary Alert

Period: In common language, the word *period* can mean how long a thing lasts. For example, we might talk about the "period of time spent at the grocery store." Astronomers use this word only when talking about repeating intervals, such as the time it takes for an object to orbit once.

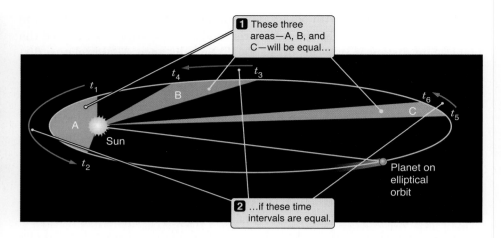

Figure 2.28 Kepler's second law: An imaginary line between a planet and the Sun sweeps out an area as the planet orbits. Kepler's second law states that if the three intervals of time shown are equal, then the three areas A, B, and C will be the same.

the Sun is 29.8 kilometers per second (km/s). When Earth is closest to the Sun, it travels at 30.3 km/s. When it is farthest from the Sun, it travels at 29.3 km/s.

Kepler found an elegant way to describe the changing speed of a planet in its orbit about the Sun. Look at **Figure 2.28**, which shows a planet at six different points in its orbit. Imagine a straight line connecting the Sun with this planet. We can think of this line as "sweeping out" an area as it moves with the planet from one point to another. **Kepler's second law**, also called Kepler's **law of equal areas**, states that the area swept out by a planet in the same amount of time is always the same, regardless of the location of the planet in its orbit. In Figure 2.28, if the three time intervals are equal (that is, $t_1 \rightarrow t_2 = t_3 \rightarrow t_4 = t_5 \rightarrow t_6$), then the three areas A, B, and C will be equal as well.

Kepler's Third Law

Planets close to the Sun travel on shorter orbits than planets that are far from the Sun. Jupiter's average distance from the Sun, for example, is 5.2 times larger than Earth's average distance from the Sun. Since an orbit's circumference is proportional to its radius, Jupiter must travel 5.2 times farther in its orbit about the Sun than Earth does in its orbit. If the two planets were traveling at the same speed, Jupiter would complete one orbit in 5.2 years. But Jupiter takes almost 12 years to complete one orbit. Jupiter not only has farther to go, but is also *moving more slowly than Earth*. The farther a planet is from the Sun, the larger the circumference of its orbit and the lower its speed.

Kepler discovered a mathematical relationship between the **period** of a planet's orbit and its average distance from the Sun. **Kepler's third law** states that *period squared is equal to distance cubed*. We explore this relationship in more detail in **Working It Out 2.2.**

Working It Out 2.2 | Kepler's Third Law

Just as *squaring* a number means that you multiply it by itself, as in $a^2 = a \times a$, *cubing* it means that you multiply it by itself again, as in $a^3 = a \times a \times a$. Kepler's third law states that the square of the period of a planet's orbit, P_{years}, measured in years, is equal to the cube of the semimajor axis of the planet's orbit, A_{AU}, measured in astronomical units. Translated into math, the law says

$$(P_{\text{years}})^2 = (A_{\text{AU}})^3$$

Here, astronomers use nonstandard units as a matter of convenience. Years are handy units for measuring the periods of orbits, and astronomical units are handy units for measuring the sizes of orbits. When we use years and astronomical units as our units, we get the relationship just shown (**Figure 2.29**). It is important to realize that *our choice of units in no way changes the physical relationship* we are studying. For example, if we instead chose seconds and meters as our units, this relationship would read

$$(3.2 \times 10^{-8} \text{ years/second} \times P_{\text{seconds}})^2$$
$$= (6.7 \times 10^{-12} \text{ AU/meter} \times A_{\text{meters}})^3$$

which simplifies to

$$(P_{\text{seconds}})^2 = 2.9 \times 10^{-19} \times (A_{\text{meters}})^3$$

Suppose that we want to know the average radius of Neptune's orbit, in AU. First, we need to find out how long Neptune's period is in Earth years, which can be determined by careful observation of Neptune's position relative to the fixed stars. Neptune's period is 165 years. Plugging this into Kepler's third law, we find that

$$(P_{\text{years}})^2 = (A_{\text{AU}})^3$$
$$(165)^2 = (A_{\text{AU}})^3$$

To solve this equation, we must first square the left side to get 27,225 and then take its cube root.

Calculator hint: A scientific calculator usually has a cube root function. It sometimes looks like $x^{1/y}$ and sometimes like $^x\sqrt{y}$. You use it by typing the base number, hitting the button, and then typing the root you are interested in (2 for square root, 3 for cube root, and so on). Occasionally, a calculator will instead have a button that looks like x^y (or y^x). In this case, you need to enter the root as a decimal. For example, if you want to take the square root, you type 0.5 because the square root is denoted by ½. For

the cube root, you type 0.333333333 (repeating) because the cube root is denoted by ⅓.

To find the length of the semimajor axis of Neptune's orbit, we might type 27,225 [$x^{1/y}$] 3. This gives

$$30.1 = A_{\text{AU}}$$

so the average distance between Neptune and the Sun is 30.1 AU.

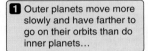
1 Outer planets move more slowly and have farther to go on their orbits than do inner planets…

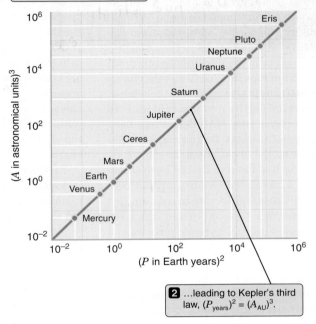
2 …leading to Kepler's third law, $(P_{\text{years}})^2 = (A_{\text{AU}})^3$.

Figure 2.29 A plot of A^3 versus P^2 for the eight classical and three of the dwarf planets in our Solar System shows that they obey Kepler's third law. (Note that by plotting powers of 10 on each axis, we are able to fit both large and small values on the same plot. We will do this frequently.)

READING ASTRONOMY News

You might think that information as fundamental as the motions of Earth and the Moon is fixed and unchanging, but given the right circumstances, these values can change in an instant. This article appeared in the March 1, 2010, online edition of Bloomberg Businessweek.

Chilean Quake Likely Shifted Earth's Axis, NASA Scientist Says

By **ALEX MORALES,** *Bloomberg Businessweek*

The earthquake that killed more than 700 people in Chile on Feb. 27 probably shifted the Earth's axis and shortened the day, a National Aeronautics and Space Administration scientist said.

Earthquakes can involve shifting hundreds of kilometers of rock by several meters, changing the distribution of mass on the planet. This affects the Earth's rotation, said Richard Gross, a geophysicist at NASA's Jet Propulsion Laboratory in Pasadena, California, who uses a computer model to calculate the effects.

"The length of the day should have gotten shorter by 1.26 microseconds (millionths of a second)," Gross, said today in an e-mailed reply to questions. "The axis about which the Earth's mass is balanced should have moved by 2.7 milliarcseconds (about 8 centimeters or 3 inches)."

The changes can be modeled, though they're difficult to physically detect given their small size, Gross said. Some changes may be more obvious, and islands may have shifted, according to Andreas Rietbrock, a professor of Earth Sciences at the U.K.'s Liverpool University who has studied the area impacted, though not since the latest temblor.

Santa Maria Island off the coast near Concepcion, Chile's second-largest city, may have been raised 2 meters (6 feet) as a result of the latest quake, Rietbrock said today in a telephone interview. He said the rocks there show evidence pointing to past earthquakes shifting the island upward in the past.

Ice Skater Effect

"It's what we call the ice-skater effect," David Kerridge, head of Earth hazards and systems at the British Geological Survey in Edinburgh, said today in a telephone interview. "As the ice skater puts her arms out when she's going around in a circle, and she pulls her arms in, she gets faster and faster. It's the same idea with the Earth going around if you change the distribution of mass, the rotation rate changes."

Rietbrock said he hasn't been able to get in touch with seismologists in Concepcion to discuss the quake, which registered 8.8 on the Richter scale.

"What definitely the earthquake has done is made the Earth ring like a bell," Rietbrock said.

The magnitude 9.1 Sumatran earthquake in 2004 that generated an Indian Ocean tsunami shortened the day by 6.8 microseconds and shifted the axis by about 2.3 milliarcseconds, Gross said.

The changes happen on the day and then carry on "forever," Benjamin Fong Chao, dean of Earth Sciences of the National Central University in Taiwan, said in an e-mail.

"This small contribution is buried in larger changes due to other causes, such as atmospheric mass moving around on Earth," Chao said.

Evaluating the News

1. What has happened to Earth's rotation rate? What caused this?

2. David Kerridge refers to something called the "ice-skater effect." What is that? Does this comparison to an everyday phenomenon make the argument that the rotation rate has changed easier or more difficult to understand? Does it make the claim more or less convincing?

3. Study the numbers in the article. Gross claims that the length of day should have gotten shorter by 1.26 microseconds. Use the list of prefixes in the inside cover to express this number in scientific notation. Is it a big or small number? Does this change meaningfully affect the amount of time you have to get your next homework assignment done?

4. Is this a unique event? Have similar incidents happened before?

5. Is the changing of Earth's rotation rate a scientific hypothesis? Recall the discussion of falsifiability in Chapter 1. Is this hypothesis falsifiable? If so, how might we test it?

6. Does this article have anything to do with business? Why might *Bloomberg Businessweek* report on this story?

7. Connect the ice-skater effect in this news story to the orbit of the Moon. The Moon is moving away from us by about 1 cm every century. Is its orbital period getting longer or shorter? Would you guess that its speed is increasing or decreasing?

SUMMARY

2.1 Earth's rotation on its axis causes the apparent daily motion of the Sun, Moon, and stars. Our location on Earth and Earth's location in its orbit determine which stars we see at night.

2.2 The tilt of Earth's axis determines the seasons, by changing the angle at which sunlight strikes the surface in different locations.

2.3 The Moon is in synchronous rotation about Earth. The relative locations of the Sun, Earth, and Moon determine the

phases of the Moon. The phase of the Moon is determined by how much of its bright side is visible from Earth.

2.4 Special alignments of the Sun, Earth, and Moon result in solar and lunar eclipses.

2.5 Copernicus introduced the heliocentric model of the Solar System, which more simply explains the motions of the planets. Kepler developed three laws that describe the motions of the planets. A frame of reference is a coordinate system within which an observer measures positions and motions.

✧ SUMMARY SELF-TEST

1. The Sun, Moon, and stars
   a. change their relative positions over time.
   b. appear to move each day because Earth rotates.
   c. rise north or south of east and set north or south of west, depending on their location on the celestial sphere.
   d. all of the above

2. The stars we see at night depend on
   a. our location on Earth.
   b. Earth's location in its orbit.
   c. the time of the observation.
   d. all of the above

3. The seasons are caused by _____.

4. You see the Moon rising, just as the Sun is setting. What phase is it in?
   a. full   b. new
   c. first quarter d. third quarter
   e. waning crescent

5. You see the first quarter Moon on the meridian. Where is the Sun?
   a. on the western horizon   b. on the eastern horizon
   c. below the horizon   d. on the meridian

6. You do not see eclipses every month because
   a. you are not very observant.
   b. the Sun, Earth, and the Moon line up only about twice a year.
   c. the Sun, Earth, and the Moon line up only about once a year.
   d. eclipses happen randomly and are unpredictable.

7. A frame of reference is a coordinate system within which an observer measures _____ and _____.

8. Place the following in order from largest to smallest semimajor axis:
   a. a planet with a period of 84 Earth days
   b. a planet with a period of 1 Earth year
   c. a planet with a period of 2 Earth years
   d. a planet with a period of 0.5 Earth years

9. A planet moves fastest when it is _____ to the Sun and slowest when it is _____ from the Sun.

QUESTIONS AND PROBLEMS

True/False and Multiple-Choice Questions

10. **T/F:** Kepler's three laws explain *why* the planets orbit the Sun as they do.

11. **T/F:** The celestial sphere is not an actual object in the sky.

12. **T/F:** Eclipses happen somewhere on Earth every month.

13. **T/F:** The phases of the Moon are caused by the relative position of Earth, the Moon, and the Sun.

14. **T/F:** If a star rises north of east, it will set south of west.

15. **T/F:** From the North Pole, all stars in the night sky are circumpolar stars.

16. The tilt of Earth's axis causes the seasons because
   a. one hemisphere of Earth is closer to the Sun in summer.
   b. the days are longer in summer.
   c. the rays of light strike the ground more directly in summer.
   d. both a and b
   e. both b and c

17. On the vernal and autumnal equinoxes,
   a. the entire Earth has 12 hours of daylight and 12 hours of darkness.
   b. the Sun rises due east and sets due west.
   c. the Sun is located on the celestial equator.
   d. all of the above
   e. none of the above

18. We always see the same side of the Moon because
    a. the Moon does not rotate on its axis.
    b. the Moon rotates once each revolution.
    c. when the other side of the Moon is facing toward us, it is unlit.
    d. when the other side of the Moon is facing Earth, it is on the opposite side of Earth.
    e. none of the above

19. You see the Moon on the meridian at sunrise. The phase of the Moon is
    a. waxing gibbous.   b. full.
    c. new.   d. first quarter.
    e. third quarter.

20. A lunar eclipse occurs when the _____ shadow falls on the _____.
    a. Earth's; Moon   b. Moon's; Earth
    c. Sun's; Moon   d. Sun's; Earth

21. Kepler's second law says that
    a. planetary orbits are ellipses with the Sun at one focus.
    b. the square of a planet's orbital period equals the cube of its semimajor axis.
    c. for every action there is an equal and opposite reaction.
    d. planets move fastest when they are closest to the Sun.

22. Suppose you read in the newspaper that a new planet has been found. Its average speed in its orbit is 33 km/s. When it is closest to its star, it moves at 31 km/s, and when it is farthest from its star, it moves at 35 km/s. This story is in error because
    a. the average speed is far too fast.
    b. Kepler's third law says the planet has to sweep out equal areas in equal times, so the speed of the planet cannot change.
    c. planets stay at a constant distance from their stars; they don't move closer or farther away.
    d. Kepler's second law says the planet must move fastest when it's closest, not when it is farthest away.
    e. using these numbers, the square of the orbital period will not be equal to the cube of the semimajor axis.

Conceptual Questions

23. In your study group, two of your fellow students are arguing about the phases of the Moon. One argues that the phases are caused by the shadow of Earth on the Moon. The other argues that the phases are caused by the orientation of Earth, the Moon, and the Sun. Explain how the photos in the chapter-opening figure falsify one of these hypotheses.

24. Why is there no "east celestial pole" or "west celestial pole"?

25. Earth has a North Pole, a South Pole, and an equator. What are their equivalents on the celestial sphere?

26. Polaris was used for navigation by seafarers such as Columbus as they sailed from Europe to the New World. When Magellan sailed the South Seas, he could not use Polaris for navigation. Explain why.

27. If you were standing at Earth's North Pole, where would you see the north celestial pole relative to your zenith?

28. If you were standing at Earth's South Pole, which stars would you see rising and setting?

29. Where on Earth can you stand and, over the course of a year, see the entire sky?

30. What do we call the group of constellations through which the Sun appears to move over the course of a year?

31. We tend to associate certain constellations with certain times of year. For example, we see the zodiacal constellation Gemini in the Northern Hemisphere's winter (Southern Hemisphere's summer) and the zodiacal constellation Sagittarius in the Northern Hemisphere's summer. Why do we not see Sagittarius in the Northern Hemisphere's winter (Southern Hemisphere's summer) or Gemini in the Northern Hemisphere's summer?

32. Assume that you are flying along in a jetliner.
    a. Define your frame of reference.
    b. What relative motions might take place within your frame of reference?

33. The tilt of Jupiter's rotational axis with respect to its orbital plane is 3°. If Earth's axis had this tilt, explain how it would affect our seasons.

34. Describe the Sun's apparent daily motion on the celestial sphere
    a. at the vernal equinox.
    b. at the summer solstice in Earth's Northern Hemisphere.

35. Why is winter solstice not the coldest time of year?

36. There is only a certain region on Earth where at some time of year the Sun can appear precisely at the zenith. Describe this region.

37. Many cities have main streets laid out in east–west and north–south alignments.
    a. Why are there frequent traffic jams on east–west streets during both morning and evening rush hours within a few weeks of the equinoxes?
    b. Considering your conclusion in (a), if you work in the city during the day, would you rather live east or west of the city?

38. Earth rotates on its axis and wobbles like a top.
    a. How long does it take to complete one rotation?
    b. How long does it take to complete one wobble?

39. Why do we always see the same side of the Moon?

40. What is the approximate time of day when you see the full Moon near the meridian? At what time is the first quarter (waxing) Moon on the eastern horizon? Use a sketch to help explain your answers.

41. Assume that the Moon's orbit is circular. Suppose you are standing on the side of the Moon that faces Earth.
    a. How would Earth appear to move in the sky as the Moon made one revolution around Earth?
    b. How would the "phases of Earth" appear to you, as compared to the phases of the Moon as seen from Earth?

42. Sometimes artists paint the horns of the waxing crescent Moon pointing toward the horizon. Is this depiction realistic? Explain.

43. Astronomers are sometimes asked to serve as expert witnesses in court cases. Suppose you are called in as an expert witness, and the defendant states that he could not see the pedestrian because the full Moon was casting long shadows across the street at midnight. Is this claim credible? Why or why not?

44. From your own home, why are you more likely to witness a partial eclipse of the Sun rather than a total eclipse?

45. Why do we not see a lunar eclipse each time the Moon is full or witness a solar eclipse each time the Moon is new?

46. Why does the fully eclipsed Moon appear reddish?

47. In the Gregorian calendar, the length of a year is not 365 days, but actually about 365.25 days. How do we handle this extra quarter day to keep our calendars from getting out of sync?

48. Does the occurrence of solar and lunar eclipses disprove the notion that the Sun and the Moon both orbit around Earth? Explain your reasoning.

49. Vampires are currently prevalent in popular fiction. These creatures have extreme responses to even a tiny amount of sunlight (the response depends on the author), but moonlight doesn't affect them at all. Is this logical? How is moonlight related to sunlight?

50. Suppose you are on a plane from the Northern to the Southern Hemisphere. On the way there, you realize something amazing. You have just experienced the longest day of the year in the Northern Hemisphere and are about to experience the shortest day of the year in the Southern Hemisphere on the same day! On what day of the year are you flying? How do you explain this amazing phenomenon to the person in the seat next to you?

51. Each ellipse has two foci. The orbits of the planets have the Sun at one focus. What is at the other focus?

52. Ellipses contain two axes, major and minor. Half the major axis is called the semimajor axis. What is especially important about the semimajor axis of a planetary orbit?

53. What is the eccentricity of a circular orbit?

54. The speed of a planet in its orbit varies in its journey around the Sun.
  a. At what point in its orbit is the planet moving the fastest?
  b. At what point is it moving the slowest?

55. The distance that Neptune has to travel in its orbit around the Sun is approximately 30 times greater than the distance that Earth must travel. Yet it takes nearly 165 years for Neptune to complete one trip around the Sun. Explain why.

Problems

56. Earth is spinning along at 1,674 km/h at the equator. Use this number to find Earth's equatorial diameter.

57. The waxing crescent Moon appears to the east of the Sun and moves farther east each day. Does this mean it rises earlier each day or later? By how much?

58. Romance novelists sometimes say that as the hero rides off into the sunset, the full Moon is overhead. Is this correct? Why or why not? Draw a picture of the Sun, Moon, and Earth at full Moon phase to explain your answer.

59. You can often make out the unlit portion of the Moon, even during crescent phase. This is because of a phenomenon called earthshine, where sunlight reflects off Earth, reflects off the Moon, and comes to your eye. Draw a picture of the Sun, Moon, and Earth during crescent phase, and draw the path the light takes to come to you.

60. Suppose you are an astronaut on the Moon. Explain the relative motion of Earth as seen from your lunar frame of reference.

61. Many people "winter over" every year in Antarctica, at the scientific bases there. One man claims to you that he was able to see the full Moon there in December. Is this possible? Draw a picture to show why or why not.

62. According to the Reading Astronomy News feature, Earth's rotation rate has shortened by 1.26 microseconds. How many days will it take for our clocks to be off by 1 second?

63. If you travel at a speed of 60 mph for a distance of 15 miles, how long does the trip take? Suppose that instead, you speed to your destination at 65 mph. How long does this speedy trip take?

64. Suppose you are on vacation in Australia, right on the Tropic of Capricorn. What is your latitude? What is the largest angle from the south celestial pole at which stars are circumpolar at your location?

65. The Moon's orbit is tilted by about 5° relative to Earth's orbit around the Sun. What is the highest altitude in the sky that the Moon can reach, as seen in Philadelphia (latitude 40° north)?

66. Imagine you are standing on the South Pole at the time of the southern summer solstice.
  a. How far above the horizon will the Sun be at noon?
  b. How far above (or below) the horizon will the Sun be at midnight?

67. Find out the latitude where you live. Draw and label a diagram showing that your latitude is the same as (a) the altitude of the north celestial pole and (b) the angle (along the meridian) between the celestial equator and your local zenith. What is the noontime altitude of the Sun as seen from your home at the times of winter solstice and summer solstice?

68. The southernmost star in the famous Southern Cross constellation lies approximately 65° south of the celestial equator. From which U.S. state(s) is the entire Southern Cross visible?

69. Let's say you use a protractor to estimate an angle of 40° between your zenith and Polaris. Are you in the continental United States or Canada?

70. Suppose the tilt of Earth's equator relative to its orbit were 10° instead of 23.5°. At what latitudes would the Arctic and Antarctic Circles and the two Tropics be located?

71. Suppose you would like to witness the "midnight Sun," when the Sun appears just above the northern horizon at midnight, but you don't want to travel any farther north than necessary.
    a. How far north (that is, to which latitude) would you have to go?
    b. When would you make this trip?

72. The vernal equinox is now in the zodiacal constellation of Pisces. Wobbling of Earth's axis will eventually cause the vernal equinox to move into Aquarius, beginning the legendary, long-awaited "Age of Aquarius." How long, on average, does the vernal equinox spend in each of the 12 traditional zodiacal constellations?

73. Referring to Figure 2.14a, estimate when Thuban will once again be the northern pole star.

74. The apparent diameter of the Moon is approximately 0.5°. About how long does it take the Moon to move its own diameter across the fixed stars on the celestial sphere?

75. If the Moon were in its same orbital plane, but twice as far from Earth, which of the following would happen?
    a. The phases that the Moon goes through would remain unchanged.
    b. Total eclipses of the Sun would not be possible.
    c. Total eclipses of the Moon would not be possible.

76. Which, if either, occurs more often: (a) total eclipses of the Moon seen from Earth or (b) total eclipses of the Sun seen from the Moon? Explain your answer.

77. Suppose you discover a new dwarf planet in our Solar System with a semimajor axis of 46.4 AU. What is its period (in Earth years)?

78. Suppose you discover a planet around a Sun-like star. From careful observation over several decades, you find that its period is 12 Earth years. Find the semimajor axis cubed and then the semimajor axis.

79. Suppose you read in a tabloid newspaper that "experts have discovered a new planet with a distance from the Sun of 1 AU, and a period of 3 years." Use Kepler's third law to argue that this is impossible nonsense.

80. The **sidereal day** is the length of time the Earth takes to rotate to the same position relative to the stars: 23^{h}56^m. The **solar day** is the length of time it takes the Earth to rotate to the same position relative to the Sun: 24^h. Perform the following steps to figure out what is going on (remember that Earth rotates around its axis counterclockwise, as viewed from North, and orbits the Sun counterclockwise).
    a. Sketch the Earth–Sun system, looking down from North. Add one observer, Bill, placed at the noon position on Earth, and another observer, Sally, placed at the midnight position on Earth.

    b. Add a second image of Earth one *sidereal* day later, so that Sally is in the same direction relative to distant stars. You will need to exaggerate the motion of the Earth around the Sun for clarity.
    c. In this second sketch, where is the Sun relative to Bill? Is it noon for Bill?
    d. What has to happen now to bring Bill back to the same position relative to the Sun?
    e. Will this make the solar day longer or shorter than the sidereal day?
    f. Draw a third image of Earth, with Bill once again in the noon position. Does Sally see the same stars in the sky as she did in the first image? How do you know?

81. Suppose there is an observer, Rita, in Ecuador at noon. As Earth spins around its axis, Rita experiences the day, moving from noon to sunset to midnight to sunrise to noon again. At the same time, she is being carried around the Sun as Earth revolves around it. Because of this, Rita has to rotate a little further than 360° to experience noon. The time for one 360° rotation is a sidereal day. The time from noon to noon is the solar day, which is 4 minutes longer than the sidereal day, because of the "extra" rotation. Imagine that Earth revolved around the Sun twice as fast.
    a. Would Earth travel farther in its orbit, or less far, in one sidereal day? By how much?
    b. Would this mean Earth has to rotate more or less to bring Rita back to noon again? By how much?
    c. If the current "extra" rotation takes 4 minutes, how long would the "extra" rotation take in this imaginary, faster system?
    d. In reality, the sidereal day is 23^{h}56^m. How long is the solar day in reality? How long is the solar day in this imaginary, faster system?

 SmartWork, Norton's online homework system, includes algorithmically generated versions of these questions, plus additional conceptual exercises. If your instructor assigns questions in SmartWork, log in at **smartwork.wwnorton.com**.

 StudySpace is a free and open website that provides a Study Plan for each chapter of **Understanding Our Universe**. Study Plans include animations, reading outlines, vocabulary flashcards, and multiple-choice quizzes, plus links to premium content in SmartWork and the ebook. Visit **wwnorton.com/studyspace**.

Exploration | Kepler's Laws

wwnorton.com/studyspace

In this exploration, we will be exploring how Kepler's Laws apply to the orbit of Mercury. Visit StudySpace, and open the Planetary Orbit Simulator applet in Chapter 2. This simulator animates the orbits of the planets, allowing you to control the simulation speed as well as a number of other parameters. Here, we focus on exploring the orbit of Mercury, but you may wish to spend some time examining the orbits of other planets as well.

Kepler's First Law

To begin exploring the simulation, in the "Orbit Settings" panel, use the drop-down menu next to "set parameters for" to select "Mercury" and then click "OK." Click the "Kepler's 1st Law" tab at the bottom of the control panel. Use the radio buttons to select "show empty focus" and "show center."

1. **How would you describe the shape of Mercury's orbit?**

Deselect "show empty focus" and "show center," and select "show semiminor axis" and "show semimajor axis." Under Visualization Options, select "show grid."

2. **Use the grid markings to estimate the ratio of the semiminor axis to the semimajor axis.**

3. **Calculate the eccentricity of Mercury's orbit from this ratio using the equation $e = \sqrt{1-(\text{ratio})^2}$.**

Kepler's Second Law

Click the "reset" button near the top of the control panel, set parameters for Mercury, and then click on the "Kepler's 2nd Law" tab at the bottom of the control panel. Slide the "adjust size" slider to the right, until the fractional sweep size is ⅛.

Click "start sweeping." The planet moves around its orbit, and the simulation fills in area until ⅛ of the ellipse is filled. Click "start sweeping" again as the planet arrives at the right-most point in its orbit (that is, at the point in its orbit farthest from the Sun). You may need to slow the animation rate using the slider under "Animation Controls." Click "show grid" under the visualization options. (If the moving planet annoys you, you can pause the animation.) One easy way to estimate an area is to count the number of squares.

4. **Count the number of squares in the yellow area and in the red area. You will need to decide what to do with fractional squares. Are the areas the same? Should they be?**

Kepler's Third Law

Click the "reset" button near the top of the control panel, set parameters for Mercury, and then click on the "Kepler's 3rd Law" tab at the bottom of the control panel. Select "show solar system orbits" in the Visualization Options panel. Study the graph in the Kepler's 3rd Law tab. Use the eccentricity slider to change the eccentricity of the simulated planet. First, make the eccentricity smaller and then larger.

5. **Did anything in the graph change?**

6. **What does this tell you about the dependence of the period on the eccentricity?**

Set parameters back to those for Mercury. Now use the semimajor axis slider to change the semimajor axis of the simulated planet.

7. **What happens to the period when you make the semimajor axis smaller?**

8. **What happens when you make it larger?**

9. **What does this tell you about the dependence of the period on the semimajor axis?**

3 Laws of Motion

We have learned that the planets, including Earth, orbit around the Sun, but we have not yet touched on the reason why. Gravity is the force that holds the planets in orbit. Because the Sun is far more massive than all the other parts of the Solar System combined, its gravity shapes the motions of every object in its vicinity, from the almost circular orbits of some planets to the extremely elongated orbits of comets.

As our understanding of the universe has expanded, we have come to realize that our Solar System is but one example of gravity at work. Gravity binds stars into the colossal groups that we call galaxies. Gravity holds the planets and stars together and keeps the thin blanket of air we breathe close to Earth's surface. Gravity caused a vast cloud of gas and dust to collapse 4.5 billion years ago to form our Solar System. Gravity shapes space and time. As we continue our journey outward through the cosmos, we will repeatedly find gravity at the center of our growing understanding.

✧ LEARNING GOALS

The figure at right shows the launch of a space shuttle. Scientists view a picture like this with pride, recognizing not only our astonishing accomplishment of sending people into space but also our understanding of the science that makes such achievements possible. When a scientist looks at this picture, she draws the arrows of force in her head, figuring out what forces are acting, in which direction, and with what strength. (An accomplished scientist will also search for things that she does *not* understand, hoping to find a new avenue of study.) By the end of this chapter, you should know how to think about and diagram forces such as gravity. You should

understand how to draw action-reaction pairs of forces. And you should be able to look at a photograph like this one and see the fundamental laws of physics at work. You should also be able to:

- List the physical laws that govern the motion of all objects

- Use proportionality to describe patterns and relationships in nature

- Combine motion and gravitation to explain planetary orbits

- Explain how theories lead to new knowledge

EXHAUST
PUSHING
ROCKET

EARTH'S
GRAVITY

Greetings from
KENNEDY SPACE CENTER

Cape Canaveral, FL

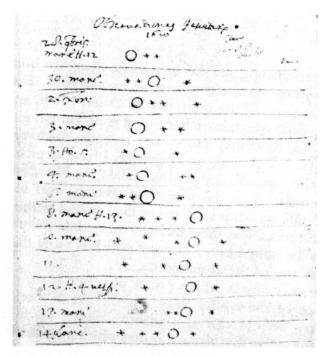

Figure 3.1 Galileo's observations show the moons of Jupiter orbiting the planet over time.

3.1 Galileo: The First Modern Scientist

Galileo Galilei is one of the heroes of astronomy. He was the first to use a telescope to make and report on significant discoveries about astronomical objects, and much has been written about the considerable danger that Galileo (as he is commonly known) found himself in as a result of these discoveries. Galileo's telescopes were relatively small, yet they were sufficient for him to observe spots on the Sun, the uneven surface and craters of the Moon, and the large numbers of stars in the band of light in the sky called the Milky Way.

But two other sets of observations made Galileo famous. When he turned his telescope to the planet Jupiter, he observed several "stars" in a line near Jupiter, and over time saw that there were actually four of these stars, and that their positions changed from night to night (**Figure 3.1**). Galileo correctly reasoned that these were moons in orbit around Jupiter, and thus he provided the first observational evidence that some objects in the sky did not orbit Earth. (They are the largest of the many moons of Jupiter and are still referred to as the Galilean moons.) He also observed that the planet Venus went through phases similar to the Moon's and that the phases correlated with the size of the image of Venus in his telescope (**Figure 3.2**). This is difficult to explain in a geocentric model, but it makes sense in the heliocentric one. These observations in particular convinced Galileo that Copernicus was correct in placing the Sun at the center of our Solar System.

Galileo's public support for Copernican heliocentricity got him into trouble with the Catholic Church. In 1632, Galileo published his bestselling book, *Dialogo sopra i due massimi sistemi del mondo (Dialogue Concerning the Two Chief World Systems)*. In the *Dialogo*, the champion of the Copernican view of the universe, Salviati, is a brilliant philosopher. The defender of an Earth-centered universe, Simplicio, uses arguments made by the classical Greek philosophers and the pope, and sounds silly and ignorant. Galileo had submitted drafts of his book to Church censors, but the censors found the final version unacceptable. The perceived attack on the pope attracted the attention of the Church,

Figure 3.2 Modern photographs of the phases of Venus show that when we see Venus more illuminated, it also appears smaller, implying that Venus is farther away at that time.

and Galileo was eventually placed under house arrest. The book was placed on the Index of Prohibited Works, along with Copernicus's *De Revolutionibus*; but it traveled across Europe, was translated into other languages, and was read by other scientists.

Galileo's work on the motion of objects was at least as fundamental a contribution as his astronomical observations. He conducted actual experiments with falling and rolling objects—a different approach from natural philosophers who had thought about objects in motion but had not actually experimented with them. As with his telescopes, Galileo also improved or developed new technologies to allow him to conduct his experiments. For example, his work on falling objects demonstrated that gravity on Earth accelerates all objects at the same rate, independent of mass. Galileo's observations and experiments with many types of moving objects, such as carts and balls, led him to disagree with the philosophers about when and why objects continued to move or came to rest. Specifically, Galileo found that *an object in motion will continue moving along a straight line with a constant speed until an unbalanced force acts on it to change its state of motion*—an idea with implications for not only the motion of carts and balls but also the orbits of planets.

3.2 The Rise of Scientific Theory: Newton's Laws Govern the Motion of All Objects

Empirical rules, like the ones that Kepler spent his life developing, are only the first step in understanding a phenomenon. Empirical rules *describe*, but they do not *explain*. Kepler described the orbits of planets as ellipses, but he did not explain *why* they should be so.

The next step is to try to understand those empirical rules in terms of more general physical principles or laws. Beginning with basic physical principles and using the tools of mathematics, a scientist works to **derive** the empirically determined rules. Or sometimes a scientist starts with physical laws and predicts relationships, which are then verified (or falsified). When these predictions are indeed borne out by experiment and observation, then we just may have learned something fundamental about how the universe works.

One of the earliest advances in theoretical science was also one of the greatest intellectual accomplishments. The work of Sir Isaac Newton on the nature of motion set the standard for what we now refer to as *scientific theory* and *physical law*. Newton proposed three beautifully elegant laws that govern the motions of all objects in the universe.

Newton's laws of motion are essential to our understanding of the motions of the planets and all other celestial bodies. These laws enabled Newton to look at the motion of a shot fired from a cannon and see the orbits of the planets around the Sun. Through Newton's laws, we see Earth's true place in the universe.

Newton's First Law: Objects at Rest Stay at Rest; Objects in Motion Stay in Motion

The law of **inertia**, which became the cornerstone of physics as **Newton's first law of motion**, was first discovered by Galileo. Newton's first law of motion states that an object in motion tends to stay in motion, and in the same direction, until an unbalanced **force** acts upon it; and an object at rest tends to stay at rest until

Vocabulary Alert

Derive: In common language, this word means to obtain something, often a chemical substance, from a specific source. For example, we might say, "Chocolate is derived from cacao beans." Astronomers, mathematicians, and physicists all use this word to mean deducing a conclusion logically (or mathematically) from a set of starting principles.

Inertia: In common language, we think of inertia as a tendency to remain motionless. Astronomers and physicists think more generally of inertia as the tendency of matter to resist a change in motion—of an object at rest to remain at rest, and of a moving object to remain in motion.

Force: In common language, the word *force* has many meanings. Astronomers specifically mean a push or a pull.

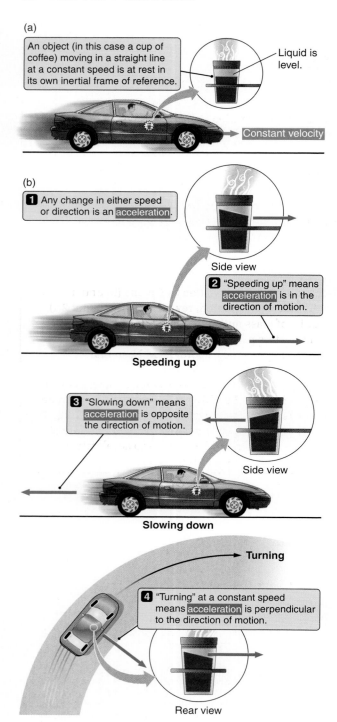

(a)

An object (in this case a cup of coffee) moving in a straight line at a constant speed is at rest in its own inertial frame of reference.

Liquid is level.

Constant velocity

(b)

1 Any change in either speed or direction is an acceleration.

Side view

2 "Speeding up" means acceleration is in the direction of motion.

Speeding up

3 "Slowing down" means acceleration is opposite the direction of motion.

Side view

Slowing down

Turning

4 "Turning" at a constant speed means acceleration is perpendicular to the direction of motion.

Rear view

Figure 3.3 (a) An object moving in a straight line at a constant speed is at rest in its own inertial frame of reference. (b) Any change in the velocity of an object is an acceleration. (Throughout the text, velocity arrows are shown as red and acceleration arrows are shown as green.)

Uniform: Here, astronomers mean that neither the speed nor the direction of travel changes.

an unbalanced force acts upon it. We can remember this law with the much shorter expression, "Things keep on keepin' on." We specify an **unbalanced force** because it is possible for two or more forces to oppose one another in such a way that they are perfectly balanced. In that case, the forces have no effect on the object's motion. For example, a ball sitting on level ground does not accelerate, because the force of gravity is exactly balanced by the force of the surface of Earth pushing up on it.

Recall from Chapter 2 the concept of a frame of reference. Within a frame of reference, only the relative motions between objects have any meaning. There is no perceptible difference between an object at rest and an object in uniform motion. The object "at rest" beside you on the front seat of your car as you drive down the highway is moving at 100 kilometers per hour (kph) according to a bystander along the side of the road—and it is moving at 200 kph according to a car traveling at 100 kph in oncoming traffic. All of these perspectives are equally valid.

A frame of reference moving in a straight line at a constant speed is referred to as an **inertial frame of reference**. Motion is meaningful only when measured relative to an inertial frame of reference, and *any* inertial frame of reference is as good as any other. As illustrated in **Figure 3.3a**, in the frame of reference of a cup of coffee, *it is at rest relative to itself* even if the car is speeding down the road.

Newton's Second Law: Motion Is Changed by Unbalanced Forces

Newton often gets credit for Galileo's insight about inertia because Newton took the crucial next step. Newton's first law says that in the absence of an unbalanced force, an object's motion does not change. **Newton's second law of motion** goes on to say that *if an unbalanced force acts on an object, then the object's motion does change*. Even more, Newton's second law tells us *how* the object's motion changes in response to that force.

In the preceding paragraphs we spoke of "changes in an object's motion," but what does that phrase really mean? When you are in the driver's seat of a car, a number of controls are at your disposal. On the floor are a gas pedal and a brake pedal. You use these to make the car speed up or slow down. A *change in speed* is one way the car's motion can change. But also remember the steering wheel in your hands. When you are moving down the road and you turn the wheel, your speed does not necessarily change, but the direction of your motion does. A *change in direction* is also a change in motion.

Together, the speed and direction of an object's motion are called the object's **velocity**. "Traveling at 50 kph" indicates speed; "traveling north at 50 kph" indicates velocity. The rate at which the velocity of an object changes is called **acceleration**. Acceleration tells you how rapidly a change in velocity happens. For example, if you go from 0 to 100 kph in 4 seconds, you feel a strong push from the seat back as it shoves your body forward, causing you to accelerate along with the car. However, if you take 2 minutes to go from 0 to 100 kph, the acceleration is so slight that you hardly notice it (**Working It Out 3.1**).

Because the gas pedal on a car is often called the accelerator, some people think *acceleration* always means that an object is speeding up. But we need to stress that, as used in physics, any change in motion is an acceleration. **Figure 3.3b** illustrates the point. Slamming on your brakes and going from 100 to 0 kph in 4 seconds is just as much acceleration as going from 0 to 100 kph in 4 seconds. Similarly, the acceleration you experience as you go through a fast, tight

Working It Out 3.1 | Finding the Acceleration

The equation for acceleration states

$$\text{Acceleration} = \frac{\text{How much velocity changes}}{\text{How long the change takes to happen}}$$

We can write this more compactly by expressing the change in velocity as $v_2 - v_1$ and the change in time as $t_2 - t_1$. Acceleration is commonly abbreviated as a. So our equation can be translated to math as

$$a = \frac{v_2 - v_1}{t_2 - t_1}$$

For example, if an object's speed goes from 5 meters per second (m/s) to 15 m/s, then the final velocity is $v_2 = 15$ m/s. The initial velocity is $v_1 = 5$ m/s. So the change in velocity $(v_2 - v_1)$ is 10 m/s. If that change happens over the course of 2 seconds, then the change in time, $t_2 - t_1$, is 2 seconds. So the acceleration is 10 m/s divided by 2 seconds:

$$a = \frac{10 \text{ m/s}}{2 \text{ s}} = 5 \text{ m/s/s}$$

This is the same as saying "5 meters per second squared," which is written as 5 m/s^2 or 5 m s^{-2}

If we want to know how an object's motion is changing, we need to know two things: what unbalanced force is acting on the object, and what is the mass of the object? We can say this formula in words as "the acceleration is equal to the strength of the unbalanced force divided by the mass." We can translate it into math as

$$\text{acceleration} = \frac{\text{strength of unbalanced force}}{\text{mass}}$$

This is a lot to write every time we want to talk about acceleration. Typically, we let a stand for acceleration, F for force, and m for mass:

$$a = \frac{F}{m}$$

Newton's second law is often written as $F = ma$ (to understand how we changed this equation, revisit Working It Out 2.1), giving force as units of mass multiplied by units of acceleration, or kilograms times meters per second squared (kg m/s^2). The units of force are named newtons, abbreviated N, so that $1 \text{ N} = 1 \text{ kg m/s}^2$.

This is the mathematical statement of Newton's second law of motion. This equation says three things: (1) when you push on an object, that object accelerates in the direction you are pushing; (2) the harder you push on an object, the more it accelerates; and (3) the more massive the object is, the harder it is to accelerate it.

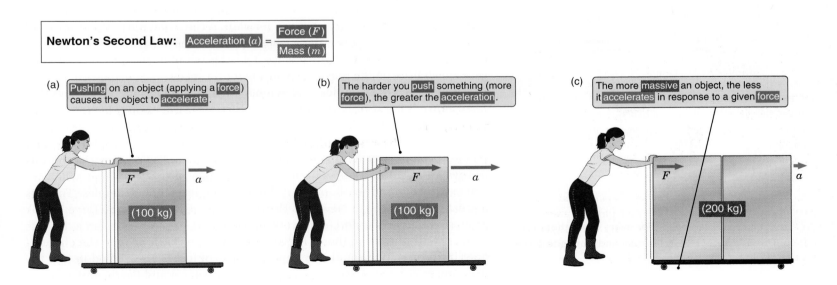

Newton's Second Law: Acceleration $(a) = \dfrac{\text{Force } (F)}{\text{Mass } (m)}$

(a) Pushing on an object (applying a force) causes the object to accelerate.

(b) The harder you push something (more force), the greater the acceleration.

(c) The more massive an object, the less it accelerates in response to a given force.

Figure 3.4 Newton's second law of motion says that the acceleration experienced by an object is determined by the force acting on the object, divided by the object's mass. (Throughout the text, force arrows are shown as blue.)

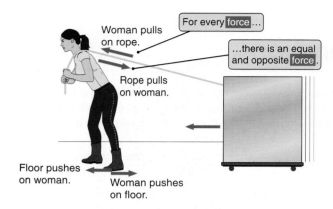

Woman pulls on rope.

For every force …

…there is an equal and opposite force.

Rope pulls on woman.

Floor pushes on woman.

Woman pushes on floor.

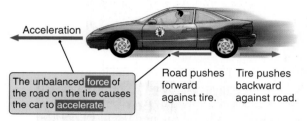

Acceleration

The unbalanced force of the road on the tire causes the car to accelerate.

Road pushes forward against tire.

Tire pushes backward against road.

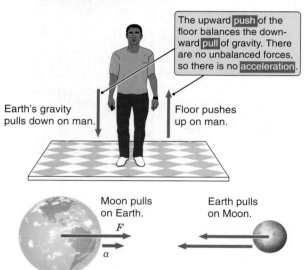

The upward push of the floor balances the downward pull of gravity. There are no unbalanced forces, so there is no acceleration.

Earth's gravity pulls down on man.

Floor pushes up on man.

Moon pulls on Earth.

Earth pulls on Moon.

F

a

Figure 3.5 Newton's third law states that for every force there is always an equal and opposite force. These opposing forces always act on the two different objects in the same pair.

For every force there is an equal and opposite force.

turn at a constant speed is every bit as real as the acceleration you feel when you slam your foot on the gas pedal or the brake pedal. Speeding up, slowing down, turning left, turning right—*if you are not moving in a straight line at a constant speed, you are experiencing an acceleration.*

Newton's second law of motion says that unbalanced forces cause accelerations. The acceleration an object experiences depends on two things. First, it depends on the strength of the unbalanced force acting on the object to change its motion (**Figure 3.4a**). Push three times as hard, and the object experiences three times the acceleration (**Figure 3.4b**). The change in motion occurs in the direction the unbalanced force points. Push an object away from you, and it will accelerate away from you.

The acceleration that an object experiences also depends on its inertia. Some objects—for example, an empty box that a refrigerator was delivered in—are easily shoved around by humans. However, a refrigerator, even though it is about the same size, is not easily shoved around. For our purposes, an object's **mass** is interchangeable with its inertia. An object has inertia because it has mass (**Figure 3.4c**). The greater the mass, the greater the inertia, and the *less* acceleration will occur in response to the same unbalanced force.

Newton's Third Law: Whatever Is Pushed, Pushes Back

Imagine you are standing on a skateboard and pushing yourself along with your foot. Each shove of your foot against the ground sends you faster along your way. But why does this happen? Your muscles flex, and your foot exerts a force on the ground. (Earth does not noticeably accelerate, because its great mass gives it great inertia.) Yet this does not explain *why* you experience an acceleration. The fact that you accelerate means that as you push on the ground, *the ground must be pushing back on you.*

Part of Newton's genius was his ability to see patterns in such everyday events. Newton realized that *every* time one object exerts a force on another, a matching force is exerted by the second object on the first. That second force is exactly as strong as the first force but is in exactly the *opposite* direction. You push backward on Earth, and Earth pushes you forward. A canoe paddle pushes backward through the water, and the water pushes forward on the paddle, sending the canoe along its way. A rocket engine pushes hot gases out of its nozzle, and those hot gases push back on the rocket, propelling it into space.

All of these are examples of **Newton's third law of motion**, which says that *forces always come in pairs, and those pairs are always equal in strength but opposite in direction.* The forces in these action-reaction pairs always act on two different objects. Your weight pushes down on the floor, and the floor pushes back on you with the same amount of force. For every force there is *always* an equal and opposite force. This is one of the few times when we can say *always* and really mean it. **Figure 3.5** gives a few examples. There is a great game hiding in Newton's third law. It is called "find the force." Look around you at all the forces at work in the world, and for each force, find its mate. It will *always* be there! For another example, turn back to the chapter-opening figure. Two equal and opposite pairs of forces are labeled in this figure. A pair of forces are operating between the rocket and Earth, and a pair of forces are operating between the rocket and the exhaust. Because the force of the exhaust on the rocket is larger than the force of gravity on the rocket, the rocket accelerates upward.

To see how Newton's three laws of motion work together, study **Figure 3.6**.

An astronaut is adrift in space, motionless with respect to the nearby space shuttle. With no tether to pull on, how can the astronaut get back to the ship? Suppose the 100-kg astronaut throws a 1-kg wrench directly away from the shuttle at a speed of 10 m/s. Newton's second law says that to cause the motion of the wrench to change, the astronaut has to apply a force to it in the direction away from the shuttle. Newton's third law says that the wrench must therefore push back on the astronaut with just as much force in the opposite direction. The force of the wrench on the astronaut causes the astronaut to begin drifting toward the shuttle. How fast will the astronaut move? Turn to Newton's second law again. A force that causes the 1-kg wrench to accelerate to 10 m/s will have much less effect on the 100-kg astronaut. Because acceleration equals force divided by mass, the 100-kg astronaut will experience only 1/100 as much acceleration as the 1-kg wrench, and so will have 1/100 the final velocity. The astronaut will drift toward the shuttle, but only at the leisurely rate of 1/100 × 10 m/s, or 0.1 m/s.

3.3 Gravity Is a Force between Any Two Objects Due to Their Masses

Having explored both Kepler's empirical description of the motions of planets about the Sun and Newton's laws of motion, we turn now to **gravity**, which unites these two pillars of empirical and theoretical science. The gravitational acceleration near the surface of Earth is usually written as g and has an average value of 9.8 m/s². Experiments show that all objects on Earth fall with this same acceleration. Whether you drop a marble or a cannonball, after 1 second it will be falling at a speed of 9.8 m/s, after 2 seconds at 19.6 m/s, and after 3 seconds at 29.4 m/s. (Air resistance becomes a factor at higher speeds, but it is negligible for dense, slow objects.)

Newton realized that if all objects fall with the same acceleration, then the gravitational *force* on an object must be determined by the object's *mass*. To see why, look back at Newton's second law (acceleration equals force divided by mass, or $a = F/m$). The only way gravitational acceleration can be the same for all objects is if the value of the force divided by the mass is the same for all objects. In other words, make an object twice as massive and you double the gravitational force acting on it. Make an object three times as massive and you triple the gravitational force acting on it.

In precise terms, the gravitational force acting on an object is commonly referred to as the object's **weight**. On the surface of Earth, weight is just mass multiplied by Earth's gravitational acceleration, g. Due to our everyday use of language, it is easy to see why people confuse mass and weight. We often say that an object with a mass of 2 kg "weighs" 2 kg, but it is more correct to express a weight in terms of newtons (N), the metric unit of force:

$$F_{\text{weight}} = m \times g$$

where F_{weight} is an object's weight in newtons, m is the object's mass in kilograms, and g is Earth's gravitational acceleration in meters per second squared. On Earth, an object with a *mass* of 2 kg has a *weight* of 2 kg × 9.8 m/s², or 19.6 N. On the Moon, where the gravitational acceleration is 1.6 m/s², the 2-kg mass would have a weight of 2 kg × 1.6 m/s², or 3.2 N. Although your mass remains the same wherever you are, your weight varies. On the Moon, your weight is about one-sixth of your weight on Earth.

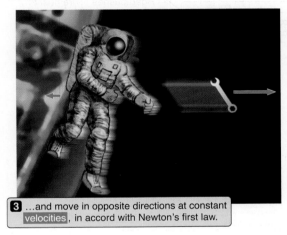

1 An astronaut adrift in space pushes on a wrench, which, according to Newton's third law, pushes back on the astronaut.

2 The wrench and the astronaut experience accelerations as dictated by Newton's second law…

3 …and move in opposite directions at constant velocities, in accord with Newton's first law.

Figure 3.6 According to Newton's laws, if an astronaut adrift in space throws a wrench, the two will move in opposite directions at speeds that are inversely proportional to their masses. (Acceleration and velocity arrows are not drawn to scale.)

All objects on Earth fall with the same acceleration, *g*.

Vocabulary Alert

Weight: In common language, we often use *weight* and *mass* interchangeably. For example, we often interchange kilograms (a unit of mass) and pounds (a unit of weight), but this is technically incorrect. Astronomers use *mass* to refer to the amount of stuff in an object, and *weight* to refer to the force exerted on that object by the planet's gravitational pull. Your weight changes with location, but your mass stays the same no matter what planet you are on.

Newton Reasons His Way to a Law of Gravity

Newton's next great insight came from applying his third law of motion to gravity. Newton's third law states that for every force there is an equal and opposite force. Therefore, if Earth exerts a force of 19.6 N on a 2-kg mass sitting on its surface, then that 2-kg mass *must* exert a force of 19.6 N on Earth as well. Drop a 10-kg frozen turkey and it falls toward Earth—but at the same time, Earth falls toward the 10-kg turkey. The reason we do not notice the motion of Earth is that Earth has a lot of inertia. In the time it takes a 10-kg turkey to fall to the ground from a height of 1 kilometer, Earth has "fallen" toward the turkey by a tiny fraction of the size of an atom.

Newton reasoned that this relationship should work with either object. If doubling the mass of an object doubles the gravitational force between the object and Earth, then doubling the mass of Earth ought to do the same thing. In short, the gravitational force between Earth and an object must be equal to the product of the two masses multiplied by something:

$$\text{Gravitational force} = \text{Something} \times \text{Mass of Earth} \times \text{Mass of object}$$

If the mass of the object were three times greater, then the force of gravity would be three times greater. Likewise, if the mass of Earth were three times what it is, the force of gravity would have to be three times greater as well. If *both* the mass of Earth *and* the mass of the object were three times greater, the gravitational force would increase by a factor of 3×3, or 9 times. Because objects fall toward the center of Earth, we know that gravity is an attractive force acting along a line between the two masses.

"And by the way," reasoned Newton, "why are we restricting our attention to Earth's gravity?" If gravity is a force that depends on mass, then there should be a gravitational force between *any* two masses. Suppose we have two masses—call them mass 1 and mass 2, or m_1 and m_2 for short. The gravitational force between them is something multiplied by the product of the masses:

$$\text{Gravitational force between two objects} = \text{Something} \times m_1 \times m_2$$

We have gotten this far just by combining Galileo's observations of falling objects with (1) Newton's laws of motion and (2) Newton's belief that Earth is a mass just like any other mass. There has been no wiggle room—there is nothing arbitrary in what we have done. But what about that "something" in the previous expression? Today we have instruments sensitive enough that we can put two masses close to each other in a laboratory, measure the force between them, and determine the value of that something directly. Yet Newton had no such instruments. He had to look elsewhere to continue his exploration of gravity.

Kepler had already thought about this question. He reasoned that because the Sun is the focal point for planetary orbits, the Sun must be responsible for exerting an influence over the motions of the planets. Kepler speculated that whatever this influence is, it must grow weaker with distance from the Sun. (After all, it must surely require a stronger influence to keep Mercury whipping around in its tight, fast orbit than it does to keep the outer planets lumbering along their paths around the Sun.) Kepler's speculation went even further. Although he did not know about forces or inertia or gravity, he did know quite a lot about geometry, and geometry alone suggested how this solar "influence" might change for planets progressively farther from the Sun.

As Kepler himself may have speculated, imagine that you have a certain amount of paint to spread over the surface of a sphere. If the sphere is small, you will get a thick coat of paint. But if the sphere is larger, the paint has to spread farther and you get a thinner coat. The surface area of a sphere depends on the square of the sphere's radius. Double the radius of a sphere, and the sphere's surface becomes four times what it was. If you paint this new, larger sphere, the paint must cover four times as much area and the thickness of the paint will be only a fourth of what it was on the smaller sphere. Triple the radius of the sphere and the sphere's surface will be nine times as large, and the thickness of the coat of paint will be only a ninth as thick.

Kepler thought the influence that the Sun exerts over the planets might be like the paint in this example. As the influence of the Sun extended farther and farther into space, it would spread out to cover the surface of a larger and larger imaginary sphere centered on the Sun. The influence of the Sun should be proportional to 1 divided by the square of the distance between the Sun and a planet. We refer to this relationship as an **inverse square law**.

Kepler had an interesting idea, but not a scientific hypothesis with testable predictions. He lacked an explanation for how the Sun influences the planets as well as the mathematical tools to calculate how an object would move. Newton had both. If gravity is a force between *any* two objects, then there should be a gravitational force between the Sun and each of the planets. Might this gravitational force be the same as Kepler's "influence"? If so, gravity might behave according to an inverse square law. Newton's expression for gravity now came to look like this:

$$\frac{\text{Gravitational force}}{\text{between two objects}} = \text{Something} \times \frac{m_1 \times m_2}{(\text{Distance between objects})^2}$$

There is still a "something" left in this expression, and that something is a constant. This constant measures the intrinsic strength of gravity between objects, and it is the same for all objects. Newton named it the **universal gravitational constant**, written as G. It was not until many years later that the actual value of G was first measured. Today, the value of G is accepted as 6.673×10^{-11} m³/(kg s²). The precise value of this constant is still an area of active research for physicists.

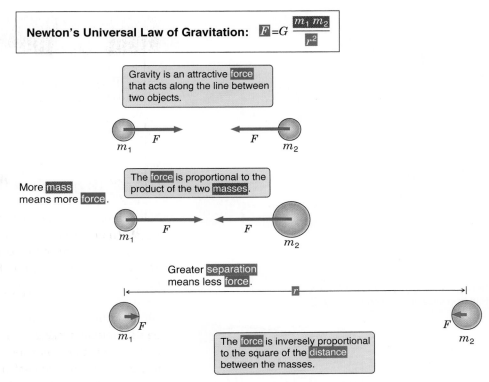

Newton's Universal Law of Gravitation: $F = G\dfrac{m_1 m_2}{r^2}$

Gravity is an attractive force that acts along the line between two objects.

More mass means more force.

The force is proportional to the product of the two masses.

Greater separation means less force.

The force is inversely proportional to the square of the distance between the masses.

Figure 3.7 Gravity is an attractive force between two objects. The force of gravity depends on the masses of the objects and the distance between them.

Putting the Pieces Together: A Universal Law for Gravitation

Newton's **universal law of gravitation**, illustrated in **Figure 3.7**, states that gravity is a force between any two objects having mass and has these properties:

Gravity is governed by an inverse square law.

1. It is an attractive force, F, acting along a straight line between the two objects.
2. It is proportional to the mass of one object, m_1, multiplied by the mass of the other object, m_2. If we double m_1, F increases by a factor of 2. Likewise, if we double m_2, F increases by a factor of 2.

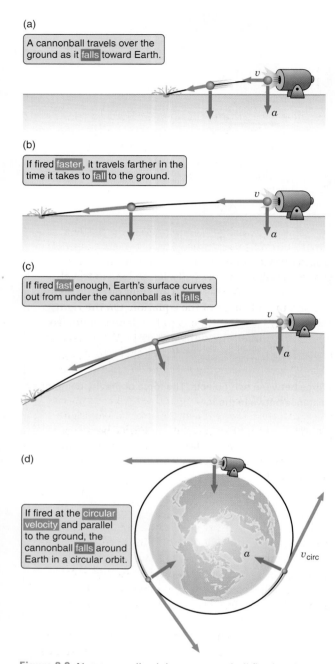

Figure 3.8 Newton realized that a cannonball fired at the right speed would fall around Earth in a circle. Velocity (*v*) is indicated by a red arrow and acceleration (*a*) by a green arrow.

▶❚❚ **AstroTour: Newton's Laws and Universal Gravitation**

3. It is inversely proportional to the square of the distance *r* between the centers of the two objects. If we double *r*, *F* decreases by a factor of 4. If we triple *r*, *F* falls by a factor of 9 (see **Working It Out 3.2**).

Translated into mathematics, and including the constant of proportionality *G*, the universal law of gravitation is

$$F = \frac{Gm_1m_2}{r^2}.$$

3.4 Orbits Are One Body "Falling Around" Another

Newton used his laws of motion and his proposed law of gravity to *calculate* the paths that planets should follow as they move around the Sun. When he did so, his calculations predicted that planetary orbits could be ellipses with the Sun at one focus, that planets should travel faster when closer to the Sun, and that the square of the period should equal the semimajor axis cubed (in appropriate units). In short, Newton's universal law of gravitation *predicted* that planets should orbit the Sun in just the way that Kepler's empirical laws described. This was the moment when it all came together. By *explaining* Kepler's laws, Newton found important corroboration for his law of gravitation. And in the process he moved the cosmological principle (recall Chapter 1) out of the realm of interesting ideas and into the realm of testable scientific theories.

Newton's laws tell us how an object's motion changes in response to forces and how objects interact with each other through gravity. To go from statements about how an object's motion is *changing* to more practical statements about where an object *is*, we have to carefully "add up" the object's motion over time. To see how we can do this, let's begin with a "thought experiment"—the same thought experiment that helped lead Newton to his understanding of planetary motions.

Newton Fires a Shot around the World

Drop a cannonball and it falls directly to the ground, just as any mass does. However, if instead we fire the cannonball from a cannon that is level with the ground, as shown in **Figure 3.8a**, it behaves differently. The cannonball still falls to the ground in the same time as before, but while falling it is also traveling *over* the ground, following a curved path that carries it some horizontal distance before it finally lands. The faster the ball is fired from the cannon (**Figure 3.8b**), the farther it will go before hitting the ground.

In the real world this experiment reaches a natural limit. To travel through air, the cannonball must push the air out of its way—an effect we normally refer to as air resistance—which slows it down. But because this is only a thought experiment, we can ignore such real-world complications. Instead imagine that, having inertia, the cannonball continues along its course until it runs into something. As the cannonball is fired faster and faster, it goes farther and farther before hitting the ground. If the cannonball flies far enough, Earth's surface "curves out from under it" (**Figure 3.8c**). Eventually we reach a point where the cannonball is flying so fast that the surface of Earth curves away from the cannonball at exactly the same rate that the cannonball is falling toward Earth. This is the case shown in **Figure 3.8d**. At this point the cannonball, which always falls *toward the center of Earth*, is literally "falling around the world."

In 1957 the Soviet Union used a rocket to lift Sputnik 1, an object about

Working It Out 3.2 | # Newton's Law of Gravity: Playing with Proportionality

If you are an artist, you play with colors and patterns of light and dark as you create new works. If you are a musician, you might play with tones and rhythms as you compose. Writers play with combinations of words. In exactly the same sense, scientists play with mathematics as they seek new insights into the world around them.

One of the most useful ways to play with equations is to study the proportionalities. We often use the phrase "goes like" to describe proportionality. For example, in Newton's law of gravity, we can see that the force goes like the mass of the objects involved:

$$F = \frac{Gm_1m_2}{r^2}$$

What does that mean? It means that if the mass of one object is doubled, the force is as well. If the mass of one object is increased by a factor of 2.63147, the force is as well. This makes for a handy shorthand way of making calculations without having to actually plug G, r, and the m_2 into the equation. Suppose that your mass, m_1, suddenly increased by a factor of 2 (doubled). How would this affect the force of gravity, F, on you (that is, your weight)? F would also increase by a factor of 2, so that your weight would double. This is somewhat intuitive. If you suddenly had twice as much stuff in your insides, you should certainly weigh twice as much.

If you study the equation, you will see that we are just using a rule we learned in Working It Out 2.1: whatever you do to one side of the equation, you also have to do to the other side. In this case, we multiplied one of the terms on the right by two, so we also had to multiply the term on the left by two.

Let's consider another example. Suppose you go to the doctor's office, and he finds that your weight, F, has been reduced by 10 percent; your new weight is 0.90 times your old weight. How much has your mass decreased? To figure this out, we multiply the term on the left by 0.90. So we also have to multiply something on the right by 0.90. G has not changed, and neither has Earth's mass or radius. All that's left is your mass, which is 0.90 times what it was the last time you were at the doctor. You have lost 10 percent of your body mass.

In these examples, we have been focusing on terms that are *directly* proportional to each other. *Inverse* proportions are slightly more complicated: in this case, when one term is doubled, the other is halved. This happens when one term is in the denominator while the other is in the numerator. We've seen something like this before, when we learned about the relationship between distance, time, and speed:

$$s = \frac{d}{t}$$

Speed and distance are directly proportional to each other. If you traveled twice as far as your friend in the same amount of time, you were going twice as fast as he was. But speed and time are inversely proportional. If it took you twice as long to travel that distance, you were going *half* as fast. Consider our rule about doing the same thing to both sides of the equation. On the right, you multiply the time by two. But time is in the denominator, so you have effectively multiplied the right side by ½. You must also do this on the left, and multiply s by ½. (Notice that two things that are proportional or inversely proportional do not have the same units.)

Let's return to Newton's law of gravity and ask what happens when we change r, the distance between the objects. We might ask, for example, "How would the gravitational force between Earth and the Moon change if the distance between them were doubled?" Newton's law of gravitation is inversely proportional to r^2; that is, it goes like $1/r^2$:

$$F = \frac{Gm_1m_2}{r^2}$$

So if we double r, we expect F to decrease; so far so good. But what do we do about the square? If we put $(2r)$ in where (r) is in the equation, we see that we would have to write $(2r)^2$. This means that the r is squared *and* the 2 is squared. So $(2r)^2 = 4r^2$. We have effectively multiplied the right-hand side by ¼. We must also do that on the left, so the force is ¼ as strong as before. Doubling the distance reduces the force by a factor of 4. This $1/r^2$ proportionality occurs in many contexts in astronomy. Take a moment to calculate how the force would change if the distance increased by a factor of 3, 5, or 10, and if it decreased by a factor of ½, ¼, or ¹⁄₁₀. Once you can do this, you will have a tool you can use again and again in the remaining chapters.

the size of a basketball, high enough above Earth's upper atmosphere that air resistance ceased to be a concern, and Newton's thought experiment became a matter of great practical importance. Sputnik 1 was moving so fast that it fell around Earth, just as the cannonball did in Newton's mind. Sputnik 1 was the first human-made object to orbit Earth. This is what an **orbit** is: one object falling freely around another.

Orbiting means "falling around" a world.

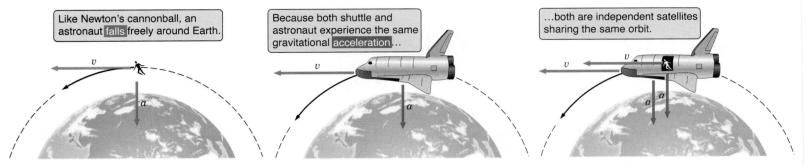

Figure 3.9 A "weightless" astronaut has not escaped Earth's gravity. Rather, an astronaut and a spacecraft share the same orbit as they fall around Earth together.

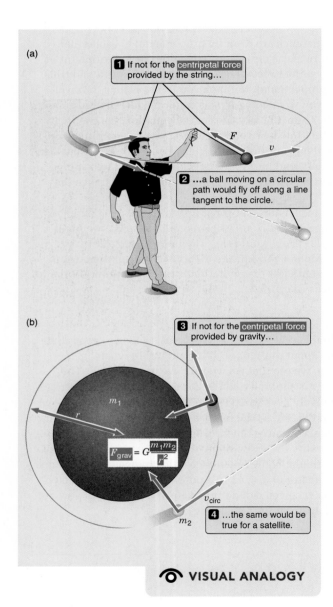

Figure 3.10 (a) A string provides the centripetal force that keeps a ball moving in a circle. (We are ignoring the smaller force of gravity that also acts on the ball.) (b) Similarly, gravity provides the centripetal force that holds a satellite in a circular orbit.

The concept of orbits also answers the question of why astronauts float freely about the cabin of a spacecraft. It is *not* because they have escaped Earth's gravity; it is Earth's gravity that holds them in their orbit. Instead the answer lies in Galileo's early observation that every object falls in just the same way, regardless of its mass. The astronauts and the spacecraft are both moving in the same direction, at the same speed, and are experiencing the same gravitational acceleration, *so they fall around Earth together*. **Figure 3.9** demonstrates this point. The astronaut is orbiting Earth just as the spacecraft is orbiting Earth. On the surface of Earth our bodies try to fall toward the center of Earth, but the ground gets in the way. We experience our weight when we are standing on Earth because the ground pushes on us hard enough to counteract the force of gravity, which is trying to pull us down. In the spacecraft, however, nothing interrupts the astronaut's fall, because the spacecraft is falling around Earth in just the same orbit. The astronaut is in **free fall**.

When one object is falling around another, much more massive object, we say that the less massive object is a **satellite** of the more massive object. Planets are satellites of the Sun, and moons are natural satellites of planets. Newton's imaginary cannonball is a satellite. The spacecraft and the astronauts are independent satellites of Earth that conveniently happen to share the same orbit.

How Fast Must Newton's Cannonball Fly?

If fired fast enough, Newton's cannonball falls around the world; but just how fast is "fast enough"? Newton's orbiting cannonball moves along a circular path at constant speed. This type of motion is referred to as **uniform circular motion**. You are probably familiar with other examples of uniform circular motion. For example, think about a ball whirling around your head on a string, as shown in **Figure 3.10a**. If you were to let go of the string, the ball would fly off in a straight line in whatever direction it was traveling at the time. It is the string that keeps this from happening, constantly changing the direction the ball is traveling. This central force is called a **centripetal force**. Using a more massive ball, speeding up its motion, or making the circle smaller so that the turn is tighter all increase the force needed to keep the ball moving in a circle.

In the case of Newton's cannonball (or a satellite), there is no string to hold the ball in its circular motion. Instead the force is provided by gravity, as illustrated in **Figure 3.10b**. For Newton's thought experiment to work, the force of gravity must be just right to keep the satellite moving on its circular path. Because this force has a specific strength, it follows that the satellite must be moving at a particular speed, which we call v_{circ}, its **circular velocity**. *If the satellite were moving at any other velocity, it would not be moving in a circular orbit.* Remember the cannonball. If the cannonball were moving too slowly, it would drop below the circular path and hit the ground. Similarly, if the cannonball were moving too fast, its motion would carry it above the circular orbit. Only a cannonball moving at just the right velocity—the circular velocity—will fall around Earth on a circular path.

Most Orbits Are Not Perfect Circles

Some Earth satellites travel a circular path at constant speed. Just like the ball on a string, satellites traveling at the circular velocity remain the same distance from Earth at all times, neither speeding up nor slowing down in orbit. But what if the satellite were in the same place in its orbit and moving in the same direction, but traveling *faster* than the circular velocity? The pull of Earth is as strong as ever, but because the satellite has greater speed, its path is not bent by Earth's gravity sharply enough to hold it in a circle. So the satellite begins to climb above a circular orbit.

As the distance between the satellite and Earth increases, the satellite slows down. Think about a ball thrown into the air, as shown in **Figure 3.11a**. As the ball climbs higher, Earth's gravity slows the ball down. The ball climbs more and more slowly until its vertical motion stops for an instant and then is reversed; the ball falls back toward Earth, speeding up along the way. Our satellite does the same thing. As the satellite moves away from Earth, Earth's gravity slows the satellite down. The farther the satellite is from Earth, the more slowly the satellite moves—just like the ball thrown into the air. And just like the ball, the satellite reaches a maximum height on its curving path and then begins falling back toward Earth. As the satellite falls back toward Earth, Earth's gravity speeds it up. The satellite's orbit has changed from circular to elliptical.

This happens for any object in an elliptical orbit, including a planet orbiting the Sun. Kepler's second law says that a planet moves fastest when it is closest to the Sun and slowest when it is farthest from the Sun. Now we know why. As shown in **Figure 3.11b**, planets lose speed as they pull away from the Sun and then gain that speed back as they fall inward toward the Sun.

Newton's laws do more than explain Kepler's laws: they predict orbits beyond Kepler's empirical experience. **Figure 3.12** shows a series of satellite orbits, each with the same point of closest approach to Earth but with different velocities at that point. The greater the speed a satellite has at its closest approach to Earth, the farther the satellite is able to pull away from Earth, and the more eccentric its orbit becomes. As long as it remains elliptical, no matter how eccentric, the orbit is called a **bound orbit** because the satellite is gravitationally bound to the object it is orbiting.

In this sequence of faster and faster satellites there comes a point of no return—a point when the satellite is moving so fast that gravity is unable to reverse its outward motion, so the satellite coasts away from Earth, never to return. This indeed is possible. The lowest speed at which this happens is called the **escape velocity**, v_{esc}. Once a satellite's velocity at closest approach equals or exceeds v_{esc}, it is in an **unbound orbit**. The object is no longer gravitationally bound to the body that it was orbiting. A comet traveling on an unbound orbit makes only a single pass around the Sun and then is back off into deep space, never to return.

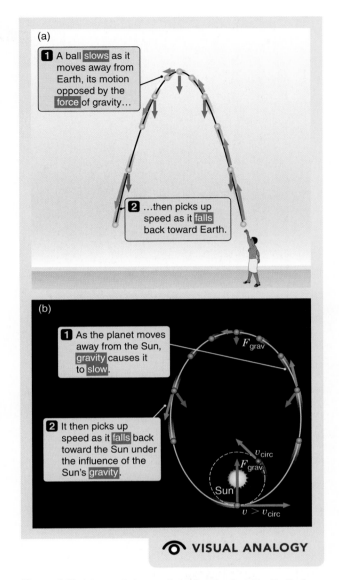

(a)

1 A ball slows as it moves away from Earth, its motion opposed by the force of gravity...

2 ...then picks up speed as it falls back toward Earth.

(b)

1 As the planet moves away from the Sun, gravity causes it to slow.

2 It then picks up speed as it falls back toward the Sun under the influence of the Sun's gravity.

F_{grav}

v_{circ}

F_{grav}

Sun

$v > v_{\text{circ}}$

VISUAL ANALOGY

Figure 3.11 (a) A ball thrown into the air slows as it climbs away from Earth and then speeds up as it heads back toward Earth. (b) A planet on an elliptical orbit around the Sun does the same thing. (Although no planet has an orbit as eccentric as the one shown here, the orbits of comets can be far more eccentric.)

▶❚❚ **AstroTour: Elliptical Orbits**

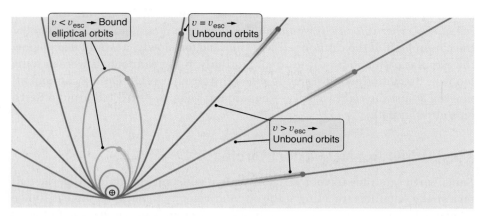

Figure 3.12 A range of different orbits that share the same point of closest approach but differ in velocity at that point. An object's velocity at closest approach determines the orbit shape and whether the orbit is bound or unbound.

Newton's Theory Is a Powerful Tool for Measuring Mass

Newton showed that the same physical laws that describe the flight of a cannonball on Earth also describe the motions of the planets through the heavens. In doing so, Newton opened up an entirely new way of investigating the universe. Copernicus may have dislodged Earth from the center of the universe and started us on the way toward the cosmological principle, but it was Newton who moved the cosmological principle out of the realm of philosophy and into the realm of testable scientific theory by extending his theory of gravity on Earth to the gravitational interaction between astronomical objects.

Besides being more philosophically satisfying than simple empiricism, Newton's method is far more powerful. For example, Newton's laws can be used to measure the masses of Earth and the Sun. This calculation could never be done with Kepler's empirical rules alone. If we can measure the size and period of an orbit—*any* orbit—then we can use Newton's law of universal gravitation to calculate the mass of the object being orbited. This is also true for the masses of other planets, distant stars, our galaxy and distant galaxies, and vast clusters of galaxies. In fact, *almost all of our knowledge about the masses of astronomical objects comes directly from the application of this one law.*

Newton found that his laws of motion and gravitation predicted elliptical orbits that agree exactly with Kepler's empirical laws. *This is how Newton tested his theory that the planets obey the same laws of motion as cannonballs and how he confirmed that his law of gravitation is correct.* Had Newton's predictions not been borne out by observation, he would have had to throw them out and go back to the drawing board.

Kepler's empirical rules for planetary motion pointed the way for Newton and provided the crucial observational test for Newton's laws of motion and gravitation. At the same time, Newton's laws of motion and gravitation provided a powerful new understanding of why planets and satellites move as they do. Theory and empirical observation work hand in hand, and our understanding of the universe strides forward.

> Newton's laws provided a physical explanation for Kepler's empirical results.

READING ASTRONOMY News

Reminder: Clear Your Car of Snow

By **Jason Schreiber**, *Union Leader*

Motorists who fail to properly clear snow and ice from their vehicles after this week's powerful nor'easter may get a cold reception from police.

Authorities are warning drivers to clean up or face possible fines under a negligent driving law.

"It basically comes down to laziness and disrespect for other drivers. Unfortunately, you have to have these laws because people just lack the common sense when driving a vehicle," Plaistow police Sgt. Pat Caggiano said.

The law was enacted in 2002 following the death of Jessica Smith. The Nashua woman was involved in a fatal accident caused by a large piece of ice that flew off a tractor-trailer in 1999.

The law has been on the books for almost a decade, but police said many drivers still don't realize that they could be ticketed for an improperly cleaned vehicle.

The problem occurs most often after a storm like the one on Wednesday, which dumped close to two feet of snow in some areas. Some drivers clear away only enough snow to be able to see through the windshield, leaving several inches covering the hood, roof, and trunk, police said.

Danville Police Chief Wade Parsons knows the dangers of falling ice and snow. A year ago his department investigated an incident in which a 40-pound chunk of ice slid off the roof of a box truck and smashed through the windshield of a car driven by 82-year-old Edo Crescentini.

The Kingston man was traveling east on Route 111 in Danville when the ice blew off the westbound-headed truck. The ice flew into the eastbound lane and struck the windshield of Crescentini's Ford Taurus. The ice just missed his head.

"Motorists can be held liable for any snow or ice that blows off their car and causes injury, property damage or results in an accident," Parsons said.

Snow left clinging to windows can create another traffic hazard as well by obstructing a driver's view.

"This is a real concern for us when motorists don't completely clean the snow and ice off their vehicles before heading out, especially following a substantial snowstorm like we experienced (Wednesday). Visibility at intersections is greatly reduced because of the high snow banks. That, coupled with a car that has not been completely cleaned off, creates a recipe for disaster," Parsons said.

The law states that anyone who "drives a vehicle negligently or causes a vehicle to be driven negligently . . . or in a manner that endangers or is likely to endanger any person or property shall be guilty of a violation."

Motorists can face fines of $250 to $500 for a first offense and up to $1,000 for a second or subsequent offense.

State Police Sgt. Gary Wood of Troop A in Epping said troopers will stop drivers if they notice snow and ice build-up.

The trooper uses his own discretion when deciding whether to give a ticket.

Police said their main concern is getting the message out to drivers.

"If you're driving your vehicle down the street with what amounts to basically two eye holes cleared out on your windshield or what looks like a mattress of snow and ice on your roof, you're likely to get stopped," Hampstead police Lt. John Frazier said.

Evaluating the News

1. In the first accident described, Jessica Smith was killed in an accident caused by a large piece of ice that flew off a tractor-trailer. This could be taken to imply that she was directly struck by the ice. Is there another way this accident could have happened?

2. In the second accident described, a 40-pound chunk of ice slid off the roof of a box truck and smashed through a windshield. About how big would a 40-pound chunk of ice be? (Hint: A gallon of water weighs 8 pounds.)

3. In this second accident, if the ice had been set gently on the windshield, would the windshield have broken?

4. Make a sketch of the second accident as it occurred. Label the direction the truck was traveling, the direction the ice was traveling through the air, and the direction the Ford Taurus was traveling.

5. Assuming that the car and the truck were each traveling at 72 km/hr (45 mph), and that the ice was going about the same speed as the truck, how fast were the ice and the windshield traveling relative to each other when the ice hit the windshield in the second accident?

6. Are you surprised at the results of your calculations? Why or why not?

SUMMARY

3.1 Galileo's experiments set the stage for Newton.

3.2 Newton's three laws (in shorthand: inertia, $F = ma$, and "every action has an equal and opposite reaction") govern the motion of all objects. Unbalanced forces cause accelerations, or changes in motion. Proportionality describes patterns and relationships in nature. Mass is the property of matter that gives it inertia. Inertia resists changes in motion.

3.3 Gravity, as one of the fundamental forces of nature, binds the universe together. Gravity is a force between any two objects due to their masses. The force of gravity is proportional to the mass of each object, and inversely proportional to the square of the distance between them.

3.4 Planets orbit the Sun in bound, elliptical orbits because of gravity. Each circular orbit has a characteristic circular velocity. Orbits become "unbound" when the object reaches escape velocity. Knowing these properties of orbits allows astronomers to measure the masses of planets or stars.

✦ SUMMARY SELF-TEST

1. Suppose that you drop the following objects off of a tall tower. Rank the objects in terms of the gravitational force on them, from smallest to largest.
    a. an apple
    b. a decorative Styrofoam "apple"
    c. a decorative gold "apple"

2. Rank the choices from question 1 in terms of the gravitational acceleration they experience, from smallest to largest.

3. Imagine you are walking along a forest path. Which of the following is *not* an action-reaction pair in this situation?
    a. The gravitational force between you and Earth; the gravitational force between Earth and you.
    b. Your shoe pushes back on Earth; Earth pushes forward on your shoe.
    c. Your foot pushes back on the inside of your shoe; your shoe pushes forward on your foot.
    d. You push down on Earth; Earth pushes you forward.

4. An unbalanced force must be acting when
    a. an object accelerates.
    b. an object changes direction but not speed.
    c. an object changes speed but not direction.
    d. an object changes speed and direction.
    e. all of the above

5. In Newton's universal law of gravitation,
    a. the force is proportional to both masses.
    b. the force is proportional to the radius.
    c. the force is proportional to the radius squared.
    d. the force is inversely proportional to the orbiting mass.

6. Suppose you are transported to a planet with twice the mass of Earth, but the same radius of Earth. Your weight would _____ by a factor of _____.
    a. increase; 2   b. increase; 4
    c. decrease; 2   d. decrease; 4

QUESTIONS AND PROBLEMS

True/False and Multiple-Choice Questions

7. T/F: The natural state of objects is to be at rest. This is why a book, given a push across a table, will eventually slow to a stop.

8. T/F: A force is required to keep an object moving at the same velocity.

9. T/F: When an object falls to Earth, Earth also falls up to the object.

10. T/F: To find the mass of a central object, such as the Sun, we only need to know the semimajor axis and period of an orbit, such as Earth's.

11. Imagine that you are pulling a small child on a sled by means of a rope. Which of the following are action-reaction pairs?
    a. You pull forward on the rope; the rope pulls backward on you.
    b. The sled pushes down on the ground; the ground pushes up on the sled.
    c. The sled pushes forward on the child; the child pushes back on the sled.
    d. The rope pulls forward on the sled; the sled pulls backward on the rope.

12. Suppose that you read about a new car that can go from 0 to 100 km/h in only 2.0 seconds. What is this car's acceleration?
    a. about 50 km/h
    b. about 14 m/s^2
    c. about 50 km/s^2
    d. about 200 km
    e. about 0.056 km/h^2

13. Imagine that you are standing at the top of a tall tower. You drop four objects, all the size of a bowling ball. Each is made of a different substance: Styrofoam, lead, Bubble Wrap, and pumpkin. Neglecting air resistance, in what order do they reach the ground?
    a. lead, pumpkin, Bubble Wrap, Styrofoam
    b. lead, Bubble Wrap, pumpkin, Styrofoam
    c. Styrofoam, lead, Bubble Wrap, pumpkin
    d. None of the above; they all reach the ground at the same time.

Conceptual Questions

14. Kepler's and Newton's laws all tell us something about the motion of the planets, but there is a fundamental difference between them. What is the difference?

15. Aristotle taught that the natural state of all objects is to be at rest. Even though this idea seems consistent with what we observe around us, explain why Aristotle's reasoning was wrong.

16. What is inertia? How is it related to mass?

17. What is the difference between speed and acceleration?

18. When riding in a car, we can sense changes in speed or direction through the forces that the car applies on us. Do we wear seat belts in cars and airplanes to protect us from speed or from acceleration? Explain your answer.

19. Imagine a planet moving in a perfectly circular orbit around the Sun. Because the orbit is circular, the planet is moving at a constant speed. Is this planet experiencing acceleration? Explain your answer.

20. An astronaut standing on Earth can easily lift a wrench having a mass of 1 kg, but not a scientific instrument with a mass of 100 kg. In the International Space Station she is quite capable of manipulating both, although the scientific instrument responds much more slowly than the wrench. Explain why.

21. In 1920, a *New York Times* editor refused to publish an article based on rocket pioneer Robert Goddard's paper that predicted spaceflight, saying that "rockets could not work in outer space because they have nothing to push against" (a statement that the *Times* did not retract until July 20, 1969, the date of the *Apollo 11* Moon landing). You, of course, know better. What was wrong with the editor's logic?

22. Explain the difference between weight and mass.

23. On the Moon, your weight is different from your weight on Earth. Why?

24. Picture a swinging pendulum. During a single swing, the bob of the pendulum first falls toward Earth and then moves away until the gravitational force between it and Earth finally stops its motion. Would a pendulum swing if it were in orbit? Explain your reasoning.

25. Describe the difference between a bound orbit and an unbound orbit.

26. Two objects are leaving the vicinity of the Sun, one traveling in a bound orbit and the other in an unbound orbit. What can you say about the future of these two objects? Would you expect either of them to eventually return?

27. Suppose astronomers discovered a object approaching the Sun in an unbound orbit. What would that say about the origin of the object?

28. What is the advantage of launching satellites from spaceports located near the equator? Why are satellites never launched toward the west?

Problems

29. A sports car accelerates from 0 kph to 100 kph in 4 seconds.
    a. What is its average acceleration?
    b. Suppose the car has a mass of 1,200 kg. How strong is the force on the car?
    c. What supplies the "push" that accelerates the car?

30. A train pulls out of a station accelerating at 0.1 m/s². What is its speed, in kilometers per hour, 2½ minutes after leaving the station?

31. Flybynite Airlines takes 3 hours to fly from Baltimore to Denver at a speed of 800 kph. To save fuel, management orders its pilots to reduce their speed to 600 kph. How long will it now take passengers on this route to reach their destination?

32. Suppose that you are pushing a small refrigerator of mass 50 kg on wheels. You push with a force of 100 N.
    a. What is the refrigerator's acceleration?
    b. Assume the refrigerator starts at rest. How long will the refrigerator accelerate at this rate before it gets away from you (that is, is moving faster than you can run—on the order of 10 m/s)?

33. You are driving down a straight road at a speed of 90 kph and see another car approaching you at a speed of 110 kph in the frame of the road.
    a. Relative to your own frame of reference, how fast is the other car approaching you?
    b. Relative to the other driver's frame of reference, how fast are you approaching the other driver's car?

34. You are riding along on your bicycle at 20 kph and eating an apple. You pass a bystander.
    a. How fast is the apple moving in your frame of reference?
    b. How fast is the apple moving in the bystander's frame of reference?
    c. Whose perspective is more valid?

35. How long does it take Newton's cannonball, moving at 7.9 km/s just above Earth's surface, to complete one orbit around Earth?

36. Two balls, one of gold with a mass of 100 kg and one of wood with a mass of 1 kg, are suspended 1 meter apart. What is the attractive force, in newtons, of
    a. the gold ball acting on the wooden ball?
    b. the wooden ball acting on the gold ball?

37. Earth's mean radius and mass are 6,371 km and 5.97×10^{24} kg, respectively. Show that the acceleration of gravity at the surface of Earth is 9.80 m/s².

38. (a) What does an acceleration of 9.8 m/s² mean? (b) If you jumped from a plane, after 1 second you would be falling at a speed of 9.8 m/s. After 2 seconds your speed would be $2 \times 9.8 = 19.6$ m/s, or slightly more than 70 kph. In the absence of any air resistance, how fast would you be falling after 20 seconds?

39. *Weight* refers to the force of gravity acting on a mass. We often calculate the weight of an object by multiplying its mass by the local acceleration due to gravity. The value of gravitational acceleration on the surface of Mars is 0.377 times that on Earth. If your mass is 85 kg, your weight on Earth is 830 N ($m \times g = 85$ kg $\times$ 9.8 m/s² = 830 N). What would be your mass and weight on Mars?

40. Suppose you go skydiving.
　a. Just as you fall out of the airplane, what is your gravitational acceleration?
　b. Would this acceleration be bigger, smaller, or the same if you were strapped to a flight instructor, and so had twice the mass?
　c. Just as you fall out of the airplane, what is the gravitational force on you (assume your mass is 70 kg)?
　d. Is the gravitational force bigger, smaller, or the same if you were strapped to a flight instructor and so had twice the mass?

41. The average distance of Uranus from the Sun is about 19 times Earth's distance from the Sun. How much stronger is the Sun's gravitational force on Earth than on Uranus?

42. Earth speeds along at 29.8 km/s in its orbit. Neptune's nearly circular orbit has a radius of 4.5×10^9 km, and the planet takes 164.8 years to make one trip around the Sun. Calculate the speed at which Neptune plods along in its orbit. (Hint: The circular velocity is the speed at which an object travels in a circle. This is the distance traveled [the circumference of the circle, $2\pi r$] divided by the time it takes.)

43. Assume that a planet just like Earth is orbiting the bright star Vega at a distance of 1 astronomical unit (AU). The mass of Vega is twice that of the Sun.
　a. How long in Earth years will it take to complete one orbit around Vega?
　b. How fast is the Earth-like planet traveling in its orbit around Vega?

44. Using 6,371 km for Earth's radius, compare the gravitational force acting on NASA's space shuttle when sitting on its launchpad with the gravitational force acting on it when orbiting 350 km above Earth's surface.

 SmartWork, Norton's online homework system, includes algorithmically generated versions of these questions, plus additional conceptual exercises. If your instructor assigns questions in SmartWork, log in at **smartwork.wwnorton.com**.

 StudySpace is a free and open website that provides a Study Plan for each chapter of **Understanding Our Universe**. Study Plans include animations, reading outlines, vocabulary flashcards, and multiple-choice quizzes, plus links to premium content in SmartWork and the ebook. Visit **wwnorton.com/studyspace**.

Exploration | Newtonian Features

wwnorton.com/studyspace

In the last Exploration, we used the Planetary Orbit Simulator to explore Kepler's laws for Mercury. Now that we know how Newton's laws explain why Kepler's laws describe orbits, we will revisit the simulator to explore the Newtonian features of Mercury's orbit. Visit StudySpace, and open the Planetary Orbit Simulator applet in Chapter 3.

Accelerations

To begin exploring the simulation, set parameters for "Mercury" in the "Orbit Settings" panel and then click "OK." Click the Newtonian Features tab at the bottom of the control panel. Select "show solar system orbits" and "show grid" under Visualization Options. Change the animation rate to 0.01, and press the "start animation" button.

Examine the graph at the bottom of the panel.

1. **Where is Mercury in its orbit when the acceleration is smallest?**

2. **Where is Mercury in its orbit when the acceleration is largest?**

3. **What are the values of the largest and smallest accelerations?**

In the Newtonian Features graph, mark the boxes for "vector" and "line" that correspond to the acceleration. These will insert an arrow that shows the direction of the acceleration, and a line that extends the arrow.

4. **To what solar system object does the arrow point?**

Think about Newton's second law.

5. **In what direction is the force on the planet?**

Velocities

Examine the graph at the bottom of the panel again.

6. **Where is Mercury in its orbit when the velocity is smallest?**

7. **Where is Mercury in its orbit when the velocity is largest?**

8. **What are the values of the largest and smallest velocities?**

Add the velocity vector and line to the simulation by clicking on the boxes in the graph window. Study the arrows carefully.

9. **Are the velocity and the acceleration always perpendicular (is the angle between them always 90 degrees)?**

10. **If the orbit were a perfect circle, what would the angle be between the velocity and the acceleration?**

Hypothetical Planet

Use the Orbit Settings to change the semimajor axis to 0.8 AU.

11. **How does this imaginary planet's orbital motion compare to Mercury's?**

Now change the semimajor axis to 0.1 AU.

12. **How does this planet's orbital period now compare to Mercury's?**

13. **Summarize your observations of the relationship between the speed of an orbiting object and the semimajor axis.**

4 Light and Telescopes

Light is fundamental to our experience because it is how we most directly perceive things out of our reach. Our knowledge of the universe beyond Earth comes primarily from light given off or reflected by astronomical objects. Fortunately, light is a *very* informative messenger. Light from an object tells us about its temperature, composition, speed, and even the nature of the material between us and the object.

Light transports energy generated in the heart of a star outward through the star and off into space. In this chapter, we begin to learn about light and how astronomers use it as a tool to investigate the universe. Studying light is how we know what stars are made of, how they form, and how they die. Studying light also teaches us, somewhat surprisingly, about the size and scale of the universe, how and when it began, and how it is changing over time. To understand these mysteries, we must first understand how astronomers investigate them.

✧ LEARNING GOALS

Unlike the physicist or the chemist, who has control over the conditions in a laboratory, the astronomer must try to learn the secrets of the universe from the light that reaches us from distant objects. But this information must first be collected and processed before it can be analyzed and converted into useful knowledge. In the image at right, a student has just acquired a new telescope. She is excited to point it at the sky, but has taken a moment to figure out the path the light takes as it comes through the telescope—an essential step toward figuring out how to use it. By the end of this chapter, you should be able to make similar diagrams for other telescopes and explain why most modern telescopes are reflectors like this one. You should understand how astronomical observations of different kinds of light give us different kinds of information about the universe. You should also be able to:

- Compare and contrast waves and particles

- Explain how light sometimes acts like a wave and sometimes acts like a particle

- Explain how astronomical observations using different kinds of light give us different kinds of information about the universe

- List the types of telescopes, and match them with the types of light they collect

- Understand how a telescope's aperture and focal length relate to resolution and image size

EYEPIECE

APERTURE

SECONDARY
MIRROR

PRIMARY MIRROR

4.1 The Speed of Light

In the 1670s, Danish astronomer Ole Rømer (1644–1710) was studying the moons of Jupiter, measuring the times when each moon disappeared behind the planet. Much to his amazement Rømer found that rather than maintaining a regular schedule, the observed times of these events would slowly drift in comparison with predictions. Sometimes the moons disappeared behind Jupiter sooner than expected, and at other times they went behind Jupiter later than expected. Rømer realized that the difference depended on where Earth was in its orbit. If he began tracking the moons when Earth was closest to Jupiter, then by the time Earth was farthest from Jupiter, the moons were a bit over 16½ minutes "late." But if he waited until Earth was once again closest to Jupiter, the moons once again passed behind Jupiter at the predicted times.

Rømer was seeing the first clear evidence that light travels at a finite speed. As shown in **Figure 4.1**, the moons appeared "late" when Earth was farther from Jupiter because the light had to travel the extra distance between the two planets. The value that Rømer announced in 1676 was a bit on the low side—2.25×10^8 meters per second (m/s)—because the size of Earth's orbit was not well known. Modern measurements of the speed of light in a **vacuum** give a value of 2.99792458×10^8 m/s. In this book, we will round up to 3×10^8 m/s. The International Space Station moves around Earth at a dazzling speed of about 28,000 kilometers per hour, orbiting Earth in about 90 minutes. Light travels almost 40,000 times faster than this and could circle Earth in only ⅐ of a second.

The travel time of light is a convenient way of expressing cosmic distances, and the basic unit is the **light-year**. Pause for a moment to consider the light-year, the distance light travels in one year. Imagine traveling around Earth in ⅐ of a second. Now try to imagine traveling at that speed for about 8 minutes. This is the distance to the Sun. Now try to imagine traveling at that speed for an entire year. This is not even ¼ of the way to the star closest to the Sun. The universe is immense.

The speed of light in a vacuum is a fundamental constant, usually written as *c*. Keep in mind, however, that light travels at this speed *only* in a vacuum. The speed of light through a medium such as air or glass is *always* less than *c*.

The speed of light in a vacuum is 300,000 km/s.

A light-year is the distance that light travels in one year.

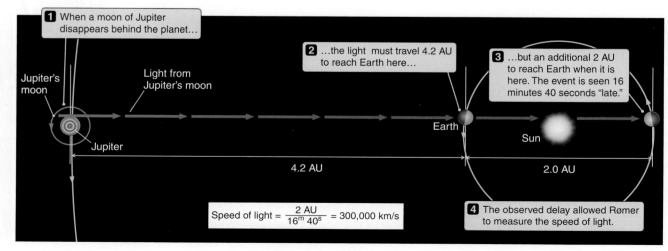

Figure 4.1 Danish astronomer Ole Rømer realized that apparent differences between the predicted and observed orbital motions of Jupiter's moons depend on the distance between Earth and Jupiter. He used these observations to measure the speed of light.

1 When a moon of Jupiter disappears behind the planet…

2 …the light must travel 4.2 AU to reach Earth here…

3 …but an additional 2 AU to reach Earth when it is here. The event is seen 16 minutes 40 seconds "late."

Jupiter's moon

Light from Jupiter's moon

Jupiter

Earth

Sun

4.2 AU

2.0 AU

$$\text{Speed of light} = \frac{2 \text{ AU}}{16^m 40^s} = 300{,}000 \text{ km/s}$$

4 The observed delay allowed Rømer to measure the speed of light.

4.2 What Is Light?

To understand what light is and how it interacts with matter, we must first understand what matter is. **Matter** is anything that occupies space and has mass. **Atoms**
are the fundamental building blocks of matter, consisting of a central massive
nucleus of **protons**, which are positively charged, and **neutrons**, which have no
charge. A cloud of negatively charged **electrons** surrounds the nucleus. Atoms
with the same number of protons are all of the same type, known as an **element**.
For example, an atom with two protons is the element helium (**Figure 4.2**). **Molecules** are groups of atoms bound together by shared electrons. We return to this
model of the atom throughout the book.

A charged particle, such as an electron or a proton, creates an **electric field**
that points away from the charge if the charge is positive, as shown in **Figure 4.3**,
or toward it if the charge is negative. This field exerts an **electric force** on any
nearby charged particle. An electric force is caused by the *presence* of charged
particles, and it points directly toward or away from those particles. If the two
particles have opposite charges, they are attracted to each other. If they have the
same charge, they repel one another. A **magnetic force**, on the other hand, is a
force arising from the *motion* of charged particles. Moving charges produce **magnetic fields** that circle around them and exert a force on nearby moving charged
particles. Motionless charges produce no magnetic force. An electromagnetic
wave consists of changing electric and magnetic fields.

Electric and magnetic fields are tied together. A changing electric field causes
a magnetic field, and a changing magnetic field causes an electric field. These
changes feed on each other. Once the process starts, it can become self-sustaining.
An accelerating charged particle creates these varying fields, producing an **electromagnetic wave** (**Figure 4.4**) that travels out in all directions from the accelerating charge. The Scottish physicist James Clerk Maxwell (1831–1879) developed
a theory of electromagnetic waves, and he predicted the speed at which such a
wave should travel. When Maxwell carried out this calculation, he discovered
that this speed should be 3×10^8 m/s—which is the speed of light. Maxwell had
shown that light is an electromagnetic wave.

Maxwell's wave description of light also explains how light is made and how
it interacts with matter. Imagine a sink full of still water. A drop falls from the

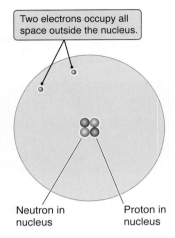

Figure 4.2 The helium atom.

Accelerating charges emit light.

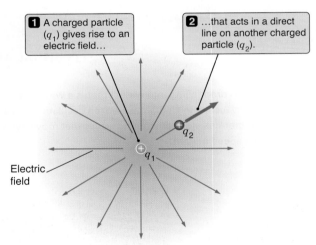

Figure 4.3 Maxwell's theory of electromagnetic radiation describes how electrically charged particles move and
interact. A positively charged particle, q_1, has an electric
field that acts outward along the field lines.

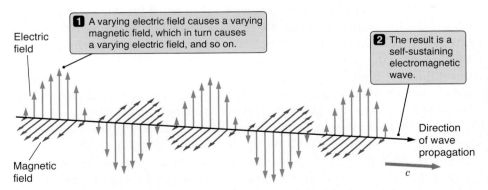

Figure 4.4 Far from its source, an electromagnetic wave consists of oscillating electric and
magnetic fields that are perpendicular both to each other and to the direction in which the
wave travels.

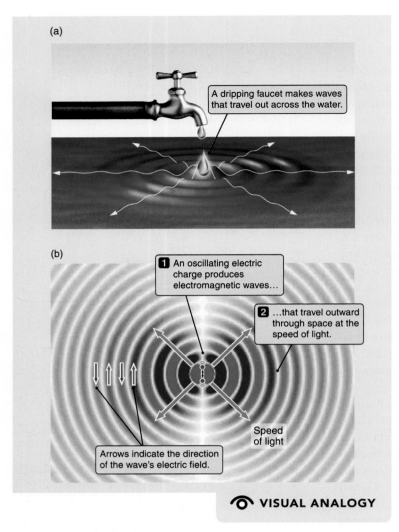

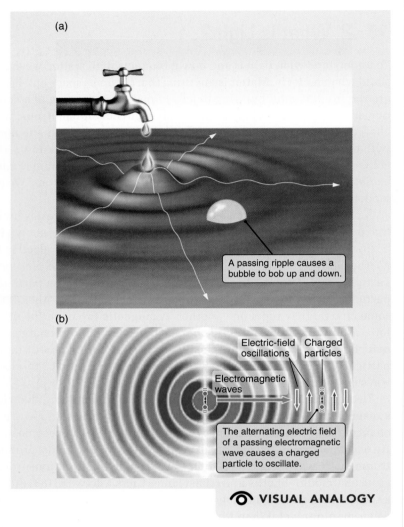

Figure 4.5 (a) A drop falling into water creates a wave that moves across the water's surface. (b) In similar fashion, an oscillating (accelerated) electric charge generates electromagnetic waves that move away from the charge at the speed of light.

Figure 4.6 (a) When waves moving across the surface of water reach a bubble, they cause the bubble to bob up and down. (b) Similarly, a passing electromagnetic wave causes an electric charge to oscillate in response to the wave.

faucet into the sink. This causes a disturbance, or wave, that moves outward as a ripple on the surface of the water (**Figure 4.5a**). In much the same way, an accelerating electric charge (**Figure 4.5b**) causes a disturbance that moves outward through space as an electromagnetic wave. However, whereas the ripples in the sink are *mechanical* distortions of the water's surface and require a medium to travel through, light waves are a fundamentally different type of wave and do not require a medium. Light waves result not from mechanical distortion, but from changes in the strength of the electric and magnetic fields.

Now imagine that a soap bubble is floating in the sink, as in **Figure 4.6a**. The bubble remains stationary until the ripple from the dripping faucet reaches it. The rising and falling water causes the bubble to rise and fall. Similarly, the electromagnetic wave causes charged particles to move about (**Figure 4.6b**). It takes energy to produce an electromagnetic wave, and that energy is carried through space by the wave. Matter far from the source of the wave can absorb this energy. In this way, some of the energy lost by the generating particles is transferred to other charged particles. The *emission* and *absorption* of light by matter are the

Working It Out 4.1 | # Wavelength and Frequency

When you tune in to a radio station at 770 AM, you are receiving an electromagnetic signal that is broadcast at a frequency of 770 kilohertz (kHz), or 7.7×10^5 Hz. We can use the relationship between wavelength and frequency to calculate the wavelength of the AM signal:

$$\lambda = \frac{c}{f} = \frac{3 \times 10^8 \text{ m/s}}{7.7 \times 10^5/\text{s}} = 390 \text{ m}$$

This AM wavelength is about four times the length of a football field.

We can compare this wavelength with that of a typical FM signal, 99.5 FM, which is broadcast at a frequency of 99.5 megahertz (MHz), or 9.95×10^7 Hz:

$$\lambda = \frac{c}{f} = \frac{3 \times 10^8 \text{ m/s}}{9.95 \times 10^7/\text{s}} = 3 \text{ m}$$

This FM wavelength is about 10 feet. FM wavelengths are much shorter than AM wavelengths.

The human eye is most sensitive to green and yellow light, which has a wavelength of about 500–550 nm. Green light with a wavelength of 520 nm has a frequency of

$$f = \frac{c}{\lambda} = \left(\frac{3 \times 10^8 \text{ m/s}}{5.2 \times 10^{-7} \text{ m}} \right) = \frac{5.8 \times 10^{14}}{\text{s}} = 5.8 \times 10^{14} \text{ Hz}$$

This frequency corresponds to 580 *trillion* (580 million million) wave crests passing by each second.

---

result of the interaction of electric and magnetic fields with electrically charged particles.

Waves are characterized by four quantities—*amplitude*, *speed*, *frequency*, and *wavelength*—as illustrated in **Figure 4.7**. The **amplitude** of a wave is the height of the wave above the equilibrium position. (In the case of light waves, the amplitude is related to the brightness of the light.) A wave travels at a particular speed, v. (In the case of light waves, as we have seen, the speed is 3×10^8 m/s and is written as c.) The number of wave crests passing a point in space each second is called the wave's **frequency**, f. Frequency is measured in cycles per second, or **hertz** (Hz). The distance from one crest to the next is the **wavelength**, λ (this is the Greek letter *lambda*). One commonly used unit for wavelengths of light is the nanometer (nm). A nanometer is one-billionth (10^{-9}) of a meter.

The speed of a wave can be found by multiplying the frequency and the wavelength. Translating this idea into math, we have $v = \lambda f$. The speed of light in a vacuum is always c, so once the wavelength of a wave of light is known, its frequency is known, and vice versa. Since its speed is constant, its wavelength and frequency are inversely proportional to each other: if you increase its wavelength, you must decrease its frequency. **Working It Out** 4.1 further explores this relationship.

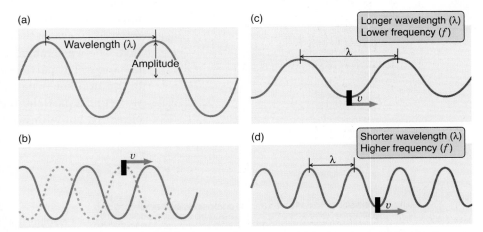

Figure 4.7 (a) A wave is characterized by the distance from one peak to the next (wavelength, λ), the maximum height above the medium's undisturbed state (amplitude), and (b) the speed (v) at which the wave pattern travels. (c) A wave with a longer wavelength has a lower frequency. (d) Conversely, a wave with a shorter wavelength has a higher frequency.

A long wavelength means low frequency, and a short wavelength means high frequency.

Light Is a Wave, but It Is Also a Particle

Though the wave theory of light successfully describes many phenomena, there are also many phenomena that it does not describe well. Scientists working in the late 19th and early 20th centuries discovered that many of the puzzling aspects of light could be better understood by thinking of light as a particle. We now know that light sometimes acts like a wave and sometimes acts like a particle. In

▶❙❙ **AstroTour:** Light as a Wave, Light as a Photon

The energy of a photon is proportional to its frequency.

the particle model, light is made up of massless particles called **photons** (*phot–* means "light," as in *photograph*, and *–on* signifies a particle). Photons always travel at the speed of light, and they carry energy.

The particle description of light is tied to the wave description of light by relating the energy of a photon and the frequency or wavelength of the wave. The two are directly proportional to each other, and the constant of proportionality is Planck's constant, h, which is equal to 6.63×10^{-34} joule-seconds (a joule is a unit of energy). Specifically, we write $E = hf$. The higher the frequency of the electromagnetic wave, the greater the energy carried by each photon. For example, photons of blue light have higher frequencies, and so carry more energy, than do photons of red light.

The total amount of energy that a beam of light carries is called the **intensity**. A beam of red light can carry just as much energy as a beam of blue light, but the red beam will have more photons than the blue beam. In this sense, light is similar to money. Ten dollars is 10 dollars, but it takes a lot more pennies (low-energy photons) than quarters (high-energy photons) to make up 10 dollars (**Figure 4.8**). When physicists speak of photons, they say that the light energy is **quantized**. The word *quantized*, which has the same root as the word *quantity*, means that something is subdivided into discrete units. A photon is a *quantum* of light.

The wave-particle description of light conflicts with everyday, commonsense ideas about the world. It is hard for us to imagine a single thing sharing the properties of a wave on the ocean *and* the properties of a beach ball, yet light does just that. The wave model of light is clearly the correct description to use in many instances. The particle description of light is also clearly the correct description to use in other instances. But how can the same thing—light—act like both a wave *and* a particle? Light is what light is. The trouble lies not with the nature of reality, but with what our brains are used to thinking about.

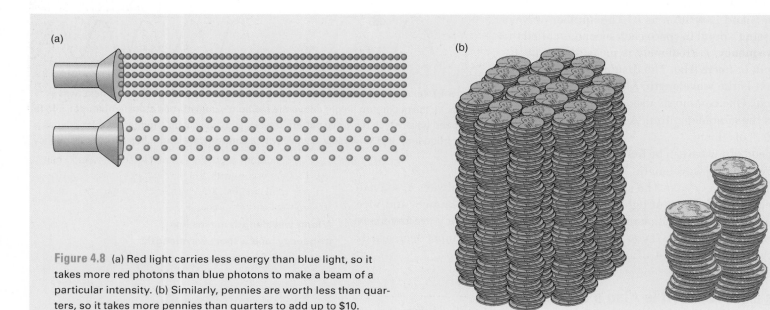

(a)

(b)

Figure 4.8 (a) Red light carries less energy than blue light, so it takes more red photons than blue photons to make a beam of a particular intensity. (b) Similarly, pennies are worth less than quarters, so it takes more pennies than quarters to add up to $10.

◉ **VISUAL ANALOGY**

4.3 How Do We Collect Light?

A telescope is a device for collecting and focusing light. Telescopes come in two primary types: **refracting telescopes**, which use lenses; and **reflecting telescopes**, which use mirrors. The size of a telescope is given by the diameter of the collecting area (the primary lens or mirror) and is called the **aperture**. When astronomers refer to a 4.5-meter telescope, they mean that the primary mirror (or lens) is 4.5 meters in diameter.

The ratio of light's speed in a vacuum, c, to its speed in a medium, v, is the medium's **index of refraction**, n—that is, $n = c/v$. For example, n is approximately 1.5 for typical glass, so the speed of light in glass is about 200,000 km/s. Because the speed of light changes as it enters a medium, light that enters glass at an angle is bent. The amount of bend depends on the angle and on the index of refraction of the glass. **Figure 4.9a** shows a schematic diagram of wavefronts striking a medium at an angle. **Figure 4.9b** shows an actual light ray passing into and out of a medium. The ray is bent each time the index of refraction changes. This change of direction, called **refraction,** is the basis for the refracting telescope (**Figure 4.10**). Because the telescope's glass lens is curved, light at the outer edges of the lens is refracted more than light near its center. This concentrates the light rays entering the telescope, bringing them to a sharp focus in the telescope's *focal plane* and creating an image. The distance between the lens and the focal plane is the **focal length** of the telescope. Increasing the focal length (by decreasing how sharply the lens is curved) increases the size and separation of objects in the image.

The larger the area of the lens, the more light-gathering power it has and the fainter the stars we can observe. There are physical limits on the size of refracting telescopes because large lenses are very heavy. There is another serious drawback to refracting telescopes: *chromatic aberration*. This is the effect that produces rainbows when sunlight passes through a piece of cut glass. Sunlight is made up of all the colors of the rainbow, and each color refracts at a slightly different angle because the index of refraction depends on the wavelength of the

(a)

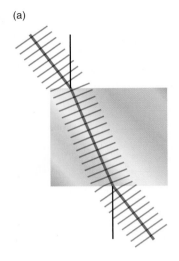

(b)

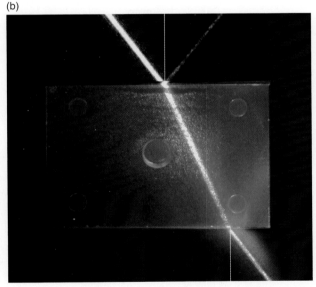

Figure 4.9 (a) A schematic of a series of wavefronts entering a new medium. (b) An actual light ray entering and leaving a medium. Light waves are refracted (bent) when entering a medium with a higher index of refraction. They are refracted again as they reenter the medium with the lower index of refraction.

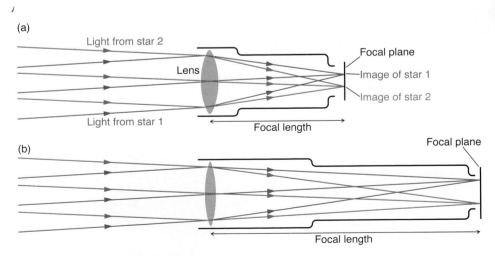

Figure 4.10 (a) A refracting telescope uses a lens to collect and focus light from two stars, forming images of the stars in its focal plane. (b) Telescopes with longer *focal length* produce larger, more widely separated images.

light. In astronomical applications, this chromatic aberration produces blurry images. An image of a star, for example, will have a colored halo around the star's location, rather than a crisp white star. Manufacturers of quality cameras and telescopes use a **compound lens**, composed of two types of glass, to correct for chromatic aberration.

Reflecting telescopes use mirrors rather than lenses to focus the light into an image (**Figure 4.11**). Light coming from a star first strikes the *primary mirror.* Typically, a *secondary mirror* then folds the light path from the primary mirror to the focal plane. The secondary mirror allows a significant reduction in the length and weight of the telescope.

Reflecting telescopes have several important advantages over refracting telescopes. Because mirrors do not spread out white light to form rainbows, chromatic aberration is no longer a problem. Primary mirrors can be supported from the back, and they can be made thinner and therefore weigh less than objective lenses. There are other advantages as well, and all large telescopes made today are of the reflector type. Look back at the chapter-opening figure, which shows a reflecting telescope. In this case, the focal plane is located at the side of the telescope instead of at the bottom. This innovation makes the telescope easier to use because the telescope doesn't need to be lifted off the ground for the user to look through it. Innovations like this one are common, but the fundamental concept of using mirrors to focus the light to a focal plane is always the same.

4.4 How Do We Detect Light?

Human eyes respond to light with wavelengths ranging from about 350 nm (deep violet) to 700 nm (far red). The retina is the light detector in the human eye (**Figure 4.12**), and the individual receptor cells that respond to light falling on the retina are called rods and cones. Cones are located near the eye's optical axis at the center of our vision, while rods are located away from the eye's optical axis and are responsible for our peripheral vision. As photons from a star enter the pupil of the eye, they strike cones at the center of vision. The cones then send a signal to the brain, which interprets this message as "I see a star." What limits the faintest stars we can see with our unaided eyes, assuming a clear, dark night and good eyesight? This limit is determined in part by two factors that are characteristic of all detectors of light: integration time and quantum efficiency.

Integration time is the time interval during which the eye can add up photons. The brain "reads out" the information gathered by the eye about every 100 mil-

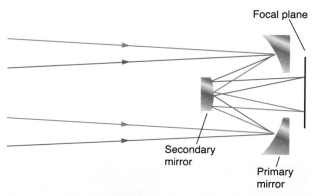

▶❙❙ **AstroTour:** Geometric Optics and Lenses

Figure 4.11 Reflecting telescopes use mirrors to collect and focus light. Large telescopes typically use a secondary mirror that directs the light back through a hole in the primary mirror to an accessible focal plane behind the primary mirror.

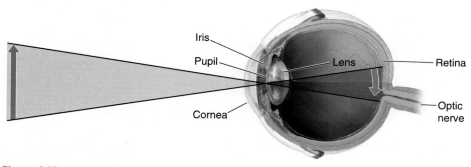

Figure 4.12 A schematic view of the human eye.

liseconds (ms). Anything that happens faster than that appears to happen all at once. If two images on a computer screen appear 30 ms apart, you will see them as a single image because your eyes will sum (or integrate) whatever they see over an interval of 100 ms. However, if the images occur 200 ms apart, you will see them as separate images. This relatively brief integration time is the biggest factor limiting our nighttime vision. Stars too faint for us to see with our unaided eyes are those that produce too few photons for our eyes to detect in 100 ms.

Quantum efficiency also restricts our nighttime vision. This effect determines how many responses occur for each photon received. For the human eye, 10 photons must strike a cone within 100 ms to activate a single response. So the quantum efficiency of our eyes is about 10 percent: for every 10 events, the eye sends one signal to the brain. Together, integration time and quantum efficiency determine the rate at which photons must arrive before the brain says, "Aha, I see something."

However, even when we receive a sufficient number of photons in a short enough time span to see a star, our vision is further limited by the eye's **resolution**, which refers to how close two points of light can be to each other before we can no longer distinguish them. Unaided, the best human eyes can resolve objects separated by 1 arcminute ($\frac{1}{60}$ of a degree), an angular distance of about $\frac{1}{30}$ the diameter of the full Moon. This may seem small; but when we look at the sky, thousands of stars and galaxies may hide within a patch of sky with this diameter.

Until 1840, the retina of the human eye was the only astronomical detector. Permanent records of astronomical observations were limited to what an experienced observer could sketch on paper while working at the eyepiece of a telescope, as illustrated in **Figure 4.13a**.

Photographic Plates

In 1840, John W. Draper (1811–1882), a New York chemistry professor, created the earliest known astronomical photograph, shown in **Figure 4.13b**. His subject was the Moon. Early photography was slow and very messy, and astronomers were reluctant to use it. In the late 1870s, a faster, simpler process was invented and astronomical photography took off. Astronomers could now create permanent

Figure 4.13 (a) A drawing of the galaxy M51 made by William Parsons (Lord Rosse, 1800–1867) in 1845. (b) A photograph of the Moon taken by John W. Draper in 1840.

images of planets, nebulae, and galaxies with ease. Thousands of photographic plates soon filled the "plate vaults" of major observatories.

The quantum efficiency of most photographic systems used in astronomy was very low—typically 1–3 percent, even poorer than that of the human eye. But unlike the eye, photography can overcome poor quantum efficiency by increasing the integration time to many hours of exposure. Photography made it possible for astronomers to record and study objects that were previously invisible.

Photography is not without its problems. Very faint objects often require long exposures that can take up much of an observing night. Each photographic plate could be used only once, and they were expensive. By the middle of the 20th century, the search was on for electronic detectors that would overcome many of the deficiencies of photographic plates.

Charge-Coupled Devices

In 1969, scientists at Bell Laboratories invented a remarkable detector called a **charge-coupled device**, or **CCD**. Astronomers soon realized that this was the tool they had been looking for, and the CCD became the detector of choice in almost all astronomical imaging applications. The output from a CCD is a digital signal that can be sent directly from the telescope to image-processing software or stored electronically for later analysis.

A CCD is an ultrathin wafer of silicon, less than the thickness of a human hair, that is divided into a two-dimensional array of picture elements, or **pixels**, as seen in **Figure 4.14a**. When a photon strikes a pixel, it creates a small electric charge within the silicon. As each CCD pixel is read out, the digital signal that flows to the computer is almost precisely proportional to the accumulated charge. The first astronomical CCDs were small arrays containing a few hundred thousand pixels. The larger CCDs used in astronomy today—like the one seen in **Figure 4.14b**—may contain as many as 100 million pixels. Still larger arrays are under development as ever-faster computing power keeps up with image-processing demands.

> The CCD is the astronomer's detector of choice.

(a)

(b)

(c)

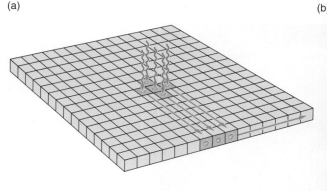
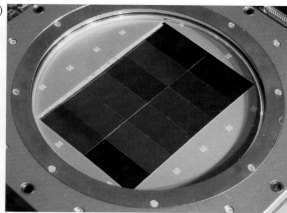

Figure 4.14 (a) A simplified diagram of a charge-coupled device (CCD). Photons from a star land on pixels (gray squares) and produce free electrons within the silicon. The electrons are moved to the collecting register at the bottom. Each row is then moved out to the right, where the charge of each pixel is converted into a digital signal. (b) This very large CCD has 12,288 × 8,192 pixels. (c) CCDs are used in many consumer electronic devices, such as camera phones.

CCDs have found their way into many devices that we now take for granted, such as digital cameras, digital video cameras, and camera phones. Your cell phone takes color pictures (**Figure 4.14c**) by using a grid of CCD pixels arranged in groups of three. Each pixel in a group is constructed to respond only to a particular range of colors—only to red light, for example. This degrades the resolution of the camera because each spot in the final image requires three pixels of information. Astronomers choose instead to use all the pixels on the camera to measure the number of photons that fall on each pixel, without regard to color. They put filters in front of the camera to allow through only light of particular wavelengths. Color pictures like those from the Hubble Space Telescope are constructed by taking multiple pictures, coloring each one, and then carefully aligning and overlapping them to produce beautiful and informative images. Sometimes the colors are "true"; that is, they are close to the colors you would see if you were actually looking at the object with your eyes. At other times, the colors represent different portions of the spectrum and tell you the temperature or composition of different parts of the object. Using changeable filters instead of designated color pixels gives astronomers greater flexibility and greater resolution.

Spectra and Spectrographs

You have almost certainly seen a rainbow like the one in **Figure 4.15** spread out across the sky. This sorting of light by colors is really a sorting by wavelength. When we talk about light being spread out according to wavelength, we are referring to the **spectrum** of the light. On the long-wavelength end of the visible spectrum is red light, between about 600 and 700 nm. At the other end of the visible spectrum is violet light, which is the bluest of blue light; the shortest-wavelength violet light that our eyes can see has a wavelength of about 350 nm. Stretched out between the two, literally in a rainbow, is the rest of the visible spectrum (**Figure 4.16**). The colors in the visible spectrum, in order of decreasing wavelength, are

Red   Orange   Yellow   Green   Blue   Indigo   Violet

Light can have wavelengths that are much shorter or much longer than our eyes can perceive. The whole range of different wavelengths of light is collectively referred to as the **electromagnetic spectrum**.

Follow along in Figure 4.16 as we take a tour of the electromagnetic spectrum, beginning with visible light and working our way out. Beyond the short-wavelength, high-frequency blue end of the visible spectrum, with wavelengths between 40 and 350 nm, is **ultraviolet (UV)** light. At a wavelength shorter than 40 nm, or 4×10^{-8} meter, we refer to the light as **X-rays**. As we continue to even shorter wavelengths, less than about 10^{-10} meter, we come to *gamma radiation*, or **gamma rays**.

We can also start at the red end of the visible spectrum and proceed in the other direction. Light with wavelengths longer than about 700 nm (7×10^{-7} meter) and shorter than 500 μm (5×10^{-4} meter) is referred to as **infrared (IR)** radiation. When the wavelength of light gets still longer than this, we start calling it **microwave** radiation. The longest-wavelength light, with wavelengths longer than a few centimeters, is called **radio waves**.

Astronomers conventionally use nanometers when referring to wavelengths at visible and shorter wavelengths, micrometers or "microns" (μm) for wavelengths

Figure 4.15 The visible part of the electromagnetic spectrum is laid out in all its glory in the colors of this rainbow.

Visible light is only one small segment of the electromagnetic spectrum.

Vocabulary Alert

Radiation: In common language, radiation is associated with emissions from nuclear bombs or radioactive substances. In some cases, this radiation actually is light in the form of gamma rays. In other cases, it's particles. Astronomers use the word to mean energy carried through space—that is, light. Astronomers often use the two words *light* and *radiation* interchangeably, especially when talking about wavelengths that are not in the visible range.

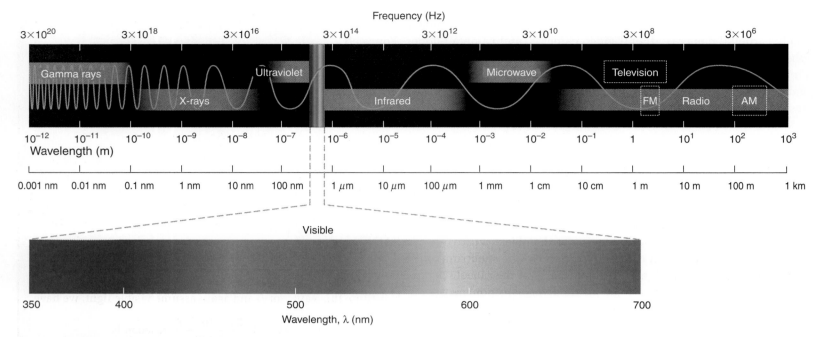

Frequency (Hz)

Wavelength (m)

Visible

Wavelength, λ (nm)

Figure 4.16 By convention, the electromagnetic spectrum is broken into loosely defined regions ranging from gamma rays to radio waves. Throughout the book, we use these labels for astronomical images, with an icon to remind you. Images taken in Gamma rays (G), X-rays (X), ultraviolet (U), visible (V), infrared (I), and radio (R) are all labeled. If more than one region is represented, multiple labels are highlighted.

in the infrared, and millimeters (mm), centimeters (cm), and meters (m) for wavelengths in the microwave and radio regions of the electromagnetic spectrum.

Spectrographs—or **spectrometers**, as they are often called—are tools that allow us to probe the composition and physical properties of distant objects by studying their spectra. Each type of atom or molecule has its own distinct impact on the spectrum of a star, either adding light at specific wavelengths or taking it away. Atoms and molecules produce **emission lines** when they add light to the spectrum, and **absorption lines** when they take light away. In either case, as we'll see in more detail in Chapter 10, each type of atom or molecule has a unique set of lines, which act like fingerprints telling astronomers what particular atoms and molecules the object is composed of. **Spectroscopy** is the study of an object's light in terms of its component wavelengths, and we'll encounter its many applications throughout the chapters to come as we continue our journey through the Solar System and beyond.

4.5 Resolution and the Atmosphere

Earlier in this chapter, we learned that increasing the focal length of a telescope increases the size and separation between the images it produces (as in Figure 4.10). Using the vocabulary of Section 4.4, we can now say that the telescope with a longer focal length has a higher *resolution*. This is the critical reason why telescopes allow us to see stars so much more clearly than we can with the naked eye: the focal length of a human eye is typically about 20 millimeters, whereas telescopes used by professional astronomers often have focal lengths of tens or even hundreds of meters. Such telescopes make images that are far larger than those formed by our eyes, and consequently they contain far more detail.

However, focal length explains only one difference between the resolution of telescopes and the unaided eye. A second difference arises from the wave nature of light. As light waves pass through the aperture of a telescope, they spread out from the edges of the lens or mirror (**Figure 4.17**). The distortion of a wavefront

Diffraction, or blurring of an image, depends on the ratio of wavelength to telescope aperture.

Working It Out 4.2 | Diffraction Limit

The ultimate limit on the angular resolution, θ, of a lens—the diffraction limit—is determined by the ratio of the wavelength of light, λ, to the aperture, D:

$$\theta = 2.06 \times 10^5 \frac{\lambda}{D} \text{ arcseconds}$$

The constant, 2.06×10^5, changes the units to arcseconds. An arcsecond is a tiny angular measure found by first dividing a degree by 60 to get arcminutes, and then by 60 again to get arcseconds. To get an idea of the size of an arcsecond, imagine that you hand a tennis ball to your friend and ask her to run 8 miles down a straight road away from you and hold up the tennis ball. If she actually does it, the angle you perceive from one side of the tennis ball to the other is approximately 1 arcsecond.

Both λ and D must be expressed in the same units (usually meters). The smaller the ratio of λ/D, the better the resolution. The size of the human pupil can change from about 2 mm in bright light to 8 mm in the dark. A typical pupil size is about 4 mm, or 0.004 meter. Visible (green) light has a wavelength of 550 nm, or 5.5×10^{-7} meters. Substituting in, we have

$$\theta = 2.06 \times 10^5 \left(\frac{5.5 \times 10^{-7} \text{ m}}{0.004 \text{ m}}\right) \text{arcseconds}$$

$$= 28.3 \text{ arcseconds}$$

or about 0.5 arcminute. But, as we learned earlier, the best resolution that the human eye can achieve is about 1 arcminute, and 2 arcminutes is actually more typical. We do not achieve the theoretical resolution with our eyes, because the optical properties of the lens and the physical properties of the retina are not perfect.

How does this compare to the resolution of a telescope? Consider Hubble Space Telescope, operating in the visible part of the spectrum. Its primary mirror has a diameter of 2.4 meters. If we substitute this value for D and again assume visible light, we have

$$\theta = 2.06 \times 10^5 \left(\frac{5.5 \times 10^{-7} \text{ m}}{2.4 \text{ m}}\right) \text{arcseconds}$$

$$= 0.047 \text{ arcseconds}$$

This is about 600 times better than the resolving power of the human eye.

(a series of parallel waves) as it passes the edge of an opaque object is called **diffraction**. Diffraction "diverts" some of the light from its path, slightly blurring the image made by the telescope. The degree of blurring depends on the wavelength of the light and the telescope's aperture. The larger the aperture, the smaller the problem posed by diffraction. The best resolution that a given telescope can achieve is known as the **diffraction limit** (see **Working It Out 4.2**).

The diffraction limit tells us that larger telescopes get better resolution. Theoretically, the 10-meter Keck telescopes have a diffraction-limited resolution of 0.0113 arcsecond in visible light, which would allow you to read newspaper headlines 60 km away. But for telescopes with apertures larger than about

Figure 4.17 (a) Light waves from a star are diffracted by the edges of a telescope's lens or mirror. (b) This diffraction causes the stellar image to be blurred, limiting a telescope's ability to resolve objects. (c) Diffraction from a circular aperture illuminated by green laser light.

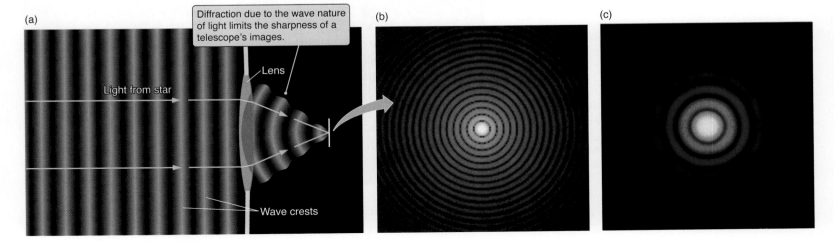

(a)

Diffraction due to the wave nature of light limits the sharpness of a telescope's images.

Light from star

Lens

Wave crests

(b)

(c)

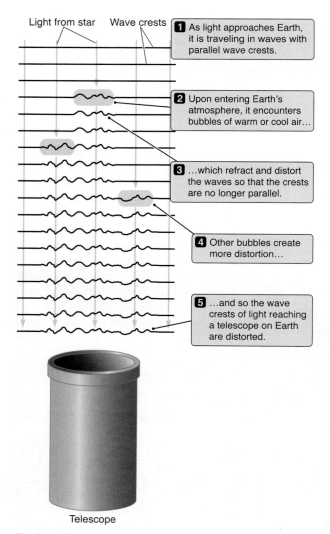

1 As light approaches Earth, it is traveling in waves with parallel wave crests.

2 Upon entering Earth's atmosphere, it encounters bubbles of warm or cool air...

3 ...which refract and distort the waves so that the crests are no longer parallel.

4 Other bubbles create more distortion...

5 ...and so the wave crests of light reaching a telescope on Earth are distorted.

Telescope

Figure 4.18 Bubbles of warmer or cooler air in Earth's atmosphere distort the wavefront of light from a distant object.

Figure 4.19 Images of (a) Neptune and (b) a star cluster taken without and with adaptive optics. The corrected images are on the right.

a meter, Earth's atmosphere stands in the way of better resolution. If you have ever looked out across a road on a summer day, you have seen the air shimmer as light is bent this way and that by turbulent bubbles of warm air rising off the hot pavement.

The problem is less pronounced when we look overhead, but the twinkling of stars in the night sky is caused by the same phenomenon. As telescopes magnify the angular diameter of a planet, they also magnify the shimmering effects of the atmosphere. The limit on the resolution of a telescope on the surface of Earth caused by this atmospheric distortion is called **astronomical seeing**. One advantage of launching telescopes such as the Hubble Space Telescope into orbit around Earth is that they are not hampered by astronomical seeing.

Modern technology has improved ground-based telescopes with computer-controlled **adaptive optics**, which compensate for much of the atmosphere's distortion. To better understand how adaptive optics work, we need to look more closely at how Earth's atmosphere smears out an otherwise perfect stellar image. Look again at Figure 4.17. Light from a distant star arrives at the top of Earth's atmosphere as a flat, parallel wavefront. If Earth's atmosphere were perfectly

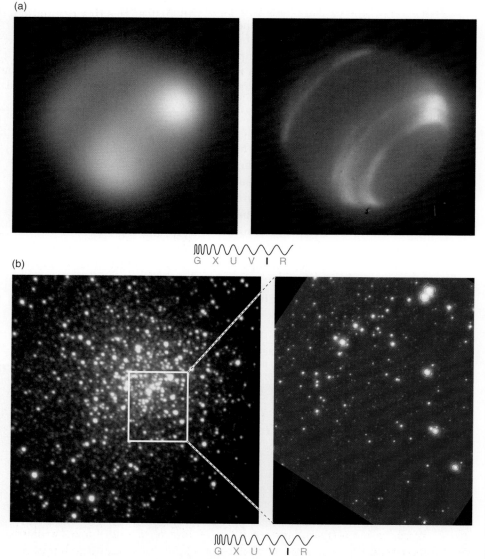

(a)

(b)

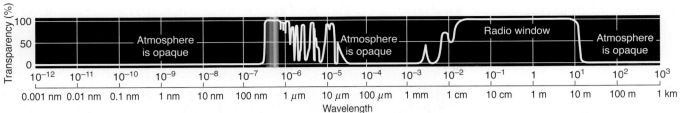

Figure 4.20 Earth's atmosphere blocks most electromagnetic radiation.

uniform, the wavefront would remain flat as it reached the objective lens or primary mirror of a ground-based telescope. After making its way through the telescope's optical system, the wavefront would produce a tiny diffraction disk in the focal plane, as shown in Figure 4.17b. But Earth's atmosphere is not uniform. It is filled with small bubbles of air that have slightly different temperatures than their surroundings. Different temperatures mean different densities, and different densities mean each bubble bends light differently.

These air bubbles act as weak lenses, and by the time the wavefront reaches the telescope it is far from flat, as shown in **Figure 4.18**. Instead of a tiny diffraction disk, the image in the telescope's focal plane is distorted and swollen. Adaptive optics flattens out this distortion. First, an optical device within the telescope measures the wavefront. Then, before reaching the telescope's focal plane, the light is reflected off yet another mirror, which has a flexible surface. A computer analyzes the wavefront and bends the flexible mirror so that it accurately corrects for the distortion of the wavefront. Examples of images corrected by adaptive optics are shown in **Figure 4.19**. The widespread use of adaptive optics has made the image quality of ground-based telescopes competitive with that of the Hubble Space Telescope.

Image distortion is not the only problem caused by Earth's atmosphere. Nearly all of the X-ray, ultraviolet, and infrared light arriving at Earth fails to reach the ground because it is partially or completely absorbed by ozone, water vapor, carbon dioxide, and other atmospheric molecules.

4.6 Observing in Wavelengths beyond the Visible

As shown in **Figure 4.20**, our atmosphere is transparent in several regions of the spectrum. The largest of these atmospheric "windows" is in the radio portion, and so we are able to build radio telescopes on the ground (instead of in space). Most **radio telescopes** are large, steerable dishes, typically tens of meters in diameter, such as the one shown in **Figure 4.21a**. The world's largest single-dish radio telescope is the 305-meter Arecibo dish, built into a natural bowl-shaped depression in Puerto Rico (**Figure 4.21b**). This huge structure is too big to steer. Arecibo can only observe sources that pass within 20° of the zenith as Earth's rotation carries them overhead.

As large as radio telescopes are, they have relatively poor angular resolution. A telescope's angular resolution is determined by the ratio λ/D, where λ is the wavelength of electromagnetic radiation and D is the telescope's aperture. A larger ratio means poorer resolution. Radio telescopes have diameters much larger than the apertures of most optical telescopes, and that helps. But

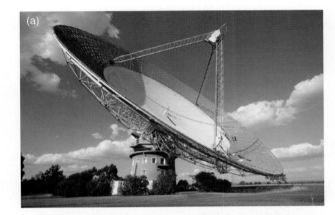

Figure 4.21 (a) A large radio telescope in Australia. (b) The Arecibo radio telescope is the world's largest single-dish telescope. The steerable receiver suspended above the dish permits limited pointing toward celestial targets as they pass close to the zenith.

Figure 4.22 The Very Large Array (VLA) in New Mexico.

the wavelengths of radio waves are much longer than the wavelengths of visible light. Radio telescopes are thus hampered by the very long wavelengths they are designed to receive. Consider the huge Arecibo dish. Its resolution is typically about 1 arcminute, no better than the unaided human eye. So radio astronomers have had to develop ways to improve the resolution, and one of the cleverest is the **interferometer**.

When we mathematically combine the signals from two radio telescopes, they act like a telescope with a diameter equal to the separation between them. For example, if two 10-meter telescopes are located 1,000 meters apart, the D in λ/D is 1,000, not 10. This is called an interferometer because it makes use of the wavelike properties of light, in which signals from the individual telescopes *interfere* with one another. Usually several telescopes are employed in an **interferometric array**. Through the use of very large arrays, astronomers can attain and even exceed the angular resolution possible with optical telescopes.

One of the larger radio interferometric arrays is the Very Large Array (VLA) in New Mexico, shown in **Figure 4.22**. The VLA is made up of 27 individual movable dishes spread out in a Y-shaped configuration up to 36 km across. At a wavelength of 10 cm, this array can achieve resolutions of less than 1 arcsecond. The Very Long Baseline Array (VLBA) employs 10 radio telescopes spread out over more than 8,000 km from the Virgin Islands in the Caribbean to Hawaii in the Pacific. At a wavelength of 10 cm, this array can reach resolutions better than 0.003 arcsecond. A radio telescope put into near-Earth orbit as part of a Space Very Long Baseline Interferometer (SVLBI) overcomes even this limit. Future SVLBI projects would extend the baseline to as much as 100,000 km, yielding resolutions far exceeding those of any existing optical telescope.

Optical telescopes can also be arrayed to yield resolutions greater than those of single telescopes, although for technical reasons the individual units cannot be spread as far apart as radio telescopes. The Very Large Telescope (VLT), operated by the European Southern Observatory (ESO) in Chile, combines the four 8-meter telescopes (**Figure 4.23**) with four movable 1.8-meter auxiliary telescopes. It has a baseline of up to 200 meters, yielding angular resolution in the milliarcsecond range.

Getting above Earth's Atmosphere: Airborne and Orbiting Observatories

Earth's atmosphere distorts telescopic images, and molecules such as water vapor in Earth's atmosphere block large parts of the electromagnetic spectrum from getting through to the ground, so astronomers try to locate their instruments above as much of the atmosphere as possible. Most of the world's larger astronomical telescopes are located 2,000 meters or more above sea level. Mauna Kea, a dormant Hawaiian volcano and home of the Mauna Kea Observatories (MKO), rises 4,200 meters above the Pacific Ocean. At this altitude the MKO telescopes sit above 40 percent of Earth's atmosphere; more important, 90 percent of Earth's atmospheric water vapor lies below. Still, for the infrared astronomer the remaining 10 percent is troublesome.

One way to solve the water vapor problem is to make use of high-flying aircraft. NASA's Kuiper Airborne Observatory (KAO), a modified C-141 cargo aircraft, carried a 90-cm telescope and was among the first of these flying observatories. It cruised at altitudes of 12–14 km, above most of Earth's water vapor. NASA retired KAO in 1995 and replaced it with the Stratospheric Observatory for Infrared Astronomy (SOFIA), which began operations in 2011. SOFIA carries a 2.5-meter telescope and works in the far-infrared region of the spectrum, from 1 to 650 micrometers (μm). It flies in the stratosphere at an altitude of about 12 km, at the low end of KAO's flight range, but still above 99 percent of the water vapor in Earth's lower atmosphere.

Airborne observatories overcome atmospheric absorption of infrared light by placing telescopes above most of the water vapor in the atmosphere. But gaining full access to the complete electromagnetic spectrum is yet another matter. That requires getting completely above Earth's atmosphere. The first astronomical satellite was the British Ariel 1, launched in 1962 to study solar ultraviolet and X-ray radiation and the energy spectrum of primary cosmic rays. Today we have a multitude of orbiting astronomical telescopes covering the electromagnetic spectrum from gamma rays to microwaves.

Optical telescopes, such as the 2.4-meter Hubble Space Telescope (HST), can operate successfully at modest altitudes in what is called low Earth orbit, 600 km above Earth's surface. Launched in 1990, HST has been collecting UV, visible, and IR data for over two decades. For other satellites, 600 km is not nearly high enough. The Chandra X-ray Observatory, NASA's X-ray telescope, cannot see through even the tiniest traces of atmosphere and therefore orbits more than 16,000 km above Earth's surface. And even this is not distant enough for some telescopes. NASA's Spitzer Space Telescope, an infrared telescope, is so sensitive that it needs to be completely free from Earth's own infrared radiation. The solution was to put it into a *solar* orbit, trailing tens of millions of kilometers behind Earth. Many future space telescopes, including the James Webb Space Telescope, NASA's replacement for HST, will orbit free of Earth, bound only to the Sun.

You can now see the significance of the various types of telescopes that astronomers use. Astronomers try valiantly to cover the entire electromagnetic spectrum, though each region presents different challenges that must be overcome if we are to see the entire universe.

Figure 4.23 The Very Large Telescope (VLT) operated by the European Southern Observatory in Chile. Movable auxiliary telescopes allow the four large telescopes to operate as an optical interferometer.

Orbiting observatories explore regions of the spectrum inaccessible from the ground.

The James Webb Space Telescope will replace the aging Hubble Space Telescope. This telescope will be much larger and much farther away, and it will be optimized in the infrared. Scientists are busy constructing both the telescope and the cameras that will be attached to it. This article describes progress on one of these instruments.

New Space Telescope Relies on Never-Before-Manufactured Material; No Problem, Says NASA

By **REBECCA BOYLE,** *Popular Science*

NASA engineers working on the James Webb Space Telescope are doing a lot of things from scratch—they've had to design new mirrors and a foldy space cocoon, for instance—but their newest work may take the cake: To survive the coldest reaches of space, they invented a brand-new composite material. They nicknamed it unobtanium.

The heart of the Webb Telescope is the car-sized Integrated Science Instrument Module, which will hold all the telescope's instruments, packed tightly together. The scope's high-precision optics need stability, so the chassis must be tough enough to avoid warping in the extreme deep freeze of 1 million miles from Earth. But it also must be tough enough to survive the stress of launch.

Engineers combed the scientific literature to find a material that could meet those standards, and came up with nothing. They decided they had to whip something up, so they used mathematical models to measure several ingredients. Finally, they found two composite materials that would yield something called a carbon fiber/cyanate-ester resin system.

It was ideal for making the chassis' 3-inch tubes, but the engineers still had to figure out how to put them together—the ISIM has about 900 components. Ultimately, they used several methods, including nickel-alloy fittings, clips,

and special composite plates joined with an adhesive that they also invented.

Once they made all this, they had to figure out how to test it, because the three-story Space Environment Simulator at Goddard Space Flight Center wasn't cold enough.

After it launches sometime in 2014, the Webb telescope will chill out at a Lagrange point, which is a special point in space where the gravitational fields of the Earth and moon are in equilibrium, allowing a satellite to sit between them. Shielded from the sun by a massive space umbrella, the telescope's instruments will be exposed to daily temperatures around 39 Kelvin, or −389.5 degrees Fahrenheit. The simulator can only reach about 100 K, which is sufficient for testing most other space equipment.

But the team wanted to chill the truss to 27 K, roughly equivalent to Pluto's surface temperature, just to make sure it wouldn't crack. So they built a shroud that resembled a tuna fish can and inserted it into the chamber. Once the air was removed from the test chamber, they pumped helium into the shroud to cool off the truss to 27 K.

During 26 days of tests, it didn't crack. As the mathematical models showed, it shrunk by 170 microns—about the width of a needle—once the temperature reached 27 K. NASA's warp limit was 500 microns, so the super-material proved its worth.

No sweat, said Jim Pontius, the ISIM lead

mechanical engineer. "The technology challenges are what attracted the people to the program," he said.

Evaluating the News

1. Sum up this article. What part of the telescope is being tested?
2. What is so unusual about the environment this telescope will inhabit? Why is that difficult to design for?
3. Using the vocabulary you learned in this chapter, explain why it is so important that the chassis of the telescope not warp. What would happen if it did?
4. Find a picture of the James Webb Space Telescope model online. Describe the "*massive* space umbrella." In what sense is this reporter probably using the word *massive*? Does she mean "massive" as a scientist would use the word, or does she mean "large"?
5. Compare the results of the test (170 microns) with the tolerance (500 microns) allowed. Would you be comfortable launching this instrument with these measured parameters? Why or why not?
6. Speculate about the value of a telescope like the James Webb Space Telescope. What might we learn that is of value? What other spin-off technologies might result from this effort?

SUMMARY

4.1 The speed of light in a vacuum is 300,000 km/s, and nothing can travel faster.

4.2 Light carries both information and energy throughout the universe. Light is simultaneously an electromagnetic wave and a stream of particles called photons.

4.3 Optical telescopes come in two basic types: refractors and reflectors. All large astronomical telescopes are reflectors.

4.4 The CCD is today's astronomical detector of choice. From gamma rays to visible light to radio waves, all radiation is an electromagnetic wave.

4.5 Large telescopes collect more light and have greater resolution. Earth's atmosphere blocks many spectral regions and distorts telescopic images.

4.6 Radio and optical telescopes can be arrayed to greatly increase angular resolution. Putting telescopes in space solves problems created by Earth's atmosphere.

✦ SUMMARY SELF-TEST

1. Light acts like
 a. a wave.
 b. a particle.
 c. both a wave and a particle.
 d. neither a wave nor a particle.

2. All large astronomical telescopes are reflectors because
 a. chromatic aberration is minimized.
 b. they are not as heavy.
 c. they can be shorter.
 d. all of the above

3. Which of the following can be observed from the ground?
 a. radio waves
 b. gamma radiation
 c. UV light
 d. X-ray light
 e. visible light

4. Rank the following in order of decreasing wavelength.
 a. gamma rays
 b. visible
 c. infrared
 d. ultraviolet
 e. radio waves

5. Match the following properties of telescopes (lettered) with their corresponding definitions (numbered).
 a. aperture
 b. resolution
 c. focal length
 d. chromatic aberration
 e. diffraction
 f. interferometer
 g. adaptive optics

 (1) several telescopes connected to act as one
 (2) distance from lens to focal plane
 (3) diameter
 (4) ability to distinguish objects that appear close together in the sky
 (5) computer-controlled active focusing
 (6) rainbow-making effect
 (7) smearing effect due to sharp edge

QUESTIONS AND PROBLEMS

True/False and Multiple-Choice Questions

6. T/F: Nothing can travel faster than 3×10^8 m/s.

7. T/F: The frequency of a wave is related to the energy of the photon.

8. T/F: Blue light has a longer wavelength than red light.

9. T/F: Blue light has more energy than red light.

10. T/F: Visible light is an electromagnetic wave.

11. If the wavelength of a beam of light were halved, how would that affect the frequency?
 a. The frequency would be four times larger.
 b. The frequency would be two times larger.
 c. The frequency would not change.
 d. The frequency would be two times smaller.
 e. The frequency would be four times smaller.

12. How does the speed of light in a medium compare to the speed in a vacuum?
 a. It's the same, since the speed of light is a constant.

 b. The speed in the medium is always faster than the speed in a vacuum.
 c. The speed in the medium is always slower than the speed in a vacuum.
 d. The speed in the medium may be faster or slower, depending on the medium.

13. Astronomers put telescopes in space to
 a. get closer to the stars.
 b. avoid atmospheric effects.
 c. look primarily at radio wavelengths.
 d. improve quantum efficiency.

14. The advantage of an interferometer is that
 a. the resolution is dramatically increased.
 b. the focal length is dramatically increased.
 c. the light-gathering power is dramatically increased.
 d. diffraction effects are dramatically decreased.
 e. chromatic aberration is dramatically decreased.

15. The angular resolution of a ground-based telescope is usually determined by
    a. diffraction.　　b. refraction.
    c. the focal length.　　d. atmospheric seeing.

Conceptual Questions

16. We know that the speed of light in a vacuum is 3×10^8 m/s. Is it possible for light to travel at a lower speed? Explain your answer.

17. Is a light-year a measure of time or distance, or both?

18. Would you describe light as a wave or a particle, or both? Explain your answer.

19. Does a stationary charged particle create an electric field? Does it create a magnetic field?

20. Name the variables that describe a wave.

21. Name the types of electromagnetic radiation that lie at the extreme ends of the electromagnetic spectrum.

22. If photons of blue light have more energy than photons of red light, how can a beam of red light carry as much energy as a beam of blue light?

23. Galileo's telescope used simple lenses rather than compound lenses. What is the primary disadvantage of using a simple lens in a refracting telescope?

24. The largest astronomical refractor has an aperture of 1 meter. Why is it impractical to build a larger refractor with, say, twice the aperture?

25. Name and explain at least two advantages that reflecting telescopes have over refractors.

26. Your camera may have a zoom lens, ranging between wide angle (short focal length) and telephoto (long focal length). How would the size of an image in the camera's focal plane differ between wide angle and telephoto?

27. What causes refraction?

28. How do manufacturers of quality refracting telescopes and cameras avoid the problem of chromatic aberration?

29. Consider two optically perfect telescopes having different diameters but the same focal length. Is the image of a star larger or smaller in the focal plane of the larger telescope? Explain your answer.

30. Name two ways in which Earth's atmosphere interferes with astronomical observations.

31. Explain *why* stars twinkle.

32. Why do we not have ground-based gamma-ray and X-ray telescopes?

33. Explain adaptive optics and how they improve a telescope's image quality.

34. Explain integration time and how it contributes to the detection of faint astronomical objects.

35. Explain quantum efficiency and how it contributes to the detection of faint astronomical objects.

36. Why are the world's largest telescopes located on high mountains?

37. Some people believe that we put astronomical telescopes in orbit because doing so gets them closer to the objects they are observing. Explain what is wrong with this common misconception.

Problems

38. The index of refraction, n, of a diamond is 2.4. What is the speed of light within a diamond?

39. You are tuned to 790 on AM radio. This station is broadcasting at a frequency of 790 kHz (7.90×10^5 Hz). What is the wavelength of the radio signal? You switch to 98.3 on FM radio. This station is broadcasting at a frequency of 98.3 MHz (9.83×10^7 Hz). What is the wavelength of this radio signal?

40. Many amateur astronomers start out with a 4-inch (aperture) telescope and then graduate to a 16-inch telescope. By what factor does the light-gathering power (which goes like the area) of the telescope increase?

41. Compare the light-gathering power of a large astronomical telescope (aperture 10 meters) with that of the dark-adapted human eye (aperture 8 mm).

42. Compare the angular resolution of the Hubble Space Telescope (aperture 2.4 meters) with that of a typical amateur telescope (aperture 20 cm).

43. Assume a telescope has an aperture of 1 meter. Calculate the telescope's resolution when observing in the near-infrared region of the spectrum ($\lambda = 1{,}000$ nm). Calculate the resolution in the violet region of the spectrum ($\lambda = 400$ nm). In which region does the telescope have better resolution?

44. Assume that the maximum aperture of the human eye, D, is approximately 8 mm and the average wavelength of visible light, λ, is 5.5×10^{-4} mm.
    a. Calculate the diffraction limit of the human eye in visible light.
    b. How does the diffraction limit compare with the actual resolution of 1–2 arcminutes (60–120 arcseconds)?
    c. To what do you attribute the difference?

45. The resolution of the human eye is about 1.5 arcminutes. What would the aperture of a radio telescope (observing at 21 cm) have to be to have this resolution? Even though the atmosphere is transparent at radio wavelengths, we do not see in the radio region of the electromagnetic spectrum. Using your calculations, explain why humans do not see in the radio.

46. One of the earliest astronomical CCDs had 160,000 pixels, each recording 8 bits (256 levels of brightness). A new generation of astronomical CCDs may contain a billion pixels, each recording 15 bits (32,768 levels of brightness). Compare the number of bits of data that each CCD type produces in a single image.

47. In problem 46, you calculated the number of bits in a new-generation CCD image. To read out a new chip in 1 minute, how many bits per second have to be transferred? (That is, how fast does the computer have to transfer the data?)

48. Consider a CCD with a quantum efficiency of 80 percent and a photographic plate with a quantum efficiency of 1 percent. If

an exposure time of 1 hour is required to photograph a celestial object with a given telescope, how much observing time would we save by substituting a CCD for the photographic plate?

49. The VLBA employs an array of radio telescopes ranging across 8,000 km of Earth's surface from the Virgin Islands to Hawaii.
   a. Calculate the angular resolution of the array when radio astronomers are observing interstellar water molecules at a microwave wavelength of 1.35 cm.
   b. How does this resolution compare with the angular resolution of two large optical telescopes separated by 100 meters and operating as an interferometer at a visible wavelength of 550 nm?

50. When operational, the SVLBI may have a baseline of 100,000 km. What will its angular resolution be when we are studying an interstellar molecule emitting at a wavelength of 17 mm from a distant galaxy?

51. The *Mars Reconnaissance Orbiter (MRO)* flies at an average altitude of 280 km above the martian surface. If its cameras have an angular resolution of 0.2 arcsecond, what is the size of the smallest objects that the MRO can detect on the martian surface?

52. As with light, the speed, wavelength, and frequency of gravitational waves are related as $c = \lambda \times f$. If we were to observe a gravitational wave from a distant cosmic event with a frequency of 10 hertz (Hz), what would be the wavelength of the gravitational wave?

 SmartWork, Norton's online homework system, includes algorithmically generated versions of these questions, plus additional conceptual exercises. If your instructor assigns questions in SmartWork, log in at **smartwork.wwnorton.com**.

 StudySpace is a free and open website that provides a Study Plan for each chapter of **Understanding Our Universe**. Study Plans include animations, reading outlines, vocabulary flashcards, and multiple-choice quizzes, plus links to premium content in SmartWork and the ebook. Visit **wwnorton.com/studyspace**.

Exploration | Light as a Wave

wwnorton.com/studyspace

Visit StudySpace, and open the "Light as a Wave, Light as a Photon" AstroTour in Chapter 4. Watch the first section, and click through, using the "Play" button until you reach Section 2 of 3.

Here we will explore the following questions: How many properties does a wave have? Are any of these related to each other?

Work your way through to the experimental section, where you can adjust the properties of the wave. Watch the simulation for a moment to see how fast the frequency counter increases.

1. **Increase the wavelength, using the arrow key. What happens to the rate of the frequency counter?**

2. **Reset the simulation and then decrease the wavelength. What happens to the rate of the frequency counter?**

3. **How are the wavelength and frequency related to each other?**

4. **Imagine that you increase the frequency instead of the wavelength. How should the wavelength change when you increase the frequency?**

5. **Reset the simulation, and increase the frequency. Did the wavelength change in the way you expected?**

6. **Reset the simulation, and increase the amplitude. What happens to the wavelength and the frequency counter?**

7. **Decrease the amplitude. What happens to the wavelength and the frequency counter?**

8. **Is the amplitude related to the wavelength or frequency?**

9. **Why can't you change the speed of this wave?**

5 The Formation of Stars and Planets

The space between the stars is far from empty. Between the stars are giant clouds of cool gas and dust. Hot gas fills the space between these clouds, pressing on them and helping to keep them together. Each of the atoms and molecules in a cloud is gravitationally attracted to every other particle. Some of these clouds will collapse under this gravity, fragmenting to form multiple stars, and sometimes further fragmenting to form planets. It is somewhat shocking to find that dust and gas, along with a few physical principles, create everything that we find both on Earth and throughout the universe.

Until the last quarter of the 20th century, every discussion of the Solar System necessarily started with an accounting of its pieces: "There are *planets* with such and such properties. There are *moons*, there are *asteroids*, and there are *comets*...." The *origin* of the Solar System remained speculative. However, in the last few decades, both stellar astronomers and planetary scientists have found themselves arriving at the *same* picture of our early Solar System from two very different directions. This unified understanding is the foundation for the way we now think about the Sun and the objects that orbit it.

✧ LEARNING GOALS

Our Solar System is an unmistakable by-product of the birth of the Sun. The discovery of planetary systems surrounding other stars has shown that our Solar System is probably not unique. The figure at right shows a montage of photographs of various steps in the formation of stars and planets. By the end of this chapter, you should be able to identify the physical principles operating at each step and fill in the details between the steps. You should also be able to:

- Understand the role that gravity and angular momentum play in the formation of stars and planets

- Diagram the process by which dust grains in the disk around a young star stick together to form larger and larger solid objects

- Know why planets orbit the Sun in a plane, and why they revolve in the same direction that the Sun rotates

- Explain how temperature in the disk affects the composition and location of planets, moons, and other bodies

- Understand how astronomers find planets around other stars

- Explain why planetary systems around other stars are common

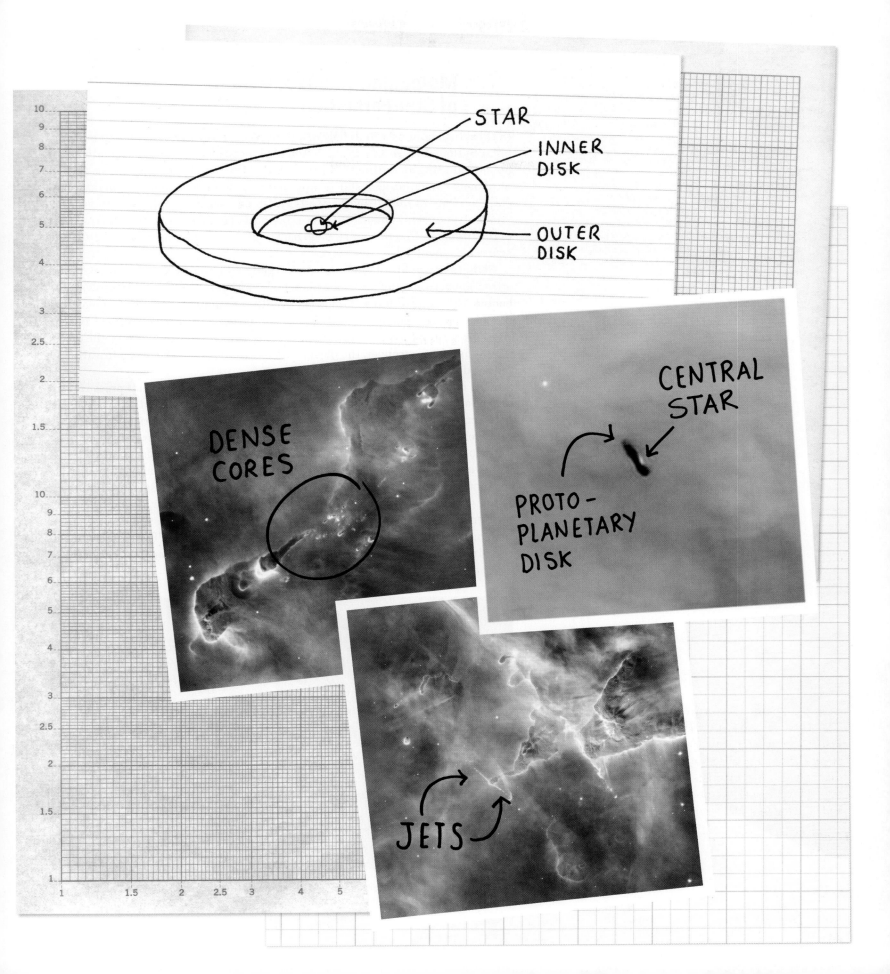

Vocabulary Alert

Pressure: In common language, we often use *pressure* interchangeably with the word *force*. Astronomers specifically use pressure to mean the force that atoms or molecules exert on each other over an area when they are zooming around quickly and colliding with each other.

Dense: In common language, we use this word in many ways, some of which are metaphorical and unkind, as in "You can't understand this? You are so dense!" Astronomers specifically use *density* to mean "the amount of mass packed into a volume"; denser material contains more mass in the same amount of space. In practical terms, you are familiar with density by how heavy an object feels for its size: a billiard ball and a tennis ball are roughly the same size, but the billiard ball has greater mass and therefore feels heavier because it is denser.

Our Solar System is tiny compared to the universe.

▶❚❚ AstroTour: **Star Formation**

Some molecular clouds collapse under their own weight.

5.1 Molecular Clouds Are the Cradles of Star Formation

In thinking about how it all began, a good place to start is right here in our own **Solar System** (**Figure 5.1**)—a collection of planets, moons, and other smaller bodies surrounding an ordinary star that we call the Sun. In general, such a system of planets surrounding a star is a **planetary system**, and there are many planetary systems in the Milky Way Galaxy. Do not confuse our Solar System with "the universe." Our Solar System is a tiny part of our galaxy, which is a tiny part of the universe. You may wish to go back to Figure 1.2 to remind yourself of the size scales involved.

Self-gravity holds planets and stars together. Self-gravity is the gravitational attraction between the parts of a planet or star that pulls all the parts toward its center. This inward force is opposed by either structural strength (such as that of rocks that make up terrestrial planets), or the outward force resulting from gas **pressure** within a star. If the outward force is weaker than self-gravity, the object contracts. If it is stronger, the object expands. In a stable object, the inward and outward forces are balanced.

This concept of the opposing forces of self-gravity and internal pressure can also be applied to the **interstellar medium**, the gas and dust that fill the vast spaces between the stars in a galaxy. As shown in **Figure 5.2**, an **interstellar cloud**—a relatively **dense** region of the interstellar medium—has self-gravity: each part of the cloud is gravitationally attracted to every other part of the cloud. In most interstellar clouds, the internal pressure is much stronger than self-gravity, so the clouds should expand. But the opposing pressure of the less dense but much hotter gas surrounding the clouds helps to hold them together. (Pressure is proportional to both density and temperature.) The densest, coolest clouds are called **molecular clouds** because the conditions within them allow hydrogen atoms to combine to form hydrogen *molecules.* Some molecular clouds are massive enough,

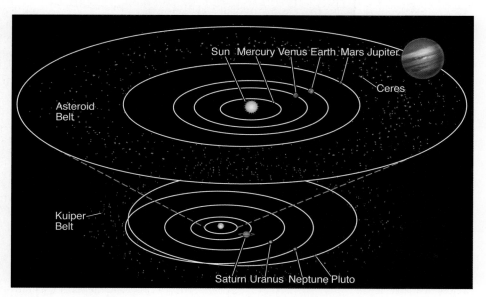

FIGURE 5.1 Our Solar System. Pluto and Ceres are the only dwarf planets shown. Planetary sizes are not to scale.

dense enough, and cool enough that their self-gravity overwhelms their internal pressure, and they collapse under their own weight.

We might expect that the collapse of a molecular cloud should happen quickly. In practice, the process goes very slowly because several other effects slow the collapse. One important effect is conservation of angular momentum, which we explore later in this chapter. (This is the "ice-skater effect" we saw in Chapter 2's Reading Astronomy News article.) Turbulence—the random motion of pockets of gas within the cloud—and magnetic fields also play a role. However, these effects are temporary, and gravity is weak but relentless. As the forces that oppose the cloud's self-gravity gradually fade away, the cloud slowly collapses.

Molecular Clouds Fragment as They Collapse

As a molecular cloud becomes smaller and denser, its self-gravity grows stronger. Suppose a cloud starts out being 4 light-years across. By the time the cloud has collapsed to 2 light-years across, the different parts of the cloud are, on average, only half as far apart as when the collapse started. As a result, the gravitational attraction they feel toward each other will be four times stronger. When the cloud is ¼ as large as when the collapse began, the force of gravity will be 16 times as strong. As a cloud collapses, the inward force of gravity increases; as gravity increases, the collapse speeds up; as the collapse speeds up, the gravitational force increases even faster.

Molecular clouds are never uniform. Some regions are denser and collapse more rapidly than surrounding regions. As these regions collapse, their self-gravity becomes stronger, so they collapse even faster. **Figure 5.3** shows this process. Slight variations in the density of the cloud become very dense concentrations of gas. The end result is that instead of collapsing into a single object, the cloud fragments into a number of very dense **molecular-cloud cores**. A single molecular cloud may form hundreds or thousands of molecular-cloud cores, each of

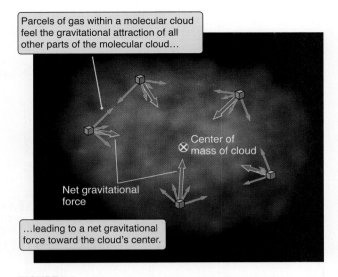

Parcels of gas within a molecular cloud feel the gravitational attraction of all other parts of the molecular cloud...

Center of mass of cloud

Net gravitational force

...leading to a net gravitational force toward the cloud's center.

FIGURE 5.2 Self-gravity causes a molecular cloud to collapse.

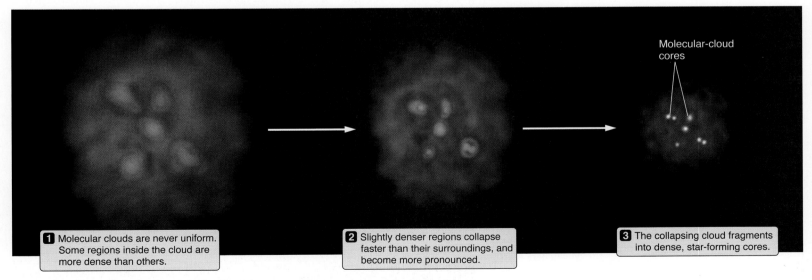

1 Molecular clouds are never uniform. Some regions inside the cloud are more dense than others.

2 Slightly denser regions collapse faster than their surroundings, and become more pronounced.

3 The collapsing cloud fragments into dense, star-forming cores.

Molecular-cloud cores

FIGURE 5.3 As a molecular cloud collapses, denser regions within the cloud collapse more rapidly than less dense regions. As this process continues, the cloud fragments into a number of very dense molecular-cloud cores that are embedded within the large cloud. These cloud cores may go on to form stars.

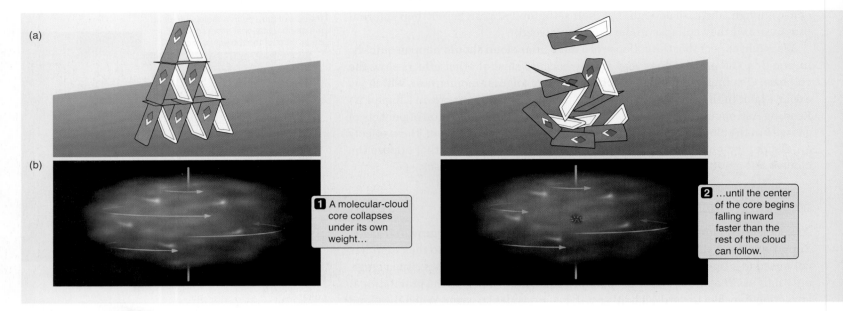

FIGURE 5.4 (a) When the bottom layer is knocked out from under a house of cards, the entire structure collapses from the ground up. (b) Similarly, when a molecular-cloud core gets very dense, it collapses from the inside out. Conservation of angular momentum causes the infalling material to form an accretion disk that feeds the growing protostar.

which is typically a few light-months in size. Some of these cores will eventually form stars.

As a molecular-cloud core collapses, the gravitational forces grow stronger still. Eventually, gravity is able to overwhelm the opposing forces caused by pressure, magnetic fields, and turbulence. This happens first near the center of the cloud core, because that's where the cloud material is most strongly concentrated. The inner parts of the cloud core start to fall rapidly inward, "pulling the bottom out" from the next outer layer. Without this support, the next outer layer begins to fall freely toward the center as well. The process continues: each layer of the cloud core falls inward in turn, thereby removing support from the layers still farther out. As shown in **Figure 5.4**, the cloud core collapses like a house of cards when the bottom layer is knocked out. The whole structure comes crashing down.

5.2 The Protostar Becomes a Star

We will now follow what happens to that innermost core and then go back to find out what happens to the "leftovers." The core is called a **protostar**: an object that is about to be a **star**. The surface of the protostar is heated to a temperature of thousands of degrees as the cloud collapses. Particles are pulled toward the center by gravity. As they fall, they move faster and faster. As they become more densely packed, they begin to crash into each other, causing random motions and raising the temperature of the core. We summarize this process by saying that gravitational energy is converted to thermal energy. The surface of the protostar is tens of thousands of times larger than the surface of the Sun, and each square meter of that enormous surface is radiating energy away. As a result, the protostar is thousands of times more **luminous** than our Sun.

Although the protostar is emitting a lot of light, astronomers often cannot see it in visible light. There are two reasons for this. First, most of the protostar's radiation is in the infrared part of the spectrum, not the visible. Second, the protostar is buried deep in the heart of a dense and dusty molecular cloud. Dust blocks visible light. However, astronomers are able to view these objects in the infrared

Vocabulary Alert

Luminous: In common language, *luminous* and *bright* are used interchangeably. Astronomers, however, observe objects at great distances. They use the word *bright* to describe how an object appears in our sky. They use the word *luminous* to describe how much light the object emits, in all parts of the spectrum. A bright star might be luminous, or it might just be very close. A faint star might be extremely luminous but very, very far away.

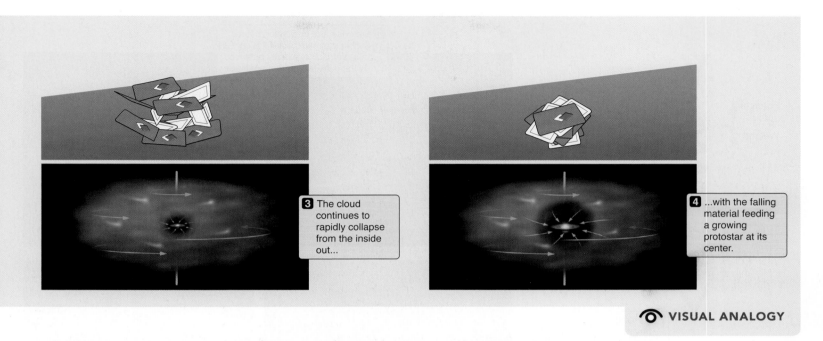

3 The cloud continues to rapidly collapse from the inside out...

4 ...with the falling material feeding a growing protostar at its center.

○ **VISUAL ANALOGY**

part of the spectrum because much of the longer-wavelength infrared light from the protostar *is* able to escape through the cloud. In addition, the radiation from the star heats the dust in the cloud core, and this dust also glows in the infrared.

Sensitive infrared instruments developed since the 1980s have revolutionized the study of protostars and other young stellar objects. Dark clouds have revealed themselves to be entire clusters of dense cloud cores, young stellar objects, and glowing dust when viewed in the infrared. **Figure 5.5** shows infrared pictures of stars forming within columns of gas and dust in the Eagle Nebula.

A Shifting Balance: The Evolving Protostar

At any given moment, the protostar is in balance, with the forces from hot gas pushing outward and gravity pulling inward exactly opposing each other. However, this balance is constantly changing. How can an object be in perfect balance and yet be changing at the same time? **Figure 5.6** shows a simple spring balance. If an object is placed on it, the spring compresses until the downward force of the weight of the object is exactly balanced by the upward force of the spring. The more the spring is compressed, the harder the spring pushes back. We measure the weight of the object by determining the point at which the pull of gravity and the push of the spring are equal.

Let's now slowly pour sand onto our spring balance. At any time, the downward weight of the sand is balanced by the upward force of the spring. As the weight of the sand increases, the spring is slowly compressed. The spring and the weight of the sand are always in balance, but this balance is *changing with time* as more sand is added. The situation is analogous to our protostar, in which the outward pressure of the gas behaves like the spring. Material falls onto the protostar, adding to its mass

FIGURE 5.5 This Hubble Space Telescope image of the Eagle Nebula shows dense columns of molecular gas and dust. Infrared images of the same field, also taken with the HST, show young stars forming within these columns.

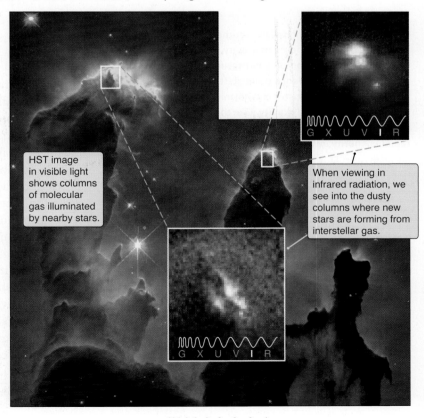

HST image in visible light shows columns of molecular gas illuminated by nearby stars.

When viewing in infrared radiation, we see into the dusty columns where new stars are forming from interstellar gas.

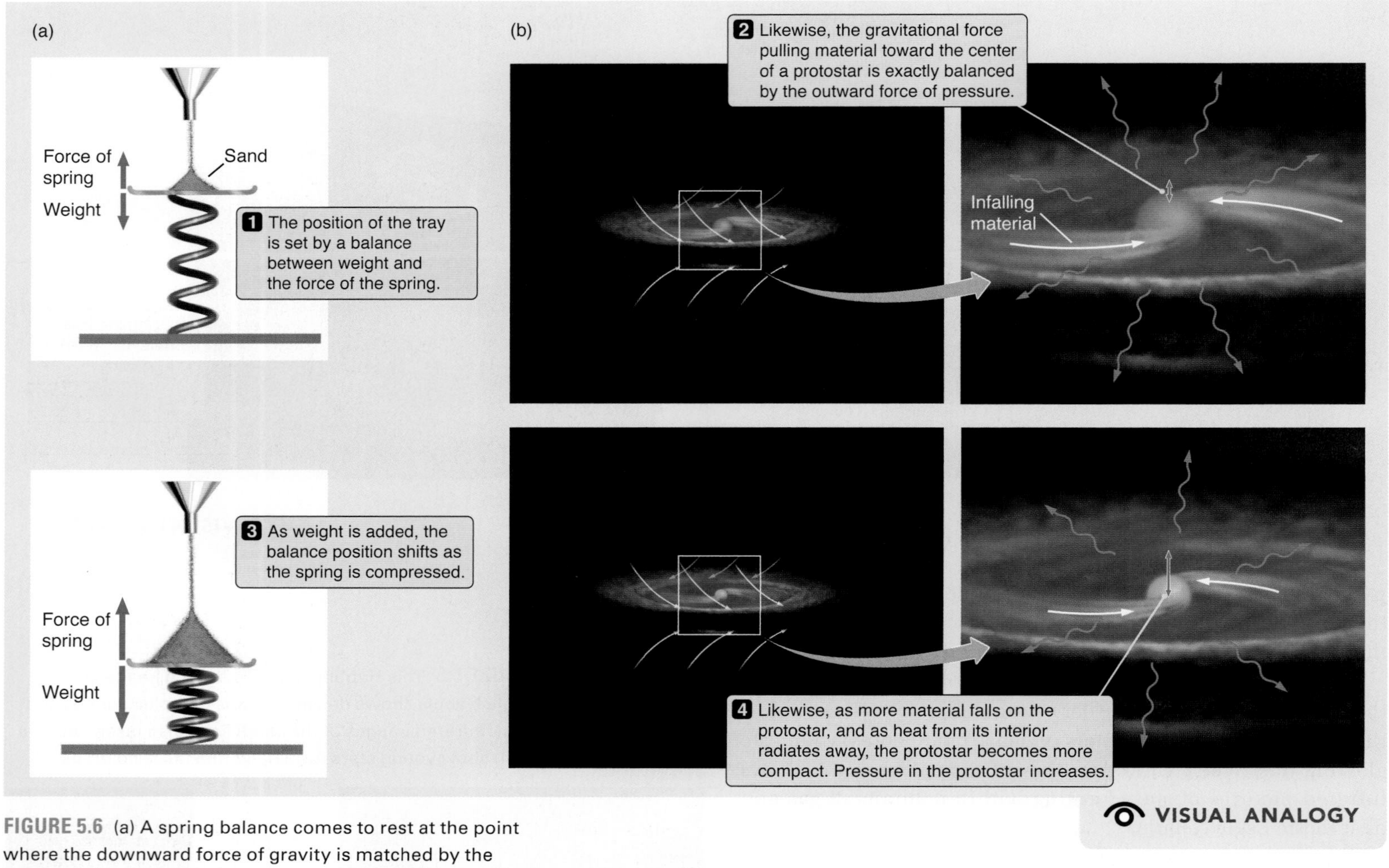

(a)

Force of
spring
Weight

Sand

1 The position of the tray
is set by a balance
between weight and
the force of the spring.

Force of
spring

Weight

3 As weight is added, the
balance position shifts as
the spring is compressed.

(b)

2 Likewise, the gravitational force
pulling material toward the center
of a protostar is exactly balanced
by the outward force of pressure.

Infalling
material

4 Likewise, as more material falls on the
protostar, and as heat from its interior
radiates away, the protostar becomes more
compact. Pressure in the protostar increases.

👁 **VISUAL ANALOGY**

FIGURE 5.6 (a) A spring balance comes to rest at the point
where the downward force of gravity is matched by the
upward force of the compressed spring. As sand is added,
the location of this balance point shifts. (b) Similarly,
the structure of a protostar is determined by a balance
between pressure and gravity. Like the spring balance, the
structure of the protostar constantly shifts as the protostar
radiates energy away and additional material falls onto its
surface.

▶‖ **AstroTour: Solar System Formation**

and gravitational pull. Additionally, the protostar slowly loses internal thermal
energy by radiating it away. But the material that has fallen onto the protostar
also compresses the protostar and heats it up. The interior becomes denser and
hotter, and the pressure rises—just enough to balance the increased weight of
the material above it. Dynamic balance is always maintained.

This dynamic balance persists as energy is radiated away and the protostar
slowly contracts. Gravitational energy is converted to thermal energy, which
heats the core, raising the pressure to oppose gravity. This process continues,
with the protostar becoming smaller and smaller and its interior growing hotter
and hotter, until the center of the protostar is finally hot enough to "ignite", or
begin turning hydrogen into helium, a process that is the main energy source for
most stars. This sequence of events is shown in **Figure 5.7**.

The protostar's mass determines whether it will actually become a star. As the
protostar slowly collapses, the temperature at its center gets higher and higher. If
the protostar's mass is greater than about 0.08 times the mass of the Sun (0.08 $M_\odot$),
the temperature in its core will eventually reach 10 million K,[1] and the nuclear
reaction that converts hydrogen into helium will begin. The newly born star will

---

[1] Here we are using the scientific temperature scale, Kelvin, which has degrees of the same size
as the Celsius scale, but has a zero point at -273°C. Water freezes at 273 K and boils at 373 K.

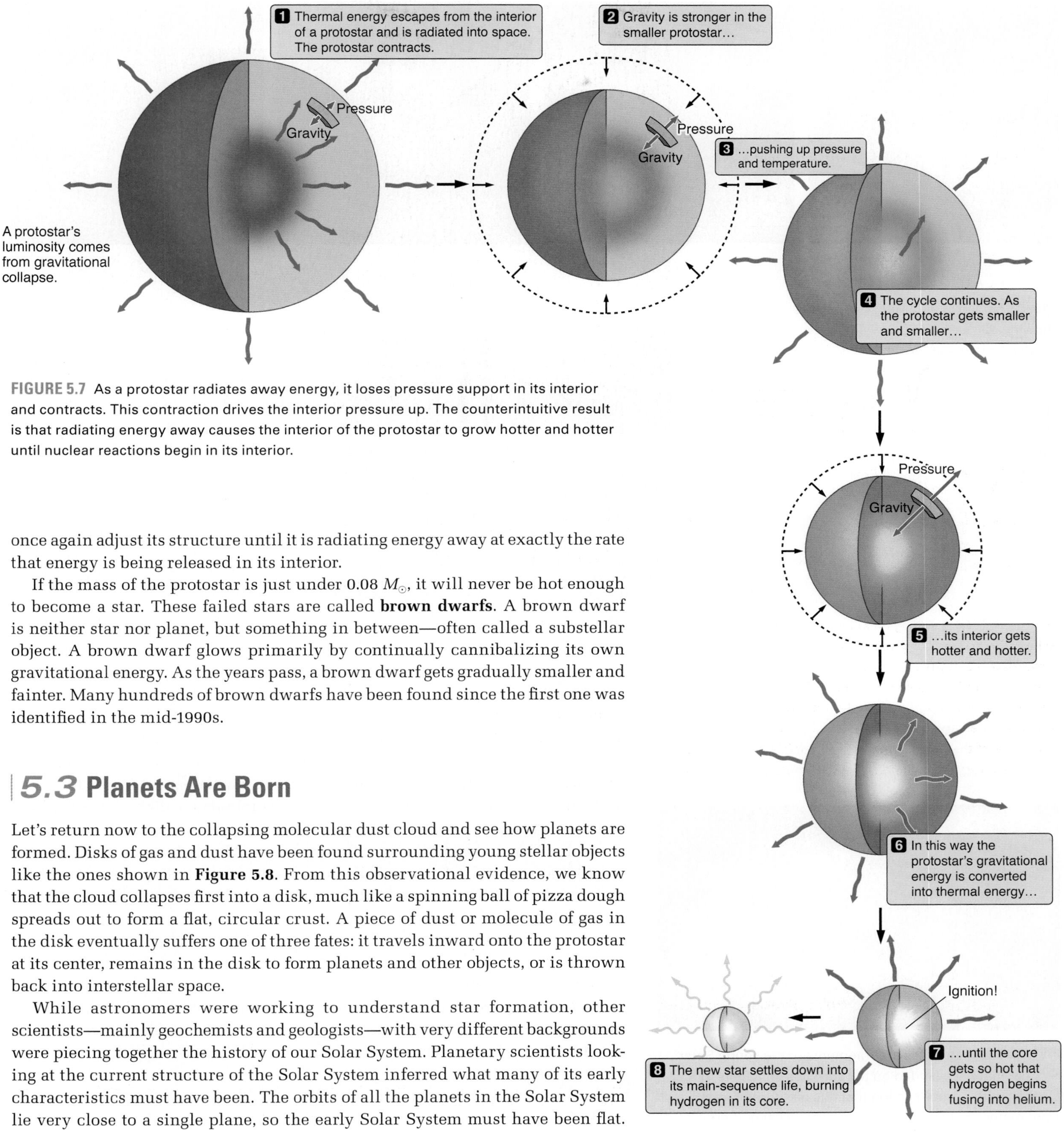

1 Thermal energy escapes from the interior of a protostar and is radiated into space. The protostar contracts.

2 Gravity is stronger in the smaller protostar…

Pressure
Gravity

A protostar's luminosity comes from gravitational collapse.

Pressure
Gravity

3 …pushing up pressure and temperature.

4 The cycle continues. As the protostar gets smaller and smaller…

FIGURE 5.7 As a protostar radiates away energy, it loses pressure support in its interior and contracts. This contraction drives the interior pressure up. The counterintuitive result is that radiating energy away causes the interior of the protostar to grow hotter and hotter until nuclear reactions begin in its interior.

Pressure
Gravity

5 …its interior gets hotter and hotter.

6 In this way the protostar's gravitational energy is converted into thermal energy…

Ignition!

8 The new star settles down into its main-sequence life, burning hydrogen in its core.

7 …until the core gets so hot that hydrogen begins fusing into helium.

once again adjust its structure until it is radiating energy away at exactly the rate that energy is being released in its interior.

If the mass of the protostar is just under $0.08\ M_\odot$, it will never be hot enough to become a star. These failed stars are called **brown dwarfs**. A brown dwarf is neither star nor planet, but something in between—often called a substellar object. A brown dwarf glows primarily by continually cannibalizing its own gravitational energy. As the years pass, a brown dwarf gets gradually smaller and fainter. Many hundreds of brown dwarfs have been found since the first one was identified in the mid-1990s.

5.3 Planets Are Born

Let's return now to the collapsing molecular dust cloud and see how planets are formed. Disks of gas and dust have been found surrounding young stellar objects like the ones shown in **Figure 5.8**. From this observational evidence, we know that the cloud collapses first into a disk, much like a spinning ball of pizza dough spreads out to form a flat, circular crust. A piece of dust or molecule of gas in the disk eventually suffers one of three fates: it travels inward onto the protostar at its center, remains in the disk to form planets and other objects, or is thrown back into interstellar space.

While astronomers were working to understand star formation, other scientists—mainly geochemists and geologists—with very different backgrounds were piecing together the history of our Solar System. Planetary scientists looking at the current structure of the Solar System inferred what many of its early characteristics must have been. The orbits of all the planets in the Solar System lie very close to a single plane, so the early Solar System must have been flat.

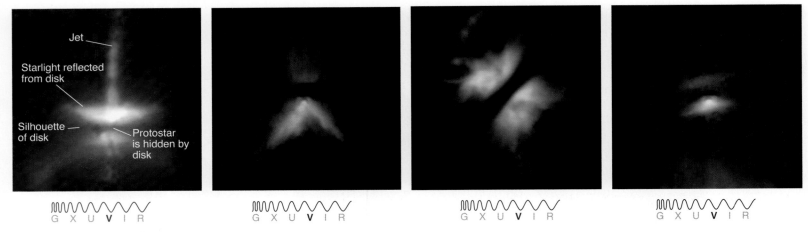

Jet

Starlight reflected from disk

Silhouette of disk

Protostar is hidden by disk

G X U V I R   G X U V I R   G X U V I R   G X U V I R

FIGURE 5.8 (a) and (b) Hubble Space Telescope images showing disks around newly formed stars. The dark bands are the shadows of the disks seen more or less edge on. Bright regions are dust illuminated by starlight. Some disk material may be expelled in a direction perpendicular to the plane of the disk in the form of violent jets. (c) and (d) Hubble Space Telescope images of dust disks around young stars. In these images, the disks shine by reflected starlight. Planets may be forming or have already formed in these disks.

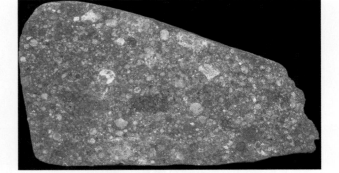

FIGURE 5.9 Meteorites are fragments of the young Solar System that have landed on the surfaces of planets. It is clear from this cross section that this meteorite formed from many smaller components that stuck together.

Additionally, all the planets orbit the Sun in the same direction, so the material from which the planets formed must have been orbiting in the same direction as well. **Meteorites** include pieces of material that are left over from the Solar System's youth. Many meteorites, such as the one in **Figure 5.9**, resemble a piece of concrete in which pebbles and sand are mixed with a much finer filler, suggesting that the larger bodies in the Solar System must have grown from the aggregation of smaller bodies. Following this chain of thought back in time, we find an early Solar System in which the young Sun was surrounded by a flattened disk of both gaseous and solid material. Our Solar System formed from this swirling disk of gas and dust.

As astronomers and planetary scientists compared notes, they realized they had arrived at the *same* picture of the early Solar System from two completely different directions. The connection between the formation of stars and the origin and evolution of the Solar System is one of the cornerstones of both astronomy and planetary science—a central theme of our understanding of our Solar System.

Figure 5.10 presents the protostar that formed the Sun roughly 5 billion years ago. Surrounding the protostellar Sun was a flat, orbiting disk of gas and dust like those seen around protostars today. Each bit of the material in this thin disk orbited according to the same laws of motion and gravitation that govern the orbits of the planets. This disk is referred to as a **protoplanetary disk**. It was 1/100 as massive as the protostar, but this amount was more than enough to account for the bodies that make up the Solar System today. What is it about the process of star formation that leads not only to a star, but to a flat, orbiting collection of gas and dust as well? The answer to this question involves something called *angular momentum*.

The Collapsing Cloud Rotates

You have likely seen a figure-skater spinning on the ice (**Figure 5.11**). Like any rotating object, the spinning skater has some amount of **angular momentum**. The amount of angular momentum depends on three things:

1. How fast the object is rotating. The faster an object is rotating, the more angular momentum it has.

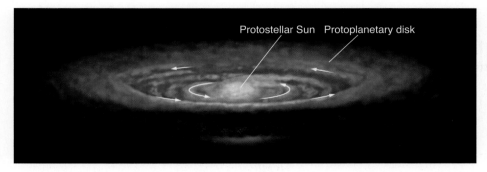

Protostellar Sun   Protoplanetary disk

FIGURE 5.10 When you consider the young Sun, think of it as being surrounded by a flat, rotating disk of gas and dust that is flared at its outer edge.

2. The mass of the object. If a billiard ball and a tennis ball are spinning at the same speed, the billiard ball has more angular momentum because it has more mass.

3. How the mass of the object is distributed relative to the spin axis—that is, how far or spread out the object is. For an object of a given mass and rate of rotation, the more spread out it is, the more angular momentum it has. A spread-out object that is rotating slowly might have more angular momentum than a compact object rotating rapidly.

In **Working It Out 5.1**, we see how to calculate the angular momentum of an orbiting body—Earth.

Angular momentum is very important in physics and astronomy because it is *conserved*; the amount of angular momentum possessed by an isolated object or group of objects does not change. This statement is referred to as the law of **conservation of angular momentum**. There are many other conservation laws in physics, including the laws of conservation of momentum, of energy, and of electric charge. Conservation laws are at the heart of many discoveries. Whenever it appears that a conservation law has been violated, we know there must be something new to understand.

Both an ice-skater and a collapsing interstellar cloud are affected by conservation of angular momentum. An ice-skater can control how rapidly she spins by pulling in or extending her arms. As she pulls in her arms to become more compact, she must spin faster to maintain the same angular momentum. When her arms are held tightly in front of her, the skater's spin becomes a blur. She finishes with a flourish by throwing her arms and leg out—an action that abruptly slows her spin by spreading out her mass. Her angular momentum, which accounts for both her speed and her compactness, remains the same throughout.

Just as the ice-skater speeds up when she pulls in her arms, the cloud that formed our Sun rotated faster and faster as it collapsed. However, there is a puzzle here. Suppose the Sun formed from a typical cloud—one that was about a light-year across and was rotating so slowly that it took a million years to complete one rotation. By the time such a cloud collapsed to the size of our Sun, it would be spinning so fast that it would complete a rotation in only 0.6 second. This is more than 3 million times faster than our Sun actually spins. At this rate of rotation, the Sun would tear itself apart. It appears that angular momentum was not conserved in the actual formation of the Sun—but that can't be right. We must be missing something. Where did the angular momentum go?

👁 **VISUAL ANALOGY**

FIGURE 5.11 A figure-skater relies on the principle of conservation of angular momentum to change the speed of her spin. In the same way, a collapsing cloud spins faster as it becomes smaller.

Working It Out 5.1 | Angular Momentum

We will encounter angular momentum throughout this book. The angular momentum, L, of an orbiting body depends on the mass of the body, m, the speed, v, and the radius of the body's orbit, r. Expressing this mathematically, we have

$$L = mvr$$

As an example, we can apply this relationship to the *orbital angular momentum*, L_o, of Earth in its orbit about the Sun. Earth's mass is 5.97×10^{24} kilograms (kg), its orbital speed is 2.98×10^4 meters per second (m/s), and its orbital radius is 1.496×10^{11} meters. This gives us

$$L_o = (5.97 \times 10^{24}\,\text{kg})\,(2.98 \times 10^4\,\text{m/s})\,(1.496 \times 10^{11}\,\text{m})$$
$$L_o = 2.66 \times 10^{40}\,\text{kg m}^2/\text{s}$$

We can compare this with the angular momentum of Earth around its own axis, its *spin angular momentum*, which is calculated differently. Since Earth is a solid object, filling the space

in which it rotates, we must calculate the angular momentum of every tiny piece and then add them up. If we calculate the spin angular momentum for a uniform sphere, we find that the spin angular momentum, L_s, is proportional to the square of the radius of the object, R, and inversely proportional to the period, P:

$$L_s = \frac{4\pi M R^2}{5P}$$

If we then assume Earth is a uniform sphere, using a radius of 6.38×10^6 meters, and a rotation period of 86,400 seconds (1 day), we find that

$$L_s = \frac{4\pi(5.97 \times 10^{24}\,\text{kg})\,(6.38 \times 10^6\,\text{m})^2}{5(86,400\,\text{s})}$$
$$L_s = 7.07 \times 10^{33}\,\text{kg m}^2/\text{s}$$

So Earth's orbital angular momentum is nearly 4 million times its spin angular momentum.

An Accretion Disk Forms

The *direction* of the collapse is important. Imagine that the ice-skater bends her knees, compressing herself downward instead of bringing her arms toward her body. As she does this, she again makes herself less spread out, but her speed does not change because no part of her body has become any closer to the axis of spin. Similarly, a clump of a molecular cloud can flatten out without speeding up (**Figure 5.12**). As the clump collapses, its self-gravity increases, and the inner parts begin to fall freely inward, raining down on the growing object at the center. The outer portions of the clump lose the support of the collapsed inner portion, and they start falling inward too. As this material makes its final inward plunge, it lands on a thin, rotating **accretion disk**.

The formation of accretion disks is common in astronomy, so it is worth taking a moment to think carefully about this process. As the material falls toward

FIGURE 5.12 A rotating clump of a molecular cloud collapses in a direction parallel to its axis of rotation, forming an accretion disk.

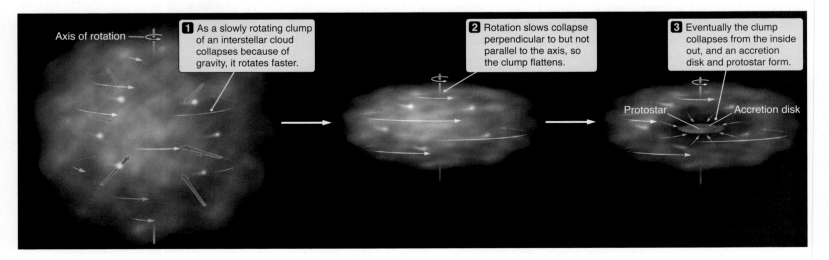

1. As a slowly rotating clump of an interstellar cloud collapses because of gravity, it rotates faster.

Axis of rotation

2. Rotation slows collapse perpendicular to but not parallel to the axis, so the clump flattens.

3. Eventually the clump collapses from the inside out, and an accretion disk and protostar form.

Protostar

Accretion disk

the protostar, it travels on curved, almost always elliptical paths. These paths are oriented randomly except for one key feature—either they all go clockwise, or they all go counterclockwise, when viewed from a direction along the axis of rotation. This is what we mean when we say the cloud rotates. Imagine yourself in such a cloud, near the edge, looking toward the center. As you watch, all the material is orbiting from left to right, but some of it is traveling upward and some downward. Some of it is on steep orbits, traveling more vertically, and some is on shallow orbits, traveling mostly to the side. Now imagine that two pieces of material collide and stick together to form a larger piece. Both are on shallow orbits, but one is traveling up and the other is traveling down. What will happen to the new, larger piece? It will still orbit from left to right, but the upward motion and the downward motion will cancel out, so the orbit will become more shallow. Imagine this same scenario for two pieces on steep orbits. Again, the orbit will become more shallow. In this way, the upward and downward motions of the material cancel each other out, and a disk is formed in which all the material has very shallow orbits, and all orbits still proceed in the same overall direction—either clockwise or counterclockwise. The angular momentum is unchanged, because all the material has maintained its distance from the axis (**Figure 5.13**).

Astronomers can now explain why the Sun does not have the same angular momentum that was present in the original clump of cloud. The radius of a rotating accretion disk is thousands of times greater than the radius of the star that will form at its center. Much of the angular momentum in the original interstellar clump ends up in its accretion disk rather than in its central protostar.

Most of the matter that lands on the accretion disk either ends up as part of the star or is ejected back into interstellar space, sometimes in the form of violent jets, as seen in **Figure 5.14**. Material swirling in these jets carries angular

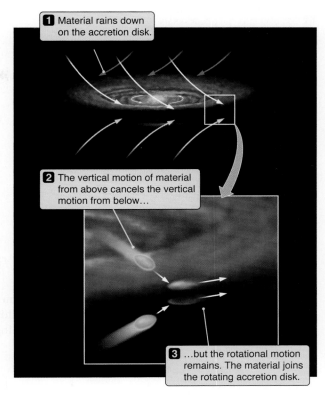

1 Material rains down on the accretion disk.

2 The vertical motion of material from above cancels the vertical motion from below…

3 …but the rotational motion remains. The material joins the rotating accretion disk.

FIGURE 5.13 Gas from a rotating molecular cloud falls inward from opposite sides, piling up onto a rotating disk.

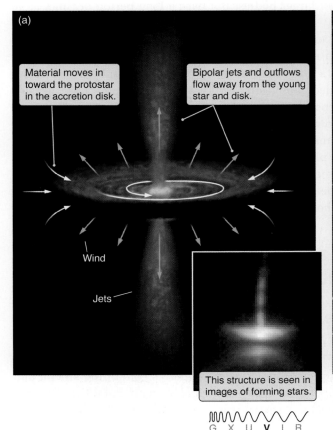

(a)

Material moves in toward the protostar in the accretion disk.

Bipolar jets and outflows flow away from the young star and disk.

Wind

Jets

This structure is seen in images of forming stars.

G X U V I R

(b)

G X U V I R

FIGURE 5.14 (a) Material falls onto an accretion disk around a protostar and then moves inward, eventually falling onto the star. In the process, some of this material is driven away in powerful jets that stream perpendicular to the disk. (b) This infrared Spitzer Space Telescope image shows jets streaming outward from a young, developing star. Note the nearly edge-on, dark accretion disk surrounding the young star.

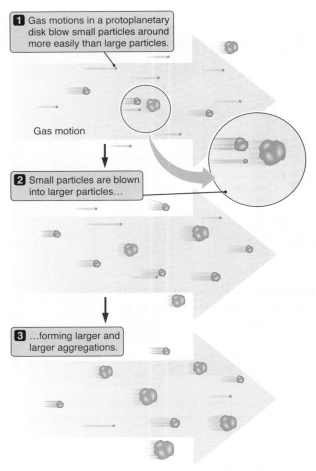

1 Gas motions in a protoplanetary disk blow small particles around more easily than large particles.

Gas motion

2 Small particles are blown into larger particles…

3 …forming larger and larger aggregations.

FIGURE 5.15 Motions of gas in a protoplanetary disk blow smaller particles of dust into larger particles, making the larger particles larger still. This process continues, eventually creating objects many meters in size.

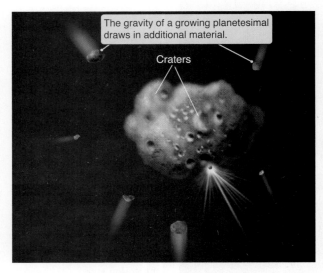

The gravity of a growing planetesimal draws in additional material.

Craters

FIGURE 5.16 The gravity of a planetesimal is strong enough to attract surrounding material, which causes the planetesimal to grow.

momentum away from the accretion disk. However, a small amount of material is left behind in the disk. It is the objects in this leftover disk—the dregs of the process of star formation—that form planets.

Theoretical calculations by astronomers long predicted that accretion disks should be found around young stars. Look back at Figure 5.8, which shows Hubble Space Telescope images of edge-on accretion disks around young stars. The dark bands are the shadows of the edge-on disks, the top and the bottom of which are illuminated by light from the forming star. Our Sun and Solar System formed from a protostar and disk much like those in these pictures.

Small Objects Stick Together to Become Large Objects

Random motions in the accretion disk push the smaller grains of solid material past larger grains, and as this happens, the smaller grains stick to the larger grains. (The "sticking" process among smaller grains is due to the same static electricity that causes dust balls to grow under your bed.) Starting out at only a few micrometers across, the slightly larger bits of dust grow to the size of pebbles and then to clumps the size of boulders, which are not as easily pushed around by gas (**Figure 5.15**). When clumps grow to about 100 meters across, the objects are so few and far between that they collide less frequently, and their growth rate slows down but does not stop.

For two large clumps to stick together rather than explode into many small pieces, they must bump into each other gently—very gently. Collision speeds must be about 0.1 m/s or less for colliding boulders to stick together. Your stride is probably about a meter, so to walk this slowly, you would take one step every 10 seconds. In an accretion disk, collisions more violent than this do happen, breaking these clumps back into smaller pieces. The process is not uniformly one of creating larger and larger bodies. But over a long period of time, large bodies do form.

Eventually measuring up to several hundred meters across, objects grow by "sweeping up" smaller objects that get in their way. But as the clumps reach the size of about a kilometer, a different process becomes important. These kilometer-sized objects, now called **planetesimals** (literally "tiny planets"), are massive enough that their gravity begins to pull on nearby bodies (**Figure 5.16**). No longer does the planetesimal grow only by chance collisions with other objects; now it can pull in and capture small objects that lie outside its direct path. The growth speeds up, and the larger planetesimals quickly consume most of the remaining bodies in the vicinity of their orbits. The final survivors of this process are large enough to be called **planets**. As with the major bodies in orbit about the Sun, some of the planets may be relatively small and others quite large.

5.4 The Inner and Outer Disk Have Different Compositions

Rock, Metal, and Ice

On a hot summer day, ice melts and water quickly evaporates; but on a cold winter night, even the water in our breath freezes into tiny ice crystals before our eyes. Some materials, such as iron, **silicates** (minerals containing silicon and oxygen), and carbon—metals and rocky materials—remain solid even at quite high temperatures. Such substances, which are capable of withstanding high

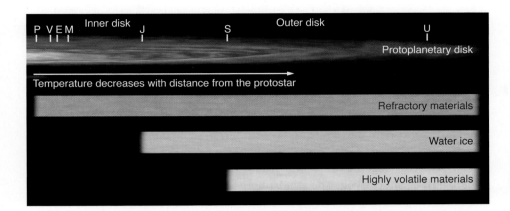

Inner disk   Outer disk

P V E M   J   S   U

Protoplanetary disk

Temperature decreases with distance from the protostar

Refractory materials

Water ice

Highly volatile materials

FIGURE 5.17 Differences in temperature within a protoplanetary disk determine the composition of dust grains that then evolve into planetesimals and planets. Shown here are the protostar (P) and the orbits of Venus (V), Earth (E), Mars (M), Jupiter (J), Saturn (S), and Uranus (U).

temperatures without melting or being vaporized, are referred to as **refractory materials**. Other materials, such as water, ammonia, and methane, can remain in a solid form only if their temperature is quite low. These less refractory substances are called **volatile materials** (or "volatiles" for short).

Differences in temperature from place to place within the protoplanetary disk significantly affect the makeup of the dust grains in the disk (**Figure 5.17**). In the hottest parts of the disk (closest to the protostar), only refractory substances can exist in solid form. In the inner disk, dust grains are composed almost entirely of refractory materials. Somewhat farther out in the disk, some hardier volatiles, such as water ice and certain organic substances, can survive in solid form, adding to the materials that make up dust grains. Highly volatile components such as methane, ammonia, and carbon monoxide ices and some simple organic molecules survive in solid form only in the coldest, outermost parts of the accretion disk, far from the central protostar. The differences in composition of dust grains within the disk are reflected in the composition of the planets formed from that dust. Planets closest to the central star are composed primarily of refractory materials such as rock and metals but are deficient in volatiles. Those that form farthest from the central star contain refractory materials, but they also contain large quantities of ices and organic materials.

Chaotic encounters can change this organization of planetary compositions. In a process called **planet migration**, gravitational scattering can force some planets to end up far from where they formed. Uranus and Neptune originally may have formed near the orbits of Jupiter and Saturn but were then driven outward to their current locations by gravitational encounters with Jupiter and Saturn. A planet can also migrate when it gives up some of its orbital angular momentum to the disk material that surrounds it. Such a loss of angular momentum causes the planet to slowly spiral inward toward the central star. We will see examples from other planetary systems when we discuss *hot Jupiters* in Section 5.6.

Solid Planets Gather Atmospheres

Once a solid planet has formed, it may continue growing by capturing gas from the protoplanetary disk. To do so, though, it must act quickly. Young stars and protostars emit fast-moving particles and intense radiation that can quickly disperse the remains of the accretion disk. Planets like Jupiter have about 10 million years to form and accumulate an atmosphere. Massive planets can capture more of the hydrogen and helium gas that makes up the bulk of the disk.

The gas that is captured by a planet at the time of its formation is the planet's

Volatile ices survive in the outer disk, but only refractory solids survive in the inner disk.

Vocabulary Alert

Organic: *Organic* often means "pertaining to life"; or, in the case of food labeled "organic" in the United States, it means that a list of rules has been obeyed regarding the use of pesticides and herbicides. To scientists, *organic* means "carbon based," as in organic chemistry. Organic substances are those with carbon in them.

Ice: In common language, *ice* refers specifically to the solid form of water, although the term *dry ice* is also common. To astronomers, ice refers to the solid form of any type of volatile material.

Chaotic (or chaos): *Chaotic* is commonly used to mean "messy or disorganized." To scientists, *chaotic* means that a small change in the initial state can lead to a large change in the final state of a system.

primary atmosphere. The primary atmosphere of a large planet can be more massive than the solid body, as in the case of giant planets such as Jupiter. Very large objects can acquire mini accretion disks as they capture gas and dust. Some of this material may grow into larger bodies in much the same way that material in the protoplanetary disk formed planets. The result is a mini "solar system"— a group of moons that orbit about the planet.

A less massive planet may also capture some gas from the protoplanetary disk, only to lose it later. The gravity of small planets may be too weak to hold low-mass gases such as hydrogen and helium. Even if a small planet is able to gather some hydrogen and helium from its surroundings, this primary atmosphere will be short-lived. The atmosphere around a small planet like our Earth is a **secondary atmosphere**, which forms later in the life of a planet. Volcanism is one important source of a secondary atmosphere because it releases carbon dioxide and other gases from the planet's interior. In addition, volatile-rich comets that formed in the outer parts of the disk fall inward toward the new star and sometimes collide with planets. Comets may provide a significant source of water, organic compounds, and other volatile materials on planets close to the central star.

5.5 A Tale of Eight Planets

Nearly 5 billion years ago, our Sun was a protostar surrounded by a protoplanetary disk of gas and dust. Over the course of a few hundred thousand years, much of the dust collected into planetesimals—clumps of rock and metal near the Sun and aggregates of rock, metal, ice, and organic materials farther from the Sun. Within the inner few astronomical units (AU) of the disk, several rock and metal planetesimals—probably less than a half dozen—quickly grew in size to become the dominant masses in their orbits. These few either captured most of the remaining planetesimals or ejected them from the inner part of the disk. These dominant planetesimals had now become planet-sized bodies with masses ranging between 1/20 and one Earth mass. They became the **terrestrial planets**. Mercury, Venus, Earth, and Mars are the surviving terrestrial planets. One or two others may have formed in the young Solar System but were later destroyed.

For several hundred million years following the formation of the four surviving terrestrial planets, leftover pieces of debris still in orbit around the Sun continued to rain down on their surfaces. Much of this barrage may have originated in the outer Solar System. Today we can still see the scars of these early impacts on the cratered surfaces of some of the terrestrial planets, such as the surface of Mercury shown in **Figure 5.18**. This rain of debris continues today, but at a much lower rate.

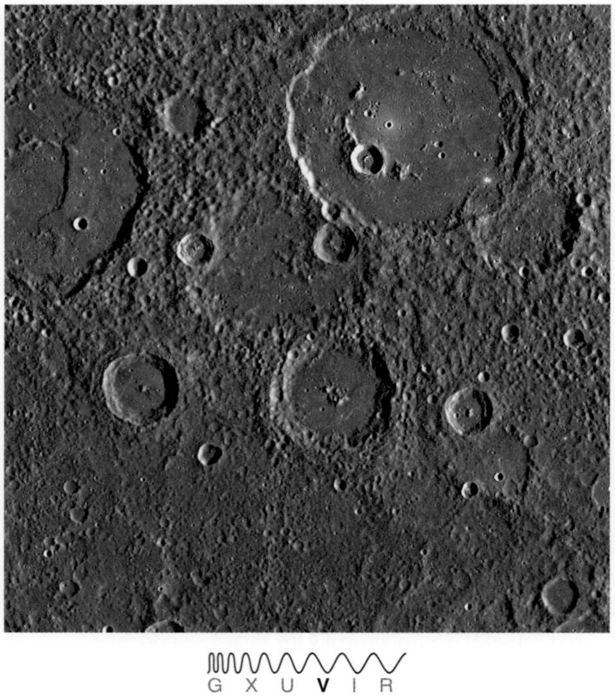

Before the proto-Sun became a true star, gas in the inner part of the protoplanetary disk was still plentiful. During this early period, Earth and Venus held on to weak primary atmospheres of hydrogen and helium, but these thin atmospheres were lost to space. The terrestrial planets had no thick atmospheres until the formation of the secondary atmospheres that now surround Venus, Earth, and Mars. Mercury's proximity to the Sun and the Moon's small mass prevented these bodies from retaining significant secondary atmospheres.

Five AU from the Sun and beyond, planetesimals combined to form a number of bodies with masses about 5–10 times that of Earth. These planet-sized objects

G X U V I R

FIGURE 5.18 Large impact craters on Mercury (and on solid bodies throughout the Solar System) record the final days of the Solar System's youth, when planets and planetesimals grew as smaller planetesimals rained down on their surfaces.

formed from planetesimals containing volatile ices and organic compounds in addition to rock and metal. Four such massive bodies later became the cores of the **giant planets**: Jupiter, Saturn, Uranus, and Neptune. Mini accretion disks formed around these planetary cores, capturing large amounts of hydrogen and helium and funneling this material onto the planets' surfaces.

Jupiter's massive solid core captured and retained the most gas—roughly 300 times the mass of Earth. The other planetary cores captured lesser amounts of hydrogen and helium, perhaps because their cores were less massive or because less gas was available to them. Saturn has less than 100 Earth masses of gas, whereas Uranus and Neptune captured only a few Earth masses' worth of gas.

This model indicates that it could take up to 10 million years for a Jupiter-like planet to form. Some scientists do not think that our protoplanetary disk could have survived long enough to form gas giants such as Jupiter through this process. All the gas may have dispersed in roughly half that time, cutting off Jupiter's supply of hydrogen and helium. There must be more to the story. A process in which the protoplanetary disk fragments into massive clumps equivalent to that of a large planet may resolve this potential conflict. It is possible that both processes played a role in the formation of our own and other planetary systems.

During the formation of the planets, gravitational energy was converted into thermal energy as the individual atoms and molecules moved faster. This conversion warmed the gas surrounding the cores of the giant planets. Proto-Jupiter and proto-Saturn probably became so hot that they actually glowed a deep red color, similar to the heating element on an electric stove. Their internal temperatures may have reached as high as 50,000 K. However, they were never close to becoming stars. As we saw in Section 5.2, a ball of gas must have a mass at least 0.08 times the mass of the Sun for it to become a star. This is about 80 times the mass of Jupiter.

The composition of the moons of the giant planets followed the same trend as the planets that formed around the Sun: the innermost moons formed under the hottest conditions and therefore contained the smallest amounts of volatile material. When Io formed, Jupiter was glowing so intensely that it rivaled the distant Sun. The high temperatures created by the glowing planet evaporated most of the volatile substances nearby. Io today contains no water at all. However, water is probably relatively plentiful on Europa, Ganymede, and Callisto because these moons formed farther from warm, glowing Jupiter.

Moons formed from the mini accretion disks around the giant planets.

Not all planetesimals in the protoplanetary disk went on to become planets. Jupiter is a true giant of a planet. Its gravity kept the region between it and Mars so stirred up that most planetesimals there never formed a large planet. (The one exception is Ceres, which is large enough to be classified as a **dwarf planet**.) This region, now referred to as the **asteroid belt**, contains many planetesimals that remain from this early time.

Planetesimals persist to this day in the outermost part of the Solar System as well. Formed in a deep freeze, these objects have retained most of the highly volatile materials found in the grains present at the formation of the protoplanetary disk. Unlike the crowded inner part of the disk, the outermost parts of the disk had planetesimals that were too sparsely distributed for large planets to grow. Icy planetesimals in the outer Solar System remain today as **comet nuclei**—relatively pristine samples of the material from which our planetary system formed. The frozen, distant dwarf planets Pluto and Eris are especially large examples of these residents of the outer Solar System.

Asteroids and comet nuclei are planetesimals that survive to this day.

The early Solar System was a remarkably violent and chaotic place. Many objects in the Solar System show evidence of cataclysmic impacts that reshaped worlds. A dramatic difference between the terrains of the northern and southern hemispheres on Mars, for example, has been interpreted by some planetary scientists as the result of one or more colossal collisions. Mercury has a crater on its surface from an impact so devastating that it caused the crust to buckle on the opposite side of the planet. In the outer Solar System, one of Saturn's moons, Mimas, sports a crater roughly one-third the diameter of the moon itself. Uranus suffered a collision that was violent enough to literally knock the planet on its side. Today its axis of rotation is tilted almost at a right angle to its orbital plane.

Not even our own Earth escaped devastation by these cataclysmic events. Our best current hypothesis for the formation of the Moon proposes that, in addition to the four terrestrial planets we know today, the early Solar System included at least one other terrestrial planet—a planet about the same size and mass as Mars. As the newly formed planets were settling into their present-day orbits, this fifth planet suffered a grazing collision with Earth and was completely destroyed. The remains of the planet, together with material knocked from Earth's outer layers, formed a huge cloud of debris encircling Earth. For a brief period Earth may have displayed a magnificent group of rings like those of Saturn. In time, this debris coalesced into the single body we know as our Moon.

Look back to the chapter-opening figure, which shows some of the steps involved in the formation of stars and planets. Now that we have explored both a generic example and our own Solar System, you can fill in the gaps between the steps shown to tell the entire story of the formation of stars and planets.

5.6 Planetary Systems Are Common

We began this chapter by discussing the formation of a generic planetary system around a generic star. Only later did we turn to the specifics of our own Solar System. We did this to make an important point: according to our current understanding, *there is absolutely nothing special about the conditions that led to the formation of our Sun and our Solar System*. When astronomers turn their telescopes to young stars around us, they see just the sorts of disks from which our own Solar System formed (**Figure 5.19**). From what we know, the physical processes that led to the formation of our Solar System should be commonplace wherever new stars are being born.

In 1995, astronomers announced the first confirmed **extrasolar planet** around a solar-type star, a Jupiter-sized body orbiting surprisingly close to 51 Pegasi. Today the number of known extrasolar planets, sometimes called

FIGURE 5.19 An edge-on circumstellar dust disk is seen extending outward to 60 AU from the young (12-million-year-old) star AU Microscopii. The star itself, whose brilliance would otherwise overpower the disk, is hidden behind an occulting disk placed in the telescope's focal plane. Its position is represented by the dot.

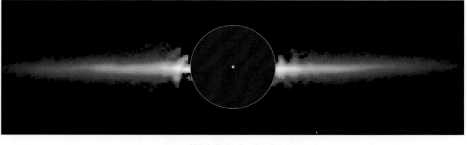

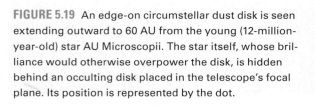

exoplanets, has grown to many hundreds, and new discoveries are occurring almost daily.

The discovery of extrasolar planets raises the question of what we mean by the term *planet*. Within our own Solar System, we feel reasonably comfortable with our definition of a planet. But what about those extrasolar bodies? The International Astronomical Union defines an extrasolar planet as a body that orbits a star other than our Sun and has a mass less than 13 Jupiters. Objects more massive than this but less than $0.08\ M_\odot$ are brown dwarfs. Objects more massive than $0.08\ M_\odot$ are defined as stars.

The Search for Extrasolar Planets

As a planet orbits a star, the planet's gravity tugs the star around ever so slightly. We can sometimes detect this wobble and infer the properties of the planet—its mass and its distance from the star. To see how this works, we must understand the Doppler effect.

Have you ever stood outside and listened as a fire truck sped by with sirens blaring? As the fire truck came toward you, its siren had a certain high pitch; but as it passed by, the pitch of the siren dropped noticeably. If you were to close your eyes and listen, you would have no trouble knowing when the fire truck passed you, just from the change in pitch of its siren. You do not even need a fire engine to hear this effect. The sound of normal traffic behaves in the same way. As a car drives past, the pitch of the sound that it makes suddenly drops. This effect is known as the **Doppler effect**.

The pitch of a sound is like the color of light: it is determined by the wavelength of the wave. What we perceive as higher pitch corresponds to sound waves with shorter wavelengths. Sounds that we perceive as lower in pitch are waves with longer wavelengths. When an object is moving toward us, the waves that it emits, whether light or sound, "crowd together" in front of the object. You can see how this works by looking at **Figure 5.20**, which shows the locations of successive wave crests emitted by a moving object.

The wavelength of light as measured from the source's frame of reference is called the **rest wavelength**, as shown in **Figure 5.21a**. If an object such as a star is moving toward you, the light reaching you from the object has a shorter wavelength than its rest wavelength—it is bluer than if it were stationary. We say that the light from an object moving toward us is **blueshifted (Figure 5.21b)**. In contrast, light from a source that is moving away from you is shifted to longer wavelengths. The light that you see is redder than if the source were not moving away from you, so we say that the light is **redshifted**, as seen in **Figure 5.21c**. The amount by which the wavelength of light is shifted by the Doppler effect is called the **Doppler shift** of the light, and it depends on the speed of the object emitting the light. The faster the object is moving with respect to you, the larger the shift.

The Doppler shift provides information only about the **radial velocity** of an object—that is, whether an object is moving toward you or away from you along the line of sight. The technique of examining Doppler shifts in the

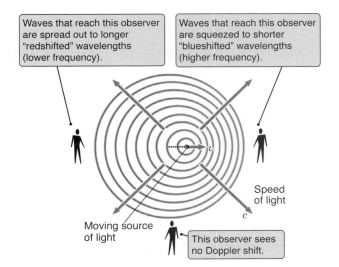

FIGURE 5.20 Motion of a light or sound source relative to an observer may cause waves to be spread out (*redshifted*, or made lower in pitch) or squeezed together (*blueshifted*, or made higher in pitch). Such a change in the wavelength of light or the frequency of sound is called a Doppler shift.

▶❙❙ **AstroTour: The Doppler Effect**

FIGURE 5.21 Light from astronomical objects will be blueshifted or redshifted (a), depending on whether they are moving toward us (b) or away from us (c).

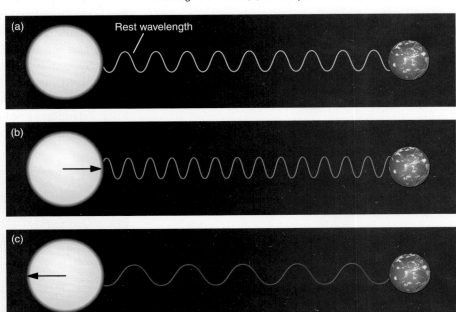

Working It Out 5.2 | Making Use of the Doppler Shift

We noted in Section 4.4 that atoms and molecules emit and absorb light only at certain wavelengths. The spectrum of an atom or molecule has absorption or emission lines that look something like a bar code instead of a rainbow, and each type of atom or molecule has a unique set of lines. These lines are called *spectral lines*. A prominent spectral line of hydrogen atoms has a rest wavelength λ_{rest} of 656.3 nanometers (nm). Suppose that, using a telescope, you measure the wavelength of this line in the spectrum of a distant object and find that instead of seeing the line at 656.3 nm, you see it at a wavelength λ_{obs} of 659.0 nm. You could then infer that the object is moving at a radial velocity of

$$v_r = \frac{\lambda_{obs} - \lambda_{rest}}{\lambda_{rest}} \times c$$

$$v_r = \frac{659.0\,\text{nm} - 656.3\,\text{nm}}{656.3\,\text{nm}} \times (3 \times 10^8\,\text{m/s})$$

$$v_r = 1.2 \times 10^6\,\text{m/s}$$

The object is moving away from you (because the wavelength became longer) with a speed of 1.2×10^6 m/s, or 1,200 kilometers per second (km/s).

Now consider our stellar neighbor, Alpha Centauri, which is moving toward us with a radial velocity of –21.6 km/s (-2.16×10^4 m/s). (Negative velocity means the object is moving toward us.)

What is the observed wavelength λ_{obs} of a magnesium line in Alpha Centauri's spectrum having a rest wavelength λ_{rest} of 517.27 nm? First, we need to manipulate the Doppler equation to get λ_{obs} all by itself. Then we can plug in all the numbers.

$$v_r = \frac{\lambda_{obs} - \lambda_{rest}}{\lambda_{rest}} \times c$$

Solve this equation for λ_{obs} to get

$$\lambda_{obs} = \lambda_{rest} + \frac{v_r}{c} \lambda_{rest}$$

Both terms on the right contain λ_{rest}. Factor it out to make the equation a little more convenient.

$$\lambda_{obs} = \left(1 + \frac{v_r}{c}\right) \lambda_{rest}$$

We are ready to plug in some numbers to solve for the observed wave length.

$$\lambda_{obs} = \left(1 + \frac{-2.16 \times 10^4\,\text{m/s}}{3 \times 10^8\,\text{m/s}}\right) \times 517.27\,\text{nm}$$

$$\lambda_{obs} = 517.23\,\text{nm}$$

Although the observed Doppler blueshift (517.23 – 517.27) is only –0.04 nm, it is easily measurable with modern instrumentation.

light from stars to detect extrasolar planets is therefore called the **spectroscopic radial velocity method.** If an object is moving across your line of sight, then the light is neither redshifted nor blueshifted. In **Working It Out 5.2**, we explore how to use the Doppler shift to calculate the radial velocity of an astronomical object, or the amount by which its light has shifted from our standpoint.

We can see how the Doppler shift helps us find planets by using our own Solar System as an example. Imagine that an alien astronomer points its spectrograph toward the Sun. Both the Sun and Jupiter orbit a common center of gravity that lies just outside the surface of the Sun, as shown in **Figure 5.22**. Our alien astronomer would find that the Sun's radial velocity varies by ±12 m/s, with a period equal to Jupiter's orbital period of 11.86 years. From this informa-

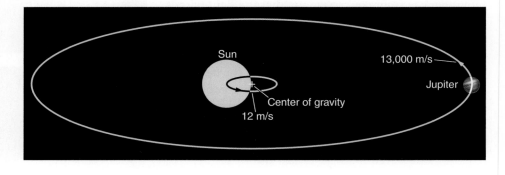

Figure 5.22 Both the Sun and Jupiter orbit around a common center of gravity, which lies just outside the Sun's surface. Spectroscopic measurements made by an extrasolar astronomer would reveal the Sun's radial velocity varying by ±12 m/s over an interval of 11.86 years, Jupiter's orbital period.

tion, the alien would rightly conclude that the Sun has at least one planet with a mass comparable to Jupiter's. Without greater precision, it would be unaware of the other less massive major planets. But, spurred on by the excitement of its discovery, it would improve the sensitivity of its instruments. If the alien could measure radial velocities as small as 2.7 m/s, it would be able to detect Saturn, and if the precision extended to radial velocities as small as 0.09 m/s, it would be able to detect Earth.

Current technology limits the precision of our own radial velocity instruments to about 1 m/s, but it has still been the most successful ground-based approach to finding extrasolar planets thus far. This technique enables astronomers to detect giant planets around solar-type stars, but we are still far from being able to use it to reveal bodies with masses similar to those of our own terrestrial planets.

Another technique for finding extrasolar planets is the **transit method**, in which we observe the effect of a planet passing in front of its parent star. When this happens, the light from the star diminishes by a tiny amount, as illustrated in **Figure 5.23**. Try picturing the geometry required to observe a transit, however, and you may see a major limitation. For a planet to pass in front of a star from our perspective, Earth must lie nearly in the orbital plane of the planet. There is another important difference between the radial velocity and transit methods: Whereas the radial velocity method gives us the mass of the planet and its orbital distance from a star, the transit method provides the *size* of a planet. Current ground-based technology limits the sensitivity of the transit method to about 0.1 percent of a star's brightness.

Using the transit method, our aforementioned alien astronomer could infer the existence of Earth only if it were located somewhere in the plane of Earth's orbit and could detect a 0.009 percent drop in the Sun's brightness. Our own astronomers now have space instruments with that capability. Using the transit method, the French CoRoT satellite discovered a planet with a diameter only 1.7 times that of Earth. In 2009, NASA launched a solar-orbiting telescope called Kepler with even greater capability. Its instruments are able to detect transits of Earth-sized planets. This telescope has more than doubled the number of extra-solar planetary candidates in just a few short years.

Still another technique involves a relativistic effect called **gravitational lensing** (see Chapter 13). The gravitational field of an unseen planet bends light from a distant star in such a way that it causes the star to brighten temporarily while the planet is passing in front of it. Gravitational lensing, like the radial velocity method, provides an estimate of the mass of the planet. Like the transit method used in space, lensing is also capable of detecting Earth-sized planets.

A fourth method is **direct imaging**. This technique is straightforward conceptually, but it is the most technically difficult technique because it involves searching for a relatively faint planet in the overpowering glare of a bright star—a challenge far more difficult than looking for a star in the dazzling brilliance of a clear daytime sky. Even when an object is detected by direct imaging, there remains the major problem of determining whether the observed object is indeed a planet. Suppose we detect a faint object near a bright star. Might it be a distant star that just happens to be in the line of sight? Future observations could tell if it shares the bright star's motion through space. Could it be a brown dwarf rather than a true planet? Further observations to determine the object's mass would be required.

Some planets have been discovered by this method using large, ground-based telescopes operating in the infrared region of the spectrum (**Figure 5.24**). The first

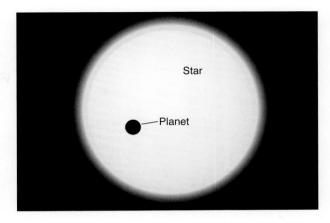

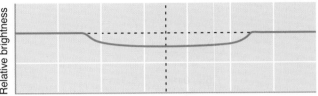

FIGURE 5.23 As a planet passes in front of a star, it blocks some of the light coming from the star's surface, causing the brightness of the star to decrease slightly. (The brightness decrease has been exaggerated in this illustration.)

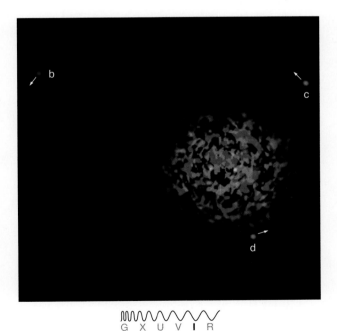

FIGURE 5.24 An infrared image shows three planets (labeled "b," "c," and "d"), each with a mass several times that of Jupiter, orbiting the star HR 8799 (hidden behind a mask.)

visible-light discovery was made from space while the Hubble Space Telescope was observing Fomalhaut, a bright naked-eye star only 25 light-years away. The planet, Fomalhaut b, has a mass no more than three times that of Jupiter and orbits within a dusty debris ring some 17 billion km from the central star (**Figure 5.25**).

The most exciting discoveries, however, could come with future missions. Future observatories will not only detect Earth-like planets around nearby stars; they will also measure the planets' physical and chemical characteristics.

Other Planetary Systems Are Not Like Ours

Our searches for extrasolar planets have been remarkably successful. Since the first extrasolar planet around a solar-type star was confirmed in 1995, astronomers have found hundreds more, many with multiple planets. A representative few are shown in **Figure 5.26**. NASA's Kepler mission is currently searching for terrestrial-type planets. The field is changing so fast that the most up-to-date information can only be found online.

As the number of known planetary systems grows, we can compare own Solar System with other planetary systems. Most of them *do not look like our own.* Many contain **hot Jupiters**, which are Jupiter-type planets orbiting solar-type stars in tight circular orbits that are closer to their parent stars than Mercury is to our own Sun. Others have planets in highly eccentric orbits, unlike the relatively circular planetary orbits in our Solar System. Still others contain planets far more massive than Jupiter. Does this mean that our Solar System is an oddball? Not necessarily—but maybe. We await further results from the Kepler spacecraft.

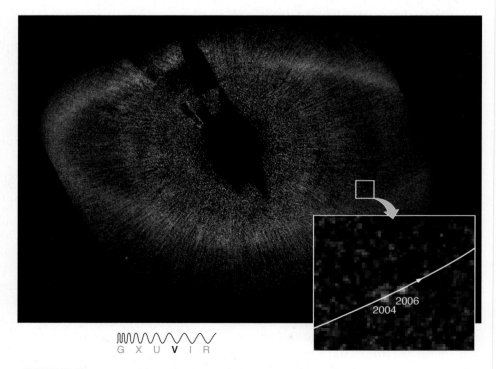

G X U V I R

FIGURE 5.25 Fomalhaut b is seen here moving in its orbit around Fomalhaut, a nearby star easily visible to the naked eye. The parent star, hidden by an obscuring mask, is about a billion times brighter than the planet, which is located within a dusty debris ring that surrounds the star.

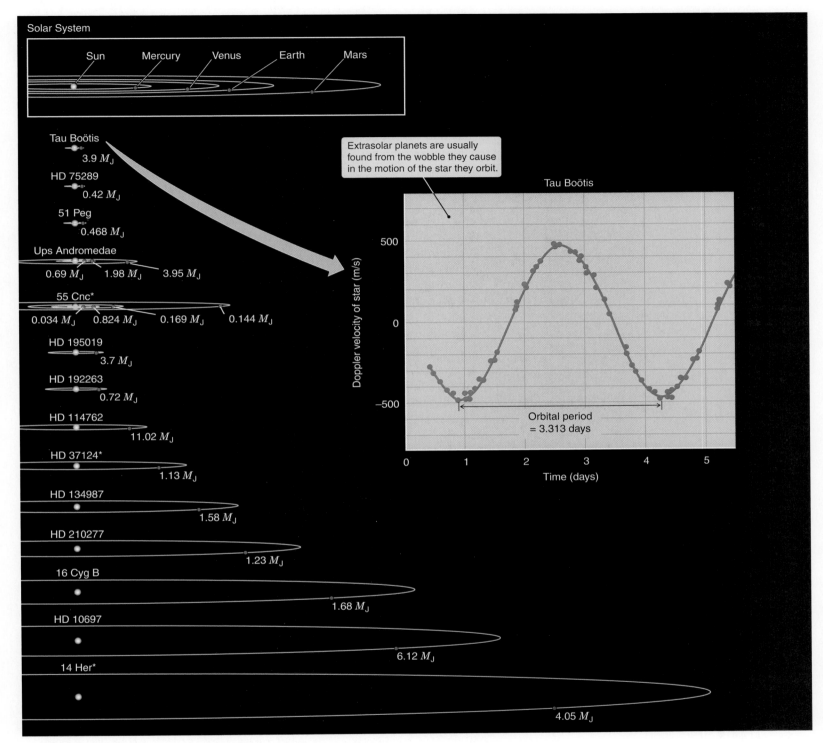

Solar System

Sun Mercury Venus Earth Mars

Tau Boötis
3.9 M_J

HD 75289
0.42 M_J

51 Peg
0.468 M_J

Ups Andromedae
0.69 M_J 1.98 M_J 3.95 M_J

55 Cnc*
0.034 M_J 0.824 M_J 0.169 M_J 0.144 M_J

HD 195019
3.7 M_J

HD 192263
0.72 M_J

HD 114762
11.02 M_J

HD 37124*
1.13 M_J

HD 134987
1.58 M_J

HD 210277
1.23 M_J

16 Cyg B
1.68 M_J

HD 10697
6.12 M_J

14 Her*
4.05 M_J

Extrasolar planets are usually found from the wobble they cause in the motion of the star they orbit.

Tau Boötis

Orbital period = 3.313 days

Time (days)

Doppler velocity of star (m/s)

* Multiple planet system with other planets not shown due to size constraints.
  55 Cnc: 3.835 M_J at 5.8 AU
  HD 37124: 0.61 M_J at 0.53 AU, 0.68 M_J at 3.2 AU, and 0.6 M_J at 1.6 AU
  14 Her: 2.1 M_J at 6.9 AU

FIGURE 5.26 Planetary systems have been discovered around hundreds of stars other than the Sun, confirming what astronomers have long suspected—that planets are a natural and common by-product of star formation. A few of these planetary systems are represented here, along with an example of the radial velocity method. (Masses of planets are given in units of Jupiter masses, M_J.) The inner Solar System is shown to scale.

As an example, let's consider the hot Jupiters. A massive planet orbiting very close to its parent star will tug the star very hard, creating large radial velocity variations in the star. This means that hot Jupiters are relatively easy targets when we are using the spectroscopic radial velocity method. In addition, large planets orbiting close to their parent stars are more likely to move in front of the star periodically and reveal themselves with the transit method. Therefore, these hot Jupiter systems may *not* be representative of *most* planetary systems. They may just be easier to find. Scientists refer to this bias as a *selection effect*.

Many astronomers were surprised by the hot Jupiters because, according to the formation theory they had at the time (based only on our Solar System), these giant, volatile-rich planets should not have been able to form so close to their parent stars. Jupiter-type planets should form in the more distant, cooler regions of the protoplanetary disk, where the volatiles that make up much of their composition are able to survive. Perhaps hot Jupiters formed much farther away from their parent star and subsequently migrated inward. The mechanism by which a planet could migrate so far is not well understood, but it must involve an interaction with gas or planetesimals in which orbital angular momentum is somehow transferred from the planet to its surroundings, allowing it to spiral inward.

Our studies of planetary systems, many unlike our own, challenge some aspects of our understanding of planet formation. Yet the message conveyed by our discoveries is clear: the formation of planets frequently, and perhaps always, accompanies the formation of stars. The implications of this conclusion are profound. In a galaxy of 100 billion stars, and a universe of hundreds of billions of galaxies, how many planets (or even moons) with Earth-like conditions might exist? And with all of these Earth-like worlds in the universe, how many might play host to the particular category of chemical reactions that we refer to as "life"? In Chapter 18, we explore the idea of what kind of planet life might choose for its home.

The insights we have gained in this chapter complete the revolution that Copernicus started long ago when he had the audacity to challenge authority and suggest that Earth was not the center of all things. Now we know that planetary systems are common, and that the rules of their formation are in effect both on Earth and in distant molecular clouds. All we have to do to see our own history is to look at the stars that continue to form around us today. The chapter opening figure shows regions where such processes are occurring. Now you know how to look at that figure and see where stars are being formed and what will happen next. Keep an eye out for more figures like this in the news, so you can sharpen your skills at finding the newly forming stars.

Astronomy is not without its controversies! In a field that is heavily dominated by public funding, there is an ethic of "science for everyone": making data from publicly owned telescopes openly available within a year, so that everyone can benefit from new discoveries. However, tension exists between this ethic and the scientists' need to thoroughly evaluate the data as in this 2010 news article.

Orbiting Telescope Spots Possible Planets

By **Thomas H. Maugh II**, *Los Angeles Times*

In only its first 43 days of operation, NASA's Kepler orbiting telescope has identified 706 potential planets circling other stars, the agency said Tuesday.

Although some are likely to be false positives that will be weeded out by further observation, the announcement could still double the known number of so-called extrasolar planets, a figure that now stands at 461.

Significantly, many of the planets are the size of Earth or slightly larger, suggesting that they might be rocky, water-bearing locations where life could have evolved.

"This is the single largest announcement of planets that's ever happened," said Caltech astrophysicist John Johnson, who was not part of the team that made the announcement. "It's definitely momentous."

The announcement was accompanied by some controversy, however. The Kepler team, which is headquartered at NASA's Ames Research Center in Northern California, made public the data on only 306 of the potential planets. That information was presented in a paper posted on the astrophysics pre-publication website arXiv.org.

The data about the best 400 candidates were kept confidential by the team members, giving them a chance to confirm that each sighting actually is a planet, a process that requires ground-based observations. Other teams cannot verify the potential planets without knowing their location.

Normally, NASA requires that data collected from its orbiting observatories or other missions be released quickly. But the Kepler team argued that delays in the launch and other problems have limited its ability to follow up on the sightings. In particular, the

region of space occupied by the potential new planets is visible only from ground-based observatories during the summer, so team members have not yet had a chance to verify the data.

NASA had earlier extended the period during which the data could be kept confidential while the team worked out data processing problems. All of the withheld data will be made public in six months, NASA said, with additional data released quarterly.

"I'm somewhat ambivalent about that," said astrophysicist B. Scott Gaudi of Ohio State University, who was not part of the team. "The sooner the data gets out there, the more eyes that will see it. On the other hand, the discovery of an Earth-sized planet is a really big deal, not the kind of thing you want to have reported incorrectly."

Added Johnson: "Nobody likes to wait, particularly when you could get your hands on history-changing data."

Kepler, launched from Cape Canaveral, Fla., in March 2009, carries a 55-inch mirror, whose field of view is about 33,000 times wider than that of the Hubble Space Telescope. From its orbit above Earth, it focuses constantly on an area of the sky near the constellations Cygnus and Lyra, near the plane of the Milky Way. That area encompasses about 105 square degrees—a small fraction of the sky.

The team is studying about 150,000 sun-like stars in this region out of a total of 4.5 million; those stars are from tens of light-years to 3,000 light-years away. Kepler photographs them every 30 minutes, looking for the minute variations in light output that might indicate an orbiting planet has passed between the individual stars and Earth. The telescope will monitor the same region for at least three years.

In doing so, it will provide multiple observations of planets with short orbital periods, or years, and it will detect those with longer years that don't often pass between their stars and Earth.

The telescope will detect only those planets whose orbits put them between the targeted stars and Earth. Because these systems account for but a small segment of the heavens, the ultimate data could suggest that the actual number of planets in the cosmos is immense.

Most of the smaller potential planets observed so far have short orbital periods and are relatively close to their stars.

"That's a good sign if you are an observer," Johnson said, "but perhaps a bad sign if you are a theorist." The consensus among theorists has been that most planets should be farther out from their stars. "There is supposed to be a planet desert in this region."

Evaluating the News

1. Summarize this article. What are the Kepler scientists doing and why?
2. What planet-finding method does Kepler use?
3. List the arguments in favor of an extension to the Kepler team's ownership of the Kepler data.
4. List the arguments against such an extension.
5. In general, why might scientists hesitate to publish or release data?
6. What kinds of extraordinary claims might come from this data? How would these claims affect you as a member of the public?
7. Look online for more recent news of the Kepler mission. How did this issue turn out? In February 2011, how many planets were announced?

SUMMARY

5.1 Stars and their planetary systems form from collapsing interstellar clouds of gas and dust.

5.2 A protostar becomes a star when the central temperature and pressure are high enough for nuclear reactions to begin.

5.3 Planets are a common by-product of star formation, and many stars are surrounded by planetary systems. The Solar System grew from a protoplanetary disk of gas and dust that surrounded the forming Sun.

5.4 Solid terrestrial planets (Mercury, Venus, Earth, and Mars)

formed in the inner disk, where temperatures were high. Giant gaseous planets (Jupiter, Saturn, Uranus, and Neptune) formed in the outer disk, where temperatures were low.

5.5 Our Solar System formed about 5 billion years ago, nearly 9 billion years after the birth of the universe. Dwarf planets formed in the asteroid belt and in the region beyond the orbit of Neptune. Asteroids and comet nuclei remain today as leftover debris.

5.6 Hundreds of extrasolar planets have been found orbiting other stars within our galaxy.

✧ SUMMARY SELF-TEST

1. Place the following in order of their occurrence, as a star forms:
   a. Nuclear reactions begin and a star is born.
   b. A cloud contracts under gravity.
   c. A disk forms because angular momentum is conserved.
   d. Clumps form from static electricity.
   e. Planetesimals form from collisions.
   f. A wind blows from the central star.

2. The terrestrial planets are different from the giant planets because when they formed,
   a. the inner solar system was richer in heavy elements.
   b. the inner solar system was hotter than the outer solar system.
   c. the outer solar system took up a bigger volume, so there was more material to form planets.
   d. the inner solar system was moving faster, so centrifugal force was more important.

3. Planetary systems are probably
   a. exceedingly common—every star has planets.
   b. common—many stars in our galaxy have planets.
   c. rare—few stars have planets.
   d. exceedingly rare—only one star has planets.

4. Nuclear reactions require very high _____ and _____.
   a. temperature; pressure
   b. volume; pressure
   c. density; volume
   d. mass; area
   e. temperature; mass

5. The spectroscopic radial velocity method preferentially detects
   a. large planets close to the central star.
   b. small planets close to the central star.
   c. large planets far from the central star.
   d. small planets far from the central star.
   e. none of the above. The method detects all of these equally well.

QUESTIONS AND PROBLEMS

True/False and Multiple-Choice Questions

6. **T/F:** A molecular cloud is held together by gravity.

7. **T/F:** Gravity and angular momentum are both important in the formation of planetary systems.

8. **T/F:** A protostar has nuclear reactions inside.

9. **T/F:** Volatile materials are solid only at low temperatures.

10. **T/F:** The Solar System formed from a giant cloud of dust and gas that collapsed under gravity.

11. **T/F:** Gravitational lensing is similar to the transit method in that both require the planet to pass in front of a bright object.

12. **T/F:** Objects that move toward us appear redder than if they are at rest.

13. Molecular clouds collapse because of
    a. gravity.
    b. angular momentum.

    c. static electricity.
    d. nuclear reactions.

14. Since angular momentum is conserved, an ice-skater who throws her arms out will
    a. rotate more slowly.
    b. rotate more quickly.
    c. rotate at the same rate.
    d. stop rotating entirely.

15. Clumps grow into planetesimals by
    a. gravitationally pulling in other clumps.
    b. colliding with other clumps.
    c. attracting other clumps with opposite charge.
    d. conserving angular momentum.

16. The terrestrial planets and the giant planets have different compositions because
    a. the giant planets are much larger.
    b. the terrestrial planets are closer to the Sun.

   c. the giant planets are mostly made of solids.

   d. the terrestrial planets have few moons.

17. Of the following, which planet still has its primary atmosphere?

   a. Mercury

   b. Earth

   c. Mars

   d. Jupiter

18. Extrasolar planets have been detected by the

   a. spectroscopic radial velocity method.

   b. transit method.

   c. gravitational lensing method.

   d. direct imaging method.

   e. all of the above

Conceptual Questions

19. Compare the size of our Solar System with the size of the universe.

20. What is the source of the material that now makes up the Sun and the rest of the Solar System?

21. Build a house of cards. Watch carefully as you knock it down. Which layer of cards reaches the table first? How does this compare to the collapsing dust cloud that you learned about in the chapter?

22. Figure 5.5 shows a molecular cloud in which stars are currently forming. Of the two visible pillars in this image, one is colored and one is black. What does this tell you about the relative positions of these pillars and the source that is illuminating them? Can you find anything in the image that might be responsible for illuminating the pillars?

23. In a poetic sense, planets are sometimes referred to as children of the Sun. A more accurate description, though, would be siblings of the Sun. Explain why.

24. Describe the different ways that stellar astronomers and planetary scientists each came to the same conclusion about how planetary systems form.

25. Describe the formation of the Solar System in a few sentences of your own.

26. Examine Figure 5.7. Suppose that the universe were different, and in step 2, the shrinking star caused the gravity to decrease, instead of increase, so that the pressure arrow became longer than the gravity arrow. What would this mean for the formation of stars?

27. What is represented by the little gray box in steps 1, 2, and 5 of Figure 5.7?

28. What is a protoplanetary disk?

29. Physicists describe certain properties, such as angular momentum and energy, as being "conserved." What does this mean? Do these conservation laws imply that an individual object can never lose or gain angular momentum or energy? Explain your reasoning.

30. How does the law of conservation of angular momentum control a figure-skater's rate of spin?

31. Explain why the law of conservation of angular momentum

created problems for early versions of the theory of solar system formation.

32. What is an accretion disk?

33. Describe the process by which tiny grains of dust grow to become massive planets.

34. Look under your bed for "dust bunnies." If there aren't any, look under your roommate's bed, the refrigerator, or any similar place that might have some. Once you find them, blow one toward another. Watch carefully and describe what happens as they meet. What happens if you repeat this with another dust bunny? Will these dust bunnies ever have enough gravity to begin pulling themselves together? If they were in space, instead of on the floor, might that happen? What force prevents their mutual gravity from drawing them together into a "bunny-tesimal" under your bed?

35. Study the image of Mercury in Figure 5.18. How does this image support the idea that many, many planetesimals were once zooming around the early Solar System? The largest crater in the upper right has several smaller craters inside it. Which happened first, the larger crater or the smaller ones? How do you know? What does this tell you about their relative ages? This reasoning becomes extremely important in the next chapter.

36. There are two reasons that the inner part of a protoplanetary disk is hotter than the outer disk. What are they?

37. During the process in which a clump of a molecular cloud collapses to form an accretion disk, energy exists in three forms. Name and describe them.

38. Why do we find rocky material everywhere in the Solar System, but large amounts of volatile material only in the outer regions? Would you expect the same to be true of other planetary systems? Explain your answer.

39. When we think of ice, we tend to think of water ice; but among scientists, this definition is very restrictive. Explain why.

40. Why were the four giant planets able to collect massive gaseous atmospheres, whereas the terrestrial planets could not?

41. What is the meaning of the term *organic*?

42. Explain the source of the atmospheres now surrounding three of the terrestrial planets.

43. What happened to all the leftover Solar System debris after the last of the planets formed?

44. Name the terrestrial and the giant planets in our Solar System.

45. Describe four methods that astronomers use to search for extrasolar planets.

46. Examine Figure 5.22. Redraw this figure looking straight down on the system. Now draw a series of pictures from that same orientation, showing one complete orbit of Jupiter around the Sun. Label the motions of the Sun and Jupiter (toward, away, neither) as they would be viewed by an observer off the page to the right. Are the Sun and Jupiter ever on the same side of the center of mass?

47. Examine Figure 5.23. Redraw this figure, paying close attention to where the line on the graph drops in brightness. Now, add three more graphs. In the first, show what happens to the light curve if the planet crosses much closer to the bottom of the star. In the second, show what happens to the light curve if the planet is much larger than the one in Figure 5.23. In the third, show what happens to the light curve if the planet crosses the precise middle of the star, but from top to bottom instead of side to side.

48. Why is the transit method so restrictive as a method for finding extrasolar planets?

49. Why has it been so difficult for astronomers to take a picture of an extrasolar planet?

50. Many of the planets that astronomers have found orbiting other stars have been giant planets with masses more like that of Jupiter than of Earth, and with orbits located very close to their parent stars. Does this prove that our Solar System is unusual? Explain your answer.

51. Step outside and look at the nighttime sky. Depending on the darkness of the sky, you may see dozens or hundreds of stars. Would you expect many or very few of those stars to be orbited by planets? Explain your answer.

Problems

52. Use information about the planets given in the back cover of the book to answer these questions:
   a. What is the total mass of all the planets in our Solar System, expressed in Earth masses ($M_\oplus$)?
   b. What fraction of this total planetary mass does Jupiter represent?
   c. What fraction does Earth represent?

53. Compare Jupiter's orbital angular momentum with its spin angular momentum using the following values: $m = 1.9 \times 10^{27}$ kg, $v = 13.1$ km/s, $r = 5.2$ AU, $R = 7.15 \times 10^4$ km, and $P = 9$ hours 56 minutes. Assume Jupiter to be a uniform body. What fraction does each component (orbital and spin) contribute to Jupiter's total angular momentum? Refer to Working It Out 5.1 for help.

54. Assume that the Sun is a uniform sphere with a radius of 700,000 km and a rotation period of 27 days. Near the end of its life, the remnant of the Sun will be a type of object called a white dwarf with a radius of only 5,000 km. Assuming that the mass remains unchanged, what will the Sun's new rotation period be as a white dwarf?

55. The mass of the Sun is 2×10^{30} kg. Using information provided in the previous two questions, compare Jupiter's orbital angular momentum with the Sun's spin angular momentum. What does this comparison tell you about the distribution of angular momentum in the Solar System?

56. Venus has a radius 0.949 times that of Earth and a mass 0.815 times that of Earth. Its rotation period is 243 days. What is the ratio of Venus's spin angular momentum to that of Earth? Assume that Venus and Earth are uniform spheres.

57. Jupiter has a mass equal to 318 Earth masses, an orbital radius of 5.2 AU, and an orbital velocity of 13.1 km/s. Earth's orbital velocity is 29.8 km/s. What is the ratio of Jupiter's orbital angular momentum to that of Earth?

58. In the text we give an example of a molecular cloud clump having a diameter of 10^{16} meters and a rotation period of 10^6 years collapsing to a sphere the size of the Sun (1.4×10^9 meters in

diameter). We point out that if all the cloud's angular momentum went into that sphere, the sphere would have a rotation period of only 0.6 second. Do the calculation to confirm this result.

59. The asteroid Vesta has a diameter of 530 km and a mass of 2.7×10^{20} kg.
   a. Calculate the density of Vesta.
   b. The density of water is 1,000 kilograms per cubic meter (kg/m³), and that of rock is about 2,500 kg/m³. What does this difference tell you about the composition of this primitive body?

60. You observe a spectral line of hydrogen at a wavelength of 502.3 nm in a distant star. The rest wavelength of this line is 486.1 nm. What is the radial velocity of this star? Is it moving toward or away from Earth?

61. The best current technology can measure radial velocities larger than about 1 m/s. Suppose that you are observing an iodine line with a wavelength of 575 nm. How large a shift in wavelength would a radial velocity of 1 m/s produce?

62. Earth tugs the Sun around as it orbits, but it causes a much smaller effect than any known extrasolar planet, only 0.09 m/s. How large a shift in wavelength does this cause in the Sun's spectrum at 575 nm?

63. If an alien astronomer observed a plot of the light curve as Jupiter passed in front of the Sun, by how much would the Sun's brightness drop during the transit?

64. A hot Jupiter nicknamed "Osiris" has been found around a solar-mass star, HD 209458. It orbits the star in only 3.525 days.
   a. What is the orbital radius of this extrasolar planet?
   b. Compare its orbit with that of Mercury around our own Sun. What environmental conditions must Osiris experience?

65. The French CoRoT satellite has detected a planet with a diameter of 1.7 Earth diameters.
   a. How much larger is the volume of this planet than Earth's?
   b. Assume that the density of the planet is the same as Earth's. How much more massive is this planet than Earth?

66. Consider the planet CoRoT-11b. It was discovered using the transit method, and astronomers have followed up with radial velocity measurements. Because of this, we know *both* its size and its mass. From this, we can find the density, which gives us a clue about whether the object is gaseous or rocky. This planet has a radius of 1.43 R_J, and a mass of 2.33 M_J.
   a. What is the mass of this planet in kg?
   b. What is the radius in meters?
   c. What is the volume of the planet?
   d. What is its density? How does this density compare to the density of water (1,000 kg/m³)? Is the planet likely to be rocky or gaseous?

67. The extrasolar planet Osiris passes directly in front of its solar-type parent star, HD 209458 (diameter = 1.7×10^6 km), every 3.525 days, decreasing the brightness of the star by about 1.7 percent (0.017).
   a. What is the diameter of Osiris?
   b. Compare the diameter of this extrasolar planet with that of Jupiter (mean diameter = 139,800 km).

SmartWork, Norton's online homework system, includes algorithmically generated versions of these questions, plus additional conceptual exercises. If your instructor assigns questions in SmartWork, log in at **smartwork.wwnorton.com**.

 StudySpace is a free and open website that provides a Study Plan for each chapter of **Understanding Our Universe**. Study Plans include animations, reading outlines, vocabulary flashcards, and multiple-choice quizzes, plus links to premium content in SmartWork and the ebook. Visit **wwnorton.com/ studyspace**.

Exploration | Doppler Shift

wwnorton.com/studyspace

Visit the StudySpace Interactive Simulation for Chapter 5: Doppler Shift Demonstrator. This simple applet has only a few "moving parts." First, start the emission from the source (the sphere labeled "S" in the lower panel). Watch as the waves emanate from the source and travel to the observer (the sphere labeled "O"). Compare these to the waves as emitted from the source and the waves as detected by the observer in the top left panel.

Grab the source with your mouse, and move it toward the observer.

1. What happens to the waves as emitted from the source?

2. What happens to the waves as detected by the observer?

Grab the source with your mouse, and move it away from the observer.

3. What happens to the waves as emitted from the source?

4. What happens to the waves as detected by the observer?

Now, adjust the rate to be as fast as possible, and then grab the source and move it slowly in a large circle.

5. What happens to the waves as detected by the observer?

Now, relate this experience to the method of extrasolar planet detection known as the spectroscopic radial velocity method.

6. Which of the three situations above corresponds to the motion of the star when a planet is detected by this method?

7. What do we, as observers, see happening to the light waves over time as the star makes this motion?

6 Terrestrial Worlds in the Inner Solar System

The past five decades have been a wonderfully exciting time of planetary exploration and discovery. Unmanned probes visited all eight major planets, and astronauts walked on the surface of the Moon. Today, new missions continue to revolutionize our understanding of the Solar System, but the vast quantity of information about the planets returned by space probes can be hard to digest. One way to organize this information is to compare the four innermost planets in our Solar System: Mercury, Venus, Earth, and Mars, collectively known as the terrestrial planets. Although the Moon is Earth's lone natural satellite, and not a planet, we also discuss it here because of its similarities to the terrestrial planets.

These worlds' similarities and differences highlight our fundamental questions about them. For example, when we explain why the Moon is covered with craters, we must also explain why preserved craters on Earth are rare. An explanation for why Venus has a very dense atmosphere should also explain that Earth and Mars do not. This approach is called **comparative planetology**. Comparing worlds with one another teaches us what shapes a planet's fate.

✧ LEARNING GOALS

The figure at right shows a photo of the Moon, taken with a large telescope, and a hand-drawn sketch of the surrounding area. From these images, we can find the relative ages of different features, and if we compare them to the ages of Moon rocks, we can find the actual ages of the features. By the end of this chapter, you should be able to look at a similar image and identify which geologic features occurred early in the history of that world and which occurred late. You should also be able to:

- Explain the roles of the four processes that shape a terrestrial planet's surface

- Identify the age of a planet's surface from the concentration of craters

- Explain how radiometric dating tells us the ages of rocks

- Diagram Earth's interior using information gained from seismic waves generated by earthquakes

- List the forms that tectonism takes on different planets

- List the many ways in which erosion modifies and wears down surface features

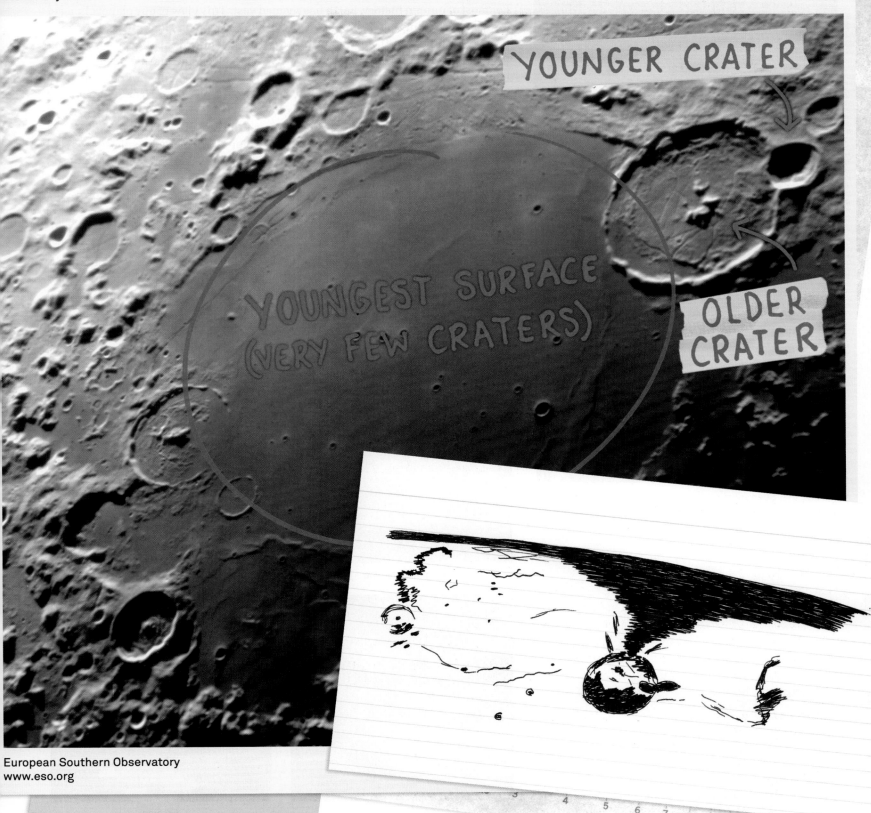

Lunar Surface: Mare Humorum
15 January 1999

YOUNGER CRATER

OLDER CRATER

YOUNGEST SURFACE
(VERY FEW CRATERS)

European Southern Observatory
www.eso.org

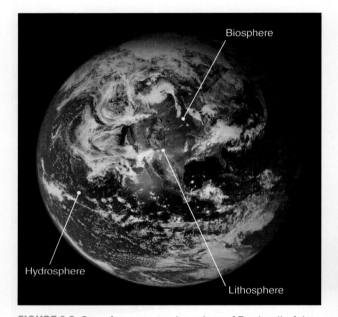

FIGURE 6.1 In December 1968, *Apollo 8* astronauts photographed our planet rising above the Moon.

FIGURE 6.2 Seen from space, the colors of Earth tell of the diversity of our planet's features.

6.1 Four Main Processes Shape Our Planet

For most of human history, Earth has seemed vast. This view of our planet changed forever with a single snapshot taken by *Apollo 8* astronauts looking back at Earth from space (**Figure 6.1**). For many, this was a philosophical turning point, compelling us to view the world for what it truly is: a "small blue marble." The psychological impact of this realization is rivaled only by the change in our scientific perspective: we now see Earth in the context of the larger universe. In a closer view from space (**Figure 6.2**), Earth is awash with color. White clouds drift in our atmosphere, and white snow and ice cover the planet's frozen poles. The blue of oceans, seas, lakes, and rivers of liquid water—Earth's **hydrosphere**—covers most of the planet. Brown shows us the outer rocky shell of Earth, referred to as Earth's **lithosphere**, the outer part of which is called the **crust**. And green is the telltale sign of vegetation, part of Earth's **biosphere**—the most extraordinary among Earth's many distinctions.

Four principal geological processes constantly reshape our planet (**Figure 6.3**). **Impact cratering** leaves distinctive scars in Earth's crust that tell of large collisions in our past. **Tectonism** folds and breaks Earth's crust, forming mountain ranges, valleys, and deep ocean trenches, and causing earthquakes. **Volcanism** can spill sheets of lava and ash over vast areas, forming mountains or plains in the process. **Erosion** slowly but persistently levels Earth's surface.

Impact cratering

Tectonism

Volcanism

Erosion

FIGURE 6.3 Examples of impact cratering, tectonism, volcanism, and erosion can all be found on Earth.

TABLE | 6.1

Comparison of Physical Properties of the Terrestrial Planets and the Moon

	MERCURY	VENUS	EARTH	MARS	MOON*
Orbital semi-major axis (AU)	0.387	0.723	1.000	1.524	384,000 km
Orbital period (years)†	0.241	0.615	1.000	1.881	27.32^d
Orbital velocity (km/s)	47.4	35.0	29.8	24.1	1.02
Mass ($M_\oplus = 1$)	0.055	0.815	1.000	0.107	0.012
Equatorial radius (km)	2,440	6,052	6,378	3,397	1,738
Equatorial radius ($K_\oplus = 1$)	0.383	0.949	1.000	0.533	0.272
Density (water = 1)	5.43	5.24	5.51	3.93	3.34
Rotation period†	58.65^d	243.02^d	$23^h\ 56^m$	$24^h\ 37^m$	27.32^d
Obliquity (degrees)‡	0.01	177.36	23.44	25.19	6.68
Surface gravity (m/s²)	3.70	8.87	9.81	3.71	1.62
Escape velocity (km/s)	4.25	10.36	11.19	5.03	2.38

*The Moon's orbital radius and orbital period are given in kilometers and days, respectively. The Moon's orbital radius and velocity are given relative to the Earth.
†The superscript letters *d*, *h*, and *m* stand for days, hours, and minutes of time, respectively.
‡An obliquity greater than 90° indicates that the planet rotates in a retrograde, or backward, direction.

Tectonism, volcanism, and impact cratering affect Earth's surface by forming mountains, valleys, and ocean basins, creating **topographic relief**. Erosion by running water and wind wears down hills, mountains, and continents; the resulting debris fills in valleys, lakes, and ocean basins. If erosion were the only geological process operating, it would eventually smooth out the surface of the planet completely. Because Earth is a geologically and biologically active world, however, its surface is an ever-changing battleground between processes that build up topography and those that tear it down. Each of the four types of geological process shaping the surface of Earth leaves its own distinctive signature. These signatures mark the surfaces of other planets as well.

Table 6.1 compares the basic physical properties of the terrestrial planets and the Moon.

> Tectonism, volcanism, and impact cratering roughen up planetary surfaces; erosion smooths them down.

▶‖ AstroTour: **Processes That Shape the Planets**

6.2 Impacts Help Shape the Evolution of the Planets

Although all terrestrial planets are subject to tectonism, volcanism, impact cratering, and erosion, the relative importance of each of these processes varies. Large impacts cause the most concentrated and sudden release of energy. Planets and other objects orbit the Sun at very high speeds; Earth, for example, orbits at an average speed of around 30 km/s, equivalent to 67,000 miles per hour (mph). Because the kinetic energy of an object is proportional to its speed squared, collisions between orbiting bodies release huge amounts of energy. When an object hits a planet, its kinetic energy heats and compresses the surface and throws material far from the resulting **impact crater** (**Figure 6.4**). Sometimes this material, called **ejecta**, falls back to the surface of the planet with enough energy to cause **secondary craters**. The rebound of heated and compressed material can

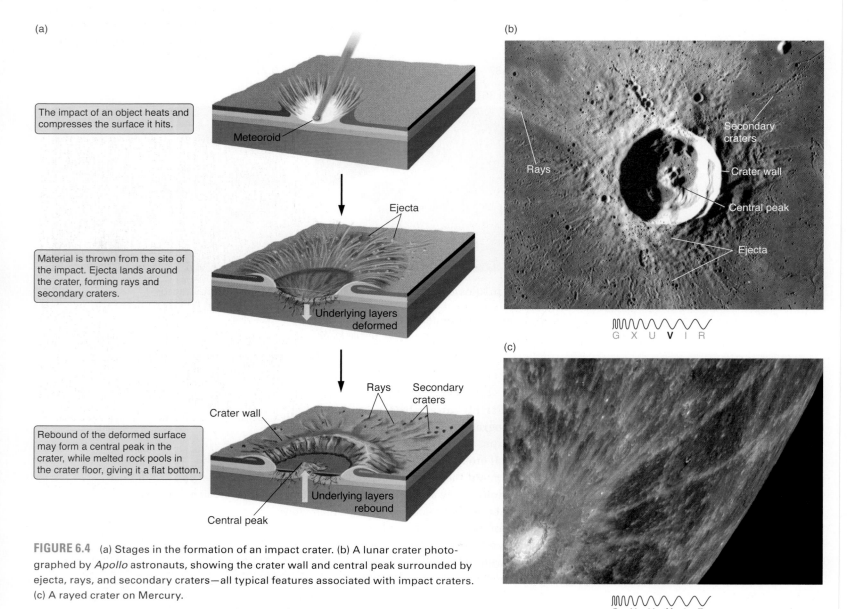

(a)

The impact of an object heats and compresses the surface it hits.

Meteoroid

Ejecta

Material is thrown from the site of the impact. Ejecta lands around the crater, forming rays and secondary craters.

Underlying layers deformed

Rebound of the deformed surface may form a central peak in the crater, while melted rock pools in the crater floor, giving it a flat bottom.

Crater wall   Rays   Secondary craters

Central peak   Underlying layers rebound

(b)

Secondary craters

Rays

Crater wall

Central peak

Ejecta

G X U **V** I R

(c)

G X U **V** I R

FIGURE 6.4 (a) Stages in the formation of an impact crater. (b) A lunar crater photographed by *Apollo* astronauts, showing the crater wall and central peak surrounded by ejecta, rays, and secondary craters—all typical features associated with impact craters. (c) A rayed crater on Mercury.

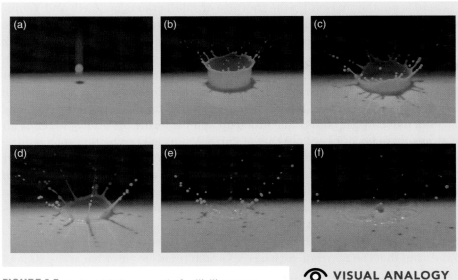

FIGURE 6.5   A drop hitting a pool of milk illustrates the formation of features in an impact crater, including crater walls (b and c), secondary craters (d and e), and a central peak (f).

👁 **VISUAL ANALOGY**

FIGURE 6.6   Meteor Crater (also known as Barringer Crater), located in northern Arizona, is an impact crater 1.2 km in diameter. It was formed some 50,000 years ago by a nickel-iron meteoroid's collision with Earth.

also lead to the formation of a central peak or a ring of mountains on the crater floor. These processes are similar, on a much larger scale, to what happens when a drop lands in a glass of milk, as shown in **Figure 6.5**.

The energy of an impact can be great enough to melt or even **vaporize** rock, and the floors of some craters are the cooled surfaces of pools of rock melted by the impact. The energy released in an impact can also lead to the formation of new minerals. Some minerals, such as shock-modified quartz, form *only* during an impact. These distinctive minerals are evidence of ancient impacts on Earth's surface. The space rocks that cause these impacts are referred to by three closely related terms: **meteoroid**, **meteor**, and **meteorite**. *Meteoroid* refers to the objects while they are in space; they become *meteors* when they pass through the atmosphere; and they end up as *meteorites* if they survive to hit the ground.

Meteor Crater in Arizona is one of the best-preserved impact craters on our planet (**Figure 6.6**). This crater was caused by a fragment of an asteroid impacting Earth about 50,000 years ago. From the crater's size and shape, and remaining pieces of the impacting body, we know that the nickel-iron asteroid fragment was about 50 meters across, had a mass of about 300 million kilograms (kg), and traveled at 13 kilometers per second (km/s). Approximately half of the original mass was vaporized in the atmosphere before hitting the ground. This collision released about 300 times as much total energy as the first atomic bomb. Yet, at only 1,200 meters in diameter, Meteor Crater is tiny compared with impact craters seen on the Moon or with ancient impact scars on Earth.

Impact craters cover the surfaces of Mercury, Mars, and the Moon; for example, the Moon has millions of craters of all different sizes, one on top of another (**Figure 6.7**). On Earth and Venus, by comparison, most impact craters have been destroyed. Fewer than 200 impact craters have been identified on Earth, and about 1,000 have been located on Venus. Earth's crater shortage is primarily due to plate tectonics in Earth's ocean basins and erosion on land, while lava flows on Venus have destroyed its craters.

G X U V I R

FIGURE 6.7   The Moon's terrain, shown here in a photograph taken during the *Apollo 17* mission, has been heavily cratered by impacts.

Vocabulary Alert

Vaporize: In common language, this often means "to destroy completely," with connotations that even the stuff the object was made of is not there anymore. Astronomers use this word specifically to mean "turn into vapor." Here, the rocks are being turned into gas by the energy of the collision.

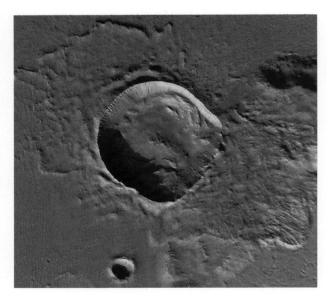

FIGURE 6.8 Some craters on Mars look like those formed by rocks thrown into mud, suggesting that material ejected from the crater contained large amounts of water. This crater is about 20 km across.

> The number of craters on a surface indicates the age of the surface.

There is another reason for the shortage of small craters on Earth and Venus. Whereas the surface of the Moon is directly exposed to cosmic bombardment, the surfaces of Earth and Venus are partially protected by their atmospheres. Rock samples from the Moon show craters smaller than a pinhead, formed by micrometeoroids. In contrast, most meteoroids smaller than 100 meters in diameter that enter Earth's atmosphere are either burned up or broken up by friction before they reach Earth's surface. (Small meteorites found on the ground are probably pieces of much larger bodies that broke up upon entering the atmosphere.) With an atmosphere far thicker than that of Earth, Venus is even better protected.

We can tell a lot about the surface of a planet by studying its craters. For example, craters on the Moon's surface are often surrounded by strings of smaller secondary craters formed from material thrown out by the impact, like those shown in Figure 6.4b. Some craters on Mars have a very different appearance. These craters are surrounded by structures that look much like the pattern you might see if you threw a rock into mud (**Figure 6.8**). The flows appear to indicate that the martian surface rocks contained water or ice at the time of the impact. Not all martian craters have this feature, so the water or ice must have been concentrated in only some areas, and these icy locations might have changed with time.

At the time these craters formed, there may have been liquid water on the surface of Mars. Features resembling canyons and dry riverbeds are further evidence of this hypothesis. But there is another possibility. Today the surface of Mars is dry or frozen, suggesting that water that might once have been on the surface has soaked into the ground, much like water frozen in the ground in Earth's polar regions. The energy released by the impact of a meteoroid could melt this ice, turning the surface material into slurry with a consistency much like wet concrete. When thrown from the crater by the force of the impact, this slurry would hit the surrounding terrain and slide across the surface, forming the mudlike craters we see today.

Calibrating a Cosmic Clock

Early in its history, every planet experienced a period of heavy bombardment because lots of planetesimals were roaming around the early Solar System. The number of *visible* craters on a planet is determined by the rate at which those craters are destroyed. Geologically active planets such as Earth, Mars, and Venus retain scars of only the more recent impacts. By contrast, the Moon's surface still preserves the scars of craters dating from the early years of the Solar System. With no atmosphere or surface water and a cold, geologically dead interior, the lunar surface has remained essentially unchanged for more than a billion years. Mercury's well-preserved craters suggest that it, like the Moon, has been geologically inactive for a long time. Planetary scientists use this cratering record to estimate the **ages** and geological histories of planetary surfaces—extensive cratering signifies an older planetary surface and minimal geological activity.

We can use the amount of cratering as a clock to measure the relative ages of surfaces. But to say that a surface with *this many* craters is *this* old, but a surface with *that many* craters is *that* old, we need to know how fast the clock runs. In other words, we need to "calibrate the cratering clock."

We use the Moon to perform this calibration. The surface of the Moon is not uniform. Parts of it are so heavily cratered that some craters overlap their neighbors. Other parts are much smoother, telling of more recent geological activity.

Between 1969 and 1976, *Apollo* astronauts and Soviet unmanned probes visited the Moon and brought back samples from nine different locations on the lunar surface. By measuring relative amounts of various radioactive elements and the elements into which they decay—a process called **radiometric dating**—scientists were able to find the ages of different lunar regions (see **Working It Out 6.1**). The results of that work were surprising. Although smooth areas on the Moon were indeed younger than heavily cratered areas, they were still very old. The oldest, most heavily cratered regions on the Moon date back to about 4.4 billion years ago, whereas most of the smoother parts of the lunar surface are typically 3.1 billion to 3.9 billion years old. This means that almost all of the cratering in the Solar System took place within its first billion years (**Figure 6.9**). Compare this figure to the chapter-opening figure, which shows this analysis using ground-based images of a crater in Mare Humorum.

Craters and radiometric dating together tell us the age of most of the surface of the Moon.

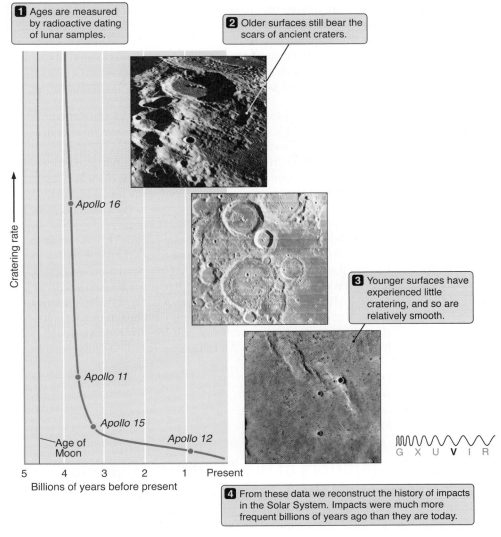

1 Ages are measured by radioactive dating of lunar samples.

2 Older surfaces still bear the scars of ancient craters.

3 Younger surfaces have experienced little cratering, and so are relatively smooth.

Apollo 16

Apollo 11

Apollo 15

Apollo 12

Age of Moon

G X U **V** I R

5 4 3 2 1 Present
Billions of years before present

Cratering rate

4 From these data we reconstruct the history of impacts in the Solar System. Impacts were much more frequent billions of years ago than they are today.

FIGURE 6.9 Radiometric dating of lunar samples returned from specific sites by *Apollo* astronauts is used to determine how the cratering rate has changed over time. Cratering records can then be used to tell us the age of other parts of the lunar surface.

Working It Out 6.1 | How to Read Cosmic Clocks

A geologist can find the age of a mineral by measuring the relative amounts of a radioactive element (also called a **radioisotope**) and the decay products it turns into. An isotope is an atom with the same number of protons but a different number of neutrons as other atoms of the same element. The radioactive element is known as the **parent**, and the products are called **daughters**. The time interval over which a radioisotope decays to half its original amount is called its **half-life**. With every half-life that passes, the remaining amount will decrease by a factor of 2. For example, over three half-lives, the final amount, P_F, of a parent radioisotope will be $\frac{1}{2} \times \frac{1}{2} \times \frac{1}{2} = \frac{1}{8}$ of its original amount, P_O. If we express the number of half-lives more generally as n, then we can translate this relationship into math:

$$\frac{P_F}{P_O} = \left(\frac{1}{2}\right)^n$$

This equation has an exponential expression on the right-hand side. This means that we are raising one-half to the power of n. Try to remember some of the rules of exponents. For example, if $n = 0$, the right-hand side of this equation will be 1. That makes sense. If no half-lives have gone by, then the ratio of the final amount to the initial amount should be 1. If $n = 1$, then the right-hand side of the equation is one-half, and the ratio of the final amount to the initial amount is one-half. One half-life has gone by, and we have one-half of the material left.

Calculator alert: This kind of exponent is not the same as the "times 10 to the" operation that we learned about in Chapter 1. Since the base of the exponent in this case is not 10, you cannot use the EE or EXP key on your calculator. Instead, you need to use the x^y (sometimes y^x) key. First, type the base (0.5) into your calculator, then hit x^y, and then type in your value for n. For example, if you are calculating the fraction of material that is left after three half-lives, you would type [0][.][5][x^y][3][=] to find the answer (try it: you should get $\frac{1}{8}$, or 0.125).

The most abundant form of the element uranium is ^{238}U (pronounced "uranium two-thirty-eight"), which decays through a series of reactions to ^{206}Pb (pronounced "lead two-oh-six"). It takes 4.5 billion years for half of a sample of ^{238}U to decay through these processes to ^{206}Pb. This means that a sample of pure ^{238}U would contain equal amounts of uranium and lead after 4.5 billion years had passed. If we were to find such a half-and-half sample, we would know that half the uranium atoms had turned to lead, so that

$$\frac{P_F}{P_O} = \frac{1}{2}$$

Compare this to the previous equation, and convince yourself that if $(\frac{1}{2})^n = (\frac{1}{2})$, then $n = 1$. So we find that the mineral formed one half-life, or 4.5 billion years, ago.

When analyzing the chemical composition of rocks to find their age, you have the ratio on the left, not the number of half-lives. The number of half-lives is what you are looking for! Solving for this in the general case requires a nontrivial mathematical manipulation involving logarithms, so it may be easiest to try reasonable numbers in your calculator until you find approximately the right one.

Let's look at an example of finding the age from the ratio of material that has decayed, this time with a different form of uranium (^{235}U) that decays to a different form of lead (^{207}Pb) with a half-life of 700 million years. Suppose that a lunar mineral brought back by astronauts has 15 times as much ^{207}Pb as ^{235}U. Assuming that the sample was pure uranium when it formed, this means that $^{15}/_{16}$ of the ^{235}U has decayed to ^{207}Pb, leaving only $^{1}/_{16}$ of the parent remaining in the mineral sample. Now we know that the left side of the equation is $^{1}/_{16}$, or 0.0625. This is less than $\frac{1}{2}$, so we know that n is bigger than one (more than one half-life has passed). Let's try 2. Putting that into the calculator gives 0.25—that's too big. Let's try 5. That gives 0.03125—too small. Try 4. Aha! After $n = 4$ half-lives, the ratio of the remaining material to the original material is 0.0625. To find out how much time has passed, we multiply the number of half-lives by the length of a half-life: 4×700 million = 2.8 billion years old.

6.3 The Interiors of the Terrestrial Planets Tell Their Own Tales

To understand the processes responsible for all but wiping away the visible record of impacts in Earth's past and for continually remaking the surface of our planet, we begin by looking below its surface. What lies hundreds and thousands of kilometers below our feet, and how do we know?

We Can Probe the Interior of Earth

Humans have literally only scratched the surface of our planet. The deepest holes ever drilled have gone only about 12 km below the surface, and Earth's center is 6,350 km deep. Even so, scientists know a lot about the interior of Earth. The strength of Earth's gravity tells us its mass.

Dividing the mass by the volume gives an average density of 5,500 kilograms per cubic meter (kg/m³). But surface material averages only 2,900 kg/m³. Since the density of the whole planet is greater than the density of the surface, the interior must be much denser than the surface and should contain large amounts of iron, whose density is nearly 8,000 kg/m³. We can also consider meteorites. Because meteorites and Earth formed at the same time out of similar materials, we can reason that the overall composition of Earth should resemble the composition of meteorite material, which includes minerals with large amounts of iron.

Vibrations from earthquakes provide more information about our planet's interior. When an earthquake occurs, vibrations spread out through and across the planet as **seismic waves**. There are different kinds of seismic waves. **Surface waves** travel across the surface of a planet, much like waves on the ocean. Surface waves from earthquakes can sometimes be seen rolling across the countryside like ripples on water. These waves cause Earth's surface to move during an earthquake, buckling roadways, for example.

Other waves probe the interior of our planet. **Primary waves** (P waves) are *longitudinal* waves, which compress and then rarefy material. Imagine a spring, stretched out along the floor. A quick push along its length (**Figure 6.10a**) will

Seismic waves provide information about Earth's interior.

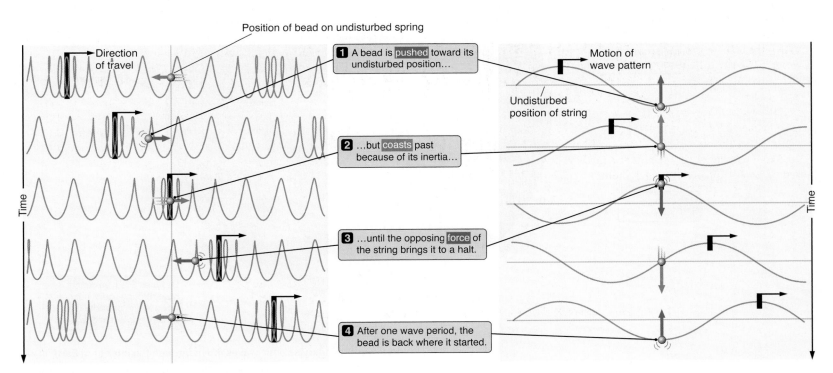

FIGURE 6.10 Mechanical waves result from forces that act to even out disturbances. (a) A longitudinal wave involves oscillations along the direction of travel of the wave. (b) A transverse wave involves oscillations that are perpendicular to the direction in which the wave travels. Primary seismic waves are longitudinal, and secondary seismic waves are transverse.

make a longitudinal wave. Sound waves are longitudinal waves. **Secondary waves** (S waves) are *transverse* waves that move material sideways (**Figure 6.10b**).

The progress of seismic waves through Earth's interior depends on the characteristics of the material they are moving through (**Figure 6.11**). P waves can travel through either solids or liquids, but S waves travel only through solids. Additionally, all seismic waves travel at different speeds depending on the density, temperature, and composition of the rock. When the density of rock varies abruptly, waves can be refracted (bent) or even reflected at the boundary, just as light is refracted or reflected by a pane of glass. This behavior creates "shadows" of the

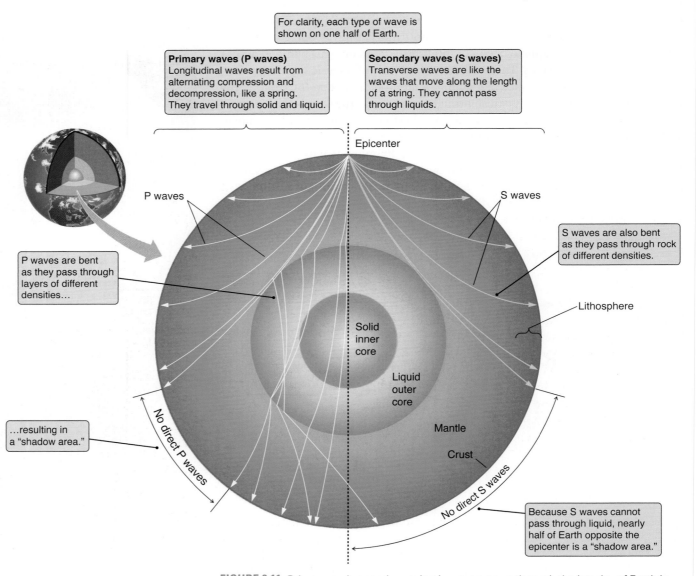

For clarity, each type of wave is shown on one half of Earth.

Primary waves (P waves)
Longitudinal waves result from alternating compression and decompression, like a spring. They travel through solid and liquid.

Secondary waves (S waves)
Transverse waves are like the waves that move along the length of a string. They cannot pass through liquids.

Epicenter

P waves

S waves

P waves are bent as they pass through layers of different densities…

S waves are also bent as they pass through rock of different densities.

Lithosphere

Solid inner core

Liquid outer core

…resulting in a "shadow area."

No direct P waves

Mantle

Crust

No direct S waves

Because S waves cannot pass through liquid, nearly half of Earth opposite the epicenter is a "shadow area."

FIGURE 6.11 Primary and secondary seismic waves move through the interior of Earth in distinctive ways. Measurements of when and where different types of seismic waves arrive after an earthquake enable us to test predictions of detailed models of Earth's interior. Note the "shadow areas" caused by the refraction of primary waves (yellow) at the outer boundary of the liquid outer core and the inability of secondary waves (blue) to pass through the liquid outer core.

liquid core on the opposite side of Earth, as shown in Figure 6.11. Much of our understanding of Earth's liquid outer core is due to studies of these shadow areas.

Seismometers measure seismic waves. For more than a hundred years, thousands of seismometers scattered around the globe have measured the vibrations from countless earthquakes and other seismic events, such as volcanic eruptions and nuclear explosions. When we combine all the measurements together, we build a comprehensive picture of the interior of our planet.

Building a Model of Earth's Interior

To go beyond this basic sketch, geologists turn to the laws of physics, combined with knowledge of the properties of materials and how they behave at different temperatures and pressures. For example, scientists take into account the fact that the pressure at any point in Earth's interior must be just high enough that the outward forces balance the weight of all the material above that point. With such considerations in mind, scientists then construct a model of Earth's interior that follows the rough outline of their sketch but that is also physically consistent. They next introduce earthquakes in their model, calculate how seismic waves would propagate through a model Earth with that structure, and predict what those seismic waves would look like at seismometer stations around the globe. They then test their model by comparing these predictions with actual observations of seismic waves from real earthquakes. The extent to which the predictions agree with the observations points out both strengths and weaknesses of the model. The structure of the model is adjusted (always remaining consistent with the known physical properties of materials) until a good match is found between prediction and observation.

This method is how geologists arrived at our current picture of the interior of Earth, shown in **Figure 6.12**. The major subdivisions of Earth's interior include a two-component **core**, a thick **mantle**, and a crust. Earth's core consists primarily of iron, nickel, and other dense metals. The mantle is made of medium-density materials. The crust is made of lower-density materials.

As Figure 6.12 illustrates, Earth's interior is not uniform. The materials have been separated by density, a process known as **differentiation**. When rocks of different types are mixed together, they tend to stay mixed. Once this rock melts, however, the denser materials sink to the bottom and the less dense materials float to the top. Today, little of Earth's interior is molten; but the differentiated structure shows that Earth was once much hotter, and its interior was liquid throughout. All of the terrestrial planets and the Moon were once molten. The cross sections in Figure 6.12 show the differentiated structure of each of the terrestrial planets and Earth's Moon.

The Moon Was Formed from Earth

As Figure 6.12 shows, the Moon has only a tiny core, composed of material similar to Earth's mantle. The best model explaining the Moon's composition involves a catastrophic impact. A Mars-sized protoplanet may have collided with Earth when it was very young, blasting off and vaporizing some of Earth's partly differentiated crust and mantle (**Figure 6.13**). This debris condensed to form the Moon. During the vaporization stage of the collision, most gases were lost to space. This model explains the similar composition of the Moon and Earth's mantle, and it accounts for the Moon's relative lack of water and other volatiles while Earth, Mars, and Venus are volatile rich.

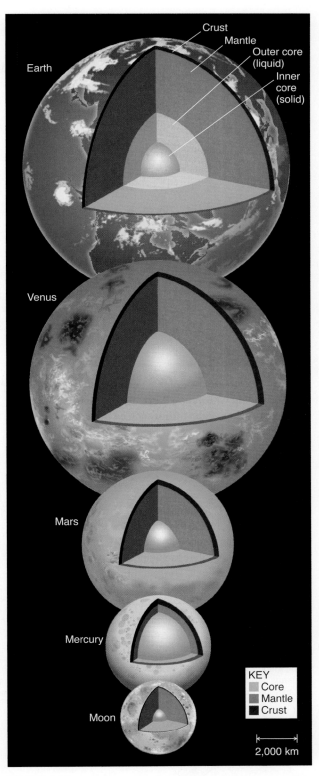

FIGURE 6.12 A comparison of the interiors of the terrestrial planets and Earth's Moon. Some fractions of the cores of Mercury, Venus, and Mars are probably liquid.

FIGURE 6.13 An artist's rendition of a Mars-sized proto-planet colliding with young Earth. The leading hypothesis proposes that debris from such an impact coalesced around Earth and formed our Moon.

FIGURE 6.14 (a) Tidal stresses pull Earth and its oceans into a tidal bulge. (b) Earth's rotation pulls its tidal bulge slightly out of alignment with the Moon. (c) As Earth's rotation carries us through these bulges, we experience the well-known ocean tides. The magnitude of the tides has been exaggerated in these diagrams for clarity. In these figures, the observer is looking down from above Earth's North Pole. Sizes and distances are not to scale.

6.4 The Evolution of Planetary Interiors Depends on Heating and Cooling

The deeper we go within a planet, the higher the temperature climbs. This thermal energy drives the planet's geological activity. The balance between energy received and energy produced and emitted governs the heating and cooling within a planet.

Part of the thermal energy in the interior of Earth is left over from when Earth formed. The tremendous energy of collisions, together with energy from short-lived radioactive elements, melted the planet. The surface of Earth radiated energy into space, cooling rapidly and forming a solid crust on top of a molten interior. Because a solid crust does not conduct thermal energy well, it kept in the remaining heat. Over the eons, energy from the interior of the planet continued to leak through the crust and radiate into space. The interior of the planet slowly cooled, and the mantle and the inner core solidified.

However, if this residual thermal energy were the only source of heating in Earth's interior, Earth's outer core would long ago have solidified completely. Additional sources of energy must be continuing to heat the interior of Earth if we are to account for today's high internal temperatures.

One source of continued heating is friction from tidal effects of the Moon and Sun. The gravitational pull of the Moon causes Earth to stretch along a line that points approximately in the direction of the Moon. The resulting tides are called **lunar tides**. The gravitational pull of the Sun causes Earth to stretch along a line that points approximately in the direction of the Sun. The resulting tides, which are about half as strong as lunar tides, are called **solar tides**. In both cases, the rotation of Earth drags the tides slightly ahead of the position of the Moon or Sun in the sky (**Figure 6.14**). When the Moon, Earth and Sun are all in a line, at new and full Moon, the lunar and solar tides overlap, creating more extreme tides ranging from extra-high high tides to extra-low low tides (**Figure 6.15**). This is called **spring tide**—not because of the season, but because the water appears to leap out of the sea. When the Moon, Earth and Sun make a right angle, at the Moon's first and third quarter, the lunar and solar tides stretch Earth in different directions, creating less extreme tides. This is called **neap tide** from the Saxon word *neafte*, which means "scarcity"; at these times of the month, shellfish and other food gathered in the tidal region is less accessible because the low tide is higher than at other times.

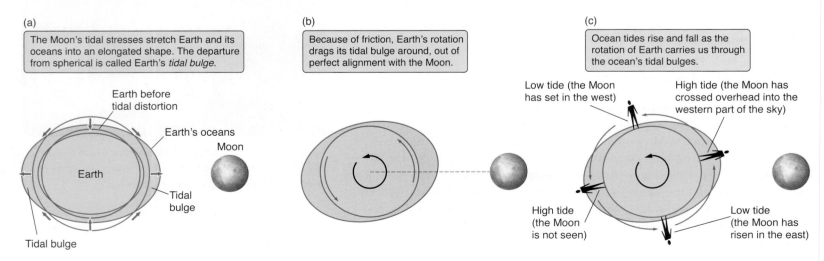

(a)

The Moon's tidal stresses stretch Earth and its oceans into an elongated shape. The departure from spherical is called Earth's *tidal bulge*.

Earth before tidal distortion

Earth's oceans

Moon

Earth

Tidal bulge

Tidal bulge

(b)

Because of friction, Earth's rotation drags its tidal bulge around, out of perfect alignment with the Moon.

(c)

Ocean tides rise and fall as the rotation of Earth carries us through the ocean's tidal bulges.

Low tide (the Moon has set in the west)

High tide (the Moon has crossed overhead into the western part of the sky)

High tide (the Moon is not seen)

Low tide (the Moon has risen in the east)

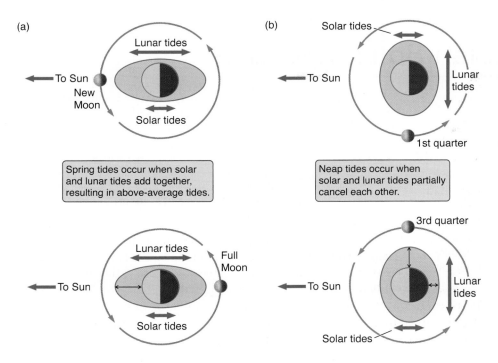

(a)

Lunar tides

To Sun
New
Moon

Solar tides

Spring tides occur when solar
and lunar tides add together,
resulting in above-average tides.

Lunar tides

To Sun

Full
Moon

Solar tides

(b)

Solar tides

To Sun

Lunar
tides

1st quarter

Neap tides occur when
solar and lunar tides partially
cancel each other.

3rd quarter

To Sun

Lunar
tides

Solar tides

FIGURE 6.15 Earth experiences solar tides that are about half as strong as tides due to the Moon. The interactions of solar and lunar tides result in either (a) spring tides when they are added together or (b) neap tides when they partially cancel one another.

Tidal effects are caused by the change in the strength of gravity across a solid object (**Figure 6.16**). For example, the Moon's pull on Earth is strongest on the part of Earth closest to the Moon and weakest on the part farthest from the Moon. This stretches Earth. While the oceans respond dramatically to the tidal stresses from the Moon and Sun, these tidal stresses also affect the *solid* body of Earth. Earth is somewhat elastic (like a rubber ball), and tidal stresses cause the ground to rise under you by about 30 cm twice each day. This takes a lot of energy, which ends up dissipated as heat. Similarly, a ball of silly putty will heat up if you stretch it and squish it repeatedly.

But tidal heating is not enough to explain the elevated temperature of Earth's interior. Instead, most of the "extra" energy in Earth's interior comes from long-lived radioactive elements trapped below the surface. As these radioactive elements decay, they release energy, heating the planet's interior. Equilibrium between heating of the interior and the loss of energy to space determines Earth's interior temperature. As radioactive elements decay, the amount of thermal energy generated declines, and Earth's interior becomes cooler as it ages.

Temperature plays an important role in a planet's interior structure, but it's not the only influence—whether a material is solid or liquid depends on pressure as well. Higher pressure forces atoms and molecules closer together and makes the material more likely to become a solid. Toward the center of Earth, the effects of temperature and pressure oppose each other: The higher temperatures make it more likely that material will melt, but the higher pressures favor a solid form. In the outer core of Earth the high temperature wins, allowing the material to exist in a molten state. At the center of Earth, even though the temperature is higher, the pressure is so great that the inner core of Earth is solid.

Planets lose their internal thermal energy through a combination of convection, conduction, and radiation. You are familiar with radiation in the form of heat, which is actually infrared light. Your skin is an infrared detector; you hold a hand over the stove to find out whether it's hot. All objects radiate light, and the hotter they are, the more light they radiate (see **Working It Out 6.2**). The type of light radiated (infrared, optical, ultraviolet, etc.) also depends on the tempera-

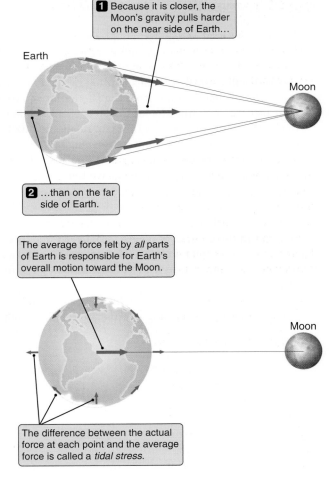

1 Because it is closer, the Moon's gravity pulls harder on the near side of Earth...

Earth

Moon

2 ...than on the far side of Earth.

The average force felt by *all* parts of Earth is responsible for Earth's overall motion toward the Moon.

Moon

The difference between the actual force at each point and the average force is called a *tidal stress*.

FIGURE 6.16 Tidal stresses stretch Earth along the line between Earth and the Moon but compress Earth perpendicular to this line.

Working It Out 6.2 | The Stefan-Boltzmann Law

Look at **Figure 6.17**, which shows the spectra of a light source at several different temperatures. This source is assumed to emit electromagnetic radiation only because of its temperature, not its composition. This kind of source is called a **blackbody**, and if we graph the intensity of its emitted radiation across all wavelengths (as in Figure 6.17), we obtain a characteristic curve called a **blackbody spectrum**. As the object's temperature increases, it emits more radiation at every wavelength, so each increase in temperature raises the curve. The **luminosity** of the object (the total amount of light emitted) increases. In fact, it increases quite fast as the temperature rises. Adding up all the energy in a spectrum shows that the increase in luminosity, L, is proportional to the *fourth power* of the temperature, T:

$$L \propto T^4$$

This result is known as the **Stefan-Boltzmann law** because it was discovered in the laboratory by Austrian physicist Josef Stefan (1835–1893) and derived by his student, Ludwig Boltzmann (1844–1906).

The amount of energy radiated *by each square meter* of the surface of an object each second is called the flux, $\mathcal{F}$. This is proportional to the luminosity, but easier to measure. (Think about trying to catch all the photons emitted by the Earth in all directions, versus just the ones from the particular square meter under your feet. You can find the luminosity from the flux by multiplying by the total surface area.) We can relate the flux to the temperature by using the **Stefan-Boltzmann constant**, σ (the Greek letter sigma). The value of σ is 5.67×10^{-8} W/(m²K⁴), where W stands for watts. Expressing all this in math, we find:

$$\mathcal{F} = \sigma T^4$$

Even modest changes in temperature can result in large changes in the amount of power radiated by an object. If the temperature of an object doubles, the flux increases by a factor of 2^4, or 16. If the temperature triples, then the flux goes up by a factor of 3^4, or 81.

Suppose we want to find the flux and luminosity of Earth. Earth's average temperature is 288 K, so the flux from its surface is

$$\mathcal{F} = \sigma T^4$$
$$\mathcal{F} = (5.67 \times 10^{-8} \text{ W/(m}^2\text{K}^4)) (288 \text{ K})^4$$
$$\mathcal{F} = 390 \text{ W/m}^2$$

But if we want to know the total energy radiated every second, we must multiply by the surface area (A) of Earth. Surface area is given by $4\pi R^2$, and the radius of Earth is 6,378,000 meters, or 6.378×10^6 meters. So the luminosity is

$$L = \mathcal{F} \cdot A$$
$$L = \mathcal{F} \cdot 4\pi R^2$$
$$L = (390 \text{ W/m}^2) \cdot (4\pi \cdot (6.378 \times 10^6 \text{ m})^2)$$
$$L \approx 2 \times 10^{17} \text{ W}$$

This means that Earth emits the equivalent of 2,000,000,000,000,000 hundred-watt light bulbs. This is an enormous amount of energy, but not anywhere close to the amount emitted by the Sun.

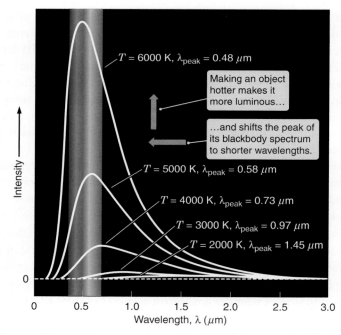

FIGURE 6.17 Blackbody spectra emitted by sources with temperatures ranging from 2000 K to 6000 K. At higher temperatures the peak of the spectrum shifts toward shorter wavelengths, and the amount of energy radiated per second from each square meter of the source increases.

ture of the object. Hotter objects emit more of their light at shorter wavelengths (see **Working It Out 6.3**). We describe the mechanism of *convection* as it relates to plate tectonics in Section 6.5. We cover *thermal conduction* when we talk about energy transfer within the Sun's interior in Chapter 11. For now, we'll simply say that liquids transfer heat by convection, and solids transfer heat by conduction,

Working It Out 6.3 | Wien's Law

Look again at Figure 6.17. Notice where the peak of each curve lines up along the horizontal axis. As the temperature, T, increases, the *peak* of the spectrum shifts toward shorter wavelengths. This is an inverse proportion. Translating this into math, and inserting the constant of 2,900 μm·K (derived from measurements of this relationship) to fix the units, we find:

$$\lambda_{peak} = \frac{2,900\ \mu m \cdot K}{T}$$

This is **Wien's law** (pronounced "Veen's law"), named for German physicist Wilhelm Wien (1864–1928), who discovered the relationship. In this equation, λ_{peak} (pronounced "lambda peak") is the wavelength where the spectrum is at its peak, where the electromagnetic radiation from an object is greatest.

Suppose we want to calculate the peak wavelength at which Earth radiates. If we insert Earth's average temperature of 288 K into Wien's law, we get

$$\lambda_{peak} = \frac{2,900\ \mu m \cdot K}{T}$$

$$\lambda_{peak} = \frac{2,900\ \mu m \cdot K}{288\ K}$$

$$\lambda_{peak} = 10.1\ \mu m$$

or slightly more than 10 μm. Earth's radiation peaks in the infrared region of the spectrum.

---

noting that some substances may act as either solids or liquids, depending on the time scale considered. For example, Earth's mantle behaves like a solid on short time scales, such as during an earthquake, but on geologic time scales it behaves like an extremely thick fluid and convects.

The internal temperature of a planet depends on its size. A planet's volume determines the amount of interior heating. The planet's surface area determines how much energy is lost, because thermal energy has to escape through the planet's surface. Smaller planets have more surface area in comparison to their small volumes, so they cool off faster, whereas larger planets have a smaller surface-area-to-volume ratio and cool off more slowly. It should be no surprise to learn that the smaller objects (Mercury and the Moon) are not geologically active in comparison to the larger terrestrial planets (Venus, Earth, and Mars).

Most Planets Generate Their Own Magnetic Fields

A compass needle lines up with Earth's magnetic field and points "north" and "south." The north-pointing end does not point at the *geographic* North Pole (about which Earth spins), but rather at a location in the Arctic Ocean off the coast of northern Canada. This is Earth's north *magnetic* pole. Earth's south magnetic pole is off the coast of Antarctica, 2,800 km from Earth's geographic South Pole. Earth behaves as if it contained a giant bar magnet that was slightly tilted with respect to Earth's rotation axis and had its two endpoints near the two magnetic poles (**Figure 6.18**).

Earth's magnetic field is constantly changing. At the moment, the north magnetic pole is traveling several tens of kilometers per year toward the northwest. If this rate continues, the north magnetic pole could be in Siberia before the end of the century.

The geological record shows that much more dramatic changes in the magnetic field have occurred over the history of our planet. When a magnet made of material such as iron gets hot enough, it loses its magnetization. As the material cools, it again becomes magnetized by any magnetic field surrounding it. Thus, iron-bearing minerals record the Earth's magnetic field at the time that they cooled.

> Earth's magnetic field behaves like a giant bar magnet, but the orientation of the poles changes with time.

(a)

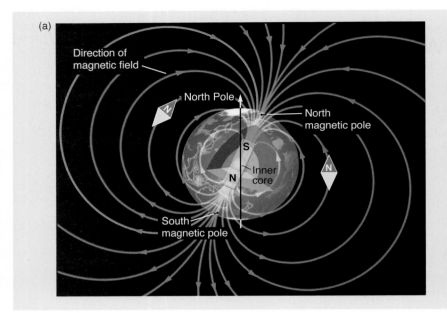

(b)

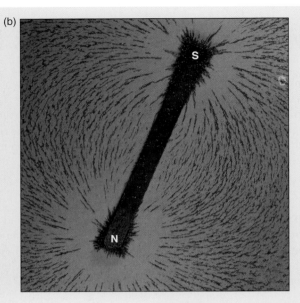

VISUAL ANALOGY

FIGURE 6.18 (a) Earth's magnetic field can be visualized as though it were a giant bar magnet tilted relative to Earth's axis of rotation. Compass needles line up along magnetic field lines and point toward Earth's north magnetic pole. Note that because bar magnets are attracted to each other's opposite poles, Earth's hypothetical bar magnet has its *south* pole coincident with Earth's *north* magnetic pole. (b) Iron filings sprinkled around a bar magnet help us visualize such a magnetic field.

The lack of magnetic fields on Venus and Mars is puzzling.

From these minerals, geologists learn how Earth's magnetic field has changed over time. Although Earth's magnetic field has existed for billions of years, the north and south magnetic poles switch from time to time. On average, these reversals in Earth's magnetic field take place about every 500,000 years.

Earth's magnetic field is *not* actually due to a bar magnet buried within the planet. Magnetic fields result from moving electric charges, and Earth's magnetic field is created by the combination of Earth's rotation about its axis and a liquid, electrically conducting, moving outer core. From this combination, Earth converts mechanical energy into magnetic energy.

During the *Apollo* program, astronauts measured the Moon's local magnetic field, and small satellites have searched for global magnetism. The Moon has a very weak field, possibly none at all, because the Moon is very small and therefore has a cool interior. It also has a very small core. However, remnant magnetism is preserved in lunar rocks from an earlier time when the Moon likely had a molten core and a magnetic field.

Other than Earth, Mercury is the only terrestrial planet with a significant magnetic field today. Rotation and a large iron core, parts of which are molten and circulating, cause Mercury's magnetic field. Venus should have a magnetic field because it should have an iron-rich core and partly molten interior like Earth's. Its lack of a magnetic field might be attributed to its extremely slow rotation. On the other hand, Mercury also rotates slowly (once every 58.6 Earth days). Venus's lack of a magnetic field is a puzzle.

Mars has a weak magnetic field, presumably frozen in place early in its history. The magnetic signature occurs only in the ancient crustal rocks. Geologically younger rocks lack this residual magnetism, so the planet's magnetic field has long since disappeared. The lack of a strong magnetic field today on Mars might be the result of its small core. However, given that Mars is expected to have a partly molten interior and rotates rapidly, the lack of a field is still surprising.

6.5 Tectonism: How Planetary Surfaces Evolve

With a better understanding of planetary interiors, we now turn to their surfaces. If you have been on a drive through mountainous or hilly terrain, you may have noticed places like the one shown in **Figure 6.19**, where the roadway has been cut through rock. These cuts show layers of rock that have been bent, broken, or fractured into pieces. These layers tell the story of Earth through the vast expanse of geologic time.

Continents Drift Apart and Come Back Together

Early in the 20th century, it became apparent that parts of Earth's continents could be fitted together like pieces of a giant jigsaw puzzle. In addition, the layers in the rock and the fossil record on the east coast of South America match those on the west coast of Africa. Based on evidence like this, Alfred Wegener (1880–1930) proposed **continental drift**: his hypothesis that the continents were originally joined in one large landmass that broke apart as the continents began to "drift" away from each other over millions of years.

This idea was met with great skepticism because geologists could not imagine a mechanism that could move such huge landmasses. In the 1960s, however, studies of the ocean floor provided compelling evidence for continental drift (**Figure 6.20**). These surveys showed surprising characteristics in bands of basalt, a type of rock formed from cooled lava, that were found on both sides

FIGURE 6.19 Tectonic processes fold and warp Earth's crust, as seen in the rocks along this roadside cut.

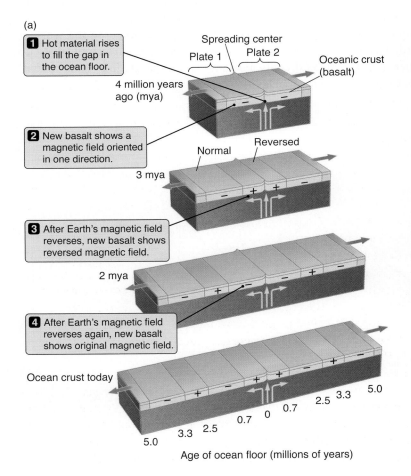

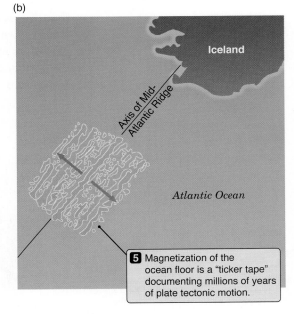

FIGURE 6.20 (a) As new seafloor is formed at a spreading center, the cooling rock becomes magnetized. The magnetized rock is then carried away by tectonic motions. (b) Maps like this one of banded magnetic structures in the seafloor near Iceland provide support for the theory of plate tectonics.

▶❙❙ **AstroTour:** Continental Drift

The theory explaining motions of Earth's crust is called *plate tectonics*.

of the ocean rifts. In these rifts, hot material rises toward Earth's surface, creating new ocean floor. When this hot material cools, it becomes magnetized along the direction of Earth's magnetic field, thus recording the direction of Earth's magnetic field at that time. Greater distance from the rift indicates an older ocean floor and an earlier time. Combined with radiometric dates for the rocks, this magnetic record showed that the continental plates have moved over long geological time spans.

More recently, precise surveying techniques and global positioning system (GPS) methods have confirmed these results more directly. Some areas are being pulled apart by about the length of a pencil each year. Over millions of years of geological time, such motions add up. Over 10 million years—a short time by geological standards—15 cm/year becomes 1,500 km, and maps definitely need to be redrawn.

Today, geologists recognize that Earth's crust is composed of a number of relatively brittle segments, or **lithospheric plates**, and that motion of these plates is constantly changing the surface of Earth. This theory, perhaps the greatest advance in 20th-century geology, is called **plate tectonics**. Plate tectonics is responsible for a wide variety of geological features on our planet.

Plate Tectonics Is Driven by Convection

Moving lithospheric plates requires immense forces. These forces are the result of thermal energy escaping from the interior of Earth through the process of **convection**. If you have ever watched water in a heated pot on a stovetop, then you have seen convection (**Figure 6.21a**). Thermal energy from the stove warms water at the bottom of the pot. The warm water expands slightly, becoming less dense than the cooler water above it. The cooler water sinks, displacing the warmer water upward. When the lower-density water reaches the surface, it gives up part of its energy to the air and cools, becomes denser, and sinks back toward the bottom of the pot. Water rises in some locations and sinks in others, forming convection "cells."

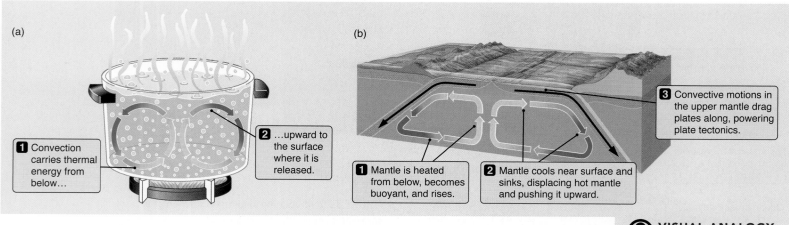

(a)

1 Convection carries thermal energy from below…

2 …upward to the surface where it is released.

(b)

1 Mantle is heated from below, becomes buoyant, and rises.

2 Mantle cools near surface and sinks, displacing hot mantle and pushing it upward.

3 Convective motions in the upper mantle drag plates along, powering plate tectonics.

👁 **VISUAL ANALOGY**

FIGURE 6.21 (a) Convection occurs when a fluid is heated from below. (b) Similarly, convection in Earth's mantle drives plate tectonics, although the time scale and velocities involved are very different from those in a pot boiling on your stovetop.

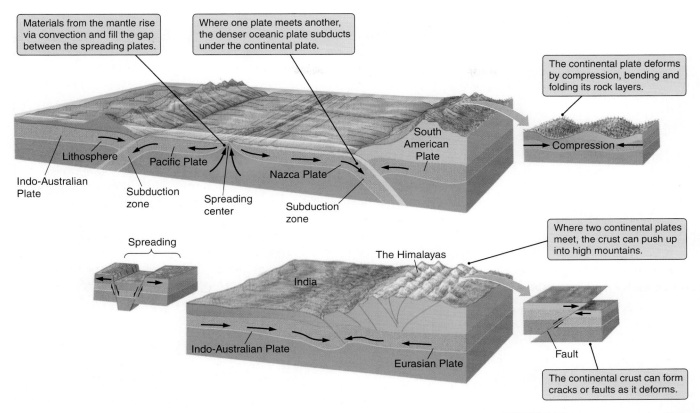

Materials from the mantle rise via convection and fill the gap between the spreading plates.

Where one plate meets another, the denser oceanic plate subducts under the continental plate.

The continental plate deforms by compression, bending and folding its rock layers.

Lithosphere

Pacific Plate

South American Plate

Compression

Indo-Australian Plate

Subduction zone

Spreading center

Nazca Plate

Subduction zone

Spreading

The Himalayas

India

Where two continental plates meet, the crust can push up into high mountains.

Indo-Australian Plate

Eurasian Plate

Fault

The continental crust can form cracks or faults as it deforms.

FIGURE 6.22 Divergence and collisions of tectonic plates create a wide variety of geological features.

Radioactive decay provides the heat source to drive convection in the mantle (**Figure 6.21b**). Earth's mantle is not molten, but it is somewhat mobile and so allows convection to take place *very slowly*. Earth's crust is divided into seven major plates and a half-dozen smaller ones floating on top of the mantle. Convection cells in Earth's mantle drive the plates, carrying both continents and ocean crust along with them. Convection also creates new crust along rift zones in the ocean basins, where mantle material rises up, cools, and slowly spreads out.

Figure 6.22 illustrates plate tectonics and some of its consequences. If material rises and spreads in one location, then it must converge and sink in another. In sinking regions, one plate slides beneath the other and convection drags the crust material down into the mantle. The Mariana Trench—the deepest part of Earth's ocean floor—is such a place. Much of the ocean floor lies between rising and sinking zones, and so the ocean floor is the youngest portion of Earth's crust. In fact, the *oldest* seafloor rocks are less than 200 million years old.

In some places, however, the plates are not sinking; they are colliding with each other and being shoved upward. The highest mountains on Earth, the Himalayas, grow 0.5 meter per century as the Indo-Australian Plate collides with the Eurasian Plate. In still other places, plates meet at oblique angles and slide along past each other. One such place is the San Andreas Fault in California, where the Pacific Plate slides past the North American Plate. A **fault** is a fracture in Earth's crust.

Locations where plates meet tend to be very active geologically. One of the best ways to see the outline of Earth's plates is to look at a map of where earthquakes and volcanism occur, like the map in **Figure 6.23**. Where plates run into each other, enormous stresses build up. Earthquakes occur when two plates finally and

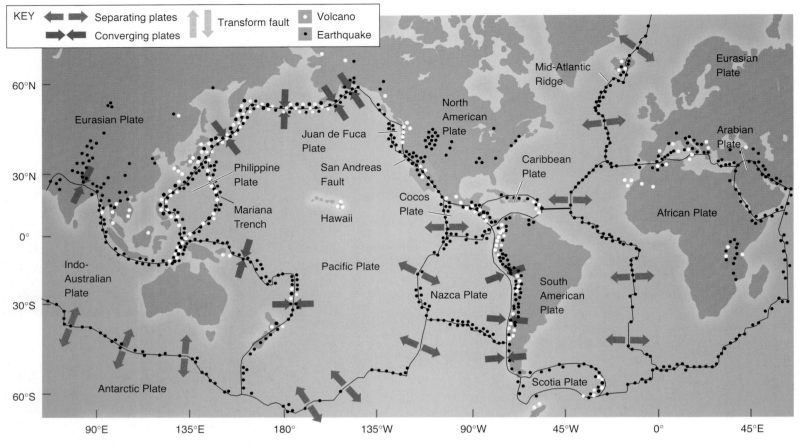

FIGURE 6.23 Major earthquakes and volcanic activity are often concentrated along the boundaries of Earth's principal tectonic plates.

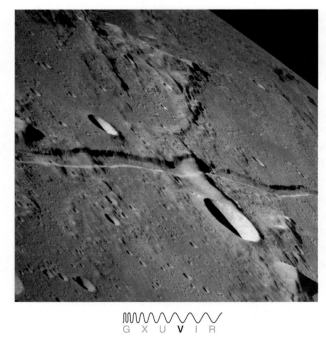

FIGURE 6.24 An *Apollo 10* photograph of Rima Ariadaeus, a 2-km-wide valley between two tectonic faults on the Moon.

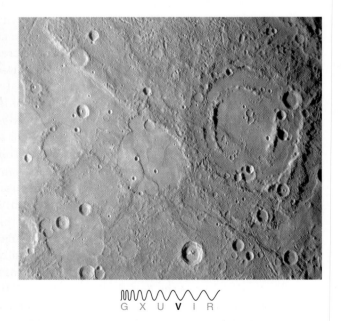

FIGURE 6.25 Faults on Mercury create cliffs hundreds of kilometers long, such as the one seen here running from upper left to lower right and alongside a double-ringed crater. Such features are consistent with compression of Mercury's crust as the planet cooled and contracted in size.

suddenly slip past each other, relieving the stress. Volcanoes are created when friction between plates melts rock, which is then pushed up through cracks to the surface. As plates shift, some parts move more rapidly than others, causing the plates to stretch, buckle, or fracture. These effects are seen on the surface as folded and faulted rocks. Mountain chains also are common near converging plate boundaries, where plates buckle and break.

Tectonism on Other Planets

Somewhat surprisingly, we have observed *plate* tectonics only on Earth. However, all of the terrestrial planets and some moons show evidence of tectonic disruptions. Fractures have cut the crust of the Moon in many areas, leaving fault valleys such as the one shown in **Figure 6.24**. Most of these features are the result of large impacts that have cracked and distorted the lunar crust.

Mercury also has fractures and faults. In addition, numerous cliffs on Mercury are hundreds of kilometers long (**Figure 6.25**). Like the other terrestrial planets, Mercury was once molten. After the surface of the planet cooled and its crust formed, the interior of the planet continued to cool and shrink. As the planet shrank, Mercury's crust cracked and buckled in much the same way that a grape skin wrinkles as it shrinks to become a raisin. To explain the faults seen on the planet's surface, the volume of Mercury must have shrunk by about 5 percent after the planet's crust formed.

Possibly the most impressive tectonic feature in the Solar System is the Valles Marineris, on Mars (**Figure 6.26**). Stretching nearly 4,000 km and nearly four times as deep as the Grand Canyon, this chasm, if located on Earth, would link San Francisco with New York. Valles Marineris includes a series of massive cracks in the crust of Mars that are thought to have formed as local forces, perhaps related to mantle convection, pushed the crust upward from below. Once formed, the cracks were eroded by wind, water, and landslides, resulting in the structure we see today. Other parts of Mars have faults similar to those on the Moon, but cliffs like those on Mercury are absent.

Venus's mass is only 20% less than Earth's, and its radius is only 5% smaller. Because of these similarities, many scientists speculated that Venus might also show evidence of plate tectonics. However, their speculations were not borne out by the *Magellan* spacecraft, which used radar to peer through the thick, dense layers of clouds enshrouding the planet. *Magellan* mapped about 98 percent of the surface of Venus, providing the first high-resolution views of the surface of the planet. *Magellan*'s view of one face of Venus is shown in **Figure 6.27**. Although Venus has a wealth of volcanic features and tectonic fractures, there is no evidence of lithospheric plates or plate motion of the sort seen on Earth. Yet the relative scarcity of impact craters on Venus suggests that most of its surface is less than 1 billion years old.

The absence of plate tectonics on Venus is a puzzle. The interior of Venus should be very much like the interior of Earth, and convection should be occurring in its mantle. On Earth, mantle convection and plate tectonics release thermal energy from the interior. Earth also has a few **hot spots**, like the Hawaiian Islands, where hot deep-mantle material rises, releasing thermal energy. On Venus, hot spots may be the principal way that thermal energy escapes from the planet's interior. Circular fractures called coronae on the surface of Venus, ranging from a few hundred kilometers to more than 1,000 km across, may be the result of upwelling plumes of hot mantle that have fractured Venus's lithosphere.

Alternatively, a radically different form of tectonism may be at work on Venus.

(a)

MMMMWWW
G X U **V** I R

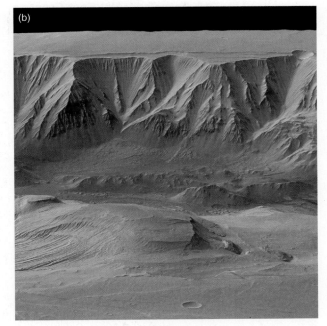

(b)

MMMMWWW
G X U **V** I R

FIGURE 6.26 (a) A mosaic of *Viking Orbiter* images shows Valles Marineris, the major tectonic feature on Mars, stretching across the center of the image from left to right. This canyon system is more than 4,000 km long. The dark spots on the left are huge shield volcanoes in the Tharsis region. (b) This close-up perspective view of the canyon wall was photographed by the European Space Agency's *Mars Express* spacecraft.

Energy may build up in the interior until large chunks of the lithosphere melt and overturn, releasing an enormous amount of energy. Then the surface cools and solidifies. This idea remains highly controversial, but it may explain the relatively young surface of Venus. It also drives home the point of how different the geological histories of the various planets appear to be. Our understanding of why Venus and Earth are so different with regard to plate tectonics is still very uncertain.

> Earth's tectonic plates are unique in the Solar System.

6.6 Volcanism: A Sign of a Geologically Active Planet

The molten rock inside Earth, called **magma**, does not come to Earth's surface from its molten core as is sometimes believed. Seismic waves show that magma originates in the lower crust and upper mantle, where sources of thermal energy combine. These sources include rising convection cells in the mantle, frictional heating generated by movement in the crust, and concentrations of radioactive elements.

Terrestrial Volcanism Is Related to Tectonism

▶❚❚ **AstroTour: Hot Spot Creating a Chain of Islands**

Because Earth's thermal energy sources are not uniformly distributed, volcanoes are usually located along plate boundaries and over hot spots. Maps such as the one shown in Figure 6.23 leave little doubt that most terrestrial volcanism is linked to the same forces responsible for plate motions. A tremendous amount of friction is generated as plates slide under each other. This friction raises the temperature of rock toward its melting point.

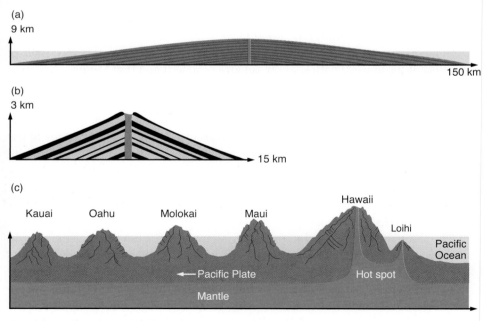

FIGURE 6.28 Magma reaching Earth's surface commonly forms (a) shield volcanoes, such as Mauna Loa, which have gently sloped sides built up by fluid lava flows; and (b) composite volcanoes, such as Vesuvius, which have steeply symmetrical sides built up by viscous lava flows. (c) Hot spots are convective plumes of lava that form a series of volcanoes as the plate above them slides by.

G X U V I **R**

FIGURE 6.27 The atmosphere of Venus blocks our view of the surface in visible light. This false-color view of Venus is a radar image made by the *Magellan* spacecraft. Bright yellow and white areas are mostly fractures and ridges in the crust. Some circular features seen in the image may be hot spots—regions of mantle upwelling. Most of the surface is formed by lava flows, shown in orange.

Material at the base of a lithospheric plate is under a great deal of pressure due to the weight of the plate pushing down on it. This pressure increases the melting point of the material, forcing it to remain solid even at high temperature. As this material is forced up through the crust, the pressure drops; as the pressure drops, so does the material's melting point. Material that was solid at the base of a plate may melt as it nears the surface. Places where convection carries hot mantle material toward the surface are frequent sites of eruptions. Iceland, which is one of the most volcanically active regions in the world, sits astride one such place—the Mid-Atlantic Ridge (see Figure 6.23).

Once lava reaches the surface of Earth, it can form many types of structures (**Figure 6.28**). Flows often form vast sheets, especially if the eruptions come from long fractures called **fissures**. If very **fluid** lava flows from a single "point source," it can spread out over the surrounding terrain or ocean floor, forming what is known as a **shield volcano** (Figure 6.28a). Thick lava flows alternating with explosively generated rock deposits can form a steep-sided structure called a **composite volcano** (Figure 6.28b) or smaller circular mounds called volcanic domes.

A third setting for terrestrial volcanism occurs where convective plumes rise toward the surface in the interiors of lithospheric plates, creating local hot spots (Figure 6.28c). Volcanism over hot spots works much like volcanism elsewhere, except that the convective upwelling occurs at a single spot rather than in a line along the edge of a plate. These hot spots melt mantle and lithospheric material and force it toward the surface of Earth.

Earth has numerous hot spots, including the regions around Yellowstone Park and the Hawaiian Islands. The Hawaiian Islands are a chain of shield volcanoes that formed as their lithospheric plate moved across a hot spot. The island ceases to grow as the plate motion carries the island away from the hot spot. Erosion, occurring since the island's inception, continues to wear the island away. Meanwhile, a new island grows over the hot spot. Today the Hawaiian hot spot is located off the southeast coast of the Big Island of Hawaii, where it continues to power the active volcanoes. On top of the hot spot, the newest Hawaiian island, Loihi, is forming. Loihi is already a massive shield volcano, rising more than 3 km above the ocean floor. Loihi will eventually break the surface of the ocean and merge with the Big Island of Hawaii—but not for another 100,000 years.

Volcanism Occurs Elsewhere in the Solar System

Even before the *Apollo* astronauts, photographs showed flowlike features in the dark regions of the Moon. Some of the first observers to use telescopes thought that these dark areas looked like bodies of water—thus the name **maria** (singular: *mare*), Latin for "seas." The maria are actually vast hardened lava flows, similar to basalts on Earth. Because the maria contain relatively few craters, these volcanic flows must have occurred after the period of heavy bombardment ceased.

When the *Apollo* astronauts returned rock samples from the lunar maria, many of them were found to contain gas bubbles typical of volcanic materials (**Figure 6.29**). The lava that flowed across the lunar surface must have been relatively fluid (**Figure 6.30**). The fluidity of the lava, due partly to its chemical composition, explains why lunar basalts form vast sheets that fill low-lying areas such as impact basins. It also explains the Moon's lack of classic volcanoes: the lava was too fluid to pile up.

The samples also showed that most of the lunar lava flows are older than 3 billion years. Only in a few limited areas of the Moon are younger lavas thought to exist; most of these have not been sampled directly. Samples from the heavily

Vocabulary Alert

Fluid: In common language, this is used mainly as a noun. Here, it is used as an adjective, describing how easily a substance flows. If it is very fluid, it flows like water. If it is not very fluid, it flows like molasses or tar. Another adjective often used in this context is *viscous*. A viscous substance does not flow easily.

FIGURE 6.29 This rock sample from the Moon, collected by the *Apollo 15* astronauts from a lunar lava flow, shows gas bubbles typical of gas-rich volcanic materials. This rock is about 6 cm by 12 cm.

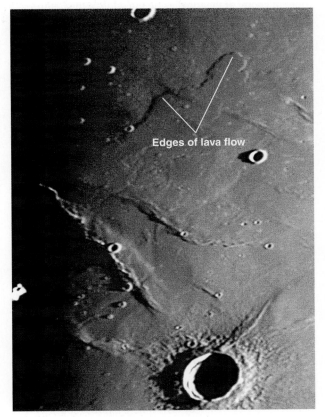

FIGURE 6.30 Lava flowing across the surface of Mare Imbrium on the Moon must have been relatively fluid to have spread out for hundreds of kilometers in sheets that are only tens of meters thick.

cratered terrain of the Moon also originated from magma, so the young Moon must have gone through a molten stage. These rocks cooled from a "magma ocean" and are more than 4 billion years old, preserving the early history of the Solar System. Most of the sources of heating and volcanic activity on the Moon shut down some 3 billion years ago—unlike on Earth, where volcanism continues. This conclusion is consistent with our earlier argument that smaller planets should cool more efficiently and thus be less active than larger planets.

Mercury also shows evidence for past volcanism. *Mariner 10* and *Messenger* missions revealed smooth plains similar in appearance to lunar maria. These sparsely cratered plains are the youngest areas on Mercury, created when fluid lava flowed into and filled huge impact basins. *Messenger* also found a number of volcanoes (**Figure 6.31**).

More than half the surface of Mars is covered with volcanic rocks. Lava covered huge regions of Mars, flooding the older, cratered terrain. Most of the vents or fissures that created these flows are buried under the lava that poured forth from them (**Figure 6.32**). Among the most impressive features on Mars are its enormous shield volcanoes. These volcanoes are the largest mountains in the Solar System. Olympus Mons, standing 27 km high at its peak and 550 km wide at its base (**Figure 6.33**), would tower over Earth's largest mountains. Despite the difference in size, most of the very large volcanoes of Mars are shield volcanoes, just like their Hawaiian cousins. Olympus Mons and its neighbors grew as the result of hundreds of thousands of individual eruptions. These volcanoes have remained over their hot spots for billions of years, growing ever taller and broader with each successive eruption.

Lava flows and other volcanic landforms span nearly the entire history of Mars, estimated to extend from the formation of crust some 4.4 billion years ago to geologically recent times, and to cover more than half of the red planet's surface. But remember, "recent" in this sense could still be more than 100 million years ago. Although some "fresh-appearing" lava flows have been identified on Mars, until rock samples are radiometrically dated we will not know the age of these latest eruptions. Mars could, in principle, experience eruptions today.

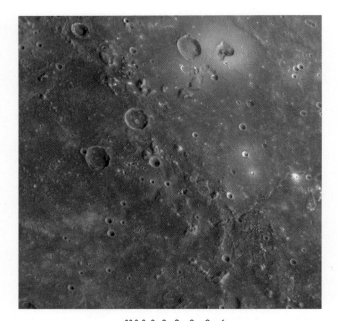

FIGURE 6.31 The largest volcano yet seen on Mercury (right of center at the top) lies on the edge of the Caloris Basin. The volcano is about 50 km across.

FIGURE 6.32 A Mars crater flooded by lava is seen in this perspective view taken by ESA's *Mars Express* spacecraft.

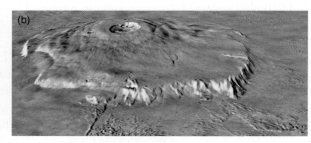

FIGURE 6.33 The largest known volcano in the Solar System, Olympus Mons on Mars is a 27-km-high shield-type volcano, similar to but much larger than Hawaii's Mauna Loa. (a) A partial view of Olympus Mons taken by the *Mars Global Surveyor*. (b) This oblique view was created from an overhead *Viking* image and topographic data provided by the *Mars Orbiter* laser altimeter.

Venus has the most volcanoes of all the terrestrial planets. Radar images reveal a wide variety of volcanic landforms. These include highly fluid flood lavas covering thousands of square kilometers, enormous shield volcanoes, dome volcanoes, and lava channels thousands of kilometers long. These lavas must have been extremely hot and fluid to flow for such long distances.

6.7 Erosion: Wearing Down the High Spots and Filling In the Low

Erosion levels planetary surfaces, wearing down high spots and filling in low ones. The term *erosion* covers a wide variety of processes. The first step in the process of erosion is called "weathering," in which rocks are broken into smaller pieces and may be chemically altered. For example, rocks on Earth are physically weathered along shorelines, where the pounding waves break them into beach sand. Other weathering processes include chemical reactions, such as the combining of oxygen in the air with iron in rocks to form a type of rust. One of the most efficient forms of weathering is freeze-thaw cycles, during which liquid water runs into crevices and then freezes, expanding and shattering the rock.

After weathering, the resulting debris can be carried away by flowing water, glacial ice, or blowing wind and deposited in other areas as sediment. Where material is eroded, we can see features such as river valleys, wind-sculpted hills, or mountains carved by glaciers. Where eroded material is deposited, we see features such as river deltas, sand dunes, or piles of rock at the bases of mountains and cliffs. Erosion is most efficient on planets with water and wind. On Earth, where water and wind are so dominant, most impact craters are worn down and filled in even before they are turned under by tectonic activity.

Even though the Moon and Mercury have almost no atmosphere and no running water, a type of erosion is still at work. Radiation from the Sun and from deep space very slowly decomposes some types of minerals, effectively weathering the rock. Such effects are only a few millimeters deep at most. Impacts of micrometeoroids also chip away at rocks. In addition, landslides can occur wherever gravity and differences in elevation are present. Although water enhances landslide activity, landslides are also seen on Mercury and the Moon.

Earth, Mars, and Venus all show the effects of windstorms. Images of Mars (**Figure 6.34**) and Venus returned by spacecraft landers show surfaces that have been subjected to the forces of wind. Sand dunes are common on Earth and Mars, and some have been identified on Venus. Orbiting spacecraft have also found wind-eroded hills, and in surface patterns called wind streaks. These surface patterns appear, disappear, and change in response to winds blowing sediments around hills, craters, and cliffs. They serve as local "wind vanes," telling planetary scientists about the direction of local prevailing surface winds. Planet-encompassing dust storms occur on Mars.

Today Earth is the only planet where the temperature and atmospheric conditions allow extensive liquid surface water to exist. Water is an extremely powerful agent of erosion and dominates erosion on Earth. Every year, rivers and streams on Earth deliver about 10 billion metric tons of sediment into the oceans. Even though today there is no liquid water on the surface of Mars, at one time water likely flowed across its surface in vast quantities. Features resembling water-carved channels on Earth such as those shown in **Figure 6.35** imply the past presence of liquid water. In addition, many regions on Mars show small networks

FIGURE 6.34 This panoramic view of the surface of Mars, taken by the *Opportunity* lander, shows windblown sand dunes.

Venus has more volcanoes than any other terrestrial planet.

FIGURE 6.35 Gully channels in a crater on Mars, taken by *Mars Reconnaissance Orbiter*. The gullies near the crater's rim (upper left) show meandering and braided patterns similar to those of water-carved channels on Earth.

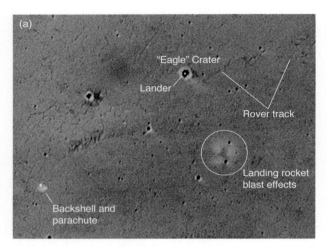

FIGURE 6.36 (a) An orbital view of the Mars rover *Opportunity*'s landing site. After landing, the protected package bounced along the surface and ended up in the "Eagle" crater—a landing that mission scientists called a "hole in one." The white dot in the middle of the crater is the lander. Tracks leading out of the crater were made by *Opportunity* as it began its exploration of Mars. (b) Layered rock seen near the landing site may represent sediments deposited by water.

of valleys that are thought to have been carved by flowing water. Some parts of Mars may once have contained even oceans and glaciers.

The search for water is central to the exploration of Mars. In 2004, NASA sent two instrument-equipped roving vehicles, *Opportunity* and *Spirit*, to search for evidence of water on Mars. *Opportunity*'s landing site was chosen because orbital data suggested the presence of hematite, an iron-rich mineral that forms in the presence of water, in that area. *Opportunity* was to test the validity of that orbital information. The rover landed inside a small crater (**Figure 6.36a**). For the first time, martian rocks were available for study in the original order in which they were laid down. Previously, the only rocks that landers and rovers had come across were those that had been dislodged from their original settings by either impacts or river floods.

The layered rocks at the *Opportunity* site (**Figure 6.36b**) revealed that they had once been soaked in or transported by water. The form of the layers was typical of sandy sediments laid down by gentle currents of water. (Some geologists think that this layering is deposits of volcanic ash rather than water sediments. This is an example of honest disagreement among scientists, all looking at the same data.) Rover instruments found a mineral so rich in sulfur that it had almost certainly formed by precipitation from water. Magnified images of the rocks showed "blueberries," small spherical grains a few millimeters across that probably formed in place among the layered rocks. These are similar to terrestrial features that form by the percolation of water through sediments. Analysis of these blueberries revealed abundant hematite, confirming the orbital data. Further observations by the European Space Agency's *Mars Express* and NASA's *Mars Odyssey* orbiters have shown the hematite signature and the presence of sulfur-rich compounds in a vast area surrounding the *Opportunity* landing site. These observations suggest an ancient martian sea larger than the combined area of Earth's Great Lakes and as much as 500 meters deep.

Spirit landed in Gusev, a 170-km impact crater. This site was chosen because it showed signs of ancient flooding by a now-dry river. It was hoped that surface deposits would provide further evidence of past liquid water. Surprise, and perhaps some disappointment, followed when *Spirit* revealed that the flat floor of

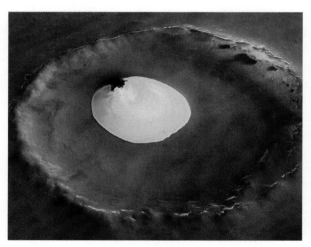

FIGURE 6.37 This image, taken by ESA's *Mars Express* mission, shows water ice in a crater in the north polar region of Mars. The crater is about 35 km in diameter, and the ice is estimated to be about 200 meters thick. The 300-meter-high crater wall blocks much of the sunlight at this high latitude, keeping the ice from vaporizing. The colors are close to natural in this image, but the relief has been exaggerated by a factor of 3.

Gusev consisted primarily of basaltic rock. Only when the rover ventured cross-country to some low hills located 2.5 km from the landing site did it find basaltic rocks showing clear signs of having been chemically altered by liquid water.

Where did the water go? Some escaped into the thin atmosphere of Mars, and at least some of it is locked up as ice in the polar regions, just as the ice caps on Earth hold much of our planet's water. But, unlike our own polar caps, those on Mars are a mixture of frozen carbon dioxide and frozen water, so water must be hiding elsewhere on Mars. Small amounts of water are found on the surface, as seen in **Figure 6.37**, and NASA's *Phoenix* lander found water ice just a centimeter or so beneath surface soils at high northern latitudes (**Figure 6.38**). However, most of the water on Mars appears to be trapped well below the surface of the planet. Recent radar imaging by NASA's *Mars Reconnaissance Orbiter* indicates huge quantities of subsurface water ice, not only in the polar areas as expected but also at lower latitudes.

Although Earth and Mars are the only terrestrial planets that show evidence for liquid water at any time in their histories, water ice could exist on Mercury and the Moon today. Some deep craters in the polar regions of both Mercury and the Moon have floors in perpetual shadow. Temperatures in these permanently shadowed areas remain below 180 K. For many years, planetary scientists speculated that ice—perhaps from comets—could be found in these craters. In the early 1990s, radar measurements of Mercury's north pole and infrared measurements of the Moon's polar areas seemed to support this possibility. In 2009, NASA's *Lunar Reconnaissance Orbiter* (*LRO*) and a companion spacecraft known as *LCROSS* were put into lunar orbit to continue the search for possible sources of subsurface water ice. *LCROSS* intentionally crashed the second stage of its Centaur rocket into a lunar polar-region crater, Cabeus. *LRO* collected data from the resulting plume, finding that more than 150 kg of water ice and water vapor were blown out of the crater by the impact. These data revealed the presence of large amounts of water buried beneath the crater's floor. The existence of ice on the Moon could make future lunar colonization much more practical.

FIGURE 6.38 Water ice appears a few centimeters below the surface of Mars in this trench dug by a robotic arm on the *Phoenix* lander. The trench measures about 20 by 30 cm.

Water ice could exist on Mercury as it does on Mars and the Moon.

READING ASTRONOMY News

Nearly all of what people know about the terrestrial planets (except Earth) and our Moon comes from observations of the surface properties of these worlds. Just below Earth's surface are pockets of elements that are rare on the surface; other Solar System objects may hide similar treasures.

Moon Is Wetter, Chemically More Complex than Thought, NASA Says

By **Amina Khan**, *Los Angeles Times*

The moon is a much wetter—and more chemically complicated—place than scientists had believed, according to new data released Thursday by NASA.

Last year, after the space agency hurtled a rocket into a frozen crater at the moon's south pole and measured the stuff kicked up by the crash, scientists calculated that the plume contained about 25 gallons of water. But further analysis over the last 11 months indicates that the amount of water vapor and ice was more like 41 gallons—an increase of 64%.

"It's twice as wet as the Sahara desert," said Anthony Colaprete, the lead scientist for the Lunar Crater Observation and Sensing Satellite, or LCROSS, mission at NASA Ames Research Center in Northern California.

The instruments aboard the satellite, including near-infrared and visible light spectrometers, scanned the lunar debris cloud and identified the compounds it contained. They determined that about 5.6% of the plume was made of water, give or take 2.9%. It also included a surprising array of chemicals, including mercury, methane, silver, calcium, magnesium, pure hydrogen and carbon monoxide.

The findings were reported in six related papers published online Thursday by the journal *Science*.

"The lunar closet is really at the poles, and I think there's a lot of stuff crammed into the closet that we really haven't investigated yet," said Peter Schultz, a planetary geologist at Brown University in Providence, R.I., and one of the LCROSS team members.

The new measurements allowed Colaprete to estimate that the entire Cabeus crater could hold as much as 1 billion gallons of water.

Potentially, that would be really handy for future space explorers who might use the moon as an interplanetary way station. Along with providing water to drink, it could be mined for breathable oxygen and used to make hydrogen fuel for long-distance spacecraft.

"You can't take a lot of big things to the moon and you can't take much water, so we're learning to live off the land," said Lawrence Taylor, a planetary geochemist at the University of Tennessee in Knoxville, who was not involved in the studies.

However, plans to develop a space colony on the moon were put on hold by the Obama administration this year.

Evaluating the News

1. In early analysis of the data, scientists discovered 25 gallons of water. In this article, that number has been revised upward to 41 gallons. What astronomical observation methods did they use to discover this?
2. The article states that 5.6 percent of the plume was made of water, give or take 2.9 percent. What is the smallest percentage of the plume that could be water? What is the largest? Is this a large range?
3. Consider the list of other chemicals that were found in this experiment. Name one way in which each of these elements is useful or hazardous to humans.
4. Average daily urban water use in the United States is roughly 100 gallons per household. If Cabeus crater does have 1 billion gallons of water, how many American households could this crater support for one day? For one month? For a year? For 10 years?
5. What do your answers to question 4 imply for the long-term survival of a colony on the Moon? How does this prediction change if intense methods of water reuse and recycling are implemented in the colony? How does this prediction change if the water is also being used to make oxygen to breathe and hydrogen for fuel?

SUMMARY

6.1 Comparative planetology is the key to understanding the planets. Impacts, volcanism, and tectonism create topographic relief on the terrestrial planets.

6.2 Relative concentrations of impact craters divulge the ages of planetary surfaces. Radiometric dating tells us the ages of rocks. Planets protected by atmospheres, like Earth and Venus, have fewer impact craters.

6.3 Scientists use seismic waves to investigate Earth's interior structure. The Moon was most probably created when a Mars-sized protoplanet collided with Earth.

6.4 Earth's interior is heated by both radioactive decay and tidal effects. The Moon and the Sun both cause tides on Earth; the tides sometimes add, and sometimes cancel each other. Earth has a strong magnetic field, but Venus and Mars do

not. The cause for this difference between the terrestrial planets is uncertain.

6.5 The theory explaining the change in the Earth's surface is called plate tectonics. While all the terrestrial planets show tectonic disruptions, only Earth has tectonic plates. Smooth areas on the Moon and Mercury are ancient lava flows.

6.6 Volcanism links Earth's surface to the molten interior. While Venus has the most volcanoes, the largest mountains in the Solar System are volcanoes on Mars.

6.7 Erosion wears down the surfaces of Venus and Mars by wind and the surface of Earth by wind and water. Mars today has large amounts of subsurface water ice and once had liquid water on its surface.

✧ SUMMARY SELF-TEST

1. _____, _____, and _____ build up structures on the terrestrial planets, while _____ tears them down.
   a. Impacts, erosion, volcanism; tectonism
   b. Impacts, tectonism, volcanism; erosion
   c. Tectonism, volcanism, erosion; impacts
   d. Tectonism, impacts, erosion; volcanism

2. If crater A is inside of crater B, we know that
   a. crater A was formed before crater B.
   b. crater B was formed before crater A.
   c. both craters were formed at about the same time.
   d. crater B formed crater A.
   e. crater A formed crater B.

3. If a radioactive element A decays into radioactive element B with a half-life of 20 seconds, then after 40 seconds,
   a. none of element A will remain.
   b. none of element B will remain.
   c. half of element A will remain.
   d. one-quarter of element A will remain.

4. We find out about the interiors of the terrestrial planets from
   a. seismic waves.
   b. satellite observations of gravitational fields.
   c. physical arguments about cooling.

   d. satellite observations of magnetic fields.
   e. all of the above.

5. Earth's interior is heated by _____.
   a. angular momentum and gravity
   b. radioactive decay and gravity
   c. radioactive decay and tidal effects
   d. angular momentum and tidal effects
   e. gravity and tidal effects

6. Lava flows on the Moon and Mercury created large, smooth plains. We don't see similar features on Earth because
   a. Earth has less lava.
   b. Earth had fewer large impacts in the past.
   c. Earth has plate tectonics that recycle the surface.
   d. Earth is large compared to the size of these plains, so they are not as noticeable.
   e. Earth's rotation rate is much faster than that of either of these other worlds.

7. Rank the following worlds by the strength of wind erosion currently operating on their surfaces, from greatest to least:
   a. Mercury   b. Venus
   c. Earth   d. Moon
   e. Mars

QUESTIONS AND PROBLEMS

True/False and Multiple-Choice Questions

8. **T/F:** Volcanism has been present on all the terrestrial planets.

9. **T/F:** Impact cratering no longer affects the terrestrial planets.

10. **T/F:** The geologic processes that shape a planet have ceased on Earth.

11. **T/F:** Seismic waves reveal the structure of the interior of Earth.

12. **T/F:** Large worlds remain geologically active longer than small ones.

13. **T/F:** Mercury has plate tectonics.

14. **T/F:** Mercury has no volcanoes.

15. **T/F:** Wind erosion is an important process on Venus.

16. Of the four processes that shape the surface of a terrestrial world, the one with the greatest potential for catastrophic rearrangement is
    a. impacts.   **b.** volcanism.
    c. tectonism.   **d.** erosion.

17. We can find the relative age of features on a world because
    a. the ones on top must be older.
    b. the ones on top must be younger.
    c. the larger ones must be older.
    d. the larger ones must be younger.
    e. all the features we can see are the same age.

18. We can find the actual age of features on a world by
    a. radioactive dating of rocks retrieved from the world.
    b. comparing cratering rates on one world to those on another.
    c. assuming that all features on a planetary surface are the same age.
    d. both a and b.
    e. both b and c.

19. Impacts on the terrestrial worlds
    a. are more common than they used to be.
    b. have occurred at approximately the same rate since the formation of the Solar System.
    c. are less common than they used to be.
    d. periodically become more common and then are less common for a while.
    e. Never occur any more.

20. _____ waves are compressional, while _____ waves oscillate perpendicular to line of travel.
    a. Longitudinal; transverse
    b. Transverse; longitudinal
    c. P waves; S waves
    d. S waves; P waves
    e. both a and d
    f. both b and c

21. The terrestrial worlds that may still be geologically active are
    a. Earth, Moon and Mercury.
    b. Earth, Mars, and Venus.
    c. Earth, Venus, and Mercury.
    d. Earth only.
    e. Earth and Venus only.

22. Spring tides occur only when
    a. the Sun is near the vernal equinox in the sky.
    b. the Moon is in first or third quarter.
    c. the Moon, Earth, and Sun form a right angle.
    d. the Moon is in new or full phase.
    e. the Sun is in full phase.

23. If an object becomes hotter, it
    a. becomes bluer.
    b. becomes redder.
    c. becomes brighter.
    d. becomes fainter.
    e. both a and c
    f. both b and d

24. Scientists know the history of Earth's magnetic field because
    a. the magnetic field hasn't changed since the formation of Earth.
    b. they see how it's changing today, and project that back in time.
    c. the magnetic field gets frozen into rocks, and plate tectonics spreads them out.
    d. they compare the magnetic fields on other planets to Earth's.
    e. there are written documents of magnetic field measurements since the beginning of Earth.

25. Water erosion is an important process on
    a. all the terrestrial worlds.
    b. Earth only.
    c. Earth and Mars only.
    d. Earth, Mars, and the Moon only.
    e. Earth, Mars, and Venus only.

Conceptual Questions

26. Imagine that you were alive in the early 20th century. How would you think about Earth differently than we do today?

27. List evidence of the four geologic processes that shape the terrestrial planets: tectonism, volcanism, impact cratering, and erosion.

28. In discussing the terrestrial planets, why do we include our Moon?

29. What is meant by *comparative planetology*, and why is it important?

30. One region on the Moon is covered with craters, while another is a smooth volcanic plain. Which is older? How do we know?

31. Can all rocks be dated with radiometric methods? Explain.

32. Suppose that you have two rocks, each containing a radioactive isotope and its decay products. In rock A, there is an equal amount of the parent and the daughter. In rock B, there is twice as much daughter as parent. Which rock is older? Explain how you know this from the information given.

33. Explain how scientists know that rock layers at the bottom of Arizona's Grand Canyon are older than those found on the rim.

34. Make a sketch of an imaginary planetary surface with three craters of distinctly different ages. Label them from youngest to oldest.

35. Describe the sources of heating that are responsible for the generation of Earth's magma.

36. Explain why the Moon's core is cooler than Earth's.

37. Name the three components that make up Earth's interior.

38. Compare the cores of the terrestrial worlds, as shown in Figure 6.12. What is unique about Earth's core?

39. What is meant by *differentiation* of a planet, and what causes it?

40. Explain the difference between longitudinal waves and transverse waves.

41. Why are S waves useless for probing Earth's solid inner core?

42. How do scientists know that Earth's core includes a liquid zone?

43. Compare and contrast tectonism on Venus, Earth, and Mercury.

44. Volcanoes have been found on all of the terrestrial planets. Where are the largest volcanoes in the inner Solar System?

45. Describe the collision theory of the formation of the Moon.

46. Examine Figure 6.14a. This figure shows two tidal bulges on Earth, both caused by the Moon. Compare this figure to Figure 6.16. In your own words, describe why the Moon's gravity causes two tides on Earth—one on the side closest to the Moon, and one on the side farthest away.

47. If you see two objects of the same type, and one is bluer and brighter than the other, what conclusion can you draw about the temperature of the bluer object?

48. A favorite object for amateur astronomers is the double star Albireo; one of its components is a golden yellow and the other a bright blue. What do these colors tell you about the relative temperatures of the two stars?

49. Describe plate tectonics, and identify the planet(s) on which this process has been observed.

50. Describe and explain the evidence for reversals in the polarity of Earth's magnetic field.

51. Why do earthquakes and volcanoes tend to occur near plate boundaries?

52. What is meant by *erosion*? What processes contribute to erosion?

53. Does the age of a planetary surface tell you the age of the planet? Why or why not?

54. Explain some of the evidence that Mars once had liquid water on its surface.

55. Why is there no liquid water on Mars today?

56. Why does the discovery of water on the Moon affect our thinking about eventual lunar colonization?

Problems

57. Earth has a radius of 6,378 km.
    a. What is its volume? (Hint: The volume of a sphere of radius r is $(4/3)\pi r^3$.)
    b. What is its surface area? (Hint: The surface area of a sphere of radius r is $4\pi r^2$.)
    c. Suppose Earth's radius suddenly became twice as big. What would happen to the volume? What would happen to the surface area?

58. Suppose you find a piece of ancient pottery and take it to the laboratory of a physicist friend. He finds that the glaze contains radium, a radioactive element that decays to radon and has a half-life of 1,620 years. He tells you that the glaze couldn't have contained any radon when the pottery was being fired, but that it now contains three atoms of radon for each atom of radium. How old is the pottery?

59. Archaeological samples are often dated by radiocarbon dating. The half-life of carbon-14 is 5,700 years.

    a. After how many half-lives will the sample have only 1/64 as much carbon-14 as it originally contained?
    b. How much time will have passed?
    c. Suppose that the daughter of carbon-14 is present in the sample when it forms (even before any radioactive decay happens). Then you cannot assume that every daughter you see is the result of carbon-14 decay. If you did make this assumption, would you overestimate or underestimate the age of a sample?

60. Different radioisotopes have different half-lives. For example, the half-life of carbon-14 is 5,700 years, the half-life of uranium-235 is 704 million years, the half-life of potassium-40 is 1.3 billion years, and the half-life of rubidium-87 is 49 billion years.
    a. Explain why you would not use an isotope with a half-life similar to that of carbon-14 to determine the age of the Solar System.
    b. The age of the universe is approximately 14 billion years. Does that mean that no rubidium-87 has decayed yet?

61. The average temperature of Mars is about 210 K. What is the blackbody flux from a square meter of Mars?

62. A panel with an area of 1 square meter (m²) is heated to a temperature of 500 K. How many watts is it radiating into its surroundings?

63. The average temperature of Venus is about 773 K. What is the peak wavelength of radiation from a blackbody the same temperature as Venus?

64. Suppose that the flux from a planet is measured to be 350 W/m². What is its average temperature?

65. A blackbody's peak emission occurs at 500 nm (in the middle of the visible range). What is the temperature of this object?

66. Some of the hottest stars known have a blackbody temperature of 100,000 K.
    a. What is the peak wavelength of their radiation?
    b. Can this wavelength be observed from Earth's surface? Explain your answer.

67. Your body, at a temperature of about 37°C (98.6°F), emits radiation in the infrared region of the spectrum.
    a. What is the peak wavelength, in micrometers, of your emitted radiation?
    b. Assuming an exposed body-surface area of 0.25 m², how many watts of power do you radiate?

68. Assume that Earth and Mars are perfect spheres with radii of 6,378 km and 3,390 km, respectively.
    a. Calculate the surface area of Earth.
    b. Calculate the surface area of Mars.
    c. Seventy-two percent of Earth's surface is covered with water. Compare the amount of Earth's land area with the total surface area of Mars.

69. The object that created Arizona's Meteor Crater was estimated to have a radius of 25 meters and a mass of 300 million kg. Calculate the density of the impacting object, and explain what that may tell you about its composition. (Recall that the density of an object is its mass divided by its volume.)

70. Earth's mean radius is 6,378 km, and its mass is 5.97×10^{24} kg.
   a. Calculate Earth's average density. Show your work. Do not look up this value.
   b. The average density of Earth's crust is 2,600 kg/m³. What does this tell you about Earth's interior?

71. Assume that the east coast of South America and the west coast of Africa are separated by an average distance of 4,500 km. Assume also that GPS measurements indicate that these continents are now moving apart at a rate of 3.75 cm/year. If this rate has been constant over geological time, how long ago were these two continents joined together as part of a supercontinent?

 SmartWork, Norton's online homework system, includes algorithmically generated versions of these questions, plus additional conceptual exercises. If your instructor assigns questions in SmartWork, log in at **smartwork.wwnorton.com**.

 StudySpace is a free and open website that provides a Study Plan for each chapter of **Understanding Our Universe**. Study Plans include animations, reading outlines, vocabulary flashcards, and multiple-choice quizzes, plus links to premium content in SmartWork and the ebook. Visit **wwnorton.com/ studyspace**.

Exploration | Earth's Tides

wwnorton.com/studyspace

Open the Chapter 6 Interactive Simulation in StudySpace called "Tidal Bulge Simulator."

Before you start the simulation, examine the setup. You are looking down on Earth from the North Pole.

1. **Are the sizes of the Earth and Moon approximately to scale in this image?**

2. **Is the distance between them to scale in this image?**

3. **Are the tides shown to scale?**

4. **Explain why the authors of the simulation made the scaling choices they did.**

5. **In this position, is the east coast of North America experiencing high or low tide?**

Recall from Chapter 2 that Earth rotates counterclockwise when viewed from this vantage point. Click the box that says "Include Effects of Earth's Rotation."

6. **What happened to the tidal bulges?**

7. **Why does Earth's rotation have this effect?**

Click the box that says "Run."

8. **Over the course of one day, how many high tides does the east coast of North America experience?**

9. **How many low tides does it experience in one day?**

10. **Why are there two high tides? That is, what causes the tide on the side of Earth away from the Moon?**

11. **Do the tides change as the Moon orbits?**

Once the Moon has orbited back to the right side of the window, stop the simulation by unchecking "Run." Click the box that says "Include Sun."

12. **What happened to the tides when you added the Sun to the simulation?**

Now, run the simulation again. Stop the simulation when the Moon is at first quarter. Remove the check mark from the "Include Sun" box.

13. **What changed about the tides when you removed the Sun?**

Run the simulation until the Moon is full. Stop the simulation by unchecking "Run." Click the box that says "Include Sun."

14. **What changed about the tides when you added the Sun back in?**

15. **Which astronomical body dominates the tides on Earth, the Moon or the Sun?**

7 Atmospheres of Venus, Earth, and Mars

Earth's atmosphere surrounds us like an ocean of air. It is responsible for all of our weather, be it pleasant or stormy. Without our atmosphere, there would be neither clouds nor rain; nor would there be streams, lakes, or oceans. There would be no living creatures. Without an atmosphere, Earth would look something like the Moon, and we would not exist.

Among the five terrestrial bodies that we discussed in Chapter 6, only Venus and Earth have dense atmospheres. Mars has a very low-density atmosphere, and the atmospheres of Mercury and the Moon are so sparse that they can hardly be detected. Why should some of the terrestrial planets have dense atmospheres while others have little or essentially none? Are atmospheres created right along with the planets they envelop, or do they appear later? For answers to these questions, we need to look back nearly 5 billion years to a time when the planets were just completing their growth.

✧ **LEARNING GOALS**

A thick blanket of atmosphere warms and sustains Earth's temperate climate. The figure at right shows a photograph of an aurora, which originates from the interaction of particles from the Sun with the atmosphere, as indicated in the sketch. By the end of the chapter, you should be able to explain what conditions need to be present to create an aurora, and to sketch Earth's magnetic field and its interaction with the solar wind. You should also be able to:

- Diagram the layers of atmospheres, and explain the existence of these layers

- Compare the strength of the greenhouse effect on Earth, Venus, and Mars

- List the differences in the atmospheres of Earth, Venus, and Mars

- Describe how Earth's atmosphere has been reshaped by life

- Evaluate the evidence that shows Earth's climate is changing

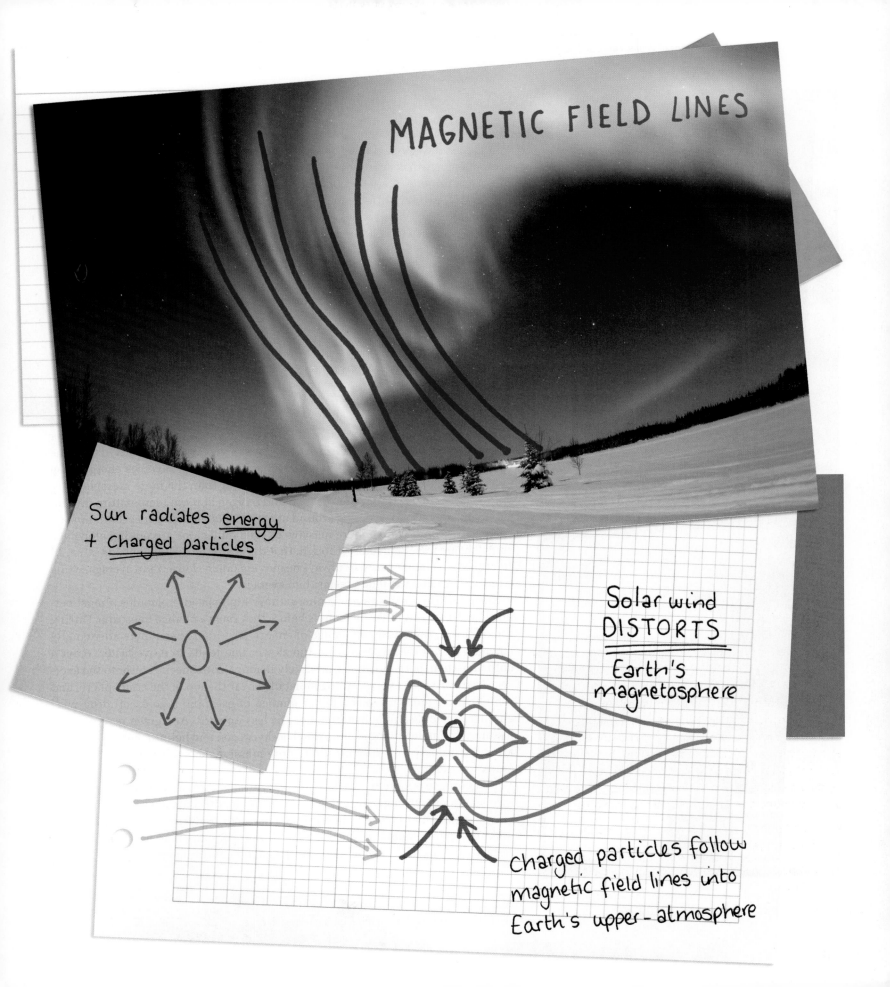

MAGNETIC FIELD LINES

Sun radiates energy
+ charged particles

Solar wind
DISTORTS

Earth's
magnetosphere

Charged particles follow
magnetic field lines into
Earth's upper-atmosphere

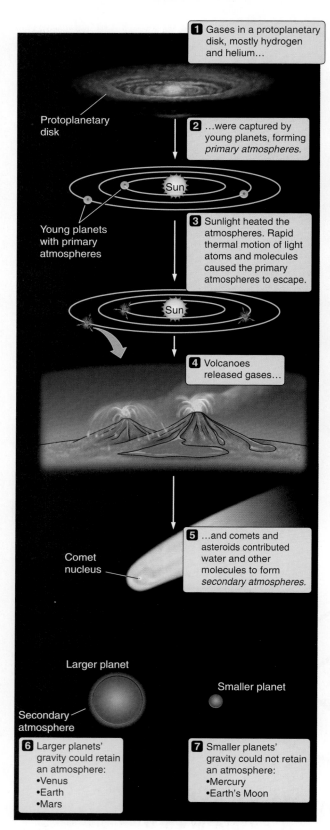

1 Gases in a protoplanetary disk, mostly hydrogen and helium…

Protoplanetary disk

2 …were captured by young planets, forming *primary atmospheres.*

Sun

Young planets with primary atmospheres

3 Sunlight heated the atmospheres. Rapid thermal motion of light atoms and molecules caused the primary atmospheres to escape.

Sun

4 Volcanoes released gases…

5 …and comets and asteroids contributed water and other molecules to form *secondary atmospheres.*

Comet nucleus

Larger planet

Smaller planet

Secondary atmosphere

6 Larger planets' gravity could retain an atmosphere:
•Venus
•Earth
•Mars

7 Smaller planets' gravity could not retain an atmosphere:
•Mercury
•Earth's Moon

FIGURE 7.1 Phases in the formation of the atmosphere of a terrestrial planet.

7.1 Atmospheres Are Oceans of Air

The terrestrial planets' atmospheres formed in phases, as illustrated in **Figure 7.1**. Young planets captured some of the residual hydrogen and helium that filled the protoplanetary disk surrounding the Sun. Gas capture continued until the supply of gas ran out. The gaseous atmosphere collected by a newly formed planet is called its **primary atmosphere**.

The terrestrial planets, with their small masses and therefore weak gravity, lacked the ability to hold on to light gases such as hydrogen and helium. When the supply of gas in the protoplanetary disk ran out, the planets' primary atmospheres began leaking back into space.[1] How can gas molecules escape from a planet? Objects—whether molecules or rockets—escape from planets when their vertical speed exceeds the escape velocity, as explained in Chapter 3. Given molecules of the same mass, hotter ones move faster than cooler ones. Intense radiation from the Sun, the primary source of thermal energy in the atmospheres of the terrestrial planets, raises the kinetic energy of atmospheric atoms and molecules enough that some may escape their planet's gravitational field.

Let's look more closely at how molecules move within a planetary atmosphere. The kinetic energy of a molecule (or any object) is determined by its mass and its speed. When a volume of air contains molecules with different masses, the *average* kinetic energy tends to be distributed equally among the different types. In other words, each type of molecule, from the lightest to the most massive, will have the same average kinetic energy. But if each type has the same average energy, then the less massive molecules must be moving faster than the more massive ones. For example, in a mixture of hydrogen and oxygen at room temperature, hydrogen molecules will be rushing around at about 2,000 meters per second (m/s) on average, while the much more massive oxygen molecules are poking along at a sluggish 500 m/s. Remember, though, that these are the *average* speeds. A small fraction of the molecules will always be moving much slower than average, while a few will be moving much faster than average.

Deep within a planet's atmosphere, these high-speed molecules almost certainly collide with other molecules before they have a chance to escape. During the collision, there is an exchange of energy. The high-speed molecule usually emerges with a lower speed, and the slower one tends to move faster. After a collision, then, both molecules are likely to move with speeds closer to the average. There are fewer surrounding molecules near the top of the atmosphere, and a high-speed molecule has a better chance of escaping before colliding with another molecule if it is heading more or less upward. At a given temperature, less massive molecules and atoms such as hydrogen and helium move faster, and are more quickly lost to space, than more massive molecules such as nitrogen or carbon dioxide.

Returning to our picture of the early Solar System, we can now see why the terrestrial planets, strongly heated by the Sun and having only a weak gravitational grasp, lost the hydrogen and helium they had acquired from the protoplanetary disk. The giant planets, on the other hand, were far more massive than their terrestrial cousins and were situated in the cooler environment of the outer Solar

---

[1] Not all atmospheric loss came from slow leakage. Impacts by large planetesimals may have blasted away substantial amounts of the terrestrial planets' primary atmospheres.

System. Stronger gravity and lower temperatures enabled them to retain nearly all of their massive primary atmospheres, which we explore further in Chapter 8.

Some Atmospheres Developed Later

Accretion, volcanism, and impacts are likely responsible for Earth's atmosphere today, which we call its **secondary atmosphere**. During the planetary accretion process, minerals containing water, carbon dioxide, and other volatile matter collected in Earth's interior. Later, as the interior heated up, these gases were released from the minerals that had held them. Volcanism then brought the gases to the surface, where they accumulated and created our secondary atmosphere.

Impacts by huge numbers of comets and asteroids may have been another important source of gases. As the giant planets of the outer Solar System grew to maturity, they perturbed the orbits of comets and asteroids. Some comets and asteroids were scattered into the inner parts of the Solar System, where they could strike the terrestrial planets, in the process bringing water, carbon monoxide, methane, and ammonia. Water from these impacts mixed with the water that had been released into the atmospheres by volcanism. On Earth, and perhaps on Mars as well, most of the water vapor then condensed as rain and flowed into the lower areas to form the earliest oceans.

Other molecules did not survive in their original form. Ultraviolet (UV) light from the Sun easily fragments molecules such as ammonia and methane. Ammonia, for example, is broken down into hydrogen and nitrogen. When this happens, the lighter hydrogen atoms quickly escape to space, leaving behind the much heavier nitrogen atoms. The nitrogen atoms then combine to form more massive nitrogen molecules, making these molecules even less likely to escape into space. Decomposition of ammonia by sunlight became the primary source of molecular nitrogen in the atmospheres of the terrestrial planets and on one of Saturn's moons, Titan. Molecular nitrogen now makes up most of the atmosphere on Earth and on Titan.

Mercury's relatively small mass and its proximity to the Sun caused it to lose nearly all of its secondary atmosphere to space, just as it had earlier lost its primary atmosphere. Even molecules as massive as carbon dioxide can escape from a small planet if the temperature is high enough. Furthermore, intense UV radiation from the Sun can break molecules into less massive fragments, which are lost to space even more quickly. Because the Moon is much farther from the Sun than Mercury is, the Moon is much cooler than Mercury. But the Moon's mass is so small that molecules can easily escape, even at low temperatures. Small mass and relative proximity to the Sun doomed both Mercury and the Moon to remain almost totally "airless."

7.2 A Tale of Three Planets—The Evolution of Secondary Atmospheres

Venus, Earth, and Mars are volcanically active today or have been volcanically active in their geological past, and all must have shared the intense cometary showers of the early Solar System. Their similar geological histories suggest that their early secondary atmospheres might also have been quite similar. Venus and Earth are similar in both mass and composition, and they have orbits that are less than 0.3 astronomical units (AU) apart (this is about one-third of Earth's

▶❚❚ **AstroTour:** Atmospheres: Formation and Escape

Primary atmospheres consist of captured gas.

Secondary atmospheres are a result of accretion, volcanism, and impacts.

Vocabulary Alert

Perturb: This word is uncommon in everyday usage, although the form *perturbed* is sometimes used to indicate that someone is upset. Astronomers use it to mean changes to an object's orbit, often caused by gravitational interactions.

average distance from the Sun). Mars is also similar in composition, but it has a mass only about a tenth that of Earth or Venus. The atmospheres of Venus and Mars today are nearly identical in composition: mostly carbon dioxide, with much smaller amounts of nitrogen.

Despite the similarities in atmospheric composition, Mars and Venus have vastly different *amounts* of atmosphere. The atmospheric pressure on the surface of Venus is nearly a hundred times greater than Earth's. By contrast, the average surface pressure on Mars is less than a hundredth of that on our own planet. Earth differs in another important respect in that, alone among these planets, its atmosphere is made up primarily of nitrogen and oxygen, with only a trace of carbon dioxide. Although these planets most likely started out with atmospheres of similar composition and comparable quantity, they ended up being very different from one another. Why did they evolve so differently?

How Mass Affects a Planet's Atmosphere

The major distinction between Venus and Mars cannot be explained in terms of planetary mass alone. Venus is nearly eight times as massive as Mars, so we can assume that it probably had about eight times as much carbon within its interior to produce carbon dioxide, the principal secondary-atmosphere component of both planets. Even allowing for the differences in planetary mass, however, Venus today has over 2,500 times more atmospheric mass than Mars. Why such a large difference? We can find the answer by considering the relative strengths of their surface gravity, which involves both the mass and radius of a planet. Venus has the gravitational pull necessary to hang on to its atmosphere; Mars has less gravitational attraction to keep its atmosphere. Furthermore, when a planet such as Mars loses so much of its atmosphere to space, the process begins to take on a *runaway* behavior. With less atmosphere, there are fewer intervening molecules to keep breakaway molecules from escaping, and the rate of escape increases. This process in turn leads to even less atmosphere and still greater escape rates.

> **Stronger surface gravity caused Venus to retain its atmosphere more effectively than Mars.**

The Atmospheric Greenhouse Effect

Differences in the present-day masses of the atmospheres of Venus, Earth, and Mars have a large effect on their surface temperatures, and thus on the evolution of their atmospheres. In **Working It Out 7.1**, we calculate the temperature of a planet by finding the equilibrium between the amount of energy it radiates and the amount of energy it receives. We find that this gives a good result for planets without atmospheres; but for Earth, and especially Venus, the calculations are far from measured values. This is excellent news—it means there is something new to learn. In this case the "something" is the **atmospheric greenhouse effect**, which traps solar radiation.

The atmospheric greenhouse effect in planetary atmospheres and the **conventional greenhouse effect** operate in different ways, although the end results are much the same. Planetary atmospheres and the interiors of greenhouses are both heated by trapping the Sun's energy, but here the similarities end. The conventional greenhouse effect is what happens in a car on a sunny day when you leave the windows closed. Sunlight pours through the car's windows, heating the interior and raising the internal air temperature. With the windows closed, hot air is trapped and temperatures can climb as high as 180°F (82°C).

Working It Out 7.1 | How Can We Find the Temperature of a Planet?

The temperature of a planet is determined by a balance between the amount of sunlight being absorbed and the amount of energy being radiated back into space. If the planet's temperature remains constant, then it must radiate as much energy as it absorbs, just as the bucket in **Figure 7.2** fills and empties at the same rate, keeping the water level constant. We now have the tools we need to turn this qualitative idea into a real prediction of the temperatures of the planets.

Begin with the amount of sunlight being absorbed. From the Sun's perspective, a planet looks like a circular disk with a radius equal to the radius of the planet, R. The area of the planet that is lit by the Sun is

$$(\text{Absorbing area of planet}) = \pi R^2$$

The amount of energy striking a planet also depends on the intensity of sunlight at the distance at which the planet orbits. Imagine a light bulb on a stage in a large theater. If you are standing on the stage with the light bulb, the light is intense—every part of your body receives a lot of energy each second from the bulb. At the back row of the theater, the light is less intense: at this greater distance, the bulb's energy is spread out over a much larger area. The intensity falls off as the area of a sphere, $4\pi d^2$,

where d is the distance from you to the bulb. Similarly, the intensity of sunlight at a planet's orbit is the luminosity of the Sun, L, divided by $4\pi d^2$.

$$(\text{Intensity of sunlight}) = \frac{L}{4\pi d^2}$$

A planet does not absorb all the sunlight that falls on it. **Albedo**, a, is the fraction of light that *reflects* from a planet. The fraction of the sunlight that is *absorbed* by the planet is 1 minus the albedo. A planet covered entirely in snow would have a high albedo, close to 1, while a planet covered entirely by black rocks would have a low albedo, close to 0.

$$(\text{Fraction of sunlight absorbed}) = 1 - a$$

We can now calculate the energy absorbed by the planet each second. Writing this relationship as an equation, we say that

$$\begin{pmatrix} \text{Energy} \\ \text{absorbed} \\ \text{each second} \end{pmatrix} = \begin{pmatrix} \text{Absorbing} \\ \text{area of} \\ \text{planet} \end{pmatrix} \times \begin{pmatrix} \text{Intensity} \\ \text{of} \\ \text{sunlight} \end{pmatrix} \times \begin{pmatrix} \text{Fraction of} \\ \text{sunlight} \\ \text{absorbed} \end{pmatrix}$$

$$= \pi R^2 \times \frac{L}{4\pi d^2} \times (1 - a)$$

The amount of energy that the planet radiates away depends on the surface area of the planet, $4\pi R^2$. The Stefan-Boltzmann law tells us that each square meter radiates energy equal to σT^4 every second. Putting these two pieces together gives

$$\begin{pmatrix} \text{Energy} \\ \text{radiated} \\ \text{each second} \end{pmatrix} = \begin{pmatrix} \text{Surface} \\ \text{area of} \\ \text{planet} \end{pmatrix} \times \begin{pmatrix} \text{Energy radiated} \\ \text{per square meter} \\ \text{per second} \end{pmatrix}$$

$$= 4\pi R^2 \times \sigma T^4$$

For a planet at constant temperature, each second the "Energy radiated" must be equal to "Energy absorbed." When we set these two quantities equal to each other, we arrive at the expression

$$\begin{pmatrix} \text{Energy radiated} \\ \text{each second} \end{pmatrix} = \begin{pmatrix} \text{Energy absorbed} \\ \text{each second} \end{pmatrix}$$

$$4\pi R^2 \, \sigma T^4 = \pi R^2 \frac{L}{4\pi d^2} (1 - a)$$

or

$$T^4 = \frac{L}{16\sigma\pi d^2} (1 - a)$$

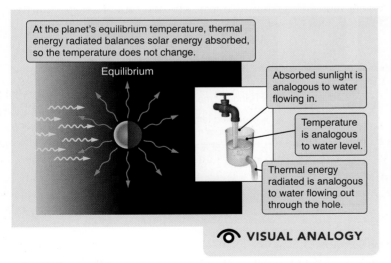

At the planet's equilibrium temperature, thermal energy radiated balances solar energy absorbed, so the temperature does not change.

Equilibrium

Absorbed sunlight is analogous to water flowing in.

Temperature is analogous to water level.

Thermal energy radiated is analogous to water flowing out through the hole.

VISUAL ANALOGY

FIGURE 7.2 Planets are heated by absorbing sunlight (and also sometimes by internal heat sources) and cooled by emitting thermal radiation into space. If there are no other sources of heating or means of cooling, then the equilibrium between these two processes determines the temperature of the planet.

CONTINUED

How Can We Find the Temperature of a Planet? CONTINUED

This gives us a T^4, but we want just T. If we take the fourth root of each side, we get

$$T = \left(\frac{L(1-a)}{16\sigma\pi d^2} \right)^{1/4}$$

Fortunately, L, σ, and π are known. If we express the distance from the Sun in AU, the equation simplifies to:

$$T = 279 \text{ K} \times \left(\frac{1-a}{d^2_{\text{AU}}} \right)^{1/4}$$

For a blackbody ($a = 0$) at 1 AU from the Sun, the temperature is 279 K. For Earth, with an albedo of 0.306, and a distance from the Sun of 1 AU, the temperature is

$$T = 279 \text{ K} \times \left(\frac{1-0.306}{1^2} \right)^{1/4}$$

$$T = 255 \text{ K}$$

Calculator hint: To take the fourth root of a number, you can either use the x^y (or sometimes y^x) button on your calculator and put in 0.25 for the exponent, or you may have a button labeled $x^{1/y}$, which allows you to put in 4 for the y. For example, if you are calculating $3^{1/4}$, you can either type [3][x^y][0][.][2][5] or [3][$x^{1/y}$][4].

Figure 7.3 plots the predicted and actual temperatures of the planets of the Solar System. The vertical bars show the range of temperatures measured for the surface of each planet (or, for the giant planets, at the tops of their clouds). The large black dots show our predictions using the equation above. Overall, there is fairly good agreement, so our basic understanding of *why* planets have the temperatures they do is probably close to the mark. The data and predictions for Mercury and Mars agree particularly well.

For Earth and the giant planets, the actual temperatures are a bit higher than the predicted temperatures. For Venus, the actual temperature is much higher than our prediction. As we built our mathematical model for the equilibrium temperatures of planets, we made a number of assumptions. For example, we assumed that the temperature of the planet was the same everywhere. This is clearly not true; for example, we might expect planets to be hotter on the day side than on the night side. We also assumed that a planet's only source of energy is the sunlight falling on it. Finally, we assumed that a planet is able to radiate energy into space freely as a blackbody.

The discrepancies between our model and the measured temperatures tell us that for some of these planets, some or all of these assumptions must be incorrect. The question of *why* these planets are hotter than expected leads us to a number of new and interesting insights. Scientific ideas sometimes succeed and sometimes fail, but even when they fail they can teach us a lot about the universe.

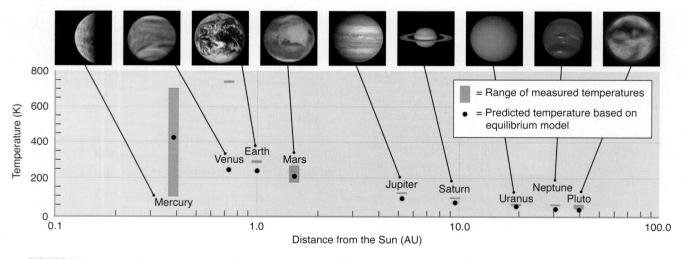

FIGURE 7.3 Predicted temperatures for the major planets and Pluto, based on the equilibrium between absorbed sunlight and thermal radiation into space, are compared with ranges of observed surface temperatures. Some predictions are correct. Interestingly, others are not.

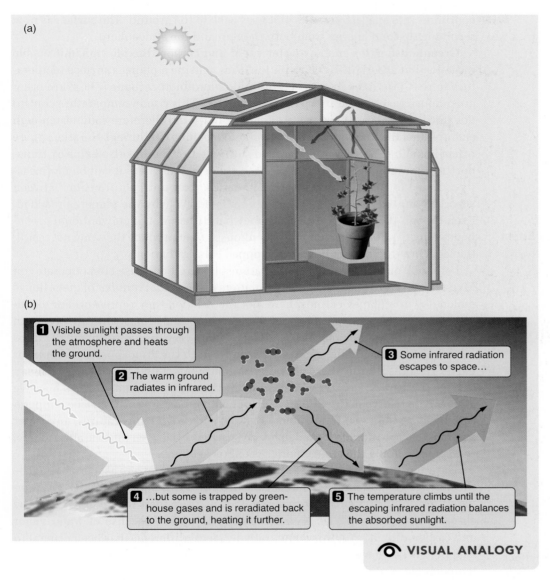

(a)

(b)

1 Visible sunlight passes through the atmosphere and heats the ground.

2 The warm ground radiates in infrared.

3 Some infrared radiation escapes to space...

4 ...but some is trapped by greenhouse gases and is reradiated back to the ground, heating it further.

5 The temperature climbs until the escaping infrared radiation balances the absorbed sunlight.

👁 **VISUAL ANALOGY**

FIGURE 7.4 (a) In an actual greenhouse, infrared radiation is trapped by glass, which is transparent to visible wavelengths but opaque to infrared. The trapped infrared radiation causes the temperature of the greenhouse to rise to a new equilibrium temperature. (b) In the atmospheric greenhouse effect, greenhouse gases such as water and carbon dioxide absorb infrared radiation and reradiate it in all directions, slowing the transport of energy out of the atmosphere. This causes the temperature of the planet to rise to a new equilibrium.

The atmospheric greenhouse effect is illustrated in **Figure 7.4**. Atmospheric gases such as nitrogen, oxygen, carbon dioxide, and water vapor freely transmit visible light, allowing the Sun to warm the planet's surface. The warmed surface now radiates the excess energy in the infrared. But carbon dioxide and water vapor—among other kinds of atmospheric molecules—strongly absorb infrared radiation, converting it to thermal energy. These same molecules subsequently reradiate this thermal energy in all directions—some of the thermal energy

▶❙❙ **AstroTour:** **Greenhouse Effect**

Without the greenhouse effect, Earth would freeze.

continues into space, but much of it goes back to the ground. The surface is now receiving thermal energy from both the Sun and the atmosphere.

Greenhouse gases such as water vapor and carbon dioxide transmit visible radiation but absorb infrared radiation, thus causing a planet's surface temperature to rise. (Methane, nitrous oxide, and chlorofluorocarbons [CFCs] are other greenhouse gases found in Earth's atmosphere.) This rise in temperature continues until the surface becomes sufficiently hot—and therefore radiates enough energy—that the fraction of infrared radiation leaking out through the atmosphere is just enough to balance the absorbed sunlight. Convection also helps by transporting thermal energy to the top of the atmosphere, where it can be more easily radiated to space. In short, the temperature rises until equilibrium between absorbed sunlight and thermal energy radiated away by the planet is reached. Even though the mechanisms are somewhat different, the conventional greenhouse effect and the atmospheric greenhouse effect produce the same net result: the local environment is heated by trapped solar radiation.

Let's look more closely at how the atmospheric greenhouse effect operates on Mars, Earth, and Venus. What really matters is the actual *number* of greenhouse molecules in a planet's atmosphere, not the *fraction* they represent. For example, even though the atmosphere of Mars is composed almost entirely of carbon dioxide—an effective greenhouse molecule—its very thin atmosphere contains relatively few greenhouse molecules compared to the atmospheres of Venus or Earth. As a result, the atmospheric greenhouse effect on Mars raises the mean surface temperature by only about 5 K. At the other extreme, Venus's massive atmosphere of carbon dioxide and sulfur compounds raises its surface temperature by more than 400 K, to about 737 K. At such high temperatures, any water and most carbon dioxide locked up in surface rocks is driven into the atmosphere, further enhancing the atmospheric greenhouse effect.

The atmospheric greenhouse effect on Earth is not as severe as it is on Venus—the average global temperature near Earth's surface is about 288 K (15°C). Temperatures on Earth's surface are about 33 K warmer than they would be in the absence of an atmospheric greenhouse effect, mainly because of water vapor and carbon dioxide. Yet this comparatively small difference has been crucial in shaping the Earth we know. Without the greenhouse effect, Earth's average global temperature would be well below the freezing point of water, leaving us with a world of frozen oceans and ice-covered continents.

How has the atmospheric greenhouse effect made the composition of Earth's atmosphere so different from the high-carbon-dioxide atmospheres of Venus and Mars? We may find the answer in Earth's particular location in the Solar System. Consider early Earth and early Venus, each having about the same mass, but with Venus orbiting somewhat closer to the Sun than Earth does. Volcanism poured out large amounts of carbon dioxide and water vapor to form early secondary atmospheres on both planets. Most of Earth's water quickly rained out of the atmosphere to fill vast ocean basins. But Venus was closer to the Sun, and its surface temperatures were higher than Earth's. Most of the rainwater on Venus immediately re-evaporated, much as it does today in Earth's desert regions. Venus was left with a planet-wide surface containing very little liquid water and an atmosphere filled with water vapor. The continuing buildup of both water vapor and carbon dioxide in Venus's atmosphere led to a runaway atmospheric greenhouse effect that drove up the surface temperature of the planet. Ultimately, the surface of Venus became so hot that no liquid water could exist there.

The early difference between a watery Earth and an arid Venus changed the ways that their atmospheres and surfaces evolved. On Earth, water erosion caused by rain and rivers continually exposed fresh minerals, which then reacted chemically with atmospheric carbon dioxide to form solid carbonates. This reaction removed some of the atmospheric carbon dioxide, burying it within Earth's crust as a component of a rock called limestone. Later, the development of life in Earth's oceans accelerated the removal of atmospheric carbon dioxide. Tiny sea creatures built their protective shells of carbonates, and as they died they built up massive beds of limestone on the ocean floors. As a result of water erosion and the chemistry of life, all but a trace of Earth's total inventory of carbon dioxide is now tied up in limestone beds. Earth's particular location in the Solar System seems to have spared it from the runaway atmospheric greenhouse effect. If all the carbon dioxide now in limestone beds had not been locked up by these reactions, Earth's atmospheric composition would resemble that of Venus or Mars.

Differences in the amounts of water on Venus, Earth, and Mars are not so well understood. Geological evidence indicates that liquid water was once plentiful on the surface of Mars, and the *Mars Odyssey* spacecraft has found evidence that significant amounts of water still exist in the form of subsurface ice. Earth's liquid and solid water supply is even greater: about 10^{21} kilograms (kg), or 0.02 percent of its total mass. More than 97 percent of Earth's water is in the oceans, which have an average depth of about 4 kilometers (km). Earth today has 100,000 times more water than Venus. What happened to all the water on Venus? One possibility is that water molecules high in its atmosphere were broken apart into hydrogen and oxygen by solar UV radiation. Hydrogen atoms, being of very low mass, were quickly lost to space. Oxygen, however, would eventually have migrated downward to the planet's surface, where it would have been removed from the atmosphere by bonding with surface minerals.

7.3 Earth's Atmosphere— The One We Know Best

Now that we understand some of the overall processes that have influenced the evolution of the terrestrial planet atmospheres, we will look in depth at each of them. We begin with Earth's atmosphere, not only because we know it best but also because it provides an introduction to atmospheric structure and weather phenomena that will help us better understand the atmospheres of Venus and Mars, and even those of Titan and the giant planets.

We can describe Earth's atmosphere as a blanket of gas that is several hundred kilometers deep and has a total mass of approximately 5×10^{18} kg (about 5,000 trillion metric tons), which is less than one-millionth of Earth's total mass. Earth's atmosphere presses with 100,000 newtons on each square meter of our planet's surface. We express this amount of pressure with a unit called a **bar**: Earth's average atmospheric pressure at sea level is approximately 1 bar. This is equivalent to the increase in pressure we might experience underwater at a depth of 10 meters (33 feet). We are largely unaware of Earth's atmospheric pressure, because the same pressure exists both within and outside of our bodies. The two precisely balance one another. As we learned in Chapter 5, the pressure at any point within a star must be great enough to balance the weight of the overlying layers. The same principle holds true in a planetary atmosphere: the atmospheric pressure on a planet's surface must be great enough to hold up the weight of the overlying atmosphere.

Vocabulary Alert

Bar: This common word has many everyday meanings, but scientists often use the bar as a unit of pressure equal to the average atmospheric pressure at sea level on Earth. A millibar (mb) is one thousandth of a bar.

Life Controls the Composition of Earth's Atmosphere

Earth's atmosphere is relatively uniform on a global scale. Two principal gases make up our atmosphere: about four-fifths of it is nitrogen, and one-fifth is oxygen. There are also many important minor constituents such as water vapor and carbon dioxide, the amounts of which vary somewhat depending on global location and season.

The composition of Earth's atmosphere is very different from that of Venus and Mars. We have already discussed the differences in carbon dioxide content, but what truly sets Earth apart from all other known planets is its oxygen. Earth's atmosphere contains abundant amounts of oxygen, while other planets' atmospheres do not. Why should this be so? Oxygen is a highly reactive gas. It chemically combines with, or oxidizes, almost any material it touches. The rust (iron oxide) that forms on steel is an example. A planet that has significant amounts of oxygen in its atmosphere requires a means of replacing what is lost through oxidation. On Earth, plants perform this task.

The oxygen concentration in Earth's atmosphere has changed over the history of the planet, as shown in **Figure 7.5**. When Earth's secondary atmosphere first appeared about 4 billion years ago, it was almost totally free of oxygen. About 2.8 billion years ago, an ancestral form of cyanobacteria (single-celled organisms that contain chlorophyll, which enables them to obtain energy from sunlight) began releasing oxygen into Earth's atmosphere as a waste product of its own metabolism. At first, this oxygen combined readily with exposed metals and minerals in surface rocks and soils and so was removed from the atmosphere as quickly as it formed. Ultimately, the explosive growth of plant life accelerated the production of oxygen, building up atmospheric concentrations that approached today's levels only about 250 million years ago. All true plants, from tiny green algae to giant redwoods, use the energy of sunlight to build carbon compounds out of carbon dioxide and produce oxygen as a waste product in the process called photosynthesis. Earth's atmospheric oxygen content is held in a delicate balance primarily by plants. If plant life disappeared, so would nearly all of Earth's atmospheric oxygen, and therefore all animal life—including humans.

> Life is responsible for the oxygen in Earth's atmosphere.

FIGURE 7.5 The amount of oxygen in Earth's atmosphere has built up over time as a result of plant life on the planet.

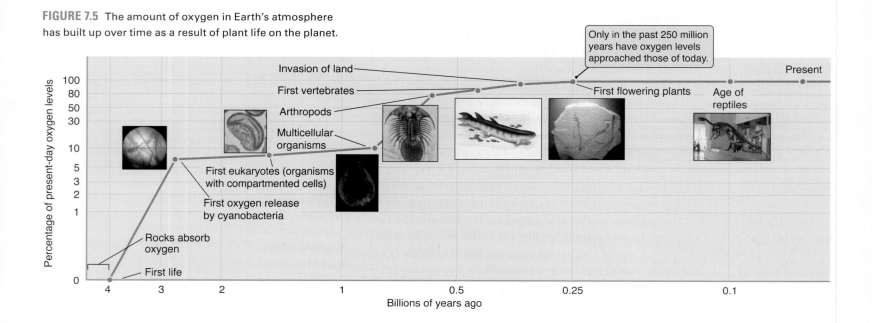

Earth's Atmosphere Is Layered Like an Onion

Our atmosphere is made up of several distinct layers. Earth's lowermost atmospheric layer, the one in which we live and breathe, is called the **troposphere** (**Figure 7.6a**). It contains 90 percent of Earth's atmospheric mass and is the

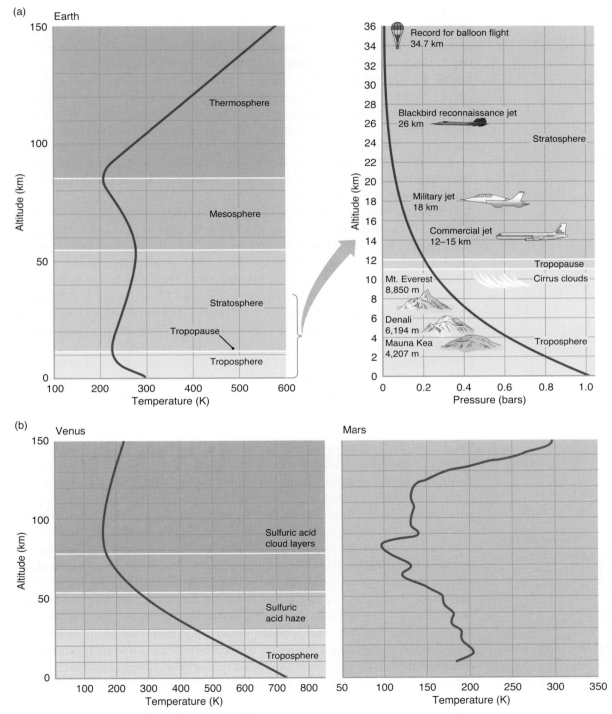

FIGURE 7.6 (a) Temperature and pressure are plotted for Earth's atmospheric layers as a function of height. (b) Atmospheric temperatures for Venus and Mars are shown for comparison.

source of all of our weather. Within the troposphere, atmospheric pressure, density, and temperature all decrease with increasing altitude. Convection circulates air between the lower and upper levels, and such circulation tends to diminish the temperature extremes caused by heating at the bottom and cooling at the top.

Convection also affects the vertical distribution of atmospheric water vapor. The ability of air to hold water in the form of vapor depends very strongly on the air temperature: the warmer the air, the more water vapor it can hold. As air is convected upward, it cools. When the air temperature decreases to the point at which the air can no longer hold all of its water vapor, water begins to condense to tiny droplets or ice crystals. In large numbers these become visible to us as clouds. When these droplets combine to form large drops, they fall as rain or snow. For this reason, most of the water vapor in Earth's atmosphere stays within 2 km of the surface. At an altitude of 4 km, the Mauna Kea Observatories are higher than approximately one-third of Earth's atmosphere, but they lie above nine-tenths of the atmospheric water vapor. This is important for astronomers who observe the sky in the infrared region of the spectrum, because water vapor strongly absorbs infrared light.

Above the troposphere and extending upward to an altitude of 50 km above sea level is the **stratosphere**. This is a region in which little convection takes place, because the temperature-altitude relationship reverses and the temperature begins to *increase* with altitude. This reversal is caused by ozone, which warms the stratosphere by absorbing UV radiation from the Sun. The ozone layer protects terrestrial life from these high-energy UV photons. The region above the stratosphere is called the **mesosphere**. It extends from an altitude of 50 km to about 90 km. In the mesosphere there is no ozone to absorb sunlight, so temperatures once again decrease with altitude. The base of the stratosphere and the upper boundary of the mesosphere are two of the coldest levels in Earth's atmosphere.

At altitudes above 90 km, solar UV radiation and the high-energy particles of the **solar wind** ionize atoms and molecules, once again causing the temperature to increase with altitude. This region is called the **thermosphere**, and it is the hottest part of the atmosphere. Near the top of the thermosphere, at an altitude of 600 km, the temperature can reach 1000 K. The gases within and beyond the thermosphere are ionized by ultraviolet photons and high-energy particles from the Sun to form a **plasma** layer known as the **ionosphere**. Some frequencies of radio waves reflect off this layer back to Earth's surface, enabling amateur radio operators to communicate with each other around the world.

Even farther out is a large region filled with charged particles from the Sun that have been captured by Earth's magnetic field. This is Earth's **magnetosphere**. It has a radius approximately 10 times the radius of Earth, and a volume over 1,000 times the volume of the planet. Magnetic fields have no effect on charged particles unless the particles are moving. Charged particles are free to move *along* the direction of the magnetic field, but if they try to move *across* the direction of the field they experience a force that is perpendicular both to their motion and to the magnetic-field direction. This force causes them to loop around the direction of the magnetic field, as illustrated in **Figure 7.7a**.

If the field is pinched together at some point, particles moving into the pinch will feel a magnetic force that reflects them back along the direction they came from. If charged particles are located in a region where the field is pinched on

Vocabulary Alert

Solar wind: In common language, *wind* means a tangible flow of air caused by temperature differences between one place and another. You can feel wind with your skin, and sometimes wind can even blow down trees or rip the roofs off of houses. The *solar wind* is much more tenuous than ordinary wind and is not detectable in this way. However, in the sense that it is a large number of high-energy charged particles all moving more or less in the same direction, it is analogous to atmospheric winds.

Plasma: In common usage, this word almost always refers to a component of blood (or a type of television). But it has another scientific meaning, referring to a gas in which most of the atoms and molecules have lost some electrons, leaving them with a net positive charge. Because these particles are charged, they interact with electric and magnetic fields.

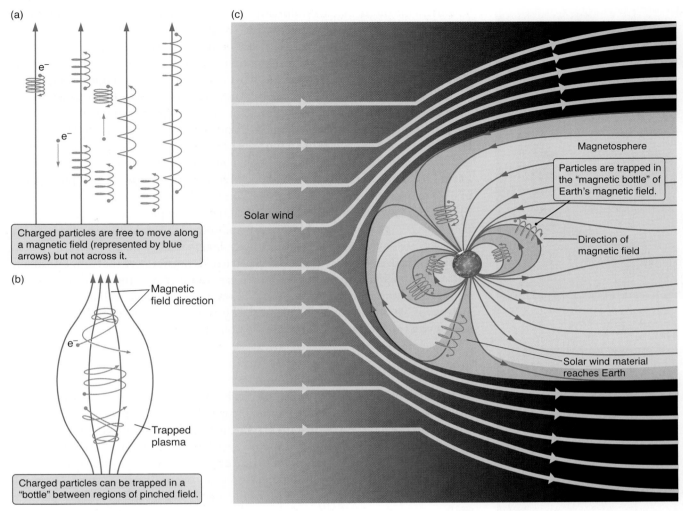

(a)

e⁻

e⁻

Charged particles are free to move along a magnetic field (represented by blue arrows) but not across it.

(b)

Magnetic field direction

e⁻

Trapped plasma

Charged particles can be trapped in a "bottle" between regions of pinched field.

(c)

Solar wind

Magnetosphere

Particles are trapped in the "magnetic bottle" of Earth's magnetic field.

Direction of magnetic field

Solar wind material reaches Earth

FIGURE 7.7 (a) The motion of charged particles, in this case electrons, in a uniform magnetic field. (b) When the field is pinched, charged particles can be trapped in a "magnetic bottle." (c) Earth's magnetic field acts like a bundle of magnetic bottles, trapping particles in Earth's magnetosphere. In all these images, the radius of the helix the charged particle follows is greatly exaggerated.

both ends, as shown in **Figure 7.7b**, then they may bounce back and forth many times. This is called a magnetic bottle. Earth's magnetic field is pinched together at the two magnetic poles and spreads out around the planet, creating a gigantic magnetic bottle.

Earth and its magnetic field are immersed in the solar wind. When these particles first encounter Earth's magnetic field, the smooth flow is interrupted and their speed suddenly drops—they are diverted by Earth's magnetic field like a river is diverted around a boulder. As they flow past, some of these charged particles become trapped by Earth's magnetic field, where they bounce back and forth between Earth's magnetic poles as illustrated in **Figure 7.7c**.

An understanding of Earth's magnetosphere is of great practical importance. Some regions in the magnetosphere contain especially strong concentrations of energetic charged particles, which can be very damaging to both electronic equipment and astronauts. Disturbances in Earth's magnetosphere can cause changes in Earth's magnetic field that are large enough to trip power grids, cause blackouts, and wreak havoc with communications.

Earth's magnetic field traps charged particles from the Sun.

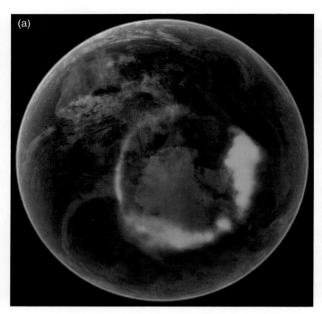

(a)

G X U V I R

(b)

G X U V I R

FIGURE 7.8 Auroras result when particles trapped in Earth's magnetosphere collide with molecules in the upper atmosphere. (a) An auroral ring around Earth's south magnetic pole, as seen from space. (b) Aurora borealis—the "northern lights"—viewed from the ground.

Nonuniform solar heating creates our weather.

Earth's magnetic field also funnels energetic charged particles down into the ionosphere in two rings located around the magnetic poles. These charged particles (mostly electrons) collide with atoms such as oxygen, nitrogen, and hydrogen in the upper atmosphere, causing them to glow like the gas in a neon sign. These glowing rings, called **auroras**, can be seen from space (**Figure 7.8a**). When viewed from the ground (**Figure 7.8b**), they look like eerie, shifting curtains of multicolored light. People living far from the equator are often treated to spectacular displays of the aurora borealis ("northern lights") in the Northern Hemisphere or the aurora australis in the Southern Hemisphere. When the solar wind is particularly strong, auroras can be seen at lower latitudes far from their usual zone. Look back at the chapter-opening figure, which shows both a photograph of an aurora and a handmade sketch. You should now be able to describe how this aurora formed, and sketch the interaction of the particles and the magnetic field that gave rise to them. Auroras are not unique to Earth. They have also been observed on Venus, Mars, all of the giant planets, and some moons.

The general structure we have described here is not unique to Earth's atmosphere either. The major components—troposphere, stratosphere, and ionosphere—also exist in the atmospheres of Venus and Mars (Figure 7.6b) as well as in the atmospheres of Titan and the giant planets. The magnetospheres of the giant planets are among the largest structures in the Solar System.

Why the Winds Blow

Winds are the natural movement of air, both locally and on a global scale, in response to variations in temperature from place to place. The air is usually warmer in the daytime than at night, warmer in the summer than in winter, and warmer at the equator than in the polar regions. Large bodies of water, such as oceans, also affect atmospheric temperatures. Heating a gas increases its pressure, which in turn causes it to push into its surroundings. These pressure differences cause winds. The strength of the winds is governed by the magnitude of the temperature difference from place to place.

As air in Earth's equatorial regions is heated by the warm surface, it begins to rise because of convection. The warmed surface air displaces the air above it, which then has no place to go but toward the poles. This air becomes cooler and denser as it moves toward the poles, and so it sinks back down through the atmosphere. It displaces the surface polar air, which is forced back toward the equator, completing the circulation. As a result, the equatorial regions remain cooler and the polar regions remain warmer than they otherwise would be. This planet-wide flow of air between the equator and the poles is called **Hadley circulation (Figure 7.9a)**. Other factors break up the planet-wide flow into a series of smaller Hadley cells. Planet rotation is a major factor here. Most planets and their atmospheres are rotating rapidly, and the effects produced by this rotation strongly interfere with Hadley circulation by redirecting the horizontal flow (**Figure 7.9b**).

The effect of Earth's rotation on winds—and on the motion of any object—is called the **Coriolis effect**, an effect of relative motion. This occurs because as the Earth rotates, it carries objects around with it. Objects near the equator move in the direction of rotation (toward the east) faster than objects closer to the poles do. If you were to fire a gun directly north from a point in the Northern Hemisphere, its path would appear to curve to the east. This is because the bullet is already traveling to the east faster than other objects farther north—including the

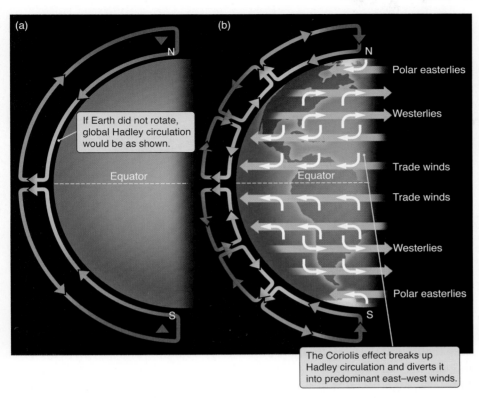

FIGURE 7.9 (a) The classic Hadley circulation. (b) Hadley flow often breaks up into smaller circulation cells. The poleward-equatorward flow is diverted into zonal flow by the Coriolis effect.

ground (**Figure 7.10**). The opposite is also true. If you fire from a northern point toward the equator, the bullet will appear to curve toward the west, because it is traveling east more slowly than the ground at the equator. This effect is important in understanding atmospheric circulation: when a volume of air starts to move directly toward or away from the poles, the Coriolis effect diverts it in a direction that is more or less parallel to the planet's equator.

This change in motion creates winds that blow predominantly in an east–west direction (see Figure 7.9b). Meteorologists call these **zonal winds**. More rapid rotations produce a stronger Coriolis effect and stronger zonal winds. Zonal winds are often confined to relatively narrow bands of latitude. Between the equator and the poles in most planetary atmospheres, the zonal winds alternate between easterlies (those blowing *from the east* and toward the west) and westerlies (those blowing *from the west* and toward the east). This unfortunate terminology is a historical carryover from early terrestrial meteorology, in which winds are labeled not by the direction they are blowing toward but by the direction they come from.

In Earth's atmosphere, several bands of alternating zonal winds lie between the equator and each hemisphere's pole. This zonal pattern is called Earth's **global circulation**. The best-known zonal currents are the subtropical trade winds—more or less easterly winds that once carried sailing ships from Europe westward to the Americas—and the midlatitude prevailing westerlies that carried them home again (see Figure 7.9b).

Nonuniform heating, together with the Coriolis effect, causes east–west zonal winds.

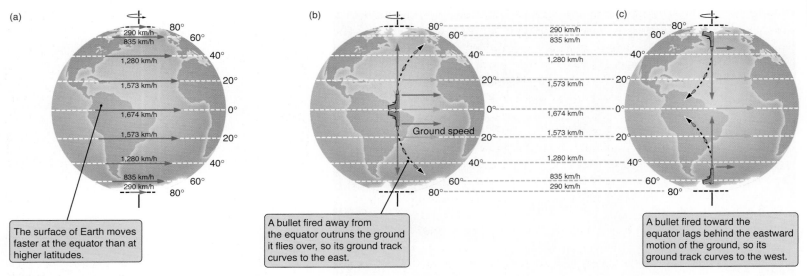

The surface of Earth moves faster at the equator than at higher latitudes.

A bullet fired away from the equator outruns the ground it flies over, so its ground track curves to the east.

A bullet fired toward the equator lags behind the eastward motion of the ground, so its ground track curves to the west.

FIGURE 7.10 The Coriolis effect causes objects to appear to be deflected as they move across the surface of Earth.

Embedded within Earth's global circulation pattern are systems of winds associated with large high- and low-pressure regions. A combination of a low-pressure region and the Coriolis effect produces a circulating pattern called **cyclonic motion**. Cyclonic motion is associated with stormy weather, including hurricanes. Similarly, high-pressure systems are localized regions where the air pressure is higher than average. We can think of these regions of greater-than-average air concentration as "mountains" of air. Owing to the Coriolis effect, high-pressure regions rotate in a direction opposite to that of low-pressure regions. These high-pressure circulating systems experience **anticyclonic motion** and are generally associated with fair weather.

7.4 Venus Has a Hot, Dense Atmosphere

Venus and Earth are similar in many ways—so similar that they might be thought of as sister planets. Indeed, when we used the laws of radiation to predict temperatures for the two planets, we concluded that they should be very close. But that was before we considered the greenhouse effect and the role of carbon dioxide in blocking the infrared radiation that a planetary surface typically emits. Venus's atmosphere is 96 percent carbon dioxide, with a small amount (3.5 percent) of nitrogen and still lesser amounts of other gases. This thick blanket of carbon dioxide traps the infrared radiation from Venus, driving the temperature at the surface of the planet to a sizzling 737 K, which is hot enough to melt lead.

Whereas Earth is a lush paradise, the runaway greenhouse effect has turned Venus into a convincing likeness of hell—an analogy made complete by the presence of choking amounts of sulfurous gases. Large variations in the observed amounts of sulfurous compounds in the high atmosphere of Venus suggest to planetary scientists that the source of sulfur may be sporadic episodes of volcanic activity. Additionally, the atmospheric pressure at the surface of Venus is 92 times greater than in your classroom—equal to the water pressure at an ocean depth of 900 meters on Earth, which would crush the hull of a World War II–era submarine. Venus may be our sister planet in many respects, but it's unlikely humans will ever visit its surface.

With its massive carbon dioxide atmosphere, Venus is a "poster child" for the greenhouse effect.

As on Earth, the atmospheric temperature of Venus decreases continuously throughout the planet's troposphere, dropping to a low of about 160 K at the tropopause. At an altitude of approximately 50 km, Venus's atmosphere has an average temperature and pressure similar to our own atmosphere at sea level. At altitudes between 50 and 80 km (see Figure 7.6b), the atmosphere is cool enough for sulfurous oxide vapors to react with water vapor to form clouds of concentrated sulfuric acid droplets (H_2SO_4). These dense clouds completely block our view of the surface of Venus, as **Figure 7.11** shows. In the 1960s, spacecraft with cloud-penetrating radar provided low-resolution views of the surface of Venus. But it was not until 1975, when the Soviet Union succeeded in landing cameras there, that people got a clear picture of the surface. Radar images taken by the *Magellan* spacecraft in the early 1990s (see Figure 6.27) produced a global map of the surface of Venus.

Unlike the other planets, Venus rotates on its axis in a direction opposite to its motion around the Sun. Relative to the stars, Venus spins once every 243 Earth days; but a solar day on Venus—the time it takes for the Sun to return to the same place in the sky—is only 117 Earth days. Venus has no seasons, because its axis of rotation is nearly perpendicular to the orbital plane. Its extremely slow rotation means that Coriolis effects on its atmosphere are small, resulting in a global circulation that is quite close to a classic Hadley pattern (see Figure 7.9a). Venus is the only planet known to behave in this way.

Imagine yourself standing (and surviving) on the surface of Venus. Since sunlight cannot easily penetrate the dense clouds above you, noontime on the surface of Venus is no brighter than a very cloudy day on Earth. The Sun is setting, which takes a long time because the planet rotates so slowly. Just when you think you'll get some relief from the scorching heat of the setting Sun, you find that the thick atmosphere efficiently transfers heat from the day side to the night side. The temperature does not drop at night, nor does it drop near the poles. Such small temperature variations also mean there is almost no wind at the surface. High in the atmosphere, however, you would find winds of 110 m/s. High temperatures and very light winds keep the lower atmosphere of Venus free of clouds and hazes, and strong scattering of light by the dense atmosphere turns any view you might have of distant scenes hazy and bluish. (We see the same effect, but to a lesser extent, in our own atmosphere.)

G X U V I R

FIGURE 7.11 Thick clouds obscure our view of the surface of Venus.

7.5 Mars Has a Cold, Thin Atmosphere

Compared to Venus, the surface of Mars is almost hospitable. For this reason we can confidently expect that humans will eventually set foot on the red planet, quite likely before the end of this century and possibly much sooner. They will find a stark landscape, colored reddish by the oxidation of iron-bearing surface minerals (**Figure 7.12**). The sky will sometimes be a dark blue, but more often it will have a pinkish color caused by windblown dust. The lower density of the Mars atmosphere makes it more responsive than Earth's to heating and cooling, so its temperature extremes are greater. Near the equator at noontime, future astronauts may experience a comfortable 20°C—a cool room temperature. However, nighttime temperatures typically drop to a frigid −100°C, and during the polar night the air temperature can reach −150°C, cold enough to freeze carbon dioxide out of the air in the form of a dry-ice frost.

For human visitors, the low surface pressure will certainly be uncomfortable

The surface of Mars is cold and the air is thin.

G X U V I R

FIGURE 7.12 A view of the surface of Mars as seen in this panorama from the rover *Spirit*. In the absence of dust, the sky's thin atmosphere would appear deep blue.

Seasonal changes affect climate more on Mars than on Earth.

as well. The average atmospheric surface pressure on Mars is equivalent to the pressure at an altitude of 35 km above sea level on Earth, far higher than our highest mountain. Pressures range from a high of 11.5 millibars (1.1 percent of Earth's pressure at sea level) in the lowest impact basins of Mars to a mere 0.3 millibars at the summit of Olympus Mons.

The inclination of the Mars equator to its orbital plane is similar to Earth's, so both planets have similar seasons. But the effects on Mars are larger for two reasons: Mars varies more in its annual orbital distance from the Sun than does Earth, and the low density of the Mars atmosphere makes it more responsive to seasonal change. The large daily, seasonal, and latitudinal surface temperature differences on Mars often create locally strong winds, some estimated to be higher than 100 m/s. High winds can stir up huge quantities of dust (**Figure 7.13**) and distribute it around the planet's surface. For more than a century, astronomers have watched the seasonal development of springtime dust storms on Mars. The stronger ones spread quickly and within a few weeks can envelop the entire planet in a shroud of dust (**Figure 7.14**). Such large amounts of windblown dust can take many months to settle out of the atmosphere. Seasonal movement of

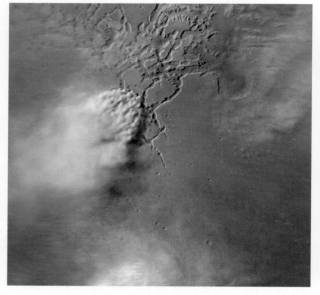

G X U V I R

FIGURE 7.13 A dust storm raging in the canyon lands of Mars.

June 26, 2001 September 4, 2001

G X U V I R

FIGURE 7.14 Hubble Space Telescope images showing the development of a global dust storm that enshrouded Mars in September 2001.

dust from one area to another alternately exposes and covers large areas of dark, rocky surface. This phenomenon led some astronomers of the late 19th and early 20th centuries to believe they were witnessing the seasonal growth and decay of vegetation on Mars. Public imagination carried the astronomers' interpretations a step further—to stories about advanced civilizations on Mars and invasions of Earth by warlike Martians (a theme found in some movies).

7.6 Climate Change: Is Earth Getting Warmer?

To understand observations of climate change, we must first clearly understand the distinction between *climate* and *weather*. The state of Earth's atmosphere at any given time and place is **weather**. Weather is small scale and short term. **Climate** is the term used to define the *average* state of Earth's atmosphere, including temperature, humidity, winds, and so on. Climate describes the planet as a whole.

Earth's climate has lengthy temperature cycles, usually lasting hundreds of thousands of years and occasionally producing shorter cold periods called ice ages (**Figure 7.15**). These changes in the mean global temperature are far smaller than typical geographic or seasonal temperature changes, but Earth's atmosphere is so sensitive to mean global temperature that it takes a drop of only a few degrees to plunge our climate into an ice age. Scientists still do not understand all the mechanisms controlling these climate-changing temperature swings. An external influence, such as small changes in the Sun's energy output, may be the cause. Changes in Earth's orbit or the inclination of its rotation axis have also been suggested. Or these temperature changes may be triggered internally by volcanic eruptions (which can produce global sunlight-blocking clouds or hazes) or long-term interactions between Earth's oceans and its atmosphere.

As mentioned in Section 7.2, the greenhouse effect clearly is responsible for warming Earth and Venus above the temperatures that would be implied by their distance from the Sun. Even Mars is warmer than it would be without an atmosphere, although its atmosphere is much thinner so the effect is much weaker. In our own Solar System, we have astronomical evidence of how the greenhouse effect can influence planetary atmospheres, including Earth's. Astronomers view these different planets as a cautionary tale, showing the results of varied "doses" of greenhouse gases.

For example, imagine an extreme hypothetical scenario in which large quantities of carbon dioxide or another greenhouse gas suddenly appeared in Earth's atmosphere. The increased warming would raise surface temperatures, driving more carbon dioxide and water vapor into the atmosphere, which in turn would increase the strength of the atmospheric greenhouse effect. The result would be even greater warming and still larger amounts of atmospheric water vapor, ultimately creating a surface pressure at least 300 times as great as it is now. This pressure is even higher than that of present-day Venus, due to the enormous amount of terrestrial water available. Earth's surface temperature would rise, and it might exceed 800 K. Long before reaching this stage, our planet would have become devoid of all life.

The physics associated with real, smaller changes to our planet's atmosphere is much more complicated than in this extreme scenario. An increase in cloud cover, caused by increased water in the atmosphere, might decrease the amount

Climate is the average state of Earth's atmosphere.

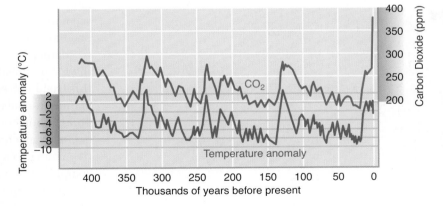

FIGURE 7.15 Global temperature and CO_2 variations over the last 450,000 years of Earth's history. Notice the two y-axes on this graph. The y-axis on the left goes with the temperature data, and the y-axis on the right goes with the CO_2 data. These two data sets have been plotted on the same graph to make the similarities and differences easier to see. *Temperature anomaly* is the departure from the long-term average. A positive anomaly means the global temperature was warmer than the average. A negative anomaly means the temperature was cooler than the average.

FIGURE 7.16 Global temperature variations of Earth over the last 130 years, shown in the form of the temperature anomaly.

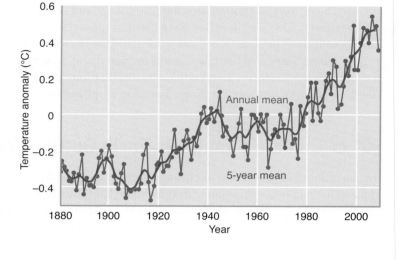

Increases in Earth's greenhouse gases can lead to significant climate change.

of sunlight reaching Earth's surface. Ocean currents are critical in transporting energy from one part of Earth to another, but no one knows how increased temperatures may affect those systems. The process is so complex that it is still not possible to accurately predict the long-term outcome of the small changes that humans are now making to the composition of Earth's atmosphere. In a real sense, we are *experimenting* with the atmosphere of Earth. We are asking the question, "What happens to Earth's climate if we steadily increase the number of greenhouse molecules in its atmosphere?" We do not yet know the answer, but we are already seeing some results. Figure 7.15 shows the carbon dioxide levels and temperature of Earth's atmosphere over the last 450,000 years. Notice that temperature and carbon dioxide are clearly correlated. When one rises, so does the other. Since the industrial revolution, the carbon dioxide level has risen to previously unknown levels. At the scale of this figure, it's difficult to see what's happening to temperature in the past few hundred years, although it does not seem to be following the sudden decline seen after other temperature peaks. We need to "zoom in" to see this more clearly.

Temperature measurements over the past 130 years show a steady increase in the mean global temperature (**Figure 7.16**). Most computer models suggest that this trend represents the beginning of a long-term change caused by the buildup of human-made greenhouse gases. A few climatologists argue that this is a temporary short-term cycle. However, we know that our atmosphere is a delicately balanced mechanism. Earth's climate is a complex, chaotic system within which tiny changes can produce enormous and often unexpected results. To add to the complexity, Earth's climate is intimately tied to ocean temperatures and currents. We see examples of this connection in the periodic El Niño and La Niña conditions, small shifts in ocean temperature that cause much larger global changes in air temperature and rainfall. Recent studies suggest that changes in the flow of the Gulf Stream current in the North Atlantic have very large effects on the climates of North America and northern Europe—and that these changes may take place not over centuries, but within a matter of decades. The best models show that a small addition of greenhouse gases to the atmosphere is extremely likely to have large effects over the relatively short period of about 50–100 years. By continuing to pump significant amounts of carbon dioxide into the atmosphere, we are effectively conducting a planet-wide experiment to find out just how sensitive our climate actually is to small changes in current conditions.

Climate change is quite possibly the most contentious and polarizing scientific issue in the United States today. The debate creates a lot of confusion among the public that stems from not having a fundamental understanding of the physical principles, and from the difficulty of determining which sources are credible.

Climate Change Skeptics Are Less "Credible" Scientists, Finds Study

By **James Dacey**, *physicsworld.com*

A survey of 1,372 climate scientists has concluded that the overwhelming majority support the basic idea that humans are significantly affecting the Earth's climate. The study also claims that the scientists, who are skeptical of anthropogenic climate change (ACC), tend to hold less "credible" publication records.

The survey, led by William Anderegg at Stanford University, included only researchers who have at some point written scientific assessments or signed public documents in relation to ACC. Participants were asked whether they were either convinced or unconvinced by the basic tenets of ACC as outlined by the Intergovernmental Panel on Climate Change.

Anderegg's team finds that 97% of the respondents are convinced by the idea. What is more, the surveyors rank all the participants by the number of climate science publications, and find that only one of the top 50 scientists, and three of the top one hundred, remain unconvinced by the arguments of ACC.

Despite the clear-cut results, the survey has already received criticism with regard to its methods.

Lorraine Whitmarsh, a social science researcher at Cardiff University, welcomes the study as the first attempt to rate the "credibility" of climate scientists with different views about climate change. She is a bit concerned, however, about the selection process for the survey's participants.

"The [survey] deliberately selects scientists who have signed high-profile public documents about their views, and so exclude those researchers 'behind the scenes,' perhaps with less extreme views one way or the other," she says.

Indeed, Whitmarsh points out that the survey excludes the 26% of researchers who are neither convinced nor unconvinced by the ACC arguments.

Andrew Russell, a climate researcher at the University of Manchester, says that the findings are "interesting," but is not sure how they will help in the communication of climate science. "The science can and should win the argument on its own," he says.

Russell believes that the media is often to blame when it comes to over-emphasising the scale of skepticism towards ACC within the climate science community. "There are valid and truly skeptical questions that need asking of climate science but they don't fit the narrative that parts of the media have constructed so they don't get aired," he says.

The research is published in *Proceedings of the National Academy of Sciences*.

Evaluating the News

1. Consider the standard of quality: the number of papers published on climate research. How do the two groups (the convinced and the unconvinced) compare when measured this way? Is the number of papers published a convincing metric to you?

2. Out of the top 100 climate scientists, what percentage of scientists affirmed the role of human activity in climate change? Do you find such consensus convincing? If not, what percentage would convince you?

3. Consider Whitmarsh's point about scientists who were excluded from the study. In your opinion, does leaving these scientists out make the findings of the study weaker, stronger, or leave them unaffected? Why or why not?

4. Scientists sometimes make the point that presenting all opinions as equally credible is misleading, and that the media should highlight the qualifications and merits of scholars on opposing sides of an issue. How could the media do that in a way that was clear to you? How could they do it in a way that was clear to someone not educated in science?

SUMMARY

7.1 Primary atmospheres consist mainly of hydrogen and helium captured from the protoplanetary disk. The terrestrial planets lost their primary atmospheres to space soon after the planets formed. Secondary atmospheres are created by volcanic gases and from volatile matter brought in by comets and asteroids striking a planet.

7.2 The atmospheric greenhouse effect keeps Earth from freezing, but it turns the atmosphere of Venus into an inferno.

7.3 Plant life is responsible for the oxygen in Earth's atmosphere. Earth's magnetosphere shields us from the solar wind. The Coriolis effect causes zonal and cyclonic patterns of atmospheric circulation.

7.4 Venus has a massive, hot carbon dioxide atmosphere.

7.5 Mars has a thin, cold carbon dioxide atmosphere.

7.6 Humans are modifying Earth's atmosphere and affecting global climate. This may lead to unintended consequences.

✦ SUMMARY SELF-TEST

1. Place in chronological order the following steps in the formation and evolution of Earth's atmosphere:
   a. Plant life converts CO_2 to oxygen.
   b. Hydrogen and helium are lost from the atmosphere.
   c. Volcanoes, comets, and asteroids increase the inventory of volatile matter.
   d. Hydrogen and helium are captured from the protoplanetary disk
   e. Oxygen enables the growth of new life-forms.
   f. Life releases CO_2 from the subsurface into the atmosphere.

2. The atmospheric greenhouse effect is present on
   a. Venus.
   b. Earth.
   c. Mars.
   d. all of the above.

3. The difference in climate between Venus, Earth, and Mars is primarily caused by
   a. the composition of their atmospheres.
   b. their relative distances from the Sun.
   c. the thickness of their atmospheres.
   d. the time at which their atmospheres formed.

4. Earth's magnetosphere
   a. shields us from the solar wind.
   b. is essential to the formation of auroras.
   c. extends far beyond Earth's atmosphere.
   d. all of the above

5. The words *weather* and *climate*
   a. mean essentially the same thing.
   b. refer to very different time scales.
   c. refer to very different size scales.
   d. both b and c

QUESTIONS AND PROBLEMS

True/False and Multiple-Choice Questions

6. **T/F:** The current atmospheres of the terrestrial planets were formed when the planets formed.

7. **T/F:** At the time of planet formation, the atmospheres of the planets consisted primarily of hydrogen and helium.

8. **T/F:** Life is responsible for the presence of oxygen in Earth's atmosphere.

9. **T/F:** Comets and asteroids are likely the source of most of the water on Earth.

10. **T/F:** The atmospheric greenhouse effect occurs on Earth, Venus, and Mars.

11. **T/F:** The atmospheric greenhouse effect occurs because greenhouse gases slow the escape of infrared radiation.

12. **T/F:** The stratosphere is where most weather happens.

13. **T/F:** All other things being equal, a planet with a high albedo will have a lower temperature than a planet with a low albedo.

14. **T/F:** The temperature of Earth has measurably risen over the last 100 years.

15. **T/F:** The fraction of CO_2 in Earth's atmosphere is at the highest level measured in the last 450,000 years.

16. Venus is hot and Mars is cold primarily because
    a. Venus is closer to the Sun.
    b. Venus has a much thicker atmosphere.
    c. the atmosphere of Venus is dominated by CO_2, but the atmosphere of Mars is not.
    d. Venus has stronger winds.

17. The atmosphere of Mars is often pink-orange because
    a. the atmosphere is dominated by carbon dioxide.
    b. the Sun is at a low angle in the sky.
    c. Mars has no oceans to reflect blue light to the sky.
    d. winds lift dust into the atmosphere.

18. Convection in the _____ causes weather on Earth.
    a. stratosphere   **b.** mesosphere
    c. troposphere   **d.** ionosphere

19. Auroras are the result of
    a. the interaction of particles from the Sun and Earth's atmosphere.
    b. upper-atmosphere lightning strikes.
    c. destruction of stratospheric ozone, which leaves a hole.
    d. the interaction of Earth's magnetic field with Earth's atmosphere.

20. The ozone layer protects life on Earth from
    a. high-energy particles from the solar wind.
    b. micrometeorites.
    c. ultraviolet radiation.
    d. charged particles trapped in Earth's magnetic field.

21. Hadley circulation is broken into zonal winds by
    a. convection from solar heating.
    b. hurricanes and other storms.
    c. interactions with the solar wind.
    d. the planet's rapid rotation.

22. Earth experiences long-term climate cycles spanning
    a. about 140,000,000 years.
    b. about 140,000 years.
    c. about 1,400 years.
    d. about 140 years.

23. Over the last 450,000 years, Earth's temperature has closely tracked
    a. solar luminosity.
    b. oxygen levels in the atmosphere.
    c. the size of the ozone hole.
    d. carbon dioxide levels in the atmosphere.

24. Uncertainties in climate science are dominated by
    a. uncertainty about physical causes.
    b. uncertainty about current effects.
    c. uncertainty about past effects.
    d. uncertainty about future effects.

25. _____ percent of climate scientists favor a human-based explanation for climate change on Earth.
    a. Ten   b. Fifty
    c. Seventy-five d. Ninety-seven

Conceptual Questions

26. Describe the origin and fate of primary atmospheres.

27. Primary atmospheres of the terrestrial planets were composed almost entirely of hydrogen and helium. Explain why they contained only these gases and not others.

28. How were the secondary atmospheres of the terrestrial planets created?

29. Mercury and the Moon have extremely tenuous atmospheres. What are two possible sources of these atmospheres?

30. Nitrogen, the principal gas in Earth's atmosphere, was not a significant component of the protoplanetary disk from which the Sun and planets formed. Where did Earth's nitrogen come from?

31. Some of Earth's water was released aboveground by volcanism. What is another likely source of Earth's water?

32. The force of gravity holds objects tightly to the surfaces of the terrestrial planets. Yet atmospheric molecules are constantly escaping into space. Explain how these molecules are able to overcome gravity's grip. How does the mass of a molecule affect its ability to break free?

33. Explain what is meant by a runaway atmosphere, and describe how it works.

34. We attribute the warming of a planet's surface to the atmospheric greenhouse effect. How does this mechanism differ from the warming of a conventional greenhouse?

35. Name at least two greenhouse molecules other than carbon dioxide.

36. In what way is the atmospheric greenhouse effect beneficial to terrestrial life?

37. Examine Figure 7.3. Why is the range of temperatures on Mercury so much larger than for any other planet?

38. Why is Venus very hot and Mars very cold if both of their atmospheres are dominated by carbon dioxide, an effective "greenhouse" molecule?

39. Study Figure 7.4. Redraw part (b) of this figure as is. Then draw the corresponding figure for a planet with no greenhouse gases, and draw it again for a planet with a continuously increasing amount of greenhouse gases. Be sure to label each drawing.

40. In what ways does plant life affect the composition of Earth's atmosphere?

41. Astronomers have developed the technical capability to detect Earth-like planets around other stars. What observations would indicate that these planets harbor some form of life as we know it?

42. You check the barometric pressure and find that it is reading only 920 millibars. Two possible effects could be responsible for this lower-than-average reading. What are they?

43. What mechanism produces the aurora borealis (northern lights)?

44. What is the principal cause of winds in the atmospheres of the terrestrial planets?

45. How does the solar wind affect Earth's upper atmosphere, and what effects can it have on society?

46. Why are humans unable to get a clear view of the surface of Venus, as we have so successfully done with Mars?

47. Assume you are somehow able to survive on the surface of Venus. Describe your environment.

48. Explain why surface temperatures on Venus hardly vary between day and night and between the equator and the poles.

49. In 1975 the Soviet Union landed two camera-equipped spacecraft on Venus, giving planetary scientists their first (and only) close-up views of the planet's surface. Both cameras ceased to function after only an hour. What environmental conditions most likely led to their demise?

50. Humans may eventually travel to the surface of Mars. Describe the environment they will experience.

51. Mars has seasons similar to those of Earth, but more extreme. Explain why.

52. Explain the difference between climate and weather.

53. What is the evidence that the greenhouse effect exists on Earth, Venus, and Mars?

54. Study Figure 7.15. How far back in time do these data go? From this graph, what can you say about the relationship between CO_2 concentrations and temperature?

55. Describe the overall trend in Figure 7.16.

Problems

56. Figure 7.5 shows the development of oxygen in Earth's atmosphere over the history of the planet. This graph is logarithmic on both axes. Scientists use logarithmic graphs to study exponential behavior—a type of process that goes increasingly faster over time. Study the y-axis to see how the numbers increase with height above the x-axis. Is the distance on the page between 0 and 1 the same as the distance between 1 and 2? How about the distance between 0 and 10 versus the distance between 90 and 100? Redraw the y-axis of this graph on a linear scale by making a graph with evenly spaced intervals every 10 percent. Compare your new graph with the one in the text. How are they different?

57. Exponentials can be a difficult mathematical concept to master, but they are critical in modeling much of the behavior of the natural world. To help you understand exponentials, think about a checkerboard. A checkerboard has 8 rows of 8 squares, so it has 64 squares on it.
   a. Imagine that you put one penny on the first square, two pennies on the second, three pennies on the third, four on the fourth, and so forth. This is linear behavior: each step is the same size; each time you go to a new space, you add the same number of pennies—1. How many pennies will be on the 64th square?
   b. Now imagine that you do this in an exponential fashion, so that you put one penny on the first space, two pennies on the second, *four* pennies on the third, *eight* pennies on the fourth, and so on. Predict how many pennies will be on the 64th square. Now, calculate this the long way on your calculator, multiplying by 2 for each square, all the way to the last square. (It's a pain, but it's important to do it just once so that you really understand.) How many pennies will be on the 64th square? Were you close with your prediction? You may notice two things. One: figuring out the linear case was much easier to visualize, and to see the pattern. That's perfectly normal. Humans are notoriously bad at understanding exponential behavior. Two: the exponential case seems to go along fine, increasing by reasonable amounts, until the numbers suddenly become absurd. It's important to know which things in nature behave linearly and which behave exponentially. The runaway atmospheres discussed in this chapter are exponential behaviors because of the feedback loops involved.

58. The total mass of Earth's atmosphere is 5×10^{18} kg. Carbon dioxide (CO_2) makes up about 0.06 percent of Earth's atmospheric mass.
   a. What is the mass of CO_2 (in kilograms) in Earth's atmosphere?

   b. The annual global production of CO_2 is now estimated to be nearly 3×10^{13} kg. What annual fractional increase does this represent?

59. Figure 7.15 shows two sets of data on the same graph. The x-axis is time, which is the same for both data sets. The y-axis on the right is the amount of CO_2 in the atmosphere. The y-axis on the left is the temperature anomaly, the difference between the temperature that year and the average temperature over the entire period of time. Do both these lines show the same trend? Does this *necessarily* mean that one causes the other? What other information do we have that leads us to believe the increasing CO_2 levels are driving the increasing temperatures?

60. Water pressure in Earth's oceans increases by 1 bar for every 10 meters of depth. Calculate how deep you would have to go to experience pressure equal to the atmospheric surface pressure on Venus.

61. A planet with no atmosphere at 1 AU from the Sun would have an average blackbody surface temperature of 279 K if it absorbed all the Sun's electromagnetic energy falling on it (albedo = 0).
   a. What would the average temperature on this planet be if its albedo were 0.1, typical of a rock-covered surface?
   b. What would the average temperature be if its albedo were 0.9, typical of a snow-covered surface?

62. Consider a hypothetical planet named Vulcan. If such a planet were in an orbit one-fourth the size of Mercury's and had the same albedo as Mercury, what would be the average temperature on Vulcan's surface? Assume that the average temperature on Mercury's surface is 450 K.

63. Earth's average albedo is approximately 0.31.
   a. Calculate Earth's average temperature in degrees Celsius.
   b. Does this temperature meet your expectations? Explain why or why not.

64. The orbit of Eris, a dwarf planet, carries it out to a maximum distance of 97.7 AU from the Sun. Assuming an albedo of 0.8, what is the average temperature of Eris when it is farthest from the Sun?

 SmartWork, Norton's online homework system, includes algorithmically generated versions of these questions, plus additional conceptual exercises. If your instructor assigns questions in SmartWork, log in at **smartwork.wwnorton.com**.

 StudySpace is a free and open website that provides a Study Plan for each chapter of **Understanding Our Universe**. Study Plans include animations, reading outlines, vocabulary flashcards, and multiple-choice quizzes, plus links to premium content in SmartWork and the ebook. Visit **wwnorton.com/studyspace**.

Exploration | Climate Change

One prediction about climate change is that as the planet warms, ice in the polar caps and in glaciers will melt. This certainly seems to be occurring in the vast majority of glaciers and ice sheets around the planet. It is reasonable to ask whether this actually matters, and why. Here you will explore several consequences of the melting ice on Earth.

Experiment 1: Floating Ice

For this experiment, you will need a permanent marker, a translucent plastic cup, water, and ice. Place a few ice cubes in the cup, and add water until the ice cubes float (that is, they don't touch the bottom). Mark the water level on the outside of the cup with the marker, and label this mark so that later you will know it was the initial water level.

1. **As the ice melts, what do you expect to happen to the water level in the cup?**

Wait for the ice to melt completely; then mark the cup again.

2. **What happened to the water level in the cup when the ice melted?**

3. **Given the results of your experiment, what can you predict will happen to global sea levels when the Arctic ice sheet, which floats on the ocean, melts?**

Experiment 2: Ice on Land

For this experiment, you will need the same materials as in experiment 1, plus a paper or plastic bowl. Fill the cup about halfway with water and then mark the water level, labeling it so you know it's the initial level. Poke a hole in the bottom of the bowl, and set the bowl over the cup. Add some ice cubes to the bowl.

4. **As the ice melts, what do you expect to happen to the water level in the cup?**

Wait for the ice to melt completely; then mark the cup again.

5. **What happened to the water level in the cup when the ice melted?**

6. **In this experiment, the water in the cup is analogous to the ocean, and the ice in the bowl is analogous to ice on land. Given the results of your experiment, what can you predict will happen to global sea levels when the Antarctic ice sheet, which sits on land, melts?**

Experiment 3: Why Does It Matter?

Search online using the phrase "Earth at night" to find a satellite picture of Earth, taken at night. The bright spots on the image trace out population centers. In general, the brighter they are, the more populous the area (although there is a confounding factor relating to technological advancement).

7. **Where do humans tend to live—near coasts, or inland?**

Coastal regions are, by definition, near sea level. If both the Arctic and Antarctic ice sheets melted completely, sea levels would rise by 80 meters.

8. **How would a sea-level rise of 60–80 meters (the range of reasonable predictions) over the next few decades affect the global population? (To help you think about this, remember that a meter is about 3 feet, and one story of a building is about 10 feet.)**

8

The Giant Planets

Although they are very far away, the giant planets—Jupiter, Saturn, Uranus, and Neptune—are so large that with the unaided eye, we can see all but Neptune. The brilliance of Jupiter and Saturn is comparable to the brightest stars. In contrast, Uranus is only slightly brighter than the faintest stars visible on a dark night, and Neptune cannot be seen without the aid of binoculars.

In 1781, William Herschel was producing a catalog of the sky when he noticed a tiny disk that he thought was a comet. The object's slow nightly motion soon convinced him that it was a new planet—Uranus. Astronomers in the 19th century found that Uranus was straying from its predicted path in the sky, perhaps because of the gravitational pull of an unknown planet. Armed with mathematical predictions, Johann Gottfried Galle began a search at the Berlin Observatory. He found Neptune on his first observing night, in the predicted position. Neptune is the outermost classical planet in our Solar System.

The Hubble Space Telescope has provided new insight on the giant planets, but our greatest leaps of knowledge have come from space probes: *Pioneer*, *Voyager*, *Galileo*, and *Cassini*.

✧ LEARNING GOALS

Unlike the solid, rocky planets of the inner Solar System, four worlds in the outer Solar System were able to capture and retain gases and volatile materials from the Sun's protoplanetary disk and swell to enormous size and mass. The image at right shows two observations of the Great Red Spot on Jupiter; a student has used this information to calculate the rotation period of the giant planet. By the end of this chapter, you should be able to compare the size and origin of the Great Red Spot to Earth-based phenomena, and you should understand how it could be used to find the rotation of the planet. You should also be able to:

- Differentiate the giant planets from each other and from the terrestrial planets

- Describe why the giant planets look the way they do

- Determine the sizes of the giant planets and their wind speeds

- Describe how gravitational energy turns into thermal energy and how that affects the atmospheres of the giant planets

- Explain the extreme conditions deep within the interiors of the giant planets

- Describe the origin and general structure of the rings of the giant planets

```
~Pat/Desktop/jupiter.fits
                    DIA: 315px
|--------------------------------------|

                                + x_0 = 70px

TIME: 23:07
```

~Pat/Desktop/jupiter.fits
Diameter = 315px
Circumference = 990px
x_0 = 70px
t = 23:07

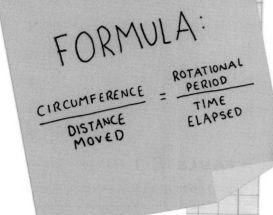

FORMULA:

$$\frac{CIRCUMFERENCE}{DISTANCE\ MOVED} = \frac{ROTATIONAL\ PERIOD}{TIME\ ELAPSED}$$

```
~Pat/Desktop/jupiter_1.fits
                    DIA: 315px
|--------------------------------------|

                   + x_0        + x_1 = 240px

TIME: 00:47                        Δx = 170px
```

~Pat/Desktop/jupiter_1.fits
Diameter = 315px
Circumference = 990px
x_0 = 70px
x_1 = 240px
Δx = 170px
t = 00:47
Δt = 1.67h

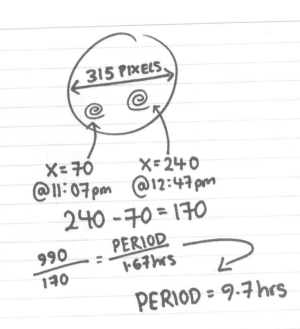

315 PIXELS

X = 70
@11:07pm

X = 240
@12:47pm

$$240 - 70 = 170$$

$$\frac{990}{170} = \frac{PERIOD}{1.67\,hrs}$$

$$PERIOD = 9.7\,hrs$$

8.1 Giant Planets Are Large, Cold, and Massive

Jupiter, the closest giant planet, is more than 5 astronomical units (AU) from the Sun (see **Table 8.1**). The Sun shines only dimly and provides very little warmth in these remote parts of the Solar System. From Jupiter, the Sun is but a tiny disk, just $1/27$ as bright as it appears from Earth. At the distance of Neptune, the Sun no longer looks like a disk at all; it appears as a brilliant star about 500 times brighter than the full Moon in our own sky. Daytime on Neptune is equivalent to a perpetual twilight here on our own planet. With so little sunlight available for warmth, daytime temperatures hover around 123 K at the cloud tops on Jupiter, and they can dip to just 37 K on Neptune's moon Triton.

Jupiter, Saturn, Uranus, and Neptune are all enormous in comparison to their small, rocky terrestrial cousins. Jupiter is the largest of the eight planets; it is more

---

TABLE 8.1

Comparison of Physical Properties of the Giant Planets

	JUPITER	SATURN	URANUS	NEPTUNE
Orbital semi-major axis (AU)	5.203	9.54	19.2	30
Orbital period (Earth years)	11.86	29.45	84.0	164.8
Orbital velocity (km/s)	13.06	9.64	6.80	5.43
Mass ($M_\oplus = 1$)	317.8	95.16	14.5	17.1
Equatorial radius (km)	71,490	60,270	25,560	24,760
Equatorial radius ($R_\oplus = 1$)	11.21	9.45	4.01	3.88
Density (water = 1)	1.326	0.687	1.270	1.638
Rotation period (Earth days)	0.414	0.444	0.718	0.671
Obliquity (degrees)[*]	3.13	26.73	97.77	28.32
Surface gravity (m/s²)	24.79	10.44	8.69	11.15
Escape velocity (km/s)	59.5	35.5	21.3	23.5

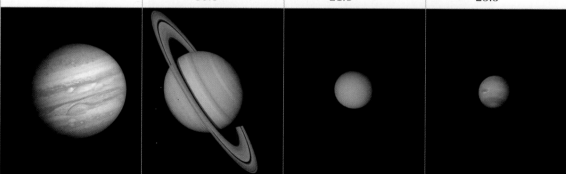

[*]An obliquity greater than 90° indicates that the planet rotates in a retrograde, or backward, direction.

Working It Out 8.1 | Finding the Diameter of a Giant Planet

Suppose that Newton's laws tell us a particular planet is moving at a speed of precisely 25 kilometers per second (km/s) relative to Earth's own motion. As the planet moves across the sky toward a background star, you begin to make careful observations. You note the moment when the planet first eclipses the star, and you wait while the star passes behind the center of the planet and emerges from the opposite side. You find that this event takes exactly 2,000 seconds. The planet has traveled a specific distance (its diameter), at a specific speed, during this time. We can use the equation from Working It Out 2.1 that relates speed, distance, and time to find this distance:

$$\text{speed} = \frac{\text{distance}}{\text{time}}$$

We can rearrange this equation to solve for distance, first multiplying both sides by time and then exchanging the right and left sides of the equation:

$$\text{distance} = \text{speed} \times \text{time}$$

Plugging in our speed of 25 km/s and our time of 2,000 seconds, we find the distance the planet traveled:

$$\text{distance} = 25 \text{ km/s} \times 2,000 \text{ s}$$
$$\text{distance} = 50,000 \text{ km}$$

The planet's diameter is equal to the distance it traveled during the 2,000 seconds, or 50,000 km.

The exact center of a planet rarely passes directly in front of a star, but observations of occultations from several widely separated observatories can provide the geometry necessary to calculate both the planet's size and its shape. Occultations of radio signals transmitted from orbiting spacecraft have also provided accurate measures of the sizes and shapes of planets and their moons, as have images taken by spacecraft cameras.

---

than one-tenth the size of the Sun itself and more than 11 times the size of Earth. Saturn is only slightly smaller than Jupiter, with a diameter of 9.5 Earths. Uranus and Neptune are each about 4 Earth diameters across; Neptune is the slightly smaller of the two. The most accurate measurements of planet sizes have come from observing the length of time it takes a planet to eclipse, or "occult," a star (**Working It Out 8.1**). Scientists call these events **stellar occultations** (**Figure 8.1**).

Not counting the mass of the Sun, the giant planets contain 99.5 percent of all the mass in the Solar System. All other Solar System objects—terrestrial planets, dwarf planets, moons, asteroids, and comets—are included in the remaining 0.5 percent. Jupiter is more than twice as massive as all the other planets in the Solar System combined: Jupiter is 318 times as massive as Earth and 3½ times as massive as Saturn, its closest rival. Even so, its mass is only about a thousandth that of the Sun. Uranus and Neptune are the lightweights among the giant planets, but each is still more than 15 times as massive as Earth.

Before the space age, scientists measured a planet's mass by observing the motions of its moons. In Chapter 3 we learned that we can use Newton's law of gravitation and Kepler's third law to predict the motion of a moon if we know the planet's mass. Conversely, if we know the motion of the moon, we can find the planet's mass. This technique necessarily worked only with planets that have moons, and the accuracy of those early calculations was limited by how precisely astronomers could measure the positions of the moons with ground-based telescopes. Planetary spacecraft now make it possible to measure the masses of planets much more accurately. As a spacecraft flies by, the planet's gravity deflects it. By tracking and comparing the spacecraft's radio signals using several antennae here on Earth, we can detect tiny changes in the spacecraft's path and accurately measure the planet's mass.

Gravitational tugs on spacecraft are used to find the masses of the giant planets.

FIGURE 8.1 Occultations occur when a planet, moon, or ring passes in front of a star. Careful measurements of changes in the starlight's brightness and the duration of these changes give information about the size and properties of the occulting object.

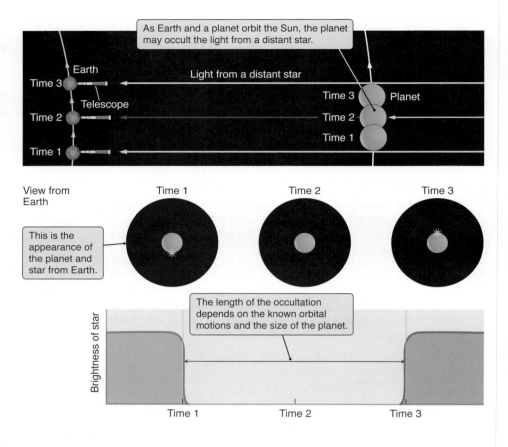

We See Only Atmospheres, Not Surfaces

The giant planets are made up primarily of gases and liquids. Jupiter and Saturn are composed of hydrogen and helium and are known as **gas giants**. Uranus and Neptune are known as **ice giants** because they both contain much larger amounts of water and other ices than do Jupiter and Saturn. On a giant planet, a relatively shallow atmosphere merges seamlessly into a deep liquid **ocean**, which in turn merges smoothly into a denser liquid or solid core. Although the atmospheres of the giant planets are shallow compared with the depth of the liquid layers below, they are still much thicker than those of the terrestrial planets—thousands of kilometers rather than hundreds. Only the very highest levels of these atmospheres are visible to us. In the case of Jupiter or Saturn, we see the tops of a layer of thick **clouds**, the highest of many other layers that lie below. Although a few thin clouds are visible on Uranus, we mostly find ourselves looking into a clear, seemingly bottomless atmosphere. Atmospheric models tell us that thick cloud layers must lie below, but strong scattering of sunlight by molecules (see Chapter 7) in the clear part of the atmosphere prevents us from seeing these lower cloud layers. Neptune displays a few high clouds with a deep, clear atmosphere showing between them.

Jupiter's Chemical Composition Is Much Like the Sun's

In Chapter 6 we learned that the terrestrial planets are composed mostly of rocky minerals, such as silicates, along with various amounts of iron and other metals. While the atmospheres of the terrestrial planets contain lighter materials,

Vocabulary Alert

Ocean: Here on Earth, this word commonly and specifically means a vast expanse of salty liquid water. Astronomers have expanded the definition of *ocean* to mean a vast expanse of any liquid—not necessarily water.

Clouds: Just as oceans on other worlds are not necessarily bodies of water, the *clouds* that an astronomer discusses are not necessarily clouds of water vapor—remember the clouds of sulfuric acid we saw on Venus.

the masses of these atmospheres—and even of Earth's oceans—are insignificant when compared with the total planetary masses. The terrestrial planets are the densest objects in the Solar System, ranging from 3.9 (Mars) to 5.5 (Earth) times the density of water.

In contrast, the giant planets have lower densities because they are composed mainly of lighter materials such as hydrogen, helium, and water. Among giant planets, Neptune has the highest density, about 1.6 times that of water. Saturn has the lowest density, only 0.7 times the density of water. This means that Saturn would actually float in water with 70 percent of its volume submerged, if you could find a large enough lake. Jupiter and Uranus have densities between those of Neptune and Saturn.

The chemical compositions of the giant planets are not all the same. Astronomers use the relative amounts of the elements in the Sun as a standard reference, termed solar **abundance**. As illustrated in **Figure 8.2**, hydrogen (H) is the most abundant element, followed by helium (He). Jupiter's chemical composition is quite similar to that of the Sun. Jupiter has about a dozen hydrogen atoms for every atom of helium, which is typical of the Sun and the universe as a whole. Only 2 percent of its mass is made up of **heavy elements** (atoms more massive

Vocabulary Alert

Abundance: In common language, this word means a more than adequate supply, or simply "enough." At times, it carries the connotation of having *much* more than is needed. Astronomers use this word very specifically to refer to percentages of chemical composition. If you take a sample of an object, and a certain fraction of it is hydrogen, then that fraction is the *abundance* of hydrogen in your sample. This means that if you are more abundant in hydrogen than your friend is, you must be less abundant in everything else, even if you have more atoms overall. This is because all the fractions must necessarily add together to equal one.

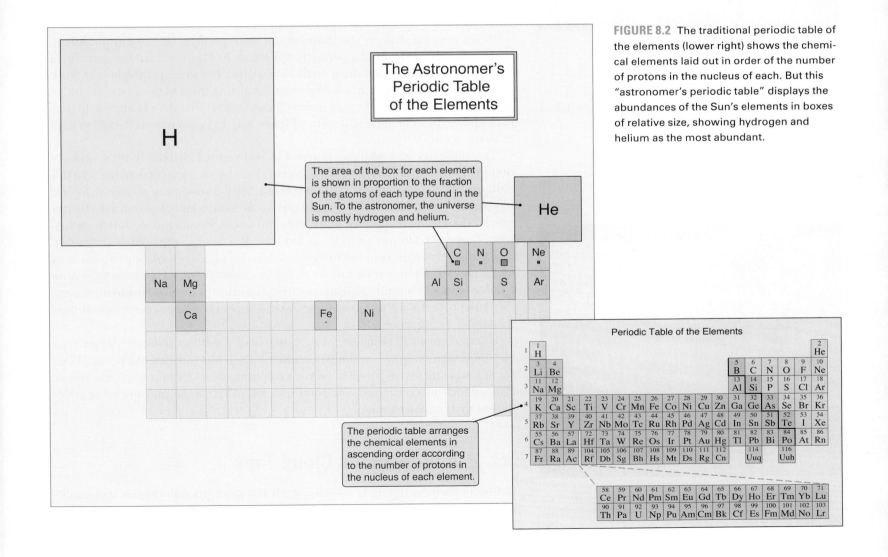

FIGURE 8.2 The traditional periodic table of the elements (lower right) shows the chemical elements laid out in order of the number of protons in the nucleus of each. But this "astronomer's periodic table" displays the abundances of the Sun's elements in boxes of relative size, showing hydrogen and helium as the most abundant.

FIGURE 8.3 Hubble Space Telescope image of Saturn, taken in 1999. The oblateness of the planet is apparent. The large orange moon Titan appears near the top of the disk of Saturn, along with its black shadow.

FIGURE 8.4 Jupiter imaged by *Cassini*.

than helium). Atoms of oxygen (O), carbon (C), nitrogen (N), and sulfur (S) have combined with hydrogen to form molecules of water (H_2O), methane (CH_4), ammonia (NH_3), and hydrogen sulfide (H_2S), respectively. More complex combinations are also common. Helium and certain other gases, such as neon and argon, are **inert** gases that do not combine with other elements to make molecules in the atmosphere. Jupiter's liquid core, which contains most of the planet's iron and silicate and much of its water, is left over from the original rocky planetesimal around which Jupiter grew.

The principal compositional differences among the four giant planets lie in their abundances of hydrogen and helium. Saturn is more abundant in heavy elements than Jupiter (and so less abundant in hydrogen and helium). In Uranus and Neptune, heavy elements are so abundant that they are major components of these two planets. This evidence supports the formation hypothesis discussed in Chapter 5.

Rotation of the Giant Planets

Giant planets rotate rapidly, so their days are short. A day on Jupiter is just under 10 hours long, and Saturn's is only a little longer. Neptune and Uranus have rotation periods of 16 and 17 hours, respectively, so their days are intermediate in length between those of Jupiter and Earth.

Rapid rotation distorts the shape of the giant planets. If they did not rotate, these **fluid** bodies would be perfectly spherical. In Chapter 5 we learned why a collapsing, rotating cloud must settle into a disk. The same principle is at work in the rapidly rotating giant planets as well, causing them to be **oblate**—to bulge at their equators. Saturn is very oblate; its equatorial diameter is almost 10 percent greater than its polar diameter (**Figure 8.3**). In comparison, the oblateness of Earth is only 0.3 percent.

The intensity of a planet's seasons is determined primarily by a planet's **obliquity**—the inclination of its equatorial plane to its orbital plane. Earth's obliquity of 23.5° causes our distinct seasons. With an obliquity of only 3°, Jupiter has almost no seasons at all. The obliquities of Saturn and Neptune are slightly greater than those of Earth or Mars, thus causing moderate but well-defined seasons. Curiously, Uranus spins on an axis that lies nearly in the plane of its orbit. This causes its seasons to be extreme; each polar region alternately experiences 42 years of continuous sunshine, followed by 42 years of total darkness. Why does Uranus have an obliquity so different from the other planets? Many astronomers think the planet was "knocked over" by the impact of a huge planetesimal near the end of its accretion phase.

Uranus is one of five major Solar System bodies with an obliquity larger than 90°. Its obliquity is 98°, and a value greater than 90° indicates that the planet rotates in a clockwise direction when seen from above its orbital plane. Venus, Pluto, Pluto's moon Charon, and Neptune's moon Triton are the only other major bodies that behave this way.

8.2 A View of the Cloud Tops

Jupiter is perhaps the most colorful of all the giant planets (**Figure 8.4**). Parallel bands, ranging in hue from bluish gray to various shades of orange, reddish brown, and pink, stretch out across its large, pale yellow disk. The darker bands are called belts and the lighter ones are called zones. Many small clouds appear along the edges of, or within, the belts. The most prominent feature is a large, often brick-red oval in Jupiter's southern hemisphere known as the **Great Red Spot**, seen at the lower right in Figure 8.4. With a length of 25,000 km and a width of 12,000 km, the Great Red Spot could comfortably hold two Earths side by side within its boundaries (**Figure 8.5**).

The Great Red Spot has been circulating in Jupiter's atmosphere for at least 300 years. It was first seen shortly after the invention of the telescope, and it has since varied unpredictably in size, shape, color, and motion as it drifts among Jupiter's clouds. Observations of small clouds distributed within the Great Red Spot show that it is an enormous atmospheric whirlpool, swirling in a counterclockwise direction with a period of about a week. Its cloud pattern looks a lot like that of a terrestrial hurricane, but it rotates in the opposite direction, exhibiting anticyclonic rather than cyclonic flow. (Anticyclonic flow indicates a high-pressure system.) Comparable whirlpool-like behavior is observed in many of the smaller oval-shaped clouds found elsewhere in Jupiter's atmosphere as well as in similar clouds observed in the atmospheres of Saturn and Neptune.

Jupiter is a roiling, swirling giant with atmospheric currents and vortices so complex that scientists still do not fully understand the details of how they interact with one another, even after decades of analysis. The Great Red Spot alone displays more structure than was visible over all of Jupiter before the space age. Dynamically, it also reveals some rather bizarre behavior, such as "cloud cannibalism." In a series of time-lapse images, *Voyager* observed a number of Alaska-sized clouds being swept into the Great Red Spot. Some of these clouds were carried around the vortex a few times and then ejected, while others were swallowed up and never seen again (**Figure 8.6**). Other smaller clouds with structure and behavior similar to that of the Great Red Spot are seen in Jupiter's middle latitudes.

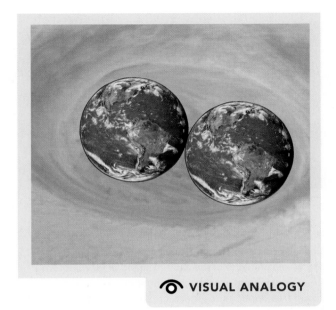

👁 **VISUAL ANALOGY**

FIGURE 8.5 The Great Red Spot on Jupiter is twice the size of Earth.

FIGURE 8.6 This sequence of images, obtained by the *Voyager* spacecraft during its encounter with Jupiter, shows the swirling, anticyclonic motion of Jupiter's Great Red Spot.

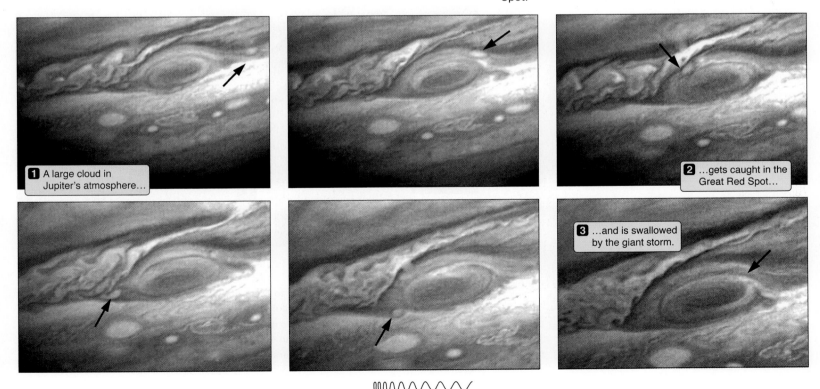

1 A large cloud in Jupiter's atmosphere…

2 …gets caught in the Great Red Spot…

3 …and is swallowed by the giant storm.

FIGURE 8.7 *Cassini* image of Saturn, taken in 2004 as the spacecraft approached the planet. This was the last time the whole planet fit in the camera's field of view. Saturn's colors are vibrant in this image. Sunlight is scattered by the cloud-free upper atmosphere, creating a blue sliver of light in the northern hemisphere.

Saturn is adorned by a magnificent system of rings, unmatched by any other planet in the Solar System (**Figure 8.7**). Saturn is both farther away than Jupiter and somewhat smaller in actual size, so from Earth it appears less than half as large as Jupiter. Like Jupiter, Saturn displays atmospheric bands, but they tend to be wider and their colors and contrasts much more subdued than those on Jupiter. A relatively narrow, meandering band in the mid-northern latitudes encircles the planet in a manner similar to that of our own terrestrial jet stream. Individual clouds on Saturn have been seen only rarely from Earth. On these infrequent occasions, large, white, cloudlike features suddenly erupt in the tropics, spread out in longitude, and then fade away over a period of a few months. The largest clouds are larger than the continental United States, but many that we see are smaller than terrestrial hurricanes. Close-up views from the *Cassini* spacecraft (**Figure 8.8**) show immense lightning-producing storms in a region of Saturn's southern hemisphere known to mission scientists as "storm alley."

From Earth, even through the largest telescopes, Uranus and Neptune look like tiny, featureless pale bluish-green disks. Infrared imaging reveals individual clouds and belts, giving these distant planets considerably more character (**Figure 8.9**). The strong absorption of reflected sunlight by methane causes the atmospheres of Uranus and Neptune to appear dark in the near infrared, allowing the highest clouds and bands to stand out in contrast against the dark background (**Figure 8.10**).

A large, dark, oval feature in Neptune's southern hemisphere, first observed in images taken by *Voyager 2* in 1989, reminded astronomers of Jupiter's Great Red Spot, so they called it the Great Dark Spot. However, the Neptune feature was gray rather than red, and it changed its length and shape more rapidly than the Great

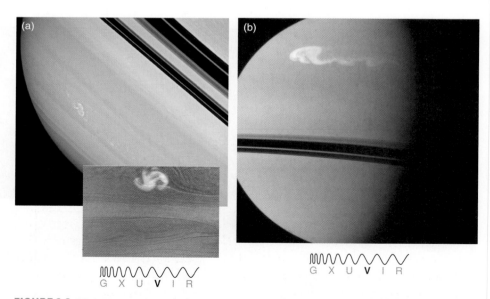

FIGURE 8.8 Violent storms are known to erupt on Saturn. (a) An enhanced *Cassini* image of an intense lightning-producing storm (left of center) located in Saturn's "storm alley." The inset shows a similar storm on Saturn's night side, illuminated by sunlight reflecting off Saturn's rings. (b) In December 2010, an enormous storm in Saturn's northern hemisphere was discovered by amateur astronomers and subsequently imaged by *Cassini*, as shown here.

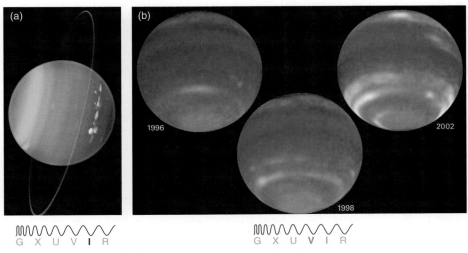

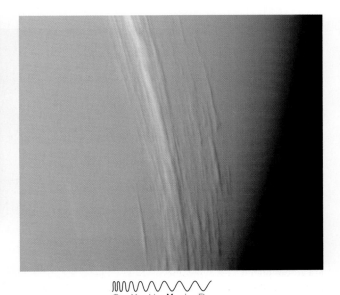

FIGURE 8.9 Ground-based Keck telescope image of Uranus (a) and Hubble Space Telescope images of Neptune (b), taken at a wavelength of light that is strongly absorbed by methane. The visible clouds are high in the atmosphere. The rings of Uranus show prominently because the brightness of the planet has been subdued by methane absorption. Seasonal changes in cloud formation on Neptune are evident over a 6-year interval.

FIGURE 8.10 Neptune's clouds cast shadows through a clear atmosphere onto the cloud deck 50 km below in this *Voyager 2* image.

Red Spot. By 1994, it had disappeared—but a different dark spot of comparable size had appeared briefly in Neptune's northern hemisphere.

8.3 A Journey into the Clouds

Our visual impression of the giant planets is based on our two-dimensional view of their cloud tops. Atmospheres, though, are three-dimensional structures whose temperature, density, pressure, and even chemical composition vary with height and over horizontal distances. As a rule, atmospheric temperature, density, and pressure all increase with decreasing altitude, although temperature is sometimes higher at very high altitudes, as in Earth's thermosphere. A thin haze above the cloud tops is visible in profile above the **limbs** of the planets. The composition of the haze particles remains unknown, but they may be smog-like products created when ultraviolet sunlight acts on hydrocarbon gases such as methane.

Water is the only substance in Earth's lower atmosphere that can condense into clouds, but the atmospheres of the giant planets contain a variety of volatile materials that can form clouds (**Figure 8.11**). Beneath those planets' cloud tops are dense layers of cloud separated by regions of relatively clear atmosphere. Because each kind of volatile condenses at a particular temperature and pressure, each therefore forms clouds at a different altitude. Convection carries volatile materials upward along with all other atmospheric gases, and when a particular volatile reaches an altitude with its condensation temperature, most of the volatile condenses and separates from the other gases, so very little of it is carried higher aloft.

The farther a planet is from the Sun, the colder its troposphere will be. Distance from the Sun thus determines the altitude at which a particular volatile, such as ammonia or water, will condense to form a cloud layer on each of the planets (see Figure 8.11). If temperatures are too high, some volatiles may not condense at all.

Different volatiles produce different clouds at different heights.

Vocabulary Alert

Limb: In common language, we use this word to refer to something that sticks out—an arm, a leg, or a tree branch. Astronomers use it to refer to the outer edge of the visible disk of a planet, moon or the Sun.

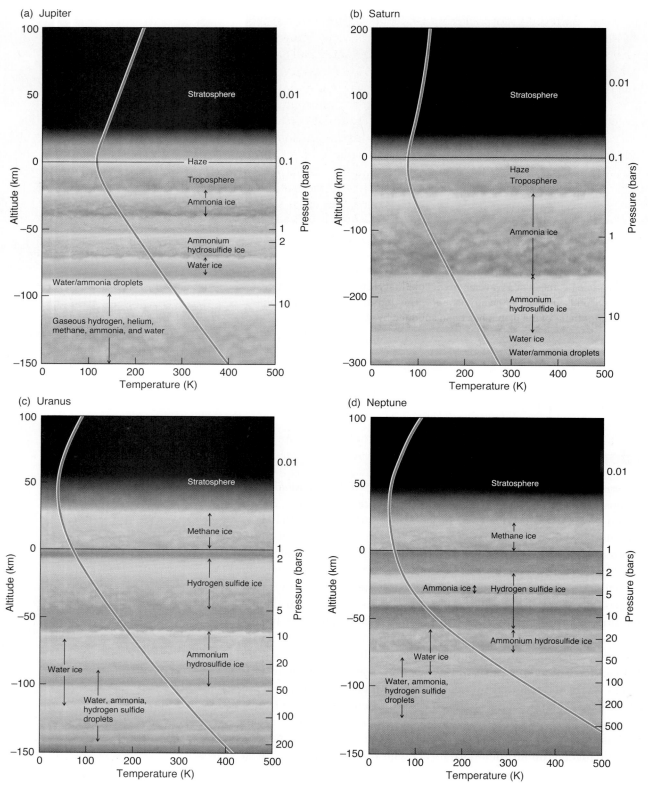

FIGURE 8.11 Volatile materials condense at different levels in the atmospheres of the giant planets, leading to chemically different types of clouds at different depths in the atmosphere. The red line in each diagram shows how atmospheric temperature changes with height. The arbitrary zero points of altitude are at 0.1 bar for Jupiter (a) and Saturn (b) and 1.0 bar for Uranus (c) and Neptune (d). Recall that 1.0 bar corresponds approximately to the atmospheric pressure at sea level on Earth. Note that Saturn's vertical scale is compressed to show the layered structure better.

The highest clouds in the frigid atmospheres of Uranus and Neptune are crystals of methane ice. The highest clouds on Jupiter and Saturn are made up of ammonia ice. Methane never makes ice in the warmer atmospheres of Jupiter and Saturn.

Why are some clouds so colorful, especially Jupiter's? These tints and hues must come from impurities in the ice crystals, similar to the way that syrups color snow cones. These impurities are probably elemental sulfur and phosphorus, as well as various organic materials produced when ultraviolet sunlight breaks up hydrocarbons and the fragments recombine to form complex organic compounds.

Uranus and Neptune are bluish green for much the same reason that Earth's oceans are blue. Methane gas is much more abundant in the atmospheres of Uranus and Neptune than in Jupiter and Saturn. Like water, methane gas absorbs the longer wavelengths of light—yellow, orange, and red. Absorption of the longer wavelengths leaves only the shorter wavelengths—green and blue—to be scattered from the relatively cloud-free atmospheres of Uranus and Neptune.

8.4 Winds and Storms—Violent Weather on the Giant Planets

The rapid rotation of the giant planets and resulting strong Coriolis effects (see Chapter 7) in their atmospheres create much stronger zonal winds than we see in the atmospheres of the terrestrial planets, even though less thermal energy is available. On Jupiter the strongest winds are equatorial westerlies, which have been clocked at speeds of 550 kilometers per hour (km/h), as seen in **Figure 8.12a**. (Remember from Chapter 7 that westerly winds are those that blow *from*, not toward, the west.) At higher latitudes, the winds alternate between easterly and westerly in a pattern that seems to be related to Jupiter's banded structure. Near a latitude of 20° south, the Great Red Spot vortex lies between a pair of easterly and westerly currents with opposing speeds of more than 200 km/h. If you think this might imply something about the relationship between zonal flow and vortices, you are right.

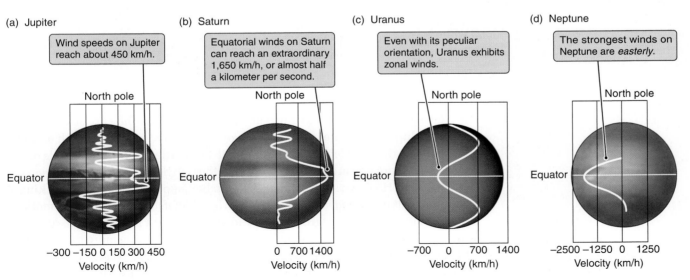

(a) Jupiter — Wind speeds on Jupiter reach about 450 km/h. — North pole — Equator — −300 −150 0 150 300 450 Velocity (km/h)

(b) Saturn — Equatorial winds on Saturn can reach an extraordinary 1,650 km/h, or almost half a kilometer per second. — North pole — Equator — 0 700 1400 Velocity (km/h)

(c) Uranus — Even with its peculiar orientation, Uranus exhibits zonal winds. — North pole — Equator — −700 0 700 1400 Velocity (km/h)

(d) Neptune — The strongest winds on Neptune are *easterly*. — North pole — Equator — −2500 −1250 0 1250 Velocity (km/h)

FIGURE 8.12 Strong winds blow in the atmospheres of the outer planets, driven by powerful convection and the Coriolis effect on these rapidly rotating worlds.

Working It Out 8.2 | How Do Astronomers Measure Wind Speeds on Distant Planets?

How do astronomers measure wind speeds on planets that are so far away? If we can see individual clouds in their atmospheres, we can measure their winds. As on Earth, clouds are carried by the local winds. By measuring the positions of individual clouds and noting how much they move during an interval of a day or so, we can calculate the local wind speed. We need to know one important additional piece of information, though: how fast the planet itself is rotating. The reason is that we want to measure the speed of the winds with respect to the planet's rotating surface.

In the case of the giant planets, of course, there is no solid surface against which to measure the winds. We must instead assume a hypothetical surface—one that rotates as though it were somehow "connected" to the planet's deep interior. As we will see near the end of this chapter, periodic bursts of radio energy caused by the rotation of a planet's magnetic field tell us how fast the interior of the planet is rotating. Interestingly, radio bursts from Saturn have, at different times, implied different rotation periods. This is puzzling, because there can be only one true rotation period. Until this phenomenon is understood, astronomers have adopted an average value.

Look back at the chapter-opening figure. In that figure, a student has used two images of Jupiter to work out the rotation period of Jupiter, assuming the whole planet rotates at the same speed that the Great Red Spot travels. This turns out to be quite a good assumption. Notice that to work out the rotation period, the student did not need to know the actual radius of Jupiter, in kilometers, say. Instead, she was able to take a ratio of sizes in the image. That's helpful in many situations. On the other hand, if we happen to know the radius, we can turn the question around and ask how fast the features are moving.

Let's see an example of how this works using a small white cloud in Neptune's atmosphere. The cloud, on Neptune's equator, is observed to be at longitude 73.0° west on a given day. (This longitude system is anchored in the planet's deep interior.) The spot is then seen at longitude 153.0° west exactly 24 hours later. Neptune's equatorial winds have carried the white spot 153.0° – 73.0° = 80.0° in longitude in 24 hours.

The circumference, C, of a planet is given by $2\pi r$, where r is the equatorial radius. The equatorial radius of Neptune is 24,760 km. So the circumference is

$$C = 2\pi r$$
$$C = 2\pi (24{,}760 \text{ km})$$
$$C = 155{,}600 \text{ km}$$

There are also 360° of longitude in the full circle represented by the circumference, so each degree of longitude stretches across 432 km (155,600 divided by 360). Because Neptune's equatorial winds have carried the spot 80°, we conclude that it has traveled 432 × 80 = 34,560 km in one day. This means that the speed is

$$\text{speed} = \frac{\text{distance}}{\text{time}}$$

$$\text{speed} = \frac{34{,}560 \text{ km}}{1 \text{ day}}$$

Converting days to hours gives more familiar units:

$$\text{speed} = \frac{34{,}560 \text{ km}}{1 \text{ day}} \times \frac{1 \text{ day}}{24 \text{ hours}}$$

$$\text{speed} = 1{,}440 \text{ km/h}$$

The wind speed is 1,440 km/h.

The equatorial winds on Saturn are also westerly, but they are stronger than those on Jupiter. In the early 1980s, *Voyager* measured speeds as high as 1,690 km/h. Later, the Hubble Space Telescope (HST) recorded maximum speeds of 990 km/h, and more recently *Cassini* has found them to be intermediate between the *Voyager* and HST measurements. What can be happening here? As **Working It Out 8.2** explains, we measure winds on other planets by observing the motions of their clouds. Saturn's winds appear to decrease with height, so the *apparent* time variability of Saturn's equatorial winds may be nothing more than changes in the height of the cloud tops. Alternating easterly and westerly winds also occur at higher latitudes; but unlike Jupiter's case, this alternation seems to bear no clear association with Saturn's atmospheric bands (**Figure 8.12b**). This is one example of the many unexplained differences among the giant planets.

Saturn's jet stream, at latitude 45° north, is a narrow, meandering river of atmosphere with alternating crests and troughs (**Figure 8.13**). It is similar to Earth's jet streams, where high-speed winds blow generally from west to east but wander toward and away from the poles. Nested within the crests and troughs of Saturn's jet stream are anticyclonic and cyclonic vortices. These are similar in both form and size to terrestrial high- and low-pressure systems, which bring us alternating periods of fair and stormy weather. Observing and analyzing similar atmospheric systems on other planets often helps us understand how our own weather works.

Our knowledge of global winds on Uranus (**Figure 8.12c**) is poorer than that of the other giant planets. When *Voyager 2* flew by Uranus in 1986, the few clouds we saw were all in its southern hemisphere because its northern hemisphere was in complete darkness at the time. The strongest winds observed were 650-km/h westerlies in the middle to high southern latitudes, and no easterly winds were detected. Because its peculiar orientation makes Uranus's poles warmer than its equator, some astronomers had predicted that the global wind system of Uranus might be very different from that of the other giant planets. But *Voyager 2* observed that the dominant winds on Uranus were zonal, just as they are on the other giant planets. This tells us that the Coriolis effect is more important than atmospheric temperature patterns in determining the structure of the global winds on all the giant planets.

As Uranus has continued in its orbit, previously hidden regions have become visible (**Figure 8.14**). Observations by HST and ground-based telescopes have shown bright cloud bands in the far north extending over 18,000 km in length and have revealed wind speeds of up to 900 km/h. As Uranus's long year passes, we will continue to learn much more about the northern hemisphere of Uranus.

As on Jupiter and Saturn, the strongest winds on Neptune occur in the tropics (**Figure 8.12d**). The surprise is that they are easterly rather than westerly, with speeds in excess of 2,000 km/h. Westerly winds with speeds higher than 900 km/h have been seen in Neptune's south polar regions. With wind speeds five times greater than those of the fiercest hurricanes on Earth, Neptune and Saturn are the windiest planets known. On Neptune, the southern hemisphere's summer solstice occurred in 2005, so much of the north is still in darkness. We'll have to wait a while before we can get a good look at its northern hemisphere. Each season on Neptune lasts 40 Earth years.

On the giant planets, the thermal energy that drives convection comes in part from the Sun, but primarily from the hot interiors of the planets themselves (**Working It Out 8.3**). The Coriolis effect shapes that convection into atmospheric vortices, familiar to us on Earth as high- and low-pressure systems, hurricanes, and supercell thunderstorms. On the giant planets, convective vortices are visible as isolated circular or oval cloud structures, such as the Great Red Spot on Jupiter and the Great Dark Spot on Neptune. As the atmosphere ascends near the centers of the vortices, it expands and cools. Cooling condenses certain volatile materials into liquid droplets, which then fall as rain. As they fall, the raindrops collide with surrounding air molecules, stripping electrons from the molecules and thereby developing tiny electric charges in the air. The cumulative effect of countless falling raindrops can generate an electric charge and resulting electric field so great that they create a surge of current and a flash of lightning. A single observation of Jupiter's night side by *Voyager 1* revealed several dozen lightning bolts within an interval of 3 minutes. The strength of these bolts is estimated to be equal to or greater than the "superbolts" that occur in the tops of high convec-

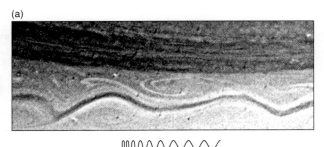
(a)

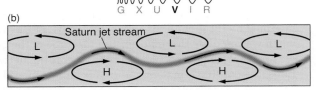

(b)

FIGURE 8.13 (a) *Voyager* image of a jet stream in Saturn's northern hemisphere, similar to jet streams in our terrestrial atmosphere. (b) The jet stream dips equatorward around regions of low pressure and is forced poleward around regions of high pressure.

FIGURE 8.14 Uranus is approaching equinox in this 2006 HST image. Much of its northern hemisphere is becoming visible. The dark spot in the northern hemisphere (to the right) is similar to but smaller than the Great Dark Spot seen on Neptune in 1989.

Working It Out 8.3 | Internal Thermal Energy Heats the Giant Planets

In Chapter 7 we learned about the equilibrium between the absorption of sunlight and the radiation of infrared light into space, and we saw how the resulting equilibrium temperature is modified by the greenhouse effect on Venus, Earth, and Mars. Yet when we calculate this equilibrium for three of the giant planets, we find that something seems amiss. According to these calculations, the equilibrium temperature for Jupiter should be 109 K; but when it is measured, we find instead an average temperature of about 124 K. A difference of 15 K might not seem like much, but remember that according to the Stefan-Boltzmann law, the energy radiated by an object depends on its temperature raised to the fourth power. Applying this relationship to Jupiter, we find

$$\left(\frac{T_{actual}}{T_{expected}}\right)^4 = \left(\frac{124}{109}\right)^4 = 1.67$$

This result is somewhat startling: Jupiter is radiating roughly two-thirds more energy into space than it absorbs in the form of sunlight. Similarly, the energy escaping from Saturn is about 1.8 times greater than the sunlight that it absorbs. Neptune emits 2.6 times as much energy as it absorbs from the Sun. Strangely, whatever internal energy may be escaping from Uranus is negligible compared with the absorbed solar energy.

With energy continually escaping from the interiors of the giant planets, we might wonder how they have maintained their high internal temperatures over the past 4½ billion years. The answer is that they have been and are still shrinking in size, thereby converting gravitational energy into thermal energy. This continual production of thermal energy replaces the energy that is escaping from their interiors. This is the primary energy source for replacing the internal energy that leaks out of the interior of Jupiter, and it is probably an important source for the other giant planets as well. Jupiter is contracting by only 1 millimeter or so per year. If it were to continue at this rate, in a billion years, Jupiter would shrink by only 1,000 km, a little more than 1 percent of its radius.

Thermal energy drives powerful convection on the giant planets.

tive clouds in the terrestrial tropics. *Cassini* has also imaged lightning flashes in Saturn's atmosphere, and radio receivers on *Voyager 1* picked up lightning static in the atmospheres of both Uranus and Neptune.

8.5 The Interiors of the Giant Planets Are Hot and Dense

At depths of a few thousand kilometers, the atmospheric gases of Jupiter and Saturn are so compressed by the weight of the overlying atmosphere that they liquefy. At depths of about 20,000 km in Jupiter's atmosphere and 30,000 km in Saturn's, the pressure climbs to 2 million bars and the temperature reaches 10,000 K. Under these conditions, hydrogen molecules are battered so violently that their electrons are stripped free, and the hydrogen acts like a liquid metal. In this state, it is called metallic hydrogen. The difference between a liquid and a highly compressed, very dense gas is subtle, so on Jupiter and Saturn there is no clear boundary between the atmosphere and the ocean of liquid hydrogen and helium that lies below. These oceans are tens of thousands of kilometers deep. Uranus and Neptune are less massive than Jupiter and Saturn, have lower interior pressures, and contain a smaller fraction of hydrogen—their interiors probably contain only a small amount of liquid hydrogen, with little or none of it in a metallic state.

Hydrogen-helium oceans lie at depths of a few thousand kilometers on Jupiter and Saturn.

Figure 8.15 shows the structure of the giant planets. At the center is a dense, liquid core consisting of a very hot mixture of heavier materials such as water, rock, and metals. The temperature at Jupiter's center is thought to be as high as 35,000 K, and the pressure may reach 45 million bars. Central temperatures and pressures of the other, less massive giant planets are correspondingly lower than

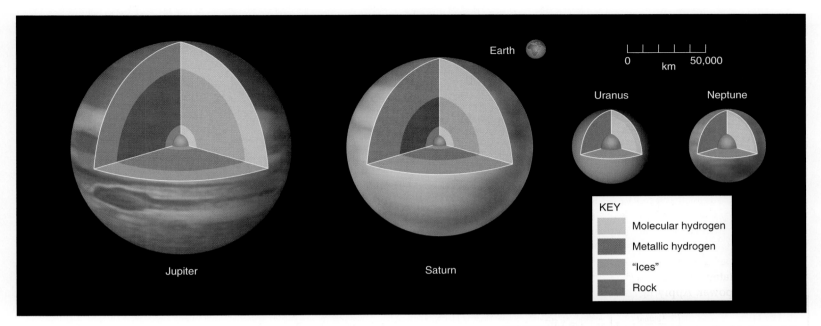

FIGURE 8.15 The central cores and outer liquid shells of the interiors of the giant planets. Only Jupiter and Saturn have significant amounts of the molecular and metallic forms of liquid hydrogen surrounding their cores.

those of Jupiter. It may seem strange that water is still liquid at temperatures of tens of thousands of degrees. Like an enormous version of a pressure cooker, the extremely high pressures at the centers of the giant planets prevent water from turning to steam.

Differentiation has occurred and is still occurring in Saturn, and perhaps in Jupiter too. On Saturn, helium condenses out of the hydrogen-helium oceans. (Helium can also be compressed to a metal, but it does not reach this metallic state under the physical conditions existing in the interiors of the giant planets.) Because these droplets of helium are denser than the hydrogen-helium liquid in which they condense, they sink toward the center of the planet, converting gravitational energy to thermal energy. This process heats the planet and enriches helium in the core while depleting it in the upper layers. In Jupiter's hotter interior, by contrast, the liquid helium is mostly dissolved along with the liquid hydrogen.

The Cores of Uranus and Neptune Are Different from the Cores of Jupiter and Saturn

The heavy-element components of the cores of Jupiter and Saturn have masses of about 5–10 Earth masses. Jupiter and Saturn have total masses of 318 and 95 Earth masses, respectively. The heavy materials in their cores contribute little to their average chemical composition. This means we can think of both Jupiter and Saturn as having approximately the same composition as the Sun and the rest of the universe: about 98 percent hydrogen and helium, leaving only 2 percent for everything else.

Uranus and Neptune are about twice as dense as Saturn, so they must be made of denser material than Saturn and Jupiter. Neptune, the densest of the giant planets, is about 1½ times as dense as uncompressed water and only about half as dense as uncompressed rock. Uranus is less dense than Neptune. These observations tell us that water and other low-density ices, such as ammonia and methane, must be the major compositional components of Uranus and Neptune, along with lesser amounts of silicates and metals. The total amount of hydrogen and helium in these planets is probably limited to no more than 1 or 2 Earth masses, and most of these gases reside in the planets' relatively shallow atmospheres.

Why do Jupiter and Saturn have so much hydrogen and helium compared with Uranus and Neptune? Why is Jupiter so much more massive than Saturn? The answers may lie both in the time that it took for these planets to form and in the distribution of material from which they formed. The cores of Uranus and Neptune were smaller and formed much later than those of Jupiter and Saturn, at a time when most of the gas in the protoplanetary disk had been blown away by the emerging Sun. Why did the cores of Uranus and Neptune form so late? Probably because the icy planetesimals from which they formed were more widely dispersed at their greater distances from the Sun. With more space between planetesimals, their cores would have taken longer to build up. Saturn may have captured less gas than Jupiter, both because its core formed somewhat later and because less gas was available at its greater distance from the Sun.

8.6 The Giant Planets Are Magnetic Powerhouses

All of the giant planets have magnetic fields that are much stronger than Earth's: their field strengths range from 50 to 20,000 times as strong. However, since field strength falls off with distance, fields at the cloud tops of Saturn, Uranus, and Neptune are comparable in strength to Earth's surface field. Even in the case of Jupiter's exceptionally strong field, the field strength at the cloud tops is only about 15 times that of Earth's surface field. In Jupiter and Saturn, magnetic fields are generated within deep layers of metallic hydrogen. In Uranus and Neptune, magnetic fields arise within deep oceans of liquid water and ammonia made electrically conductive by dissolved salts. We can illustrate the geometry of these magnetic fields as if they came from bar magnets, as shown in **Figure 8.16**.

The orientations of the magnetic field axes provide a mystery. Jupiter's magnetic axis is inclined 10° to its rotation axis—an orientation similar to Earth's—but it is offset about a tenth of a radius from the planet's center (Figure 8.16a). Saturn's magnetic axis is located almost precisely at the planet's center and is almost perfectly aligned with the rotation axis (Figure 8.16b). *Voyager 2* found that Uranus's magnetic axis is inclined nearly 60° to its rotation axis and is offset by a third of a radius from the planet's center (Figure 8.16c), but the really big surprise came when *Voyager 2* reached Neptune. The orientation of Neptune's rotation axis is similar to that of Earth, Mars, and Saturn. But Neptune's magnetic axis is inclined 47° to its rotation axis, and the center of this magnetic field is displaced from the planet's center by more than half the radius—an offset even greater than that of Uranus (Figure 8.16d). The displacement of the field is primarily toward Neptune's southern hemisphere, thereby creating a field 20 times stronger at the southern cloud tops than at the northern cloud tops. The reason for the unusual geometry of the magnetic fields of Uranus and Neptune remains unknown, but it is not related to the orientations of their rotation axes.

Giant Planets Have Giant Magnetospheres

Just as Earth's magnetic field traps energetic charged particles to form Earth's magnetosphere, the magnetic fields of the giant planets also trap energetic particles to form magnetospheres of their own. Our magnetosphere is tiny in comparison with those of the giant planets. By far the most colossal of these is Jupiter's mag-

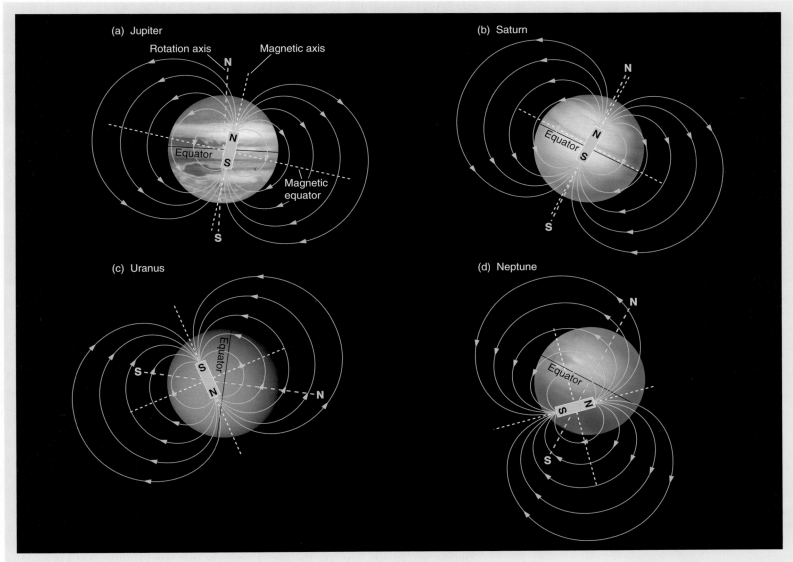

FIGURE 8.16 The magnetic fields of the giant planets can be approximated by the fields from bar magnets offset and tilted with respect to the planets' rotation axes. Compare these with Earth's magnetic field, shown in Figure 6.18.

⊙ **VISUAL ANALOGY**

netosphere. Its radius is 100 times that of the planet itself, roughly 10 times the radius of the Sun. Even the relatively weak magnetic fields of Uranus and Neptune form magnetospheres that are comparable in size to the Sun.

The solar wind (see Chapter 7) does more than supply some of the particles for a magnetosphere. The pressure of the solar wind also pushes on and compresses a magnetosphere, so the size and shape of a planet's magnetosphere depends on how the solar wind is blowing at any particular time. The tail of Jupiter's magnetosphere (**Figure 8.17**) extends well past the orbit of Saturn. The magnetic tails of Uranus and Neptune have a curious structure. Because of the tilt and the large displacements of their magnetic fields from the centers of these planets, their magnetospheres wobble as the planets rotate. This wobble causes the tails to twist like corkscrews as they stretch away from the planets.

The magnetospheres of the giant planets are enormous.

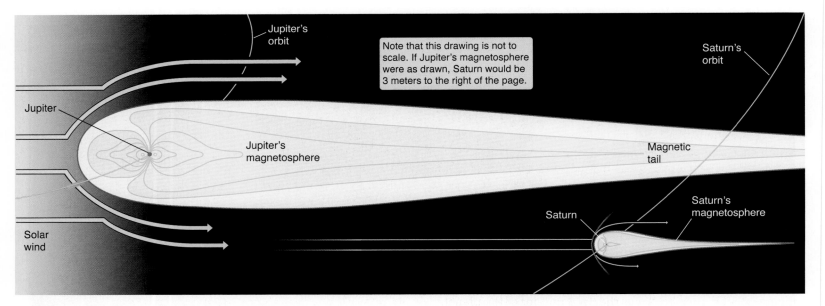

FIGURE 8.17 The solar wind compresses Jupiter's (or any other) magnetosphere in the direction toward the Sun and draws it out into a magnetic tail in the direction away from the Sun. Jupiter's tail stretches beyond the orbit of Saturn.

Jupiter has intense radiation belts.

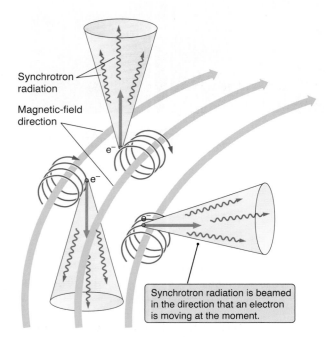

FIGURE 8.18 Rapidly moving charged particles loop around the direction of a magnetic field, giving off electromagnetic radiation as a result of the acceleration they experience. The radiation, called synchrotron radiation, is beamed in the direction of particle motion.

Radiation Belts and Auroras of the Giant Planets

As Jupiter rotates, it drags its magnetosphere around with it, and charged particles are swept around at high speeds. These fast-moving charged particles slam into neutral atoms, and the energy released in the resulting high-speed collisions heats the plasma to extreme temperatures. In 1979, while passing through Jupiter's magnetosphere, *Voyager 1* encountered a region of tenuous plasma with a temperature of over 300 million K; this is 20 times the temperature at the center of the Sun. If you're wondering why *Voyager 1* did not melt when passing through this region, remember that by *tenuous*, we mean that the plasma's particles were very far apart in space. Although each particle was extraordinarily energetic, there were so few of them that the probe passed unscathed through the plasma.

Charged particles trapped in planetary magnetospheres are concentrated in radiation belts, as we saw in Chapter 7. Earth's radiation belts are so severe as to worry astronauts, and yet the radiation belts that surround Jupiter are searing in comparison. In 1974 the *Pioneer 11* spacecraft passed through the radiation belts of Jupiter. During its brief encounter, *Pioneer 11* picked up a radiation dose of 400,000 rads, or about 1,000 times the lethal dose for humans. Several of the instruments on board were permanently damaged as a result, and the spacecraft itself barely survived to continue its journey to Saturn.

In Chapter 4 we found that anytime a charged particle experiences an acceleration, the particle emits electromagnetic radiation. Charged particles moving in a magnetic field experience a force, and this force produces an acceleration, causing the particles to spiral around the direction of the magnetic field. The acceleration causes the particles to radiate.

If the particles are traveling at a significant fraction of the speed of light, then *relativistic* effects cause the radiation they emit to be beamed in the direction they are traveling, as illustrated in **Figure 8.18**. The resulting radiation is called **synchrotron radiation**. The amount of radiation from a particle depends on the amount of acceleration the particle experiences. Electrons experience the greatest acceleration because they have the lowest masses, so they produce the majority of synchrotron radiation.

The magnetospheres of the giant planets contain energetic electrons moving in strong magnetic fields, so they satisfy the requirements for synchrotron radiation. The spectrum of synchrotron radiation is determined by the strength of the magnetic field and how energetic the radiating particles are. Synchrotron radiation from planetary magnetospheres is concentrated in the low-energy radio part of the spectrum.

Within this radio part of the spectrum, the second-brightest object in our sky is Jupiter's magnetosphere—only the Sun is brighter. And even though it is much farther away, Jupiter's magnetosphere appears much larger than the Sun in the sky. Saturn's magnetosphere is also large but much fainter than Jupiter's. Saturn does have a strong magnetic field, but pieces of rock, ice, and dust in Saturn's spectacular rings absorb magnetospheric particles. With far fewer magnetospheric electrons, there is much less radio emission from Saturn.

Except in the case of Saturn (see Working It Out 8.2), precise measurement of periodic variations in the radio signals from the giant planets tells us the planets' true rotation periods. The magnetic field of each planet is locked to the conducting liquid layers deep within the planet's interior, so the magnetic field rotates with exactly the same period as the deep interior of the planet.

This is our first encounter with synchrotron radiation, but it will not be our last. As we move outward into the universe we will find many objects—from quasars to the remnants of supernovae—that emit synchrotron radiation throughout the electromagnetic spectrum, from radio waves to X-rays.

In addition to protons and electrons from the solar wind, the magnetospheres of the giant planets contain large amounts of various elements, including sodium, sulfur, oxygen, nitrogen, and carbon from several sources, including the planets' extended atmospheres and the moons that orbit within them. The most intense radiation belt in the Solar System is a **torus** associated with Io, the innermost of Jupiter's four Galilean moons. Because of its low surface gravity and the violence of its volcanism, some of the gases erupting from Io's interior can escape the moon and become part of Jupiter's radiation belt. As charged particles are slammed into the moon by the rotation of Jupiter's magnetosphere, even more material is knocked free of Io's surface and ejected into space. Images of the region around Jupiter, taken in the light of emission lines from atoms of sulfur or sodium, show a faintly glowing torus of material supplied by the moon (**Figure 8.19**).

Other moons also influence the magnetospheres of the planets they orbit. The atmosphere of Saturn's largest moon, Titan, is rich in nitrogen. Leakage of this gas into space is the major source of a torus that forms in Titan's wake. The density of this rather remote radiation belt is highly variable because the orbit of Titan is sometimes inside and sometimes outside Saturn's magnetosphere, depending

Vocabulary Alert

Torus: Astronomers use this uncommon word because it's so much easier to write than "doughnut-shaped ring," especially if it's used as an adjective: *toroidal* versus "like a doughnut-shaped ring." A *torus* is different from a ring because it's "puffy"—geometrically, rings can be two- or even one-dimensional, but a torus is always distinctly three-dimensional.

Radio signals broadcast a planet's true rotation period.

Io is a source of magnetospheric particles.

Io's plasma torus

FIGURE 8.19 A faintly glowing torus of plasma surrounds Jupiter (center). The torus is made up of atoms knocked free from the surface of Io by charged particles. Io is the innermost of Jupiter's Galilean moons. (A semitransparent mask was placed over the disk of Jupiter to prevent its light from overwhelming the much fainter torus.)

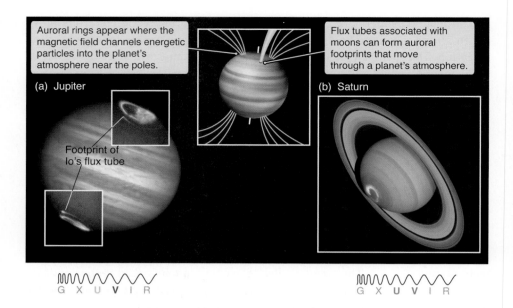

Auroral rings appear where the magnetic field channels energetic particles into the planet's atmosphere near the poles.

Flux tubes associated with moons can form auroral footprints that move through a planet's atmosphere.

(a) Jupiter

Footprint of Io's flux tube

(b) Saturn

G X U V I R

G X U V I R

Figure 8.20 Hubble Space Telescope images of auroral rings around the poles of Jupiter (a) and Saturn (b). The auroral images were taken in ultraviolet light and then superimposed on visible-light images. (High-level haze obscures the ultraviolet views of the underlying cloud layers, as the insets show.) The bright spot and trail outside the main ring of Jupiter's auroras are the footprint and wake of Io's *flux tube*.

on the strength of the solar wind. When Titan is outside Saturn's magnetosphere, any nitrogen molecules lost from the moon's atmosphere are carried away by the solar wind.

Charged particles spiral along the magnetic-field lines of the giant planets, bouncing back and forth between the two magnetic poles just like they do around Earth. The results are bright auroral rings (**Figure 8.20**) that surround the magnetic poles of the giant planets—just as the aurora borealis and aurora australis ring the north and south magnetic poles of Earth.

Jupiter's auroras have an added twist that we do not see on Earth, however. As Jupiter's magnetic field sweeps past Io, it generates an enormous electric field with an electric potential of 400,000 volts. This field accelerates electrons that then spiral along the direction of Jupiter's magnetic field. The result is a magnetic channel called a **flux tube** that connects Io with Jupiter's atmosphere near the

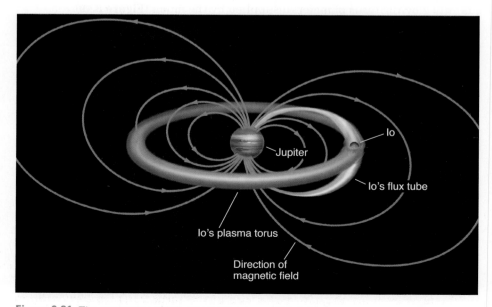

Io

Jupiter

Io's flux tube

Io's plasma torus

Direction of magnetic field

Figure 8.21 The geometry of Io's plasma torus and flux tube.

planet's magnetic poles (**Figure 8.21**). This flux tube carries an electric current of 5 million amperes, amounting to 2 trillion watts of power, or about one-sixth of the total electric power produced by generating stations on Earth. Much of the power generated within Io's flux tube is radiated away as radio energy, and these radio signals are received on Earth as intense bursts of synchrotron radiation. However, a substantial fraction of the energy of particles in the flux tube is deposited into Jupiter's atmosphere. At the very location where the flux tube intercepts the atmosphere, there is a spot of intense auroral activity. As Jupiter rotates, this spot leaves behind an auroral trail in Jupiter's atmosphere, and the footprint of Io's plasma torus, along with its wake, can be seen outside the main auroral ring in Figure 8.20a.

8.7 Rings Surround the Giant Planets

All of the giant planets have ring systems (**Figure 8.22**); none of the terrestrial planets do. Ring systems consist of countless numbers of small particles—varying in size from tiny grains to house-sized boulders—that orbit individually around the giant planets like so many tiny moons. Kepler's laws dictate that the speeds and orbital periods of all ring particles must vary with their distance from the planet, and that the closest particles move the fastest and have the shortest orbital periods. The orbital periods of particles in Saturn's bright rings, for example, range from 5 hours 45 minutes at their inner edge to 14 hours 20 minutes at their outer one. In the densely packed rings of Saturn, collisions between these particles circularize their orbits and also force them into the same plane. Any particle moving on a noncircular or inclined orbit will bump (very gently, because the particles have low relative speeds) into other particles and be unable to go any farther.

The orbits of ring particles can also be influenced by their planet's larger moons. If the moon is massive enough, it exerts a gravitational tug on the ring particles as it passes by. If this happens over and over through many orbits, the particles are pulled out of the area, leaving a lower-density gap. Such is the case with Saturn's moon Mimas, which causes the famous gap in the rings around Saturn called the Cassini Division (Figure 8.22b). Other effects are also possible. For example, most narrow rings are caught up in a tug-of-war with nearby moons, known as **shepherd moons** because of the way they "herd" the flock of ring particles. A shepherd moon just outside a ring robs orbital energy from any particles that drift outward beyond the edge of the ring, causing them to move back inward. A shepherd moon just inside a ring gives up orbital energy to a particle that has drifted too far in, nudging it back outward.

Tidal stresses are thought to be responsible for much of the material found in planetary rings. If a moon (or other planetesimal) orbits a large planet, the force of gravity will be stronger on the side of the moon close to the planet, and weaker on the side farther away. This stretches the moon out. If the tidal stresses are greater than the self-gravity that holds the moon together, the moon will be torn apart. The distance at which the tidal stresses exactly equal the self-gravity is known as the **Roche limit**. If a moon or planetesimal comes within the Roche limit of a giant planet, it is pulled apart by tidal stresses, leaving many small particles to orbit the planet. These particles gradually spread out. Collisions circularize and flatten out the orbits, as described earlier, and rings are formed.

The most complex ring system belongs to spectacular Saturn. Figure 8.22b shows the individual components of Saturn's ring system and its major divisions

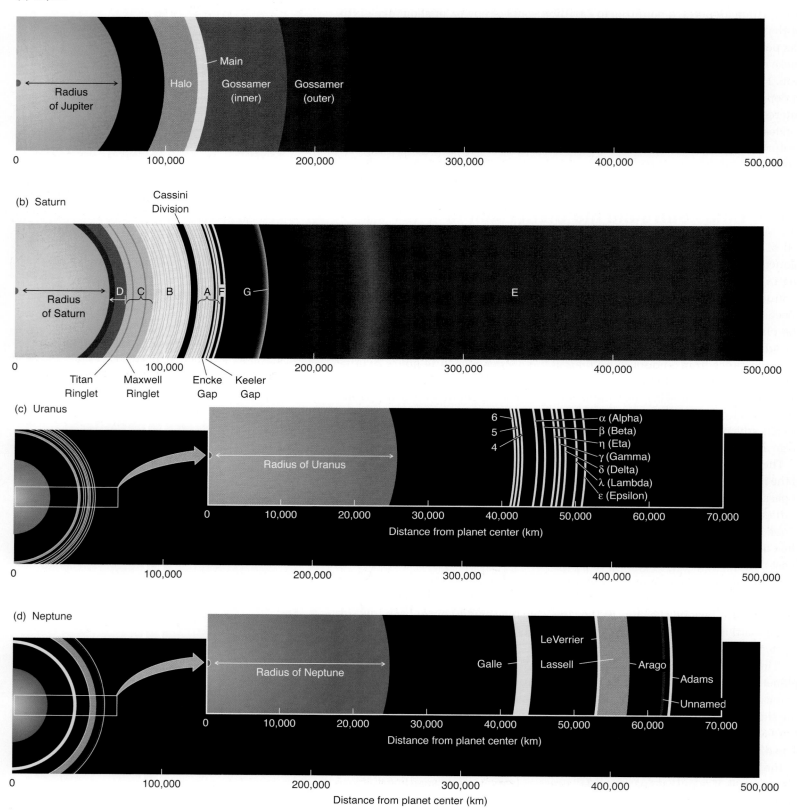

Figure 8.22 A comparison of the ring systems of the four giant planets.

and gaps. The most conspicuous features are its expansive bright rings, which dominate all photographs of Saturn. Among the four giant planets, only Saturn has rings so wide and so bright. The outermost bright ring, the A Ring, is the narrowest of the three bright rings. It has a sharp outer edge and contains several narrow gaps. On the A Ring's inner edge, the conspicuous Cassini Division is so wide (4,700 km) that the planet Mercury would almost fit within it. Astronomers once thought that it was completely empty, but images taken by *Voyager 1* show the Cassini Division is filled with material, although it is less dense than the material in bright rings.

The B Ring, whose width is roughly twice Earth's diameter, is the brightest of Saturn's rings and has no internal gaps on the scale of those seen in the other bright rings. The C Ring is so much fainter than neighboring rings that it often fails to show up in normally exposed photographs. Through the eyepiece of a telescope, though, this ring appears like delicate gauze. There is no known gap between the C Ring and either of the adjacent rings; only an abrupt change in brightness marks the boundary between them. The cause of this sharp change in the amount of ring material remains an unanswered question. Too dim to be seen next to Saturn's bright disk, the D Ring is a fourth wide ring that was unknown until imaged by *Voyager 1*. It shows less structure than any of the bright rings, and it does not appear to have a definable inner edge. The D Ring may extend all the way down to the top of Saturn's atmosphere, where its ring particles would burn up as meteors.

Saturn's bright rings are not uniform. The A and C rings contain hundreds, and the B Ring thousands, of individual **ringlets**, some only a few kilometers wide (**Figure 8.23**). Each of these ringlets is a narrowly confined concentration of ring particles bounded on both sides by regions of relatively little material.

About every 15 years, the plane of Saturn's rings lines up with Earth, and we view them edge on. The rings all but vanish for a day or so in even the largest telescopes. While the glare of the rings was absent, astronomers searched for undiscovered moons or other faint objects close to Saturn. In 1966, American astronomer Walter A. Feibelman (1930–2004) was looking for moons when he found weak but compelling evidence for a faint ring near the orbit of Saturn's moon Enceladus. In 1980, *Voyager 1* confirmed the existence of this faint ring, now called the E Ring, and found another closer one known as the G Ring. The E and G rings are examples of what astronomers call a **diffuse ring**. The particles in diffuse rings are far apart, and the rare collisions between particles can cause their individual orbits to become eccentric, inclined, or both. Because collisions are rare, the particles tend to remain in their disturbed orbits. Diffuse rings spread out horizontally and thicken vertically, sometimes without any obvious boundaries. Diffuse rings are difficult to detect except when lit from behind, when they appear bright because of the strong forward scattering of sunlight by very small ring particles.

Although Saturn's bright rings are very wide—more than 62,000 km from the inner edge of the C Ring to the outer edge of the A Ring—they are extremely thin. Saturn's bright rings are no more than a hundred meters and probably only a few tens of meters from their lower to upper surfaces. The diameter of Saturn's bright ring system is 10 million times the thickness of the rings. If the bright rings of Saturn were the thickness of the pages in this textbook, six football fields laid end to end would stretch across them.

Ring structure among the other giant planets is not as diverse as Saturn's. Most rings other than Saturn's are quite narrow, although a few are diffuse (see

Figure 8.23 This *Cassini* image of the rings of Saturn shows so many ringlets and minigaps that it looks like a close-up of an old-fashioned phonograph record. The cause of most of this structure has yet to be explained in detail.

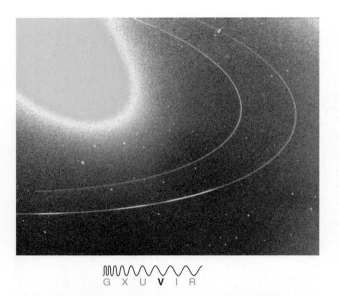

G X U V I R

Figure 8.24 *Voyager 2* image of the three brightest arcs in Neptune's Adams Ring. Neptune itself is very much overexposed in this image.

Figure 8.22a, c, and d). Jupiter's rings are made of fine dust dislodged by meteoritic impacts on the surfaces of Jupiter's small inner moons, which orbit among the rings. Nine of the 13 rings of Uranus are very narrow and widely spaced relative to their widths. Most are only a few kilometers wide, but they are many hundreds of kilometers apart. The space between these rings is filled with dust, which appears dark from our perspective. The dust is the right size to scatter sunlight forward instead of reflecting it backward. *Voyager 2* imaged this forward scattering after it passed Uranus, and showed the gaps were not empty. Four of Neptune's six rings are very narrow. One of these rings, the Adams Ring, is interesting because the material is clumped together in several **ring arcs**, high-density segments of the narrow ring (**Figure 8.24**). When first discovered, these ring arcs were a puzzle because collisions should have spread them uniformly around their orbit. Most astronomers now attribute this clumping to gravitational interactions with the moon Galatea. Some parts of Neptune's rings may be unstable: ground-based images taken 13 years after *Voyager 2* show decay in the ring arcs. One of the ring arcs may disappear entirely before the end of the century.

The Composition of Ring Material

Saturn's rings probably formed when a moon or planetesimal came within the Roche limit of Saturn. These rings reflect about 60 percent of the sunlight falling on them. They are made of water ice, though a slight reddish tint tells us they are not made of pure ice but must contain small amounts of other materials, such as silicates. The icy moons around Saturn or the frozen comets of the outer Solar System could easily provide this material.

Saturn's rings are the brightest in the Solar System and are the only ones that we know are composed of water ice. In stark contrast, the rings of Uranus and Neptune are among the darkest objects known in our Solar System. Only 2 percent of the sunlight falling on them is reflected back into space, thus making the ring particles blacker than coal or soot. No silicates or similar rocky materials are this dark, so these rings are likely composed of organic materials and ices that have been radiation-darkened by high-energy charged particles from the planets' magnetospheres. (Radiation blackens organic ices such as methane by releasing carbon from the ice's molecules.) Jupiter's rings are of intermediate brightness, suggesting that they may be rich in dark silicate materials, like the innermost of Jupiter's small moons.

Moons can contribute material to rings in other ways as well, as we see in the case of Jupiter's ring system. The brightest of Jupiter's rings is a narrow strand only 6,500 km across, consisting of material from the moons Metis and Adrastea (see **Figure 8.25**). These two moons orbit in Jupiter's equatorial plane, and the ring they form is thin. Beyond the main ring, however, are the very different **gossamer rings**, so called because they are extremely tenuous. The gossamer rings are supplied with dust by the moons Amalthea and Thebe. The innermost ring in Jupiter's system, called the Halo Ring, consists mostly of material from the main ring. As the dust particles in the main ring drift slowly inward toward the planet, they pick up electric charges and are pulled into this rather thick torus by electromagnetic forces associated with Jupiter's powerful magnetic field.

Finally, moons may contribute ring material through volcanism. Volcanoes on Jupiter's moon Io continually eject sulfur particles into space, many of which are pushed inward by sunlight and find their way into a ring. The particles in

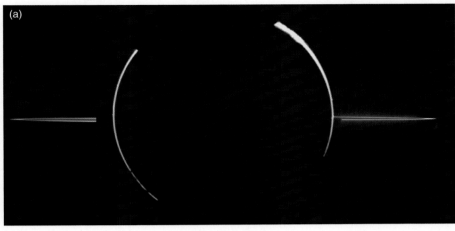

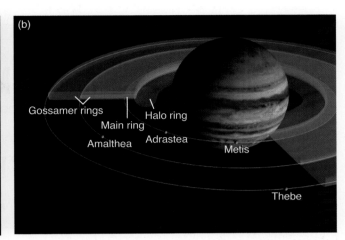

Figure 8.25 (a) A backlit *Galileo* view of Jupiter's rings. Note also the forward scattering of sunlight by tiny particles in the upper layers of Jupiter's atmosphere. (b) A diagram of Jupiter's ring system and the small moons that form the rings.

Saturn's E Ring are ice crystals ejected from icy geysers on the moon Enceladus, which is located in the very densest part of the E Ring.

Planetary Rings Are Short Lived

Planetary rings do not have the long-term stability of most Solar System objects. Ring particles are constantly colliding with one another in their tightly packed environment, either gaining or losing orbital energy. This redistribution of orbital energy can cause particles at the ring edges to leave the rings and drift away, aided by nongravitational influences such as the pressure of sunlight. Although moons may help guide the orbits of ring particles and delay the dissipation of the rings themselves, at best this can be only a temporary holding action. Most planetary rings eventually dissipate. At least one ring, however, seems immune from this eventual demise: because the volcanic emissions from Saturn's moon Enceladus are constantly supplying icy particles to Saturn's E Ring, replacing those that drift away, the E Ring will survive for as long as Enceladus remains geologically active.

It is quite unlikely that any of the planetary rings we see today have existed in their current form since the Solar System's beginning. It is far more likely that many ring systems have come and gone over the history of the Solar System. Even our own planet has probably had several short-lived rings at various times during its long history. Any number of comets or asteroids must have passed close enough to Earth to disintegrate catastrophically into a swarm of small fragments, thereby creating a temporary ring. However, Earth lacks shepherd moons to provide orbital stability.

Planetary rings in the outer Solar System will continue to come and go as long as the giant planets maintain the small moons that provide temporary ring stability. Whether any will rival the splendor of Saturn's bright ring system that we see today is, of course, unknown. We may be living in a fortuitous time when Saturn is putting on its best face.

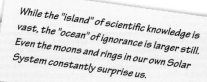
 READING ASTRONOMY News

Giant Propeller Structures Seen in Saturn's Rings

By **Denise Chow, MSNBC**

Giant propeller-shaped structures have been discovered in the rings of Saturn and appear to be created by a new class of hidden moons, NASA announced Thursday.

NASA's *Cassini* spacecraft spotted the distinctive structures inside some of Saturn's rings, marking the first time scientists have managed to track the orbits of individual objects from within a debris disk like the one that makes up Saturn's complicated ring system.

"Observing the motions of these disk-embedded objects provides a rare opportunity to gauge how the planets grew from, and interacted with, the disk of material surrounding the early sun," said the study's co-author Carolyn Porco, one of the lead researchers on the *Cassini* imaging team based at the Space Science Institute in Boulder, Colo. "It allows us a glimpse into how the solar system ended up looking the way it does."

Photos of the propellers taken by *Cassini* show them to be huge structures several thousands of miles long. By understanding how they form, astronomers hope to glean insight into the debris disks around other stars as well, researchers said.

The results of the study are detailed in the July 8 issue of the journal *Astrophysical Journal Letters*.

Cassini scientists have seen double-armed propeller structures in Saturn's rings before, but on a smaller scale than the larger, new-found features. They were first spotted in 2006 in an area now known as the "propeller belt," which is located in the middle of Saturn's outermost dense ring the A ring.

The propellers are actually gaps in the ring material [that] were created by a new class of objects, called moonlets, that are smaller than known moons but larger than the particles making up Saturn's rings. It is estimated that these moonlets could number in the millions, according to *Cassini* scientists.

The moonlets clear the space immediately around them to generate the propeller-like features, but are not large enough to sweep clear their entire orbit around Saturn, as seen with the moons Pan and Daphnis.

But in the new study, researchers [saw] a new legion of larger and rarer moons in a separate part of the A ring, farther out from Saturn. These much larger moons create propellers that are hundreds of times larger than those previously described, and these objects have been tracked for about four years.

The study was led by *Cassini* imaging team associate Matthew Tiscareno at Cornell University in Ithaca, New York.

The propeller features for these larger moons are up to thousands of miles long and several miles wide. The moons embedded in Saturn's rings appear to kick up ring material as high as 1,600 feet above and below the ring plane.

This is much greater than the typical ring thickness of about 30 feet, researchers said.

Still, the *Cassini* spacecraft is too far away to see the moons amid the swirling ring material that surrounds them. Yet, scientists estimate that the moons measure approximately half a mile in diameter, based on the size of the propellers.

According to their research, Tiscareno and

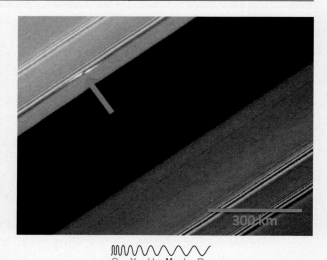
G X U V I R

Figure 8.26 *Cassini* took this close-up image of a propeller-shaped structure in the rings of Saturn.

his colleagues estimate that there are dozens of these giant propellers. In fact, 11 of them were imaged multiple times between 2005 and 2009.

One such propeller, nicknamed Bleriot after the famous aviator Louis Bleriot, has shown up in more than 100 separate *Cassini* images and one ultraviolet imaging spectrograph observation during this time.

"Scientists have never tracked disk-embedded objects anywhere in the universe before now," said Tiscareno. "All the moons and planets we knew about before orbit in empty space. In the propeller belts, we saw a swarm in one image and then had no idea later on if we were seeing the same individual objects. With this new discovery, we can now track disk-embedded moons individually over many years."

SEE **GIANT PROPELLER**

Over their four years of observation, the researchers noticed shifts in the orbits of the giant propellers as they travel around Saturn, but the cause of these disturbances has not yet been determined.

The shifting orbits could be caused by collisions with other smaller ring particles, or could be responses to these particles' gravity, the researchers said. The orbital paths of these moonlets could also be altered due to the gravitational attraction of large moons outside of Saturn's rings.

Scientists will continue to monitor the moons to see if the disk itself is driving the changes, similar to the interactions that occur in young solar systems. If so, Tiscareno said, this would be the first time such a measurement has been made directly.

"Propellers give us unexpected insight into the larger objects in the rings," said Linda Spilker, *Cassini* project scientist based at NASA's Jet Propulsion Laboratory in Pasadena, California. "Over the next seven years, *Cassini* will have the opportunity to watch the evolution of these objects and to figure out why their orbits are changing."

NASA launched the *Cassini* probe in 1997 and it arrived at Saturn in 2004, where it dropped the European *Huygens* probe on the cloudy surface of Titan, Saturn's largest moon. *Cassini* was slated to be decommissioned in September of this year, but received a life extension that now runs through 2017.

Evaluating the News

1. What new type of structure was observed in Saturn's rings?

2. Why had these structures never been observed before?

3. In the story, Carolyn Porco states: "Observing the motions of these disk-embedded objects provides a rare opportunity to gauge how the planets grew from, and interacted with, the disk of material surrounding the early sun." Explain how the rings of Saturn might tell us something about the protoplanetary disk.

4. What is "new" about the "new class of moons" that are driving the formation of propeller shapes?

5. Put this story in context of all the information that you know about Saturn's rings. How does this discovery demonstrate the ongoing nature of the scientific process and provisional nature of scientific knowledge?

SUMMARY

8.1 The giant planets are large and massive, and less dense than the terrestrial planets. Jupiter and Saturn are made up mostly of hydrogen and helium—a composition similar to that of the Sun. Uranus and Neptune contain much larger amounts of "ices" such as water, ammonia, and methane.

8.2 We see only atmospheres, not solid surfaces, on the giant planets because solid surfaces, if they exist, are deep below the cloud and liquid layers.

8.3 Clouds on Jupiter and Saturn are composed of various kinds of ice crystals colored by impurities. Uranus and Neptune have relatively few clouds.

8.4 Powerful convection and the Coriolis effect drive high-speed winds in the upper atmospheres of the giant planets. Jupiter, Saturn, and Neptune have large internal heat sources, while Uranus seems to have little or none.

8.5 The interiors of the giant planets are very hot and very dense.

8.6 The giant planets have enormous magnetospheres that emit synchrotron radiation.

8.7 All four giant planets are surrounded by rings. Planetary rings are short lived.

✦ SUMMARY SELF-TEST

1. Jupiter and Saturn have compositions similar to the Sun and are called _____ giants. Uranus and Neptune, on the other hand, are called _____ giants.

2. **T/F:** There is an abrupt change to a solid surface in the interior of a giant planet.

3. As depth increases, _____ and _____ increase, which causes changes in the chemical composition of clouds in giant planet atmospheres.

4. The following steps lead to convection in giant planet atmospheres. Place the last five in order, following the first one.
   a. Gravity pulls particles toward the center.
   b. Warm material rises and expands.

   c. Particles fall toward the center, converting gravitational energy to kinetic energy.
   d. Expanding material cools.
   e. Thermal energy heats the material.
   f. Friction converts kinetic energy to thermal energy.

5. Zonal winds on the giant planets are stronger than those on the terrestrial planets because
   a. the giant planets have more thermal energy.
   b. the giant planets rotate faster.
   c. the moons of the giant planets provide additional pull.
   d. the moons of the giant planets feed energy to their planet through the magnetosphere.

6. Deep in the interiors of the giant planets, water is still a liquid even though the temperatures are tens of thousands of degrees above the boiling point of water. This can happen because
   a. the density inside the giant planets is so high.
   b. the pressure inside the giant planets is so high.
   c. the outer solar system is so cold.
   d. space has very low pressure.

7. Imagine a giant planet, very similar to Jupiter, that was ejected from its solar system at formation. (These objects exist, and are probably numerous, although their total number is still uncer- tain.) This planet would almost certainly still have (choose all that apply)
   a. a magnetosphere.   b. thermal energy.
   c. auroras.   d. rings.

8. Saturn's bright rings are located within the Roche limit of Saturn. This supports the theory that these rings (choose all that apply)
   a. are formed of moons torn apart by tidal stresses.
   b. formed at the same time that Saturn formed.
   c. are relatively recent.
   d. are temporary.

QUESTIONS AND PROBLEMS

True/False and Multiple-Choice Questions

9. **T/F:** Uranus has extreme seasons because its pole is in the plane of the Solar System.

10. **T/F:** All the giant planets have clouds and belts.

11. **T/F:** Water never forms clouds in giant planets' atmospheres.

12. **T/F:** Storms on giant planets last much longer than storms on Earth.

13. **T/F:** The cores of the giant planets are all similar.

14. **T/F:** Auroras on Jupiter are created in part by particles from Io.

15. **T/F:** The Roche limit is the place where a giant planet's gravity equals the self-gravity of a moon.

16. The chemical compositions of Jupiter and Saturn are most similar to those of
    a. Uranus and Neptune.
    b. the terrestrial planets.
    c. their moons.
    d. the Sun.

17. The Great Red Spot on Jupiter is
    a. a surface feature.
    b. a "storm" that has been raging for more than 300 years.
    c. caused by the interaction between the magnetosphere and Io.
    d. about the size of North America.

18. The different colors of clouds on Jupiter are primarily a result of
    a. temperature.
    b. composition.
    c. motion of the clouds toward or away from us.
    d. all of the above.

19. Consider a northward-traveling particle of gas on Jupiter. According to the Coriolis effect, which way will this particle of gas turn?
    a. in the direction of rotation
    b. opposite the direction of rotation
    c. north
    d. south

20. Metallic hydrogen is not
    a. a metal that acts like hydrogen.
    b. hydrogen that acts like a metal.
    c. common in the cores of giant planets.
    d. a result of high temperatures and pressures.

21. The magnetic fields of the giant planets
    a. align closely with the rotation axis.
    b. extend far into space.
    c. are thousands of times stronger at the cloud tops than at Earth's surface field.
    d. have an axis that passes through the planet's center.

22. _____ has a ring system.
    a. Saturn   b. Jupiter
    c. Uranus   d. Neptune
    e. all of the above

Conceptual Questions

23. Describe the ways in which the giant planets differ from the terrestrial planets.

24. In what way are Uranus and Neptune not the same kind of giants as Jupiter and Saturn?

25. Jupiter's chemical composition is more like the Sun's than Earth's is. Yet both planets formed from the same protoplanetary disk. Explain why they are different today.

26. What can we learn about a Solar System object when it occults a star?

27. Explain two methods used to determine the mass of a giant planet.

28. Why can we not see the surfaces of the giant planets?

29. Jupiter is sometimes described as "a star that failed." Explain why this is a gross exaggeration.

30. Astronomers take the unusual position of lumping together all atomic elements other than hydrogen and helium into a single category, which they call heavy elements. Why is this a reasonable position for astronomers to take?

31. None of the giant planets are truly round. Explain why they have a flattened appearance.

32. The Great Red Spot is a long-lasting atmospheric vortex in Jupiter's southern hemisphere. Winds rotate counterclockwise around its center. Is the Great Red Spot cyclonic or anticyclonic? Is it a region of high or low pressure? Explain.

33. Why do the individual cloud layers in the atmospheres of the giant planets have different chemical compositions?

34. What is the source of color in Jupiter's clouds?

35. Lightning has been detected in the atmospheres of all the giant planets. How is it detected?

36. Uranus and Neptune, when viewed through a telescope, appear distinctly bluish green in color. What are the two reasons for their striking appearance?

37. Which of the giant planets have seasons similar to Earth's, and which one experiences extreme seasons?

38. Explain how astronomers measure wind speeds in the atmospheres of the giant planets.

39. What drives the zonal winds in the atmospheres of the giant planets?

40. Jupiter, Saturn, and Neptune all radiate more energy into space than they receive from the Sun. Does this violate the law of conservation of energy? What is the source of the additional energy?

41. What creates metallic hydrogen in the interiors of Jupiter and Saturn, and why do we call it metallic?

42. Jupiter's core is thought to consist of rocky material and ices, all in a liquid state at a temperature of 35,000 K. How can materials such as water be liquid at such high temperatures?

43. Saturn has a source of internal thermal energy that Jupiter may not have. What is it, and how does it work?

44. When viewed by radio telescopes, Jupiter is the second-brightest object in the sky. What is the source of this radiation?

45. What creates auroras in the polar regions of Jupiter and Saturn?

46. Identify and explain two possible mechanisms that can produce planetary ring material.

47. Will the particles in Saturn's bright rings eventually stick together to form one solid moon orbiting at the mean distance of all the ring particles? Explain your answer.

48. How does the mass of a planet's rings compare with the mass of its individual moons?

49. Explain two mechanisms that create gaps in Saturn's bright ring system.

50. Explain ways in which diffuse rings differ from other planetary rings.

51. Describe and explain a mechanism that keeps planetary rings from dissipating.

52. Astronomers believe that most planetary rings eventually dissipate. Explain why they do not last forever.

53. Name one ring that might continue to exist indefinitely, and explain why it could survive when others might not.

54. Why does Earth not have a ring?

Problems

55. The Sun appears 400,000 times brighter than the full Moon in our sky. How far from the Sun (in astronomical units) would you have to go for the Sun to appear only as bright as the full Moon appears in our nighttime sky? Compare your answer with the semimajor axis of Neptune's orbit.

56. Uranus occults a star at a time when the relative motion between Uranus and Earth is 23.0 km/s. An observer on Earth sees the star disappear for 37 minutes and 2 seconds and notes that the center of Uranus passes directly in front of the star.
   a. Based on these observations, what value would the observer calculate for the diameter of Uranus?
   b. What could you conclude about the planet's diameter if its center did not pass directly in front of the star?

57. Jupiter is an oblate planet with an average radius of 69,900 km, compared to Earth's average radius of 6,370 km.
   a. Remembering that volume is proportional to the cube of the radius, how many Earth volumes could fit inside Jupiter?
   b. Jupiter is 318 times as massive as Earth. Show that Jupiter's average density is about one-fourth that of Earth's.

58. The obliquity of Uranus is 98°. If you were located at one of the planet's poles, how far from the zenith would the Sun appear at the time of summer solstice?

59. Compare the graphs in Figure 8.11a and b. Does atmospheric pressure increase more rapidly with depth on Jupiter, or on Saturn? Compare the graphs in Figure 8.11c and d. Does pressure increase more rapidly with depth on Uranus, or on Neptune? Of the four giant planets, which has the fastest pressure rise with depth? Which has the slowest?

60. Using Figure 8.11, find the temperature at an altitude of 100 km on each of the four giant planets.

61. A small cloud in Jupiter's equatorial region is observed to be at a longitude of 122.0° west in a coordinate system that rotates at the same rate as the deep interior of the planet. (West longitude is measured along a planet's equator toward the west.) Another observation made exactly 10 Earth hours later finds the cloud at a longitude of 118.0° west. Jupiter's equatorial radius is 71,500 km. What is the observed equatorial wind speed in kilometers per hour? Is this an easterly or a westerly wind?

62. The equilibrium temperature for Saturn should be 82 K, but instead we find an average temperature of 95 K. How much more energy is Saturn radiating into space than it absorbs from the Sun?

63. Neptune radiates into space 2.6 times as much energy as it absorbs from the Sun. Its equilibrium temperature is 47 K. What is its true temperature?

 SmartWork, Norton's online homework system, includes algorithmically generated versions of these questions, plus additional conceptual exercises. If your instructor assigns questions in SmartWork, log in at **smartwork.wwnorton.com**.

 StudySpace is a free and open website that provides a Study Plan for each chapter of **Understanding Our Universe**. Study Plans include animations, reading outlines, vocabulary flashcards, and multiple-choice quizzes, plus links to premium content in SmartWork and the ebook. Visit **wwnorton.com/ studyspace**.

Exploration | Estimating Rotation Periods of Giant Planets

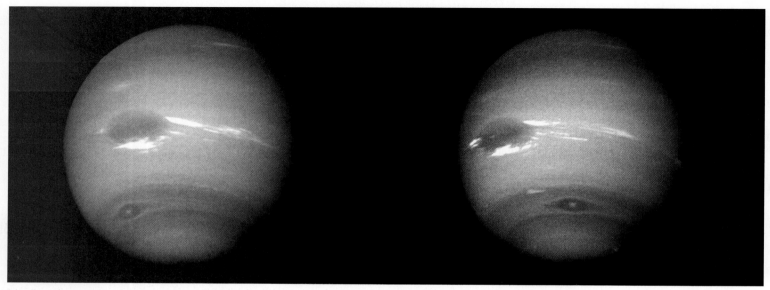

Figure 8.27 Two images of Neptune, taken 17.6 hours apart by *Voyager 2*.

G X U V I R

Study the two images of Neptune in **Figure 8.27**. The image on the left was taken first, and the image on the right was taken 17.6 hours later; during this time, the Great Dark Spot completed nearly one full rotation. The small storm at the bottom of the image of the planet completed slightly more than one rotation. You would be very surprised to see this result for, say, Mars.

1. **What does this tell you about the rotation of visible cloud tops of the planet?**

We can find the rotation period of the smaller storm in Figure 8.27 by equating two ratios, as shown in the chapter-opening figure. First, use a ruler to find the distance (in millimeters) from the left "edge" of the planet to the small storm (as shown in **Figure 8.28a**). Do this for both the left and right images in Figure 8.27. The right edge of the planet is not illuminated, so we will have to estimate the radius of the circle traveled by the storm. We do this by measuring from the limb to a line through the center of the planet (as shown in **Figure 8.28b**). Since the small storm travels along a line of latitude close to a pole, the distance it travels is significantly less than the circumference of the planet.

2. **Estimate the radius of the circle the small storm makes around Neptune (in mm) by making the appropriate measurements on Figure 8.27.**

3. **Find the circumference of this circle (in mm).**

Because the small storm has rotated *more than* one time, the total distance it has traveled is the circumference of the circle plus the distance between its location in the two images.

4. **Add those two numbers together to get the total distance traveled (in mm) between these images.**

(a)

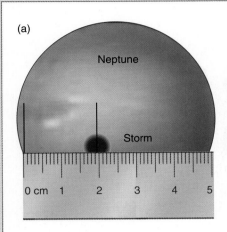

(b)

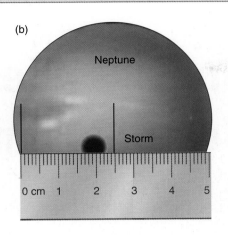

Figure 8.28 (a) How to measure the position of the storm. (b) How to measure the radius of the circle the storm traveled.

Now we are ready to take a ratio and find the rotation period. The ratio of the rotation period, T, to the time elapsed, t, must be equal to the ratio of the circumference of the circle around which it travels, C (in mm), to the total distance traveled, D (in mm).

$$\frac{T}{t} = \frac{C}{D}$$

5. **You have all the numbers you need to solve for T. What value do you calculate for the small storm's rotation period? (To check your work, note that your answer should be less than 17.6 hours. Why?)**

You may be wondering why this works at all. Clearly, the actual distance the small storm traveled is not a small number of millimeters, nor is the circumference of the circle around which it travels. To find the actual distance or circumference, however, you would multiply both values by the same constant of proportionality. Since we are taking a ratio, that constant cancels out, so we might as well leave it out from the beginning.

6. **Perform the corresponding measurements and calculations for the Great Dark Spot in Figure 8.27. What is its rotation period? Think carefully about how to find the total distance traveled, since the Great Dark Spot has rotated *less than* one time around the planet. (To check your work, note that your answer should be more than 17.6 hours. Why?)**

7. **How similar are the rotation periods for these two storms?**

8. **What does this tell you about determining the rotation periods of the giant planets using this method?**

9. **What method do astronomers use instead?**

9 Small Bodies of the Solar System

We began our discussion of the Solar System with its history. We learned that very early on—at the same time our Sun was becoming a star—tiny grains of primitive material were sticking together to produce swarms of small bodies called planetesimals. Those that formed in the hotter, inner part of the Solar System were composed mostly of rock and metal, while those in the colder, outer parts were made up of ice, organic compounds, and rock. We also learned what happened to many of these planetesimals. Some were consumed during the era of large-body building to become planets and moons, and others were ejected from the Solar System by gravitational encounters with larger bodies. However, not all of these primitive bodies suffered the same fate. Many of them are still around, and they represent a small but scientifically important component of our present-day Solar System. As we now conclude our discussion of the Solar System, we find that we have come full circle. These remaining planetesimals, and the fragments they continually create, provide planetary scientists with the opportunity to look back to the physical and chemical conditions of the earliest moments in the history of the Solar System.

✧ LEARNING GOALS

Dwarf planets are among the lesser worlds in the Solar System. Asteroids and comets are even smaller, yet these objects—and their fragments that survive the atmosphere to fall to Earth as meteorites—have told us much of what we know about the early history of the Solar System. With the dawn of the space age, robotic explorers traveling through the Solar System have shown us these small bodies. In the image at right, a student has imaged an asteroid while studying the stars in this field, and confirmed her finding using a catalog of asteroids. By the end of the chapter, you should understand what an asteroid is, what they tell us about our Solar System, and why scientists are intently studying asteroids in a special class known as near-Earth objects. You should also be able to:

- List the categories of small bodies

- Classify moons by their geologic activity

- Explain why some asteroids differentiated while others did not

- Describe the origin of different types of meteorites

- Sketch the appearance of a comet at different locations in its orbit

- Explain the importance of meteorites, asteroids, and comets to the history of the Solar System

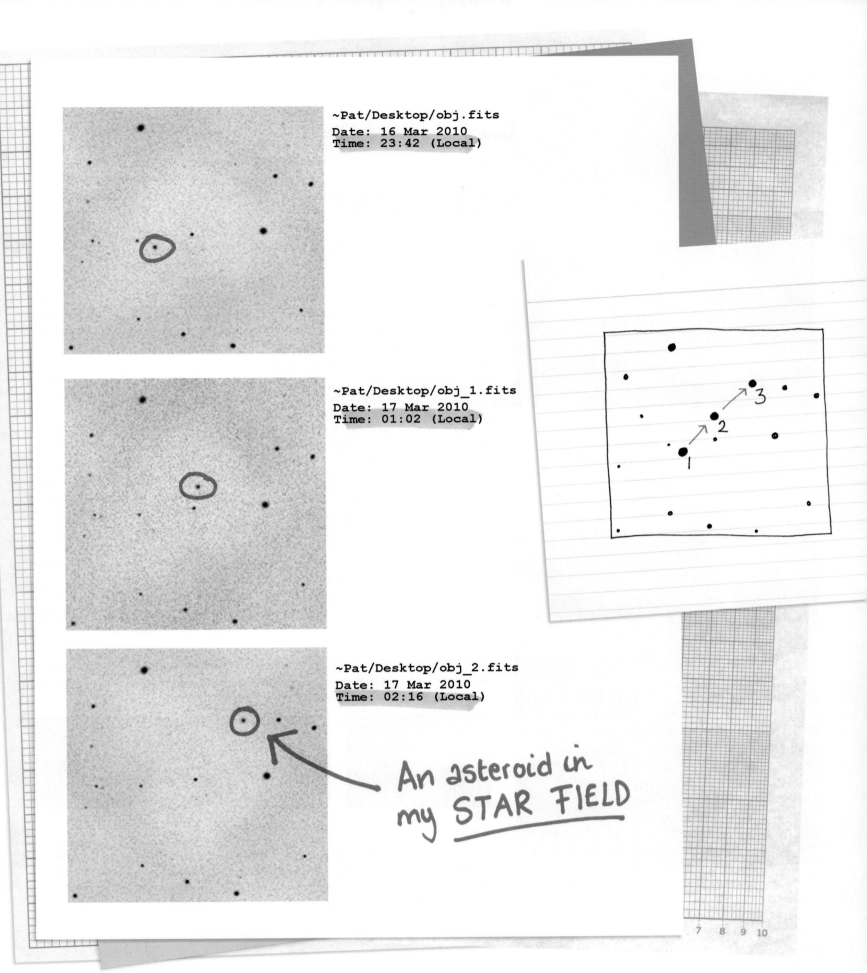

~Pat/Desktop/obj.fits
Date: 16 Mar 2010
Time: 23:42 (Local)

~Pat/Desktop/obj_1.fits
Date: 17 Mar 2010
Time: 01:02 (Local)

~Pat/Desktop/obj_2.fits
Date: 17 Mar 2010
Time: 02:16 (Local)

An asteroid in my STAR FIELD

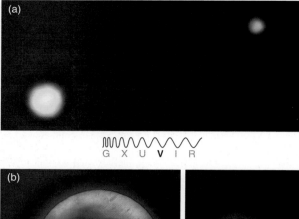

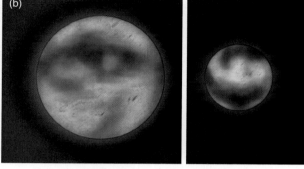

Figure 9.1 The dwarf planet Pluto and its largest moon, Charon, in (a) a Hubble Space Telescope image and (b) an artist's rendition based on analysis of HST images. Pluto and Charon are so similar in size that they could be considered a "double planet."

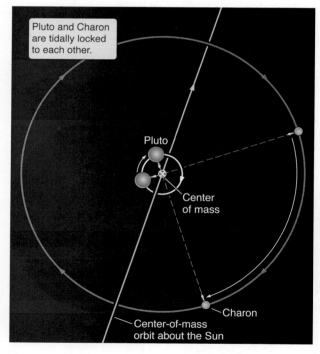

Pluto and Charon are tidally locked to each other.

Pluto

Center of mass

Charon

Center-of-mass orbit about the Sun

Figure 9.2 A diagram showing the doubly synchronous rotation and revolution in the Pluto-Charon system. The two bodies permanently show the same face to one another.

9.1 Dwarf Planets—Pluto and Others

Planets are objects that orbit the Sun and are large enough to (1) pull themselves into a round shape and (2) clear the area around their orbits. The dwarf planets meet criterion 1 by being round, but because they have considerable company in their orbits, they fail to meet criterion 2. As of this writing, there are five recognized dwarf planets in the Solar System: Pluto, Eris, Haumea, Makemake, and Ceres. Ceres is an especially large object in the main asteroid belt (a region between Mars and Jupiter that contains most of the Solar System's asteroids); the others are all found in the outer Solar System.

Pluto's orbit is 248 Earth years long and quite elliptical, periodically bringing the dwarf planet inside Neptune's nearly circular orbit—from 1979 to 1999, Pluto was closer to the Sun than Neptune. Pluto has four known moons, the largest of which is Charon, about half the size of Pluto, which in turn is only two-thirds as large as our Moon (**Figure 9.1a and b**). The total mass of the Pluto-Charon system is 1/400 that of Earth. Pluto and Charon each have a rocky core surrounded by a water-ice mantle. Pluto's surface is an icy mixture of frozen water, carbon dioxide, nitrogen, methane, and carbon monoxide, but Charon's surface is primarily dirty water ice. Pluto has a thin atmosphere of nitrogen, methane, and carbon monoxide; these gases freeze out of the atmosphere when Pluto is more distant from the Sun and therefore cooler.

Pluto and Charon are a tidally locked pair: each has one hemisphere that always faces the other (**Figure 9.2**). We know little else about these worlds' surface features, and nothing about their geological history. We hope to know more soon—the NASA spacecraft *New Horizons* will reach Pluto and Charon in 2015.

Eris (**Figure 9.3**) is about the same size as Pluto but is more massive. Eris also has a relatively large moon, called Dysnomia. Eris's highly eccentric orbit carries it from 37.8 astronomical units (AU) out to 97.6 AU, with an orbital period of 562 years. By chance, Eris was found near the most distant point in its orbit, making it currently the most remote known object in the Solar System. (The eccentric orbits of other Solar System bodies, such as Sedna, will eventually carry them farther from the Sun.) Eris is highly reflective, so it must have a coating of pristine ice. Spectra indicate this ice is probably mostly methane, which will form an atmosphere when Eris comes closest to the Sun in the year 2256.

Haumea and Makemake are both smaller and have slightly larger orbits than Pluto. Haumea has two known moons—Hi'iaka and Namaka. Haumea spins so rapidly on its axis that its shape is distorted into a flattened ellipsoid with an

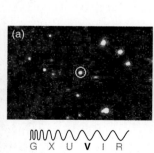

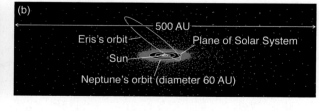

500 AU

Eris's orbit

Plane of Solar System

Sun

Neptune's orbit (diameter 60 AU)

Figure 9.3 (a) The distant dwarf planet Eris (shown within the white circle) is about the same size as Pluto and has similar physical characteristics. Small bodies of the Solar System are typically indistinguishable from stars in individual images and are found by comparing images of the same field over time. Solar System objects change position from one image to the next, but the stars remain in the same place. (b) Eris's orbit is both highly eccentric and highly inclined to the rest of the Solar System.

equatorial radius that is approximately twice its polar radius. This is the most distorted shape of any known planet, dwarf or otherwise. No moons have been discovered orbiting around Makemake, and we know less about this dwarf planet than about Pluto, Haumea, or Eris.

With a diameter of about 975 km, Ceres (**Figure 9.4**) is larger than most moons but smaller than any planet. It contains about a third of the total mass in the main asteroid belt. Ceres rotates on its axis with a period of about 9 hours, typical of many asteroids. As a large planetesimal, Ceres seems to have survived largely intact, although indications are that it underwent differentiation at some point in its early history. Spectra show that the surface of Ceres contains hydrated minerals such as clays and carbonates, indicating the presence of significant amounts of water in its interior. Perhaps as much as a quarter of its mass exists in the form of a water-ice mantle that surrounds a rocky inner core. NASA's *Dawn* mission began a one-year exploration of asteroid Vesta in 2011 and will go on to reach Ceres in 2015.

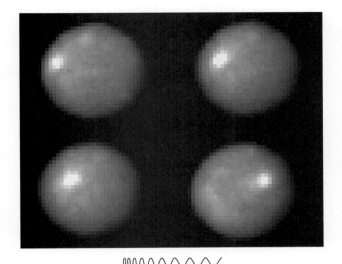

Figure 9.4 Ceres, once known as the largest asteroid, is now called a dwarf planet. Note its spherical shape, one of the criteria giving Ceres its dwarf planet status. The nature of the bright spot is unknown; but it reveals the rotation of Ceres, as seen in this set of four images taken by HST.

9.2 **Moons as Small Worlds**

The giant planets have many moons in orbit around them, ranging in size from small boulders to planet-sized objects. The major moons of the Solar System are shown in **Figure 9.5**. We will organize our tour of the moons of the Solar

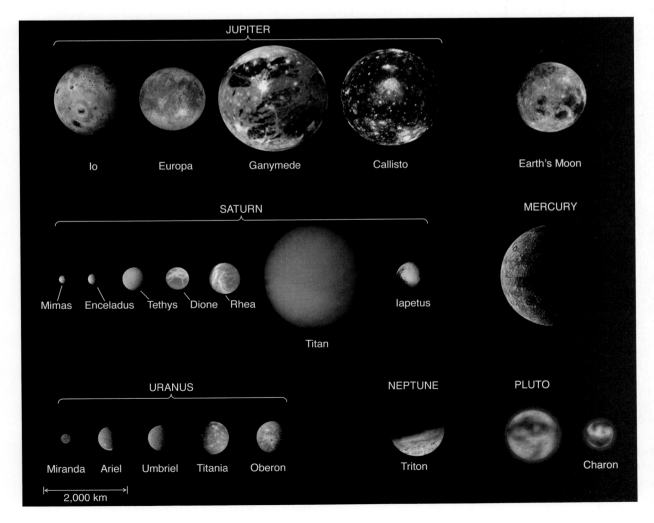

Figure 9.5 The major moons of the Solar System. Mercury and Pluto are also shown here, for comparison.

System by their geological histories as shown by surface features. Comparing these worlds in this way allows us to draw conclusions about how they formed and the physical or geological principles governing their evolution. Some moons have been frozen in time since their formation during the early development of the Solar System, while others are even more geologically active than Earth. We identify four categories of geological activity: (1) definitely active today, (2) probably or possibly active today, (3) active in the past but not today, and (4) apparently not active at any time since their formation. The first three categories are the most instructive for our comparative approach. The fourth category includes Jupiter's Callisto, Uranus's Umbriel, and a large assortment of irregular moons. (Irregular moons have orbits that are highly inclined, very elliptical, or revolve in the opposite direction from the planet's spin.) Their surfaces are heavily cratered and show no modifications other than those caused by a long history of impacts.

Geologically Active Moons

Volcanic features abound on Io, a moon of Jupiter (**Figure 9.6**). Jupiter's gravitational pull flexes Io's crust and generates enough thermal energy to melt parts of it, powering the most active volcanism known. Lava flows and volcanic ash bury impact craters as quickly as they form, so no impact craters have been observed on the surface. Explosive eruptions from more than 150 active volcanoes send debris hundreds of kilometers above Io's surface. The moon is so active that several huge eruptions are often occurring at the same time.

Mixtures of sulfur, sulfur dioxide frost, and sulfurous salts of sodium and potassium likely cause the wide variety of colors on Io's surface. Bright patches may be fields of sulfur dioxide snow. Images taken by *Voyager* and *Galileo* reveal

Io is the Solar System's most volcanically active body.

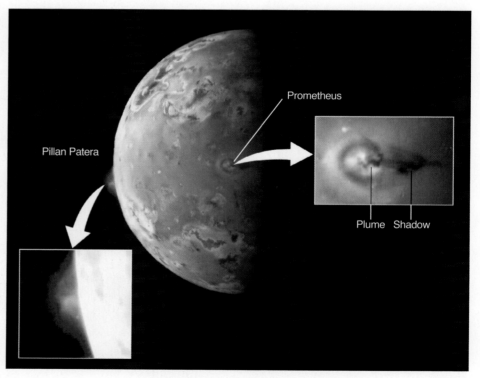

Prometheus

Pillan Patera

Plume   Shadow

Figure 9.6 When this image of Io was obtained by *Galileo*, two volcanic eruptions could be seen at once. The plume of Pillan Patera rises 140 km above the limb of the moon on the left, while the shadow of a 75-km-high plume can be seen to the right of the vent of Prometheus, near the moon's center.

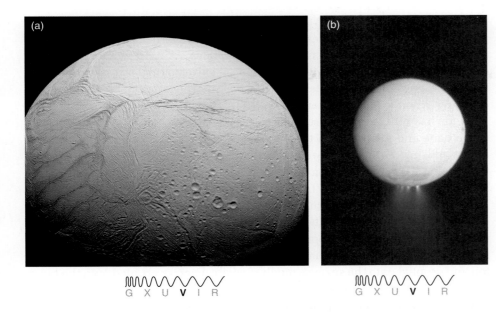

Figure 9.7 *Cassini* images of Enceladus. (a) The twisted and folded surface of deformed ice cracks near the moon's south pole (shown blue in false color) is warmer than the surrounding terrain and has been found to be the source of cryovolcanism. (b) Cryovolcanic plumes in the south polar region spew ice particles into space. In this image, there are two light sources. The surface of the moon is illuminated by sunlight reflected from Saturn while the plumes are backlit by the Sun itself, located almost directly behind Enceladus.

plains, irregular craters, and flows, all related to eruption of mostly silicate magmas onto the surface of the moon. They also show tall mountains, some nearly twice the height of Mount Everest. Huge structures have multiple **calderas** (summit craters of volcanoes), showing a long history of repeated eruptions followed by collapse of the partially emptied magma chambers. Volcanoes on Io are spread around the moon much more randomly than on Earth, implying a lack of plate tectonics.

Io's entire mass has turned inside out more than once in the past, leading to chemical differentiation. Volatiles such as water and carbon dioxide escaped into space long ago. Heavier materials sank to the interior to form a core. Sulfur and various sulfur compounds, as well as silicate magmas, are constantly being recycled to form the complex surface we see today.

Farther out in the Solar System, we find volcanism of a different type. **Cryovolcanism** is similar to normal volcanism but is driven by subsurface low-temperature volatiles rather than molten rock. Enceladus, one of Saturn's icy moons, shows a wide variety of ridges, faults, and smooth plains. This evidence of tectonic processes is unexpected for a small (500-kilometer) body. Some impact craters appear softened, perhaps by the viscous flow of ice, like the flow that occurs in the bottom layers of glaciers on Earth. Parts of the moon have no craters, indicating recent resurfacing.

Terrain near Enceladus's south pole is cracked and twisted (**Figure 9.7a**). The cracks are warmer than their surroundings, implying that tidal heating and radioactive decay within the moon's rocky core heat ice and drive it to the surface. Active cryovolcanic plumes, like those seen in **Figure 9.7b**, are energetic enough to overcome the moon's low gravity, sending tiny ice crystals into space to replace particles continually lost from Saturn's E Ring. It is a mystery that Enceladus is so active while Mimas, a neighboring moon of about the same size and also subject to tidal heating, appears to be quite dead.

Cryovolcanism also occurs on Neptune's largest moon, Triton. Triton orbits Neptune in the opposite direction from Neptune's spin, implying that Triton

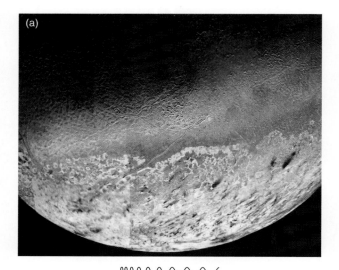

G X U **V** I R

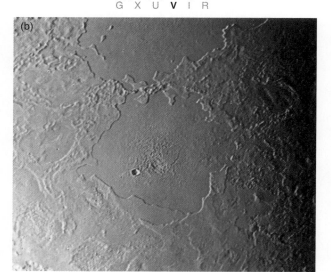

G X U **V** I R

Figure 9.8 *Voyager* images of Triton. (a) This mosaic shows various terrains on the Neptune-facing hemisphere of Triton. "Cantaloupe terrain" is visible at the top; its lack of impact craters indicates that this area is geologically younger than the bright, cratered terrain at the bottom. (b) This irregular basin on Triton has been partly filled with frozen water, forming a relatively smooth ice surface. The state of New Jersey could just fit within the basin's boundary.

Figure 9.9 A high-resolution *Galileo* image of Jupiter's moon Europa, showing where the icy crust has been broken into slabs that, in turn, have been rafted into new positions. These areas of chaotic terrain are characteristic of a thin, brittle crust of ice floating atop a liquid or slushy ocean. The area shown in this image measures approximately 25 by 50 km.

must have been captured by Neptune after the planet's formation. Conservation of angular momentum requires that moons that were formed with their planets all orbit in the same direction that the planet rotates. As Triton achieved its present circular, synchronous orbit, it experienced extreme tidal stresses from Neptune, generating large amounts of thermal energy. The interior may have melted, allowing Triton to become chemically differentiated.

Triton has a thin atmosphere and a surface composed mostly of ices and frosts of methane and nitrogen at a temperature of about 38 K. A relative absence of craters tells us the surface is geologically young. Part of Triton is covered with terrain that looks like the skin of a cantaloupe (**Figure 9.8a**), with irregular pits and hills that may be caused by slushy ice emerging onto the surface from the interior. Veinlike features include grooves and ridges that could result from ice oozing out along fractures. The rest of Triton is covered with smooth volcanic plains. Irregularly shaped depressions as wide as 200 km (**Figure 9.8b**) formed when mixtures of water, methane, and nitrogen ice melted in the interior of Triton and erupted onto the surface, much as rocky magmas erupted onto the lunar surface and filled impact basins on Earth's Moon.

Clear nitrogen ice creates a localized greenhouse effect, in which solar energy trapped beneath the ice raises the temperature at the base of the ice layer. A temperature increase of only 4°C vaporizes the nitrogen ice. As this gas is formed, the expanding vapor exerts very high pressures beneath the ice cap. Eventually the ice ruptures and vents the gas explosively into the low-density atmosphere. *Voyager 2* found four of these active cryovolcanoes on Triton. Each consisted of a plume of gas and dust as much as a kilometer wide rising 8 km above the surface, where the plume was caught by upper atmospheric winds and carried for hundreds of kilometers downwind. Dark material, perhaps silicate dust or radiation-darkened methane ice grains, is carried along with the expanding vapor into the atmosphere, from which it subsequently settles to the surface and forms dark patches streaked out by local winds, as seen near the bottom of Figure 9.8a.

Possibly Active Moons

Jupiter's moon Europa is slightly smaller than our Moon but has an outer shell of water ice. Like Io, Europa experiences continually changing tidal stresses from Jupiter that generate heat and possibly volcanism. Regions of chaotic terrain, as shown in **Figure 9.9**, are places where the icy crust has been broken into slabs

G X U **V** I R

that have shifted into new positions. In other areas the crust has split apart and the gaps have filled in with new dark material rising from the interior.

When these features were formed, Europa's crust consisted of a thin, brittle shell overlying either liquid water or slushy ice. The lightly cratered and thus geologically young surface indicates that Europa may still possess a global ocean 100 km deep that contains more water than all of Earth's oceans. Such an ocean would be salty with dissolved minerals. This reasoning is supported by *Galileo*'s magnetometer data, which show that Europa's magnetic field is variable, indicating an internal electrically conducting fluid. Europa's ocean may also contain an abundance of organic material. In fact, Europa is not so different from some places on Earth that support extreme forms of life. The conditions that create and support life—liquid water, heat, and organic material—could all be present in Europa's oceans, making Europa a high-priority target in the search for extraterrestrial life.

Europa is a target in the search for extraterrestrial life.

Saturn's largest moon, Titan, is bigger than Mercury and has a composition of about 45 percent water ice and 55 percent rocky material. What makes Titan especially remarkable is its thick atmosphere. Titan's mass and distance from the Sun have allowed it to retain an atmosphere that is 30 percent denser than Earth's. Like Earth's atmosphere, Titan's is mostly nitrogen. As Titan differentiated, various ices, including methane (CH_4) and ammonia (NH_3), emerged from the interior to form an early atmosphere. Ultraviolet photons from the Sun have enough energy to break apart ammonia and methane molecules—a process called **photodissociation**. Photodissociation of ammonia is the likely source of Titan's atmospheric nitrogen. Methane breaks into fragments that recombine to form organic compounds including complex hydrocarbons such as ethane. These compounds tend to cluster in tiny particles, creating organic smog much like the air over Los Angeles on a bad day and giving Titan's atmosphere its characteristic orange hue (**Figure 9.10a**).

The *Cassini* spacecraft, which began orbiting Saturn in 2004, gave astronomers the first close-up views of Titan's surface (**Figure 9.10b**). Haze-penetrating

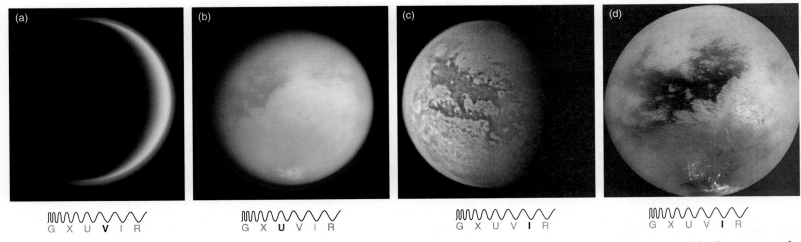

Figure 9.10 *Cassini* images of Saturn's largest moon, Titan. (a) Visible-light imaging shows Titan's orange atmosphere, which is caused by the presence of organic smoglike particles. (b) This infrared- and ultraviolet-light combined image shows surface features and a bluish atmospheric haze caused by scattering of ultraviolet sunlight by small atmospheric particles. Bright clouds in Titan's lower atmosphere are seen near the moon's south pole. (c) Infrared imaging penetrates Titan's smoggy atmosphere and reveals surface features. (d) This infrared image covers the same general area of Titan as seen in (b). The large dark area, called Xanadu, is about the size of the contiguous United States.

Figure 9.11 Radar imaging (false color) near Titan's north pole shows what appear to be lakes of liquid hydrocarbons covering 100,000 square kilometers of the moon's surface. Features commonly associated with terrestrial lakes, such as islands, bays, and inlets, are clearly visible in many of these radar images.

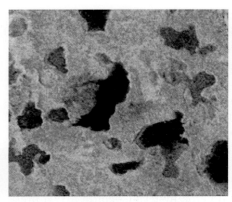

G X U V I **R**

G X U V I R

Figure 9.12 The surface of Titan viewed from the *Huygens* probe during its descent to the surface. The dark drainage patterns resemble river systems on Earth.

G X U V I R

Figure 9.13 View of the surface of Titan obtained from *Huygens*, showing a relatively flat surface littered with water-ice "rocks." The two rocks just below the center of the image are about 85 cm from the camera and roughly 15 and 4 cm across, respectively. The dark "soil" is probably made up of hydrocarbon ices.

infrared imaging shows broad regions of dark and bright terrain (**Figure 9.10c and d**). Radar imaging shows irregularly shaped features in Titan's northern hemisphere that appear to be widespread lakes of methane, ethane, and other hydrocarbons (**Figure 9.11**). Photodissociation should have destroyed all atmospheric methane within about 50 million years, so there must be some process for renewing the methane that is being destroyed by solar radiation. Heat supplied by radioactive decay could cause cryovolcanism that releases "new" methane from underground. So far, there is no direct evidence of active cryovolcanism on Titan, but the presence of abundant atmospheric methane and of methane lakes strongly suggests that Titan is indeed geologically active.

Cassini carried a probe, *Huygens*, that plunged through Titan's atmosphere measuring composition, temperature, pressure, and winds and taking pictures as it descended. *Huygens* confirmed the presence of nitrogen-bearing organic compounds in the clouds. These are key components in the production of proteins found in terrestrial life. Titan has terrains reminiscent of those on Earth (**Figure 9.12**), with networks of channels, ridges, hills, and flat areas that may be dry lake basins. The near absence of impact craters on the surface indicates recent erosion. These features suggest a cycle in which methane rain falls to the surface, washes the ridges free of dark hydrocarbons, and then collects in drainage systems that empty into low-lying liquid methane pools. An infrared camera photographed a reflection of the Sun from such a lake surface.

Once on Titan's surface, *Huygens* took pictures and made physical and compositional measurements. The surface was wet with liquid methane, which evaporated as the probe—heated to 2000 K during its passage through the atmosphere—landed in the frigid soil. The surface was also rich with other organic compounds, such as cyanogen and ethane. As shown in **Figure 9.13**, the surface around the landing site is relatively flat and littered with rounded "rocks" of water ice. The dark "soil" is probably a mixture of water and hydrocarbon ices.

In many ways, Titan resembles a primordial Earth, albeit at much lower temperatures. The presence of organic compounds that could be biological precursors in the right environment makes Titan another high-priority target for continued exploration.

Formerly Active Moons

Some moons show clear evidence of past ice volcanism and tectonic deformation, but no current geological activity. For example, Jupiter's moon Ganymede is the largest moon in the Solar System, larger than the planet Mercury. The surface is composed of two prominent terrains: a dark, heavily cratered (and therefore ancient) terrain and a bright terrain characterized by ridges and grooves. The most extensive region of ancient dark terrain includes a semicircular area about the size of the contiguous United States on the leading hemisphere. Parallel ripples occurring in many dark areas are among Ganymede's oldest surface features. They may represent surface deformation from internal processes, or they may be relics of impact cratering.

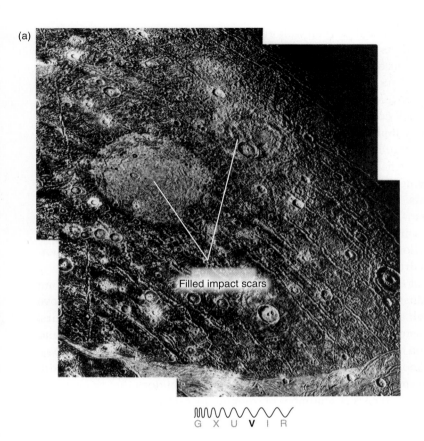

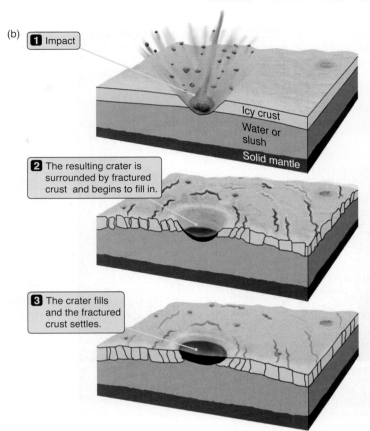

Figure 9.14 (a) A *Voyager* image of filled impact scars on Jupiter's moon Ganymede. (b) This type of impact scar forms as viscous flow smooths out structures left by impacts on icy surfaces.

Impact craters on Ganymede range up to hundreds of kilometers in diameter, and the larger craters are proportionately shallower. The icy crater rims slowly slump, like a lump of soft clay. They are seen as bright, flat, circular patches found principally in the moon's dark terrain and are thought to be scars left by early impacts onto a thin, icy crust overlying water or slush (**Figure 9.14**).

In Chapter 6 we discussed how planetary surfaces can be fractured by faults or folded by compression resulting from movements initiated in the mantle. On Ganymede, fracturing and faulting may have completely deformed the icy crust, destroying all signs of older features such as impact craters and creating the bright terrain. The energy that powered Ganymede's early activity was liberated during a period of differentiation when the moon was very young. Once the differentiation process was complete, that source of internal energy ran out and geological activity ceased.

Ganymede is not the only formerly active moon. Several other large moons in the outer Solar System have also been active in the past.

9.3 Asteroids—Pieces of the Past

The planetesimals that formed our Solar System's planets and moons have been so severely modified by planetary processes that nearly all information about their original physical condition and chemical composition has been lost. By contrast, **asteroids** constitute an ancient and far more pristine record of what the early Solar System was like. Asteroids continue to collide with one another today and produce small fragments of rock and metal, some of which crash to Earth's surface—and if you visit a planetarium or science museum, you may find a

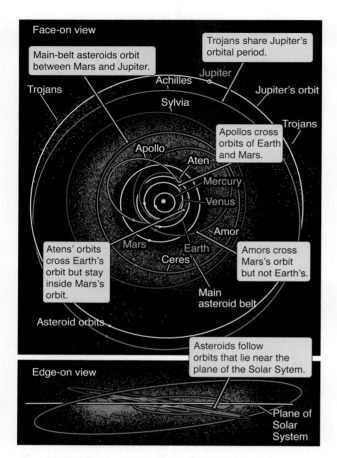

Figure 9.15 Blue dots show the locations of known asteroids at a single point in time. Most families of asteroids take their names from prototype members of the family. For example, the Apollo-family asteroids cross the orbits of Earth and Mars, like the asteroid Apollo. Most asteroids, such as Sylvia, are main-belt asteroids. Achilles was the first Trojan asteroid to be discovered.

Asteroids are relics of rocky planetesimals that formed between the orbits of Mars and Jupiter.

meteorite on display that you can walk right up to and touch. Do so with respect. The meteorite under your hand may be older than Earth itself, and might even contain tiny grains of material that predate the formation of our Solar System.

Asteroids are found throughout the Solar System (**Figure 9.15**), but most of them are located in the main asteroid belt between the orbits of Mars and Jupiter. The main asteroid belt contains several regions, known as Kirkwood gaps, that have been emptied by Jupiter's gravity. There are several groups of asteroids, divided according to their orbital characteristics. Trojan asteroids, for example, share Jupiter's orbit and are held in place by interactions with Jupiter's gravitational field.

Asteroids whose orbits bring them within 1.3 AU of the Sun are called **near-Earth asteroids**. These asteroids, along with a few comet nuclei, are known collectively as **near-Earth objects** (**NEOs**) and occasionally collide with Earth or the Moon. Astronomers estimate that between 500 and 1,000 NEOs have diameters larger than a kilometer. Collisions with these large objects are geologically important and have dramatically altered life on Earth.

Look back at the chapter-opening figure, in which a student has "found" an asteroid. Because these objects are in the Solar System, many of them move quickly enough across the sky that their motion is noticeable over a few hours. Detecting an asteroid, particularly if you weren't looking for it, is startling and exciting. It often leads amateur astronomers to many years of observing as a hobby. Amateur astronomers are a critical piece of the asteroid observation puzzle, adding orbital parameters, rotation periods, and colors to catalogs that are used by professional astronomers in a number of research projects.

Asteroids Are Fractured Rock

Most asteroids are relics of rocky or metallic planetesimals that formed in the region between Mars and Jupiter. Although early collisions between these planetesimals created several bodies large enough to differentiate, Jupiter's tidal disruption prevented them from forming a single planet.

There are many more small asteroids than large ones—indeed, most are too small for their self-gravity to have pulled them into a spherical shape. Some are very elongated, suggesting they are either fragments of larger bodies or the result of haphazard collisions between smaller bodies. Asteroids account for only a tiny fraction of the mass in the Solar System. If all of the asteroids were combined into a single body, it would be only about one-third the mass of Earth's Moon.

Astronomers have measured the masses of some asteroids by noting the effect of their gravity on spacecraft passing nearby. Knowing the mass and the size of asteroids enables us to determine their densities, which range between 1.3 and 3.5 times the density of water. Those at the lower end of this range are "rubble piles" with large gaps between the fragments, and they are considerably less dense than the meteorite fragments they create. This is what we would expect of objects that were assembled from smaller objects and then suffered a history of violent collisions.

Asteroids Viewed Up Close

In 1991, the *Galileo* spacecraft passed by the asteroid Gaspra and obtained good images of its surface. Gaspra is cratered and irregular in shape, about 9 × 10 × 20 km in size. Faint, groovelike patterns may be fractures from the impact that chipped Gaspra from a larger planetesimal. Distinctive colors imply that Gaspra is covered with a variety of rock types.

Later in its mission, *Galileo* returned to the main belt, passing close to the asteroid Ida (**Figure 9.16**). *Galileo* flew so close to Ida that it could see details as small as 10 meters across—about the size of a small house. Ida is crescent-shaped and 54 km long, ranging from 15 to 24 km in diameter. Ida's craters indicate that the surface is about a billion years old, at least twice the age estimated for Gaspra. Like Gaspra, Ida is fractured. The fractures indicate that these asteroids must be made of relatively solid rock. (You can't "crack" a loose pile of rubble.) This supports the idea that some asteroids are chips from larger, solid objects.

The *Galileo* images also revealed a tiny moon orbiting the asteroid. Ida's moon, called Dactyl, is only 1.4 km across and cratered from impacts. Moons have now been found around nearly 200 asteroids, and at least five asteroids are known to have two moons.

In early 2000, the *NEAR Shoemaker* spacecraft entered orbit around the asteroid Eros to begin long-term observations (**Figure 9.17**). Eros is one of more than 5,000 known objects whose orbits bring them within 1.3 AU of the Sun. Eros measures about 34 × 11 × 11 km and, like Gaspra and Ida, shows a surface with grooves, rubble, and impact craters, including a crater 8.5 km across. The scarcity of smaller craters suggests that its surface could be younger than Ida's. After a year of observing, the spacecraft landed on the asteroid's surface. Chemical analyses confirmed that Eros's composition is like that of primitive meteorites.

On its way to Eros, *NEAR Shoemaker* flew past the asteroid Mathilde. Mathilde was found to measure 66 × 48 × 46 km and to have a surface only half as reflective as charcoal. The overall density of Mathilde is about 1,300 kilograms per cubic meter (kg/m³; this is 1.3 times the density of water), which implies that Mathilde is a rubble pile, composed of chunks of rocky material with open spaces between them. Mathilde is covered with craters, the largest being more than 33 km across. The great number of craters suggests that Mathilde likely dates back to the very early history of the Solar System.

In November 2005, the Japanese spacecraft *Hayabusa* made contact with asteroid Itokawa. It collected samples and then returned to Earth in June of 2010. By fall of 2010, the Japan Aerospace Exploration Agency confirmed that the *Hayabusa* spacecraft had completed the first sample return mission from an asteroid—it had brought back particles from Itokawa. In addition to the significant engineering accomplishment, this mission was important because the dust was a pristine sample from a specific, well-characterized asteroid. Usually, scientists cannot match a meteorite sample to a specific asteroid.

The spectra of Mars's tiny moons, Deimos and Phobos (**Figure 9.18**), are similar to those of some asteroids. Many scientists think these moons may be asteroids

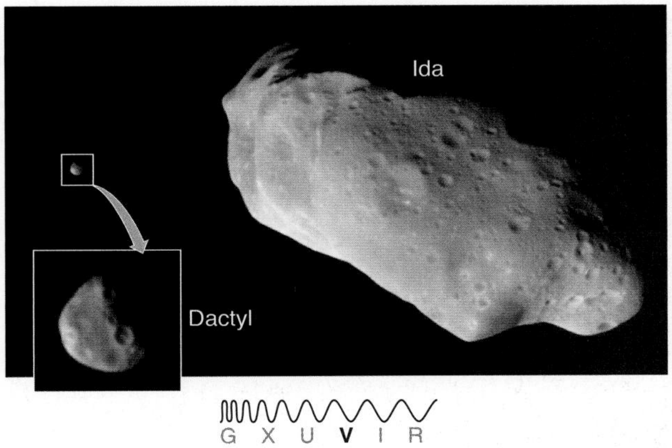

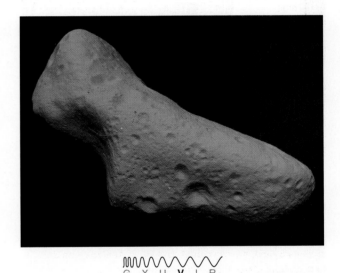

Figure 9.16 A *Galileo* spacecraft image of the asteroid Ida with its tiny moon, Dactyl (also shown enlarged in the inset).

Figure 9.17 Asteroid Eros viewed by the *NEAR Shoemaker* spacecraft. This image was produced via high-resolution scanning of the asteroid's surface by the spacecraft's laser range finder. *NEAR Shoemaker* became the first spacecraft to land on an asteroid when it was gently crashed onto the surface of Eros.

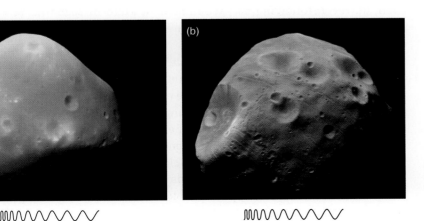

Figure 9.18 *Mars Reconnaissance Orbiter* and *Mars Express* images of Mars's two tiny moons: (a) Deimos and (b) Phobos.

that have been captured by Mars's gravity. But are Deimos and Phobos really asteroids? Controversy about this question has never been put to rest. Some scientists argue that it is unlikely Mars could have captured two asteroids, proposing that Deimos and Phobos must somehow have evolved together with Mars. A third possibility is that Mars and its moons were all parts of a much larger body that was fragmented by a collision early in the planet's history. A Russian mission scheduled for launch in November 2011 may answer this question.

9.4 Comets—Clumps of Ice

Icy planetesimals that formed from primordial material, **comets** spend most of their lives adrift in the frigid outer reaches of the Solar System. Most are much too small and far away to be seen, so no one really knows how many there are. Estimates for our Solar System range as high as a trillion (10^{12}) comets—more than all the stars in the Milky Way Galaxy. They put on a show only when they come deep enough into the inner Solar System to suffer destructive heating from the Sun. When they are close enough to show the effects of solar heating, we call them **active comets**.

The Homes of the Comets

We know where comets come from by observing their orbits as they pass through the inner Solar System. Comets fall into two distinct groups: the *Kuiper Belt* comets and the *Oort Cloud* comets. These two populations of comets are named for scientists Gerard Kuiper (1905–1973) and Jan Oort (1900–1992), who first proposed the comets' existence in the mid-20th century.

Comets from the **Kuiper Belt** orbit the Sun in a disk-shaped region aligned with the Solar System. The Kuiper Belt begins abruptly at about 50 AU, beyond the orbit of Neptune, and extends outward to perhaps a thousand AU from the Sun (**Figure 9.19**). The innermost part of the Kuiper Belt contains tens of thousands of icy planetesimals known as **Kuiper Belt objects (KBOs)**. The largest KBOs are similar in size to Pluto. With a few exceptions, the sizes of KBOs are difficult to determine. Although brightness and approximate distance are known, their albedos are uncertain. Reasonable limits for the albedos can set maximum and minimum values for their size. Some KBOs have moons, and at least one has three moons. More than 1,000 KBOs have been discovered, but many smaller ones must also be there. We know very little of the chemical and physical properties of KBOs because of their great distance. This lack of information may change soon, however: following its encounter with Pluto in 2015, the *New Horizons* spacecraft will continue outward to the Kuiper Belt, where it will attempt to fly close to one or more KBOs.

Unlike the flat disk of the Kuiper Belt, the **Oort Cloud** is a spherical distribution of planetesimals that are too distant to be seen by even the most powerful telescopes. We know the size and shape of the Oort Cloud from the orbits of this region's comets, which approach the Sun from all directions and from as far as 100,000 AU away—nearly halfway to the nearest stars.

The Kuiper Belt and Oort Cloud are enormous reservoirs of icy planetesimals that now and then fall into the inner Solar System. Gravitational perturbations by objects beyond the Solar System may kick an Oort Cloud planetesimal in toward the Sun. About every 5 million to 10 million years, a star passes within

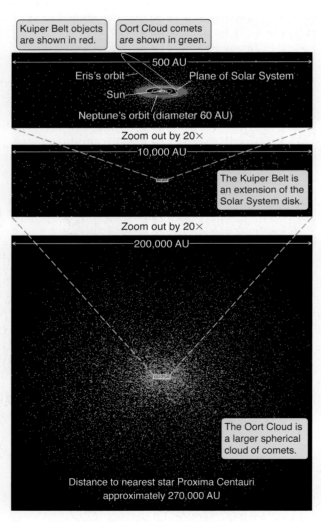

Figure 9.19 Most comets near the inner Solar System populate an extension to the disk of the Solar System called the Kuiper Belt. The spherical Oort Cloud is far larger and contains many more comets.

about 100,000 AU of the Sun, perturbing the orbits of these planetesimals. The gravitational attraction from huge clouds of dense interstellar gas can also stir up the Oort Cloud.

Unlike the icy planetesimals in the Oort Cloud, those in the Kuiper Belt are packed closely enough to interact gravitationally from time to time. In such events, one object gains energy while the other loses it. The "winner" may gain enough energy to be sent into an orbit that reaches far beyond the boundary of the Kuiper Belt. The "loser" falls inward toward the Sun.

The Orbits of Comets

The lifetime of a comet nucleus depends on how frequently it passes by the Sun and how close it comes. There are about 400 known **short-period comets**, which by definition have periods less than 200 years. Additionally, each year we discover about six new **long-period comets**, whose orbital periods are longer than 200 years. The total number of long-period comets observed to date is about 3,000.

Figure 9.20 shows the orbits of a number of comets. Long-period comets were scattered to the outer solar system by gravitational interactions, so their orbits are random. They come into the inner Solar System from all directions, some orbiting the Sun in the same direction that the planets orbit (prograde) and some orbiting in the opposite direction (retrograde). Short-period comets tend to be prograde and to have orbits in the ecliptic plane, and they frequently pass close

The Kuiper Belt and the Oort Cloud are reservoirs of comets.

▶‖ **AstroTour:** Cometary Orbits

Figure 9.20 Orbits of a number of comets (colored lines) in face-on and edge-on views of the Solar System. Populations of (a) short-period comets and (b) long-period comets have very different orbital properties. Comet Halley, which appears in both diagrams for comparison, is a short-period comet.

(a)

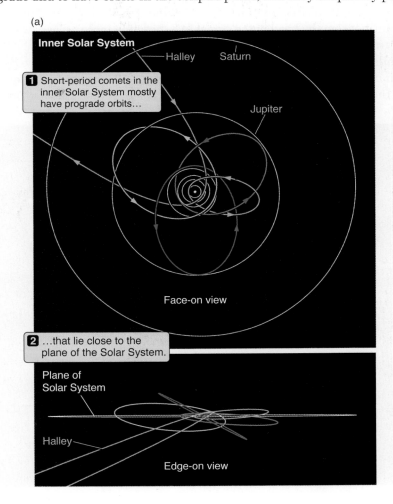

(b)

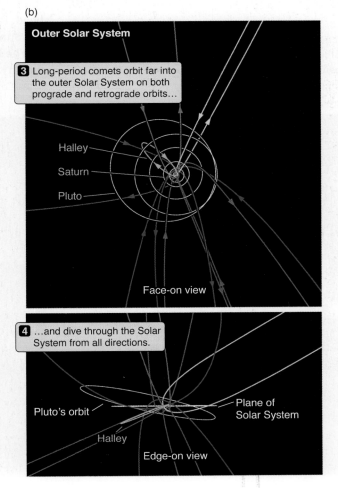

The coma is an extended cometary atmosphere surrounding the nucleus of a comet.

enough to a planet for its gravity to change the comet's orbit about the Sun. Short-period comets presumably originated in the Kuiper Belt, but as they fell in toward the Sun, they were forced into their present short-period orbits by gravitational encounters with planets.

More than 600 long-period comets have well-determined orbits. Some have orbital periods of hundreds of thousands or even millions of years. Almost all their time is spent in the Oort Cloud in the frigid, outermost regions of the Solar System. Because of their very long orbital periods, these comets have made at most one appearance throughout the course of recorded history.

Anatomy of an Active Comet

The small object at the center of the comet—the icy planetesimal itself—is the comet **nucleus**. This is by far the smallest component of a comet, but it is the source of all the material that we see stretched across the sky as a comet nears the Sun (**Figure 9.21**). Comet nuclei range in size from a few dozen meters to several hundred kilometers across. These "dirty snowballs" are composed of ice, organic compounds, and dust grains.

As a comet nucleus nears the Sun, sunlight heats its surface, vaporizing ices that stream away from the nucleus and carrying dust particles along with them. The gases and dust driven from the nucleus of an active comet form a nearly spherical atmospheric cloud around the nucleus called the **coma**. The nucleus and the inner part of the coma are sometimes referred to collectively as the comet's **head**. Pointing from the head of the comet in a direction more or less away from the Sun are long streamers of dust, gas, and ions called the **tails**.

The tails are the largest and most spectacular part of a comet. Active comets have two different types of tails, as shown in Figure 9.21. One is the **ion tail**. Many of the atoms and molecules that make up a comet's coma are ions. Because

Figure 9.21 (a) The principal components of a fully developed active comet are the nucleus, the coma, and two types of tails called the dust tail and the ion tail. Together, the nucleus and the coma are called the head. (b) Comet Hale-Bopp, the great comet of 1997.

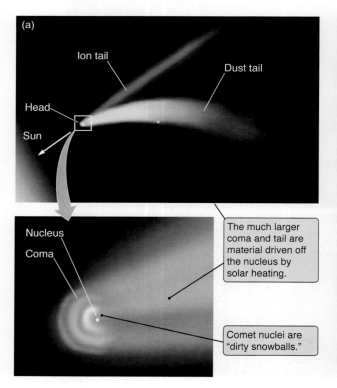

they are electrically charged, ions in the coma feel the effect of the solar wind, the stream of charged particles that blows continually away from the Sun. The solar wind pushes on these ions, rapidly accelerating them to speeds of more than 100 kilometers per second (km/s)—far greater than the orbital velocity of the comet itself—and sweeps them out into a long, wispy structure. Because the particles that make up the ion tail are so quickly picked up by the solar wind, ion tails are usually very straight and point from the head of the comet directly away from the Sun.

Dust particles also have a net electric charge and feel the force of the solar wind. In addition, sunlight itself exerts a force on cometary dust. But dust particles are much more massive than individual ions, so they are accelerated more gently and do not reach such high relative speeds as the ions do. As a result, the **dust tail** often curves gently away from the head of the comet as the dust particles are gradually pushed from the comet's orbit in the direction away from the Sun.

Figure 9.22 shows the tails of a comet at various points in its orbit. Remember that both types of tails always point *away from the Sun*, regardless of which direction the comet is moving. As the comet approaches the Sun, its two tails trail behind its nucleus. But the tails extend *ahead* of the nucleus as the comet moves outward from the Sun.

Tails vary greatly from one comet to another. Some comets display both types of tails simultaneously; others, for reasons that we do not understand, produce no tails at all. A tail often forms as a comet crosses the orbit of Mars, where the increase in solar heating drives gas and dust away from the nucleus.

The gas in a comet's tail is even more tenuous than the gas in its coma; its density is no more than a few hundred molecules per cubic centimeter. Compare this to the density of Earth's atmosphere, which at sea level contains more than 10^{19} molecules per cubic centimeter. Dust particles in the tail are typically about a micrometer (μm) in diameter, roughly the size of smoke particles.

Most naked-eye comets develop first a coma and then an extended tail as they approach the inner Solar System. Comet McNaught in 2007 was such a comet (**Figure 9.23a**). But there are exceptions. Comet Holmes was a very faint telescopic object when it reached its closest point to the Sun just beyond the orbit of Mars on May 4, 2007. Then, several months later, on October 21, as it was well on its way outward toward Jupiter's orbit, the comet suddenly became a half-million times brighter in just 42 hours. Comet Holmes became a bright naked-eye comet that graced Northern Hemisphere skies for several months (**Figure 9.23b**). Astronomers

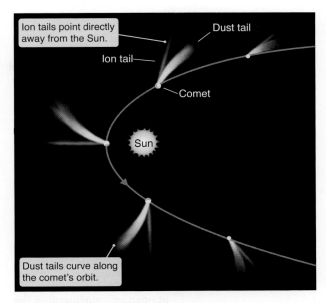

Figure 9.22 The orientation of the dust and ion tails at several points in a comet's orbit. The ion tail points directly away from the Sun while the dust tail curves along the comet's orbit.

Comet tails *always* point away from the Sun.

Figure 9.23 (a) Comet McNaught, in 2007, was the brightest comet to appear in decades, but its true splendor was visible only to observers in the Southern Hemisphere. (b) Months after its closest approach to the Sun in 2007, Comet Holmes suddenly became a half-million times brighter within hours, turning into a naked-eye comet with an angular diameter larger than that of the full Moon. Comet Holmes favored Northern Hemisphere observers.

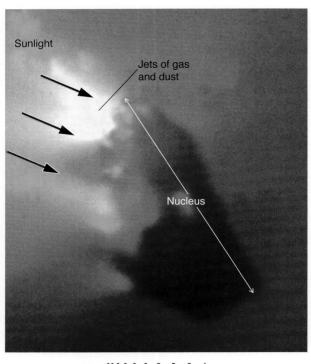

Figure 9.24 The nucleus of Comet Halley as imaged in false color by the *Giotto* spacecraft in 1986.

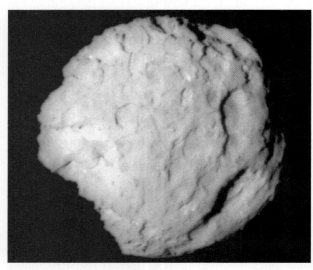

Figure 9.25 The nucleus of Comet Wild 2, imaged by the *Stardust* spacecraft.

remain puzzled over the cause of this dramatic eruption. Explanations range from a meteoroid impact to a sudden (but unexplained) buildup of subsurface gas.

Generally, long-period comets provide a more spectacular display than short-period comets. The nuclei of short-period comets are eroded from repeated heating by the Sun, and as the volatile ices are driven away, some of the dust and organic compounds are left behind on the surface of the nucleus. The buildup of this covering slows down cometary activity and makes their solar approaches less spectacular. In contrast, long-period comets are relatively pristine. More of their supply of volatile ices remains close to the surface of the nucleus, and they can produce a truly magnificent show. However, most pass too far from Earth or the Sun and never become bright enough to attract much public attention.

When will the next bright comet come along? On average, a spectacular comet appears about once per decade, but it is all a matter of chance. It might be many years from now—or it could happen tomorrow.

Visits to Comets

Comets provide a nice engineering challenge for spacecraft designers. We seldom have enough advance knowledge of a comet's visit or its orbit to mount a successful mission to intercept it. The closing speed between an Earth-launched spacecraft and a comet can be extremely high. Observations must be made very quickly, and there is a danger of high-speed collisions with debris from the nucleus. Still, nearly a dozen spacecraft have been sent to rendezvous with comets, including a five-spacecraft armada sent to Comet Halley by the Soviet, European, and Japanese space agencies in 1986. Much of what we know about comet nuclei and the innermost parts of the coma comes from data sent back by these missions.

The Soviet *VEGA* and the European *Giotto* spacecraft entered the coma of Comet Halley when they were still nearly 300,000 km from the nucleus. We learned that the dust from Comet Halley was a mixture of light organic substances and heavier rocky material, and the gas was about 80 percent water and 10 percent carbon monoxide with smaller amounts of other organic molecules. The surface of the comet's nucleus is among the darkest-known objects in the Solar System, which means that it is rich in complex organic matter that must have been present as dust in the disk around the young Sun—perhaps even in the interstellar cloud from which the Solar System formed. As *VEGA* and *Giotto* passed close by the nucleus (**Figure 9.24**), they observed jets of gas and dust moving away from its surface at speeds of up to 1,000 meters per second (m/s), far above the escape velocity. Material was coming from several small fissures that covered only about a tenth of the surface.

In 2001, NASA's *Deep Space 1* spacecraft flew within 2,200 km of Comet Borrelly's nucleus. Its tar-black surface is also among the darkest seen on any Solar System object and, much to the surprise of scientists, showed no evidence of water ice.

In 2004, NASA's *Stardust* spacecraft flew within 235 km of the nucleus of Comet Wild 2 (pronounced "vilt 2"). This comet is a newcomer to the inner Solar System. A close encounter with Jupiter in 1974 perturbed its orbit, bringing this relatively pristine body in from its orbit between Jupiter and Uranus. Wild 2's nearly spherical nucleus is about 5 km across. At least 10 gas jets are active,

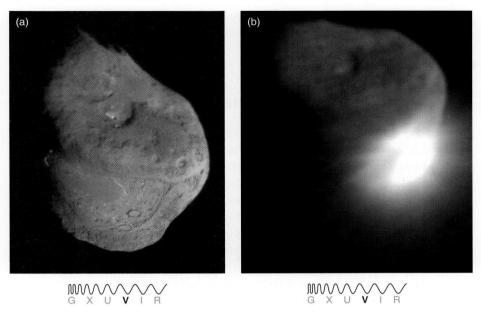

$\text{G}\ \text{X}\ \text{U}\ \textbf{V}\ \text{I}\ \text{R}$ $\text{G}\ \text{X}\ \text{U}\ \textbf{V}\ \text{I}\ \text{R}$

Figure 9.26 (a) The surface of the nucleus of Comet Tempel 1, seen just before impact by the *Deep Impact* projectile. The impact occurred between the two 370-meter-diameter craters located near the bottom of the image. The smallest features appearing in this image are about 5 meters across. (b) Sixteen seconds after the impactor struck the comet, the parent spacecraft took this image of the initial ejecta.

some of which carry surprisingly large chunks of surface material. (A few particles as large as a bullet penetrated the outer layer of the spacecraft's protective shield.) The surface of Wild 2 is covered with features that may be impact craters modified by ice sublimation, small landslides, and erosion by jetting gas (**Figure 9.25**). Some show flat floors, suggesting a relatively solid interior beneath a porous surface layer.

In 2005, NASA's *Deep Impact* spacecraft launched a 370-kg impacting projectile into the nucleus of Comet Tempel 1 at a speed of more than 10 km/s. The impact sent 10,000 tons of water and dust flying off into space at speeds of 50 m/s (**Figure 9.26**). A camera mounted on the projectile snapped photos of its target until it was vaporized by the impact. Observations of the event were made both locally by *Deep Impact* and back on Earth by a multitude of orbiting and ground-based telescopes.

Water, carbon dioxide, hydrogen cyanide, iron-bearing minerals, and a host of complex organic molecules were identified. The comet's outer layer is composed of fine dust with a consistency of talcum powder. Beneath the dust are layers made up of water ice and organic materials. One surprise for scientists was the presence of well-formed impact craters, which had been absent in close-up images of Comets Borrelly and Wild 2. Why some comet nuclei have fresh impact craters and others none remains a question.

In late 2010, the *EPOXI* spacecraft flew past Comet Hartley (**Figure 9.27**), imaging not only jets of dust and gas, indicating a remarkably active surface, but also an unusual separation of rough and smooth areas that have very different natures. Water evaporates and percolates out through the dust from the smooth area at the "waist" of the comet's nucleus, while carbon dioxide jets shoot out

$\text{G}\ \text{X}\ \text{U}\ \textbf{V}\ \text{I}\ \text{R}$

Figure 9.27 In this image of Comet Hartley taken by the *EPOXI* spacecraft, we see two distinct surface types. Water seeps through the dust at the smooth "waist" of the comet's nucleus, whereas carbon dioxide jets shoot from the rough areas.

from the rough areas. It is unclear yet whether this unusual shape is a result of how Comet Hartley formed 4.5 billion years ago or is due to more recent evolution of the comet.

9.5 Collisions Still Happen Today

Early in the 20th century, the orbit of a comet nucleus from the Kuiper Belt was perturbed, and the comet's new orbit carried it close to Jupiter. This comet, known as Shoemaker-Levy 9, passed so close to Jupiter in 1992 that tidal stresses broke it into two dozen major fragments, which subsequently spread out along its line of orbit. The fragments took one more two-year orbit about the planet, and in July 1994, the entire string of fragments crashed into Jupiter.

Over a week's time, the fragments, each traveling at 60 km/s, plunged through Jupiter's stratosphere. The impacts occurred just behind the limb of the planet, where they could not be observed from Earth, but the *Galileo* spacecraft was able to image some of the impacts. Astronomers using ground-based telescopes and the Hubble Space Telescope could see immense plumes rising from the impacts to heights of more than 3,000 km above the cloud tops at the limb. The debris in these plumes then rained back onto Jupiter's stratosphere, causing ripples like pebbles thrown into a pond. **Figure 9.28** shows the sequence of events that

Figure 9.28 Events following the impact of a fragment of Comet Shoemaker-Levy 9 on Jupiter.

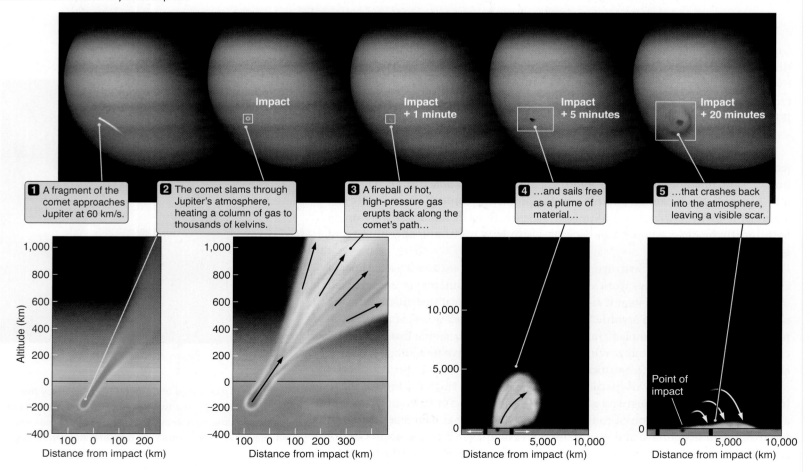

Impact

Impact + 1 minute

Impact + 5 minutes

Impact + 20 minutes

1 A fragment of the comet approaches Jupiter at 60 km/s.

2 The comet slams through Jupiter's atmosphere, heating a column of gas to thousands of kelvins.

3 A fireball of hot, high-pressure gas erupts back along the comet's path...

4 ...and sails free as a plume of material...

5 ...that crashes back into the atmosphere, leaving a visible scar.

Point of impact

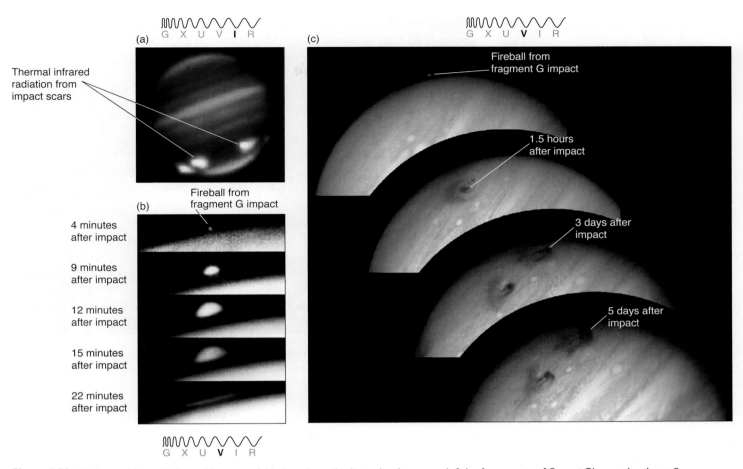

Figure 9.29 (a) Ground-based infrared images of Jupiter show the hot, glowing scars left by fragments of Comet Shoemaker-Levy 9. (b) Although the fragments struck Jupiter on the back side, these HST images show the fireball from one fragment rising above the limb of the planet. (c) HST images of the evolution of the scar left by one fragment of the comet.

took place as each fragment of the comet slammed into the giant planet. Sulfur and carbon compounds released by the impacts formed Earth-sized scars in the atmosphere that persisted for months (**Figure 9.29**) and were visible even through small amateur telescopes.

Almost all hard-surfaced objects in the Solar System still bear the scars of a time when tremendous impact events were common. The collision of Comet Shoemaker-Levy 9 with Jupiter in 1994 reminded us that although such impacts are far less frequent today than they once were, they still happen.

The Tunguska River flows through a remote region of western Siberia. In the summer of 1908, the region was blasted with the energy equivalent of 2,000 times the atomic bomb dropped over Hiroshima. **Figure 9.30** shows a map of the region, along with a photo of the devastation caused by the blast. Eyewitness accounts detailed the destruction of dwellings, the incineration of reindeer (including one herd of 700), and the deaths of at least five people. Although trees were burned or flattened for over 2,150 square kilometers—an area greater than metropolitan New York City—no crater was left behind! The Tunguska event was the result of a tremendous high-altitude explosion that occurred when a small body hit Earth's atmosphere, ripped apart, and formed a fireball before reaching Earth's

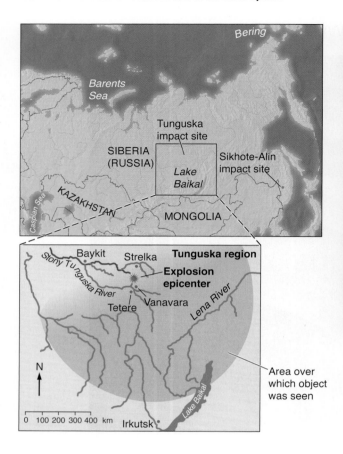

Figure 9.30 A large region of forest near the Tunguska River in Siberia was flattened in 1908 by an atmospheric explosion caused by the impact of a small asteroid or comet.

Comet or asteroid impacts are infrequent but devastating.

surface. Recent expeditions to the Tunguska area have recovered resin from the trees blasted by the event. Chemical traces in the resin suggest that the impacting object may have been a stony asteroid.

On February 12, 1947, yet another planetesimal struck Earth, this time in the Sikhote-Alin region of eastern Siberia. Composed mostly of iron, the object had an estimated diameter of about 100 meters and broke into a number of fragments before hitting the ground, leaving a cluster of craters and widespread devastation. Witnesses reported a fireball brighter than the Sun and sound that was heard 300 km away.

These impacts are sobering events. It is highly improbable that a populated area on Earth will experience a collision with a large asteroid within our lifetimes. Comets and smaller asteroids, however, are less predictable. There may be as many as 10 million asteroids larger than a kilometer across, but only about 130,000 have well-known orbits, and most of the unknowns are too small to see until they come very close to Earth. The U.S. government—along with the governments of several other nations—is aware of the risk posed by NEOs. Although the probability of a collision between a small asteroid and Earth is quite small, the consequences could be catastrophic, so NASA has been given a congressional mandate to catalog all NEOs and to scan the skies for those that remain undiscovered.

Comets present a more serious problem. Half a dozen unknown long-period comets enter the inner Solar System each year. If one happens to be on a collision course with Earth, we might not notice it until just a few weeks or months before impact. Although this has become a favorite theme of science fiction disaster stories, Earth's geological and historical record suggests that impacts by large bodies are infrequent events.

9.6 Solar System Debris

Comet nuclei that enter the inner Solar System generally disintegrate within a few hundred thousand years as a result of their repeated passages close to the Sun. Asteroids have much longer lives but still are slowly broken into pieces from occasional collisions with each other. Disintegration of comet nuclei and asteroid collisions create most of the debris that fills the inner part of the Solar System. As Earth and other planets move along in their orbits, they continually sweep up this fine debris.

This cometary and asteroidal debris is the source of most of the **meteoroids** that Earth encounters. When a meteoroid enters Earth's atmosphere, frictional heat causes the air to glow, producing an atmospheric phenomenon called a **meteor**. Some 100,000 kg of meteoritic debris is swept up by Earth every day, and what does not burn up (mostly particles smaller than 100 μm) eventually settles to the ground as fine dust. A meteoroid that survives to a planet's surface is called a **meteorite**.

Meteor showers happen when Earth's orbit crosses the orbit of a comet. Bits of dust and other debris from a comet nucleus remain in orbits that are similar to the orbit of the nucleus itself. When Earth passes through this concentration of cometary debris, the result is a meteor shower. More than a dozen comets have orbits that come close enough to Earth's to produce annual meteor showers. Because the meteoroids that are being swept up are all in similar orbits, they all enter our atmosphere moving in the same direction. As a result, the paths of all the meteors in a particular shower are parallel to one another. From our perspective all the meteors appear to originate from the same point in the sky (**Figure 9.31a**), just as the parallel rails of a railroad track appear to vanish to a single point in the distance (**Figure 9.31b**). This point is the shower's **radiant**.

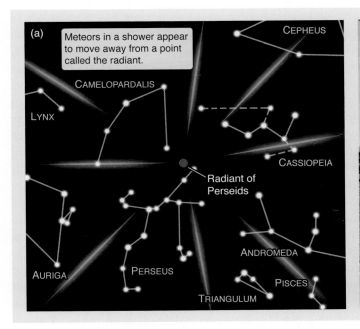

VISUAL ANALOGY

Figure 9.31 (a) Meteors appear to stream away from the radiant of the Perseid meteor shower. (b) Such streaks are actually parallel paths that appear to emerge from a vanishing point, as in our view of these railroad tracks.

Figure 9.32 Cross sections of several kinds of meteorites: (a) a chondrite (a stony meteorite with chondrules), (b) an achondrite (a stony meteorite without chondrules), (c) an iron meteorite, and (d) a stony-iron meteorite.

Fragments of asteroids are much denser than cometary meteoroids. If an asteroid fragment is large enough—about the size of your fist—it can survive all the way to the ground to become a meteorite. The fall of a 10-kg meteoroid can produce a fireball so bright that it lights up the night sky more brilliantly than the full Moon. Such a large meteoroid may create a sonic boom heard hundreds of kilometers away, or explode into multiple fragments as it nears the end of its flight. Some glow with a brilliant green color caused by metals in the original meteoroid.

Thousands of meteorites reach the surface of Earth every day, but only a tiny fraction of these are ever found and identified. Antarctica offers the best meteorite hunting in the world. Meteorites are no more likely to fall in Antarctica than anywhere else, but in Antarctica they are far easier to distinguish from their surroundings because in many places the *only* stones to be found on the ice are meteorites. Since Antarctica is actually very dry, Antarctic meteorites also tend to show little weathering or contamination from terrestrial dust or organic compounds, making them excellent specimens for study. Scientists compare meteorites to rocks found on Earth and the Moon and contrast their structure and chemical makeup with those of rocks studied by spacecraft that have landed on Mars and Venus. Meteorites are also compared with asteroids and other objects based on the colors of sunlight they reflect and absorb.

Meteorites are grouped into three categories, depending on the kinds of materials they are made of and the degree of differentiation they experienced within their parent bodies. Over 90 percent of meteorites are **stony meteorites** (**Figure 9.32a and b**), which are similar to terrestrial silicate rocks. A stony meteorite is characterized by the thin coating of melted rock that forms as it passes through the atmosphere. Many stony meteorites contain **chondrules** (see Figure 9.32a), once-molten droplets that rapidly cooled to form crystallized spheres ranging in size from sand grains to marbles. Stony meteorites containing chondrules are called **chondrites** (and those without chondrules are called **achondrites**). **Carbonaceous chondrites**, chondrites that are rich in carbon, are thought to be the very building blocks of the Solar System. Indirect measurements suggest that these meteorites are about 4.56 billion years old—consistent with all other measurements of the time that has passed since the Solar System was formed.

The second major category of meteorites, **iron meteorites (Figure 9.32c)**, is the easiest to recognize. The surface of an iron meteorite has a melted and pitted appearance generated by frictional heating as it streaked through the atmosphere. Even so, many are never found, either because they land in water or simply because no one happens to recognize them. Mars rover *Opportunity* has discovered a handful of iron meteorites on the martian surface (**Figure 9.33**). Both their appearance—typical of iron meteorites found on Earth—and their position on the smooth, featureless plains made them instantly recognizable. The final category is the **stony-iron meteorites (Figure 9.32d)**, which consist of a mixture of rocky material and iron-nickel alloys. Stony-iron meteorites are relatively rare.

Meteorites come from asteroids, which in turn come from stony-iron planetesimals. During the growth of the terrestrial planets, large amounts of thermal energy were released as larger planetesimals accreted smaller objects, and these bodies were heated further as radioactive elements inside them decayed. Despite this heating, some planetesimals never reached the high temperatures needed to melt their interiors—instead, they simply cooled off, and they have since remained pretty much as they were when they formed. These planetesimals are known as **C-type asteroids**, which are composed of primitive material that astronomers believe is essentially unmodified since the origin of the Solar System almost 4.6 billion years ago.

Some planetesimals, however, did melt and differentiate, with denser matter such as iron sinking to their center. Lower-density material—such as compounds of calcium, silicon, and oxygen—floated toward the surface and combined to form a mantle and crust of silicate rock. **S-type asteroids** may be pieces of the

Meteorites tell us the age of the Solar System.

Figure 9.33 A basketball-sized iron meteorite lying on the surface of Mars, imaged by the Mars exploration rover *Opportunity*.

mantles and crusts of such differentiated planetesimals. They are chemically more similar to igneous rocks found on Earth than to C-type asteroids, because they were hot enough at some point to lose their carbon compounds and other volatile materials to space. Similarly, **M-type asteroids** (from which iron meteorites come) are fragments of the iron- and nickel-rich cores of one or more differentiated planetesimals that shattered into small pieces during collisions with other planetesimals. Slabs cut from iron meteorites show large, interlocking crystals characteristic of iron that cooled very slowly from molten metal. Rare stony-iron meteorites may come from the transition zone between the stony mantle and the metallic core of such a planetesimal.

Vesta (the largest known asteroid, with a diameter of 525 km) has a spectrum strikingly similar to that of a peculiar group of meteorites. These meteorites, which are probably pieces of Vesta, are also like rocks taken from iron-rich lava flows on Earth and the Moon. Radiometric dating shows that they are younger than most meteorites—*only* 4.4 billion to 4.5 billion years old. From this, we deduce that early in the history of the Solar System, some planetesimals were large enough to become differentiated and to develop volcanism. But rather than going on to form planets, they collided with other planetesimals and were broken up.

The story of how planetesimals, asteroids, and meteorites are related is one of the great successes of planetary science. A wealth of information about this diverse collection of objects has come together into a self-consistent story of planetesimals growing and differentiating and then shattering in subsequent collisions. The story is even more satisfying because it fits so well with the even grander story of how most planetesimals formed into the planets and their moons.

As we stress throughout this book, patterns are extremely important in science—when they are followed as well as when they are broken. Some types of meteorites fail to follow the patterns just discussed. Whereas most achondrites have ages in the range of 4.5 billion to 4.6 billion years, some members of one group are less than 1.3 billion years old and are chemically and physically similar to the soil and atmospheric gases that our lander instruments have measured on Mars. The similarities are so strong that most planetary scientists believe these meteorites are pieces of Mars that were knocked into space by asteroidal impacts. This means researchers have pieces of another planet that they can study in laboratories here on Earth. The martian meteorites support the general belief that much of the surface of Mars is covered with iron-rich volcanic materials. Sometimes the findings are controversial. In 1996, a NASA research team announced that the meteorite ALH84001 showed possible physical and chemical evidence of past life on Mars. This was an extraordinary claim, so it required extraordinary evidence. The team's conclusions were challenged, and the debate continues even today.

If pieces of Mars have reached Earth from its orbit almost 80 million kilometers beyond our own, we might expect that pieces of our companion Moon would have found their way to Earth as well. Indeed, meteorites of another group bear striking similarities to samples returned from the Moon. Like the meteorites from Mars, these are chunks of the Moon that were blasted into space by impacts and later fell to Earth.

> To hold certain meteorites is to hold a piece of Mars.

READING ASTRONOMY News

Before science "news" gets to be "news," often someone in the scientific community must recognize it as interesting and issue a press release. These are typically very short pieces to let members of the press know something is happening, and if they want more information, they should call and ask about it. NASA issued one such press release in January 2010.

Small Asteroid 2010 AL30 Will Fly Past the Earth

By **DON YEOMANS, PAUL CHODAS, STEVE CHESLEY, AND JON GIORGINI, NASA/JPL Near-Earth Object Program Office**

Asteroid 2010 AL30, discovered by the LINEAR survey of MIT's Lincoln Laboratories on Jan. 10, will make a close approach to the Earth's surface to within 76,000 miles on Wednesday January 13 at 12:46 P.M. Greenwich time (7:46 EST, 4:46 PST). Because its orbital period is nearly identical to the Earth's one-year period, some have suggested it may be a manmade rocket stage in orbit about the Sun. However, this object's orbit reaches the orbit of Venus at its closest point to the Sun and nearly out to the orbit of Mars at its furthest point, crossing the Earth's orbit at a very steep angle, and this actually makes it very unlikely that 2010 AL30 is a rocket stage. Furthermore, our trajectory extrapolations show that this object cannot be associated with any recent launch and it has not made any close approaches to the Earth since well before the Space Age began.

It seems more likely that this is a near-Earth asteroid about 10–15 meters across, one of approximately 2 million such objects in near-Earth space. One would expect a near-Earth asteroid of this size to pass within the Moon's distance about once every week on average.

To take advantage of this close approach, there are plans to observe it with the Goldstone planetary radar on Wednesday evening, Jan. 13 beginning at 6:20 PST. The radar data could dramatically improve the object's orbit and provide additional information on its size and shape.

Evaluating the News

1. Examine **Figure 9.34**. Did this asteroid pass close to Earth?
2. According to the press release, 2010 AL30 has an orbital period close to Earth's one-year period. *If* its orbit were circular, or nearly so, what would this period tell you about the average distance of 2010 AL30 from the Sun? Why would this lead to speculation that it might be a human-made object?
3. What evidence is presented that it probably is *not* a human-made object? Is this argument convincing?
4. Compare the size of this asteroid to those mentioned in the text. If it hit Earth, would the results be catastrophic?
5. How often do similar types of asteroids pass by?
6. The final sentence states, "The radar data could dramatically improve the object's orbit." Does this mean that the radar could change the orbit of the asteroid? If so, how could that happen? If not, what did the authors really mean to say?

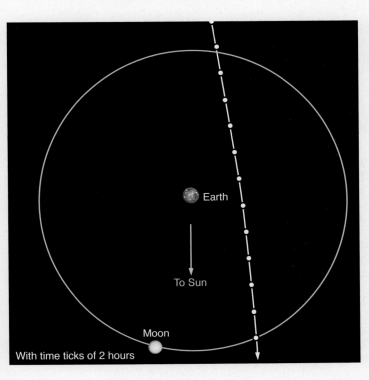

Figure 9.34 The trajectory of asteroid 2010 AL30 past Earth in mid-January 2010.

SUMMARY

9.1 Pluto, Eris, Haumea, Makemake, and Ceres are classified as dwarf planets. They are round, but they have not cleared their orbits.

9.2 The moons of the outer Solar System are composed of rock and ice. A few moons are geologically active; most are dead.

9.3 Asteroids are small Solar System bodies made of rock and metal. Some asteroids cross Earth's orbit and are potentially dangerous.

9.4 Comets are small, icy planetesimals that reside in the frigid regions beyond the planets. Comets that venture into the inner Solar System are warmed by the Sun, often producing an atmospheric coma and a long tail.

9.5 Very large asteroids or comets striking Earth create enormous explosions that can dramatically impact terrestrial life.

9.6 Meteoroids are small fragments of asteroids and comets. When a meteoroid enters Earth's atmosphere, frictional heat causes the air to glow, producing a phenomenon called a meteor. A meteoroid that survives to a planet's surface is called a meteorite.

✧ SUMMARY SELF-TEST

1. _____, _____, _____, and _____ are the four categories of small bodies in the Solar System.

2. Io is an example of a geologically _____ moon.

3. A differentiated asteroid was once _____ enough to be molten.

4. The three types of meteorites come from different parts of their parent bodies. Stony-iron meteorites are rare because
   a. they are hard to find.
   b. only a small amount of a parent body has *both* stone and iron.
   c. there is very little iron in the Solar System.
   d. the magnetic field of the Sun attracts the iron.

5. As a comet leaves the inner Solar System, its ion tail points
   a. back along the orbit.
   b. forward along the orbit.
   c. toward the Sun.
   d. away from the Sun.

6. Congress tasked NASA with searching for near-Earth objects because
   a. they might impact Earth, as others have in the past.
   b. they are close by and easy to study.
   c. they are moving fast.
   d. Congress was trying to spend some money.

QUESTIONS AND PROBLEMS

True/False and Multiple-Choice Questions

7. **T/F** There are no dwarf planets interior to Jupiter's orbit.

8. **T/F** Most moons of the outer Solar System are, or once were, geologically active.

9. **T/F** Asteroids are mostly rock and metal; comets are mostly ice.

10. **T/F** Comet tails always point directly away from the Sun.

11. **T/F** Major impacts on Earth don't happen any more.

12. **T/F** Meteors are typically caused by small pieces of dust falling through the atmosphere.

13. Pluto is classified as a dwarf planet because
    a. it is round.
    b. it orbits the Sun.
    c. it has company in its orbit.
    d. all of the above

14. Classifying moons according to their geology allows us to
    a. compare their features.
    b. explain the formation mechanisms.
    c. identify physical mechanisms responsible for their evolution.
    d. all of the above

15. Asteroids are
    a. small rock and metal objects orbiting the Sun.
    b. small icy objects orbiting the Sun.
    c. small rock and metal objects found only between Mars and Jupiter.
    d. small icy bodies found only in the outer Solar System.

16. Aside from their periods, short- and long-period comets differ because
    a. short-period comets orbit prograde, while long-period comets have either prograde or retrograde orbits.
    b. short-period comets contain less ice, while long-period comets contain more.
    c. short-period comets do not develop ion tails, while long-period comets do.
    d. short-period comets come closer to the Sun at closest approach than long-period comets do.

17. Large impacts of asteroids and comets with Earth are
    a. impossible.
    b. infrequent.
    c. unimportant.
    d. unknown.

18. Meteorites can tell us about
    a. the early composition of the Solar System.
    b. the composition of asteroids.
    c. other planets.
    d. all of the above.

Conceptual Questions

19. This chapter deals with leftover planetesimals. What happened to the others?

20. List the types of small bodies in the Solar System. Identify one distinguishing characteristic of each type.

21. Make a table of information about the dwarf planets discussed in this chapter. Include their size, number of moons, location in the Solar System, whether they are icy or rocky, and any other information you can think of that compares and contrasts these bodies. Check online to see if any new dwarf planets have been discovered, and if so, include information about those as well.

22. Describe ways in which Pluto differs significantly from the classical Solar System planets.

23. Use a search engine to find information about a moon that was not discussed in this chapter. To which of the geological activity categories do you think it belongs? Explain why.

24. Explain the process that drives volcanism on Jupiter's moon Io.

25. Describe cryovolcanism, and explain its similarities and differences with respect to terrestrial volcanism.

26. Europa and Titan may both be geologically active. What is the evidence for this?

27. Discuss evidence supporting the idea that Europa might have a subsurface ocean of liquid water.

28. Titan contains abundant amounts of methane. Why does this require an explanation? What process destroys methane in this moon's atmosphere?

29. In certain ways, Titan resembles a frigid version of early Earth. Explain the similarities.

30. Some moons display signs of geological activity in the past. Identify some of the evidence for past activity.

31. Explain why it is unlikely that the main belt asteroids will coalesce to form a planet.

32. If an asteroid is not spherical, what does that tell you about its mass?

33. Describe differences between the Kuiper Belt and the Oort Cloud as sources of comets.

34. Name the three parts of a comet. Which part is the smallest? Which is the most massive?

35. Kuiper Belt objects (KBOs) are actually comet nuclei. Why do they not display tails?

36. Comets contain substances closely associated with the development of life, such as water (H_2O), ammonia (NH_3), methane (CH_4), carbon monoxide (CO), and hydrogen cyanide (HCN). Which of these are organic compounds?

37. Sketch the appearance of a long-period comet at different locations in its orbit.

38. Picture a comet leaving the vicinity of the Sun and racing toward the outer Solar System. Does its tail point backward toward the Sun or forward in the direction the comet is moving? Explain your answer.

39. How does the composition of an asteroid differ from that of a comet nucleus?

40. Explain the importance of meteorites, asteroids, and comets to the history of life on Earth.

41. Most asteroids are found between the orbits of Mars and Jupiter, but astronomers are especially interested in the relative few whose orbits cross that of Earth. Why?

42. If collisions of comet nuclei and asteroids with Earth are rare events, why should we be concerned about the possibility of such a collision?

43. What is the source of meteors we see during a meteor shower?

44. Define meteoroid, meteor, and meteorite.

45. Make a table of the types of meteorites, a distinguishing characteristic of each type, and the origin of each type.

46. What are the differences between a comet and a meteor in terms of their size, distance, and how long they remain visible?

47. How could you and a friend, armed only with your cell phones and a knowledge of the night sky, prove conclusively that meteors are an atmospheric phenomenon?

48. During a meteor shower, all meteors trace back to a single region in the sky known as the radiant. Explain why.

Problems

49. If the dwarf planet Eris is about the same size as Pluto, why is it nearly 100 times fainter?

50. Suppose we discover a new dwarf planet that is twice as far from the Sun as Pluto, but similar in all other respects. How much fainter would it be than Pluto?

51. Io has a mass, M, of 8.9×10^{22} kg and a radius, R, of 1,820 km.
    a. The speed needed to escape from the surface of a world is called the escape velocity, v_{esc}. It is given by

    $$v_{esc} = \sqrt{\frac{2GM}{R}}$$

    Calculate the escape velocity at Io's surface.
    b. How does Io's escape velocity compare with the 1-km/s vent velocities from this moon's volcanoes?

52. Enceladus has a mass of 1.1×10^{20} kg and a radius of 250 km. Calculate the escape velocity (see previous question) at which ice crystals from this moon's cryovolcanoes must be traveling in order to escape to Saturn's E Ring.

53. Planetary scientists have estimated that Io's extensive volcanism could be covering this moon's surface with lava and ash to an average depth of up to 3 millimeters (mm) per year.
    a. Io's radius is 1,815 km. If we model Io as a sphere, what are its surface area and volume?
    b. What is the volume of volcanic material deposited on Io's surface each year?
    c. How many years would it take for volcanism to perform the equivalent of depositing Io's entire volume on its surface?
    d. How many times might Io have "turned inside out" during the life of the Solar System?

54. The inner and outer diameters of Saturn's B Ring are 184,000 and 235,000 km, respectively. If the average thickness of the ring is 10 meters and the average density is 150 kilograms per cubic meter (kg/m^3), what is the mass of Saturn's B Ring?

55. The mass of Saturn's small, icy moon Mimas is 3.8×10^{19} kg. How does this mass compare with the mass of Saturn's B Ring, as calculated in the previous question? Why is this comparison meaningful?

56. Electra is a 182-km-diameter asteroid accompanied by a small moon orbiting at a distance of 1,350 km in a circular orbit with a period of 3.92 days. Refer to Chapter 3 to answer the following questions.
    a. What is the mass of Electra?
    b. What is Electra's density?

57. Just like the engines on a jet airplane, the jets of material streaming away from the nucleus of a comet push on it, subtly altering its orbit. Scientists applied Newton's laws of motion to Comet Halley, finding the mass of the nucleus to be about 2.2×10^{14} kg. Assume the nucleus is a rectangular box with dimensions of $8 \times 8 \times 14$ km (it is actually peanut-shaped, but this is a close approximation).
    a. Calculate the volume.
    b. Find the density of the comet. Compare this to the density of water.
    c. Explain why this density confirms that the nucleus is a rubble pile.

58. At the time that Giotto and VEGA visited Comet Halley, the nucleus was losing 20,000 kg of gas and 10,000 kg of dust each second. The period of this comet is 76 years.
    a. If the comet constantly lost mass at this rate, how much mass would it lose in one year?
    b. What percentage is this of its total mass (2.2×10^{14} kg)?
    c. How many trips like this around the Sun would Halley make before it was completely destroyed?
    d. How long would that take?

59. The orbital periods of Comets Encke, Halley, and Hale-Bopp are 3.3 years, 76 years, and 2,530 years, respectively.
    a. What are the semimajor axes (in astronomical units) of the orbits of these comets?
    b. Assuming negligible perihelion distances, what are the maximum distances from the Sun (in astronomical units) reached by Comets Halley and Hale-Bopp in their respective orbits?
    c. Which would you guess is the most pristine comet among the three? Which is the least? Explain your reasoning.

60. Comet Halley has a mass of approximately 2.2×10^{14} kg. It loses about 3×10^{11} kg each time it passes the Sun.
    a. The first confirmed observation of the comet was made in 230 B.C. Assuming a constant period of 76.4 years, how many times has it reappeared since that early sighting?
    b. How much mass has the comet lost since 230 B.C.?
    c. What percentage of its total mass does this amount represent?

61. A cubic centimeter of the air you breathe contains about 10^{19} molecules. A cubic centimeter of a comet's tail may typically contain 10 molecules. Calculate the size of a cubic volume of comet tail material that would hold 10^{19} molecules.

62. The total number of asteroids larger than 1 kilometer in diameter that cross Earth's orbit (Aten and Apollo asteroids) is currently estimated to be about 3,500. Five times as many asteroids that cross Earth's orbit have diameters larger than 500 meters. Assuming this progression of number versus size remains constant for still smaller asteroids, how many would have diameters larger than 125 meters? (Note that the impact on Earth of any asteroid larger than 100 meters in diameter could cause major, widespread damage.)

63. A 1-megaton hydrogen bomb releases 4.2×10^{15} joules of energy. Compare this amount of energy with that released by a 10-km-diameter comet nucleus ($m = 5 \times 10^{14}$ kg) hitting Earth at a speed (v) of 20 km/s. You will need to use the fact that $E_K = \frac{1}{2}mv^2$ (where E_K is the kinetic energy in joules, m is the mass in kilograms, and v is the speed in meters per second).

64. One recent estimate concludes that nearly 800 meteorites with mass greater than 100 grams (massive enough to cause personal injury) strike the surface of Earth each day. Assuming that you present a target of 0.25 square meters (m^2) to a falling meteorite, what is the probability that you will be struck by a meteorite during your 100-year lifetime? (Note that the surface area of Earth is approximately 5×10^{14} m^2.)

65. Assuming that 800 meteorites with mass greater than 100 grams strike the surface of Earth each day, how much material from meteorites of this size falls on Earth each day? How much each year? If this were the only transfer of mass onto or off of Earth, how long would it take for the mass of Earth to double?

 SmartWork, Norton's online homework system, includes algorithmically generated versions of these questions, plus additional conceptual exercises. If your instructor assigns questions in SmartWork, log in at **smartwork.wwnorton.com**.

 StudySpace is a free and open website that provides a Study Plan for each chapter of **Understanding Our Universe**. Study Plans include animations, reading outlines, vocabulary flashcards, and multiple-choice quizzes, plus links to premium content in SmartWork and the ebook. Visit **wwnorton.com/studyspace**.

Exploration | Asteroid Discovery

Astronomers often discover asteroids and other small solar systems by comparing two (or more) images of the same star field, and looking for bright spots that have moved between the images. The four images below (**Figure 9.35**) are all "negative images": every dark spot would actually be bright on the sky, and all the white space is dark sky. A negative image sometimes helps the observer pick out faint details, and it is definitely preferable for printing or photocopying.

Study these four images. Can you find an asteroid that moves across the field in these images?

That's the hard way to do it. A much easier method is to use a "blink comparison," which lets you look at one image and then another very quickly. Make three photocopies of these images, cut out each image, and align all of them carefully on top of one another, so the stars overlap. You should have 12 pieces of paper, in this order: Image 1, Image 2, Image 3, Image 4; Image 1, Image 2, . . . , and so on. Staple the edge, and flip the pages with your

thumb, looking carefully at the images. Can you find the asteroid now?

1. Circle the asteroid in each image.

This method takes advantage of a feature of the human eye-brain connection. Humans are much better at noticing things that move than things that do not.

2. Why might this feature be a helpful evolutionary adaptation?

There's a third method, which is sometimes easier to see but requires high-quality digital images. In that method, one image is subtracted from another.

3. If you did that with two of these images, what would you expect to see in the resulting image?

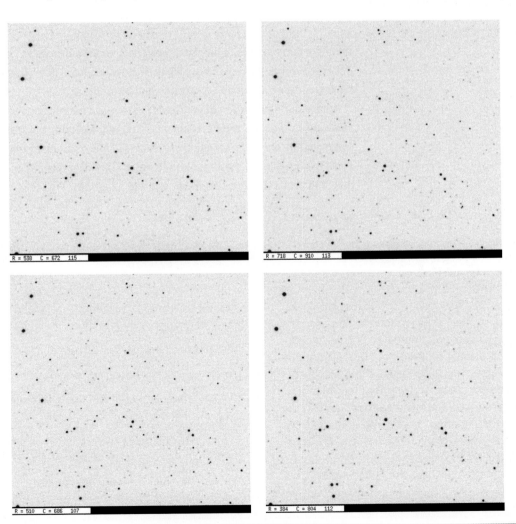

Figure 9.35

10 **Measuring the Stars**

An important part of the human experience, and the process of science, is asking questions. How far away is that star? How big is it? How luminous is it? These are questions that people have wondered about and worked to understand.

Unlike our exploration of the Solar System, we cannot send space probes to a star to take close-up pictures or land on its surface. Instead, we study the stars by observing their light, by using our current understanding of the laws of physics, and by finding patterns in subgroups of stars that enable us to extrapolate to other stars. We use our knowledge of geometry, radiation, and orbits to begin to answer age-old questions.

⟡ *LEARNING GOALS*

To all but the most powerful of telescopes, a star is just a point of light in the night sky. But by applying our understanding of light, matter, and motion to what we see, we are able to build a remarkably detailed picture of the physical properties of stars. At right, a student is using the H-R diagram to learn more about Vega, the bright star in her photo of the constellation Lyra. By the end of this chapter, you should understand why some stars are red and some are blue—and how we know this about them. You should know what the main sequence is, and why most stars are main-sequence stars. You should be able to find the luminosity and temperature of a main-sequence star from its mass and the H-R diagram. You should also be able to:

- Use the brightness of nearby stars and their distances from Earth to discover how luminous they are

- Infer the temperatures and sizes of stars from their colors

- Determine the composition and mass of stars

- Classify stars, and organize this information on a Hertzsprung-Russell diagram

- Explain why the mass and composition of a main-sequence star determine its luminosity, temperature, and size

10.1 The First Step: Measuring the Brightness, Distance, and Luminosity of Stars

Our two eyes have different views that depend on the distance to the object we are viewing. Hold up your finger in front of you, quite close to your nose. View it with your right eye only and then with your left eye only. Each eye sends a slightly different image to your brain, and so your finger *appears* to move back and forth relative to the background behind it. Now hold up your finger at arm's length, and blink your right eye, then your left. Your finger appears to move much less. This difference in perspective at different distances is the basis of our **stereoscopic vision**. This is the primary way we perceive distances. (In **Figure 10.1**, we trick your brain into perceiving distance where none exists by giving each of your eyes a slightly different view.)

Our stereoscopic vision allows us to judge the distances of objects as far away as ten meters, but beyond that it is of little use. Your right eye's view of a mountain several kilometers away is indistinguishable from the view seen by your left eye—all you know is that the mountain is too far away for you to judge its distance. The distance over which our stereoscopic vision works is limited by the separation between our two eyes, about 6 centimeters (cm). If you could separate your eyes by several meters, you could judge the distances to objects that were about half a kilometer away.

Of course, we cannot literally take our eyes out of our heads and hold them apart at arm's length, but we can compare pictures taken from two widely separated locations. The greatest separation we can get without leaving Earth is to let Earth's orbital motion carry us from one side of the Sun to the other. If we take a picture of the sky tonight and then wait 6 months and take another picture, the distance between the two locations is the diameter of Earth's orbit (2 astronomical units—AU), which should give us very powerful stereoscopic vision indeed. **Figure 10.2** shows how our view of a field of stars changes as our perspective changes during the year. This change in perspective is what enables us to measure the distances to nearby stars.

We measure distances to nearby stars by comparing the view from opposite sides of Earth's orbit.

Figure 10.1 Your brain uses the slightly different views offered by your two eyes to "see" the distances and three-dimensional character of the world around you. This stereoscopic pair shows the stars in the neighborhood of the Sun as viewed from the direction of the north celestial pole. The field shown is 40 *light-years* on a side. The Sun is at the center, marked with a green cross. The observer is 400 light-years away and has "eyes" separated by about 30 light-years. To get the stereoscopic view, hold a card between the two images and look at them from about a foot and half away. Relax and look straight at the page until the images merge into one.

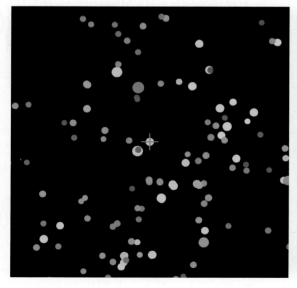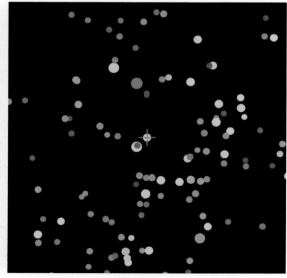

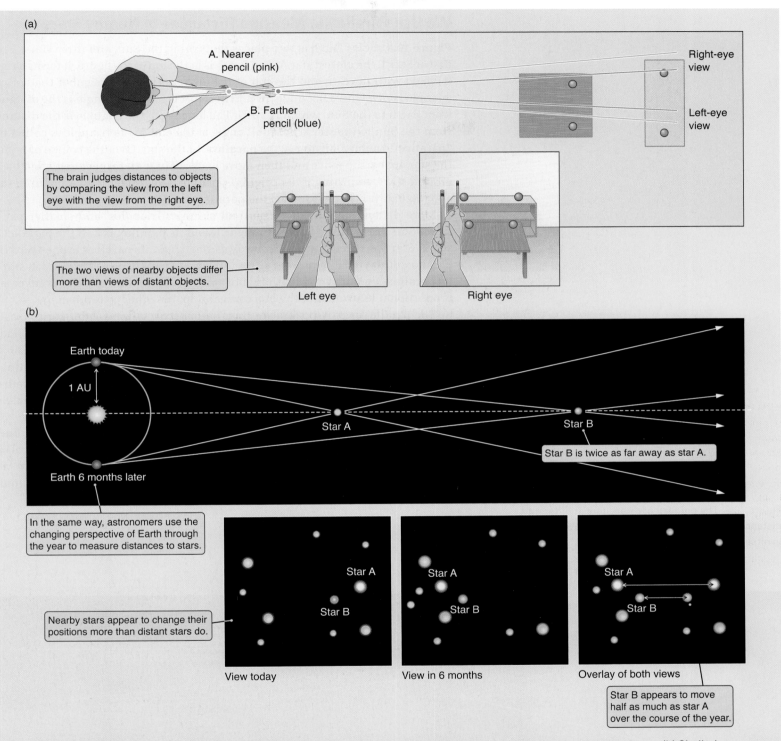

(a)

A. Nearer pencil (pink)

B. Farther pencil (blue)

Right-eye view

Left-eye view

The brain judges distances to objects by comparing the view from the left eye with the view from the right eye.

The two views of nearby objects differ more than views of distant objects.

Left eye

Right eye

(b)

Earth today

1 AU

Star A

Star B

Star B is twice as far away as star A.

Earth 6 months later

In the same way, astronomers use the changing perspective of Earth through the year to measure distances to stars.

Nearby stars appear to change their positions more than distant stars do.

Star A

Star B

View today

Star A

Star B

View in 6 months

Star A

Star B

Overlay of both views

Star B appears to move half as much as star A over the course of the year.

Figure 10.2 (a) Stereoscopic vision allows you to determine the distance to an object by comparing the view from each eye. (b) Similarly, comparing views from different places in Earth's orbit allows us to determine the distance to stars. As we move around the Sun, the apparent positions of nearby stars change more than the apparent positions of more distant stars. (The diagram is not to scale.) This is the starting point for measuring the distances to stars.

◉ VISUAL ANALOGY

We Use Parallax to Measure Distances to Nearby Stars

Figure 10.3 shows Earth at two places in its orbit, the Sun, and three stars. Look first at Star 1, the closest star. When Earth is at the top of the figure, it forms a right triangle with the Sun and the star at the other corners. (Remember that a right triangle is one with a 90° angle in it.) The short leg of the triangle is the distance from Earth to the Sun, which is 1 AU. The long leg of the triangle is the distance from the Sun to the star. The small angle at the end of the triangle is called the "parallactic angle," or simply the **parallax**, of the star. Over the course of 1 year, the star appears to move and then move back again with respect to distant background stars, returning to its original position 1 year later. The amount of this apparent shift is equal to twice the parallax.

More distant stars make longer and skinnier triangles, and smaller parallaxes. Star 2 is twice as far away as star 1, and its parallax is half the parallax of star 1. Star 3 is three times as far away as star 1, and its parallax is one-third the parallax of star 1. If there were a "star 10" located 10 times as far away as star 1, its parallax would be one-tenth that of star 1. The parallax of a star is inversely proportional to its distance: when one goes up, the other goes down.

The parallaxes of real stars are tiny. Rather than talking about parallaxes of 0.0000028°, astronomers normally measure parallaxes in units of arcseconds. Just as an hour on the clock is divided into minutes and seconds, a degree can be divided into arcminutes and arcseconds. An **arcminute** (abbreviated **arcmin**) is one-sixtieth of a degree, and an **arcsecond** (abbreviated **arcsec**) is one-sixtieth of an arcminute. An arcsecond is about equal to the angle formed by the diameter of a golf ball at the distance of 5 miles.

The distances to real stars are large, and in this book we normally use units of **light-years** to describe them. One light-year is the distance that light travels in 1 year—about 9.5 trillion kilometers. We use this unit because it is the unit you are most likely to see in a newspaper article or a popular book about astronomy.

Figure 10.3 The parallax of three stars at different distances. (The diagram is not to scale.) Parallax is inversely proportional to distance.

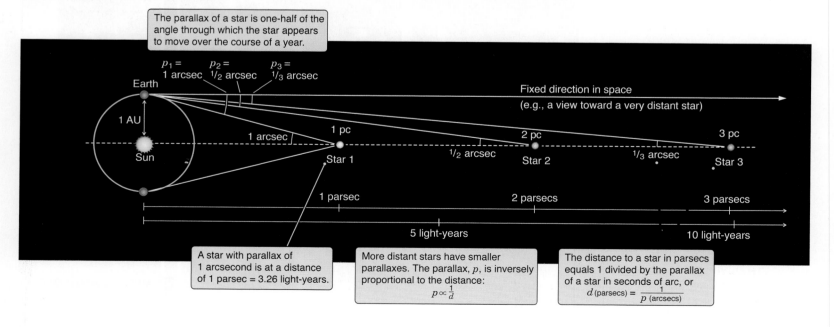

The parallax of a star is one-half of the angle through which the star appears to move over the course of a year.

$p_1 =$ 1 arcsec $p_2 =$ 1/2 arcsec $p_3 =$ 1/3 arcsec

Earth

Fixed direction in space (e.g., a view toward a very distant star)

1 AU

1 arcsec 1 pc 2 pc 3 pc

Sun

Star 1 1/2 arcsec Star 2 1/3 arcsec Star 3

1 parsec 2 parsecs 3 parsecs

5 light-years 10 light-years

A star with parallax of 1 arcsecond is at a distance of 1 parsec = 3.26 light-years.

More distant stars have smaller parallaxes. The parallax, p, is inversely proportional to the distance:
$$p \propto \frac{1}{d}$$

The distance to a star in parsecs equals 1 divided by the parallax of a star in seconds of arc, or
$$d \text{ (parsecs)} = \frac{1}{p \text{ (arcsecs)}}$$

Working It Out 10.1 | Parallax and Distance

Recall from earlier chapters that "inversely proportional" means that on one side of the equation a variable is in the numerator, while on the other side, a different variable is in the denominator. The relationship between distance (d) and parallax (p) is an inverse proportion:

$$p \propto \frac{1}{d} \text{ or } d \propto \frac{1}{p}$$

The parsec has been adopted by astronomers because it makes the relationship between distance and parallax particularly nice:

$$\left(\begin{array}{c}\text{Distance measured} \\ \text{in parsecs}\end{array}\right) = \dfrac{1}{\left(\begin{array}{c}\text{Parallax measured} \\ \text{in arseconds}\end{array}\right)}$$

or

$$d \text{ (pc)} = \frac{1}{p \text{ (arcsec)}}$$

Notice that the proportionality sign has turned into an equals sign. You don't have to remember any constants if the distance is in parsecs and the parallax is in arcseconds.

Suppose that the parallax of a star is measured to be 0.5 arcsec. To find the distance to the star, we substitute that into the parallax equation for p:

$$d = \frac{1}{0.5} = \frac{1}{1/2} = 2 \text{ pc}$$

Suppose that the parallax of a star is measured to be 0.01 arcsec. What is its distance in light-years? First we find its distance in parsecs:

$$d = \frac{1}{0.01} = \frac{1}{1/100} = 100 \text{ pc}$$

Then we convert to light-years by remembering that a parsec is 3.26 light-years.

$$d = 100 \text{ pc} \times \frac{3.26 \text{ light-years}}{1 \text{ pc}}$$

$$d = 100 \times 3.26 \text{ light-years}$$

$$d = 326 \text{ light-years}$$

The star closest to us (other than the Sun) is Proxima Centauri. Located at a distance of 4.22 light-years, Proxima Centauri is a faint member of a system of three stars called Alpha Centauri. What is this star's parallax? First, we must convert from light-years back to parsecs:

$$d = 4.22 \text{ light-years} \times \frac{1 \text{ pc}}{3.26 \text{ light-years}}$$

$$d = 1.29 \text{ pc}$$

Then we find the parallax from the distance:

$$d = \frac{1}{p}$$

solve for p to get:

$$p = \frac{1}{d}$$

Then insert our value for the distance in parsecs:

$$p = \frac{1}{1.29}$$

$$p = 0.77 \text{ arcsec}$$

This star has a parallax of only 0.77 arcsec. It is no wonder that ancient astronomers were unable to detect the apparent motions of the stars over the course of a year.

When astronomers discuss distances to stars and galaxies, however, the unit they generally use is the **parsec** (which is short for *parallax second* and is abbreviated **pc**), which is equal to 3.26 light-years.

When astronomers began to apply parallax to stars in the sky, they discovered that stars are very distant objects (an example is given in **Working It Out 10.1**). The first successful parallax measurement was made by the German astronomer F. W. Bessel (1784–1846), who in 1838 reported a parallax of 0.314 arcsec for the star 61 Cygni. This finding implied that 61 Cygni was 3.2 pc away, or 660,000 times as far away as the Sun. With this one measurement, Bessel increased the known size of the universe 10,000-fold. Today only about 60 stars within 15 light-years of the Sun are known. In the neighborhood of the Sun, each star (or star system) has about 360 cubic light-years of space all to itself.

Vocabulary Alert

Uncertainty: An uncertain distance does not mean that the distance is completely unknown. Uncertainty is simply a way of expressing how well the distance is known. For scientists, the uncertainty is sometimes the most important number, and they often get very excited when a new experiment reduces the uncertainty, even when it doesn't change the value. When astronomers discovered that the age of the universe was 13.7 billion years old, with an uncertainty of 0.1 billion years, many astronomers were more excited about the 0.1 than the 13.7. This is because this measurement was so much more certain than any that came before it.

Deep: In common language, this word has many meanings, such as how far down in the ocean an object is, or how profound an idea is. Astronomers use this word to refer to how far away an object is—how *deep* it is in space. Since distant objects are typically fainter, *deep* and *faint* are closely related and sometimes used somewhat interchangeably.

Brightness: In common language, the word *brightness* can be used to describe either how much light an object emits (its intrinsic brightness) or how much gets to you (its apparent brightness). These two things may be different due to the distance of the object. Astronomers try very hard to always use *brightness* to refer to how much light an observer receives and *luminosity* to refer to how much light is emitted.

Brightness depends on the observer's distance; luminosity does not.

Knowledge of our stellar neighborhood took a tremendous step forward during the 1990s, when the Hipparcos satellite measured the positions and parallaxes of 120,000 stars. Even this catalog has its limits. The accuracy of any given Hipparcos parallax measurement is about ±0.001 arcsec. Because of this observational **uncertainty**, our measurements of the distances to stars are not perfect. For example, a star with a Hipparcos parallax of 0.004 arcsec really has a parallax within a range of arcseconds centered on 0.004. Instead of knowing that the distance to the star is exactly 250 pc, we know only that the star is between about 200 pc and 330 pc. As an analogy, consider your speed while driving down the road. If your digital speedometer says 10 kilometers per hour (km/h), you might actually be traveling 10.4 km/h, or 9.6 km/h. The precision of your speedometer is limited to the ones place, but that doesn't mean you don't know your speed *at all*. You are certainly not traveling 100 km/h, for example. With current technology, we cannot reliably measure stellar distances of more than a few hundred parsecs using parallax. Other methods—to be discussed later—are used for more distant stars.

The Brightness of a Star

Two thousand years ago, the Greek astronomer Hipparchus classified the brightest stars he could see as being "of the first **magnitude**" and the faintest as being "of the sixth magnitude." Note that this means a brighter object has a smaller magnitude. It is also an example of logarithmic behavior, so that an object of magnitude 2 is more than twice as bright as an object of magnitude 4. Hipparchus must have had typical eyesight, because an average person under dark skies can see stars only as faint as 6th magnitude. Modern telescopes can see much "**deeper**" than this. Hubble Space Telescope can detect stars as faint as 30th magnitude—4 billion times fainter than what the naked eye can see.

Objects that are brighter than 1st magnitude in this system have magnitudes of less than one, and the magnitude can even be negative. For example, Sirius, the brightest star in the sky in the visible wavelengths, has a magnitude of −1.46. Venus can be bright enough to cast shadows, at magnitude −4.4. The magnitude of the full Moon is −12.6, and that of the Sun is −26.7. Thus, the Sun is 14.1 magnitudes (nearly half a million times) brighter than the full Moon.

The magnitude of a star, as we have discussed it, is called the star's **apparent magnitude** because it is the **brightness** of the star as it *appears* to us in our sky. Stars are found at different distances from us, so a star's apparent magnitude does not tell us how much light it actually emits. To find its *intrinsic* brightness—its **luminosity**—we must know the distance. Then we can put it on a scale with all other stars of known distances, and we can calculate how bright they *would* be if they were all located 10 pc from us. This is the **absolute magnitude**: the brightness of each star if it were located at a distance of 10.0 pc (32.6 light-years).

The brightness of astronomical objects generally varies with wavelength region (color), so astronomers use special symbols to represent magnitudes at certain colors. For example, they use *V* and *B*, respectively, to represent magnitudes in the visual (yellow-green) and blue regions of the spectrum. The term *visual* is used because yellow-green light roughly corresponds to the range of wavelengths our eyes are most sensitive to.

Finding Luminosity

Although the brightness of a star is directly measurable, it does not immediately tell us much about the star itself. As illustrated in **Figure 10.4**, an apparently bright star in the night sky may in fact be intrinsically dim but close by. Conversely, a faint star may actually be very luminous, but because it is very far away, it appears faint to us. We can specifically say that its apparent brightness, which corresponds to the intensity of the starlight that reaches us, is inversely proportional to the square of our distance from the star (see Working It Out 7.1). If we know this distance, we can then use our measurement of the star's apparent brightness to find its luminosity.

The range of possible luminosities for stars is very large. The Sun provides a convenient yardstick for measuring the properties of stars, including their luminosity. The most luminous stars are more than a million times the luminosity of the Sun. The least luminous stars have luminosities less than 1/10,000 that of the Sun. The most luminous stars are therefore over 10 billion (10^{10}) times more luminous than the least luminous stars. Very few stars are near the upper end of this range of luminosities, and the vast majority of stars are far less luminous than our Sun. **Figure 10.5** shows the relative number of stars compared to their luminosity in solar units.

There are many more low-luminosity stars than high-luminosity stars.

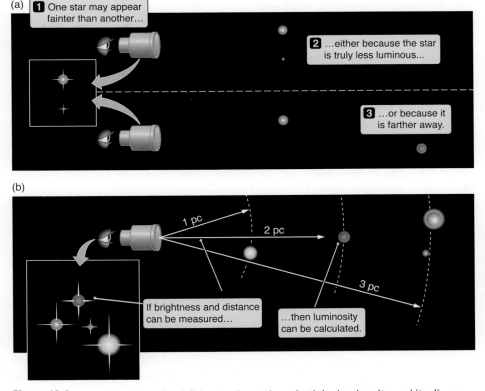

(a)
1 One star may appear fainter than another…

2 …either because the star is truly less luminous…

3 …or because it is farther away.

(b)

1 pc

2 pc

3 pc

If brightness and distance can be measured…

…then luminosity can be calculated.

Figure 10.4 The brightness of a visible star depends on both its luminosity and its distance. When brightness and distance are measured, luminosity can be calculated.

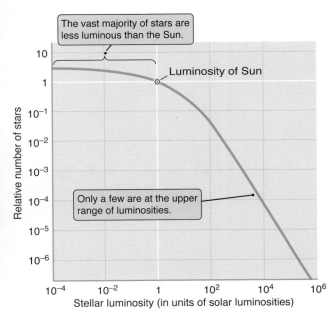

The vast majority of stars are less luminous than the Sun.

Luminosity of Sun

Relative number of stars

Only a few are at the upper range of luminosities.

Stellar luminosity (in units of solar luminosities)

Figure 10.5 The distribution of the luminosities of stars. Both axes are plotted logarithmically.

Vocabulary Alert

Surface: In common language, we don't use the word *surface* to refer to a layer within a gaseous body. But here astronomers mean the part of the star that gives off the radiation that we see. A star's surface is not solid like the surface of a terrestrial planet, and stars can certainly have more layers outside of this "surface."

10.2 Radiation Tells Us the Temperature, Size, and Composition of Stars

Two everyday concepts—stereoscopic vision, and the fact that closer objects appear brighter—have given us the tools we need to measure the distance and luminosity of the closest stars. In these two steps, stars have gone from being merely faint points of light in the night sky to being extraordinarily powerful beacons located at distances almost impossible for the mind to comprehend.

Stars are gaseous, but they are fairly dense—dense enough that the radiation from a star comes close to obeying the same laws as the radiation from objects like the heating element on an electric stove. That means we can use our understanding of blackbody radiation—results such as the Stefan-Boltzmann law (hotter at same size means more luminous) and Wien's law (hotter means bluer)—to understand the radiation from stars. In particular, these two laws enable us to measure the temperatures and sizes of our stellar neighbors.

Wien's Law Revisited: The Color and Surface Temperature of Stars

Wien's law (see Working It Out 6.3) shows that the temperature of an object determines the peak wavelength of its spectrum. Hotter objects emit bluer light. Stars with especially hot surfaces are blue, stars with especially cool surfaces are red, and our Sun is a middle-of-the-road yellow. If you obtain a spectrum of a star and measure the wavelength at which the spectrum peaks, then Wien's law will tell you the temperature of the star's **surface**. The color of a star tells us only about the temperature at the surface, because this layer is giving off the radiation that we see. Stellar interiors are far hotter than this, as we will discover.

Measuring the color of a star tells us its surface temperature.

Look back at the chapter-opening figure. The student identified that Vega was blue in her image, and immediately determined that it must be about 10,000 K or hotter. Other stars in the image have other colors. From these colors alone, you should be able to identify several stars that are definitely cooler than Vega.

In practice, it is usually not necessary to obtain a complete spectrum of a star to determine its temperature. Instead, astronomers often measure the colors of stars by comparing the brightness at two different wavelengths. The brightness of a star is usually measured through a **filter**—sometimes just a piece of colored glass—that lets through only a small range of wavelengths. Two of the most common filters are a blue filter and a "visual" (again, yellow-green) filter. From a pair of pictures of a group of stars, each taken through a different filter, we can measure the surface temperature of every star in the picture—perhaps hundreds or even thousands—all at once. When we do, we find there are many more cool stars than hot stars. We also discover that most stars have surface temperatures lower than that of the Sun.

▶‖ AstroTour: **Stellar Spectrum**

The Interaction of Light and Matter

So far, we have concentrated on what we can learn about stars by applying our understanding of thermal radiation. However, the spectra of stars are not smooth, continuous blackbody spectra. Instead, when we pass the spectra of stars through a prism, we see dark and bright lines at specific wavelengths. To understand these lines (which tell us much of what we know about the universe), we must

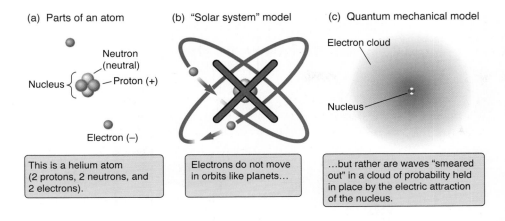

(a) Parts of an atom

Neutron (neutral)

Nucleus — Proton (+)

Electron (−)

This is a helium atom (2 protons, 2 neutrons, and 2 electrons).

(b) "Solar system" model

Electrons do not move in orbits like planets…

(c) Quantum mechanical model

Electron cloud

Nucleus

…but rather are waves "smeared out" in a cloud of probability held in place by the electric attraction of the nucleus.

FIGURE 10.6 (a) An atom (in this case, helium) is made up of a nucleus consisting of positively charged protons and electrically neutral neutrons and surrounded by less massive negatively charged electrons. (b) Atoms are often drawn as miniature "solar systems," but this model is incorrect, as the red *X* indicates. (c) Electrons are actually smeared out around the nucleus in quantum mechanical clouds of probability.

know how light interacts with matter. We learned a little about light and matter in Chapter 4. Now we need to learn more about how they interact together.

Matter occupies space and has mass. Everyday matter is made of atoms (**Figure 10.6a**). The electrons in the atom are much less massive than protons or neutrons, so almost all the mass of an atom is found in its nucleus. This description can lead to a mental picture of an atom as a tiny "solar system," with the massive nucleus sitting in the center and the smaller electrons orbiting about it, much as planets orbit about the Sun (**Figure 10.6b**). We refer to this as the Bohr model, after the Danish physicist Niels Bohr (1885–1962), who proposed it in 1913. Unfortunately, although it is easy to visualize, this model is incorrect.

Just as waves of light have particle-like properties, particles of matter have wavelike properties. A positively charged nucleus is not surrounded by planet-like electrons moving in well-behaved orbits, but by electron "clouds" or electron "waves" (**Figure 10.6c**). This is why we use a featureless cloud to represent electrons in orbit around an atomic nucleus.

Atoms Have Discrete Energy Levels

Electron "waves" in an atom can assume only certain specific forms, which depend on the energy state of the atom. We can imagine the energy states of an atom as a set of shelves in a bookcase, as shown in **Figure 10.7a**. The energy of an atom might correspond to the energy of one shelf or to the energy of the next shelf; but the energy of the atom will never be found between the two shelves.

The lowest possible energy state of an atom is called the **ground state** (**Figure 10.7b**). When the atom is in this state, the electron has its minimum energy. It can't give up any more energy to move to a lower state, because there isn't one. An atom will remain in its ground state forever unless it gets energy from outside.

Energy levels above the ground state are called **excited states**. An atom in an excited state might transition to the ground state by getting rid of the "extra" energy all at once. The atom goes from one energy state to another, but it never has an amount of energy in between. Think about money, which is similarly *quantized*. If you have a penny, a nickel, and a dime, you have 16 cents. Now imagine that you give away the nickel. You are left with 11 cents. But you never had exactly 13 cents, and certainly you never had 13.6 cents. You had 16, and

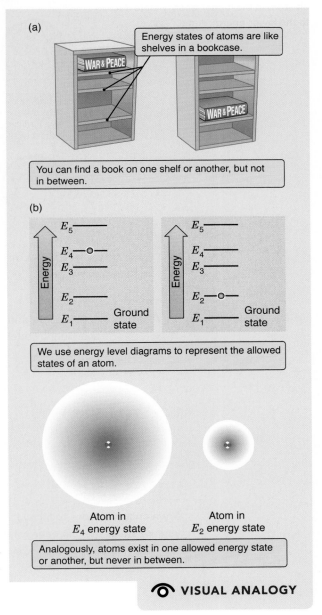

(a)

Energy states of atoms are like shelves in a bookcase.

WAR & PEACE

WAR & PEACE

You can find a book on one shelf or another, but not in between.

(b)

E_5
E_4
E_3
E_2
E_1 Ground state

E_5
E_4
E_3
E_2
E_1 Ground state

We use energy level diagrams to represent the allowed states of an atom.

Atom in E_4 energy state

Atom in E_2 energy state

Analogously, atoms exist in one allowed energy state or another, but never in between.

⊙ VISUAL ANALOGY

FIGURE 10.7 (a) Energy states of an atom are analogous to shelves in a bookcase. You can move a book from one shelf to another, but books can never be placed between shelves. (b) Atoms exist in one allowed energy state or another but never in between. There is no level below the ground state.

FIGURE 10.8 (a) The energy associated with transitions between energy states is analogous to individual coins in a handful. If you begin with a dime, a nickel, and a penny, you might give someone the nickel, leaving you with 11 cents. But you never had exactly 13 cents. First you had 16, then you had 11. (b) Similarly, an atom can give up photons with only specific energies. A photon with energy $E_2 - E_1$ is emitted when an atom in the higher-energy state decays to the lower-energy state.

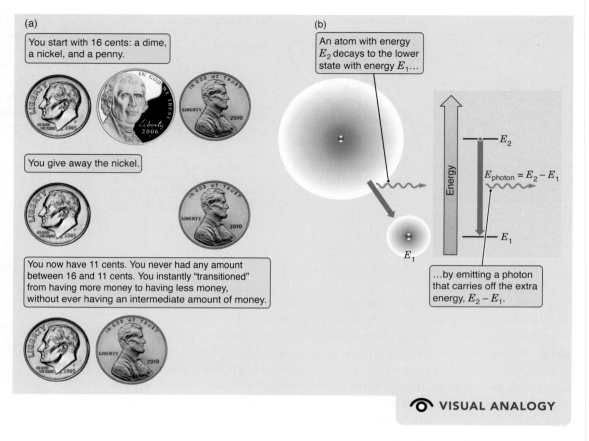

(a)

You start with 16 cents: a dime, a nickel, and a penny.

You give away the nickel.

You now have 11 cents. You never had any amount between 16 and 11 cents. You instantly "transitioned" from having more money to having less money, without ever having an intermediate amount of money.

(b)

An atom with energy E_2 decays to the lower state with energy E_1...

E_2

Energy

$E_{\text{photon}} = E_2 - E_1$

E_1

E_1

...by emitting a photon that carries off the extra energy, $E_2 - E_1$.

VISUAL ANALOGY

▶∥ **AstroTour: Atomic Energy Levels and the Bohr Model**

When an atom drops to a lower-energy state, the lost energy is carried away as a photon.

then 11 (**Figure 10.8a**). Atoms don't accept and give away money to change energy states, but they do accept and give away photons. Atoms falling from a higher-energy state with energy E_2 to a lower-energy state, E_1, lose an amount of energy exactly equal to the difference in energy levels, $E_2 - E_1$. Therefore, the energy of the photon emitted must be $E_{\text{photon}} = E_2 - E_1$. This change is illustrated in **Figure 10.8b**, where the downward arrow indicates that the atom went from the higher state to the lower state.

The energy level structure of an atom determines the wavelengths of the photons emitted by it—the color of the light that the atom gives off. An atom can emit photons with energies corresponding *only* to the difference between two of its allowed energy states. Recall from Chapter 4 that the energy, wavelength, and frequency of photons are all related. We say that a photon of energy $E_{\text{photon}} = E_2 - E_1$ has a specific wavelength $\lambda_{2 \rightarrow 1}$ and a specific frequency $f_{2 \rightarrow 1}$. This means that these are photons with a very specific color, and every photon emitted in any transition from E_2 to E_1 will have this same color.

Why was the atom in the excited state E_2 in the first place? An atom sitting in its ground state will remain there forever unless it absorbs just the right amount of energy to kick it up to an excited state. In general, the atom either absorbs the energy of a photon or it collides with another atom, or perhaps an unattached electron, and absorbs some of the other particle's energy. Atoms moving from a lower-energy state E_1 to a higher-energy state E_2 can *absorb* only energy $E_2 - E_1$, whether it comes in the form of photons or collisions.

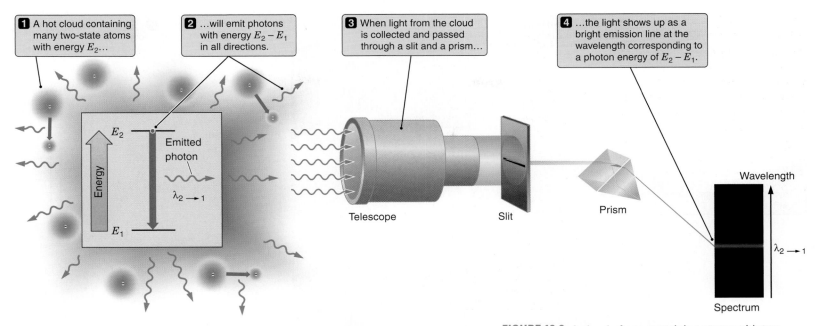

1 A hot cloud containing many two-state atoms with energy E_2...

2 ...will emit photons with energy $E_2 - E_1$ in all directions.

3 When light from the cloud is collected and passed through a slit and a prism...

4 ...the light shows up as a bright emission line at the wavelength corresponding to a photon energy of $E_2 - E_1$.

E_2

Energy

Emitted photon

$\lambda_2 \longrightarrow 1$

E_1

Telescope

Slit

Prism

Wavelength

$\lambda_2 \longrightarrow 1$

Spectrum

FIGURE 10.9 A cloud of gas containing atoms with two energy states, E_1 and E_2, emits photons with an energy $E_2 - E_1$, which appear in the spectrogram (right) as a single *emission line*.

Imagine a cloud of hot gas consisting of atoms with only two energy states, E_2 and E_1, as shown in **Figure 10.9**. Because the gas is hot, the atoms are continuously zooming around and colliding, thus getting kicked up from the ground state, E_1, into the higher-energy state, E_2. Any atom in the higher-energy state quickly decays and emits a photon in a random direction. This emitted light contains only photons with the specific energy $E_2 - E_1$. In other words, all of the light coming from the cloud is the same color. If passed through a slit and a prism, it forms a single bright line of one color, called an **emission line**. This is how some neon signs work: Each color in a "neon" sign comes from a different gas (not necessarily neon) trapped inside the glass tubes.

Now imagine that we view a white light (one with all wavelengths of photons in it) through a cool cloud of gas (**Figure 10.10a**). Almost all of these photons will pass through the cloud of gas unaffected, because they do not have the right amount of energy ($E_2 - E_1$) to be absorbed by atoms of the gas. However, photons with just the right amount of energy will be absorbed; as a result, they will be *missing* from the spectrum. We will see a sharp, dark line at the wavelength corresponding to this energy. This process is called absorption, and the dark feature is called an **absorption line**. **Figure 10.10b** shows such absorption lines in the spectrum of a star. For any element, the absorption lines occur at exactly the same wavelength as the emission lines. The energy difference between the two levels is the same whether the atom is emitting a photon or absorbing one, so the energy of the photon involved will be the same in either case.

When an atom absorbs a photon, it may quickly return to its previous energy state, emitting a photon with the same energy as the photon it just absorbed. If the atom emits a photon just like the one it absorbed, you might reasonably ask why the absorption matters at all, since the photon taken out of the spectrum was replaced by an identical one. The photon was replaced, it's true, but all of the absorbed photons were originally traveling in the *same direction*, whereas the

▶❙❙ **AstroTour: Atomic Energy Levels and Light Emission and Absorption**

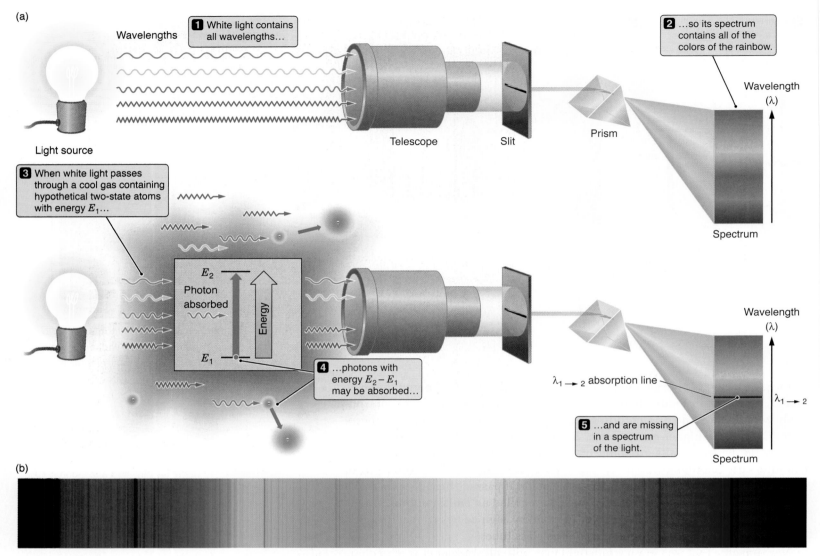

FIGURE 10.10 (a) When passed through a prism, white light produces a spectrum containing all colors. When light of all colors passes through a cloud of hypothetical two-state atoms, photons with energy $E_2 - E_1$ may be absorbed, leading to the dark absorption line in the spectrogram. (b) Absorption lines in the spectrum of a star.

emitted photons are now traveling in *random directions*. In other words, most of the photons have been diverted from their original paths. If you look at a white light *through* the cloud, you will observe an absorption line at a wavelength of $\lambda_{1 \to 2}$, but if you look at the cloud from another direction, you will observe an emission line at this wavelength.

Emission and Absorption Lines Are the Spectral Fingerprints of Atoms

Real atoms can occupy many more than just two possible energy states, so an atom of a given element is capable of emitting and absorbing photons at many different wavelengths. An atom with three energy states, for example, might jump from

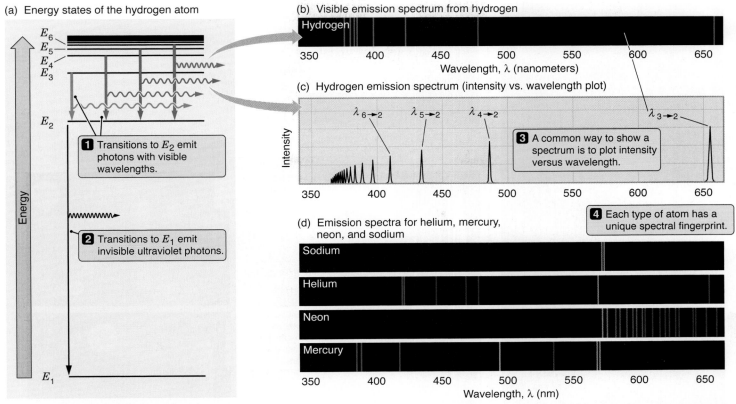

(a) Energy states of the hydrogen atom

1 Transitions to E_2 emit photons with visible wavelengths.

2 Transitions to E_1 emit invisible ultraviolet photons.

(b) Visible emission spectrum from hydrogen

Hydrogen

Wavelength, λ (nanometers)

(c) Hydrogen emission spectrum (intensity vs. wavelength plot)

$\lambda_{6\rightarrow2}$ $\lambda_{5\rightarrow2}$ $\lambda_{4\rightarrow2}$ $\lambda_{3\rightarrow2}$

3 A common way to show a spectrum is to plot intensity versus wavelength.

4 Each type of atom has a unique spectral fingerprint.

(d) Emission spectra for helium, mercury, neon, and sodium

Sodium

Helium

Neon

Mercury

Wavelength, λ (nm)

FIGURE 10.11 (a) The energy states of the hydrogen atom. Transitions to level E_2 emit photons in the visible part of the spectrum. (b) This spectrum is what you might see if you looked at the light from a hydrogen lamp projected through a prism onto a screen. (c) This graph of the brightness (intensity) of spectral lines versus their wavelength illustrates how spectra are traditionally plotted. (d) Emission spectra from several other types of gases.

state 3 to state 2, or from state 3 to state 1, or from state 2 to state 1. Its spectrum will have three distinct emission lines.

Every hydrogen atom has the same energy states available to it, and all hydrogen atoms have the same emission and absorption lines. **Figure 10.11a** shows the energy level diagram of hydrogen, along with the spectrum of emission lines for hydrogen in the visible part of the spectrum (**Figure 10.11b and c**).

Each different element has a unique set of available energy states and therefore a unique set of wavelengths at which it can emit or absorb radiation. **Figure 10.11d** shows the set of emission lines from different kinds of atoms. These unique sets of wavelengths serve as unmistakable spectral "fingerprints" for each element.

If we see the spectral lines of hydrogen, helium, carbon, oxygen, or any other element in the light from a distant object, then we know that element is present in that object. The **strength** of a line is determined in part by how many atoms of that type are present in the source. By measuring the strength of the lines from an element, astronomers can often infer the abundance of the element in the object—and sometimes the temperature, density, and pressure of the material also.

Stars Are Classified According to Their Surface Temperature

Although the hot surface of a star emits radiation with a spectrum very close to a smooth blackbody curve, this light must then escape through the outer layers of the star's atmosphere. The atoms and molecules in the cooler layers of the star's atmosphere leave their absorption line fingerprints in this light, as shown

Vocabulary Alert

Strength: In this context, *strength* means how bright the emission line is, or how faint the absorption line is. More atoms either add more light (in the case of emission) or take more light away (in the case of absorption), making a stronger line.

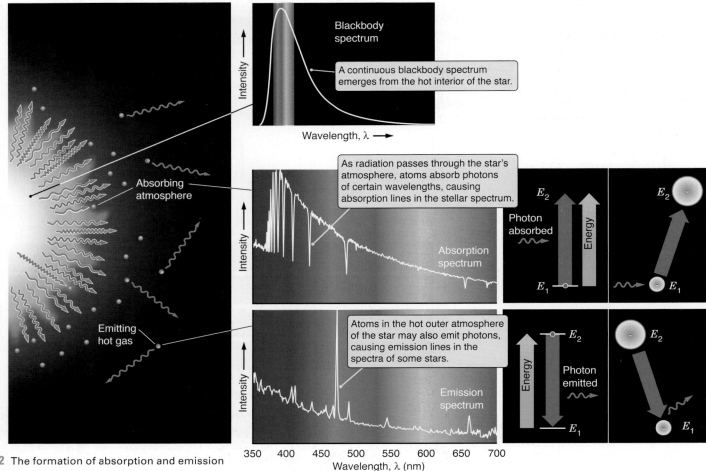

FIGURE 10.12 The formation of absorption and emission lines in the spectra of stars.

Stars are classified by the appearance of their spectra.

---

in **Figure 10.12**. These atoms and molecules, along with any gas that might be found in the vicinity of the star, can also produce emission lines in stellar spectra. Although absorption and emission lines complicate how we use the laws of blackbody radiation to interpret light from stars, spectral lines more than make up for this trouble by providing a wealth of information about the state of the gas in a star's atmosphere.

The spectra of stars were first classified during the late 1800s, long before stars, atoms, or radiation were well understood. The classification we use today was based on the prominence of particular absorption lines seen in those spectra. Stars with the strongest hydrogen lines were labeled "A stars," stars with somewhat weaker lines were labeled "B stars," and so on.

Annie Jump Cannon (1863–1941) led an effort at the Harvard College Observatory to systematically examine and classify the spectra of hundreds of thousands of stars. She dropped many of the earlier spectral types, keeping only seven that were subsequently reordered based on surface temperatures. Spectra of stars of different types are shown in **Figure 10.13**. The hottest stars, with surface temperatures over 30,000 K, are labeled "O stars." O stars have only weak absorption lines from hydrogen and helium. The coolest stars—"M stars"—have temperatures as low as about 2800 K. M stars show myriad lines from many different types of atoms and molecules. The complete sequence of **spectral types**

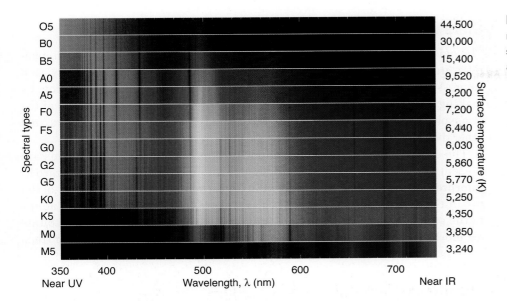

of stars, from hottest to coolest, is O, B, A, F, G, K, M. The boundaries between spectral types are not precise. A hotter-than-average G star is very similar to a cooler-than-average F star.

Astronomers divide the main spectral types into subclasses by adding numbers to the letter designations. For example, the hottest B stars are called B0 stars, slightly cooler B stars are called B1 stars, and so on. The coolest B stars are B9 stars, which are only slightly hotter than A0 stars. The Sun is a G2 star.

Returning to Figure 10.13, we can see that not only are hot stars bluer than cool stars, but the absorption lines in their spectra are quite different as well. The temperature of the gas in the atmosphere of a star affects the state of the atoms in that gas, which in turn affects the energy level transitions available to absorb radiation. In O stars, the temperature is so high that most atoms have had one or more electrons stripped from them by energetic collisions within the gas. Few transitions are available in the visible part of the electromagnetic spectrum, so the visible spectrum of an O star is relatively featureless. At lower temperatures there are more atoms that can absorb light in the visible part of the spectrum, so the visible spectra of cooler stars are more complex than are the spectra of O stars.

Most absorption lines have a temperature at which they are strongest. For example, absorption lines from hydrogen are most prominent at temperatures of about 10,000 K, which is the surface temperature of an A star. (This makes sense, because spectral-type A stars were so named because they are the stars with the strongest lines of hydrogen in their spectra.)

At the very lowest stellar temperatures, atoms in the atmosphere of the star form molecules. Molecules such as titanium oxide (TiO) are responsible for much of the absorption in the atmospheres of cool M stars.

Because different spectral lines are formed at different temperatures, we can use these absorption lines to measure a star's temperature directly. The temperatures of stars measured in this way agree extremely well with the temperatures of stars measured using Wien's law, again confirming that the physical laws that apply on Earth apply to stars also.

▶❚❚ AstroTour: **Stellar Spectrum**

Stars Consist Mostly of Hydrogen and Helium

The most obvious differences in the lines seen in stellar spectra are due to temperature, but the lines also tell us about other physical properties of stars such as pressure, chemical composition, and magnetic-field strength. In addition, by making use of the Doppler shift, we can measure rotation rates, motions of the atmosphere, expansion and contraction, "winds" driven away from stars, and other dynamic properties of stars.

The strengths of various absorption lines tell us what kinds of atoms are present in the gas and in what abundance, although we must take great care in interpreting spectra to properly account for the temperature and density of the gas in the atmosphere of a star. Typically, more than 90 percent of the atoms in the atmosphere of a star are identified as hydrogen, while helium accounts for most of what remains. All of the other elements are present only in trace amounts.

The Stefan-Boltzmann Law Revisited: Finding the Sizes of Stars

We can determine the temperature of a star directly, either from Wien's law (**Figure 10.14a**) or from the strength of its spectral lines. Once we know the temperature of a star, we also know how much radiation each square meter of the star is giving off each second. As shown in **Figure 10.14b**, each square meter of the surface of a hot blue star gives off more radiation per second than a square meter of the surface of a cool red star. Therefore, a hot star will be more luminous than a cool star of the same size. A small hot star might even be more luminous than

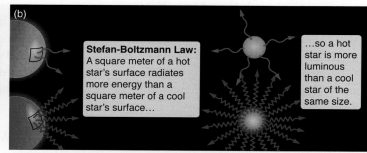

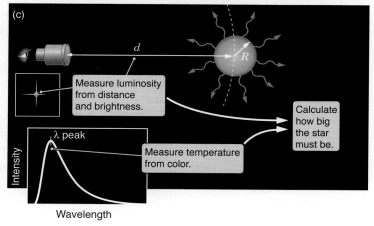

FIGURE 10.14 We measure the temperature and size of stars by applying our understanding of blackbody radiation.

a larger cool star. As noted in **Figure 10.14c**, putting the temperature and luminosity together allows us to infer the size of the star. We can calculate its radius from the Stefan-Boltzmann law and the relationship between flux and luminosity (see Working It Out 6.2).

The luminosity-temperature-radius relationship has been used to measure the sizes of thousands of stars. When we talk about the sizes of stars, we again use our Sun as a yardstick. The radius of the Sun, written as $R_\odot$, is about 700,000 km. When we look at stars around us, we find that the smallest stars we see, called white dwarfs, have radii that are only about 1 percent of the radius of the Sun ($R = 0.01\ R_\odot$). The largest stars that we see, called red supergiants, can have radii more than 1,000 times that of the Sun. There are many more stars toward the small end of this range, smaller than our Sun, than there are giant stars.

10.3 Measuring Stellar Masses

Determining the mass of an object can be a tricky business. We certainly cannot rely on the amount of light from an object or the object's size as a measure of its mass. Massive objects can be large or small, faint or luminous. The only thing that *always* goes with mass is gravity. When astronomers are trying to determine the masses of astronomical objects, they almost always wind up looking for the effects of gravity.

In Chapter 3, we found that Kepler's laws of planetary motion are the result of gravity, and we showed that the orbit of a planet can be used to measure the mass of the Sun. We can also study two *stars* that orbit about each other. About half of the higher-mass stars in the sky are actually systems consisting of several stars moving about under the influence of their mutual gravity. Most of these are **binary stars** in which two stars orbit each other as predicted by Newton's version of Kepler's laws. (However, most low-mass stars are single, and low-mass stars far outnumber their higher-mass brethren. This means that *most stars are single.*)

Binary Stars Orbit a Common Center of Mass

If the two stars in a binary system were sitting on a seesaw in a gravitational field, the support of the seesaw would have to be directly under the **center of mass** for the objects to balance, as shown in **Figure 10.15**. The two stars orbit this center of mass, a point in space that is seldom located inside either star, but usually somewhere in between.

When Newton applied his laws of motion to the problem of orbits, he found that two objects must move in elliptical orbits around each other, and that their common center of mass lies at one focus shared by both of the ellipses, as shown in **Figure 10.16**. The center of mass, which lies along the line between the two

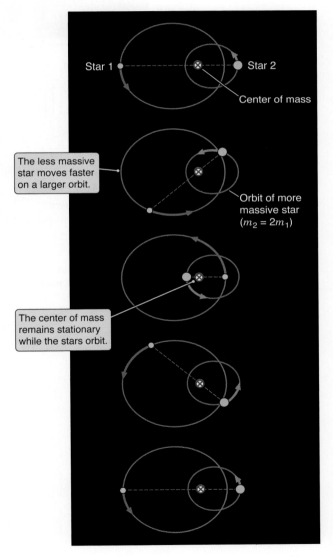

FIGURE 10.16 In a binary star system, the two stars orbit on elliptical paths about their common center of mass. In this case, star 2 has twice the mass of star 1. The eccentricity of the orbits is 0.5. There are equal time steps between the frames.

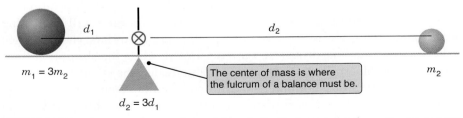

FIGURE 10.15 The center of mass of two objects is the "balance" point on a line joining the centers of two masses.

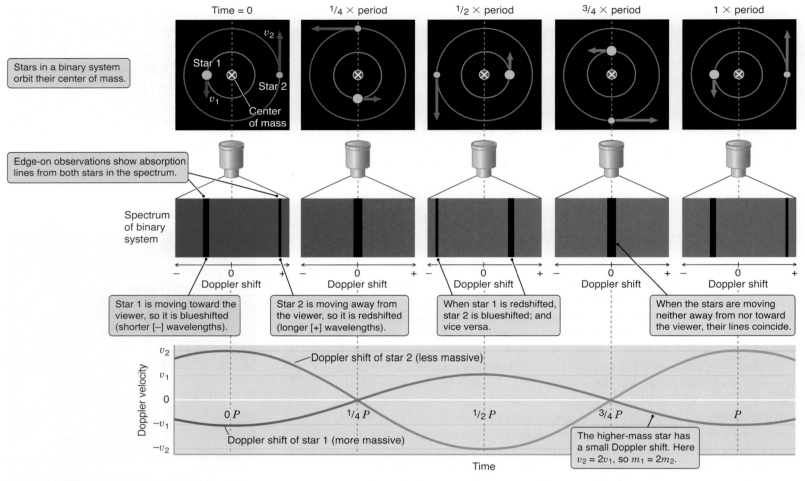

FIGURE 10.17 The orbits of two stars in a binary system are shown here, along with the Doppler shifts of the absorption lines in the spectrum of each star. Star 1 is twice as massive as star 2 and so has half the Doppler shift. *P* is the period of the orbit.

objects, remains stationary. The two objects will always be found on exactly opposite sides of the center of mass.

Because the orbit of the less massive star is larger than the more massive star's orbit, the less massive star has farther to go than the more massive star. But it must cover that distance in the same amount of time, so the less massive star must be moving *faster* than the more massive star. The velocity of a star in a binary system is inversely proportional to its mass.

Imagine that you are watching a binary star from an edge-on position, as shown in **Figure 10.17**. As star 1 is moving toward you, star 2 will be moving away from you. If you were to look at the spectra of the two stars, you might see that the absorption lines from star 1 are redshifted, while the absorption lines from star 2 are blueshifted. Half an orbital period later, the situation would be reversed: lines from star 2 would be redshifted, and lines from star 1 would be blueshifted. Comparing the size of the Doppler shift for star 1 with the size of the Doppler shift for star 2 gives the *ratio* of the masses of the two stars. That is, we can find that star 1 is two times as massive as star 2. But we can't find the actual mass of either star from these observations alone.

Kepler's Third Law Gives the Total Mass of a Binary System

In Chapter 3 we ignored the complexity of the motion of two objects about their common center of mass. Now, however, this very complexity enables us to measure the masses of the two stars in a binary system. In his derivation of Kepler's laws, Newton showed that if we can measure the period of the binary system and the average separation between the two stars, then Kepler's third law gives us the total mass. Referring to the previous subsection, if we can measure the sizes of the orbits of the two stars independently, or independently measure their velocities, then we can determine the ratios of the masses of the two stars. If we know the total mass of the system as well as the ratio of the two masses, we have all we need to determine the mass of each star separately. In other words, if we know that star 1 is two times as massive as star 2, and we know that star 1 and star 2 together are three times as massive as the Sun, then we can calculate separate values for the masses of star 1 and star 2.

There are two ways to measure the average distance and period that we need to determine the masses of stars. If a binary system is a **visual binary** (**Figure 10.18**)—that is, if we can take pictures that show the two stars separately—then we can watch over time as the stars orbit each other. We can directly measure the shapes and period of the orbits of the two stars. In many binary systems, however, the two stars are so close together and so far away from us that we cannot actually see the stars separately. We know these stars belong to binary systems only because we see the spectral lines of the two stars as they are Doppler-shifted away from each other first in one direction and then in the other; these are called **spectroscopic binary** stars. Yet even those Doppler shifts alone can sometimes be used to find the masses of the stars. If a spectroscopic binary system is an **eclipsing binary**, in which we see a dip in brightness as one star passes in front of the other (**Figure 10.19**), then the system is viewed nearly edge on. This special case is described in **Working It Out 10.2**.

The range of stellar masses is not nearly as great as the range of stellar luminosities. The least massive stars have masses of about 0.08 $M_\odot$; the most massive stars appear to have masses greater than 200 $M_\odot$. You might wonder why the mass of a star should have any limits. These limits are determined solely by

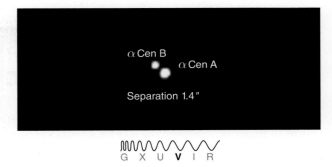

FIGURE 10.18 The two stars of a visual binary are resolved.

Newton's version of Kepler's third law gives the total mass of a binary system.

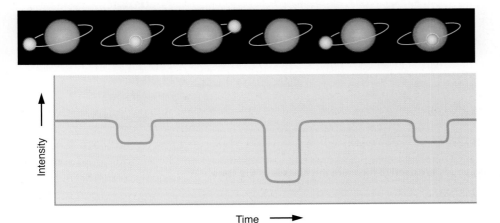

FIGURE 10.19 The shape of the light curve of an eclipsing binary can reveal information about the relative size and surface brightness of the two stars. Even though the blue star here is smaller, it is significantly more luminous because its temperature is higher.

Working It Out 10.2 | Measuring the Masses of an Eclipsing Binary Pair

You have been studying a binary star system like the one in Figure 10.16. After observing the star for several years, you accumulate the following information about the binary system:

1. The star is an eclipsing binary, so that one star passes directly in front of the other from your vantage point.
2. The observed period of the orbit is 2.63 years.
3. Star 1 has a Doppler velocity that varies between $+20.4$ and -20.4 kilometers per second (km/s).
4. Star 2 has a Doppler velocity that varies between $+6.8$ and -6.8 km/s.
5. The stars are in circular orbits. You know this because the Doppler velocities about the star are symmetric.

These data are summarized in **Figure 10.20**. You begin your analysis by noting that the star is an eclipsing binary, which tells you that the orbit of the star is edge on to your line of sight. This means the Doppler velocities tell you the total orbital velocity of each star, and you can determine the size of the orbits by using the relationship distance = speed × time. In one orbital period, star 1 travels around a circle—a distance of

$$d = v \times t$$

$$d = (20.4 \text{ km/s}) \times (2.63 \text{ years})$$

$$d = 53.7 \ \frac{\text{km} \cdot \text{years}}{\text{s}}$$

Those units are a mess, so let's multiply by the number of seconds in a year to fix them:

$$d = 53.7 \ \frac{\text{km} \cdot \text{years}}{\text{s}} \times \frac{3.156 \times 10^7 \text{s}}{\text{year}}$$

Both the seconds and the years divide out, to leave

$$d = 1.69 \times 10^9 \text{ km}$$

This distance is the circumference of the star's orbit, or 2π times the radius of the star's orbit, so star 1 is following an orbit with a radius of

$$r = \frac{d}{2\pi}$$

(Check to make sure you see how $d = 2\pi r$ turned into this!)

$$r = \frac{1.69 \times 10^9 \text{ km}}{2\pi}$$

$$r = 2.7 \times 10^8 \text{ km}$$

But there are 1.50×10^8 km in 1 AU, so

$$r = 2.7 \times 10^8 \text{ km} \times \frac{1 \text{ AU}}{1.50 \times 10^8 \text{ km}}$$

$$r = 1.8 \text{ AU}$$

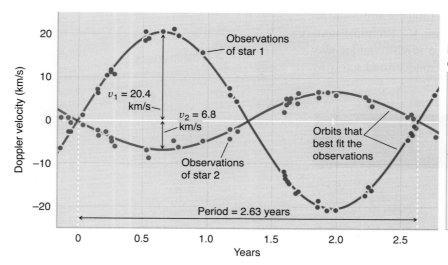

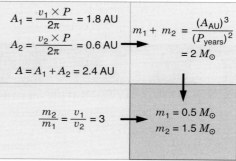

Masses are calculated from observations of stars in a binary system.

$$A_1 = \frac{v_1 \times P}{2\pi} = 1.8 \text{ AU}$$

$$A_2 = \frac{v_2 \times P}{2\pi} = 0.6 \text{ AU} \longrightarrow$$

$$A = A_1 + A_2 = 2.4 \text{ AU}$$

$$m_1 + m_2 = \frac{(A_{\text{AU}})^3}{(P_{\text{years}})^2} = 2 \ M_\odot$$

$$\frac{m_2}{m_1} = \frac{v_1}{v_2} = 3 \longrightarrow$$

$$m_1 = 0.5 \ M_\odot$$
$$m_2 = 1.5 \ M_\odot$$

FIGURE 10.20 Doppler velocities of the stars in an eclipsing binary are used to measure the masses of the stars.

CONTINUED

A similar analysis of star 2 shows that its orbit has a radius of 0.6 AU. Try it yourself to make sure you understand how to do it.

Next you apply *Newton's version of Kepler's third law*, which says that the masses of the stars in solar masses, m_1 and m_2, are related to the average distance between them in AU, A_{AU}, and the period of their orbit in years, P_{years}:

$$m_1 + m_2 = \frac{A^3_{AU}}{P^2_{years}}$$

Since the stars are always on opposite sides of the center of mass, $A_{AU} = 1.8\ AU + 0.6\ AU = 2.4\ AU$. So you can calculate the total mass of the system:

$$m_1 + m_2 = \frac{(2.4)^3}{(2.63)^2} M_\odot$$

$$m_1 + m_2 = 2.0\ M_\odot$$

You have learned that the combined mass of the two stars is twice the mass of the Sun.

To sort out the individual masses of the stars, you use the fact that the mass and velocity are inversely proportional to each other:

$$\frac{m_2}{m_1} = \frac{v_1}{v_2}$$

$$\frac{m_2}{m_1} = \frac{20.4\ \text{km/s}}{6.8\ \text{km/s}}$$

$$\frac{m_2}{m_1} = 3.0$$

Star 2 is three times as massive as star 1. Translating that statement into math gives $m_2 = 3 \times m_1$. Substituting that into the equation $m_1 + m_2 = 2.0\ M_\odot$ gives:

$$m_1 + 3m_1 = 2.0\ M_\odot$$

$$4m_1 = 2.0\ M_\odot$$

$$m_1 = \frac{2.0}{4.0} M_\odot$$

$$m_1 = 0.5\ M_\odot$$

Since $m_2 = 3 \times m_1$, $m_2 = 1.5\ M_\odot$.
Star 1 has a mass of 0.5 $M_\odot$, and star 2 has a mass of 1.5 $M_\odot$. You have just found the masses of two distant stars.

the physical processes that go on deep in the interior of the star. We will learn in the chapters ahead that a minimum stellar mass is necessary to ignite the nuclear furnace that keeps a star shining, but the furnace can run out of control if the stellar mass is too great. Thus, although the most luminous stars are 10^{10}, or 10 billion, times more luminous than the least luminous ones, the most massive stars are only about 10^3, or a thousand, times more massive than the least massive stars.

10.4 The H-R Diagram Is the Key to Understanding Stars

We have come a long way in our effort to measure the physical properties of stars. However, just knowing some of the basic properties of stars does not mean that we understand stars. The next step involves looking for patterns in the properties we have determined. The first astronomers to take this step were Ejnar Hertzsprung (1873–1967) and Henry Norris Russell (1877–1957). In the early part of the 20th century, Hertzsprung and Russell were independently studying the properties of stars. Each scientist plotted the luminosities of stars versus their surface temperatures. The resulting plot is referred to as the Hertzsprung-Russell diagram, or simply the **H-R diagram**. *The H-R diagram is one of the most used and useful diagrams in astronomy.* Everything about stars, from their formation to their old age and eventual death, is discussed using the H-R diagram.

The H-R Diagram

We begin with the layout of the H-R diagram itself, shown in **Figure 10.21**. The surface temperature is plotted on the horizontal axis (the *x*-axis), but it is plotted backward: temperature is high on the left and low on the right. Remember that. Hot blue stars are on the left side of the H-R diagram; cool red stars are on the right. Temperature is plotted logarithmically, which is to say that a step along the axis from a point representing a star with a surface temperature of 30,000 K to one with a surface temperature of 10,000 K (that is, a temperature change by a factor of 3) is the same as a step between points representing a star with a temperature of 9000 K and a star with a temperature of 3000 K (also a factor-of-3 change). The temperature axis is sometimes labeled with some other characteristic that corresponds to temperature, such as the spectral type.

Along the vertical axis (the *y*-axis), we plot the luminosity of stars—the total amount of energy that a star radiates each second. More luminous stars are toward the top of the diagram, and less luminous stars are toward the bottom.

FIGURE 10.21 The layout of the Hertzsprung-Russell, or H-R, diagram used to plot the properties of stars. More luminous stars are at the top of the diagram. Hotter stars are on the left. Stars of the same radius lie along the dotted lines moving from upper left to lower right.

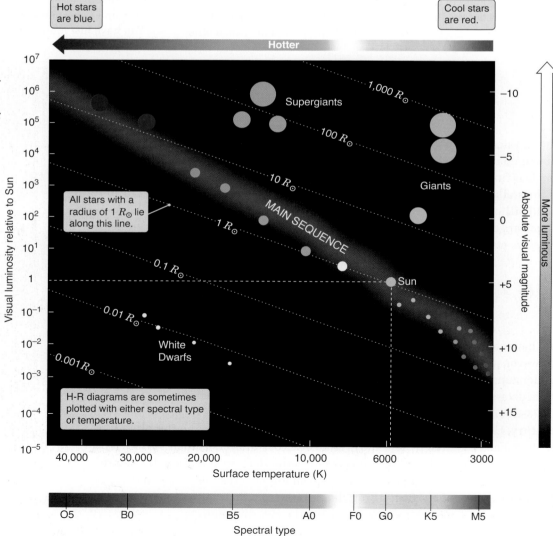

As with the temperature axis, luminosities are plotted logarithmically, in this case with each step along the axis corresponding to a multiplicative factor of 10. To understand why the plotting is done this way, recall that the most luminous stars are 10 billion times more luminous than the least luminous stars, yet all of these stars must fit on the same plot. Sometimes the luminosity axis is labeled with the absolute visual magnitude instead of luminosity.

Because each point on the H-R diagram is specified by a surface temperature and a luminosity, we can use the luminosity-temperature-radius relationship to find the radius of a star at that point as well. A star in the upper right corner of the H-R diagram is very cool, so each square meter of its surface radiates a small amount of energy. But this star is also extremely luminous. It must be huge to account for its high luminosity despite the feeble radiation coming from each square meter of its surface. Conversely, a star in the lower left corner of the H-R diagram is very hot, which means that a large amount of energy is coming from each square meter of its surface. However, this star has a very low luminosity, so it must be very small. Stars in the lower left corner of the H-R diagram are small. Moving up and to the right takes you to larger and larger stars. Moving down and to the left takes you to smaller and smaller stars. All stars of the same radius lie along slanted lines across the H-R diagram, as shown in Figure 10.21.

▶❙❙ AstroTour: H-R Diagram

The Main Sequence Is a Grand Pattern in Stellar Properties

Figure 10.22 shows 16,600 nearby stars plotted on an H-R diagram. The data are based on observations obtained by the Hipparcos satellite. A quick look at this diagram immediately reveals a remarkable fact. About 90 percent of the stars in the sky lie in a well-defined region running across the H-R diagram from lower right to upper left. This region is called the **main sequence**. On the left end of the main sequence are the O stars: hotter, larger, and more luminous than the Sun. On the right end of the main sequence are the M stars: cooler, smaller, and fainter than the Sun. If you know where a star lies on the main sequence, then you know its approximate luminosity, surface temperature, and size.

We can determine whether a star is a main-sequence star by looking at the absorption lines in its spectrum. We can also determine its temperature. Thus we can find its location on the main sequence above the location on the *x*-axis that gives its temperature. We can then read across to the *y*-axis to find its luminosity. If we know its luminosity and its apparent brightness, we can find its distance. This method of determining distances to stars is called **spectroscopic parallax**. Despite the similarity between the names, this method is very different from the *parallax* method using geometry, discussed earlier in the chapter.

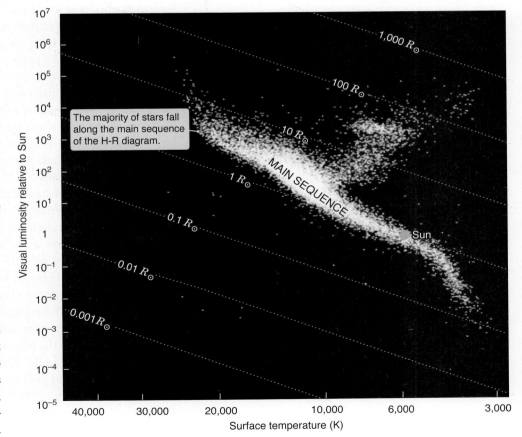

FIGURE 10.22 An H-R diagram for 16,600 stars obtained by the Hipparcos satellite. Most of the stars lie along the main sequence, a band running from the upper left of the diagram toward the lower right.

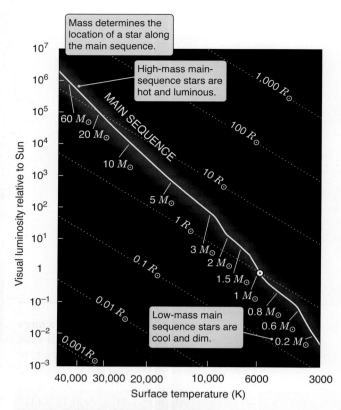

FIGURE 10.23 The main sequence of the H-R diagram is a sequence of masses.

The mass of a main-sequence star determines its fate.

If a main-sequence star is less massive than the Sun, it will be smaller, cooler, redder, and less luminous than the Sun; it will be located to the lower right of the Sun on the main sequence. On the other hand, if a star is more massive than the Sun, it will be larger, hotter, bluer, and more luminous than the Sun; it will be located to the upper left of the Sun on the main sequence. *The mass of a star determines where on the main sequence the star will lie.* In the chapter-opening figure, the student used this feature of the H-R diagram to find the luminosity and temperature of Vega.

It is difficult to overemphasize the importance of this result, so we state it again for clarity. For stars of similar chemical composition, *the mass of a main-sequence star alone determines all of its other characteristics.* Knowing the mass (and chemical composition) of a main-sequence star tells us how large it is, what its surface temperature is, how bright it is, what its internal structure is, how long it will live, how it will evolve, and what its final fate will be. Therefore, its mass determines its position on the H-R diagram (**Figure 10.23**).

Not All Stars Are Main-Sequence Stars

Although 90 percent of stars are main-sequence stars, some are not. Some stars are found in the upper right portion of the H-R diagram, well above the main sequence, so they must be large, cool giants with radii hundreds or thousands of times the radius of the Sun. At the other extreme are stars found in the far lower left corner of the H-R diagram. These stars are tiny, comparable to the size of Earth. Their small surface areas explain why they have such low luminosities despite having such high temperatures.

Stars that lie off the main sequence on the H-R diagram can be identified by their luminosities (determined by their distance) or by slight differences in their spectral lines. The density and surface pressure of gas in a star's atmosphere affect the width of a star's spectral lines. Puffed-up stars above the main sequence have lower densities and lower surface pressures compared to main-sequence stars.

When using the H-R diagram to estimate the distance to a star by the spectroscopic parallax method, astronomers must know whether the star is on, above, or below the main sequence in order to find the star's luminosity. Stars both on and off the main sequence have a property called **luminosity class**, which tells us the *size* of the star. Supergiant stars are luminosity class I, bright giants are class II, giants are class III, subgiants are class IV, main-sequence stars are class V, and white dwarfs are class WD. Luminosity classes I though IV lie above the main sequence, while class WD falls below and to the left of the main sequence. Thus, the complete spectral classification of a star includes both its spectral type, which tells us temperature and color; and its luminosity class, which indicates size.

The existence of the main sequence, together with the fact that the mass of a main-sequence star determines where on the main sequence it will lie, is a grand pattern that points to the possibility of a deep understanding of what stars are and what makes them tick. By the same token, the existence of stars that do *not* follow this pattern raises yet more questions. In the coming chapters, we will learn that the main sequence tells us what stars are and how they work, and that stars off the main sequence tell us how stars form, how they evolve, and how they die.

READING ASTRONOMY News

Unraveling the Mystery of Massive Star Birth: Scientists' Observations Show That Formation Works the Same for All Stars, Regardless of Mass

By **ESO, Garching, Germany, astronomy.com**

Astronomers have obtained the first image of a dusty disk closely encircling a massive baby star, providing direct evidence that massive stars form in the same way as smaller stars.

"Our observations show a disk surrounding a young, massive star, which is now fully formed," said Stefan Kraus from the University of Michigan.

The team of astronomers looked at an object named IRAS 13481-6124. About 20 times the mass of our Sun and 5 times its radius, the young, central star, which is still surrounded by its prenatal cocoon, is located in the constellation Centaurus and lies about 10,000 light-years away.

From archival images obtained by the NASA Spitzer Space Telescope as well as from observations done with the APEX 12-meter submillimeter telescope, astronomers discovered the presence of a jet.

"Such jets are commonly observed around young, low-mass stars and generally indicate the presence of a disk," said Kraus.

Circumstellar disks are essential in the formation process of low-mass stars such as our Sun. However, it is not known whether such disks are also present during the formation of stars more massive than about 10 solar masses where the strong light emitted might prevent mass falling onto the star. For instance, it has

been proposed that massive stars might form when smaller stars merge.

In order to discover and understand the properties of this disk, astronomers employed the European Southern Observatory's (ESO) Very Large Telescope Interferometer (VLTI). By combining light from three of the VLTI's 6-foot (1.8 meters) Auxiliary Telescopes with the Astronomical Multi-BEam combiner (AMBER) instrument, this facility allows astronomers to see details equivalent to those a telescope with a mirror of 280 feet (85 meters) in diameter would see. The resulting resolution is about 2.4 milliarcseconds, which is equivalent to picking out the head of a screw on the International Space Station, or more than 10 times the resolution possible with current visible-light telescopes in space.

With this unique capability, complemented by observations done with another of ESO's telescopes, the 11.7-foot (3.58 meters) New Technology Telescope at La Silla, Kraus and colleagues were able to detect a disk around IRAS 13481-6124.

"This is the first time we could image the inner regions of the disk around a massive young star," said Kraus. "Our observations show that formation works the same for all stars, regardless of mass."

The astronomers conclude that the system is about 60,000 years old, and the star has reached its final mass. Because of the

intense light of the star—30,000 times more luminous than our Sun—the disk will soon start to evaporate. The flared disk extends to about 130 times the Earth–Sun distance, or 130 astronomical units (AU), and has a mass similar to that of the star, roughly 20 times the Sun. In addition, the inner parts of the disk are shown to be devoid of dust.

"Further observations with the Atacama Large Millimeter/submillimeter Array (ALMA), currently being constructed in Chile, could provide much information on these inner parts, and allow us to better understand how baby massive stars became heavy," said Kraus.

Evaluating the News

1. What fundamental question was this research trying to address?
2. Why did astronomers think that low-mass and high-mass stars might form differently?
3. How might these astronomers know the distance to this star?
4. How might these astronomers know the size of the young central star?
5. How might they know the mass?
6. Compare the information about this star to the H-R diagram in Figure 10.23. Could this possibly be a main-sequence star? Why or why not?
7. What technological improvement allowed these astronomers to make these observations?

SUMMARY

10.1 The distances to nearby stars are measured stereoscopically by their parallax.

10.2 Radiation tells us the temperature, size, and composition of stars. Emission and absorption lines allow us to detect the presence of specific elements and compounds. Small, cool stars greatly outnumber large, hot stars.

10.3 Binary stellar systems enable us to measure the masses of stars.

10.4 The H-R diagram is one of the most useful diagrams in stellar astronomy, showing the relationship among the various physical properties of stars. When stars are too remote to measure parallax using geometry, we must use their spectra and the H-R diagram to estimate their distances. A star's luminosity class and temperature tell us its size. The mass and composition of a main-sequence star determine its luminosity, temperature, and size.

✧ SUMMARY SELF-TEST

For the following three questions, star A is blue and star B is red. They have equal luminosities, but star A is twice as far away as star B.

1. Which star appears brighter in the night sky?
    a. star A   **b.** star B

2. Which star is hotter?
    a. star A   **b.** star B

3. Which one is larger?
    a. star A   **b.** star B

4. If a star has very strong hydrogen absorption lines, that means
    a. the temperature is right for hydrogen to make lots of transitions.
    b. hydrogen is abundant in the star, because it is absorbing at these wavelengths.
    c. hydrogen is depleted in the star, because it's not emitting at these wavelengths.
    d. both (a) and (b) are true.

5. To find the masses of both stars in a binary system, you must find the _____ of each star, the _____ of the orbit, and the average _____ between the stars.

6. Suppose you are studying a star with a luminosity of 100 $L_\odot$ and a surface temperature of 4000 K. According to the H-R diagram, this star is a
    a. main-sequence star. **b.** red giant.
    c. white dwarf.   **d.** blue giant.

For the following three questions, suppose you are studying a main-sequence star of 10 solar masses. Consult the H-R diagram to answer these questions.

7. This star's luminosity is roughly
    a. 0.01 $L_\odot$.   **b.** 1 $L_\odot$.
    c. 10 $L_\odot$.   **d.** 1,000 $L_\odot$.

8. This star's temperature is roughly
    a. 1000 K.   **b.** 10,000 K.
    c. 20,000 K.   **d.** 100,000 K.

9. This star's radius is roughly
    a. 0.1 $R_{\odot v}$.   **b.** 1 $R_\odot$.
    c. 10 $R_\odot$.   **d.** 100 $R_\odot$.

QUESTIONS AND PROBLEMS

True/False and Multiple-Choice Questions

10. **T/F:** Red stars have cooler surfaces than blue ones.

11. **T/F:** The brightness of a star in the sky tells you how luminous it is.

12. **T/F:** An atom can emit or absorb a photon of any wavelength.

13. **T/F:** *Visual binary* is just another term for eclipsing binaries.

14. **T/F:** The mass of an isolated star must be inferred from the star's position on the H-R diagram.

15. Two stars have equal luminosities, but star A is much larger than star B. What can you say about these stars?
    a. Star A is hotter than star B.
    b. Star A is cooler than star B.
    c. Star A is farther away than star B.
    d. Star A is brighter in our sky than star B.

16. Most stars are
    a. cool and low-luminosity.   **b.** hot and low-luminosity.
    c. cool and high-luminosity.   **d.** hot and high-luminosity.

17. Suppose an atom has energy levels at 1, 3, 4, and 4.3 (in arbitrary units). Which of the following is not a possible energy for an emitted photon?
    a. 2   **b.** 1
    c. 1.3   **d.** 2.3

18. An eclipsing binary system has a primary eclipse (star A is eclipsed by star B) that is deeper than the secondary eclipse (star B is eclipsed by star A). What does this tell you about stars A and B?
    a. Star A is more luminous than star B.
    b. Star B is more luminous than star A.
    c. Star B is larger than star A.
    d. Star B is moving faster than star A.

19. Which of the following can be determined from the location of a main-sequence star on the H-R diagram?
   a. temperature   b. luminosity   c. radius
   d. mass   e. all of the above

Conceptual Questions

20. To know certain properties of a star, you must first determine the star's distance. For other properties, knowledge of distance is not necessary. Into which of these two categories would you place each of the following properties: size, mass, temperature, color, spectral type, and chemical composition? In each case, state your reasons.

21. Make a table from the information in this chapter, showing each property discussed and the methods of finding that property of an individual star. For example, one row might read:

   Property   **Method**
   Composition   Analyze the lines in the star's spectrum.

22. The light from stars passes through dust in our galaxy before it reaches us, making stars appear dimmer than they actually are. How does this phenomenon affect parallax? How does it affect spectroscopic parallax?

23. Distances to stars can be measured in inches, miles, kilometers, astronomical units, light-years, or parsecs. Why do many stellar astronomers prefer to use parsecs?

24. Even our best measurements always have experimental uncertainties. Using the measurement uncertainties given in the book, how accurately can we know the distance to a star using parallax?

25. What would happen to our ability to measure stellar parallax if we were on the planet Mars? What about Venus or Jupiter?

26. Albireo, a star in the constellation of Cygnus, is a visual binary system whose two components can be seen easily with even a small amateur telescope. Viewers describe the brighter star as "golden" and the fainter one as "sapphire blue."
   a. What does this description tell you about the relative temperatures of the two stars?
   b. What does it tell you about their respective sizes?

27. Very cool stars have temperatures around 2500 K and emit blackbody spectra with peak wavelengths in the red part of the spectrum. Do these stars emit any blue light? Why or why not?

28. It is possible for two stars to have different temperatures and different sizes but the same luminosity. Explain how.

29. The stars Betelgeuse and Rigel are both in the constellation Orion. Betelgeuse appears red in color and Rigel is bluish white. To the eye, the two stars appear equally bright. If you can compare the temperature, luminosity, or size from just this information, do so. If not, explain why.

30. Capella (in the constellation Auriga) is the sixth-brightest star in the sky. With a high-power telescope, Capella is seen to be actually two pairs of binary stars: the first pair is G-type giants; the second pair is M-type main-sequence stars. From this information, what color do you expect Capella to appear as, and why?

31. Two stars have the same luminosity, but one is larger. Compare their temperatures. Now suppose that the two stars have the same size but one is more luminous. Again, compare their temperatures.

32. You obtain a spectrum of an object in space. It has a number of sharp, bright lines. Is this object a cloud of hot gas, a cloud of cool gas, or a star?

33. A friend shows you a spectrum of an object in space. It has a rainbow of colors, with a few dark lines. Your friend claims this is an image of a rare star that has no atmosphere. "Aha!" you say, "You are wrong! This is a perfectly ordinary star with a cool atmosphere around a hotter interior!" How do you know this?

34. Look carefully at Figure 10.10b. In this spectrum, what tells you that there is a white light source? What tells you there is a cool cloud of gas? Suppose the cool cloud of gas were located behind the white light source. How would this spectrum be different?

35. Study Figure 10.11a. What are the differences between the ultraviolet photon and the red one? In this schematic diagram, all the photons leave the atom traveling to the right. Is that true for a real batch of hydrogen atoms in space?

36. In this chapter, we learned that the spectra of stars contain many absorption features, yet we treat them as blackbodies when discussing their properties. Why is this approach justified?

37. Explain why the stellar spectral types (O, B, A, F, G, K, M) are not in alphabetical order. Also explain the sequence of temperatures defined by these spectral types.

38. In Figure 10.13, there is an absorption line at about 410 nm that is weak for O stars and weak for G stars, but very strong in A stars. This particular line comes from the transition from the 2nd excited state of hydrogen up to the 6th excited state. Why is this line weak in O stars? Why is it weak in G stars? Why is it strongest in the middle of the range of spectral types?

39. Explain whether we would still be able to identify the spectral types of stars if their atmospheres contained no elements other than hydrogen and helium.

40. Other than the Sun, the only stars whose mass we can measure directly are those in eclipsing or visual binary systems. Explain why.

41. In Working It Out 10.2, we show how to calculate the masses of both stars in an eclipsing binary system using their orbital details. Implicit in these calculations is the assumption that the binary system is seen edge on. What would happen to our final values of the stars' masses if the system's orbital plane were tilted (or "inclined") instead?

42. Once we determine the mass of a certain spectral type of star that is located in a binary system, we assume that all other stars of the same spectral type and luminosity class also have the same mass. Why is this a safe assumption?

43. In Figure 10.16, two stars orbit a common center of mass.
   a. Explain why star 2 has a smaller orbit than star 1.
   b. Re-sketch this picture for the limiting case where star 1 has a very low mass, perhaps close to that of a planet.
   c. Re-sketch this picture for the limiting case where star 1 and star 2 have the same mass.

44. How do we estimate the mass of stars that are not in eclipsing or visual binary systems?

45. Star masses range from 0.08 $M_\odot$ to more than 200 $M_\odot$.
a. Why are there no stars with masses less than 0.08 $M_\odot$?
b. Why are there no stars much more massive than 200 $M_\odot$?

46. Although we tend to think of our Sun as an "average" main-sequence star, it is actually hotter and more luminous than average. Explain.

47. There are many more low-mass stars than high-mass stars. Why do you think this is so? (We will address the reasons in chapters to come).

48. Compare the temperature, luminosity, and radius of stars at the lower left and upper right of the H-R diagram (for example, Figure 10.21).

49. Logarithmic (log) plots show major steps along an axis scaled to represent equal factors, most often factors of 10. Why, in astronomy, do we sometimes use a log plot instead of the more conventional linear plot?

Problems

50. Look at Figure 10.2b. Suppose the figure included a third star, located four times as far away as star A. How much less would it move each year than star A? How much less than star B?

51. Look at Figure 10.2b. Suppose the figure included a third star, located half as far away as star A. How much more would it move each year than star A? How much more than star B?

52. Sketch a circle and mark an arc along the circumference equal to the radius of the circle. The size of the angle subtended by the arc, measured in radians, is given by the length of the arc divided by the radius (r) of the circle. Because the circumference of a circle is $2\pi r$, there must be $2\pi r/r$ or 2π radians in a circle.
a. How many degrees are there in 1 radian?
b. How many arcseconds are there in 1 radian? How does this value compare to the number of astronomical units in 1 pc (206,265)?
c. What angle does a round object that has a diameter of 1 AU make in our sky if we see it from a distance of 1 pc? A distance of 10 pc?
d. What do your answers to parts (b) and (c) of this question tell you about how the actual size of an object is related to its angular size measured in units of radians?

53. The average person has eyes that are 6 cm apart. Suppose your eye separation is average, and you see an object jump from side to side by half a degree as you blink back and forth between your eyes. How much farther away is an object that moves only one-third of a degree?

54. Sirius, the brightest star in the sky, has a parallax of 0.379 arcsec. What is its distance in parsecs? In light-years?

55. Proxima Centauri, the star nearest to Earth other than the Sun, has a parallax of 0.772 arcsec. How long does it take light to reach us from Proxima Centauri?

56. Betelgeuse (in Orion) has a parallax of 0.00763 ± 0.00164 arcsec, as measured by the Hipparcos satellite. What is the distance to Betelgeuse (in light-years), and the uncertainty in that measurement?

57. Rigel (also in Orion) has a Hipparcos parallax of 0.00412 arcsec. Given that Rigel and Betelgeuse appear equally bright in the sky, which star is actually more luminous? Betelgeuse appears reddish while Rigel appears bluish white. Which star would you say is larger, and why?

58. Sirius is actually a binary pair of two A-type stars. The brighter of the two stars is called the Dog Star, and the fainter is called the Pup Star because Sirius is in the constellation Canis Major (meaning "Big Dog"). The Pup Star appears about 6,800 times fainter than the Dog Star, even though both stars are at the same distance from us. Compare the temperatures, luminosities, and sizes of these two stars.

59. Examine Figure 10.5. This figure is plotted logarithmically on both axes. The luminosities are in units of solar luminosities.
a. How much more luminous than the Sun is a star on the far right side of the plot?
b. How much less luminous than the Sun is a star on the far left side of the plot?

60. Look further at Figure 10.5. The y-axis of this figure shows the "relative number of stars" at a particular luminosity. Relative to what? (To help you figure this out, think about the position of the Sun on this plot, at the intersection of $x = 1$ and $y = 1$. The $x = 1$ tells you that the Sun has one solar luminosity. Does the fact that the sun lies at $y = 1$ mean there is only one star with a luminosity equal to one solar luminosity? Or is there another way to interpret this?)

61. Look still further at Figure 10.5. What is the relative number of stars with luminosity 10,000 times that of the Sun? Suppose that 1 million stars had the luminosity of the Sun. How many stars would have a luminosity 10,000 times that of the Sun?

62. A hydrogen atom makes a transition in which it loses 13.6 eV of energy. (An eV is an electron-volt, a very tiny unit of energy.) Did the atom absorb or emit a photon? How do you know? What is the energy of that photon?

63. Find the peak wavelength of blackbody emission for a star with a temperature of about 10,000 K. In what region of the spectrum does this wavelength fall? What color will this star appear to us?

64. Sirius and its companion orbit around a common center of mass with a period of 50 years. The mass of Sirius is 2.35 times the mass of the Sun.
a. If the orbital velocity of the companion is 2.35 times greater than that of Sirius, what is the mass of the companion?
b. What is the semimajor axis of the orbit?

65. Study Figure 10.15. If $m_1 = m_2$, where would the center of mass be located? If $m_1 = 2m_2$, where would the center of mass be located?

 SmartWork, Norton's online homework system, includes algorithmically generated versions of these questions, plus additional conceptual exercises. If your instructor assigns questions in SmartWork, log in at **smartwork.wwnorton.com**.

 StudySpace is a free and open website that provides a Study Plan for each chapter of **Understanding Our Universe**. Study Plans include animations, reading outlines, vocabulary flashcards, and multiple-choice quizzes, plus links to premium content in SmartWork and the ebook. Visit **wwnorton.com/studyspace**.

Exploration | The H-R Diagram

wwnorton.com/studyspace

Open the "HR Explorer" interactive simulation on the StudySpace website for this chapter. This simulation allows you to compare stars on the H-R diagram in two ways. You can compare an individual star (marked by a red *X*) to the Sun by varying its properties in the box on the left half of the window. Or you can compare groups of the nearest and brightest stars. Play around with the controls for a few minutes to familiarize yourself with the simulation.

Let's begin by exploring how changes to the properties of the individual star change its location on the H-R diagram. First, press the reset button at the top right of the window.

Decrease the temperature of the star by dragging the temperature slider to the left. Notice that the luminosity remains the same. Since the temperature has decreased, each square meter of star surface must be emitting less light. What other property of the star changes in order to keep the total luminosity of the star constant?

Predict what will happen when you slide the temperature slider all the way to the right. Now do it. Did the star behave as you expected?

1. **As you move to the left across the H-R diagram, what happens to the radius?**

2. **What happens as you move to the right?**

Press "reset," and experiment with the luminosity slider.

3. **As you move up on the H-R diagram, what happens to the radius?**

4. **What happens as you move down?**

Press "reset" again and then predict how you would have to move the slider bars to move your star into the red giant portion of the H-R diagram (the upper right). Adjust your slider bars until the star is in that area. Were you correct?

5. **How would you have to adjust the slider bars to move the star into the white dwarf area of the H-R diagram?**

Press the reset button, and let's explore the right-hand side of the window. Add the nearest stars to the graph by clicking their radio button under "Plotted Stars." Using what you learned above, compare the temperatures and luminosities of these stars to the Sun (marked by the *X*).

6. **Are the nearest stars generally hotter or cooler than the Sun?**

7. **Are the nearest stars generally more or less luminous than the Sun?**

Press the radio button for the brightest stars. This will remove the nearest stars, and add the brightest stars in the sky to the plot. Compare these stars to the Sun.

8. **Are the brightest stars generally hotter or cooler than the Sun?**

9. **Are the brightest stars generally more or less luminous than the Sun?**

10. **How do the temperatures and luminosities of the brightest stars in the sky compare to the temperatures and luminosities of the nearest stars? Does this information support the claim in the chapter that there are more low-luminosity stars than high-luminosity stars? Explain.**

11

Our Star: The Sun

The Sun, a G2 star, is all but indistinguishable from billions of other stars in our galaxy. It is also the star against which we measure all others—its basic properties provide the yardsticks of modern astronomy.

The Sun may be run-of-the-mill as far as stars go, but that makes it no less awesome an object on a human scale. Its mass is over 300,000 times that of Earth, and its radius is over 100 times that of Earth. The Sun produces more energy in a second than all of the electric power plants on Earth could generate in a half-million years. And because the Sun is the only star we can study at close range, much of the detailed information that we know about stars has come only from studying our local star.

✧ LEARNING GOALS

You should never stare directly at the Sun, but there are many other ways to observe it. At right, a student is using a pinhole camera and comparing what she sees with her eye to a telescope image in the visible and a spacecraft image in the ultraviolet. The ultraviolet image has active regions corresponding to the locations of the sunspots in the visible image. By the end of this chapter, you should be able to explain the relationship between sunspots and active regions. You should also be able to:

- Describe the balance between the forces that determine the structure of the Sun

- Explain how mass is efficiently converted into energy in the Sun's core

- List the different ways that energy moves outward from the Sun's core toward its surface

- Sketch a physical model of the Sun's interior, and describe how observations of solar neutrinos and seismic vibrations on the surface of the Sun test these models

- Describe the solar activity cycles of 11 and 22 years, and explain how solar activity affects Earth

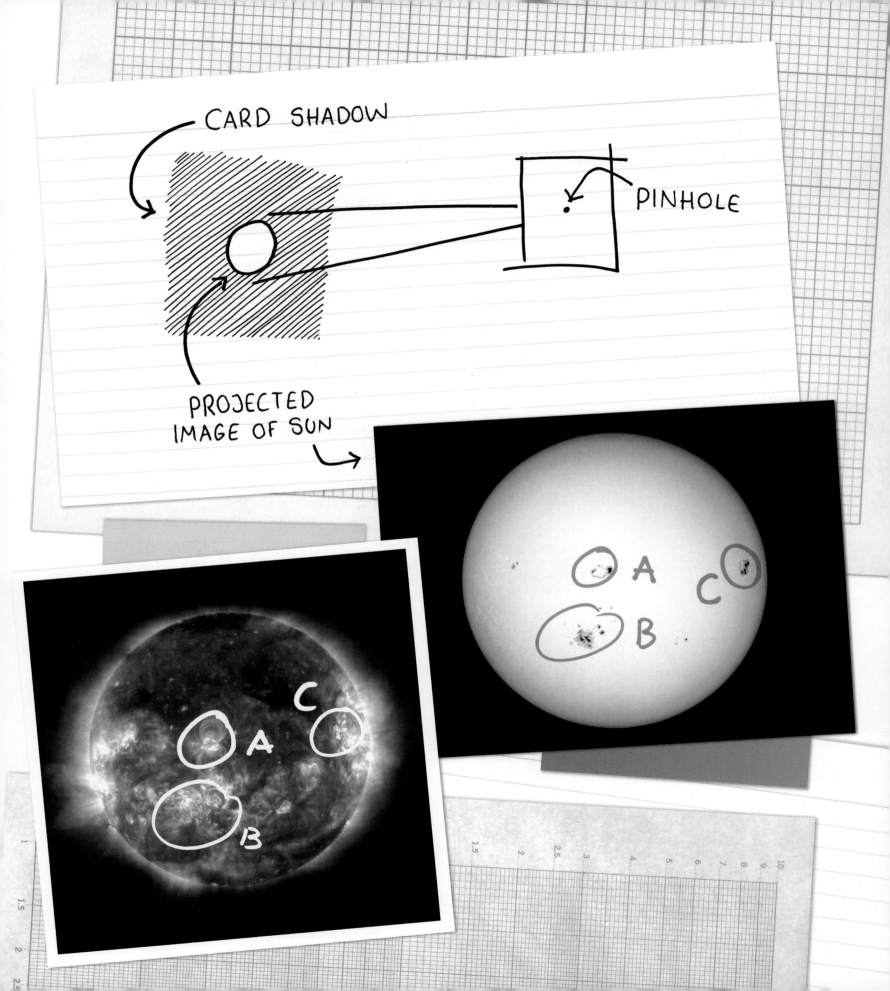

CARD SHADOW

PINHOLE

PROJECTED
IMAGE OF SUN

A

C

B

A

C

B

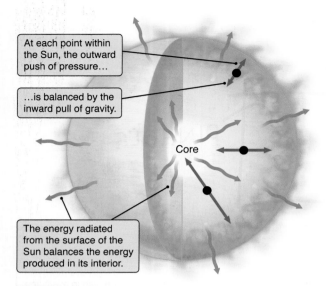

At each point within the Sun, the outward push of pressure…

…is balanced by the inward pull of gravity.

Core

The energy radiated from the surface of the Sun balances the energy produced in its interior.

FIGURE 11.1 The structure of the Sun is determined by the balance between the forces of pressure and gravity, and the balance between the energy generated in its core and energy radiated from its surface.

11.1 The Structure of the Sun Is a Matter of Balance

Our current model of the interior of the Sun is the culmination of decades of work by thousands of physicists and astronomers. Even so, the essential ideas underlying our understanding of the structure of the Sun are found in a few key insights. In turn, these insights can be summed up in a single statement: *The structure of the Sun is a matter of balance.*

The first key balance within the Sun is the balance between the forces due to pressure and gravity illustrated in **Figure 11.1**. If gravity were stronger than pressure, the Sun would collapse. If pressure were stronger than gravity, the Sun would blow itself apart. The pressure at any point within the Sun's interior must be just enough to hold up the weight of all the layers above that point. This balance between the forces of pressure and gravity is known as **hydrostatic equilibrium**. It is the same balance we saw in protostars in Chapter 5.

As we move deeper into the interior of the Sun, the weight of the material above us becomes greater, and hence the pressure must increase. In a gas, higher pressure means higher density and/or higher temperature. **Figure 11.2a** shows how conditions vary inside the Sun. As we go deeper into the Sun, the pressure climbs; and as it does, the density and temperature of the gas climb as well.

Stars like the Sun are remarkably stable objects. Models of stellar evolution

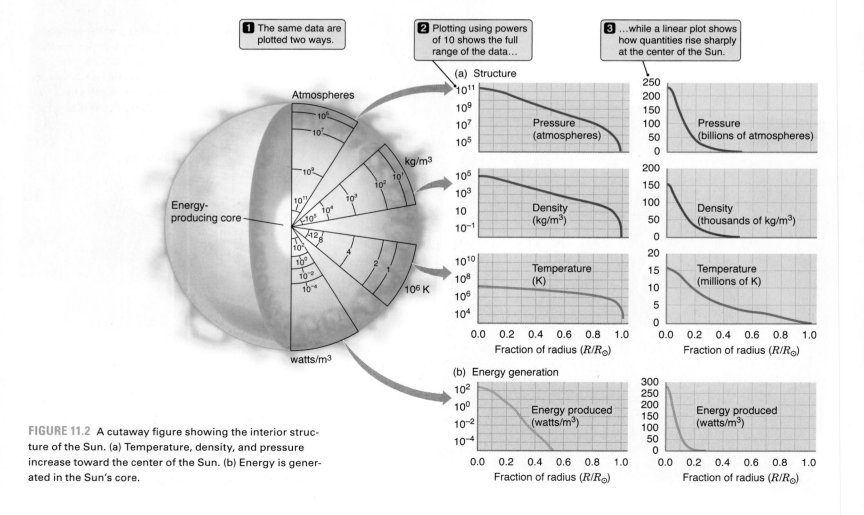

1 The same data are plotted two ways.

2 Plotting using powers of 10 shows the full range of the data…

3 …while a linear plot shows how quantities rise sharply at the center of the Sun.

(a) Structure

Atmospheres

kg/m³

Energy-producing core

10^6 K

watts/m³

Pressure (atmospheres)

Pressure (billions of atmospheres)

Density (kg/m³)

Density (thousands of kg/m³)

Temperature (K)

Temperature (millions of K)

Fraction of radius ($R/R_\odot$)

Fraction of radius ($R/R_\odot$)

(b) Energy generation

Energy produced (watts/m³)

Energy produced (watts/m³)

Fraction of radius ($R/R_\odot$)

Fraction of radius ($R/R_\odot$)

FIGURE 11.2 A cutaway figure showing the interior structure of the Sun. (a) Temperature, density, and pressure increase toward the center of the Sun. (b) Energy is generated in the Sun's core.

indicate that the luminosity of the Sun is increasing with time, but very, very slowly. The Sun's luminosity 4.5 billion years ago was about 70 percent of its current luminosity. To remain in balance, the Sun must produce just enough energy in its interior to replace the energy lost from its surface. To understand this energy balance, we must understand energy production in the interior of the Sun (**Figure 11.2b**) as well as how that energy finds its way from the interior to the Sun's surface, where it is radiated away.

The Sun Is Powered by Nuclear Fusion

The amount of energy produced by the Sun each second is truly astronomical—3.85×10^{26} watts. One of the most basic questions facing the pioneers of stellar astrophysics was how the Sun produces this energy. The answer to this question—nuclear reactions at the Sun's core—came not from astronomers' telescopes, but from theoretical physics and the laboratories of nuclear physicists.

The nucleus of most hydrogen atoms consists of a single proton. Nuclei of all other atoms are built from a mixture of protons and neutrons. Most helium nuclei, for example, consist of two protons and two neutrons. Remember that protons have a positive electric charge, and neutrons have no charge. Since like charges repel, and the closer they are the stronger the force, all of the protons in an atomic nucleus are continuously repelling each other with a tremendous force. The nuclei of atoms should fly apart due to electric repulsion—yet atoms exist. There must therefore be another force, the **strong nuclear force**, holding the nucleus together.

The strong nuclear force is very strong but acts only over very short distances, on the order of 10^{-15} meter, or about a hundred-thousandth the size of an atom. Compared to the energy required to free an electron, the amount of energy required to tear a nucleus apart is enormous. Conversely, when an atomic nucleus is formed, this same amount of energy is released. **Nuclear fusion**—the process of combining two less massive atomic nuclei into a more massive atomic nucleus—can happen only if atomic nuclei are brought close enough together for the strong nuclear force to overcome the force of electric repulsion, as illustrated in **Figure 11.3**. Many kinds of nuclear fusion can occur in stars. In the Sun, as in all other main-sequence stars, the primary process is the fusion of hydrogen into helium—a process often called **hydrogen burning**.

Mass can be converted to energy, and energy can be converted to mass. The exchange rate between the two is given by Einstein's famous equation $E = mc^2$, in which E is energy, m is mass, and c^2 is the speed of light squared. For any nuclear reaction, subtracting the mass of the outputs from the mass of the inputs gives the mass that was turned into energy. The mass of four separate hydrogen nuclei is greater than the mass of a single helium nucleus; so when hydrogen fuses to make helium, some of the mass of the hydrogen is converted to energy (see **Working It Out 11.1**).

Energy is produced in the Sun's innermost region, the **core**, where conditions are extreme. The density is about 150 times the density of water (which is 1,000 kilograms per cubic meter—kg/m³), and the temperature is about 15 million K. The atomic nuclei have tens of thousands of times more kinetic energy than atoms at room temperature. As illustrated in Figure 11.3c, under these conditions atomic nuclei slam into each other hard enough to overcome the electric repulsion and allow the strong nuclear force to act.

In hotter and denser gases, collisions happen more frequently. For this reason, the rate of nuclear fusion reactions is extremely sensitive to the temperature

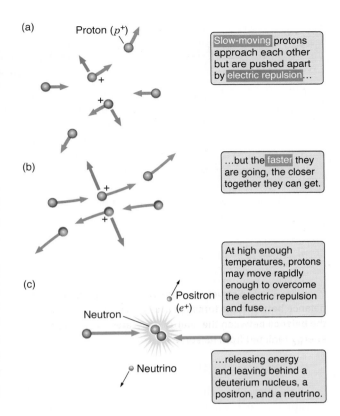

(a) Proton (p^+)

Slow-moving protons approach each other but are pushed apart by electric repulsion…

(b)

…but the faster they are going, the closer together they can get.

(c)

Positron (e^+)

Neutron

Neutrino

At high enough temperatures, protons may move rapidly enough to overcome the electric repulsion and fuse…

…releasing energy and leaving behind a deuterium nucleus, a positron, and a neutrino.

FIGURE 11.3 (a) Atomic nuclei are positively charged and electrically repel each other. (b) The faster that two nuclei are moving toward each other, the closer they will get before veering away. (c) At the temperatures and densities found in the centers of stars, nuclei are energetic enough to overcome this electric repulsion, so fusion takes place.

Solar energy production must balance what is radiated away.

Atomic nuclei are held together by the strong nuclear force.

Nuclear fusion is a very efficient source of energy.

Vocabulary Alert

Burning: In common language, this word is used to talk about fire, which is a chemical process of combining hydrocarbons with oxygen and releasing energy. Astronomers and nuclear physicists often talk about "hydrogen burning," which is *not* a chemical process because it alters the atoms themselves instead of just reassembling them into different molecules. However, hydrogen goes in, and energy comes out, so in that sense it is analogous to chemical burning.

Working It Out 11.1 | How Much Longer Will the Sun "Live"?

Like all other stars, the Sun has a limit to its lifetime, based on the amount of fuel available. By comparing the mass involved in nuclear fusion with the amount of mass available, we can calculate how long the Sun will live. Converting four hydrogen nuclei (protons) into a single helium nucleus results in a loss of mass. The mass of a hydrogen nucleus is 1.6726×10^{-27} kg. So, four hydrogen nuclei have a mass of four times that, or 6.6904×10^{-27} kg. The mass of a helium nucleus is 6.6465×10^{-27} kg, which is less than the mass of the four hydrogen nuclei. The amount of mass lost, m, is

$$m = 6.6904 \times 10^{-27} \text{ kg} - 6.6465 \times 10^{-27} \text{ kg} = 0.0439 \times 10^{-27} \text{ kg}$$

We can move the decimal point to the right and change the exponent on the 10 to rewrite the mass lost as 4.39×10^{-29} kg.

Using Einstein's equation $E = mc^2$ (and the fact that 1 kg m²/s² is 1 joule), we can see that the energy released by this mass-to-energy conversion is

$$E = mc^2 = (4.39 \times 10^{-29} \text{ kg}) (3.00 \times 10^8 \text{ m/s})^2 = 3.95 \times 10^{-12} \text{ J}$$

Each reaction that takes four hydrogen nuclei and turns them into a helium nucleus releases 3.95×10^{-12} joules of energy, which doesn't seem like very much. But consider how small an atom is, and how many there are in even a gram of material. Fusing a single gram of hydrogen into helium releases about 6×10^{11} joules, which is enough energy to boil all of the water in about 10 average backyard swimming pools. In a sense, what this number tells you is how astonishingly big the Sun really is. For the Sun to produce as much energy as it does, it must convert roughly 600 billion kilograms of hydrogen into helium every second (and 4 billion kilograms of matter is converted to energy in the process).

It's been doing this for the last 4.6 billion years. But how do we know how long the Sun will last?

Only about 10 percent of the Sun's total mass will ever be involved in fusion, because the other 90 percent will never get hot enough or dense enough for the strong nuclear force to make fusion happen. Ten percent of the mass of the Sun is

$$M_{\text{fusion}} = (0.1) (M_{\odot})$$
$$M_{\text{fusion}} = (0.1) (2 \times 10^{30} \text{ kg})$$
$$M_{\text{fusion}} = 2 \times 10^{29} \text{ kg}$$

That is the amount of fuel the Sun has available. How long will it last? The Sun consumes hydrogen at a rate of 600 billion kilograms per second, and there are 3.16×10^7 seconds in a year, so each year the Sun consumes

$$M_{\text{year}} = (600 \times 10^9 \text{ kg/s}) (3.16 \times 10^7 \text{ s/year})$$
$$M_{\text{year}} = 1.90 \times 10^{19} \text{ kg/year}$$

If we know how much fuel the Sun has (2×10^{29} kg), and we know how much the Sun burns each year (1.90×10^{19} kg/year), then we can divide the amount by the rate to find the lifetime of the Sun:

$$\text{Lifetime} = \frac{2 \times 10^{29} \text{ kg}}{1.90 \times 10^{19} \text{ kg/year}}$$

$$\text{Lifetime} = 1 \times 10^{10} \text{ years} = 10 \times 10^9 \text{ years} = 10 \text{ billion years}$$

When the Sun was formed, it had enough fuel to power it for about 10 billion years. At 4.6 billion years old, the Sun is nearly halfway through its total life span, and it will continue fusing hydrogen into helium for about 5.4 billion more years, give or take a few hundred million.

▶❙❙ AstroTour: **The Solar Core**

and density of the gas, which is why these energy-producing collisions are concentrated in the Sun's core. Half of the energy produced by the Sun is generated within the inner 9 percent of the Sun's radius. This region represents less than 0.1 percent of the volume of the Sun.

There are several reasons why hydrogen burning is the most important source of energy in main-sequence stars. Hydrogen is the most abundant element in the universe, so it is the most abundant source of nuclear fuel. Hydrogen burning is also the most efficient way to convert mass into energy. But the most important reason is that hydrogen is also the easiest type of atom to fuse. Hydrogen nuclei—protons—have an electric charge of +1. The electric barrier that must be overcome to fuse hydrogen is the repulsion of a single proton against another. To fuse carbon, the repulsion of six protons in one carbon nucleus pushing against the six protons in another carbon nucleus must be overcome. The repulsion between two carbon nuclei is therefore 36 times stronger than between two hydrogen nuclei. For this reason, hydrogen fusion occurs at a much lower temperature than does any other type of nuclear fusion.

The Proton-Proton Chain

In the cores of low-mass stars such as the Sun, hydrogen burns primarily through a process called the **proton-proton chain.** It consists of three steps, illustrated in **Figure 11.4.** In the first step, two hydrogen nuclei fuse. During this process, one of the protons turns into a neutron. To conserve energy and charge, two particles are emitted: a positively charged particle called a **positron** and a neutral particle called a **neutrino.** The new nucleus has a proton and a neutron. It is still hydrogen, because it has only one proton. But it has more than the normal number of neutrons, so it is an **isotope** of hydrogen. This particular isotope is called deuterium (^{2}H).

A positron is the **antiparticle** of an electron. It has all the same properties as the electron, except the charge is opposite. When matter (electrons) and antimatter (positrons) meet, they annihilate each other, and all their mass is converted to energy in the form of photons (light). This happens to the emitted positron inside the Sun, and the photons carry away part of the energy released when the two protons fused, heating the surrounding gas.

Neutrinos have no charge and very little mass. They interact weakly with ordinary matter, so weakly that the neutrino escapes from the Sun without further interactions with any other particles.

In the second step of the proton-proton chain, another proton slams into the deuterium nucleus, forming the nucleus of a helium isotope, ^{3}He, consisting of two protons and a neutron. The energy released in this step is carried away as a gamma-ray photon.

In the third and final step of the proton-proton chain, two ^{3}He nuclei collide and fuse, producing an ordinary ^{4}He nucleus and ejecting two protons in the process. The energy released in this step shows up as kinetic energy of the helium nucleus and ejected protons. Overall, four hydrogen nuclei have combined to form one helium nucleus.

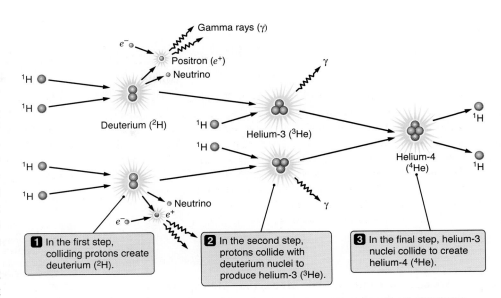

FIGURE 11.4 The Sun and all other main-sequence stars get their energy by fusing the nuclei of four hydrogen atoms to make a single helium atom. In the Sun, about 85 percent of the energy produced comes from the branch of the proton-proton chain shown here.

Hydrogen burns mostly via the proton-proton chain.

Energy Produced in the Sun's Core Must Find Its Way to the Surface

Some of the energy released by hydrogen burning in the core of the Sun escapes directly into space in the form of neutrinos, but most of the energy heats the solar interior. To understand the structure of the Sun, we must understand how thermal energy is able to move outward through the star. **Energy transport** is one of the key factors determining the Sun's structure. Energy transport can occur by conduction, convection, or radiation. In the Sun, only convection and radiation are important because it is made of gas, whereas conduction is important primarily in solids. For example, when you pick up a hot object, your fingers are heated by conduction.

Radiation transfers energy from hotter to cooler regions via photons, which carry the energy with them. Imagine a hotter region of the Sun located next to a cooler region, as shown in **Figure 11.5.** The hotter region will contain more (and more energetic) photons than the cooler region. More photons will move by

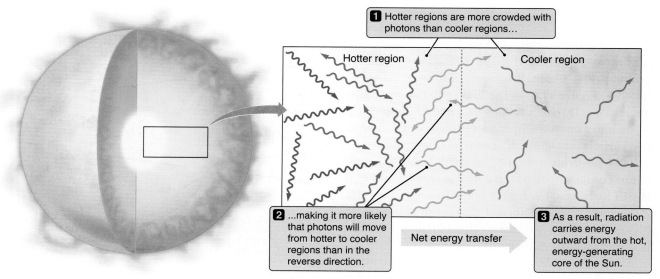

1 Hotter regions are more crowded with photons than cooler regions…

Hotter region

Cooler region

2 …making it more likely that photons will move from hotter to cooler regions than in the reverse direction.

Net energy transfer

3 As a result, radiation carries energy outward from the hot, energy-generating core of the Sun.

FIGURE 11.5 Higher-temperature regions deep within the Sun produce more radiation than do lower-temperature regions farther out. Although radiation flows in both directions, more radiation flows from the hotter regions to the cooler regions than from the cooler regions to the hotter regions. Therefore, radiation carries energy outward from the inner parts of the Sun.

chance from the hotter ("more crowded") region to the cooler ("less crowded") region than in the reverse direction. Thus, there is a net transfer of photons and photon energy from the hotter region to the cooler region, and in this way radiation carries energy outward from the Sun's core.

The transfer of energy from one point to another by radiation also depends on how freely photons can move from one point to another within a star. The degree to which matter impedes the flow of photons through it is referred to as **opacity**. The opacity of a material depends on many things, including the density of the material, its composition, its temperature, and the wavelength of the photons moving through it.

Radiative transfer is efficient in regions with low opacity. In the inner part of the Sun, opacity is relatively low, and radiation carries the energy produced in the core outward through the star. This **radiative zone** extends about 70 percent of the way out toward the surface of the Sun (**Figure 11.6**). Even though this region's opacity is relatively low, photons still travel only a short distance within the region before being absorbed, emitted, or deflected by matter, much like a beach ball being batted about by a crowd of people (**Figure 11.7**). Each interaction sends the photon in an unpredictable direction—not necessarily toward the surface of the star. The distances between interactions are so short that, on average, it takes the energy of a gamma-ray photon about 100,000 years to find its way to the outer layers of the Sun. Opacity holds energy within the interior of the Sun and lets it seep away only slowly.

From a peak of 15 million K in the core of the Sun, the temperature falls to about 100,000 K at the outer margin of the radiative zone. At this cooler temperature, the opacity is higher, so radiation is less efficient in carrying energy from one place to another. The energy that is flowing outward through the Sun "piles

Opacity impedes the outward flow of radiation.

Vocabulary Alert

Opacity: This word is related to the word *opaque*. In common language, when something is opaque, you can't see through it *at all*. The term *opacity* allows for more subtle variations. A gas can have a low opacity (be mostly transparent, allowing photons to pass through) or a high opacity (be mostly opaque, impeding photons from passing through).

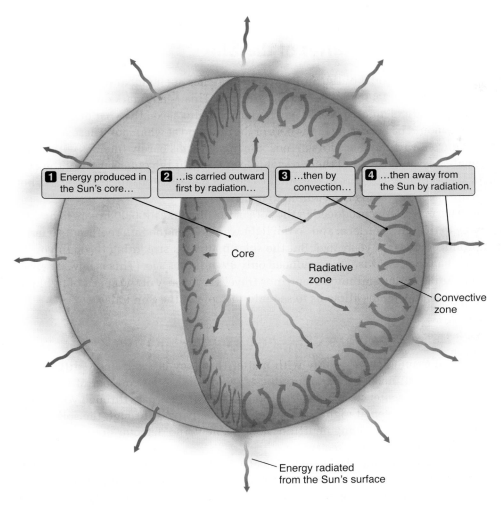

FIGURE 11.6 The interior structure of the Sun is divided into zones on the basis of where energy is produced and how it is transported outward.

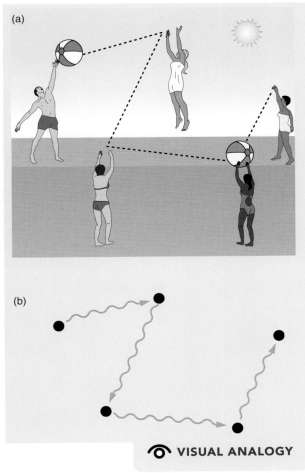

VISUAL ANALOGY

FIGURE 11.7 (a) When a crowd of people play with a beach ball, it never travels very far before someone hits it, turning it in another direction. The ball moves randomly, sometimes toward the front of the crowd, sometimes toward the back. It often takes a ball a long time to make its way from one edge of the crowd to the other. (b) When a photon travels through the Sun, it never travels very far before it interacts with an atom. The photon moves randomly, sometimes toward the center of the Sun, sometimes toward the outer edge. It takes a long time for a photon to make its way out of the Sun.

up" against this edge of the radiative zone. Beyond this region, the temperature drops off very quickly.

As we move farther from the core of the Sun, radiative transfer becomes so inefficient (and the temperature changes so quickly) that a different way of transporting energy takes over. Packets of hot gas, like hot air balloons, become buoyant and rise up through the lower-temperature gas above them, carrying energy with them. Thus, convection begins. Just as convection carries energy from the interior of planets to their surfaces (as we saw in Section 6.5), or from the Sun-heated surface of Earth upward through Earth's atmosphere, convection also plays an important role in the transport of energy outward from the interior of the Sun. The solar **convective zone** (see Figure 11.6) extends from the outer boundary of the radiative zone to just below the visible surface of the Sun, where evidence of convection can be seen in the bubbling of that surface.

In the outermost layers of the Sun, radiation again takes over as the primary mode of energy transport, and it is radiation that transports energy from its outermost layers off into space. This is the standard model of energy transport in the Sun: energy produced by fusion in the core makes its way via radiation and then convection to the outermost layers, where it is radiated away into space.

In the outer part of the Sun, energy is carried by convection.

11.2 The Interior of the Sun

The standard model of the Sun correctly matches such global properties of the Sun as its size, temperature, and luminosity. This is a remarkable feat, but the model predicts much more: in particular, it predicts exactly what nuclear reactions should be occurring in the core of the Sun, and at what rate. These reactions produce a vast number of neutrinos. Since neutrinos barely interact with other matter, essentially all of them travel into space as if the outer layers of the Sun were not there. The core of the Sun lies buried beneath 700,000 km of dense, hot matter, yet the Sun is *transparent* to neutrinos.

Thermal energy produced in the core of the Sun takes 100,000 years to find its way to the Sun's surface, and so the light we see from the Sun represents energy produced 100,000 years ago. But the solar neutrinos streaming through you as you read these words were produced only $8\frac{1}{3}$ minutes ago. (This is how far away the Sun is in light-minutes. Neutrinos travel very nearly at the speed of light.) In principle, neutrinos allow us to see into the core of the Sun today.

> Neutrinos escape freely from the core of the Sun.

We Use Neutrinos to Observe the Heart of the Sun

Neutrinos interact so weakly with matter that they are extremely difficult to observe. Fortunately, the Sun produces a truly enormous number of them. As you read this sentence, about 400 trillion solar neutrinos pass through your body. This happens even at night, since neutrinos easily pass through the solid body of Earth. With this many neutrinos about, a neutrino detector does not have to be very efficient to be effective.

The first apparatus designed to detect solar neutrinos was built underground, 1,500 meters deep, within the Homestake Mine in Lead, South Dakota. A tank there was filled with 100,000 gallons of dry-cleaning fluid. Over the course of 2 days, roughly 10^{22} solar neutrinos pass through the Homestake detector. Of these, on average only *one* neutrino interacts with a chlorine atom within the fluid to form an atom of argon. Even so, this instrument produces a measurable number of argon atoms over time.

The Homestake experiment consistently detected evidence of solar neutrinos, confirming that nuclear fusion powers the Sun. As with many good experiments, however, these results raised new questions. After astronomers' initial joy at confirming the standard solar model, they noticed that there seemed to be far fewer solar neutrinos than they had predicted. The difference between the predicted and measured number of solar neutrinos was called the **solar neutrino problem**.

One possible solution to the solar neutrino problem was that our understanding of the structure of the Sun was wrong. This seemed unlikely, however, due to the many other successes of our solar model. A second possibility was that our understanding of the neutrino itself was incomplete. The neutrino was long thought to have zero mass (like photons) and to travel at the speed of light. But if neutrinos actually do have a tiny amount of mass, then particle physics predicts that neutrinos should *oscillate* (alternate back and forth) among three different kinds, or "flavors"—the *electron*, *muon*, and *tau* neutrinos, as shown in **Figure 11.8**. Because early neutrino experiments could detect only the electron neutrino, these oscillations could explain why we observed only about a third of the expected number of neutrinos. And, as seen in Figure 11.8, electron neutrinos also change flavor as they interact with solar material during their escape from the Sun.

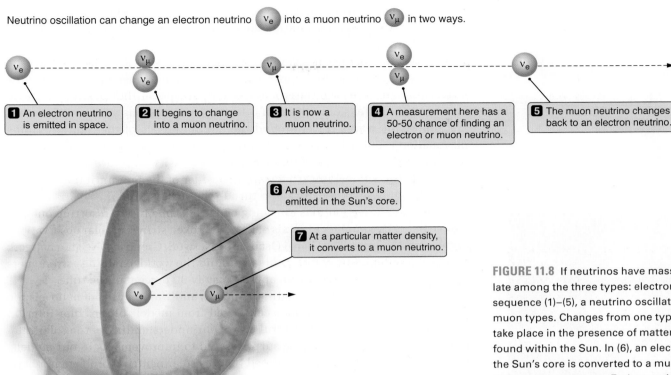

Neutrino oscillation can change an electron neutrino ν_e into a muon neutrino ν_μ in two ways.

1 An electron neutrino is emitted in space.

2 It begins to change into a muon neutrino.

3 It is now a muon neutrino.

4 A measurement here has a 50-50 chance of finding an electron or muon neutrino.

5 The muon neutrino changes back to an electron neutrino.

6 An electron neutrino is emitted in the Sun's core.

7 At a particular matter density, it converts to a muon neutrino.

FIGURE 11.8 If neutrinos have mass, they should oscillate among the three types: electron, muon, and tau. In sequence (1)–(5), a neutrino oscillates between electron and muon types. Changes from one type to another can also take place in the presence of matter at a certain density found within the Sun. In (6), an electron neutrino created in the Sun's core is converted to a muon neutrino (7), which then escapes to space. Early neutrino detectors did not recognize muon neutrinos.

Since Homestake began operating in 1965, more than two dozen additional neutrino detectors have been built, each using different reactions to detect neutrinos of different energies. Experiments at high-energy physics labs, nuclear reactors, and neutrino telescopes around the world have shown that neutrinos *do* have a nonzero mass, and this work has uncovered evidence of neutrino oscillations.

Solving the solar neutrino problem is a good example of how science works. Experiments reveal a gap in our understanding. Our known theories suggest possible resolutions, and more sophisticated experiments distinguish which of the competing hypotheses is correct. Through this process, the solar neutrino problem has led to a better understanding of basic physics. Perhaps this helps you understand why scientists are often excited to find they are wrong. Sometimes it means that completely new knowledge is right around the corner.

Helioseismology Enables Us to Observe Echoes from the Sun's Interior

In Chapter 6 we found that models of Earth's interior predict how density and temperature change from place to place within our planet. These differences affect seismic waves traveling through Earth, bending the paths of these waves. We compare measurements with model predictions to test and refine our models.

The same method can be applied to the Sun. Detailed observations of the surface of the Sun show that the Sun vibrates or "rings," something like a struck bell (**Figure 11.9a**). Unlike a well-tuned bell, however, which vibrates primarily at one frequency, the vibrations of the Sun are very complex; many different frequencies of vibrations occur simultaneously (**Figure 11.9b**). Just as geologists

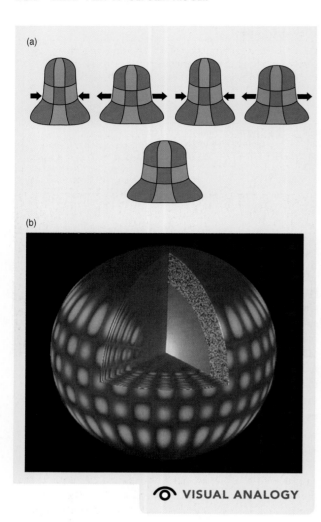

Figure 11.9 (a) A bell vibrates as waves move through it. Red shows regions where the wave is stretching the metal inward; blue shows where the wave is stretching the metal outward. If the wave is the correct wavelength, it will reinforce the vibration, causing the sound to continue for quite some time. If it is not, it will dampen the vibration, and the sound will cease. (b) The interior of the Sun rings like a bell as helioseismic waves move through it. Waves with the right wavelength amplify and sustain the vibrations, while those with the wrong wavelength are damped out and disappear. This figure shows one particular "mode" of the Sun's vibration. Red shows regions where gas is traveling inward; blue, where gas is traveling outward. We can observe these motions by using Doppler shifts.

use seismic waves from earthquakes to probe Earth's interior, solar physicists use the surface oscillations of the Sun to probe the solar interior. This science is called **helioseismology**.

To detect the disturbances of helioseismic waves on the surface of the Sun, astronomers must measure Doppler shifts of less than 0.1 meter per second while detecting changes in brightness of only a few parts per million at any given location on the Sun. Tens of millions of different wave motions are possible within the Sun. Some waves travel around the circumference of the Sun, providing information about the density of the upper convective zone. Other waves travel through the interior of the Sun, revealing the density structure near the Sun's core. Still others travel inward toward the center of the Sun until they are bent by the changing solar density and return to the surface.

All of these wave motions are going on at the same time, so sorting out this jumble requires computer analysis of long, unbroken strings of solar observations from several sources. The Global Oscillation Network Group (GONG) is a network of six solar observation stations spread around the world, enabling astronomers to observe the surface of the Sun approximately 90 percent of the time.

Scientists compare the strength, frequency, and wavelengths of helioseismic data against predicted vibrations calculated from models of the solar interior. This technique provides a powerful test of our understanding of the solar interior, and it has led both to some surprises and to improvements in our models. For example, some scientists proposed that the solar neutrino problem might be solved if we had overestimated the amount of helium in the Sun. This explanation was ruled out by analysis of the waves that penetrate to the core of the Sun. Helioseismology also showed that the value for opacity used in early solar models was too low. This realization led astronomers to recalculate the location of the bottom of the convective zone. Both theory and observation now put the base of the convective zone at 71.3 percent of the way out from the center of the Sun, and uncertainty in this number is less than half a percent.

11.3 The Atmosphere of the Sun

The Solar and Heliospheric Observatory (*SOHO*) spacecraft is a joint mission between NASA and the European Space Agency. *SOHO* moves in lockstep with Earth at a location approximately 1,500,000 km (0.01 astronomical unit; AU) from Earth and almost directly in line between Earth and the Sun. *SOHO* carries 12 scientific instruments that monitor the Sun and measure the solar wind upstream of Earth. In early 2010, NASA launched the Solar Dynamics Observatory to study the solar magnetic field with the goal of developing our understanding of the Sun to the point where we can *predict* when events will occur, rather than simply respond after they happen. These detailed observations help astronomers understand the complex nature of the solar atmosphere.

The Sun is a large ball of gas and so, unlike Earth, it has no solid surface. Instead, it has the kind of surface that a fog bank on Earth does; it is a gradual thing—an illusion, really. Imagine watching a man walking into a fog bank. After he has disappeared from view, you would say he is inside the fog bank, even though he never passed through a definite boundary. The apparent surface of the Sun is similar. Light from the Sun's surface can escape directly into space, so we can see it. Light from below the Sun's surface cannot escape directly into space, so we cannot see it.

The Sun's apparent surface is called the solar **photosphere**. (*Photo* means "light"; the photosphere is the sphere that light comes from.) There is no instant when you can say that you have suddenly crossed the surface of a fog bank, and by the same token there is no instant when we suddenly cross the photosphere of the Sun. The photosphere is about 500 km thick, across which the density and opacity of the Sun increase sharply. The reason the Sun appears to have a well-defined surface and a sharp outline when viewed from Earth (but *never* look at the Sun directly!) is that 500 km does not look very thick when viewed from a distance of 150 million km.

The Sun appears fainter near its edges than near its center (**Figure 11.10a**). This effect, called **limb darkening**, is an artifact of the structure of the Sun's photosphere. Near the edge of the Sun you are looking through the photosphere at a steep angle. As a result, you do not see as deeply into the interior of the Sun as when you are looking near the center of the Sun's disk. The light from the limb of the Sun comes from a shallower layer that is cooler and fainter (**Figure 11.10b**).

The Solar Spectrum Is Complex

In the outermost part of the Sun, the Sun's atmosphere, the density of the gas drops very rapidly with increasing altitude. The Sun's atmosphere is where all visible solar phenomena take place. It consists of three layers: the photosphere, the chromosphere, and the corona, as shown in **Figure 11.11**. Most of the radiation from below the Sun's photosphere is absorbed by matter and reemitted at the photosphere as a blackbody spectrum.

As we examine the structure of the Sun in more detail, however, we see that this simple description is incomplete. Light from the solar photosphere must

Vocabulary Alert

Limb: In common language, this word refers to an outlying body part like an arm or a leg. Astronomers use *limb* to describe the outlying area of a celestial body—the edge of its visible disk.

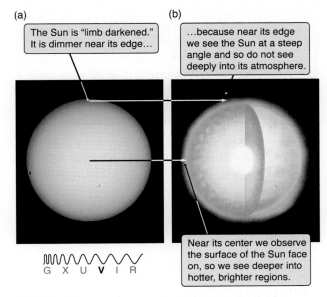

(a) The Sun is "limb darkened." It is dimmer near its edge…

(b) …because near its edge we see the Sun at a steep angle and so do not see deeply into its atmosphere.

Near its center we observe the surface of the Sun face on, so we see deeper into hotter, brighter regions.

G X U **V** I R

FIGURE 11.10 (a) When viewed in visible light, the Sun appears to have a sharp outline, even though it has no true surface. The center of the Sun appears brighter, while the limb of the Sun is darker—an effect known as limb darkening. Incidentally, the tiny black dot nearest the center of the disk is the planet Mercury, transiting the Sun. The other dots are all sunspots. (b) Looking at the middle of the Sun allows us to see deeper into the Sun's interior than looking at the edge of the Sun does. Because higher temperature means more luminous radiation, the middle of the Sun appears brighter than its limb.

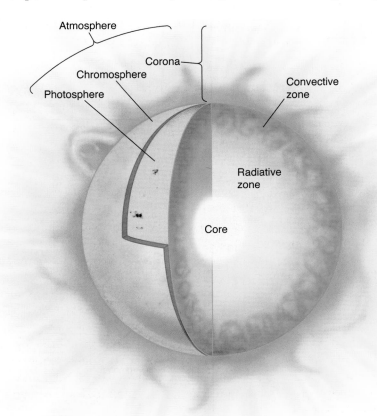

FIGURE 11.11 The components of the Sun's atmosphere.

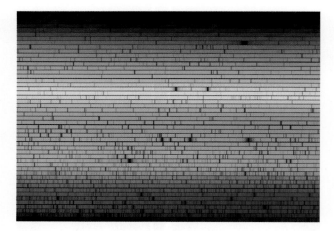

FIGURE 11.12 A high-resolution spectrum of the Sun, stretching from 400 nanometers (nm) in the lower left corner to 700 nm in the upper right corner and showing absorption lines.

escape through the upper layers of the Sun's atmosphere, which puts its fingerprints on the spectrum that we observe. As photospheric light travels upward through the solar atmosphere, atoms in the solar atmosphere absorb the light at distinct wavelengths, forming absorption lines (**Figure 11.12**). Absorption lines from more than 70 elements have been identified. Analysis of these lines forms the basis for much of our knowledge of the solar atmosphere, including the composition of the Sun, and is the starting point for our understanding of the atmospheres and spectra of other stars.

The Sun's Outer Atmosphere: Chromosphere and Corona

As we move upward through the Sun's photosphere, the temperature falls from 6600 K at the photosphere's bottom to 4400 K at its top. At this point the trend reverses and the temperature slowly begins to climb, rising to about 6000 K at a height of 1,500 km above the top of the photosphere (see **Figure 11.13**). This region above the photosphere is called the **chromosphere** (**Figure 11.14a**). The chromosphere was discovered in the 19th century during observations of total solar eclipses (**Figure 11.14b**). The chromosphere is seen most clearly at the solar limb, just above the photosphere, as a source of emission lines, especially the Hα line from hydrogen (the "hydrogen alpha line"). In fact, the deep red color of the Hα line is what gives the *chromosphere* ("the place where color comes from") its

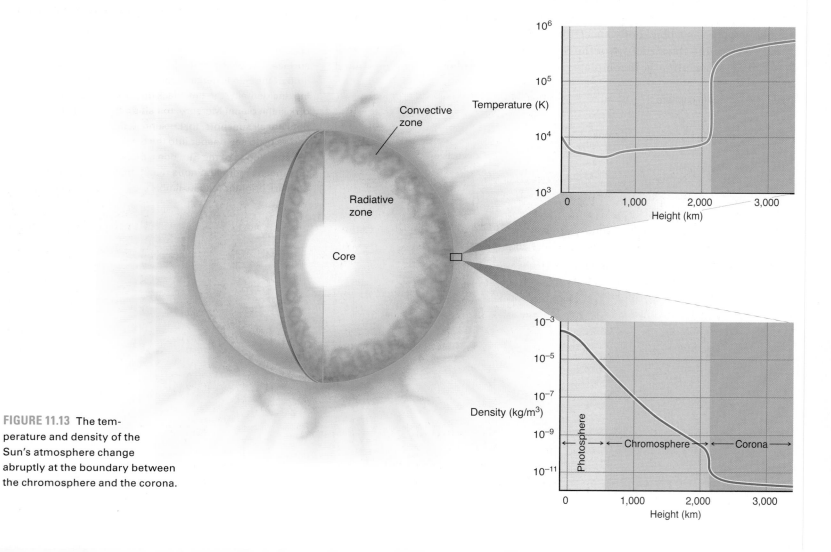

FIGURE 11.13 The temperature and density of the Sun's atmosphere change abruptly at the boundary between the chromosphere and the corona.

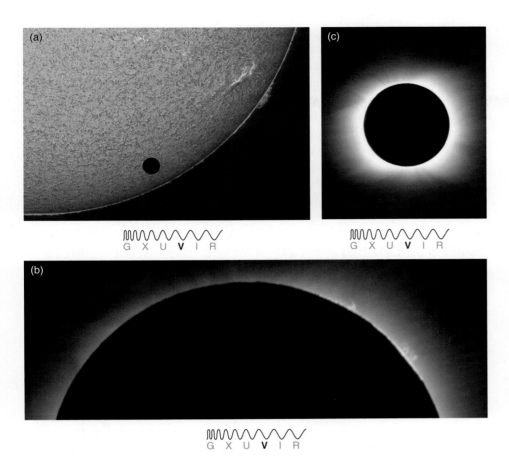

(a)

G X U **V** I R

(c)

G X U **V** I R

(b)

G X U **V** I R

FIGURE 11.14 (a) This image of the Sun, taken in Hα light during a transit of Venus, shows structure in the Sun's chromosphere. The planet Venus is seen in silhouette against the disk of the Sun. (b) The chromosphere seen during a total eclipse. (c) This eclipse image shows the Sun's corona, consisting of million-kelvin gas that extends for millions of kilometers beyond the surface of the Sun.

name. It was also from a spectrum of the chromosphere that the element helium was discovered in 1868; helium is named after *helios*, the Greek word for "Sun." The reason for the chromosphere's temperature reversal with increasing height is not well understood, but it may be caused by magnetic fields propagating through the region.

At the top of the chromosphere, across a transition region that is only about 100 km thick, the temperature suddenly soars (see Figure 11.13) while the density abruptly drops. In the **corona**, temperatures reach 1 million to 2 million K. The corona is thought to be heated by magnetic fields in much the same way the chromosphere is, but why the temperature changes so abruptly at the transition between the chromosphere and the corona is not at all clear.

The corona is visible during total solar eclipses as an eerie glow stretching several solar radii beyond the Sun's surface (**Figure 11.14c**). Because it is so hot, the corona is a strong source of X-rays, and atoms in it are highly ionized. The spectrum of the corona shows emission lines from ionic species such as Fe^{13+} (iron atoms from which 13 electrons have been stripped) and Ca^{14+} (calcium atoms from which 14 electrons have been stripped).

Solar Activity Is Caused by Magnetic Effects

The Sun's magnetic field causes virtually all of the structure in the Sun's atmosphere. High-resolution images of the Sun reveal "coronal loops" that make up much of the Sun's lower corona (**Figure 11.15**). This texture is the result of magnetic structures called flux tubes, much like the flux tube that connects Io with

The Sun's chromosphere lies above the photosphere.

G **X** U V I R

FIGURE 11.15 A close-up image of the Sun, showing the tangled structure of coronal loops.

FIGURE 11.16 X-ray images of the Sun show a very different picture of our star than do images taken in visible light. The brightest X-ray emission comes from the base of the Sun's corona, where gas is heated to temperatures of millions of kelvins. This heating is most powerful above magnetically active regions of the Sun. The dark areas are coronal holes, regions where the Sun's corona is cooler and has lower density.

Jupiter (see Chapter 8). The corona itself is far too hot to be held in by the Sun's gravity, but over most of the surface of the Sun coronal gas is confined instead by magnetic loops with both ends firmly anchored deep within the Sun. The magnetic field in the corona acts almost like a network of rubber bands that coronal gas is free to slide along but cannot cross. In contrast, about 20 percent of the surface of the Sun is covered by an ever-shifting pattern of **coronal holes**. These are apparent in X-ray images of the Sun as dark regions (**Figure 11.16**), indicating that they are cooler and lower in density than their surroundings. In coronal holes, the magnetic field points away from the Sun, and coronal material is free to stream away into interplanetary space.

We have encountered this flow of coronal material away from the Sun before. This is the same solar wind responsible for shaping the magnetospheres of planets (see Chapters 7 and 8) and for blowing the tails of comets away from the Sun (see Chapter 9). The relatively steady part of the solar wind consists of lower-speed flows with velocities of about 350 km/s, and higher-speed flows with velocities up to about 700 km/s. These higher-speed flows originate in coronal holes. Depending on their speed, particles in the solar wind take about 2–5 days to reach Earth.

Using space probes, we have been able to observe the solar wind extending out to 100 AU from the Sun. But the solar wind does not go on forever. The farther it gets from the Sun, the more it spreads out. Just like radiation, the density of the solar wind follows an inverse square law. At a distance of about 100 AU from the Sun, the solar wind is probably not powerful enough to push the interstellar medium (the gas and dust that lie between stars in a galaxy and that surround the Sun) out of the way. There the solar wind stops abruptly, "piling up" against the pressure of the interstellar medium. **Figure 11.17** shows this region of space over which the wind from the Sun holds sway. Sometime within the next 10 years, the *Voyager 1* spacecraft is expected to cross the very outer edge of this boundary and begin sending back our first "on the scene" measurements of true interstellar space.

FIGURE 11.17 The solar wind streams away from the Sun for about 100 AU, until it finally piles up against the pressure of the interstellar medium through which the Sun is traveling. The *Voyager 1* spacecraft is now passing through this boundary and is expected to cross over into the true interstellar medium soon.

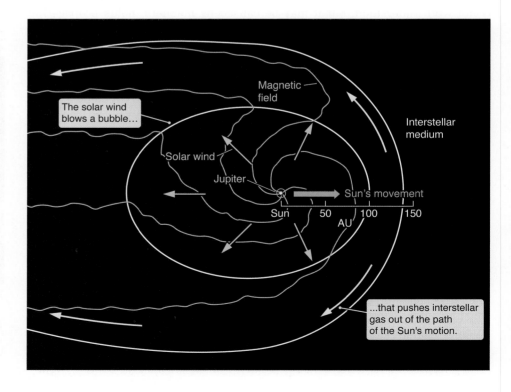

Working It Out 11.2 | Sunspots and Temperature

Sunspots are about 1500 K cooler than their surroundings. What does this tell us about their brightness? Think back to the Stefan-Boltzmann law in Chapter 6. The flux, $\mathcal{F}$, from a blackbody is proportional to the fourth power of the temperature, T. The constant of proportionality is the Stefan-Boltzmann constant, σ, which has a value of 5.67×10^{-8} W/(m²K⁴). We write this relationship as

$$\mathcal{F} = \sigma T^4$$

Remember that the flux is the amount of energy coming out of a square meter of surface every second. How much less energy comes out of a sunspot than out of the rest of the Sun? Let's take round numbers for the temperature of a typical sunspot and the surrounding photosphere: 4500 and 6000 K, respectively. We can set up two equations:

$$\mathcal{F}_{spot} = \sigma T^4_{spot}$$

and

$$\mathcal{F}_{surface} = \sigma T^4_{surface}$$

We could solve each of these separately, and then divide the value of $\mathcal{F}_{spot}$ by $\mathcal{F}_{surface}$, to find out how much fainter it is, but it's much easier to solve for the *ratio* of the fluxes. We divide the left side of one equation by the left side of the other one, and do the same for the right sides. Anything that is the same in both equations will divide out.

$$\frac{\mathcal{F}_{spot}}{\mathcal{F}_{surface}} = \left(\frac{\sigma T^4_{spot}}{\sigma T^4_{surface}}\right)$$

The constant σ divides out. Since both terms on the right are the fourth power, we can divide them first and then take them to the fourth power:

$$\frac{\mathcal{F}_{spot}}{\mathcal{F}_{surface}} = \left(\frac{T_{spot}}{T_{surface}}\right)^4$$

Plugging in our values for T_{spot} and $T_{surface}$ gives

$$\frac{\mathcal{F}_{spot}}{\mathcal{F}_{surface}} = \left(\frac{4500\ K}{6000\ K}\right)^4$$

$$\frac{\mathcal{F}_{spot}}{\mathcal{F}_{surface}} = 0.32$$

Multiplying both sides by $\mathcal{F}_{surface}$ gives

$$\mathcal{F}_{spot} = 0.32\ \mathcal{F}_{surface}$$

So the amount of energy coming from a square meter of sunspot every second is about one-third as much as the amount of energy coming from a square meter of surrounding surface every second. In other words, the sunspot is about one-third as bright as the surrounding photosphere. This is still extremely bright—if you could cut out the sunspot and place it elsewhere in the sky, it would be brighter than the full moon.

The best-known features on the surface of the Sun are relatively dark blemishes in the solar photosphere, called **sunspots**, that come and go over time. Sunspots appear dark, but only in contrast to the brighter surface of the Sun (see **Working It Out 11.2**). Early telescopic observations of sunspots made during the 17th century led to the discovery of the Sun's rotation, which has an average period of about 27 days as seen from Earth and 25 days relative to the stars. Observations of sunspots also show that the Sun, like Saturn, rotates more rapidly at its equator than it does at higher latitudes. This effect, referred to as **differential rotation**, is possible only because the Sun is a large ball of gas rather than a solid object.

Figure 11.18a is a photograph of a large sunspot group, and **Figure 11.18b** shows the remarkable structure of one of these blemishes on the surface of the Sun. Each sunspot consists of an inner dark core called the **umbra**, which is surrounded by a less dark region called the **penumbra** that shows an intricate radial pattern, reminiscent of the petals of a flower. Sunspots are caused by magnetic fields thousands of times greater than the magnetic field at Earth's surface. They occur in pairs that are connected by loops in the magnetic field. Sunspots range in size from about 1,500 km across up to complex groups that may contain several dozen individual spots and span 50,000 km or more. The largest sunspot groups are so large that they can be seen with the naked eye—*but do not look directly at the Sun!* More than one casual observer of the Sun has paid the price of blindness for a glimpse of a naked-eye sunspot. Look back at the chapter-opening figure,

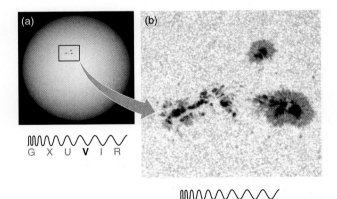

FIGURE 11.18 (a) This image from the Solar Dynamics Observatory (SDO), taken on Oct. 26, 2010, shows a large sunspot group. Sunspots are magnetically active regions that are cooler than the surrounding surface of the Sun. (b) The close-up shows a very high-resolution view of a sunspot, showing the dark *umbra* surrounded by the lighter *penumbra*. The solar surface around the sunspot bubbles with separate cells of hot gas, called granules. The smallest features are about 100 km across.

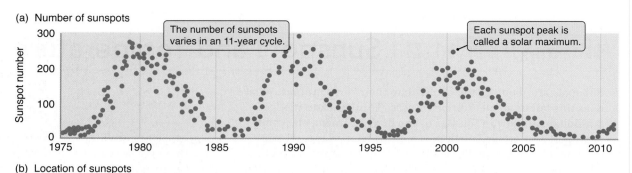

(a) Number of sunspots

The number of sunspots varies in an 11-year cycle.

Each sunspot peak is called a solar maximum.

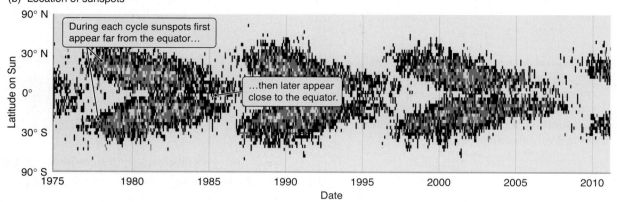

(b) Location of sunspots

During each cycle sunspots first appear far from the equator…

…then later appear close to the equator.

FIGURE 11.19 (a) The number of sunspots versus time for the last few solar cycles. (b) The solar butterfly diagram, showing the fraction of the Sun covered at each latitude.

in which a student has imaged sunspots using a pinhole camera—a tiny hole punched in a piece of cardboard and held up to the Sun. The image is projected onto a piece of paper. This is a safe way to view the Sun, and large sunspots can be imaged in this way.

Individual sunspots do not stay around for very long. Although sunspots occasionally last 100 days or longer, half of all sunspots come and go in about 2 days, and 90 percent are gone within 11 days. Sunspots have a pronounced 11-year pattern called the **sunspot cycle** (**Figure 11.19a**). Over 11 years, both the number and the location of sunspots change. At the beginning of a cycle, sunspots appear at solar latitudes about 30° north and south of the solar equator. Over the following years, sunspots are found closer to the equator as their number increases and then declines. As the last few sunspots approach the equator, sunspots begin reappearing at middle latitudes and the next cycle begins. A diagram showing the number of sunspots at a given latitude plotted against time has the appearance of a series of opposing diagonal bands and is often referred to as the sunspot "butterfly diagram" (**Figure 11.19b**).

Telescopic observations of sunspots date back 400 years, and there were naked-eye reports of sunspots by Chinese astronomers nearly 2,000 years before that. **Figure 11.20** shows the historical record of sunspot activity. Although the 11-year sunspot cycle is real, it is in fact neither perfectly periodic nor especially reliable. The time between peaks in the number of sunspots actually varies between about 9.7 and 11.8 years. The number of spots seen during a given cycle fluctuates as well, and there have been periods when sunspot activity has disappeared almost entirely. An

FIGURE 11.20 Sunspots have been observed for hundreds of years. In this plot, the 11-year cycle in the number of sunspots (half of the 22-year solar magnetic cycle) is clearly visible. Sunspot activity varies greatly over time. The period from the middle of the 17th century to the early 18th century, when almost no sunspots were seen, is called the Maunder Minimum.

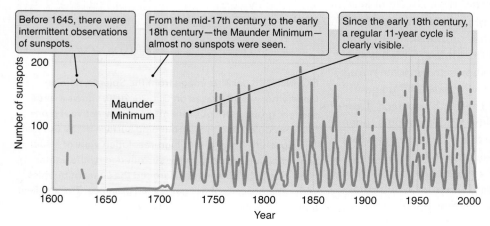

Before 1645, there were intermittent observations of sunspots.

From the mid-17th century to the early 18th century—the Maunder Minimum—almost no sunspots were seen.

Since the early 18th century, a regular 11-year cycle is clearly visible.

Maunder Minimum

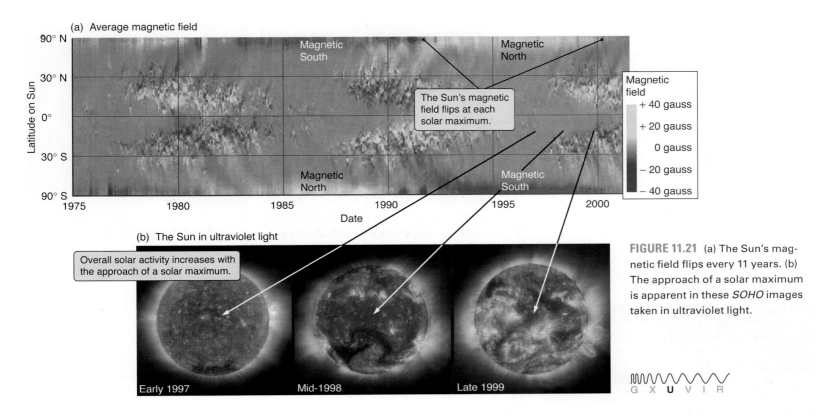

(a) Average magnetic field

Magnetic South

Magnetic North

The Sun's magnetic field flips at each solar maximum.

Magnetic North

Magnetic South

Magnetic field
+ 40 gauss
+ 20 gauss
0 gauss
− 20 gauss
− 40 gauss

Latitude on Sun

90° N
30° N
0°
30° S
90° S

1975   1980   1985   1990   1995   2000

Date

(b) The Sun in ultraviolet light

Overall solar activity increases with the approach of a solar maximum.

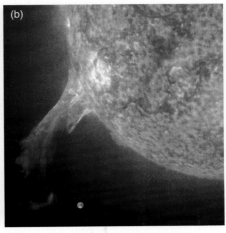

Early 1997   Mid-1998   Late 1999

G X **U** V I R

FIGURE 11.21 (a) The Sun's magnetic field flips every 11 years. (b) The approach of a solar maximum is apparent in these *SOHO* images taken in ultraviolet light.

extended lull in solar activity, called the **Maunder Minimum**, lasted from 1645 to 1715. Normally there would be six peaks in solar activity in 70 years, but virtually no sunspots were seen during the Maunder Minimum.

In the early 20th century, American solar astronomer George Ellery Hale (1868–1938) was the first to show that the 11-year sunspot cycle is actually half of a 22-year magnetic cycle during which the direction of the Sun's magnetic field reverses after each 11-year sunspot cycle. In one sunspot cycle, the leading sunspot in each pair tends to be a north magnetic pole, whereas the trailing sunspot tends to be a south magnetic pole. In the next sunspot cycle, this polarity is reversed: the leading spot in each pair is a south magnetic pole. The transition between these two magnetic polarities occurs near the peak of each sunspot cycle (**Figure 11.21a**).

Sunspots are only one of several phenomena that follow the Sun's 22-year cycle of magnetic activity. The peaks of the cycle, when the most sunspots are visible, are referred to as **solar maxima**. These are times of intense activity, as can be seen in the ultraviolet images of the Sun in **Figure 11.21b**. Sunspots are often accompanied by a brightening of the solar chromosphere that is seen most clearly in emission lines such as Hα. The magnificent loops arching through the solar corona are also anchored in these solar **active regions**. These include solar **prominences (Figure 11.22)**, magnetic flux tubes of relatively cool (5000–10,000 K) gas extending through the million-kelvin gas of the corona. Although most prominences are

FIGURE 11.22 (a) Solar prominences are magnetically supported arches of hot gas that rise high above active regions on the Sun. (b) A close-up view at the base of a large prominence. Earth is shown to scale.

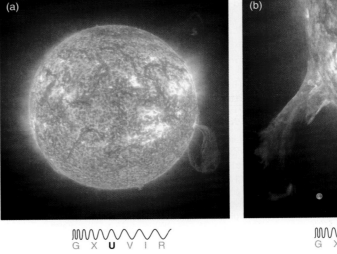

(a)

(b)

G X **U** V I R   G X **U** V I R

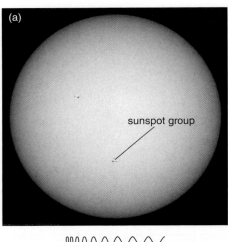

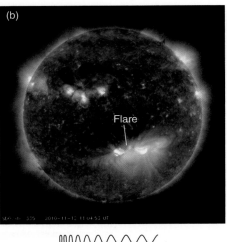

FIGURE 11.23 A large sunspot group (a) emits a powerful flare, shown (b) in ultraviolet light.

FIGURE 11.24 A *SOHO* image of a coronal mass ejection (upper right), with a simultaneously recorded ultraviolet image of the solar disk superimposed.

relatively quiet, others can erupt out through the corona, towering a million kilometers or more over the surface of the Sun and ejecting material into the corona at speeds of 1,000 km/s.

Figure 11.23 shows images of a **solar flare** erupting from a sunspot group. Solar flares are the most energetic form of solar activity, violent eruptions in which enormous amounts of magnetic energy are released over the course of a few minutes to a few hours. Solar flares can heat gas to temperatures of 20 million K, and they are the source of intense X-ray and gamma-ray radiation. Hot plasma (consisting of atoms stripped of some of their electrons) moves outward from flares at speeds that can reach 1,500 km/s. Magnetic effects can then accelerate subatomic particles to almost the speed of light. Such events, called **coronal mass ejections** (**Figure 11.24**), send powerful bursts of energetic particles outward through the Solar System. Coronal mass ejections occur about once per week during the minimum of the solar activity cycle, and as often as several times per day near the maximum of the cycle.

Solar Activity Affects Earth

The amount of solar radiation at Earth's distance from the Sun is, on average, 1.35 kilowatts per square meter. Satellite measurements of the amount of radiation coming from the Sun (**Figure 11.25**) show that this value can vary by as much as 0.2 percent over periods of a few weeks as dark sunspots and bright spots in the chromosphere move across the disk. However, the increased radiation from active regions on the Sun more than makes up for the reduction in radiation from sunspots. On average, the Sun seems to be about 0.1 percent brighter during the peak of a solar cycle than it is at its minimum.

Solar activity affects Earth in many ways. Solar active regions are the source of most of the Sun's extreme ultraviolet and X-ray emissions, energetic radiation that heats Earth's upper atmosphere and, during periods of increased solar activity, causes Earth's upper atmosphere to expand. When this happens, the swollen upper atmosphere can significantly increase the atmospheric drag on spacecraft orbiting near Earth, causing their orbits to **decay**.

Earth's magnetosphere is the result of the interaction between Earth's magnetic field and the solar wind. Increases in the solar wind, especially coronal

Vocabulary Alert

Decay: In common language, this word often describes a process of decomposition via microbes (and usually implies a bad smell). Here we mean that the satellite loses energy, and its orbit shrinks, until the satellite crashes into the atmosphere. In other contexts, we may mean *radioactive decay*, in which an atomic nucleus spontaneously emits energy and/or particles and becomes a different element.

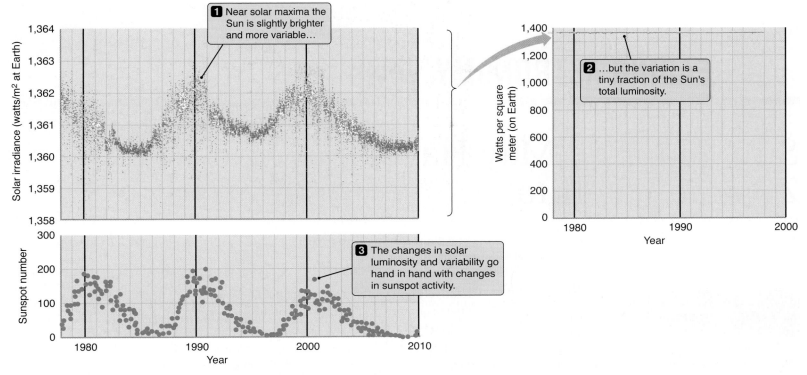

FIGURE 11.25 Measurements taken by satellites above Earth's atmosphere show that the amount of light from the Sun changes slightly over time.

mass ejections, can cause magnetic storms that disrupt power grids, causing large regional blackouts. Coronal mass ejections that are directed toward Earth also hinder radio communication and navigation, and they can damage sensitive satellite electronics, including communication satellites. In addition, energetic particles accelerated in solar flares pose one of the greatest dangers to human exploration of space.

Over the years, people have attempted to relate sunspot activity to phenomena ranging from the performance of the stock market to the mating habits of animals, but few of these supposed relationships have withstood serious scrutiny. A more serious and especially interesting idea is that past and present variations in Earth's climate might be related to solar activity. Solar activity certainly affects Earth's upper atmosphere, and we might imagine that it could affect weather patterns as well, although a mechanism to connect the two is unknown. Models suggest that observed variations in the Sun's luminosity could account for differences of only about 0.1 K in Earth's average temperature—much less than the effects due to the ongoing buildup of carbon dioxide in Earth's atmosphere. However, triggering the onset of an ice age may require a drop in global temperatures of only about 0.2–0.5 K, so the search for a definite link between solar variability and changes in Earth's climate persists.

Solar storms can disrupt electric power grids on Earth and damage satellites.

This article describes some of the very real problems that the Sun can cause.

Sun Eruption May Have Spawned Zombie Satellite

By **Zoe Macintosh, msnbc.msn.com**

Scientists have identified a massive eruption from the sun in April that reached all the way to Earth and may be responsible for knocking out a satellite, creating a so-called "zombie satellite."

The huge explosion of plasma and magnetic energy, called a coronal mass ejection, occurred on April 3 and was observed by NASA's Sun-watching *STEREO* spacecraft, according to the U.S. Naval Research Laboratory. The laboratory released new images of the solar storm last week.

The solar storm appears to have disabled the Intelsat communications satellite *Galaxy 15*, NRL officials said. *Galaxy 15* lost contact with its ground controllers on April 5 and has been drifting around Earth ever since.

Solar storms are known to put satellites at risk. The charged particles in a storm can short out electrical equipment.

The observations suggest the coronal mass ejection flung material away from the sun at a phenomenal 1,000 kilometers per second. The solar eruption was moving at 2.2 million mph while it was still close to the sun on April 3. It then slowed down to about 1.5 million mph when it reached Earth on April 5.

There is an odd twist to the *Galaxy 15* satellite failure. While the satellite has stopped communicating with its ground control center, its C-band telecommunications payload (which provided broadcast services to cus-tomers) is stuck on, earning it the "zombie satellite" nickname.

"Coronal mass ejections, or CMEs, are pow-erful eruptions of plasma and magnetic energy from the Sun's outer atmosphere, or corona," Naval Research Laboratory officials wrote in the July 7 statement. "When these sudden outbursts are directed toward Earth, they can have both breathtakingly beautiful and poten-tially damaging effects."

The study of the April 3 coronal mass ejec-tion event was performed using NASA's Solar Terrestrial Relations Observatory (*STEREO*), a set of twin spacecraft on opposite sides of Earth that continuously watch the sun in what produces a stereo view, due to the wide sepa-ration of the probes in space.

The unique lateral views provided by *STE-REO* were ideal for studying the kinemat-ics and morphology of the developing event, said Russell Howard, the *STEREO* mission's principal investigator at the Naval Research Laboratory.

Three-dimensional reconstruction of the evolving cloud of electrified gas showed its form to be a crescent-shaped "flux rope" with a shock wave driven in front.

Prior awareness that the coronal mass ejec-tion was headed straight for Earth came from the NRL-developed coronagraph aboard the *SOHO* solar observatory, which is located at a spot between the Earth and sun. That Large Angle Coronograph-Spectrograph instrument on *SOHO* observed a "halo" around the sun formed by the expanding and approaching solar eruption, NRL officials said.

Meanwhile, the now-aimless electronic signal from *Galaxy 15* has forced other com-munications satellites to conduct evasive maneuvers from time to time to avoid signal interference. But the chances of the *Galaxy 15* spacecraft hitting another satellite are so remote, they are non-existent, Intelsat officials have said. This month, *Galaxy 15* will be fly-ing near two other Intelsat satellites (*Galaxy 13* and *Galaxy 14*).

Evaluating the News

1. What does the author mean when she says the eruption "reached all the way to Earth"? It sounds like she means it was connected to the Sun and Earth at the same time. Is this possible?

2. Compare the author's description of a coro-nal mass ejection to what you have learned in the text. Is this an accurate description?

3. The author mixes her units in the descrip-tion of the speeds of the coronal mass ejec-tion. Check her math. Is 1,000 km/s the same as 2.2 million mph?

4. The coronal mass ejection slowed down as it approached Earth. What force might be acting to slow it down?

5. What was unusual about this particular satellite failure?

6. Does it seem important to monitor the Sun for such eruptions in the future?

SUMMARY

11.1 The forces due to pressure and gravity balance each other in hydrostatic equilibrium, maintaining the Sun's structure. Nuclear reactions converting hydrogen to helium are the source of the Sun's energy. Energy created in the Sun's core moves outward to the surface, first by radiation and then by convection.

11.2 As hydrogen fuses to helium in the core of the Sun, neutrinos are emitted. Neutrinos are elusive, almost massless particles that interact only very weakly with other matter. Observations of neutrinos confirm that nuclear fusion is the Sun's primary energy source.

11.3 The Sun has multiple layers, like an onion, each with a characteristic density, temperature, and pressure. The apparent surface of the Sun is called the photosphere. The temperature of the Sun's atmosphere ranges from about 6000 K near the bottom to about 1 million K at the top. Material streaming away from the Sun's corona creates the solar wind. Sunspots are photospheric regions that are cooler than their surroundings, and they reveal the 11- and 22-year cycles in solar activity. Solar storms can disrupt power grids and damage satellites.

✧ SUMMARY SELF-TEST

1. The structure of the Sun is determined by both the balance between the forces due to _____ and gravity, and the balance between energy generation and energy _____.
   a. pressure, production
   b. pressure, loss
   c. ions, loss
   d. solar wind, production

2. Place in order the following steps in the fusion of hydrogen into helium. If two or more steps happen simultaneously, use an equal sign (=).
   a. A positron is emitted.
   b. One gamma ray is emitted.
   c. Two hydrogen nuclei are emitted.
   d. Two ^{3}He collide and become ^{4}He.
   e. Two hydrogen nuclei collide and become ^{2}H.
   f. Two gamma rays are emitted.
   g. A neutrino is emitted.
   h. One deuterium nucleus and one hydrogen nucleus collide and become ^{3}He.

3. As energy moves out from the Sun's core toward its surface, it first travels by _____, then by _____, and then by _____.
   a. radiation, conduction, radiation
   b. conduction, radiation, convection
   c. radiation, convection, radiation
   d. radiation, convection, conduction

4. The physical model of the Sun's interior has been confirmed by observations of
   a. neutrinos and seismic vibrations.
   b. sunspots and solar flares.
   c. neutrinos and positrons.
   d. sample returns from spacecraft.
   e. sunspots and seismic vibrations.

5. The temperature and density change abruptly at the interface between
   a. the radiative zone and the photosphere.
   b. the photosphere and the chromosphere.
   c. the chromosphere and the corona.
   d. the corona and space.

6. The solar wind
   a. makes a perfectly spherical bubble around the Solar System.
   b. does not interact with the magnetic field of the Sun.
   c. creates a teardrop-shaped bubble around the Solar System.
   d. creates a wind pressure much stronger than a wind on Earth.

7. Sunspots, flares, prominences, and coronal mass ejections are all caused by
   a. magnetic activity on the Sun.
   b. electric activity on the Sun.
   c. the interaction of the Sun's magnetic field and the interstellar medium.
   d. the interaction of the solar wind and Earth's magnetic field.
   e. the interaction of the solar wind and the Sun's magnetic field.

8. Sunspots peak every _____ years, a consequence of the ____-year sunspot cycle.
   a. 11, 22
   b. 22, 11
   c. 5.5, 11
   d. 22, 44

QUESTIONS AND PROBLEMS

True/False and Multiple-Choice Questions

9. **T/F:** Hydrostatic equilibrium is the balance between energy production and loss.

10. **T/F:** Energy production in the Sun occurs primarily in the innermost 20 percent.

11. **T/F:** A positron is an antimatter particle produced during hydrogen fusion.

12. **T/F:** Six hydrogen nuclei are involved in the proton-proton chain, although only four of them wind up in the helium nucleus.

13. **T/F:** Photons travel outward through the Sun because the inner-most part of the Sun is more crowded with photons than the outer parts.

14. **T/F:** Energy is carried outward through the Sun only by radiation.

15. **T/F:** Neutrinos make it out of the Sun faster than photons. This means they travel faster than photons.

16. **T/F:** The chromosphere is the part of the Sun that we see.

17. **T/F:** Absorption lines in the solar spectrum are caused by cool gas between Earth and the Sun.

18. **T/F:** It is safe to stare at the Sun for the duration of a solar eclipse.

19. **T/F:** Sunspots and solar flares are related phenomena.

20. **T/F:** Sunspots can be comparable in size to Earth.

21. **T/F:** There may be a connection between solar activity and Earth's climate.

22. **T/F:** Current climate change on Earth is caused by the Sun.

23. Hydrostatic equilibrium inside the Sun means that
   a. energy produced in the core equals energy radiated from the surface.
   b. radiation pressure balances the weight of outer layers pushing down.
   c. the Sun absorbs and emits equal amounts of energy.
   d. the Sun does not change over time.

24. The energy that is emitted from the Sun is produced
   a. at the interface between the chromosphere and the photosphere.
   b. at the top of the convection zone.
   c. in the core, by nuclear fusion.
   d. at the surface.

25. In the proton-proton chain, four hydrogen nuclei are converted to a helium nucleus. This does not happen spontaneously on Earth, because the process requires
   a. vast amounts of hydrogen.
   b. very high temperatures and pressures.
   c. hydrostatic equilibrium.
   d. very strong magnetic fields.

26. The solar neutrino problem pointed to a fundamental gap in our knowledge of
   a. nuclear fusion.
   b. neutrinos.
   c. hydrostatic equilibrium.
   d. magnetic fields.

27. Sunspots appear dark because
   a. they have very low density.
   b. magnetic fields absorb most of the light that falls on them.
   c. they are cooler than their surroundings.
   d. they are regions of very high pressure.

28. Sunspots change in number and location during the solar cycle. This phenomenon is connected to
   a. the rotation rate of the Sun.
   b. the temperature of the Sun.
   c. the magnetic field of the Sun.
   d. the tilt of the axis of the Sun.

Conceptual Questions

29. The Sun's stability depends on hydrostatic equilibrium and energy balance. Describe how both of these work.

30. Explain how hydrostatic equilibrium acts as a safety valve to keep the Sun at its constant size, temperature, and luminosity.

31. In Figure 11.3, two protons are shown in a glancing collision.
   a. What is different about parts (a) and (b)?
   b. Why are the blue arrows larger in (b) than in (a)? What do these blue arrows represent?

32. What is the strong nuclear force, and how does it provide stability for atomic nuclei?

33. Radiative transfer takes photons from hot regions to cool regions, on average. Why?

34. Explain nuclear fusion and how it relates to the Sun's source of energy.

35. Two of the three atoms in a molecule of water (H_2O) are hydrogen. Why are Earth's oceans not fusing hydrogen into helium, transforming our planet into a star?

36. Some engineers and physicists have been working to solve the world's energy supply problem by constructing power plants that would convert globally plentiful hydrogen to helium. Our Sun seems to have solved this problem. On Earth, what is the major obstacle to this solution?

37. Name the part of the Sun where most of its energy is produced, and explain why the Sun's primary source of energy is located there.

38. Explain the proton-proton chain through which the Sun generates energy by converting hydrogen to helium.

39. Figure 11.4 diagrams the proton-proton chain. In this figure, a number of squiggly arrows are pointing away from the interactions. What do these squiggly arrows represent? Are they waves or particles or both?

40. The proton-proton chain is often described as "fusing four hydrogens into one helium," but actually six hydrogen nuclei are involved in the reaction. Why don't we include the other two nuclei in our description?

41. On Earth, nuclear power plants use *fission* to generate electricity. A heavy element like uranium is broken into many smaller atoms, where the total mass of the fragments is less than the original atom. Explain why fission could not be powering the Sun today.

42. In the proton-proton chain, the mass of four protons is slightly greater than the mass of a helium nucleus. Explain what happens to this "lost" mass.

43. What experiences have you had with energy transmitted to you by radiation alone? By convection alone? By conduction alone? (Hint: You are exposed to all three every day.)

44. Is gas an efficient conductor of thermal energy? Explain why or why not.

45. Suppose an abnormally large amount of hydrogen suddenly burned in the core of the Sun. What would happen to the rest of the Sun? Would we see anything as a result?

46. The Sun has a radius equal to about 2.3 light-seconds. Explain why a gamma ray produced in the Sun's core does not emerge from the Sun's surface 2.3 seconds later.

47. A high-energy photon and a neutrino are created in the Sun's core at the same instant. Which one will reach Earth first? Explain your answer.

48. Why are neutrinos so difficult to detect?

49. Discuss the "solar neutrino problem" and how this problem was solved.

50. What technique do you find in common between how we probe the internal structure of the Sun and the internal structure of Earth?

51. The Sun's visible "surface" is not a true surface, but a feature we call the photosphere. Explain why the photosphere is not a true surface.

52. The second most abundant element in the universe was not discovered here on Earth. What is it and how was it discovered?

53. Describe the solar corona. Under what circumstances can we see it without using special instruments?

54. The solar corona has a temperature of millions of degrees; the photosphere has a temperature of only about 6000 K. Why isn't the corona much, much brighter than the photosphere?

55. What is the solar wind?

56. Sunspots are dark splotches on the Sun. Why are they dark?

57. Sketch a copy of Figure 11.18b. Label the umbra and the penumbra. Draw a circle representing the size of the Earth on your sketch.

58. What have sunspots told us about the Sun's rotation?

59. The Sun rotates once every 25 days relative to the stars. The Sun rotates once every 27 days relative to Earth. Why are these two numbers different?

60. Explain the relationship between the 11-year sunspot cycle and the 22-year magnetic cycle.

61. What is a solar flare?

62. How is the fate of the Hubble Space Telescope tied to solar activity?

63. Solar flares and magnetic storms can adversely affect certain things important to our daily lives. Name some of the things that are affected.

Problems

64. Study Figure 11.2.
    a. Explain why the graphs in the left column look different from the ones on the right. Are both sets of graphs plotting the same data?
    b. In the figure, pressure, density, and temperature all go to zero at a point on the x-axis labeled 1.0. What happens at that point?
    c. Read off the y-intercept (the point where the line meets the y-axis) for the pressure graph on the left. Should this be the same as the y-intercept for the pressure graph on the right? Is it? Explain.

65. Assume that the Sun's mass is about 300,000 Earth masses and that its radius is about 100 times that of Earth. The density of Earth is about 5,500 kg/m³.
    a. What is the density of the Sun?
    b. How does this compare with the density of water?

66. The Sun rotates every 25 days relative to the stars.
    a. Assuming a sunspot is located at the equator of the Sun, how long does it take to go once around the Sun?
    b. How far has it traveled during this time?
    c. How fast is it traveling, relative to the stars?

67. The Sun shines by converting mass into energy according to Einstein's well-known relationship $E = mc^2$. Show that if the Sun produces 3.85×10^{26} joules (J) of energy per second, it must convert 4.3 billion kg of mass per second into energy. Note that 1 J/s is a watt (W), a unit name you may be more familiar with.

68. Assume that the Sun has been producing energy at a constant rate over its lifetime of 4.5 billion years (1.4×10^{17} seconds).
    a. How much mass has it lost in creating energy over its lifetime?
    b. The present mass of the Sun is 2×10^{30} kg. What fraction of its present mass has been converted into energy over the lifetime of the Sun?

69. Imagine that the source of energy in the interior of the Sun changed abruptly.
    a. How long would it take before a neutrino telescope detected the event?
    b. When would a visible-light telescope see evidence of the change?

70. On average, how long does it take particles in the solar wind to reach Earth from the Sun if they are traveling at an average speed of 400 km/s?

71. Examine Figure 11.13.
    a. What is the height on the x-axis measured relative to?
    b. In an earlier graph (Figure 11.2), the density and the temperature in the Sun's interior behaved in approximately the same way (the lines had similar shapes). Here in Figure 11.13, the solar atmosphere's density and temperature do not behave the same way. What does this tell you about the relationship between density and temperature in the atmosphere of the Sun?

72. A sunspot appears only 70 percent as bright as the surrounding photosphere. The photosphere has a temperature of approximately 5780 K. What is the temperature of the sunspot?

73. Figure 11.19 holds an enormous amount of information. To unpack this figure, we need to do a little work.
    a. Describe what is plotted in parts (a) and (b).
    b. Does the peak in sunspot number occur at the beginning or end of the sunspot cycle, or somewhere in between?

c. In part (b) of Figure 11.19, the latitude of the Sun is plotted on the y-axis. Compare this with Figure 11.21a. Do the magnetic field strengths seem to be highest in the regions where sunspots are located?

d. Compare the *SOHO* images in Figure 11.21b, first with the graph of the average magnetic field in Figure 11.21a, and then with the butterfly diagram in Figure 11.19b. When did this sunspot cycle begin? Why do the three images look so different?

74. Use Figure 11.25 to estimate the "tiny fraction" by which the Sun varies over its cycle.

75. The hydrogen bomb represents humankind's efforts to duplicate processes going on in the core of the Sun. The energy released by a 5-megaton hydrogen bomb is 2×10^{16} J.

a. How much mass did Earth lose each time a 5-megaton hydrogen bomb was exploded?

b. This textbook, *Understanding Our Universe*, has a mass of about 1.0 kg. If we converted all of its mass into energy, how many 5-megaton bombs would it take to equal that energy?

76. Verify the claim made at the start of this chapter that the Sun produces more energy per second than all the electric power plants on Earth could generate in a half-million years. Estimate or look up how many power plants there are on the planet, and how much energy an average power plant produces. Be sure to account for different kinds of power—for example, coal, nuclear, wind.

77. Let's examine how we know that the Sun cannot power itself by chemical reactions. Using Working It Out 11.1 and the fact that an average chemical reaction between two atoms releases 1.6×10^{-19} J, estimate how long the Sun could emit energy at its current luminosity. Compare that estimate to the known age of Earth.

78. How much distance must an average photon actually cover in its convoluted path from the center of the Sun to the outer layers? (Hint: consider the time it takes an average photon to leave the Sun, as well as the speed of an average photon.)

 SmartWork, Norton's online homework system, includes algorithmically generated versions of these questions, plus additional conceptual exercises. If your instructor assigns questions in SmartWork, log in at **smartwork.wwnorton.com**.

 StudySpace is a free and open website that provides a Study Plan for each chapter of **Understanding Our Universe**. Study Plans include animations, reading outlines, vocabulary flashcards, and multiple-choice quizzes, plus links to premium content in SmartWork and the ebook. Visit **wwnorton.com/ studyspace**.

Exploration | The Proton-Proton Chain

wwnorton.com/studyspace

The proton-proton chain powers the Sun by fusing hydrogen into helium. As a by-product, several different particles are produced, which eventually produce energy. The process has multiple steps, and this Exploration is designed to explore these steps in detail, hopefully to help you keep them straight.

Open the "Proton-Proton Animation" on the StudySpace website for this chapter.

Press play, and watch the animation all the way through once.

Press play again, and pause the animation after the first collision. Two hydrogen nuclei (both positively charged) have collided to produce a new nucleus with only one positive charge.

1. **Which particle carried away the other positive charge?**

2. **What is a neutrino? Did the neutrino enter the reaction, or was the neutrino produced in the reaction?**

Compare the interaction on the top with the interaction on the bottom.

3. **Did the same reaction occur in each instance?**

Press play again, and pause the animation after the second collision.

4. **What two types of nuclei entered the collision? What type of nucleus resulted?**

5. **Was charge conserved in this reaction, or was it necessary for a particle to carry charge away?**

6. **What is a gamma ray? Did the gamma ray enter the reaction, or was it produced by the reaction?**

Press play again, and allow the animation to run to the end.

7. **What nuclei enter the final collision? What nuclei are produced?**

8. **In chemistry, a catalyst is a reaction helper. It facilitates the reaction but does not get used up in the process. Are there any nuclei that act like catalysts in the proton-proton chain?**

Make a table of inputs and outputs. Which of the particles in the final frame of the animation were inputs to the reaction? Which were outputs? Fill in your table with these inputs and outputs.

9. **Which of the outputs are converted into energy that leaves the Sun as light?**

10. **Which of the outputs could become involved in another reaction immediately?**

11. **Which of the outputs is likely to stay in that form for a very long time?**

12 Evolution of Low-Mass Stars

Like all main-sequence stars, the Sun gets its energy by converting hydrogen to helium. But the Sun cannot remain a main-sequence star forever. It will eventually exhaust the hydrogen fuel source in its core. New balances must constantly be found as the Sun evolves beyond the main sequence, until at last no balance is possible.

The evolutionary course followed by every star throughout the universe is charted when the star forms, determined foremost by its mass and secondarily by its chemical composition. Relatively minor differences in the masses and chemical compositions of two stars can sometimes result in significant differences in their fates. Nevertheless, stars can be divided roughly into two broad categories: high mass and low mass. Massive, luminous O and B stars follow a course fundamentally different from that of the cooler, fainter, less massive stars found toward the lower right end of the main sequence. **Low-mass stars** have masses less than about 8 $M_\odot$. In this chapter, we examine the stages through which low-mass stars progress. Our Sun is a typical low-mass star. What better place to start than by asking about the fate of our own Sun?

✦ LEARNING GOALS

Within its core, the Sun loses over 4 billion kilograms of mass each second, as it fuses hydrogen to helium. Although the Sun may seem immortal by human standards, eventually it will run out of fuel and its time on the main sequence will come to an end. The figure at right shows an image of M57, the shroud of a dead star, taken by a small ground-based telescope. In this chapter, you should learn how these shrouds form, what they are made of, and how they are important to the chemical evolution of the galaxy. You should also be able to:

- Estimate the lifetime of a star from its mass

- Explain why the Sun will grow larger and more luminous when it runs out of fuel

- Sketch post-main-sequence evolutionary tracks on an H-R diagram

- Make a flowchart of the stages of evolution for low-mass stars

- Describe how planetary nebulae and white dwarfs form

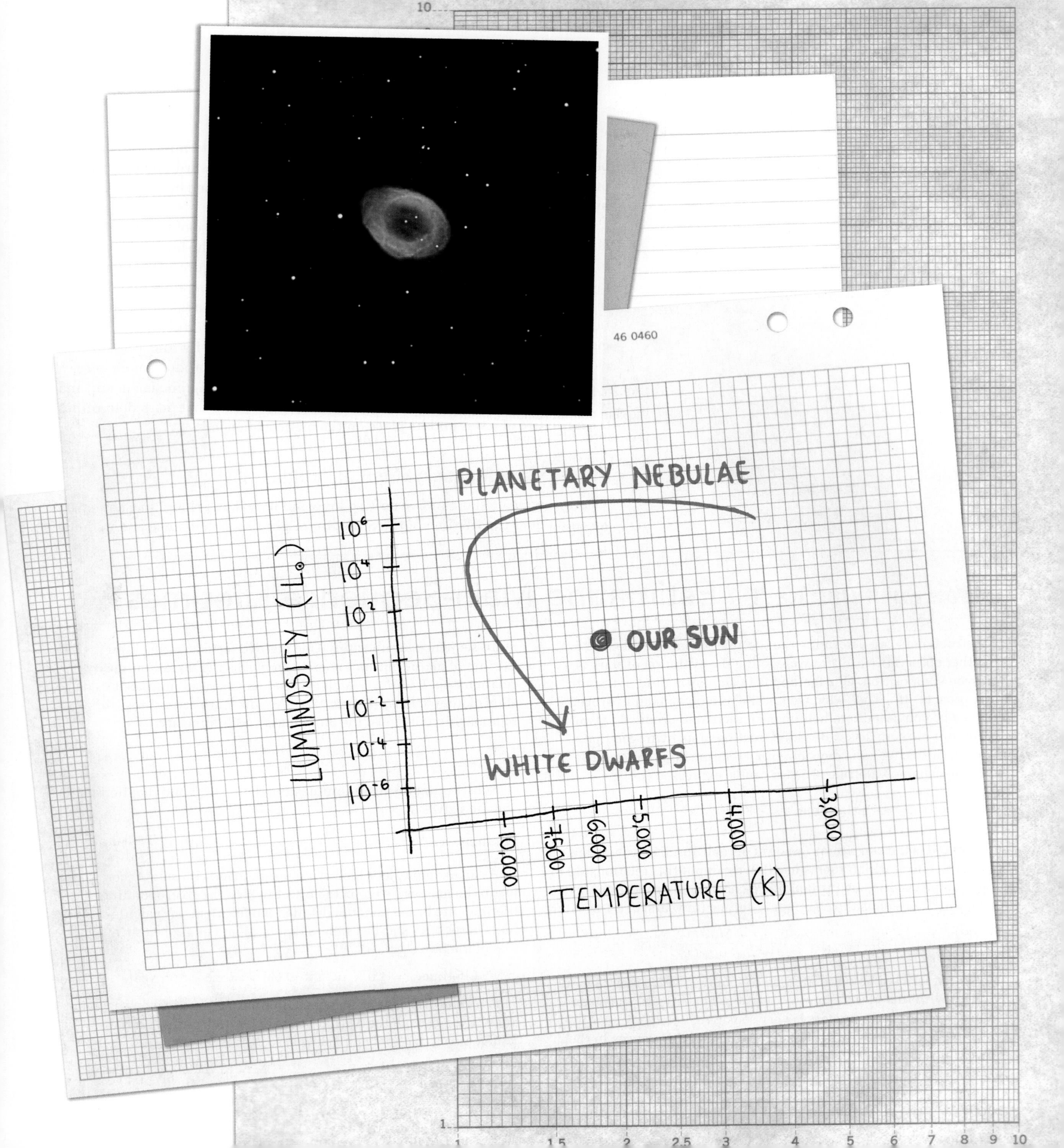
PLANETARY NEBULAE
LUMINOSITY ($L_\odot$)
10^6
10^4
10^2
1
10^{-2}
10^{-4}
10^{-6}
OUR SUN
WHITE DWARFS
10,000
7,500
6,000
5,000
4,000
3,000
TEMPERATURE (K)
46 0460

12.1 The Life and Times of a Main-Sequence Star

In Chapter 11, we learned that the structure of the Sun is determined by a balance between the inward-pushing force of gravity and the outward-pushing force of pressure. The pressure within the Sun is, in turn, maintained by energy released by nuclear fusion in the core of the star.

This is the key to understanding why the main sequence of the H-R diagram is primarily a sequence of masses with low-mass stars on the faint, cool end and high-mass stars on the luminous, hot end, as we saw in Figure 10.23. More mass means stronger gravity; stronger gravity means higher temperature and pressure in the star's interior; higher temperature and pressure mean faster nuclear reactions; faster nuclear reactions mean a more luminous star. If the Sun were more massive, it would have a different balance between gravity and pressure—it would burn its nuclear fuel more rapidly and thus would produce more energy. It would also be located at a different position on the H-R diagram: farther up and to the left on the main sequence. This reasoning tells us that mass determines the structure of a star and its place on the main sequence.

Higher Mass Means a Shorter Lifetime

A main-sequence star can live only so long. But how long is long? In Chapter 11, we calculated the lifetime of the Sun. You might think that since a more mas-

---

Working It Out 12.1 | Estimating Main-Sequence Lifetimes

Astronomers can determine the lifetime of main-sequence stars either observationally or by modeling the evolution of stars of a given composition. One method is to employ a "rule-of-thumb approach." If we use what we know about how much hydrogen must be converted into helium each second to produce a given amount of energy, as well as the fraction of its hydrogen that a star burns, we can come up with a relationship that says the main-sequence lifetime of a star can be expressed as

$$\text{Lifetime} \propto \frac{M_{\text{MS}}}{L_{\text{MS}}}$$

where M is the star's mass (amount of fuel) and L is its luminosity (the rate at which fuel is used). We can put this relationship in quantitative terms by introducing a constant of proportionality, 1.0×10^{10}, which is the computed lifetime (in years) of a 1-$M_\odot$ star:

$$\text{Lifetime} = (1.0 \times 10^{10}) \times \frac{M_{\text{MS}}/M_\odot}{L_{\text{MS}}/L_\odot} \text{ years}$$

Now, let's see how the lifetime of a more massive star compares with that of the Sun. The relationship between the mass and the luminosity of stars is very sensitive. Relatively small differences in the masses of stars result in large differences in their main-sequence luminosities. One method for estimating luminosities of main-sequence stars is known as the **mass-luminosity relationship**, $L \propto M^{3.5}$, which is based on observed luminosities of stars of known mass (Figure 12.2). As above, we can express this relationship relative to the Sun's mass and luminosity:

$$\frac{L_{\text{MS}}}{L_\odot} = \left(\frac{M_{\text{MS}}}{M_\odot}\right)^{3.5}$$

Substituting the mass-luminosity relationship into the lifetime equation gives us

$$\text{Lifetime} = (1.0 \times 10^{10}) \times \frac{M_{\text{MS}}/M_\odot}{(M_{\text{MS}}/M_\odot)^{3.5}} = (1.0 \times 10^{10}) \times \left(\frac{M_{\text{MS}}}{M_\odot}\right)^{-2.5} \text{ years}$$

As an example, let's look at a main-sequence K5 star. According to the masses listed in Table 12.1 (where the corresponding lifetimes are based on more detailed models than our calculations here), a K5 star has a mass that is about equal to 0.69 times that of the Sun:

$$\text{Lifetime}_{\text{K5}} = (1.0 \times 10^{10}) \times (0.69)^{-2.5} = 2.5 \times 10^{10} \text{ years}$$

A K5 star has a main-sequence lifetime 2.5 times as long as the Sun's. Even though the K5 star starts out with less fuel than the Sun, it burns that fuel more slowly, so it lives longer.

sive star has more mass, it will live longer. But it not only has more fuel, it burns it at a faster rate, and this is the factor that dominates. Stars with higher masses live shorter lives, not longer ones, because they burn their fuel faster. **Working It Out 12.1** develops this idea further.

The lifetimes of stars should be of more than passing interest to us. We don't yet know if life exists on planets orbiting other stars, but we think that life would be unlikely to evolve on a planet orbiting a massive star with a stable life of only a few million years (see Chapter 18). Likewise, it would be unlikely for life to develop on a planet surrounding a very low-mass star that had yet to initiate the stable nuclear burning phase.

How quickly a star runs out of fuel depends on its mass and luminosity.

The Structure of a Star Changes as It Uses Its Fuel

When the Sun formed, about 90 percent of its atoms were hydrogen atoms. Since then, the Sun has produced its energy by converting hydrogen into helium via the proton-proton chain. As the composition of a star changes, so must its structure. When we discussed the collapse of a protostar in Chapter 5, we considered the idea of a changing balance between gravity and pressure. The protostar is always in balance as it collapses, but as it radiates away thermal energy, this balance constantly shifts toward that of a smaller and denser object.

We can use the H-R diagram to help us understand how a protostar changes as it becomes a main-sequence star, and more generally to keep track of how all stars change as they evolve throughout their lifetimes. The path across an H-R diagram that a star follows as it goes through the different stages of its life is called the star's **evolutionary track.** In **Figure 12.1**, we can see typical evolutionary tracks for protostars of different masses as they collapse, growing smaller and hotter until they reach their positions as stars on the main sequence.

As a main-sequence star uses the fuel in its core, its structure must continually shift in response to the changing core composition. At any given point in its lifetime, a main-sequence star like the Sun is in balance. But the balance in the Sun today is slightly different from the balance the Sun had 4 billion years ago, and slightly different from the balance it will have 4 billion years from now. Between the time the Sun was born and the time it will leave the main sequence, its luminosity will roughly double, and most of this change will occur during the last billion years of its life on the main sequence. Stars evolve even as they "sit" on the main sequence, although this evolution is slow and modest in comparison with the events that follow. The mass and luminosity of

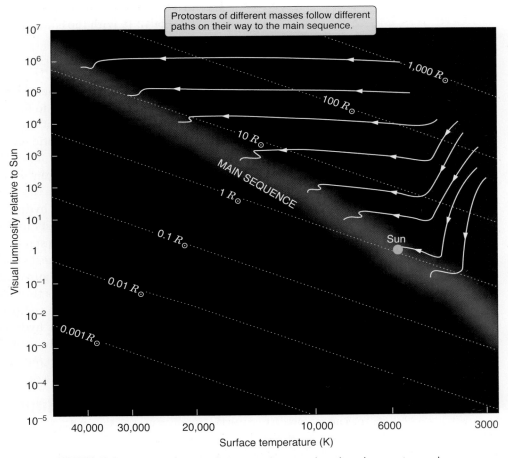

FIGURE 12.1 The H-R diagram can be used as a tool to show how a star evolves, or changes over time.

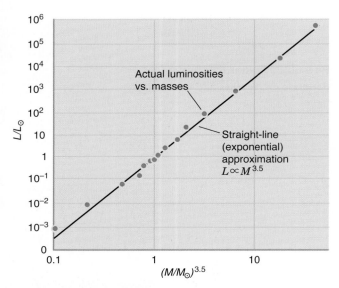

FIGURE 12.2 The mass-luminosity relationship for main-sequence stars: $L \propto M^{3.5}$. The exponent (3.5) is an average value over the wide range of main-sequence star masses. Observational data show that the deviation of stars from the average relationship depends on their mass.

TABLE 12.1

Main-Sequence Lifetimes

SPECTRAL TYPE	MASS ($M_\odot$)	LUMINOSITY ($L_\odot$)	MAIN-SEQUENCE LIFETIME (YEARS)
O5	40	500,000	8.0×10^5
B5	6.5	800	8.1×10^7
A5	2.1	20	1.05×10^9
F5	1.3	2.5	5.2×10^9
G2 (our Sun)	1.0	1.0	1.0×10^{10}
G5	0.93	0.79	1.2×10^{10}
K5	0.69	0.16	4.3×10^{10}
M5	0.21	0.0079	2.7×10^{11}

main-sequence stars are related (**Figure 12.2**) because the mass governs the rate at which nuclear reactions occur in the core. This means that the length of time a star spends on the main sequence is determined almost entirely by its mass (**Table 12.1**), with the counterintuitive result that less massive stars live longer.

Helium Ash Builds Up in the Center of the Star

We might guess that the helium produced in the core of a low-mass, main-sequence star would begin fusing to form even heavier elements. This is not the case. At the temperature found at the center of a low-mass main-sequence star, atomic collisions are not energetic enough to overcome the electric repulsion between helium nuclei.

This nonburning helium **ash** does not build up at the same rate throughout the interior of a star. Because the temperature and pressure are highest at the center of a main-sequence star, hydrogen burns most rapidly there as well. As a result, helium accumulates most rapidly at the center of the star.

If we could cut open a star and watch as it evolves, we would see its chemical composition changing most rapidly at its center and less rapidly as we move outward through the star. **Figure 12.3** shows how the chemical composition inside a star like the Sun changes throughout its main-sequence lifetime. When the Sun formed, it had a uniform composition of about 70 percent hydrogen and 30 percent helium by mass. As hydrogen was fused into helium, the helium fraction in the core of the Sun climbed. Today, roughly 5 billion years later, only about 35 percent of the mass *in the core* of the Sun is hydrogen.

Vocabulary Alert

Ash: In common language, this word specifically refers to the gray, dusty product of a fire, accumulating as the fire burns and collecting, for example, at the bottom of a fireplace. Astronomers use this word metaphorically to refer to the products of fusion, which collect in the core of the star.

12.2 A Star Runs Out of Hydrogen and Leaves the Main Sequence

Eventually, a star exhausts all of the hydrogen fuel in its core. The innermost core of the star is composed entirely of helium ash. As thermal energy leaks out of the helium core into the surrounding layers of the star, no more energy is generated within the core to replace it. The balance that has maintained the structure of the star throughout its life is now broken. The star's life on the main sequence has come to an end.

The Helium Core Is Degenerate

All of the matter we directly experience is mostly *empty space*. An atom is mostly empty except for the tiny bit of space occupied by the nucleus and the electrons. The same is true for the matter within the Sun. At the enormous temperatures within the Sun, almost all of the electrons have been stripped away from their nuclei by energetic collisions. (In other words, the gas is completely ionized.) So the gas inside the Sun is a mixture of electrons and atomic nuclei all flying about freely. Even so, the gas that makes up the Sun is still mostly empty space, and the electrons and atomic nuclei fill only a tiny fraction of the volume.

When a low-mass star like the Sun exhausts the hydrogen at its center, the situation changes. As gravity begins to win its shoving match against pressure, the helium core is crushed to an ever-smaller size and an ever-greater density, but there is a limit to how dense the core can get. The rules of quantum mechanics (the same rules that say that atoms can have only certain discrete amounts of energy and that light comes in packets called photons) limit the number of electrons that can be packed into a given volume of space at a given pressure. As the matter is compressed further and further, it finally reaches this limit. The space is now effectively "filled" with electrons that are smashed tightly together. This matter is so dense that a single cubic centimeter (about the size of a standard six-sided die) has a mass of more than 1,000 kilograms (kg). Matter that has been compressed to this point is called **electron-degenerate** matter.

Hydrogen Burns in a Shell Surrounding a Core of Helium Ash

Once a low-mass star exhausts the hydrogen at its center, nuclear burning pauses in the core—but outside the core, hydrogen continues to burn. Astronomers speak of **hydrogen shell burning** because the star's hydrogen now burns only in a shell surrounding a core of nonburning helium and degenerate electrons.

Electron-degenerate matter has a number of fascinating properties. For example, as more and more helium ash piles up on the degenerate core from the hydrogen-burning shell, the core *shrinks* in size. (This is one of the rules that degenerate matter breaks: The more massive it is, the smaller it is. This is noticeably not true for people, for example.) It does so because the added mass increases the strength of gravity and therefore the weight bearing down on the core, which means that the electrons can be smashed together into a smaller volume. The presence of the degenerate core triggers a chain of events that will dominate the evolution of our 1-$M_\odot$ star for the next 50 million years after the hydrogen runs out.

A degenerate core means stronger gravity, stronger gravity means higher pressure, and higher pressure means faster nuclear burning, producing greater and greater amounts of energy. This increase in energy generation heats the overlying

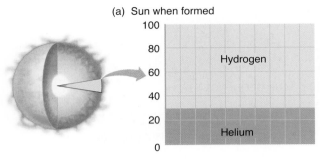

(a) Sun when formed

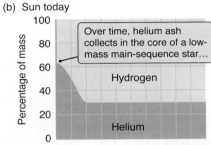

(b) Sun today

> Over time, helium ash collects in the core of a low-mass main-sequence star...

(c) Sun in 5 billion years

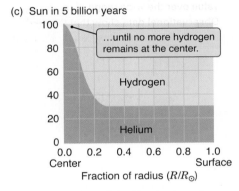

> ...until no more hydrogen remains at the center.

FIGURE 12.3 Chemical composition of the Sun is plotted here as a percentage of mass against distance from the center of the Sun. (a) When the Sun formed 5 billion years ago, about 30 percent of its mass was helium and 70 percent was hydrogen throughout. (b) Today the material at the center of the Sun is about 65 percent helium and 35 percent hydrogen. (c) The Sun's main-sequence life will end in about 5 billion years, when all of the hydrogen at the center of the Sun is gone.

Vocabulary Alert

Degenerate: In common language, this word means "to decline or deteriorate," as in "the discussion degenerated into name-calling." Astronomers use *degenerate* to mean a particular state of matter, in which matter behaves in an unfamiliar way. Electron-degenerate matter doesn't obey the rules of ordinary matter, as we will find out.

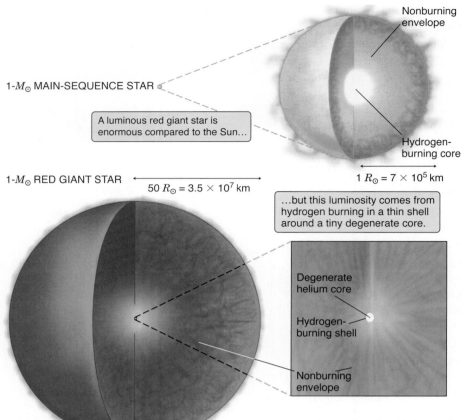

1-$M_\odot$ MAIN-SEQUENCE STAR

Nonburning envelope

Hydrogen-burning core

1 $R_\odot$ = 7 × 10⁵ km

A luminous red giant star is enormous compared to the Sun...

1-$M_\odot$ RED GIANT STAR

50 $R_\odot$ = 3.5 × 10⁷ km

...but this luminosity comes from hydrogen burning in a thin shell around a tiny degenerate core.

Degenerate helium core

Hydrogen-burning shell

Nonburning envelope

FIGURE 12.4 The structure of a star near the top of the red giant branch is compared with the structure of the Sun. Left panels compare the size of the Sun with the size of the red giant. Right panels compare the size and structure of the Sun with the core of the red giant. The panels at right are magnified about 50 times compared to the panels on the left.

An evolving low-mass star moves up and to the right on the H-R diagram.

layers of the star, causing them to expand to form a bloated, luminous giant. As illustrated in **Figure 12.4**, the internal structure of the star is now fundamentally different from the structure when the star was on the main sequence. The giant has a luminosity hundreds of times the luminosity of the Sun and a radius of over 50 solar radii (50 $R_\odot$). Yet the core is far more compact than the Sun's, and much of the star's mass becomes concentrated into a volume that is only a few times the size of Earth.

The star becomes larger and more luminous, and perhaps surprisingly, *cooler and redder* as well. The enormous expanse of the star's surface allows it to cool very efficiently. Even though its *interior* grows hotter and its luminosity increases, the *surface* temperature of the star actually begins to drop.

Tracking the Evolution of the Star on the H-R Diagram

The H-R diagram is a handy device for keeping track of the changing luminosity and surface temperature of the star as it evolves away from the main sequence. As soon as the star exhausts the hydrogen in its core, it leaves the main sequence and begins to move upward and to the right on the H-R diagram, growing more luminous but cooler. As the star continues to evolve, it grows larger and cooler. But after a time, its progress to the right on the H-R diagram ceases. The surface layers regulate how much radiation can escape from the star, thus preventing it from becoming any cooler.

Because the star can cool no further, it begins to move almost vertically upward on the H-R diagram, growing larger and more luminous but remaining about the same temperature. The star has become a **red giant**—an obvious name for a star that is now both redder and larger than it was on the main sequence. We can think of the path that a star follows on the H-R diagram as it leaves the main sequence as being a tree "branch" growing out of the "trunk" of the main sequence. Astronomers refer to this track as the **red giant branch** of the H-R diagram.

As the star leaves the main sequence, the changes in its structure occur slowly at first, but then the star moves up the red giant branch faster and faster. It takes roughly 200 million years for a star like the Sun to go from the main sequence to the top of the red giant branch. During the first half of this period, the star's luminosity increases to about 10 times the luminosity of the Sun (10 $L_\odot$). During the second half of this time, the star's luminosity skyrockets to almost 1,000 $L_\odot$. The evolution of the star, illustrated in **Figure 12.5**, is reminiscent of the growth of a snowball rolling downhill. The larger the snowball becomes, the faster it grows; and the faster it grows, the larger it becomes.

The analogy between the evolution of a red giant star and the growth of a snowball is not perfect. The helium core of the star grows in mass—but not in radius—as hydrogen is converted to helium in the hydrogen-burning shell. The increasing mass of the ever more compact helium core increases the force of gravity in the heart of the star. Stronger gravity means higher pressure, and higher pressure accelerates nuclear burning in the shell. Faster nuclear reactions

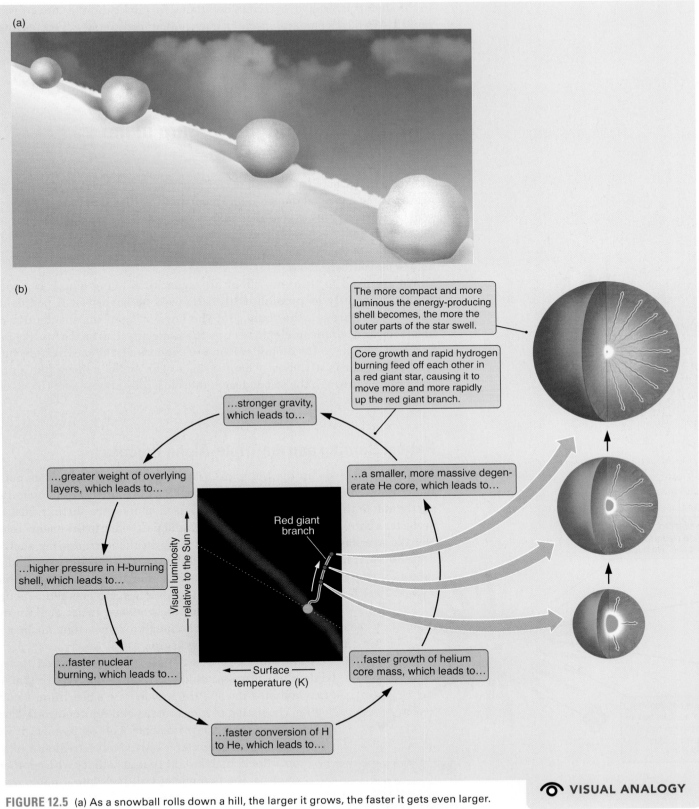

FIGURE 12.5 (a) As a snowball rolls down a hill, the larger it grows, the faster it gets even larger. (b) Similarly, as a star moves up the red giant branch, burning hydrogen to helium in a shell surrounding a degenerate helium core, its evolution feeds on itself. The luminosity of the star grows faster and faster.

in the shell convert hydrogen into helium more quickly, so the core grows more rapidly. We have come full circle in a cycle that feeds on itself. Increasing core mass leads to ever-faster burning in the shell; and the faster hydrogen burns in the shell, the faster the core mass grows. As a result, the star's luminosity climbs at an ever-higher rate.

12.3 Helium Begins to Burn in the Degenerate Core

The growth of the red giant cannot continue forever, and we find ourselves once again asking a crucial question for understanding the evolution of stars: What will be the next thing to give? The answer lies in another unusual property of the degenerate helium core: although as many *electrons* are packed into that space as the rules of quantum mechanics allow, the *atomic nuclei* in the core are still able to move freely about.

We are used to thinking about all material as being equal: if a room is packed as tightly as possible with cats, we'd be surprised if people could move freely through the room. But the laws of quantum mechanics allow electrons and atomic nuclei to occupy the same physical space (unlike people and cats). As far as the atomic nuclei are concerned, the electron-degenerate core of the star is still mostly empty space. The nuclei behave like a normal gas, moving through the sea of degenerate electrons almost as if the electrons were not there.

Helium Burning and the Triple-Alpha Process

As the star evolves up the red giant branch, its helium core grows not only smaller and more massive, but hotter as well. This increase in temperature is partly due to the gravitational energy released as the core shrinks (just like a protostar's core grows hotter as it collapses) and partly due to the energy released by the ever-faster pace of hydrogen burning in the surrounding shell. The thermal motions of the atomic nuclei in the core become more and more energetic. Eventually, at a temperature of about 10^8 K, the collisions between helium nuclei in the core become energetic enough to overcome the electric repulsion between them. Helium nuclei are slammed together hard enough for the strong nuclear force to act, and helium burning begins.

Helium burns in a two-stage process referred to as the **triple-alpha process**, which is illustrated in **Figure 12.6**. First, two helium-4 nuclei (⁴He) fuse to form a beryllium-8 nucleus (⁸Be) consisting of four protons and four neutrons. The ⁸Be nucleus is extremely unstable. Left on its own, it would break apart again after only about a trillionth of a second. But if, in that short time, it collides with another ⁴He nucleus, the two nuclei will fuse into a stable nucleus of carbon-12 (¹²C) consisting of six protons and six neutrons. The triple-alpha process takes its name from the fact that it involves the fusion of three ⁴He nuclei, which are traditionally referred to as alpha particles.

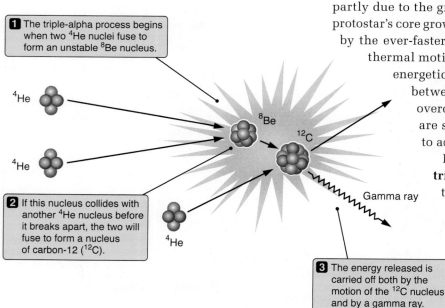

1 The triple-alpha process begins when two ⁴He nuclei fuse to form an unstable ⁸Be nucleus.

⁴He

⁴He

⁸Be

¹²C

2 If this nucleus collides with another ⁴He nucleus before it breaks apart, the two will fuse to form a nucleus of carbon-12 (¹²C).

⁴He

Gamma ray

3 The energy released is carried off both by the motion of the ¹²C nucleus and by a gamma ray.

FIGURE 12.6 The triple-alpha process produces a stable nucleus of carbon-12.

The Helium Core Ignites in a Helium Flash

In the next phase of the star's evolution, the helium in the core begins burning, as **Figure 12.7** illustrates. Degenerate material is a very good conductor of thermal energy, so any differences in temperature within the core rapidly even out. As a result, when helium burning begins at the center of the core, the energy released quickly heats the entire core. Within a few minutes the entire core is burning helium into carbon by the triple-alpha process.

In a normal gas like the air around you, the pressure of the gas comes from the random thermal motions of the atoms. Increasing the temperature of such a gas means that the motions of the atoms become more energetic, so the pressure of the gas increases. If the helium core of a red giant star were a normal gas, the increase in temperature from helium burning would increase the pressure. The core of the star would expand; the temperature, density, and pressure would decrease; nuclear reactions would slow; and the star would settle down into a new balance between gravity and pressure. These are exactly the sorts of changes that are steadily occurring within the core of a main-sequence star like the Sun as the structure of the star shifts in response to the changing composition of the core.

However, the degenerate core of a red giant is not a normal gas. The pressure in a red giant's degenerate core comes from how densely the electrons in the core are packed together. Heating the core does not change the number of electrons that can be packed into its volume, so the core's pressure does not respond to changes in temperature. And if the pressure does not increase, then the core does not expand.

Helium burns via the triple-alpha process.

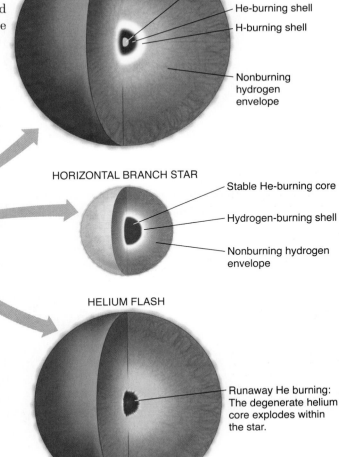

1 The degenerate helium core of the red giant ignites in a helium flash…

2 …and then the star settles onto the horizontal branch.

3 The star exhausts the helium at its center, leaves the horizontal branch, and ascends the asymptotic giant branch.

Asymptotic giant branch (AGB)

ASYMPTOTIC GIANT BRANCH STAR
- Nonburning degenerate carbon ash core
- He-burning shell
- H-burning shell
- Nonburning hydrogen envelope

HORIZONTAL BRANCH STAR
- Stable He-burning core
- Hydrogen-burning shell
- Nonburning hydrogen envelope

HELIUM FLASH
- Runaway He burning: The degenerate helium core explodes within the star.

FIGURE 12.7 A low-mass star travels a complex path on the H-R diagram at the end of its life.

The higher temperature does not change the pressure, but it does cause the helium nuclei to collide with more frequency and greater force, so the nuclear reactions become more vigorous. More vigorous reactions mean higher temperature, and higher temperature means even more vigorous reactions. Helium burning in the degenerate core runs wildly out of control as increasing temperature and increasing reaction rates feed each other. As long as the degeneracy pressure from the electrons is greater than the thermal pressure from the nuclei, this feedback loop continues.

Within seconds of helium ignition, the thermal pressure increases until it is no longer smaller than the degeneracy pressure. At this point, the core literally explodes. Because the explosion is contained deep within the star, however, we do not see it directly—although the energy released in this runaway thermonuclear reaction does lift the overlying layers of the star. This explosion is called a **helium flash**. But the drama is over within a few hours because the expanded helium-burning core is no longer degenerate, and the star is on its way toward a new equilibrium.

> A helium flash is a thermonuclear runaway—an explosion within the star.

You might imagine that helium burning in the core would cause the star to grow more luminous, but it does not. The tremendous energy released during the helium flash goes into fighting gravity and "puffing up" the core. After the helium flash, the core (which is no longer degenerate) is much larger, so the force of gravity within it and the surrounding shell is much smaller. Weaker gravity means less force pushing down on the core and the shell, which means lower pressure. Lower pressure, in turn, slows the nuclear reactions. The net result is that following the helium flash, core helium burning keeps the core of the star puffed up, and the star becomes less luminous than it was as a red giant.

The star spends the next 100,000 years or so burning helium into carbon in a nondegenerate core while hydrogen burns to helium in a surrounding shell. The star is about a hundredth as luminous as it was at the time of the helium flash. The lower luminosity means that the outer layers of the star are not puffed up as much as they were when the star was a red giant. The star shrinks, and its surface temperature climbs.

> Horizontal branch stars burn helium in the core and hydrogen in a shell.

At this point in their evolution, low-mass stars with chemical compositions similar to that of the Sun lie on the H-R diagram just to the left of the red giant branch. Stars that contain much less iron than the Sun tend to distribute themselves away from the red giant branch along a nearly horizontal line on the H-R diagram. This stage of stellar evolution takes its name from this horizontal band. The star is now referred to as a **horizontal branch** star (see Figure 12.7).

12.4 The Low-Mass Star Enters the Last Stages of Its Evolution

The evolution of a star like our Sun from the main sequence through its helium flash and on to the horizontal branch is fairly well understood. Just as our understanding of the interior of the Sun comes from computer models of the physical conditions within our local star, our understanding of the evolution of a red giant comes from computer models that look at the changes in structure as the star's degenerate helium core grows. These models show that any star with a mass of about 1 $M_\odot$ will follow the march from main sequence to helium flash and then drop down onto the horizontal branch. But when we try to use computer models to understand what happens next, the road that we follow gets a bit trickier. We just noted that differences in chemical composition between stars significantly affect where they fall on the horizontal branch. From this point on, small changes in the properties of a star—mass, chemical composition, strength of the star's magnetic field, or even the rate at which the star is rotating—can lead to noticeable differences in how the star evolves.

With this caution in mind, we continue our story of the evolution of a 1-$M_\odot$ star with solar composition, following the most likely sequence of events awaiting our Sun.

The Star Moves Up the Asymptotic Giant Branch

The behavior of a star on the horizontal branch is remarkably similar to that of a star on the main sequence. The star's life on the horizontal branch, however, is much shorter than its life on the main sequence. The star is more luminous, so it is consuming fuel more rapidly. Helium is a much less efficient nuclear fuel than hydrogen, so it has to burn fuel even faster. Even so, for 100 million years the horizontal branch star remains stable, burning helium to carbon in its core and hydrogen to helium in a shell.

The temperature at the center of a horizontal branch star is not high enough for carbon to burn, so carbon ash builds up in the heart of the star. Gravity once again begins to win as the nonburning carbon ash core is crushed by the weight of the layers of the star above it. Again, the electrons in the core are packed together as tightly as the laws of quantum mechanics allow. The carbon core is now electron-degenerate, with physical properties much like those of the degenerate helium core at the center of a red giant.

> The star forms a degenerate carbon core as it leaves the horizontal branch.

The strength of gravity in the inner parts of the star increases, which in turn drives up the pressure, which speeds up the nuclear reactions, which causes the degenerate core to grow more rapidly—we have heard this story before. The internal changes occurring within the star are similar to the changes that took place at the end of the star's main-sequence lifetime, and the path the star follows as it leaves the horizontal branch echoes that earlier phase of evolution as well. Just as the star accelerated up the red giant branch as its degenerate helium core grew, the star now leaves the horizontal branch and once again begins to grow larger, redder, and more luminous as its degenerate carbon core grows. The path that the star follows on the H-R diagram (see Figure 12.7) closely parallels the path it followed as a red giant, approaching the red giant branch as the star grows more luminous. This phase of evolution is referred to as the **asymptotic giant branch** (**AGB**) of the H-R diagram. An AGB star burns helium and hydrogen in nested concentric shells surrounding a degenerate carbon core.

Giant Stars Lose Mass

Building on our analogy between AGB stars and red giants, you might imagine that the next step in the evolution of an AGB star should be a "carbon flash" when carbon burning begins in the star's degenerate core. Yet a carbon flash never happens. Before the temperature in the carbon core becomes high enough for carbon to burn, the star loses its gravitational grip on itself and expels its outer layers into interstellar space.

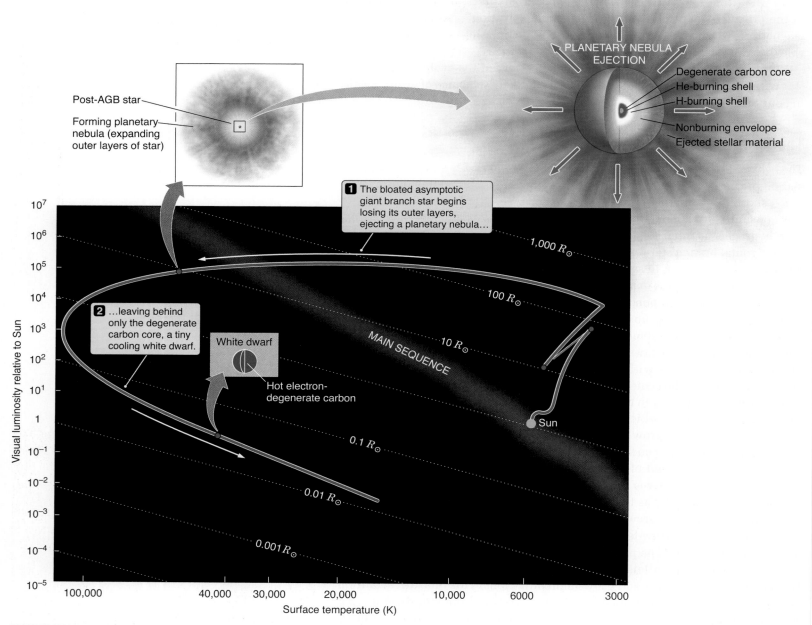

FIGURE 12.8 At the end of the AGB star's life, it ejects most of its mass in a planetary nebula, becoming a post-AGB star and finally leaving behind nothing but the star's degenerate core, which appears on the H-R diagram as a cooling *white dwarf star*.

Red giant and AGB stars are huge objects. When our Sun becomes an AGB star, its outer layers will swell to the point that they engulf the orbits of the inner planets, possibly including Earth. When a star expands to such a size, the gravitational force at its surface is only 1/10,000 as strong as the gravity at the surface of the present-day Sun. It takes little extra energy to push surface material away from the star. **Stellar mass loss** actually begins when the star is still on the red giant branch; by the time a 1-$M_\odot$ main-sequence star reaches the horizontal branch, it may have lost 10–20 percent of its total mass. As the star ascends the asymptotic giant branch, it loses another 20 percent or even more of its total mass. And by the time it is well up this branch, the star may have lost more than half of its original mass.

Mass loss on the asymptotic giant branch can be spurred on by the star's unstable interior. The extreme sensitivity of the triple-alpha process to temperature can lead to episodes of rapid energy release, which can provide the extra kick needed to expel material from the star's outer layers. Even stars that are initially quite similar can behave very differently when they reach this stage in their evolution.

Red giants and AGB stars lose mass from their outer layers.

The Post-AGB Star May Produce a Planetary Nebula

Toward the end of an AGB star's life, mass loss itself becomes a runaway process. When a star loses a bit of mass from its outer layers, the weight pushing down on the underlying layers of the star is reduced. Without this weight holding them down, the outer layers of the star puff up even larger than they were before. The star, which is now both less massive and larger, is even less tightly bound by gravity, so even less energy is needed to push its outer layers away. Much of the remaining mass of the star is ejected into space, typically at speeds of 20–30 kilometers per second (km/s).

All that is left of the low-mass star itself is a tiny, very hot electron-degenerate carbon core, surrounded by a thin envelope in which hydrogen and helium are still burning. This star is now somewhat less luminous than when it was at the top of the asymptotic giant branch, but it is still much more luminous than a horizontal branch star. The remaining hydrogen and helium in the star rapidly burn to carbon, and as more and more of the mass of the star ends up in the carbon core, the star shrinks and becomes hotter and hotter. Over the course of only 30,000 years or so following the beginning of runaway mass loss, the star moves very rapidly from right to left across the top of the H-R diagram, as shown in **Figure 12.8**.

The surface temperature of the star may eventually reach 100,000 K or hotter. Wien's law says that at such temperatures, most of the light from the star is in the high-energy ultraviolet part of the spectrum. The intense UV light heats and ionizes the expanding shell of gas that was recently ejected by the star, causing it to glow.

The mass ejected by the AGB star will pile up in a dense, expanding shell. If you were to look at such a shell through a small telescope, you would see a round or perhaps oblong patch of light, perhaps with a hole and a dot in the middle. When these glowing shells were first observed in small telescopes, they looked like fuzzy disks of planets (**Figure 12.9a**), which is why they were named **planetary nebulae**. But there is nothing planet-like about them (**Figure 12.9b**). Instead, a planetary nebula consists of the remaining outer layers of a star, ejected into space as a dying gasp at the end of the star's ascent of the asymptotic giant branch. Not all stars form planetary nebulae. Stars more massive than about 8 solar masses pass through the post-AGB stage too quickly. Stars with insufficient mass take too long, so their envelope dissipates before they can illuminate it. Some astronomers think that our own Sun will not retain enough mass during its post-AGB phase to form a planetary nebula.

Planetary nebulae can be dazzling in appearance. The most distinctive ones have earned names like Owl Nebula, Clown Nebula, Cat's Eye Nebula, and Dumbbell Nebula. This extraordinary menagerie is illustrated in **Figure 12.10**. The structure of a planetary nebula tells of eras when mass loss was slower or

Planetary nebulae may form around dying low-mass stars.

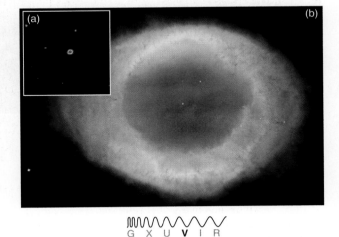

G X U **V** I R

FIGURE 12.9 At the end of its life, a low-mass star ejects its outer layers and may form a planetary nebula consisting of an expanding shell of gas surrounding the white-hot remnant of the star. (a) This picture of a famous planetary nebula called the Ring Nebula, as it appears to the eye through a small amateur telescope, shows why astronomers thought these objects looked like planets. (b) However, an HST image of the Ring shows the remarkable and complex structure of this expanding shell of gas.

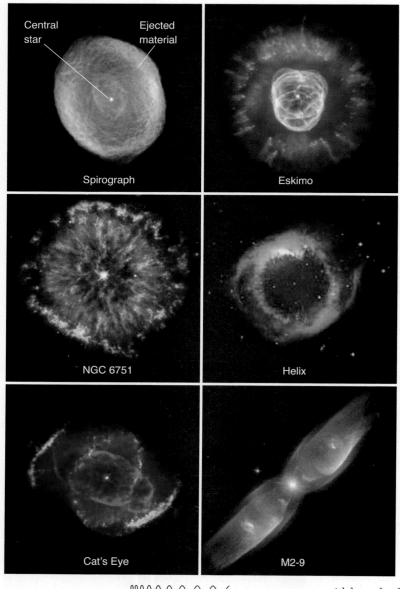

FIGURE 12.10 Planetary nebulae come in a wide variety of shapes that reveal the details of the history of mass loss from the central star.

A newly formed white dwarf is hot but tiny.

faster, and of times when mass was ejected primarily from the star's equator or its poles. The colors come from emission lines from particular atoms and ions. Look back at the chapter-opening figure, which shows another view of the Ring Nebula. From the colored rings, we know that we are seeing emission lines from different ions in different places. In this nebula, a lot of mass was lost nearly all at once. Then the mass loss ceased, resulting in a hollow shell around the central star.

Mass loss from giant stars carries the chemical elements enriching the stars' outer layers off into interstellar space. Planetary nebulae often show an overabundance of elements such as carbon, nitrogen, and oxygen, which are by-products of nuclear burning. Once this chemically enriched material leaves the star, it mixes with interstellar gas, increasing the chemical diversity of the universe.

The Star Becomes a White Dwarf

Within 50,000 years or so, a post-AGB star burns all of the fuel remaining on its surface, leaving nothing behind but a cinder—a nonburning ball of carbon. In the process, the star plummets down the left side of the H-R diagram, becoming smaller and fainter. Within a few thousand years the burned-out core shrinks to about the size of Earth, at which point it has become electron-degenerate and can shrink no further. This **white dwarf** continues to radiate energy away into space, and as it does so it cools, just like the filament of a lightbulb cools when the switch is turned off. The white dwarf moves down and to the right on the H-R diagram, following a line of constant radius. The white dwarf may remain very hot for 10 million years or so, but its tiny size means the luminosity may now be only one-thousandth of our Sun's. Many white dwarfs are known, but none can be seen without a telescope.

Figure 12.11 recaps the evolution of a solar-type 1-$M_\odot$ main-sequence star through to its final existence as a 0.6-$M_\odot$ white dwarf. This process is representative of the fate of low-mass stars. Although all low-mass stars form white dwarfs at the end points of their evolution, the exact path a low-mass star follows from core hydrogen burning on the main sequence to white dwarf depends on details particular to the star.

This is the fate of our Sun, some 6 billion or so years from now. It will become a white dwarf that will fade as it radiates its thermal energy away into space. This superdense ball—with a density of a ton per teaspoonful—actually began its life billions of years earlier as a cloud of interstellar gas billions of times more tenuous than the vacuum in the best vacuum chamber on Earth.

Once it leaves the main sequence, the Sun will travel the path from red giant to white dwarf in less than one-tenth of the time that it spent on the main sequence steadily burning hydrogen to helium in its core. Stars spend most of their luminous lifetimes on the main sequence, which is why most of the stars that we see in the sky are main-sequence stars. In the end, though, white dwarfs will carry the day. Too faint to be noticed as we look at the sky on a clear summer evening, white dwarfs still constitute the final resting place for the vast majority of stars that have been or ever will be formed.

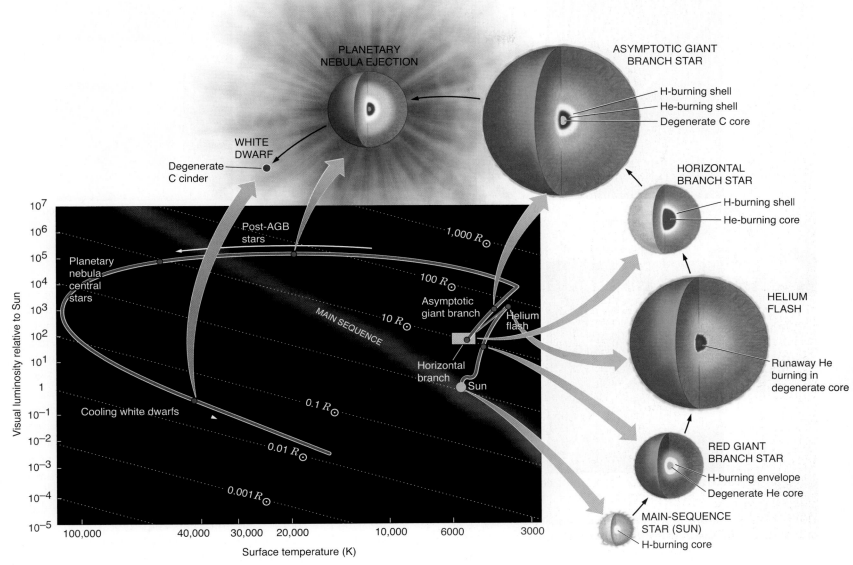

FIGURE 12.11 This H-R diagram summarizes the stages in the post-main-sequence evolution of a 1-$M_\odot$ star.

12.5 Star Clusters Are Snapshots of Stellar Evolution

As we learned in Chapter 5, when an interstellar cloud collapses, it fragments into pieces, forming a cluster of many stars of different masses. We see many such **star clusters** around us today, containing anywhere from a few dozen to millions of stars. The fact that all of the stars in a cluster formed together at nearly the same time means that clusters are snapshots of stellar evolution. A look at a cluster that is 10 million years old shows us what stars of all different masses evolve into during the first 10 million years after they are formed. A look at a cluster 10 *billion* years after it formed shows us what becomes of stars of different masses after 10 billion years pass.

Stars in clusters formed together at about the same time.

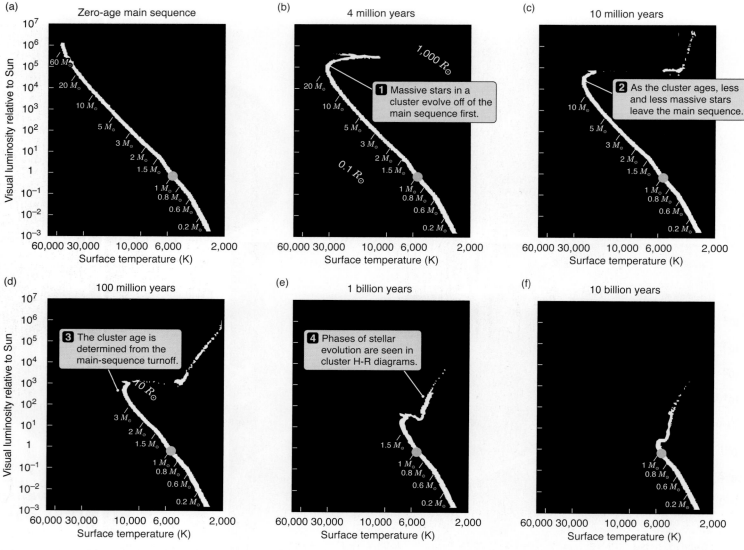

FIGURE 12.12 H-R diagrams of star clusters are snapshots of stellar evolution. These are H-R diagrams of a simulated cluster of 40,000 stars of solar composition seen at different times following the birth of the cluster. Note the progression of the main-sequence turnoff to lower and lower masses.

Figure 12.12 shows the H-R diagram of a simulated cluster of 40,000 stars as it would appear at several different ages. In Figure 12.12a stars of all masses are located on the **zero-age main sequence**, showing where they begin their lives as main-sequence stars. We would never expect to see a cluster H-R diagram that looks like the one in Figure 12.12a, however, since the stars in a cluster do not all reach the main sequence at exactly the same time. Star formation in a molecular cloud is spread out over several million years, and it takes considerable time for lower-mass stars to reach the main sequence. The H-R diagram of a very young cluster normally shows many lower-mass stars located well above the main sequence.

The more massive a star is, the shorter its life on the main sequence will be. After only 4 million years (Figure 12.12b), all stars with masses greater than about 20 $M_\odot$ have evolved off the main sequence and are now spread out across the top of the H-R diagram. (Details of the evolution of these high-mass stars are covered in Chapter 13. For now, all you need to know is that they evolve faster than low-mass stars.) As time goes on, stars of lower mass evolve off the main sequence, and the turnoff point moves toward the bottom right along the main sequence. By the time the cluster is 10 million years old (Figure 12.12c), only stars with

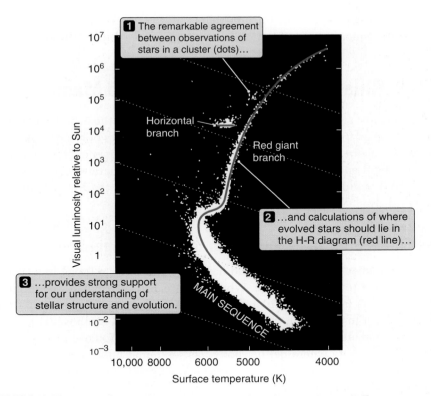

FIGURE 12.13 The observed H-R diagram of stars (dots) in the cluster 47 Tucanae agrees remarkably well with the theoretical calculation (solid line) of the H-R diagram of a 12-billion-year-old cluster.

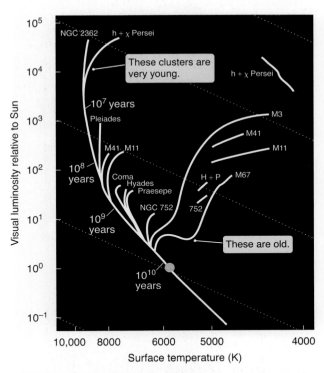

FIGURE 12.14 H-R diagrams for clusters having a range of different ages. The ages associated with the different main-sequence turnoffs are indicated.

masses less than about 15 $M_\odot$ remain on the main sequence. The location of the most massive star that is still on the main sequence is called the **main-sequence turnoff**. As the cluster ages, the main-sequence turnoff moves down the main sequence to stars of lower mass.

As a cluster ages further (Figure 12.12d and e), we see the details of all stages of stellar evolution. By the time the star cluster is 10 *billion* years old (Figure 12.12f), stars with masses of only 1 $M_\odot$ are beginning to die. Stars slightly more massive than this are seen as giants of various types. Note how few giant stars are present in any of the cluster H-R diagrams. The end phases in the evolution of a low-mass star pass so quickly in comparison with the star's main-sequence lifetime that even though the cluster started with 40,000 stars, only a handful of stars are in these phases of evolution at any given time. Similarly, even though most of the evolved stars in an old cluster are white dwarfs, all but a few of these stars will have cooled and faded into obscurity at any given time.

Figure 12.12 shows H-R diagrams as predicted by our theories of stellar evolution. **Figure 12.13** shows the observed H-R diagram for the cluster 47 Tucanae, along with a theoretical calculation of the H-R diagram for a 12-billion-year-old cluster. The fact that observed star clusters have H-R diagrams that agree so well with the predictions of models is strong support for our theories of stellar evolution.

Cluster evolution models are powerful tools for studying the history of star formation. When we observe a star cluster, the main-sequence turnoff immediately tells us its age. **Figure 12.14** traces the observed H-R diagrams for several real star clusters. NGC 2362 is a young cluster. Its complement of massive, young

The main-sequence turnoff shifts to lower-mass stars as a cluster grows older.

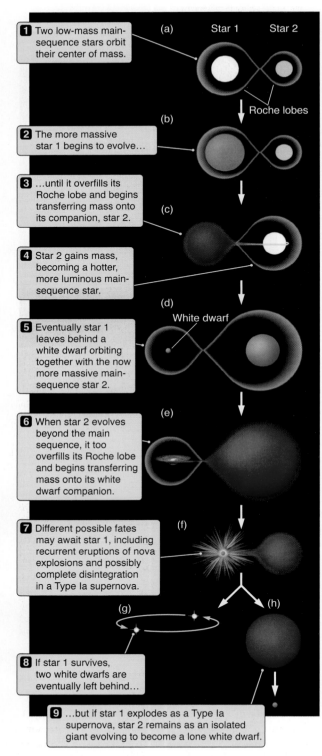

FIGURE 12.15 A compact binary system consisting of two low-mass stars passes through a sequence of stages as the stars evolve and mass is transferred back and forth. The evolution of the binary system can lead to novae or even to a Type Ia supernova that destroys the initially more massive star 1.

stars shows it to be only a few million years old. In contrast, NGC 752 is quite old; its main-sequence turnoff indicates a cluster age of about 7 billion years.

12.6 Binary Stars Sometimes Share Mass

Possibly the most significant complication in our picture of the evolution of low-mass stars arises from the fact that many stars are members of binary systems. While both members of a binary pair are on the main sequence, they usually have little effect on each other. But in some cases, if the separation between the stars is small and one star is more massive than the other, their evolution may become linked.

Mass Flows from an Evolving Star onto Its Companion

Think for a moment about what would happen if you were to travel in a spacecraft from Earth toward the Moon. When you are still near Earth, the force of Earth's gravity is far stronger than that of the Moon. As you move away from Earth and closer to the Moon, the gravitational attraction of Earth weakens, and the gravitational attraction of the Moon becomes stronger. You eventually reach an intermediate zone where neither body has the upper hand. If you continue beyond this point, the lunar gravity begins to assert itself until you find yourself firmly in the grip of the Moon.

Exactly the same situation exists between two stars. When one star swells up, its outer layers may cross that gravitational dividing line separating the star from its companion, and any material that crosses this line no longer belongs to the first star, but instead can be pulled toward the companion. A star reaches this point if it swells to fill up its portion of an invisible figure-eight-shaped volume of space set by the force of gravity exerted by each star, as shown in **Figure 12.15**. These regions surrounding the two stars—their gravitational domains—are the **Roche lobes** of the system. If the first star expands to overfill its Roche lobe, material begins to pour through the "neck" of the figure eight and fall toward the other star. Astronomers refer to this exchange of material between the two stars as **mass transfer**.

Evolution of a Close Binary System

The best way to understand how mass transfer affects the evolution of stars in a binary system is to apply what we have learned from the evolution of single low-mass stars. **Figure 12.15a** shows a close binary system consisting of two low-mass stars of somewhat different mass. The more massive of the two stars is "star 1," and the less massive of the two is "star 2." This is an ordinary binary system, and each of these stars is an ordinary main-sequence star for most of the system's lifetime.

More massive stars evolve more rapidly. Therefore, star 1 will be the first to use up the hydrogen at its center and begin to evolve off the main sequence (**Figure 12.15b**). If the stars are close enough to each other, star 1 will grow to overfill its Roche lobe, and material will transfer onto star 2 (**Figure 12.15c**). The structure of star 2 must then change to accommodate its new status as a higher-mass star. If we plotted star 2's position on the H-R diagram during this period, we would see it move up and to the left along the main sequence, becoming larger, hotter, and more luminous.

A number of interesting things can happen at this point. For example, the transfer of mass between the two stars can result in a sort of "drag" that causes the orbits of the two stars to shrink, bringing the stars closer together and further enhancing mass loss. The two stars can even reach the point where they are effectively two cores sharing the same extended envelope of material.

Because it is losing mass to star 2, star 1 never grows large enough to move to the top of the H-R diagram as a red giant. Yet it continues to evolve, burning helium in its core on the horizontal branch, proceeding through a stage of helium shell burning, and finally losing its outer layers and leaving behind a white dwarf. **Figure 12.15d** shows the binary system after star 1 has completed its evolution. All that remains of star 1 is a white dwarf orbiting about its bloated main-sequence companion, star 2.

Fireworks Occur When the Second Star Evolves

Figure 12.15e picks up the evolution of the binary system as star 2 begins to evolve off the main sequence. Like star 1 before it, star 2 grows to fill its Roche lobe; material from star 2 begins to pour through the "neck" connecting the Roche lobes of the two stars. Because the white dwarf is so small, the infalling material generally misses the star. Instead of landing directly on the white dwarf, the infalling mass forms an **accretion disk** around the white dwarf, similar in some ways to the accretion disk that forms around a protostar. As in the process of star formation, the accretion disk serves as a way station for material that is destined to find its way onto the white dwarf but starts out with too much angular momentum to hit the white dwarf directly.

A white dwarf has a mass comparable to that of the Sun, but a size comparable to that of Earth. A large mass and a small radius mean strong gravity. A kilogram of material falling from space onto the surface of a white dwarf releases 100 times more energy than a kilogram of material falling from the outer Solar System onto the surface of the Sun. All of this energy is turned into thermal energy. The spot where the stream of material from star 2 hits the accretion disk is heated to millions of kelvins, where it glows in the far ultraviolet and X-ray parts of the electromagnetic spectrum.

The infalling material accumulates on the surface of the white dwarf (**Figure 12.16a**), where it is compressed by the enormous gravitational pull of the white dwarf to a density close to that of the white dwarf itself. As more and more material builds up on the surface of the white dwarf, the white dwarf shrinks (just like the core of a red giant shrinks as it grows more massive). The density increases, and at the same time the release of gravitational energy drives up the temperature of the white dwarf. The infalling material is from the outer, unburned layers of star 2, so it is composed mostly of hydrogen.

Once the temperature at the base of the white dwarf's surface layer of hydrogen reaches about 10 million K, this hydrogen begins to burn explosively. Energy released by hydrogen burning drives up the temperature, and the higher temperature drives up the rate of hydrogen burning. This runaway reaction is much like the runaway helium burning that takes place during the helium flash, except now there are no overlying layers of the star to keep things contained. The result is a tremendous explosion—a **nova**—that blows part of the layer covering the white dwarf out into space at speeds of thousands of kilometers per second (see **Figures 12.15f and 12.16b**).

About 50 novae occur in our galaxy each year, but we can see only two or three

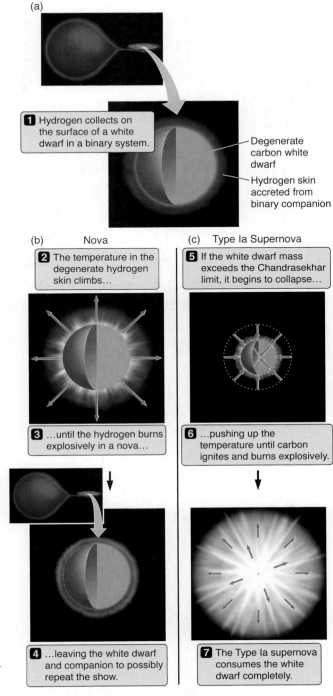

(a)

1 Hydrogen collects on the surface of a white dwarf in a binary system.

Degenerate carbon white dwarf

Hydrogen skin accreted from binary companion

(b) Nova

2 The temperature in the degenerate hydrogen skin climbs…

3 …until the hydrogen burns explosively in a nova…

4 …leaving the white dwarf and companion to possibly repeat the show.

(c) Type Ia Supernova

5 If the white dwarf mass exceeds the Chandrasekhar limit, it begins to collapse…

6 …pushing up the temperature until carbon ignites and burns explosively.

7 The Type Ia supernova consumes the white dwarf completely.

FIGURE 12.16 (a) In a binary system in which mass is transferred onto a white dwarf, a layer of hydrogen builds up on the surface of the degenerate white dwarf. (b) If hydrogen burning ignites on the surface of the white dwarf, the result is a nova. (c) If enough hydrogen accumulates to force the white dwarf to begin to collapse, carbon ignites and the result is a Type Ia supernova.

of them because dust in the disk of our galaxy blocks our view. Novae reach their peak brightness in only a few hours, and for a brief time they can be several hundred thousand times more luminous than the Sun. Although the brightness of a nova sharply drops in the weeks following the outburst, it can sometimes still be seen for years. During this time, the glow from the expanding cloud of ejected material is caused by the decay of radioactive isotopes created in the explosion.

The explosion of a nova does not destroy the underlying white dwarf star. Afterward, the binary system is in much the same configuration as before; material from star 2 is still pouring onto the white dwarf (see Figure 12.16a). This cycle can repeat itself many times, as material builds up and ignites again and again on the surface of the white dwarf. In most cases, outbursts are separated by thousands of years, so most novae have been seen only once in historical times. Some novae, however, erupt every decade or so.

Runaway burning of hydrogen on a white dwarf causes a nova.

A Stellar Cataclysm May Await the White Dwarf in a Binary System

It is possible that star 2 eventually will simply go on to form a white dwarf, leaving behind a binary system consisting of two white dwarfs, as in **Figure 12.15g**. There is another possibility, however. Through millions of years of mass transfer from star 2 onto the white dwarf, and possibly through countless nova outbursts, the white dwarf's mass slowly increases—but it cannot have a mass above 1.4 $M_\odot$ and remain a white dwarf. This value is referred to as the **Chandrasekhar limit**, named for Subrahmanyan Chandrasekhar (1910–1995), who derived it. Above this mass, even the pressure supplied by degenerate electrons is no longer enough to balance gravity, and the white dwarf will begin to collapse (**Figure 12.16c**).

As the white dwarf collapses, the conversion of gravitational energy into thermal energy quickly drives its temperature high enough to overcome the electric repulsion of carbon atoms and fuse them. Once again, the onset of nuclear burning in a degenerate gas leads to a thermonuclear runaway. But whereas the thermonuclear runaway in a nova involves only a thin shell of material on the surface of the white dwarf, runaway carbon burning involves the entire white dwarf. Within about a second, the whole white dwarf is consumed in the resulting conflagration. In this single instant, 100 times more energy is liberated than will be given off by the Sun over its entire 10-billion-year lifetime on the main sequence. Runaway fusion reactions convert a large fraction of the mass of the star into elements such as iron and nickel, and the explosion blasts the shards of the white dwarf into space at top speeds in excess of 20,000 km/s, enriching the interstellar medium with these heavier elements. The explosion completely destroys star 1, leaving star 2 behind as a lone giant to continue its evolution toward becoming a white dwarf (**Figure 12.15h**).

The mass of a white dwarf cannot exceed 1.4 $M_\odot$.

This explosion is known as a **Type Ia supernova**. Type Ia supernovae occur in a galaxy the size of the Milky Way about once a century. For a brief time they can shine with a luminosity billions of times that of our Sun, possibly outshining the galaxy itself.

The sheer number of stars in the sky, as compared to the number of astronomers on the ground, means that a lot of the time, things are happening that we are simply not aware of. Even stellar explosions sometimes are missed!

Scientists May Be Missing Many Star Explosions

By **SPACE.com staff**

Some of the brightest stellar explosions in the galaxy may be flying under astronomers' radar, a new study suggests.

Researchers using observations from a sun-studying satellite detected four novas—exploding stars not quite as bright or dramatic as supernovas. The scientists were able to follow the explosions in intricate detail over time, including before the novas reached maximum brightness.

While other astronomers had discovered all four novas before, two of them escaped detection until after they had reached peak luminosity, the study revealed. This fact suggests that many other stellar explosions, even some that are incredibly bright, may be occurring unnoticed, researchers said.

"So far, this research has shown that some novae become so bright that they could have been easily detected with the naked eye by anyone looking in the right direction at the right time, but are being missed, even in our age of sophisticated professional observatories," study lead author Rebekah Hounsell, a graduate student at Liverpool John Moores University (LJMU) in England, said in a statement.

The new observations are also allowing scientists to study nova explosions in unprecedented detail, according to researchers.

Hounsell and her colleagues analyzed measurements from an instrument aboard the U.S. Department of Defense's Coriolis satellite. The instrument, called the Solar Mass Ejection Imager (SMEI), was designed to detect disturbances in the solar wind. SMEI maps out the entire sky during its 102-minute orbit around the Earth.

The researchers found that SMEI was also detecting star explosions, or novas. Novas occur when small, extremely dense stars called white dwarfs suck up gas from a nearby companion star, igniting a runaway thermonuclear explosion.

Unlike supernovas, novas do not result in the destruction of their stars. Stars can go nova repeatedly.

SMEI detected four novas, including one confirmed repeater called RS Ophiuchi, which is found about 5,000 light-years away in the constellation Ophiuchus. RS Ophiuchi may ultimately die in a supernova explosion—one of the brightest, most dramatic events in the universe—researchers said.

Ground-based instruments missed the peak flare-up of two of these four novas, according to researchers. That suggests that space-based instruments like SMEI might be needed to pick up many novas, after which their progress can be tracked with telescopes on the ground, researchers said.

"Two of the novae observed by SMEI have confirmed that even the brightest novae may be missed by conventional ground-based observing techniques," said co-author Mike Bode, also of LJMU.

The researchers reported their results in a recent issue of the *Astrophysical Journal*.

The new observations are giving astronomers key insights into novaes' earlier days, revealing a great deal about how they start and evolve, researchers said.

"The SMEI's very even cadences and uniformly exposed images allow us to sample the sky every 102 minutes and trace the entire evolution of these explosions as they brighten and dim," said co-author Bernard Jackson of the University of California, San Diego.

The new observations have revealed, for example, that three of the explosions faltered significantly before regaining strength and proceeding. Such a "pre-maximum halt" had been theorized before, but evidence for its existence had been inconclusive, researchers said.

Since SMEI performs a survey of the entire sky every 102 minutes, the instrument could also help astronomers understand a wide variety of transient objects and phenomena, according to the research team.

"[This] work has shown how important all-sky surveys such as SMEI are and how their sets can potentially hold the key to a better understanding of many variable objects," Bode said.

Evaluating the News

1. What kind of stellar explosion is being discussed in this article?
2. Which figure in this chapter corresponds most closely to the astronomical events being discussed in this article?
3. Can this kind of explosion occur in an isolated star? Explain your reasoning.
4. In the article, the reporter states that "Such a 'pre-maximum halt' had been theorized before, but evidence for its existence had been inconclusive." Think back to the scientific method in Chapter 1. Is the reporter using the word *theory* correctly here?
5. One of the main advantages of these newly available observations is that they take a picture of the same region of the sky every 102 minutes. Why is that an advantage in studying novae?
6. Why do you think ground-based observatories missed these novae?
7. Why might the astronomers quoted in the article think that RS Ophiuchi will ultimately explode as a supernova?

SUMMARY

12.1 All stars eventually exhaust their nuclear fuel. More massive stars exhaust their fuel more quickly and have shorter lifetimes. Helium accumulates like ash in the cores of main-sequence stars.

12.2 After exhausting its hydrogen, a low-mass star leaves the main sequence and swells to become a red giant. Its helium core is made of electron-degenerate matter.

12.3 A red giant burns helium via the triple-alpha process. The core ignites in a helium flash, and the star then moves onto the horizontal branch.

12.4 A horizontal-branch star accumulates carbon ash in its core and then moves up the asymptotic giant branch. In their

dying stages, some stars form planetary nebulae. All low-mass stars eventually become white dwarfs, which are very hot but very small. These white dwarfs cool to become cold, dark carbon cinders—the stellar graveyard.

12.5 H-R diagrams of clusters give snapshots of stellar evolution. The location of the main-sequence turnoff indicates the age of the cluster.

12.6 Transfer of mass within some binary systems can lead to a nuclear explosion. A nova occurs when hydrogen collects and ignites on the surface of a white dwarf in a binary system. If the mass of the white dwarf exceeds 1.4 $M_\odot$, the entire star explodes in a Type Ia supernova.

✧ SUMMARY SELF-TEST

1. Which of the following stars will have the longest lifetime?
   a. a star 1/10 as massive as the Sun
   b. a star 1/5 as massive as the Sun
   c. the Sun
   d. a star 5 times as massive as the Sun
   e. a star 10 times as massive as the Sun

2. When the Sun runs out of hydrogen in its core, it will become larger and more luminous because
   a. it starts fusing hydrogen in a shell around a helium core.
   b. it starts fusing helium in a shell and hydrogen in the core.
   c. infalling material rebounds off the core and puffs up the star.
   d. energy balance no longer holds, and the star just drifts apart.

3. As a star leaves the main sequence, its position on the H-R diagram moves _____.
   a. up and to the left   b. up and to the right
   c. down and to the left   d. down and to the right

4. Degenerate matter is different from ordinary matter because
   a. degenerate matter doesn't interact with other particles.
   b. degenerate matter has no mass.
   c. degenerate matter doesn't interact with light.
   d. degenerate matter objects get smaller as they get more massive.

5. Place in order the following steps of the evolution of a low-mass star:
   a. The star moves onto the horizontal branch.
   b. The white dwarf cools.
   c. A clump forms in a giant molecular cloud.
   d. The star moves onto the red giant branch.
   e. The star moves onto the asymptotic giant branch.
   f. A protostar forms.
   g. The star sheds mass, producing a nebula.
   h. Hydrogen fusion begins.
   i. A helium flash occurs.

6. Place in order the following steps that lead some low-mass stars in binary systems to become novae or supernovae.
   a. Star 2 gains mass, becoming hotter and more luminous.
   b. A white dwarf orbits a more massive main-sequence star.
   c. Two low-mass main-sequence stars orbit each other.
   d. Star 1 (the more massive star) begins to evolve off the main sequence.
   e. Star 2 fills its Roche lobe and begins transferring mass to the white dwarf.
   f. Star 1 fills its Roche lobe and begins transferring mass to star 2.
   g. The white dwarf becomes either a nova or a supernova.

QUESTIONS AND PROBLEMS

True/False and Multiple-Choice Questions

7. **T/F:** The mass-luminosity relationship says that if a star is twice as massive, it is twice as bright.

8. **T/F:** When a star "runs out of fuel," that means it has fused all of its hydrogen into helium.

9. **T/F:** Even though their masses are the same, a red giant of 1 solar mass can be 50 times as large as the Sun.

10. **T/F:** Some types of material get smaller as the mass increases.

11. **T/F:** More than one kind of nuclear fusion can occur in low-mass stars.

12. **T/F:** The Sun will become a red giant and eventually a white dwarf.

13. **T/F:** A planetary nebula shines because the dust and gas are hot.

14. **T/F:** We can find the age of the stars in a cluster from its H-R diagram.

15. **T/F:** A solitary low-mass star can sometimes become a supernova.

16. As a protostar moves onto the main sequence to become a low-mass star, the evolutionary track is nearly vertical until the very end of the process. During this vertical portion,
    a. the temperature and the luminosity both fall.
    b. the luminosity falls, but the temperature remains nearly constant.
    c. the temperature falls, but the luminosity remains nearly constant.
    d. the temperature and the luminosity both remain nearly constant.

17. The helium abundance in the outer 20 percent of the Sun
    a. remains nearly the same for its entire main-sequence lifetime.
    b. gradually increases over its main-sequence lifetime.
    c. gradually decreases over its main-sequence lifetime.
    d. remains the same for a long time and then increases abruptly before it leaves the main sequence.

18. When the Sun becomes a red giant, its _____ will decrease and its _____ will increase.
    a. density, luminosity
    b. density, temperature
    c. temperature, density
    d. density, rotation

19. As a low-mass star dies, it moves across the top of the H-R diagram because
    a. it is heating up.
    b. we see deeper into the star.
    c. the dust and gas around it heat up.
    d. a new fuel source is tapped.

20. Very young star clusters have main-sequence turnoffs
    a. nowhere. All the stars have already turned off in a young cluster.
    b. at the top left of the main sequence.
    c. at the bottom right of the main sequence.
    d. in the middle of the main sequence.

21. A star cluster with a main-sequence turnoff at one solar mass is approximately
    a. 10 million years old.   b. 100 million years old.
    c. 1 billion years old.   d. 10 billion years old.

22. A low-mass star might become a nova or supernova if and only if
    a. it is in a close binary system.
    b. it is really massive.
    c. it is really dense.
    d. it has a lot of heavy elements in it.

Conceptual Questions

23. Why are most nearby stars low-mass, low-luminosity stars?

24. Is it possible for a star to skip the main sequence and immediately begin burning helium in its core? Explain your answer.

25. What is the primary reason that the most massive stars have the shortest lifetimes? (Note: The answer can be expressed in just a few words.)

26. Describe some possible ways in which a star might increase the temperature within its core while at the same time lowering its density.

27. Suppose a main-sequence star suddenly started burning hydrogen at a faster rate in its core. How would the star react? Discuss changes in size, temperature, and luminosity.

28. Astronomers typically say that the mass of a newly formed star determines its destiny from birth to death. However, there is a frequent environmental circumstance for which this statement is not true. Identify this circumstance, and explain why the birth mass of a star might not fully account for its destiny.

29. Is it fair to assume that stars do not change their structure while on the main sequence? Why or why not?

30. Suppose Jupiter was not a planet, but instead a G5 main-sequence star with a mass of 0.8 $M_\odot$.
    a. How do you think terrestrial life would be affected, if at all?
    b. How would the Sun be affected as it came to the end of its life?

31. When a star runs out of nuclear fuel in its core, why does it become more luminous? Why does the surface temperature of the star cool down?

32. Figure 12.5b shows an example of a feedback loop. Several of these occur in the end stages of stellar evolution. In your own words, describe what a feedback loop is. Give an example of a feedback loop from a field other than astronomy, and make a diagram for it like the one in Figure 12.5.

33. When a star leaves the main sequence, its luminosity increases tremendously. What does this increase in luminosity imply about the amount of time the star has left, or the amount of time the star spends on any subsequent part of its evolutionary path?

34. When it is compressed, ordinary gas heats up; but degenerate gas does not. Why, then, does a degenerate core heat up as the star continues burning the shell around it?

35. Suppose you were an astronomer making a survey of the observable stars in our galaxy. What would be your chances of seeing a star undergoing the helium flash? Explain your answer.

36. Why is a horizontal branch star (which burns helium at a high temperature) less luminous than a red giant branch star (which burns hydrogen at a lower temperature)?

37. Suppose a star is able to heat its core temperature high enough to begin fusing oxygen. Predict how the star will continue to evolve. Include how you think the star will evolve on the H-R diagram.

38. As an AGB star evolves into a white dwarf, it runs out of nuclear fuel, and one might argue that the star should cool off and move to the right on the H-R diagram. Why does the star move instead to the left?

39. Why must a white dwarf move down and to the right on the H-R diagram?

40. Why does fusion in degenerate material *always* lead to a runaway reaction?

41. The intersection of the Roche lobes in a binary system is the equilibrium point between the two stars where the gravitational attraction from both stars is equally strong and opposite in direction. Is this an example of stable or unstable equilibrium? Explain.

42. Suppose the more massive red giant star in a binary system engulfs its less massive main-sequence companion, and their nuclear cores combine. What structure do you think the new star will have? Where will the star lie along the H-R diagram?

43. In Latin, *nova* means "new." Novae, as we now know, are not new stars. Explain how novae might have gotten their name and why they are really not new stars.

44. T Corona Borealis is a well-known recurrent nova.
    a. Is it a single star or a binary system? Explain.
    b. What mechanism causes a nova to flare up?
    c. How can a nova flare-up happen more than once?

Problems

45. In the chapter, we used an approximation to relate the mass of a star to the luminosity. Examine Figure 12.2.
    a. The line is labeled "straight-line (exponential) approximation." What does this mean—how can it be straight line and exponential at the same time?
    b. In your opinion, is this approximation reasonable?

46. For most stars on the main sequence, luminosity scales with mass at a rate proportional to $M^{3.5}$ (see Working It Out 12.1). What luminosity does this relationship predict for (a) 0.50 $M_\odot$ stars, (b) 6.0 $M_\odot$ stars, and (c) 60 $M_\odot$ stars? Compare these numbers to values given in Table 12.1.

47. Compute the main-sequence lifetimes for (a) 0.50 $M_\odot$ stars, (b) 6.0 $M_\odot$ stars, and (c) 60 $M_\odot$ stars. Compare them to the values given in Table 12.1.

48. Study Figure 12.3, which shows the percentages of hydrogen and helium throughout the Sun at three different points in its lifetime.
    a. What percentage of hydrogen did the Sun have in the center of the core originally?
    b. What percentage of hydrogen is in the center of the core today?
    c. What percentage of hydrogen will be in the center of the core 5 billion years from now?
    d. What percentage of the entire Sun was hydrogen originally?
    e. What percentage of the entire Sun is hydrogen now? (Hint: Compare the number of dark blue rectangles to the total number of rectangles.)
    f. What percentage of the entire Sun will be hydrogen 5 billion years from now?

49. The escape velocity ($v_{esc} = \sqrt{2\,GM_\odot/R_\odot}$) from the Sun's surface today is 618 km/s.
    a. What will the escape velocity be when the Sun becomes a red giant with a radius 50 times greater and a mass only 0.9 times that of today?
    b. What will the escape velocity be when the Sun becomes an AGB star with a radius 200 times greater and a mass only 0.7 times that of today?
    c. How will these changes in escape velocity affect mass loss from the surface of the Sun as a red giant, and later as an AGB star?

50. After leaving the main sequence, low-mass stars "seed" the interstellar medium with heavy elements such as carbon, nitrogen, and oxygen. Assume that the universe is approximately 1.4×10^{10} (14 billion) years old and that the Sun is 4.6 billion years old. Refer to Table 12.1 for the masses and main-sequence lifetimes

of stars. How many generations of each type of star preceded the birth of our Sun and Solar System?

51. An H-R diagram shows two parameters on its axes, but they are not the only two that are interesting. For example, you might be interested to know how the radius of the star changes with temperature as the star moves off the main sequence to become a white dwarf. Using the information in Figure 12.8:
    a. Make a graph of radius versus temperature by picking points on Figure 12.8, finding the radius and temperature at that point, and then plotting a point on your new graph.
    b. Label the red giant branch, the horizontal branch, the asymptotic giant branch, the ejection of a planetary nebula, and the white dwarf phase.
    c. Compare and contrast this graph with the H-R diagram that you started from.

52. Suppose you are studying a star cluster, and you find that it has no main-sequence stars with surface temperatures hotter than 10,000 K. How old is this cluster? How do you know?

53. Can the main-sequence turnoff tell us anything besides the age of a cluster? If so, what?

54. Roughly how large does a planetary nebula grow before it disperses? Use an expansion rate of 20 km/s and a lifetime of 50,000 years.

55. In some binary systems, a red giant can transfer mass onto a white dwarf at rates of about 10^{-9} $M_\odot$ per year. Roughly how long after mass transfer begins will the white dwarf undergo a Type Ia supernova? How does this length of time compare to the typical lifetime of a low-mass star? (Hint: Assume that a typical white dwarf starts with a mass of 0.6 $M_\odot$.)

56. How fast does material in an accretion disk orbit around a white dwarf? (Hint: Use Kepler's third law.)

57. A white dwarf has a density of approximately 10^9 kg/m³. Earth has an average density of 5,500 kg/m³ and a diameter of 12,700 km. If Earth were compressed to the density of a white dwarf, what would its radius be?

58. What is the density of degenerate material? Calculate how large the Sun would be if all of its mass were degenerate.

59. Recall from Working It Out 6.2 that the luminosity of a spherical object at temperature T is given by $L = 4\pi R^2 \sigma T^4$, where R is the object's radius. If the Sun became a white dwarf with a radius of 10^7 meters, what would be its luminosity at temperatures of (a) 10^8 K, (b) 10^6 K, (c) 10^4 K, and (d) 10^2 K?

 SmartWork, Norton's online homework system, includes algorithmically generated versions of these questions, plus additional conceptual exercises. If your instructor assigns questions in SmartWork, log in at **smartwork.wwnorton.com**.

 StudySpace is a free and open website that provides a Study Plan for each chapter of **Understanding Our Universe**. Study Plans include animations, reading outlines, vocabulary flashcards, and multiple-choice quizzes, plus links to premium content in SmartWork and the ebook. Visit **wwnorton.com/ studyspace**.

Exploration | Evolution of Low-Mass Stars

wwnorton.com/studyspace

The evolution of a low-mass star, as discussed in this chapter, corresponds to many twists and turns on the H-R diagram. In this exploration, we return to the "HR Explorer" interactive simulation to investigate how these twists and turns affect the appearance of the star.

Open the H-R Explorer simulation, linked from the Chapter 12 area on StudySpace. The box labeled "Size Comparison" shows an image of both the Sun and the test star. Initially these two stars have identical properties: the same temperature, the same luminosity, and the same size.

Examine the box labeled "Cursor Properties." This box shows the temperature, luminosity, and radius of a test star located at the "X" in the H-R diagram. Before you change anything, answer these questions:

1. What is the temperature of the test star?

2. What is the luminosity of the test star?

3. What is the radius of the test star?

As a star leaves the main sequence, it moves up and to the right on the H-R diagram. Grab the cursor (the X on the H-R diagram), and move it up and to the right.

4. What changes about the image of the test star next to the Sun?

5. What is the test star's temperature? What property of the image of the test star indicates that its temperature has changed?

6. What is the test star's luminosity?

7. What is the test star's radius?

8. Ordinarily, the hotter an object is, the more luminous it is. In this case, the temperature has gone down, but the luminosity has gone up. How can this be?

The star then moves around quite a lot in that part of the H-R diagram. Look at Figure 12.11 in this chapter and then use the cursor to approximate the motion of the star as it moves up the red giant branch, back down and onto the horizontal branch, and then back to the right and up the asymptotic giant branch.

9. Are the changes you observe in the image of the star as dramatic as the ones you observed for question 4?

10. What is the most noticeable change in the star as it moves through this portion of its evolution?

Next, the star begins moving across the H-R diagram to the left, maintaining almost the same luminosity. Drag the cursor across the top of the H-R diagram to the left, and study what happens to the image of the star in the "Size Comparison" box.

11. What changed about the star as you dragged it across the H-R diagram?

12. How does its size now compare to that of the Sun?

Finally, the star drops to the bottom of the H-R diagram and then begins moving to the right. Move the cursor toward the bottom of the H-R diagram, where the star becomes a white dwarf.

13. What changed about the star as you dragged it down the H-R diagram?

14. How does its size now compare to that of the Sun?

To thoroughly cement your understanding of stellar evolution, press the reset button and then move the star from main sequence to white dwarf several times. This will help you remember how this part of a star's life appears on the H-R diagram.

13

Evolution of High-Mass Stars

So far in our discussion of the lives of stars, we have concentrated on what happens to low-mass stars like our Sun. **High-mass stars**—stars with masses greater than about 8 solar masses ($M_\odot$)—burn with luminosities thousands or even millions of times as great as the luminosity of our Sun and squander their generous allotments of nuclear fuel in less time than mammals have walked on Earth.

The differences between the ways that high-mass stars and low-mass stars evolve follow from the same relationship among gravity, pressure, and the rate of nuclear burning that determines the balance in a star like our Sun: more mass means stronger gravity and therefore more force on the inner parts of the star. The increase in force means higher pressure; higher pressure means faster reaction rates; and faster reaction rates mean greater luminosity. There are many differences in the evolution of low- and high-mass stars, but in the end they all trace back to the greater gravitational force bearing down on the interior of a high-mass star.

✧ LEARNING GOALS

Our Sun is but a pale ember when compared with the brilliance of more massive stars. These stars consume their fuel in a cosmic blink of the eye, and then in their death they dazzle us with brilliant fireworks. In the image at right, just such a dramatic explosion has occurred in the popular astrophotography target M51. This student has obtained before and after pictures of the host galaxy, so he can easily find the supernova. By the end of this chapter, you should understand what triggers such an explosion and why it happens for high-mass stars but not low-mass ones, and you should be able to determine what type of object might be left behind once the debris dissipates. You should also be able to:

- Describe how high-mass stars differ from low-mass stars

- List the stages that evolving high-mass stars experience and explain the origin of chemical elements heavier than iron

- Discuss the implications of the fact that the speed of light is a universal constant

- Describe what we know about black holes

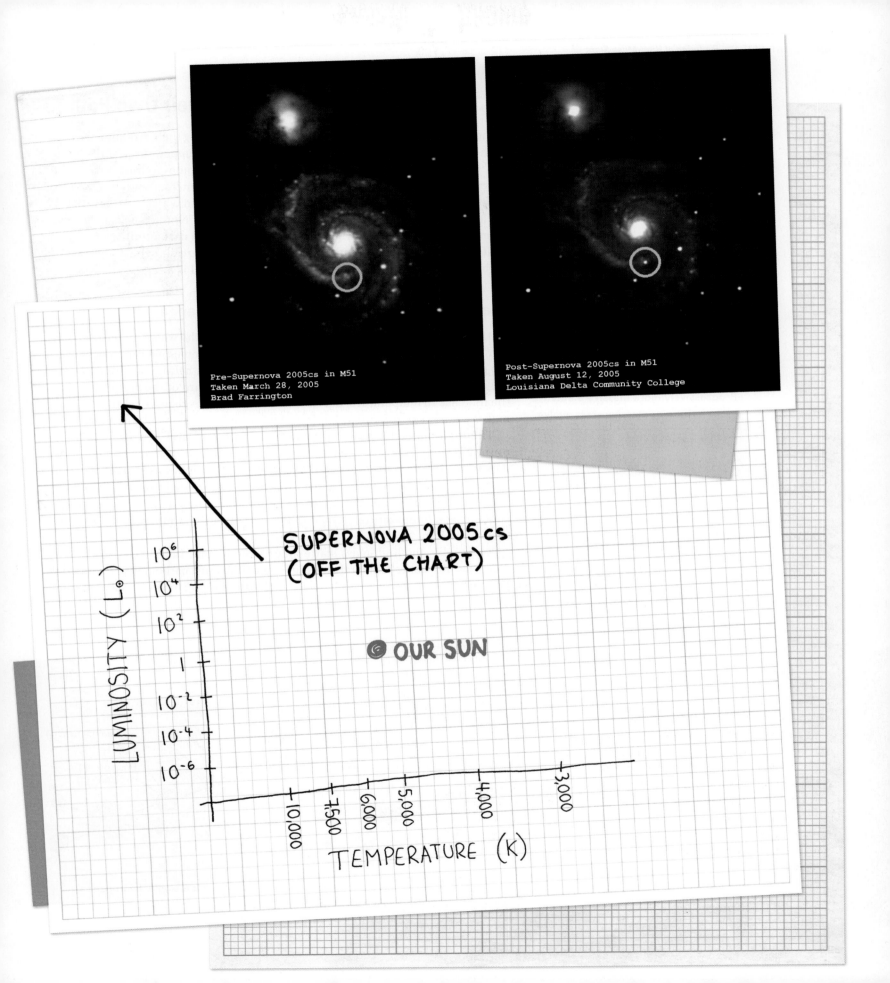

Pre-Supernova 2005cs in M51
Taken March 28, 2005
Brad Farrington

Post-Supernova 2005cs in M51
Taken August 12, 2005
Louisiana Delta Community College

SUPERNOVA 2005cs
(OFF THE CHART)

OUR SUN

LUMINOSITY ($L_\odot$)

10^6
10^4
10^2
1
10^{-2}
10^{-4}
10^{-6}

10,000
7,500
6,000
5,000
4,000
3,000

TEMPERATURE (K)

13.1 High-Mass Stars Follow Their Own Path

In the previous chapter, we followed low-mass stars along their evolutionary path. High-mass stars evolve differently. At the very high temperatures at the center of a high-mass star, nuclear reactions beyond the proton-proton chain occur. In some high-mass stars, hydrogen nuclei are able to interact with the nuclei of more massive elements such as carbon. This reaction, in which a carbon-12 (^{12}C) nucleus and a proton combine to form a nitrogen-13 (^{13}N) nucleus consisting of seven protons and six neutrons, is the first step in the **carbon-nitrogen-oxygen (CNO) cycle**. The remaining steps in the CNO cycle are shown in **Figure 13.1**. Carbon is not consumed by the CNO cycle, but instead is a catalyst.

The High-Mass Star Leaves the Main Sequence

As the high-mass star runs out of hydrogen in its core, the weight of the overlying star compresses the core. Yet long before the core becomes electron-degenerate, the pressure and temperature become high enough for helium burning. Because the core is not made of degenerate matter, the structure of the star responds to the increase in temperature, but its luminosity changes relatively little. The star makes a fairly smooth transition from hydrogen burning to helium burning.

The high-mass star now burns helium in its core and hydrogen in a surrounding shell, like a low-mass horizontal branch star. It grows in size while its surface temperature falls, so it moves off to the right on the H-R diagram, leaving the main sequence (**Figure 13.2**). Stars with more than 10 solar masses become red supergiants during their helium-burning phase. They have very cool surface temperatures (about 4000 K) and radii as much as 1,500 times that of the Sun.

Eventually, the high-mass star exhausts the helium in its core. As the core collapses, it reaches temperatures high enough to burn carbon. Carbon burning produces even more massive elements, including sodium, neon, and magnesium.

Progressively more massive elements burn as massive stars evolve.

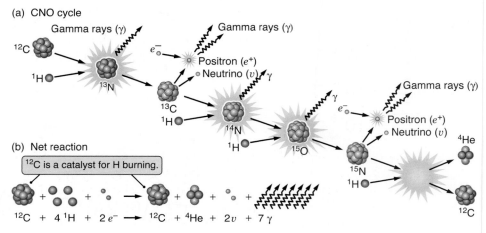

(a) CNO cycle

(b) Net reaction

^{12}C is a catalyst for H burning.

$$^{12}\text{C} + 4\,^{1}\text{H} + 2\,e^- \longrightarrow\ ^{12}\text{C} + {}^{4}\text{He} + 2v + 7\gamma$$

FIGURE 13.1 In high-mass stars, carbon serves as a catalyst for fusion of hydrogen to helium. This process is called the carbon-nitrogen-oxygen (or CNO) cycle.

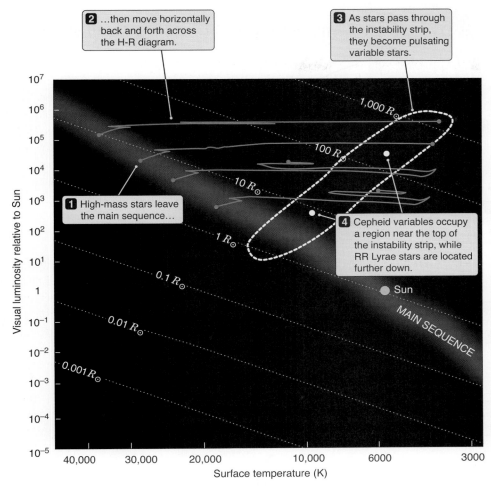

FIGURE 13.2 When massive stars leave the main sequence, they move horizontally across the H-R diagram.

The star now has a carbon-burning core surrounded by a helium-burning shell, surrounded by a hydrogen-burning shell. When carbon is exhausted, neon burning begins; and when neon is exhausted, oxygen begins to burn. The structure of the evolving high-mass star, shown in **Figure 13.3**, is like an onion; it has many concentric layers.

Not All Stars Are Stable

As a star evolves, it may make one or more passes through a region of the H-R diagram known as the **instability strip** (see Figure 13.2). Rather than finding a steady balance, stars in the instability strip pulsate, alternately growing larger and smaller. Such stars are called **pulsating variable stars**.

Thermal energy powers the pulsations. The star alternately traps and releases thermal energy, causing it to expand and contract. At each change, the star overshoots the equilibrium point where forces due to pressure and gravity

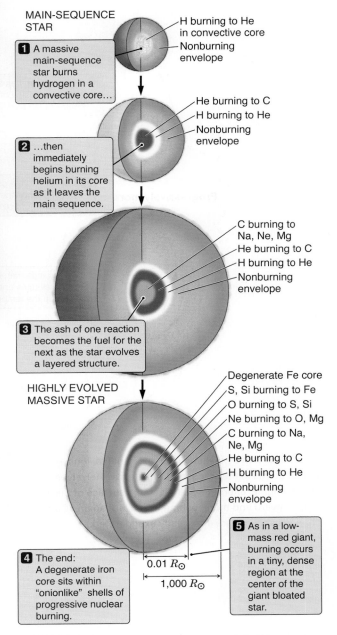

FIGURE 13.3 As a high-mass star evolves, it builds up a layered structure like that of an onion; progressively advanced stages of nuclear burning are found deeper and deeper within the star. Note the change in scale for the bottom image.

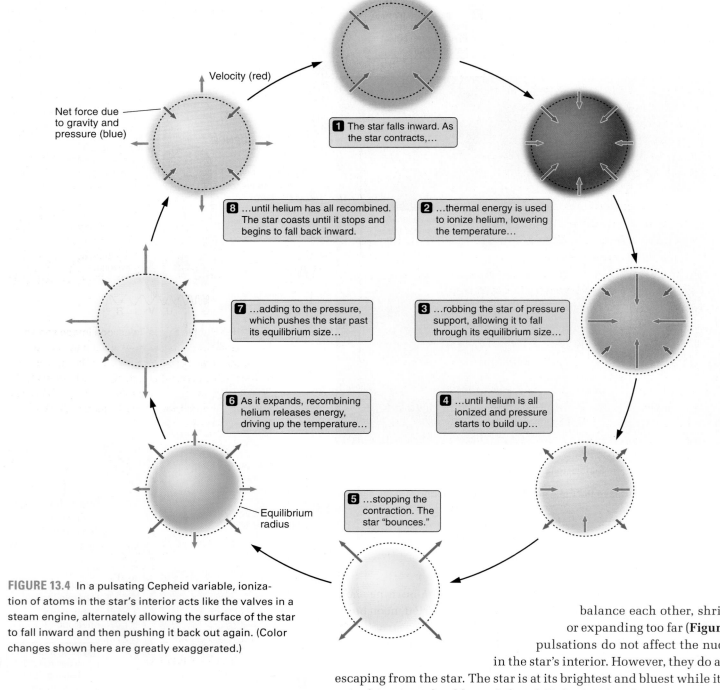

Velocity (red)

Net force due to gravity and pressure (blue)

1 The star falls inward. As the star contracts,…

2 …thermal energy is used to ionize helium, lowering the temperature…

8 …until helium has all recombined. The star coasts until it stops and begins to fall back inward.

7 …adding to the pressure, which pushes the star past its equilibrium size…

3 …robbing the star of pressure support, allowing it to fall through its equilibrium size…

6 As it expands, recombining helium releases energy, driving up the temperature…

4 …until helium is all ionized and pressure starts to build up…

5 …stopping the contraction. The star "bounces."

Equilibrium radius

FIGURE 13.4 In a pulsating Cepheid variable, ionization of atoms in the star's interior acts like the valves in a steam engine, alternately allowing the surface of the star to fall inward and then pushing it back out again. (Color changes shown here are greatly exaggerated.)

balance each other, shrinking too far or expanding too far (**Figure 13.4**). These pulsations do not affect the nuclear burning in the star's interior. However, they do affect the light escaping from the star. The star is at its brightest and bluest while it expands and at its faintest and reddest while it falls back inward.

The most luminous pulsating variable stars are the **Cepheid variables**, named after the prototype star Delta Cephei. Cepheid variables have periods ranging from 1 to 100 days, and longer-period Cepheids are more luminous. This **period-luminosity relationship** for Cepheid variables follows a consistent pattern, allowing us to use them to find the distances to galaxies beyond our own.

The instability strip intersects the low-mass horizontal branch as well. Unstable low-mass horizontal branch stars are known as **RR Lyrae variables** after their prototype star in the constellation Lyra. They are much less luminous than Cepheid variables, but they can also be used as **standard candles**. The

Vocabulary Alert

Candle: In everyday usage, a candle is simply a piece of wax containing a wick. When astronomers say that a star is a standard candle, they mean that its luminosity can be independently determined because it is a specific type of star. Comparing its apparent brightness to its intrinsic luminosity gives its distance, as we will see in Chapter 14.

instability strip also intersects the main sequence around spectral type A, and many A stars do show variability. A number of other kinds of variable stars are seen elsewhere in the H-R diagram, each driven by its own type of instability.

The pressure of the intense radiation on gas at the surface of a massive star overcomes the star's gravity and causes winds with speeds of about 3,000 kilometers per second (km/s) and mass loss ranging from about 10^{-7} to 10^{-5} $M_\odot$ per year. These numbers may sound tiny; but over millions of years, this rate of mass loss becomes significant. O stars with masses of 20 $M_\odot$ lose about 20 percent of their mass while on the main sequence, and possibly more than 50 percent over their entire lifetimes. Eta Carinae (**Figure 13.5**) is an extreme example, a star more than 100 times as massive as the Sun and as luminous as 5 million Suns. Eta Carinae is currently losing 1 $M_\odot$ every 1,000 years. However, during a 19th-century eruption when Eta Carinae became the second-brightest star in the sky, it shed 2 $M_\odot$ in only 20 years. Eta Carinae is expected to explode as a supernova eventually.

G X U V I R

FIGURE 13.5 This Hubble Space Telescope image shows an expanding cloud of dusty material ejected by the luminous blue variable star Eta Carinae. The star itself, which is largely hidden by the surrounding dust, has a luminosity 5 million times that of the Sun and a mass probably in excess of 100 $M_\odot$. Dust is created when volatile material ejected from the star condenses.

13.2 High-Mass Stars Go Out with a Bang

For a high-mass star, the end comes suddenly and amid considerable fury. An evolving high-mass star builds up its onionlike structure (see Figure 13.3) as hydrogen burns to helium, helium burns to carbon, carbon burns to sodium, neon, and magnesium, oxygen burns to sulfur and silicon, and then silicon and sulfur burn to iron. Many different types of nuclear reactions occur up to this point, forming almost all of the different stable isotopes of elements less massive than iron. But *the chain of nuclear fusion stops with iron.*

Why does the chain of nuclear fusion stop with iron? Think about gasoline, which burns because energy is released when the fuel chemically combines with oxygen. This increases the temperature, which speeds up the chemical reaction. Once it starts, the reaction is self-sustaining. The same is true of nuclear fusion reactions in the interiors of stars. When hydrogen is fused into helium, energy is released, maintaining the temperature necessary to keep the reaction going.

The energy bound up in an atomic nucleus is called the **binding energy**. A nuclear reaction that increases an atom's binding energy releases energy. Conversely, decreasing the binding energy absorbs energy. **Figure 13.6** shows the binding energy per nucleon (that is, per each proton or neutron in the nucleus) for different atomic nuclei. Moving up on the plot from helium to carbon increases the binding energy, so fusing helium to carbon releases energy. Iron is at the peak of the binding energy curve, so moving up from lighter elements to iron releases energy; conversely, moving from iron down to heavier elements *absorbs* energy. Iron fusion is not self-sustaining because it absorbs energy in the reaction.

The Final Days in the Life of a Massive Star

You can think of maintaining the balance in a star as somewhat like trying to keep a leaky ball inflated. The larger the leak, the more rapidly you have to pump air into the ball. A star that is burning hydrogen or helium is like a ball with a slow

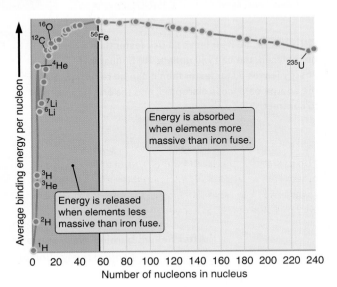

FIGURE 13.6 The binding energy per nucleon is plotted against the number of nucleons for each element. This is the energy it would take to break the atomic nucleus apart into protons and neutrons. Energy is released by nuclear fusion only if the resulting element is higher on the curve. The periodic table of elements in Appendix 1 identifies individual elements by their atomic number (the number of protons).

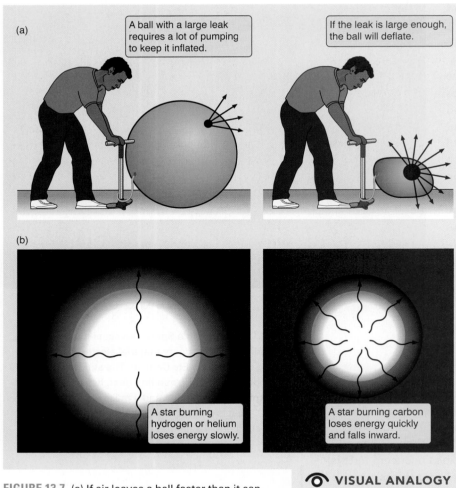

FIGURE 13.7 (a) If air leaves a ball faster than it can be replaced, the ball will deflate. (b) Similarly, if energy leaves a star faster than it can be replaced, the energy balance is disrupted, and the star begins to contract.

○ VISUAL ANALOGY

After helium burning, each stage is progressively shorter.

leak (**Figure 13.7**). At the temperatures of hydrogen or helium burning, energy leaks out of the interior of a star primarily by radiation and convection. Neither of these processes is very efficient, because the outer layers of the star act like a thick, warm blanket. Much of the energy is kept in the star, so nuclear fuels need to burn at only a relatively modest rate to support the star against gravity.

Once carbon burning begins, this balance shifts. Energy is carried primarily by neutrinos rather than radiation and convection. Neutrinos escape easily, carrying energy away from the core. The outer layers of the star fall inward, drive up the star's density and temperature, and force nuclear reactions to run faster.

Once this process of **neutrino cooling** becomes significant, the star evolves much more rapidly. Carbon burning supports the star for about a thousand years. Oxygen burning lasts about a year. Silicon burning lasts only a few days. A silicon-burning star is not much more luminous now than it was while burning helium. But because of neutrino cooling, this silicon-burning star is actually giving off about 200 million times more energy per second than its former self.

The Core Collapses and the Star Explodes

Following silicon burning, we come to the end of the line. Once a star forms its iron core, no source of nuclear energy remains to replenish the energy that is being taken away by escaping neutrinos. The star's life as a balancing act between gravity and energy production is over. No longer supported by thermonuclear fusion, the iron core of the massive star begins to collapse (**Figure 13.8**).

As the core collapses, the force of gravity increases and the density and temperature skyrocket. The core becomes electron-degenerate when it is about the size of Earth. The weight bearing down on the iron ash core is too great to be held up by electron degeneracy pressure. The collapse continues. The core reaches temperatures of more than 10 billion K, while the density exceeds 10^{10} kilograms per cubic meter (kg/m^3)—10 times the density of a white dwarf.

These conditions trigger fundamental changes in the core. At these temperatures, the nucleus of the star is filled with thermal radiation so energetic that photons can break iron nuclei apart. This process, called photodisintegration, absorbs thermal energy and begins reversing the results of nuclear fusion. At the same time, the density of the core is so great that electrons are forced into atomic nuclei, where they combine with protons to produce neutrons. The two processes together absorb much of the energy that was holding up the dying star. Neutrinos continue to pour out of the core of the dying star, taking still more energy with them. The collapse of the core accelerates, reaching a speed of 70,000 km/s, or almost one-fourth the speed of light, on its inward fall. These events take place in less than a second.

As material in the collapsing core exceeds the density of an atomic nucleus, the strong nuclear force actually becomes repulsive. About half of the collapsing core suddenly slows its inward fall. The remaining half slams into the innermost part of the star at a significant fraction of the speed of light and "bounces," sending a tremendous shock wave back out through the star.

Over the next second or so, almost 20 percent of the material in the core is converted into neutrinos. Most fly outward through the star; but at these phenomenal densities, not even neutrinos pass with complete freedom. The dense material behind the expanding shock wave traps a few tenths of a percent of the neutrinos.

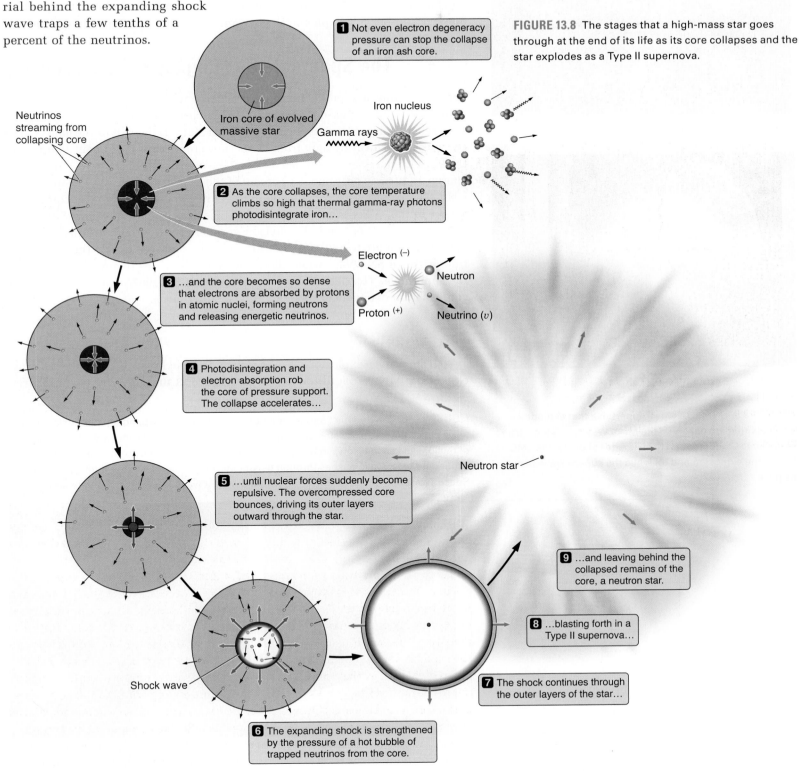

FIGURE 13.8 The stages that a high-mass star goes through at the end of its life as its core collapses and the star explodes as a Type II supernova.

Neutrinos streaming from collapsing core

Iron core of evolved massive star

Iron nucleus

Gamma rays

1 Not even electron degeneracy pressure can stop the collapse of an iron ash core.

2 As the core collapses, the core temperature climbs so high that thermal gamma-ray photons photodisintegrate iron…

Electron (−)

Neutron

Proton (+)

Neutrino (v)

3 …and the core becomes so dense that electrons are absorbed by protons in atomic nuclei, forming neutrons and releasing energetic neutrinos.

4 Photodisintegration and electron absorption rob the core of pressure support. The collapse accelerates…

5 …until nuclear forces suddenly become repulsive. The overcompressed core bounces, driving its outer layers outward through the star.

Neutron star

9 …and leaving behind the collapsed remains of the core, a neutron star.

8 …blasting forth in a Type II supernova…

Shock wave

7 The shock continues through the outer layers of the star…

6 The expanding shock is strengthened by the pressure of a hot bubble of trapped neutrinos from the core.

The energy of these trapped neutrinos drives the pressure and temperature in this region higher, inflating a bubble of extremely hot gas and intense radiation around the core of the star. The pressure of this bubble adds to the shock wave moving outward through the star. Within about a minute, the shock wave has pushed its way out through the helium shell within the star. Within a few hours, it reaches the surface of the star itself, heating the stellar surface to 500,000 K and blasting material outward at speeds of up to about 30,000 km/s. Our evolved massive star has just exploded in an event referred to as a **Type II supernova**.

13.3 The Spectacle and Legacy of Supernovae

For a brief time, a Type II supernova can shine with the light of a billion suns. Look back at the chapter-opening figure, in which a supernova shines very brightly even when compared to the dense groups of stars in the rest of the galaxy. The energy carried away by light from a supernova represents only about 1 percent of the kinetic energy being carried away by the outer parts of the star. This kinetic energy, in turn, is only about 1 percent of the energy carried away by neutrinos.

In 1987, a massive star exploded in the Large Magellanic Cloud (a companion galaxy to the Milky Way). Even at a distance of 160,000 light-years, Supernova 1987A was so bright that it dazzled sky gazers in the Southern Hemisphere (**Figure 13.9**). Scientists had already unknowingly captured one of the true scientific prizes of SN 1987A when their neutrino telescopes recorded a burst from the supernova. The detection of neutrinos from SN 1987A was a fundamental confirmation of our theories about Type II supernovae.

The Energetic and Chemical Legacy of Supernovae

Type II supernova explosions leave a rich and varied legacy to the universe. Huge expanding bubbles of million-kelvin gas (**Figure 13.10a**) glow in X-rays and drive

FIGURE 13.9 SN 1987A was a supernova that exploded in a small companion galaxy of the Milky Way called the Large Magellanic Cloud (LMC). These images show (a) the LMC before the explosion and (b) while the supernova was near its peak.

FIGURE 13.10 The Cygnus Loop is a supernova remnant—an expanding interstellar blast wave caused by the explosion of a massive star. (a) Gas in the interior with a temperature of millions of kelvins glows in X-rays, while (b) visible light comes from locations where the expanding blast wave pushes through denser gas in the interstellar medium. (c) A Hubble Space Telescope image of a location where the blast wave is hitting an interstellar cloud.

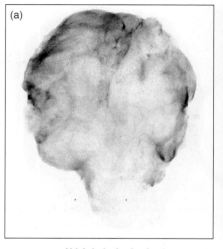

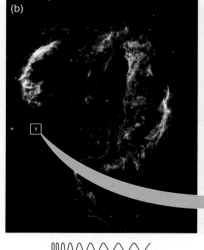

visible shock waves (**Figure 13.10b**) into the surrounding interstellar medium (the dust and gas between the stars). These bubbles are the still-powerful blast waves of supernova explosions that took place thousands of years ago. Supernova explosions compress nearby clouds (**Figure 13.10c**), triggering the initial collapse that begins star formation.

Only the least massive chemical elements were formed at the beginning of the universe: hydrogen, helium, and trace amounts of lithium, beryllium, and boron. All of the rest of the chemical elements, including a large fraction of the atoms we are made of, were formed in the stars through nuclear reactions and then returned to the interstellar medium. Low-mass stars form elements as massive as carbon and oxygen; high-mass stars produce elements as massive as iron. Yet this is not the whole story. Many naturally occurring elements are more massive than iron. If iron is the most massive element that can be formed in stars, then where do these even more massive elements come from?

Under normal circumstances, electric repulsion keeps positively charged atomic nuclei far apart. Extreme temperatures are needed to slam nuclei together hard enough to overcome this electric repulsion. Free neutrons, however, have no net electric charge, so there is no electric repulsion to prevent them from simply running into an atomic nucleus. Under normal conditions, free neutrons are rare. Under the conditions of a Type II supernova, however, free neutrons are produced in very large numbers. These are easily captured by massive atomic nuclei and later decay to become protons, forming elements more massive than iron.

Nuclear physics predicts the abundances of the elements, which can be measured on Earth, in meteorite material, and in the atmospheres of stars (**Figure 13.11**). Less massive elements are far more abundant than more massive elements, because more massive elements are progressively built up from less massive elements. An exception to this pattern is the dip in the abundances of the light elements lithium (Li), beryllium (Be), and boron (B). Nuclear burning easily destroys these elements, and they are not produced by the main reactions involved in burning hydrogen (H) and helium (He). Conversely, carbon (C), nitrogen (N),

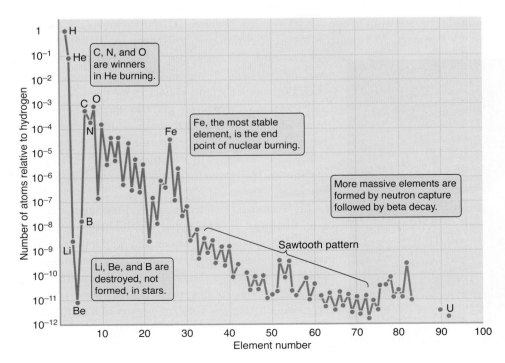

FIGURE 13.11 The relative abundances of different elements on Earth are plotted against the element number of the nucleus. This pattern is a result of the production of atomic nuclei in stars.

and oxygen (O) are big winners in the CNO cycle of hydrogen burning and the triple-alpha process of helium burning, and their high abundances reflect this fact. The spike in the abundances of elements near iron is evidence of processes that favor these tightly bound nuclei. Even the sawtooth pattern in the abundances of even- and odd-numbered elements can be understood as a consequence of the formation of atomic nuclei in stars.

Our understanding of the interiors of dying stars is confirmed by the chemical composition of the Earth under our feet. In addition, we can identify the kinds of stars that formed the atoms that make up our own bodies. Our growing understanding of the chemical evolution of the universe and our connection to it is one of the triumphs of modern astronomy.

Neutron Stars and Pulsars

Let's now return to the remains of the star. The core has collapsed to the point where it has about the same density as the nucleus of an atom. For cores less than about 3 $M_\odot$, quantum mechanical effects halt this collapse. However, instead of operating on electrons (as in white dwarfs), these effects now force *neutrons* together as tightly as the rules of quantum mechanics allow. This neutron-degenerate core is called a **neutron star**. It is roughly the size of a small city, with a radius of about 10 km. That volume is packed with a mass between 1.4 $M_\odot$ and 3 $M_\odot$. At a density of about 10^{18} kg/m^3, the neutron star is a billion times denser than a white dwarf and a thousand trillion (10^{15}) times denser than water.

The Type II supernova leaves behind a neutron-degenerate core.

If the original massive star was part of a close binary system, then the neutron star will have a companion, much like the white dwarf binary systems in Chapter 12. As the lower-mass star evolves and overfills its Roche lobe, matter falls toward the accretion disk around the neutron star, heating it to millions of kelvins and causing it to glow brightly in X-rays. This is an **X-ray binary** (**Figure 13.12**). X-ray binaries sometimes develop powerful jets, perpendicular to the accretion disk, that carry material away at speeds close to the speed of light.

X-ray binaries arise from the accretion of mass onto neutron stars.

As the core of the original massive star collapses, it spins faster, because angular momentum is conserved. A main-sequence O star rotates perhaps once every few days. As a neutron star, it might rotate tens or even hundreds of times each second. The collapsing star also concentrates its magnetic field to strengths trillions of times greater than the magnetic field at Earth's surface. A neutron star has a magnetosphere just like Earth and several other planets do, except that the neutron star's magnetosphere is vastly stronger and is whipped around many times a second by the spinning star. As in planets, in stars the magnetic axis is often not aligned with the rotation axis.

Electrons and positrons move along the magnetic-field lines and are "funneled" by the field toward the magnetic poles of the system. The particles produce radiation, which is beamed away from the magnetic poles of the neutron star as shown in **Figure 13.13**. As the neutron star rotates, these beams sweep through space much like the rotating beams of a lighthouse. The neutron star appears to flash on and off with a regular period equal to the period of rotation of the

FIGURE 13.12 X-ray binaries are systems consisting of a normal evolving star and a white dwarf, neutron star, or black hole. As the evolving star overflows its Roche lobe, mass falls toward the collapsed object. The gravitational well of the collapsed object is so deep that when the material hits the accretion disk, it is heated to such high temperatures that it radiates away most of its energy as X-rays.

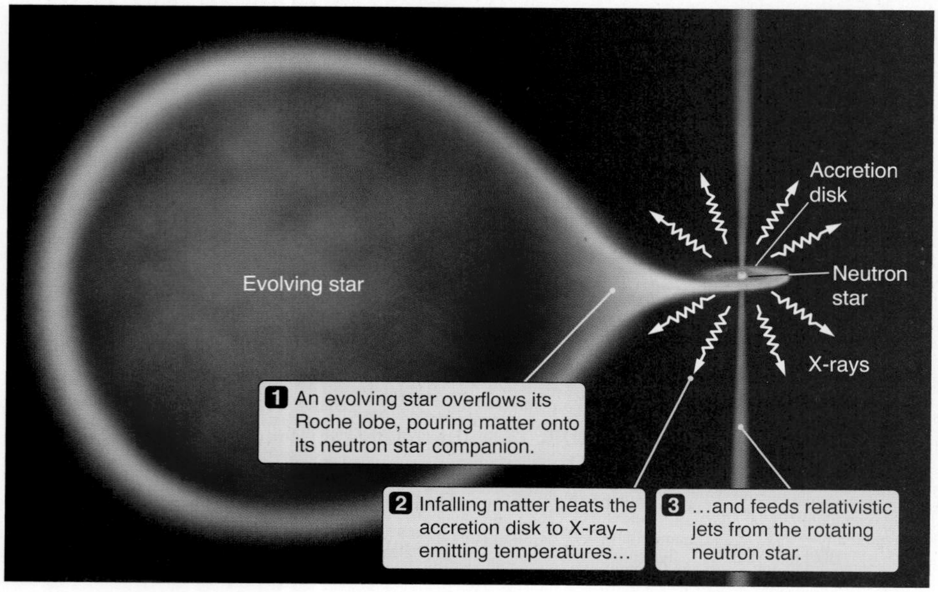

Accretion disk

Evolving star

Neutron star

X-rays

1 An evolving star overflows its Roche lobe, pouring matter onto its neutron star companion.

2 Infalling matter heats the accretion disk to X-ray–emitting temperatures…

3 …and feeds relativistic jets from the rotating neutron star.

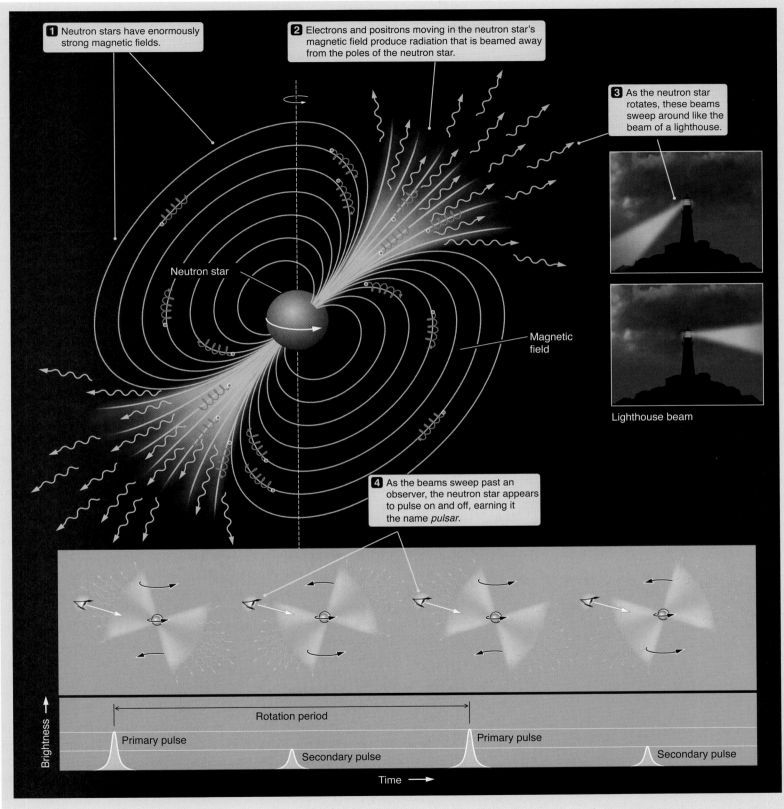

1 Neutron stars have enormously strong magnetic fields.

2 Electrons and positrons moving in the neutron star's magnetic field produce radiation that is beamed away from the poles of the neutron star.

3 As the neutron star rotates, these beams sweep around like the beam of a lighthouse.

Neutron star

Magnetic field

Lighthouse beam

4 As the beams sweep past an observer, the neutron star appears to pulse on and off, earning it the name *pulsar*.

Rotation period

Primary pulse

Secondary pulse

Primary pulse

Secondary pulse

Brightness

Time

VISUAL ANALOGY

FIGURE 13.13 As a highly magnetized neutron star rotates rapidly, light is given off, much like the beams from a rotating lighthouse lamp. From our perspective, as these beams sweep past us the star appears to pulse on and off, earning it the name *pulsar*.

star (or half the rotation period, if we see both beams). These objects are known as **pulsars**. As of this writing, more than two thousand pulsars are known, and more are being discovered all the time.

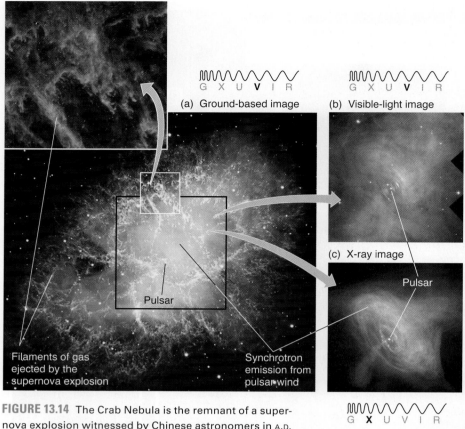

(a) Ground-based image

(b) Visible-light image

(c) X-ray image

Pulsar

Filaments of gas ejected by the supernova explosion

Synchrotron emission from pulsar wind

Pulsar

FIGURE 13.14 The Crab Nebula is the remnant of a supernova explosion witnessed by Chinese astronomers in A.D. 1054. (a) The object we see today is an expanding cloud of "shrapnel" from that earlier cataclysm. The spinning pulsar at the heart of the Crab Nebula sends off a "wind" of electrons and positrons moving at close to the speed of light. The synchrotron radiation from these particles is shown (b) in visible light and (c) in X-rays.

The Crab Nebula—Remains of a Stellar Cataclysm

In A.D. 1054, Chinese astronomers noticed a "guest star" in the direction of the constellation Taurus. The new star was so bright that it could be seen during the daytime for 3 weeks, and it did not fade away altogether for many months. This star was a fairly typical Type II supernova. Today, an expanding cloud of debris from this explosion occupies this place in the sky—an extraordinary object called the Crab Nebula (**Figure 13.14**).

The Crab Nebula has filaments of glowing gas expanding away from the central star at 1,500 km/s. These filaments contain anomalously high abundances of helium and other more massive chemical elements—the products of the nuclear reactions that took place in the supernova and its progenitor star.

The Crab pulsar at the center of the nebula flashes 60 times a second: first with a main pulse associated with one of the "lighthouse" beams; then with a fainter secondary pulse associated with the other beam. As the Crab pulsar spins 30 times a second, it whips its powerful magnetosphere around with it. At a distance from the pulsar about equal to the radius of the Moon, material in the magnetosphere must move at almost the speed of light to keep up with this rotation. Like a tremendous slingshot, the rotating pulsar magnetosphere flings particles away from the neutron star in a wind moving at nearly the speed of light. This wind fills the space between the pulsar and the expanding shell. The Crab Nebula is almost like a big balloon; but instead of being filled with hot air, it is filled with a mix of very fast particles and strong magnetic fields. Images of the Crab Nebula such as Figure 13.14b and c show this bubble as an eerie glow from synchrotron radiation released as the particles spiral around the magnetic field.

13.4 Beyond Newtonian Physics

Just as there is a Chandrasekhar limit for white dwarfs, there is an analogous limit for neutron stars. If the mass of a neutron star exceeds about 3 $M_\odot$, then gravity will be stronger than the neutron degeneracy pressure. The neutron star shrinks, and gravity increases at an ever-accelerating pace. Eventually, gravity is so strong that the escape velocity exceeds the speed of light. From this point on, nothing can escape from the collapsing object. The object is now a **black hole**. Black holes are so strange, so far from our common understanding of reality, that the laws of Newtonian physics (which we explored in Chapter 3) are inadequate

If a neutron star's mass exceeds about 3 $M_\odot$, it will collapse to a black hole.

to describe them. To do so, we must first take a step back and—as Einstein did—question our unexamined assumptions about the very nature of space and time.

The Speed of Light Is a Very Special Value

Imagine that you are sitting in a car and you throw a ball out the window at 25 miles per hour (mph), as measured in your reference frame. But if the car is moving at 50 mph down the highway, someone standing by the road will say that the ball is moving at 75 mph (25 mph from throwing it plus 50 mph from the motion of the car). To someone in oncoming traffic moving at 50 mph, the speed of the ball would be 125 mph (**Figure 13.15a**). There really is no difference among these three perspectives. The laws of physics are the same in *any* inertial reference frame.

During the closing years of the 19th century and the early years of the 20th century, physicists were conducting laboratory experiments in an attempt to better understand light. What they found puzzled them greatly. Rather than the

Experiments show that the speed of light is the same for all observers.

FIGURE 13.15 (a) The rules of motion that apply in our daily lives break down (b) when speeds approach the speed of light. The fact that light itself always travels at the same speed for any observer is the basis of special relativity. (Note that relativity also affects the relative speeds of the two spacecraft.)

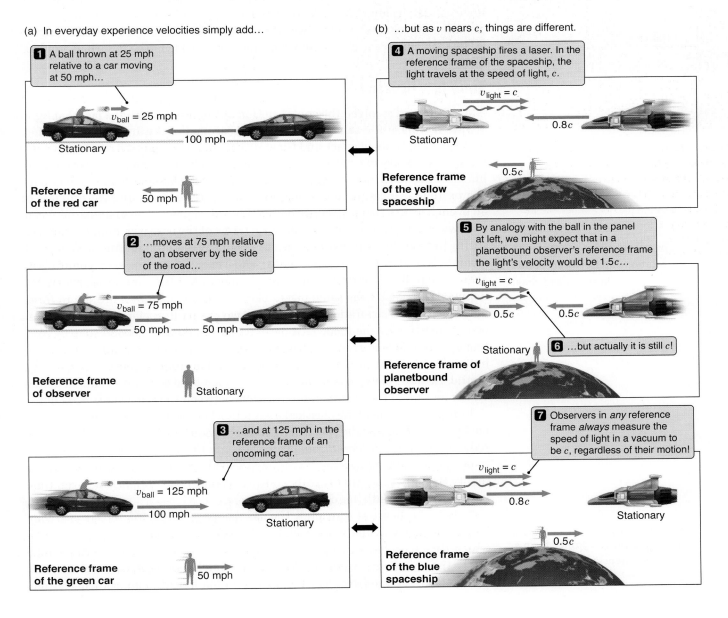

(a) In everyday experience velocities simply add…

1 A ball thrown at 25 mph relative to a car moving at 50 mph…

v_{ball} = 25 mph
100 mph
Stationary
Reference frame of the red car
50 mph

2 …moves at 75 mph relative to an observer by the side of the road…

v_{ball} = 75 mph
50 mph   50 mph
Reference frame of observer
Stationary

3 …and at 125 mph in the reference frame of an oncoming car.

v_{ball} = 125 mph
100 mph
Stationary
Reference frame of the green car
50 mph

(b) …but as v nears c, things are different.

4 A moving spaceship fires a laser. In the reference frame of the spaceship, the light travels at the speed of light, c.

$v_{light} = c$
0.8c
Stationary
0.5c
Reference frame of the yellow spaceship

5 By analogy with the ball in the panel at left, we might expect that in a planetbound observer's reference frame the light's velocity would be 1.5c…

$v_{light} = c$
0.5c 0.5c
Stationary   **6** …but actually it is still c!
Reference frame of planetbound observer

7 Observers in *any* reference frame *always* measure the speed of light in a vacuum to be c, regardless of their motion!

$v_{light} = c$
0.8c
Stationary
Reference frame of the blue spaceship
0.5c

speed of a beam of light differing from one observer to the next, as expected on the basis of Newton's physics and "common sense," they found instead that *all observers measure exactly the same value for the speed of a beam of light, regardless of their motion!*

Imagine that you are riding in a spaceship (**Figure 13.15b**) and you shine a beam of light forward. You measure the speed of the beam of light to be *c*, or 3×10^8 meters per second (m/s). That is as expected because you are holding the source of the light. But the observer on a planet you are passing *also* measures the speed of the passing beam of light to be 3×10^8 m/s. Even a passenger in an oncoming spacecraft finds that the beam from your light is traveling at exactly *c* in her own reference frame. *Every observer always finds that light in a vacuum travels at exactly the same speed,* c, *regardless of his or her own motion or the motion of the light source.*

If this discussion bothers you, imagine the reaction of those physicists. Newton's laws of motion had been the bedrock of science for 200 years, standing up to every experimental challenge that came their way. Suddenly that bedrock seemed to turn to sand. How could light have the same speed for *all* observers, regardless of their own velocity? Yet that was the inescapable experimental result. Despite all its spectacular successes, Newtonian physics seemed to be in serious trouble.

Time Is a Relative Thing

Albert Einstein resolved the contradiction by beginning with the fact that light always travels at the same speed, and he reasoned backward to find out what that must imply about space and time. He ushered in a scientific revolution with his **special theory of relativity**, which was published in 1905. Einstein focused his thinking on pairs of *events*. In relativity, an **event** is something that happens at a particular location in space at a particular time. Snapping your fingers is an event because that action has both a time and a place. The distance between any two events depends on the reference frame of the observer. Suppose you are sitting in a car that is traveling down the highway in a straight line at a constant 60 mph. You snap your fingers (event 1), and a minute later you snap your fingers again (event 2). In your reference frame *you* are stationary, and the two events happened at exactly the same place. They were separated by a minute in *time*, but there was no separation between the two events in *space*. Now imagine an observer sitting by the road. This observer agrees that the second snap of your fingers (event 2) occurred a minute after the first snap of your fingers (event 1), but to this observer the two events were separated from each other in space by a mile. In this "Newtonian" view, the *distance* between two events depends on the motion of the observer, but the *time* between the two events does not.

> Special relativity concerns the relationship between events in space and time.

Einstein questioned the classical distinction between space and time. He realized that the *only* way the speed of light can be the same for all observers is if *the passage of time is different from one observer to the next*. To Newton, and to us in our everyday lives, the march of time seems immutable and constant. But in reality the only thing that is truly constant is the speed of light, and even time itself flows differently for different observers.

> Space and time together form a four-dimensional spacetime.

This is the heart of Einstein's special theory of relativity. In our everyday Newtonian view of the world, we live in a three-dimensional space through which time marches steadily onward. Events occur in space at a certain time. By the time Einstein finished working out the implications of his insight, he had reshaped this three-dimensional universe into a four-dimensional **spacetime**.

Events occur at specific locations within this four-dimensional spacetime, but how this spacetime translates into what we perceive as "space" and what we perceive as "time" depends on our reference frame.

Einstein did not "disprove" Newtonian physics. We were not wasting our time in Chapter 3 when we studied Newton's laws of motion. Instead, Newtonian physics is *contained within* special relativity. In our everyday experience, we never encounter speeds approaching that of light. Even the fastest object ever made by humans (the *Helios* spacecraft) traveled at only about $0.00023c$. At speeds much less than the speed of light, Einstein's equations become the same equations that describe Newtonian physics. In our everyday lives, we experience a Newtonian world. Only when relative velocities approach the speed of light do things begin to depart noticeably from the predictions of Newtonian physics. When great velocities cause an effect different from what Newtonian physics predicts, we call this a **relativistic** effect.

Newtonian physics is contained within special relativity.

The Implications of Relativity Are Far-Ranging

Today, special relativity is an integral and indispensable part of all of physics, shaping our thinking about the motions of both the tiniest subatomic particles and the most distant galaxies. Here, we explore only a few of the essential insights that come from Einstein's work:

1. **What we think of as "mass" and what we think of as "energy" are actually two manifestations of the same thing.** Einstein's famous equation $E = mc^2$ says that even a *stationary* object has an intrinsic "rest" energy that equals the mass (m) of the object multiplied by the speed of light (c) squared. The speed of light is a very large number. This relationship between mass and energy says that a single tablespoon of water has a rest energy equal to the energy released in the explosion of over 300,000 tons of TNT.

2. **The speed of light is the ultimate speed limit.** All the energy in the entire universe is inadequate to accelerate a single electron to the speed of light. While getting the electron arbitrarily close to that number—$0.99999999999999999\ldots \times c$ is no problem, at least in principle—there is no getting over the hump.

3. **Time passes more slowly in a moving reference frame.** This phenomenon is referred to as **time dilation**, meaning that time is "stretched out" in the moving reference frame (**Working It Out 13.1**). Were you to compare clocks with an observer moving at 9/10 the speed of light ($0.9c$), you would find that the other observer's clock was running 0.44 times as fast as your clock. You might guess that to the other observer, your clock would be fast; but actually the other observer would find instead that *your* clock was running slow. To you, the other observer may be moving at $0.9c$; but to the other observer, *you* are moving. Either frame of reference is equally valid, so you would each find the other's clock to be slow.

A scientific observation demonstrates this effect. As illustrated in **Figure 13.16**, fast particles called cosmic-ray muons are produced 15 km up in Earth's atmosphere when high-energy cosmic rays strike atmospheric atoms or molecules. Muons at rest decay very rapidly into other particles. This happens so quickly that, even if they *could* move at the speed of light, virtually all muons would have decayed long before traveling the 15 km to reach Earth's surface. However, time dilation slows the muons' clocks, so the particles travel farther during their lifetime and reach the ground.

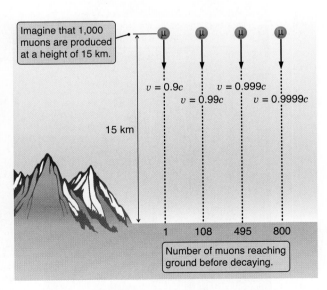

FIGURE 13.16 Muons created by cosmic rays high in Earth's atmosphere would decay long before reaching the ground if they were not traveling at nearly the speed of light. An observer on the ground sees a longer lifetime because of time dilation. Here we show what happens to 1,000 muons produced at an altitude of 15 km for a variety of speeds.

4. **"At the same time" is a relative concept.** Two events that occur at the *same* time for one observer may occur at *different* times for a different observer.

5. **An object is shorter in motion than it is at rest.** Moving objects are compressed in the direction of their motion, a phenomenon referred to as **length contraction**. A meter stick moving at 0.9c is only 0.44 meter long.

Working It Out 13.1 | The Boxcar Experiment

The special theory of relativity is a *very* counterintuitive idea, but it is so central to our modern understanding of the universe that it is worth wrestling with a bit. Consider the famous thought experiment known as the boxcar experiment. In this experiment, observer 1 is in a boxcar of a train moving to the right. Observer 1 has a lamp, a mirror, and a clock. Observer 2 is standing on the ground outside.

Figure 13.17a shows the experimental setup as seen by observer 1, who is stationary with respect to the clock. At time t_1, event 1 happens: the lamp gives off a pulse of light. The light bounces off a mirror a distance l meters away. At time t_2, event 2 happens: the light arrives at the clock. The time between events 1 and 2 is just the distance the light travels ($2l$ meters), divided by the speed of light: $t_2 - t_1 = 2l/c$.

Now let's look from the perspective of observer 2, standing on the ground outside the train (**Figure 13.17b**). In *this* observer's reference frame, he is stationary and the boxcar is moving at speed *v*. In observer 2's reference frame, the clock *moves* to the right between the two events, so the light has *farther to go*. (If you do not see this, use a ruler to measure the total length of the light path in Figure 13.17b and compare it with the total length of the light path in Figure 13.17a.) The time between the two events ($t_2 - t_1$) is longer than for observer 1, because the distance traveled is *longer* than $2l$ meters.

The two events are the *same two events*, regardless of the reference frame from which they are observed. But because the speed of light is the same for all observers, there *must* be more time between the two events when viewed by observer 2. The seconds of a moving clock are stretched. Moving clocks *must* run slow, and the passage of time *must* depend on an observer's reference frame.

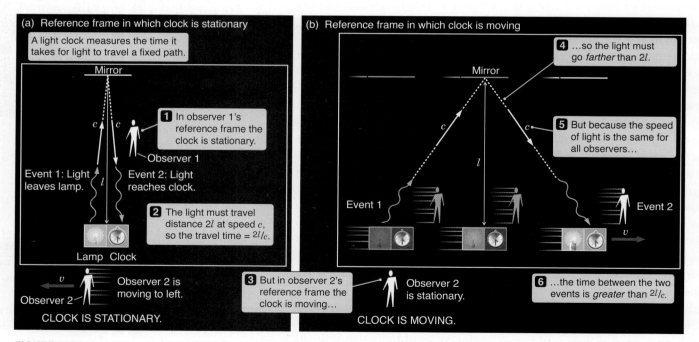

FIGURE 13.17 The "tick" of a light clock as seen in two different reference frames: (a) stationary and (b) moving. As Einstein's thought experiment demonstrates, if the speed of light is the same for every observer, then moving clocks *must* run slow.

13.5 Gravity Is a Distortion of Spacetime

Our exploration of special relativity began with the observation that the speed of light is always the same (regardless of the motion of an observer or the source) and ended by shattering our everyday notions of space and time. As we confront the properties of black holes, our concepts of space and time will be pulled even further from the comfortable absolutes of Newtonian physics. First we learned that space and time are actually just a result of our particular, limited perspective on a four-dimensional spacetime. Now we will discover that this four-dimensional spacetime is itself warped and distorted by the masses it contains. Such spacetime warping leads directly to the gravity that holds you to Earth. This realization, called the **general theory of relativity**, describes how mass distorts spacetime and is another of Einstein's great contributions to science.

A crucial clue to the fundamental connection between gravity and spacetime is that, left on their own, any two objects at the same location and moving with the same velocity will follow the same path through spacetime, regardless of their masses. A space station astronaut falls around Earth along with the space station itself. A feather dropped by an *Apollo* astronaut standing on the Moon falls at exactly the same rate as a dropped hammer. (This experiment was actually done, and videotaped, and can be found on the Internet.) Rather than thinking of gravity as a "force" that "acts on" objects, it is more accurate to think of it as a consequence of the warping of spacetime in the presence of a mass. *Gravitation is the result of the shape of spacetime that objects move through.*

Free Fall Is the Same as Free Float

The essence of special relativity is that any inertial reference frame is as good as any other. There is no way to distinguish between sitting in an enclosed spaceship floating stationary (with respect to the center of mass of the galaxy) in deep space (**Figure 13.18a**) and sitting in an enclosed spaceship traveling through our galaxy at 0.99999 times the speed of light (**Figure 13.18b**). These two cases do not *feel* any different, because there *is* no difference between them. Each is an equally valid inertial reference frame. As long as nothing accelerates either spaceship, the laws of physics are exactly the same inside both spacecraft.

General relativity begins by applying this same idea to an astronaut inside the space station orbiting Earth, as shown in **Figure 13.18c**. Our astronaut has no way to tell the difference between being inside the space station as it falls around Earth and being inside a spaceship coasting through interstellar space. If you close your eyes and jump off a diving board, for the brief time that you are falling freely through Earth's gravitational field, the sensation you feel is exactly the same as the sensation that you would feel floating in interstellar space. Even though its velocity is constantly changing as it falls, *the inside of a space station orbiting Earth is as good an inertial frame of reference as that of an object drifting along a straight line through interstellar space.* This principle—which can be simply stated as "free fall is the same as free float"—is called the **equivalence principle**.

The equivalence principle says that a falling object is simply following its "natural" path through spacetime, just like an object that drifts along a straight line at a constant speed through deep space. The natural path that an object will follow through spacetime in the absence of other forces is referred to as

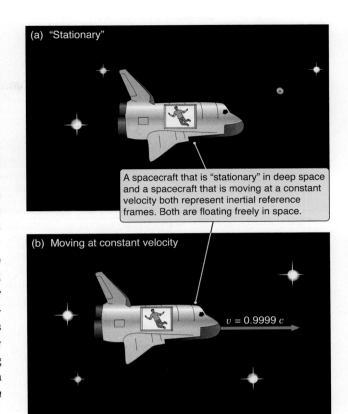

(a) "Stationary"

A spacecraft that is "stationary" in deep space and a spacecraft that is moving at a constant velocity both represent inertial reference frames. Both are floating freely in space.

(b) Moving at constant velocity

$v = 0.9999\,c$

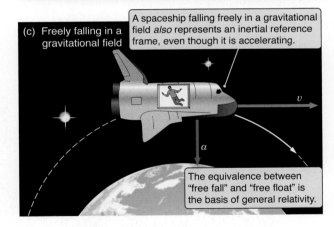

(c) Freely falling in a gravitational field

A spaceship falling freely in a gravitational field *also* represents an inertial reference frame, even though it is accelerating.

v

a

The equivalence between "free fall" and "free float" is the basis of general relativity.

FIGURE 13.18 Special relativity says that there is no difference between (a) a reference frame that is floating "stationary" in space and (b) one that is moving through the galaxy at constant velocity. General relativity adds that there is no difference between these inertial reference frames and (c) an inertial reference frame that is falling freely in a gravitational field. Free fall is the same as free float, as far as the laws of physics are concerned.

Falling objects follow curved paths through curved spacetime.

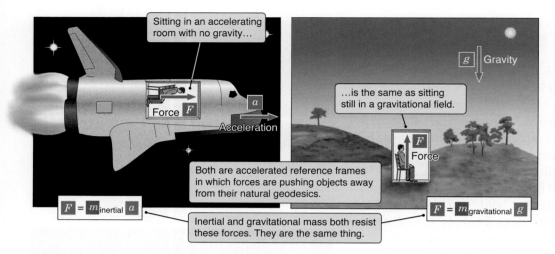

FIGURE 13.19 According to the equivalence principle, sitting in a spaceship accelerating at 9.8 m/s² feels the same as sitting still on Earth.

> **Being stationary in a gravitational field is the same as being in an accelerated reference frame.**

the object's "world line" or **geodesic**. In the absence of a gravitational field, the geodesic of an object is a straight line—hence Newton's statement of inertia that an object, unless acted on by an unbalanced external force, will move at a constant speed in a constant direction. However, in the presence of mass the shape of spacetime becomes distorted, so an object's geodesic becomes curved.

Imagine you are in a box inside a spaceship that is accelerating through deep space at a rate of 9.8 meters per second per second (m/s²) in the direction of the arrow shown in **Figure 13.19**. The floor of the box pushes on you to overcome your inertia and cause you to accelerate at 9.8 m/s², so you feel as though you are being pushed into the floor of the box. Now imagine instead that you are sitting in a closed box on the surface of Earth. The floor of the box pushes on you to keep you from following your curved geodesic. Again, you feel as though you are being pushed into the floor of the box.

According to the equivalence principle, *the two cases are identical.* There is no difference between sitting in an armchair in a spaceship with an acceleration of 9.8 m/s² and sitting in an armchair on the surface of Earth reading this book. In the first case, the force of the spaceship is pushing you away from your "floating" straight-line geodesic through spacetime. In the second case, Earth's surface is pushing you away from your curved "falling" geodesic through a spacetime that has been distorted by the mass of Earth. An acceleration is an acceleration, regardless of whether you are being accelerated off a straight-line geodesic through deep space or being accelerated off a "falling" geodesic in the gravitational field of Earth.

There is an important caveat to the equivalence principle. In an accelerated reference frame such as an accelerating spaceship, the *same* acceleration is experienced *everywhere*. In contrast, the curvature of space by a massive object changes from place to place. Tidal forces are one result of increased curvature of space closer to a gravitating body. A more careful statement of the equivalence principle is that the effects of gravity and acceleration are indistinguishable *locally*—that is, as long as we restrict our attention to small enough volumes of space that changes in gravity can be ignored.

Spacetime as a Rubber Sheet

The general theory of relativity describes how mass distorts the *geometry* of spacetime. Imagine the surface of a tightly stretched, flat rubber sheet. A marble will roll in a straight line across the sheet. Euclidean geometry, the geometry of everyday life, applies on the surface of the sheet: if you draw a circle, the circumference is equal to 2π times its radius, r; if you draw a triangle, the angles add up to 180 degrees; lines that are parallel anywhere are parallel everywhere.

Now place a bowling ball in the middle of the rubber sheet, creating a "well," as in **Figure 13.20**. The sheet will be stretched and distorted. If you roll a marble across the sheet, its path curves (Figure 13.20a). If you draw a circle around the bowling ball, the circumference is less than $2\pi r$ (Figure 13.20b). If you draw a

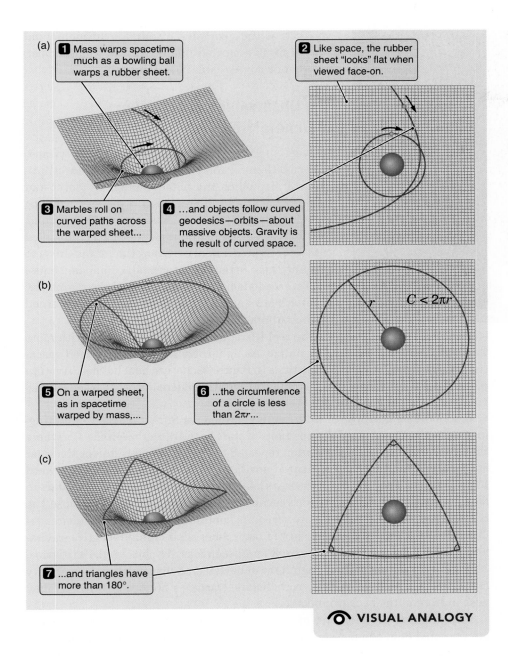

(a)

1 Mass warps spacetime much as a bowling ball warps a rubber sheet.

2 Like space, the rubber sheet "looks" flat when viewed face-on.

3 Marbles roll on curved paths across the warped sheet...

4 ...and objects follow curved geodesics—orbits—about massive objects. Gravity is the result of curved space.

(b)

5 On a warped sheet, as in spacetime warped by mass,...

6 ...the circumference of a circle is less than $2\pi r$...

r $C < 2\pi r$

(c)

7 ...and triangles have more than 180°.

🔊 **VISUAL ANALOGY**

FIGURE 13.20 Mass warps the geometry of spacetime in much the same way that a bowling ball warps the surface of a stretched rubber sheet. This distortion of spacetime has many consequences; for example, (a) objects follow curved paths or geodesics through curved spacetime, (b) the circumference of a circle around a massive object is less than 2π times the radius of the circle, and (c) angles in triangles need not add to exactly 180 degrees.

triangle, the angles add up to more than 180 degrees (Figure 13.20c). The surface of the sheet is no longer flat, and Euclidean geometry no longer applies.

Similarly, mass distorts spacetime, changing the distance between any two locations or events. You can imagine, at least in principle, stretching a rope all the way around the circumference of a circular orbit about the Sun and then comparing the length of that rope with the length of a rope taken from the orbit to the center of the Sun. You might expect to find that the circumference of the orbit is equal to 2π times the radius of the orbit, just like a circle drawn on a flat piece of paper. If you carried out this experiment, however, you would find that *the circumference of the orbit was 10 km shorter than 2π times the radius.*

We can visualize how a rubber sheet with a bowling ball on it is stretched through a third spatial dimension, but it is impossible for most people to visualize what a curved four-dimensional spacetime would "look like." Yet experiments

Mass distorts the geometry of spacetime.

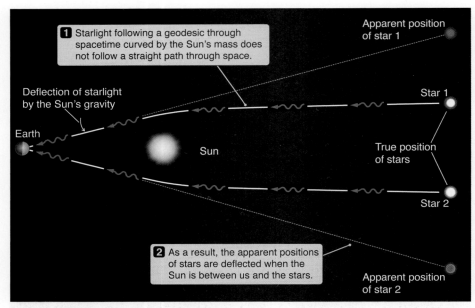

1 Starlight following a geodesic through spacetime curved by the Sun's mass does not follow a straight path through space.

Deflection of starlight by the Sun's gravity

Earth

Sun

Apparent position of star 1

Star 1

True position of stars

Star 2

2 As a result, the apparent positions of stars are deflected when the Sun is between us and the stars.

Apparent position of star 2

FIGURE 13.21 Measurements obtained by Sir Arthur Eddington during the total solar eclipse of 1919 found that the gravity of the Sun bends the light from distant stars by the amount predicted by Einstein's general theory of relativity. This is an example of gravitational lensing. Note that the "triangle" formed by Earth and the two stars contains more than 180 degrees, just like the triangle in Figure 13.20c.

Gravitational lensing can displace and distort an object's image.

verify that the geometry of our four-dimensional spacetime is distorted much like the rubber sheet, whether or not we can easily picture it.

The Observable Consequences of General Relativity

Curved spacetime does have observable consequences. General relativity predicts that an elliptical orbit, in which the planet swings in closer and then farther from the Sun, should *precess*; the long axis should slowly change its direction. For example, Newton's law of gravity predicts the major component of Mercury's precession; but even after taking this into account, there remains a very small component equal to 43 arcseconds/century that cannot be explained by Newton's laws alone. This discrepancy is explained by general relativity.

The real-life equivalent of the triangle with more than 180 degrees is easier to visualize. As light travels through distorted spacetime, its path bends, just like the lines in Figure 13.20c are bent by the curvature of the sheet. This phenomenon is called **gravitational lensing** because lenses also bend light. But how could you ever measure this?

Before the solar eclipse of 1919, English astronomer Sir Arthur Stanley Eddington (1882–1944) measured the positions of stars in the direction in the sky where the eclipse would occur. Eddington repeated his measurements during the solar eclipse and found that the measured positions of the stars had shifted outward (**Figure 13.21**). The light from the stars bent around the Sun, causing the stars to appear farther apart in Eddington's second measurement. During the eclipse, the triangle formed by Earth and any two stars contained more than 180 degrees—just like the triangle on the surface of our rubber sheet. More recently, gravitational lensing has been used to search for unseen massive objects adrift in space. These objects do not emit enough light to be detected directly, but their gravity can distort the light from background stars, causing a noticeable effect as they pass in front of the background stars.

Mass distorts the geometry of time as well. The deeper we descend into the gravitational field of a massive object, the more slowly our clocks appear to run from the perspective of a distant observer. This effect is called **general relativistic time dilation**. Suppose a light is attached to a clock sitting on the surface of a neutron star. The light is timed so that it flashes once a second. Because time near the surface of the star is dilated, an observer far from the neutron star perceives the light to be pulsing with a lower frequency—less than once a second. Now suppose there is an emission-line source on the surface of the neutron star. Because time is running slowly on the surface of the neutron star, the light that reaches the distant observer will have a lower frequency than when it was emitted. A lower frequency means a longer wavelength. So the light from the source will be seen at a longer, redder wavelength than the wavelength at which it was emitted.

This phenomenon, shown in **Figure 13.22**, is called the **gravitational redshift** because the wavelengths of light from objects deep within a gravitational well are shifted to longer, redder wavelengths. The effect of gravitational redshift is similar to the Doppler redshift. In fact, there is no way to tell the difference between light that has been redshifted by gravity and light that has been Doppler redshifted.

Bringing this phenomenon a bit closer to home, a clock on the top of Mount Everest runs faster, gaining about 80 nanoseconds (80 billionths of a second) a day compared with a clock at sea level. The difference between an object on the surface of Earth and an object in orbit is even greater. Even after allowing for slowing due to special relativity, the clocks on the satellites that make up the global positioning system (GPS) run faster than clocks on the surface of Earth. If the satellite clocks *and* your GPS receiver did not correct for this and other effects of general relativity, then the position your GPS receiver reported would be in error by up to half a kilometer within a single hour. The fact that the GPS works is actually a strong experimental confirmation of several predictions of general relativity, including general relativistic time dilation.

If you thump the surface of a rubber sheet, waves will move away from where you thumped it, something like ripples spreading out over the surface of a pond. Similarly, the equations of general relativity predict that if you "thump" the fabric of spacetime (for example, with the catastrophic collapse of a high-mass star), then ripples in spacetime, or **gravitational waves**, will move outward at the speed of light. These gravitational waves are like electromagnetic waves in some respects. Accelerating an electrically charged particle gives rise to an electromagnetic wave. Accelerating a massive object gives rise to gravitational waves.

Gravitational waves have not yet been observed, but there is strong circumstantial evidence for their existence. In 1974, astronomers discovered a binary system of two neutron stars, one of which is an observable pulsar. Using the pulsar as a precise clock, astronomers have been able to very accurately measure the orbits of both stars. The orbits are gradually decaying, which means they are losing energy *somewhere*. The energy being lost by the system is just what general relativity predicts the system should be producing in the form of gravitational waves. More recently, astronomers discovered a similar binary pair with smaller orbits, and they found once again that the orbital energy loss was consistent with the radiation of gravitational waves. These measurements very strongly suggest that gravitational waves exist.

Let's stop for a minute and think about this. Gravitational waves have *never been directly observed*. At the moment, gravitational waves are an untested prediction of a scientific theory. Is this so different from the pseudoscientific theory of "intelligent design," which claims that life on Earth was deliberately designed by an intelligent agent? Yes, because the prediction that gravitational waves exist is *falsifiable*, while intelligent design is not. The theory predicts that specific events will generate gravitational waves. Astrophysicists have developed a new kind of "telescope," called the Laser Interferometer Gravitational-Wave Observatory (LIGO), that can detect the predicted gravitational waves emanating from such events. Either we will see them or we will not. The theory that predicts gravitational waves is scientific because it can be proven wrong. Pseudoscientific theories, on the other hand, are *not* falsifiable.

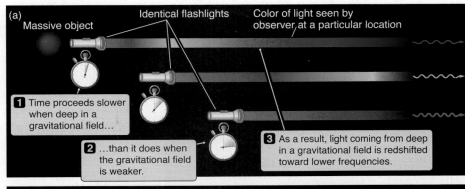

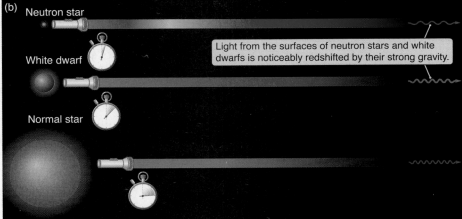

FIGURE 13.22 Time passes more slowly near massive objects due to the curvature of spacetime. As a result, to a distant observer light from near a massive object will have a lower frequency and longer wavelength. (a) The closer the source of radiation is to the object, or (b) the more massive and compact the object is, the greater the gravitational redshift will be.

Gravitational waves travel through the fabric of spacetime.

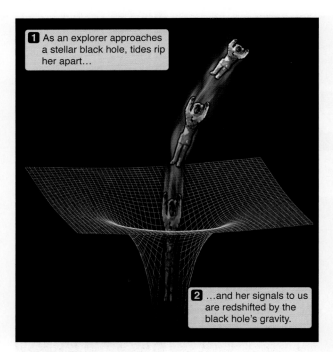

1 As an explorer approaches a stellar black hole, tides rip her apart…

2 …and her signals to us are redshifted by the black hole's gravity.

FIGURE 13.23 Journey into a black hole.

The event horizon is the boundary of no return.

Vocabulary Alert

Infinity: In common language, this word sometimes just means "really, really big" or "really, really far," as in "To infinity and beyond!" But astronomers really mean infinity in a mathematical sense—a quantity without limit or boundary. If we say infinity, there is no beyond.

13.6 Back to Black Holes

An object on the surface of a rubber sheet causes a funnel-shaped distortion that is analogous to the distortion of spacetime by a mass. Now imagine the funnel is **infinitely** deep—it gets narrower as we go deeper, but it has no bottom. This is the rubber-sheet analog to a black hole. The mathematics describing the shape of a black hole fail in the same way that the mathematical expression $1/x$ fails when $x = 0$. Such a mathematical anomaly is called a **singularity**. Black holes are singularities in spacetime.

A black hole has only three properties: mass, electric charge, and angular momentum. The amount of mass that falls into a black hole determines the extent of its distortion of spacetime. The electric charge of a black hole is the net electric charge of the matter that fell into it. The angular momentum of a black hole twists spacetime around it. Apart from these three properties, all information about the material that fell into the black hole is lost. Nothing of its former composition, structure, or history survives.

We can never actually "see" the singularity at the center of a black hole. As you approach a black hole, the escape velocity increases, until it reaches the speed of light—at this distance from the center of the black hole, even light cannot escape. For a nonrotating, spherical black hole, the radius where the escape velocity equals the speed of light is called the **Schwarzschild radius**, named for German physicist Karl Schwarzschild (1873–1916), and it is proportional to the mass of the black hole. The sphere around the black hole at this distance is called its **event horizon**. A black hole with a mass of 1 $M_\odot$ has a Schwarzschild radius of about 3 km. If Earth were squeezed into a black hole, it would have a Schwarzschild radius of only about a centimeter.

Imagine that an adventurer journeys into a black hole (**Figure 13.23**). From our perspective outside the black hole, we would see our adventurer fall toward the event horizon; but as she did, her watch would run more and more slowly and her progress toward the event horizon would slow as well. At the event horizon, the gravitational redshift becomes infinite and clocks stop altogether. Our adventurer would approach the event horizon, but she would never quite make it, *from our perspective*. Yet the adventurer's experience would be very different. From her perspective, there would be nothing special about the event horizon at all. She would fall past the event horizon and on, deeper into the black hole's gravitational well.

Actually, we have overlooked a rather crucial fact. Our intrepid explorer would have been torn to shreds long before she reached the black hole. Near the event horizon of a 3 $M_\odot$ black hole, the difference in gravitational acceleration between our explorer's feet and her head would be about a billion times her gravitational acceleration on the surface of Earth. In other words, her feet would be accelerating a billion times faster than her head. This is not an experiment we would ever want to perform. Although scientific theories *must* produce testable predictions, not *all* individual predictions have to be tested directly.

"Seeing" Black Holes

In 1974, English physicist Stephen Hawking (b. 1942) realized that black holes should be *sources* of radiation. In the ordinary vacuum of empty space, particles and their antiparticle "mates" form and then, within 10^{-21} seconds, annihilate each other. If this happens near the event horizon of a very small black hole, as shown in **Figure 13.24**, then one of the particles might fall into the black hole while the other particle escapes. Hawking showed that through this process, a black hole should emit a blackbody spectrum and that the effective temperature of this spectrum would increase as the black hole became smaller. Although this phenomenon, called **Hawking radiation**, is of considerable interest to physicists and astronomers, in a practical sense it is not a useful way to "see" a black hole.

So far, the strongest direct evidence for black holes comes from X-ray binary stars. In 1972, Cygnus X-1 was found to be rapidly flickering, changing in as little as 0.01 second. This means that the source of the X-rays must be smaller than the distance that light travels in 0.01 second, or 3,000 km. Thus, the source of X-rays in Cygnus X-1 must be smaller than Earth. Cygnus X-1 was also identified with both a radio star and with an already cataloged optical star called HD 226868. The spectrum of HD 226868 shows that it is a normal B0 supergiant star with a mass of about 30 $M_\odot$, far too cool to produce X-ray emission. But HD 226868 is part of a binary system with a period of 5.6 days. Orbit analysis shows that the mass of the unseen compact companion of HD 226868 must be at least 6 $M_\odot$. The companion to HD 226868 is too small to be a normal star, yet it is much more massive than the Chandrasekhar limit for a white dwarf or a neutron star. Such an object can only be a black hole. The X-ray emission from Cygnus X-1 arises when material from the B0 supergiant falls onto an accretion disk surrounding the black hole, as illustrated in **Figure 13.25**.

Since 1972, a number of other good candidates for stellar-mass black holes have been discovered. One such object is a rapidly varying X-ray source in our companion galaxy, the Large Magellanic Cloud. Called LMC X-3, this X-ray source orbits a B3 main-sequence star every 1.7 days, and the compact source must have a mass of at least 9 $M_\odot$. Although the evidence that these systems contain black holes is circumstantial, the arguments that lead to this conclusion seem airtight. With dozens of compelling examples of such objects on the books, the evidence is in. Black holes, once regarded as nothing more than a bizarre quirk of the mathematics describing gravitation and spacetime, exist in nature.

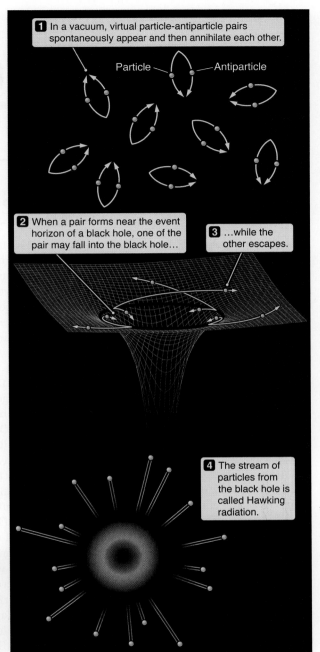

1 In a vacuum, virtual particle-antiparticle pairs spontaneously appear and then annihilate each other.

Particle — Antiparticle

2 When a pair forms near the event horizon of a black hole, one of the pair may fall into the black hole…

3 …while the other escapes.

4 The stream of particles from the black hole is called Hawking radiation.

FIGURE 13.24 In the vacuum of empty space, particles and antiparticles are constantly being created and then annihilating each other. Near the event horizon of a black hole, however, one particle may cross the horizon before it recombines with its partner. The remaining particle leaves the black hole as Hawking radiation.

FIGURE 13.25 Artist's rendering of the Cygnus X-1 binary system, showing material from the B0 supergiant being pulled off and falling onto an accretion disk surrounding the black hole, thereby producing X-ray emission.

READING ASTRONOMY News

What a Scorcher—Hotter, Heavier and Millions of Times Brighter Than the Sun

By **Ian Sample**, *The Guardian*

They are the most colossal stars ever seen and live short, bright, lives in faraway reaches of space before exploding in a blaze of glory.

One of the stars, now tagged R136a1, is estimated to weigh 265 times more than the sun and to shine millions of times more brightly. Were it to replace our own star, the intensity of its rays would sterilise the Earth, leaving it lifeless.

British astronomers spotted the stars, more massive than any others on record, using the Very Large Telescope, an aptly named observatory on a mountain top in the Atacama Desert of northern Chile.

The discovery of the stellar giants has prompted astronomers to scrap the upper limits they set on star formation, which suggested it was almost impossible for a star to grow to more than 150 times the mass of the sun.

The team, led by Paul Crowther, an astrophysicist at Sheffield University, searched two regions of space for massive stars. The first region, known as NGC 3603, is a stellar nursery 22,000 light-years away in a region of the Milky Way called the Carina spiral arm.

The second target, RMC 136a, is a cloud of gas and dust, 165,000 light-years away in the Tarantula nebula of our neighbouring galaxy, the Large Magellanic Cloud. The astronomers were able to distinguish individual stars using exquisitely sensitive infrared instruments on the telescope, and take measurements of their brightness and mass.

At least three stars examined in the first region of space weighed in at about 150 times the mass of the sun. The record-breaking star,

R136a1, was found in the second region. When born, the star could have been a staggering 320 times more massive than the sun.

Several of the stars were found to have surface temperatures above 40,000° C, which is more than seven times hotter than the sun.

"These stars are born heavy and lose weight as they age," said Crowther. "Being a little over a million years old, the most extreme star R136a1 is already middle-aged and has undergone an intense weight loss programme, shedding a fifth of its initial mass over that time. Owing to the rarity of these monsters I think it unlikely this new record will be broken any time soon."

If R136a1 were in our own solar system, it would outshine the sun as much as the sun outshines the full moon, the scientists said. The mass of the star is so great that it would reduce the length of an Earth year—the time it takes to circle the star—to just three weeks. "It would [also] bathe Earth in incredibly intense ultraviolet radiation, rendering life on our planet impossible," said Raphael Hirschi, a member of the team at Keele University.

While the latest crop of stars are the most massive and heaviest ever spotted, they are not the largest. The biggest star in the group, R136a1, is roughly 30 times as wide as the sun. Another kind of star, known as a super red giant, can grow to many hundreds of times that size—though is considerably lighter, at only 10 times the mass of the sun.

It is unlikely that any "alien" planets circle the massive stars that Crowther's team has studied. Radiation from the stars would obliterate any nearby cosmic material that

could become compact enough to be a planet. Even if some remained, planets would take longer to form than the entire life span of a massive star.

Crowther said: "We don't really know what happens when these massive stars reach the end of their lives. When some big stars die, their cores implode and they become neutron stars or black holes, but these might be different. They might blow up in a spectacular supernova and leave no remnants behind at all." The explosions could fling the weight of 10 suns worth of iron into space.

The team's observations reveal what the early universe might have looked like, when many of the first stars to be born might have been cosmic monsters like R136a1.

Before the latest discovery, the most massive star known was the Peony nebula star, which, at about 175 times the mass of the sun, could still hold the record for our own galaxy. Details of the discovery are reported in the monthly notices of the Royal Astronomical Society.

Evaluating the News

1. How much more massive is R136a1 than previous estimates of the maximum mass of a star? Is this a significant difference? If you were an astronomer, what would this tell you about your theories about maximum size? Could they be saved, or are they likely to be thrown out altogether? What if R136a1 were only 155 times the mass of the Sun? Would the same conclusions apply?

2. In the title, the reporter states that the star is "hotter" and "millions of times brighter"

SEE SCORCHER

than the Sun. Explain how this information would lead to a conclusion that the star is more massive.

3. Paul Crowther states in the article that the star is "a little over a million years old." How might he know that?

4. If Paul Crowther had stated that the star was 5 billion years old, you could immediately say, "Baloney!" Why?

5. "Stellar nurseries" are the ideal place to look for the most massive stars. Why?

6. The article states that "If R136a1 were in our solar system . . . The mass is so great that it would reduce the length of an Earth year . . . to just three weeks." What physical law are the astrophysicists using to estimate this?

SUMMARY

13.1 The CNO cycle burns hydrogen in high-mass stars. As these stars evolve, they burn and exhaust heavier elements in their core, creating concentric shells of progressive nuclear burning. If a high-mass star passes through the instability strip on the H-R diagram, it will become a pulsating variable star.

13.2 The chain of nuclear fusion consists of increasingly shorter stages of burning resulting in more massive elements, up to iron. High-mass stars eventually explode as Type II supernovae.

13.3 Type II supernovae eject newly formed massive elements into interstellar space. Some leave behind neutron stars, which contain between 1.4 $M_\odot$ and 3 $M_\odot$ of neutron-degenerate matter packed into a 10-km-diameter sphere. Accretion of mass onto neutron stars produces X-rays in some binary systems. Pulsars are rapidly spinning, magnetized neutron stars.

13.4 The speed of light is a constant. Special relativity describes the relationship between events in space and time. Time runs more slowly for moving objects.

13.5 General relativity describes how mass warps the fabric of spacetime so that objects move on the shortest path in this warped geometry. Time runs more slowly near massive objects. Objects deep in a gravitational well appear redshifted.

13.6 Black holes are singularities in spacetime. A black hole's mass determines its Schwarzschild radius, the distance from the center of the black hole to its event horizon. Within this boundary, light cannot escape.

✦ SUMMARY SELF-TEST

1. The interior of an evolved high-mass star has layers like an onion because
   a. heavier atoms sink to the bottom, because stars are not solid.
   b. before the star formed, heavier atoms accumulated in the centers of clouds, due to gravity.
   c. heavier atoms fuse closer to the center, because the temperature and pressure are higher there.
   d. different energy transport mechanisms occur at different densities.

2. Place in order the stages of nuclear burning that evolving high-mass stars experience.
   a. helium   b. neon
   c. oxygen   d. silicon
   e. hydrogen   f. carbon

3. Elements heavier than iron originate
   a. in the Big Bang.
   b. in the cores of low-mass stars.
   c. in the cores of high-mass stars.
   d. in the explosions of high-mass stars.

4. A pulsar "pulses" because
   a. its spin axis crosses our line of sight.
   b. it spins.
   c. it has a strong magnetic field.
   d. its magnetic axis crosses our line of sight.

5. The fact that the speed of light is a universal constant forces us to completely rethink classical physics. This is an example of
   a. scientists always being completely wrong.
   b. the self-correcting nature of science.
   c. a theory becoming a hypothesis and then becoming a law.
   d. the universe changing with time.

6. Which of the following are possible consequences of distortions of spacetime caused by mass?
   a. time dilation   b. gravity
   c. length contraction   d. tidal forces
   e. gravitational lensing f. precession

7. An astronaut who fell into a black hole would be stretched because
   a. the gravity is so strong.
   b. the gravity changes dramatically over a short distance.
   c. time is slower near the event horizon.
   d. black holes rotate rapidly, dragging spacetime with them.

QUESTIONS AND PROBLEMS

True/False and Multiple-Choice Questions

8. T/F: The end result of the CNO cycle is that four hydrogen nuclei become one helium nucleus.

9. T/F: Stars in the instability strip are pulsating variable stars.

10. T/F: When iron fuses into heavier elements, it produces energy.

11. T/F: Electrons and protons can combine to become neutrons.

12. T/F: Uranium is made in the core of a star.

13. T/F: A supernova can be as bright as its entire host galaxy.

14. T/F: A pulsar changes in brightness because its size pulsates.

15. T/F: The speed of light has the same value in every reference frame.

16. T/F: Inside a small, windowless room, gravitational acceleration is indistinguishable from other kinds of accelerations.

17. T/F: Euclidean geometry does not apply to space near a massive object.

18. In a high-mass star, hydrogen fusion occurs via the
a. proton-proton chain.
b. CNO cycle.
c. gravitational collapse.
d. spin-spin interaction.

19. The layers in a high-mass star occur roughly in order of
a. atomic number.
b. decay rate.
c. atomic abundance.
d. spin state.

20. Pulsations in a Cepheid variable star are controlled by
a. the spin.
b. the magnetic field.
c. the ionization state of helium.
d. the gravitational field.

21. Eta Carinae is an extreme example of
a. a massive star.
b. a rotating star.
c. a magnetized star.
d. a high-temperature star.

22. Iron fusion cannot support a star, because
a. iron oxidizes too quickly.
b. iron absorbs energy when it fuses.
c. iron emits energy when it fuses.
d. iron is not dense enough to hold up the layers.

23. When iron fusion starts in a star, a process begins that *always* results in a
a. supernova.
b. neutron star.
c. black hole.
d. pulsar.

24. Supernova remnants
a. are viewable at all wavelengths.
b. are viewable only at a few emission lines.
c. are never seen in radio waves.
d. have colors because the moving gas emits Doppler-shifted emission lines.

25. X-ray binaries are similar to another type of system we have studied. This system is
a. the Solar System.
b. progenitors of Type Ia supernovae.
c. progenitors of Type II supernovae.
d. progenitors of planetary nebulae.

26. Imagine that two spaceships travel toward each other. One spaceship travels to the right at a speed of $0.9c$, and another travels to the left at $0.9c$. The pilot of the spaceship traveling right shoots a yellow laser at the spaceship traveling left. The pilot of the spaceship traveling left observes
a. blue light traveling at c.
b. blue light traveling at $1.9c$.
c. yellow light traveling at c.
d. yellow light traveling at $1.9c$.

27. Bob is moving relative to Susie. Susie observes the explosion of two firecrackers, and after careful measuring, concludes that they exploded at the same time. Bob observes, carefully measures, and concludes that one happened before the other. Which of the following is correct?
a. Bob is wrong, but Susie is right.
b. Susie is wrong, but Bob is right.
c. Both Susie and Bob are wrong.
d. Both Susie and Bob are right.

28. Bill is standing in a small, windowless room. The force on his feet suddenly ceases. Which of the following conclusions can he draw?
a. He is in an elevator, which is now accelerating downward.
b. He is now in space, freely falling around Earth.
c. The rocket he was in has stopped accelerating upward.
d. Any of the above could be true.

29. If you could draw a very large circle in space near a black hole, its circumference
a. would equal $2\pi r$.
b. would be greater than $2\pi r$.
c. would be less than $2\pi r$.
d. can't be determined from the information given.

30. An object that passes near a black hole
a. always falls in because of gravity.
b. is sucked in because of vacuum pressure.
c. is pushed in by photon pressure.
d. is deflected by the curvature of spacetime.

31. If the Sun were replaced by a 1-solar-mass black hole,
a. the Solar System would collapse into it.
b. the planets would remain in their orbits, but smaller objects would be sucked in.
c. small objects would remain in their orbits, but larger objects would be sucked in.
d. all objects in the Solar System would remain in their orbits.

Conceptual Questions

32. Look back at the chapter-opening figure, which shows a supernova in M51. Write a brief obituary of the star that produced this supernova. Besides summarizing the important moments of its life, be sure to mention what the star leaves behind.

33. In the CNO cycle, ^{13}N emits a positron to form ^{13}C and absorbs a proton to form ^{14}N. Explain why a positron must be emitted to form ^{13}C. Is N higher or lower on the periodic table than C?

34. Explain the differences between the ways that hydrogen is converted to helium in a low-mass star (proton-proton chain) and in a high-mass star (CNO cycle). What is the catalyst in the CNO cycle, and how does it take part in the reaction?

35. How does a low-mass star begin burning helium in its core? What about a high-mass star? How are these processes different or similar?

36. Why does the core of a high-mass star not become degenerate, as in the cases of low-mass stars?

37. For what two reasons does each post-helium-burning cycle for high-mass stars (carbon, neon, oxygen, silicon, and sulfur) become shorter than the preceding cycle?

38. Cepheids are highly luminous, variable stars in which the period of variability is directly related to luminosity. Explain why Cepheids are good indicators for determining stellar distances that lie beyond the limits of accurate parallax measurements.

39. Is it possible for a high-mass star to pass through the instability strip more than once? Use Figure 13.2 to defend your answer.

40. Identify and explain two important ways in which supernovae influence the formation and evolution of new stars.

41. Is a Type II supernova really an explosion, or is it an implosion? Explain your answer.

42. Suppose that you are an astronomer studying Eta Carinae (see Figure 13.5). As a result of this observation, you determine that Eta Carinae has experienced at least two different episodes of mass loss. Defend this claim.

43. Describe what we will observe on Earth when Eta Carinae explodes.

44. Recordings show that SN 1987A was detected by neutrinos on February 23, 1987. About 3 hours later, it was detected in optical light. Why did this time delay occur?

45. Why can the accretion disk around a neutron star release so much more energy than the accretion disk around a white dwarf, even though both stars have approximately the same mass?

46. In Section 13.2, you learned that Type II supernovae blast material outward at 30,000 km/s, but the material in the Crab Nebula (Section 13.3) is expanding at only 1,500 km/s. What explains the difference?

47. What do we mean by the *binding energy* of an atomic nucleus? How does this quantity help us calculate the energy given off in nuclear fusion reactions?

48. Why is the binding energy of hydrogen zero? Would the binding energy for deuterium ("heavy hydrogen," with one proton and one neutron) also be zero? Explain.

49. An astronomer sees a redshift in the spectrum of an object. With no other information available, can she determine whether this is an extremely dense object (gravitational redshift) or one that is receding from us (Doppler redshift)? Explain your answer.

50. Einstein's special theory of relativity tells us that no object can travel faster than, or even at, the speed of light. We know that light is an electromagnetic wave, but we know that it is also a particle called a photon. If it acts as a particle, how can a photon travel at the speed of light?

51. Explain why more muons reach the ground traveling at $0.9999c$ than at $0.9c$.

52. Imagine a future cosmonaut traveling in a spaceship at 0.866 times the speed of light. Special relativity says that the length of the spaceship along the direction of flight is only half of what it was when it was at rest on Earth. The cosmonaut checks this prediction with a meter stick that he brought with him. Will his measurement confirm the contracted length of his spaceship? Explain your answer.

53. Of the four fundamental forces in nature (strong nuclear, electromagnetic, weak nuclear, and gravity), gravity is by far the weakest. Why, then, is gravity such a dominant force in stellar evolution? (Note: Although not explicitly discussed so far in this text, the weak nuclear force is involved in certain decay processes within the nucleus.)

54. Suppose astronomers discover a 3-$M_\odot$ black hole located a few light-years from Earth. Should they be concerned that its tremendous gravitational pull will lead to our planet's untimely demise?

55. Use an entire piece of paper to re-sketch Figure 13.21, adding a third star three-fourths of the way from star 1 to star 2, and a fourth star below star 2. Sketch the path of the light from these two "new" stars to Earth. Now project these paths backward into the sky to show the apparent positions of these two new stars. If you can do this, you really understand this diagram.

56. If you could watch a star falling into a black hole, how would the color of the star change as it approached the event horizon?

57. Why are we not aware of the effects of special and general relativity in our everyday lives here on Earth?

58. Many movies and TV programs (like *Star Wars, Star Trek,* or *Battlestar Galactica*) rely on faster-than-light travel. How likely is it that we will ever develop such technology in the future?

Problems

59. Make a flowchart of stellar evolution to organize your knowledge about this topic. The chart should start with a giant molecular cloud and finish with the various types of stellar corpse.

60. Our galaxy has about 50,000 stars of average mass (0.5 $M_\odot$) for every main-sequence star of 20 $M_\odot$. But 20-$M_\odot$ stars are about 10^4 times more luminous than the Sun, and 0.5-$M_\odot$ stars are only 0.08 times as luminous as the Sun.
   a. How much more luminous is a single massive star than the total luminosity of the 50,000 less massive stars?
   b. How much mass is in the lower-mass stars compared to the single high-mass star?

c. Which stars—lower-mass or higher-mass stars—contain more mass in the galaxy, and which produce more light?

61. In 1841, the 100-$M_\odot$ star Eta Carinae was losing mass at the rate of 0.1 $M_\odot$ per year. Let's put that into perspective.
   a. The mass of the Sun is 2×10^{30} kg. How much mass (in kilograms) did Eta Carinae lose each minute?
   b. The mass of the Moon is 7.35×10^{22} kg. How does Eta Carinae's mass loss per minute compare with the mass of the Moon?

62. Using values given in Section 13.1, verify that an O star can lose 20 percent of its mass during its main-sequence lifetime.

63. The approximate relationship between the luminosity and the period of Cepheid variables is L_{star} ($L_\odot$ units) $= 335\,P$ (days). Delta Cephei has a cycle period of 5.4 days and a parallax of 0.0033 arcseconds (arcsec). A more distant Cepheid variable appears 1/1,000 as bright as Delta Cephei and has a period of 54 days.
   a. How far away (in parsecs) is the more distant Cepheid variable?
   b. Could the distance of the more distant Cepheid variable be measured by parallax? Explain.

64. If the Crab Nebula has been expanding at an average velocity of 3,000 km/s since it was first observed in A.D. 1054, what was its average radius in the year 2012? (Note: There are approximately 3 $\times 10^7$ seconds in a year.)

65. Use Einstein's famous mass-energy equivalence formula ($E = mc^2$) to verify that 5.88×10^{13} joules (J) of energy is released from fusing 1.00 kg of helium.

66. According to Einstein, mass and energy are equivalent. So does a cup of coffee weigh more when it is hot, or when it is cold? Why? Do you think the difference is measurable?

67. We know that pulsars are rotating neutron stars. For a pulsar that rotates 30 times per second, at what radius in the pulsar's equatorial plane would a co-rotating satellite (rotating about the pulsar 30 times per second) have to be moving at the speed of light? Compare this to the pulsar radius of 1 km.

68. Figure 13.11 shows the relative abundance of the elements. Is this a log plot or a linear one? Explain what it means that oxygen lies on the y-axis at 10^{-3}.

69. From the information given in Section 13.3, verify the claim that Earth would be roughly the size of a football stadium if it were as dense as a neutron star.

70. If a spaceship approaching us at 0.9 times the speed of light shines a laser beam at Earth, how fast will the photons in the beam be moving when they arrive at Earth?

Exploration | The CNO Cycle

wwnorton.com/studyspace

Nuclear reactions are quite complex, and they usually involve many steps. In a previous exploration, you investigated the proton-proton chain. In this exploration, you will study the CNO cycle, which is even more complex. Visit the "CNO Cycle" interactive simulation in the Chapter 13 area on StudySpace.

First, press play and watch the animation all the way through. Press stop to clear the screen, and then press play again, allowing the animation to proceed past the first collision before pressing pause.

1. Which atomic nuclei are involved in this first collision?

2. What color is used to represent the proton (hydrogen nucleus)?

3. What is represented by the blue wiggle?

4. What atomic nucleus is created in the collision?

5. The resulting nucleus is not the same type of element as either of the two that entered the collision. Why not?

Press play again, and then pause as soon as the yellow ball and the dashed line appear.

6. Is this a collision or a spontaneous decay?

7. What is represented by the yellow ball?

8. What is represented by the dashed line?

9. The resulting nucleus has the same number of nucleons (13), but it is a different element. What happened to the proton that was in the nitrogen nucleus but is not now in the carbon nucleus?

Proceed past the next two collisions, to ^{15}O.

10. Study the pattern that's forming. When a blue ball comes in, what happens to the number of nucleons and the type of the nucleus (i.e., what happens to the "12" and the "C," or the "14" and the "N")?

11. What is emitted in these collisions?

Proceed until ^{15}N appears.

12. Is this a collision or a spontaneous decay?

13. Which previous reaction is this most like?

Now proceed to the end of the animation.

14. After this last collision, a line is drawn back to the beginning. This tells you what type of nucleus the upper red ball represents. What is this nucleus?

15. How many nucleons are not accounted for by that upper red ball? (Hint: Don't forget the ^1H that came into the collision.) These nucleons must be in the nucleus represented by the bottom red ball.

16. Carbon has six protons. Nitrogen has seven. How many protons are in the nucleus represented by the bottom red ball?

17. How many neutrons are in the nucleus represented by the bottom red ball?

18. What element is represented by the bottom red ball?

19. What is the net reaction of the CNO cycle? That is, what nuclei are combined and turned into the resulting nucleus?

20. Why do we not consider ^{12}C part of the net reaction?

14 The Expansion of Space

Unlike questions of politics and law, scientific questions are not resolved by rhetorical skills. Instead, they are settled by the results of well-crafted and carefully conducted experiments and observations. In 1920, however, the astronomers Harlow Shapley (1885–1972) and Heber D. Curtis (1872–1942) engaged in a debate about the nature of the "spiral and elliptical nebulae" that we now know to be galaxies beyond our own Milky Way Galaxy. Shapley favored the view that our galaxy was the entire universe and therefore these objects were inside the Milky Way. Curtis favored the opposite (and, we now know, correct) view that they were separate, very distant objects, much like the Milky Way. While this "Great Debate" did not resolve the issue at the time, it set the stage for Edwin P. Hubble (1889–1953), whose discoveries fundamentally changed our understanding of the universe.

✧ LEARNING GOALS

Just as stars are the building blocks of our galaxy, galaxies are the building blocks of the universe. The Hubble Space Telescope image shown at right reveals four galaxies: three large spirals and an elliptical. The student is comparing galaxies of the same type to find their relative distance from the Milky Way. By the end of this chapter, you should know how pictures like this (and follow-up observations) show that the universe contains billions upon billions of galaxies; that the universe is the same everywhere; that nearly all galaxies are moving away from us; and that the farther away they are, the faster they move. You should also be able to:

- Sketch a graph of Hubble's law, which relates the redshift of a galaxy to its distance

- Describe how Hubble's law is used to map the universe and to look back in time

- Distinguish between moving through space and being "carried along" by space

- Describe the evolution of the universe since it began approximately 13.7 billion years ago in an event called the Big Bang

- Link observations of the cosmic microwave background to the properties of the young universe

Assuming each spiral galaxy
is about the same size:

(C) IS CLOSEST

(B) IS FARTHEST

(A) IS IN BETWEEN

Hickson Compact Group 87

Hubble
Heritage

PRC99-31 • Space Telescope Science Institute • Hubble Heritage Team (AURA/STScI/NASA)

A

B

C

14.1 The Cosmological Principle Shapes Our View of the Universe

Earth and its sibling planets orbit an average star within an enormous spiral **galaxy**—a gravitationally bound grouping of stars, dust, and gas. Our galaxy contains hundreds of billions of stars. This galaxy, in turn, is but one of at least hundreds of billions of galaxies that fill a universe vast in both space and time. The **cosmological principle** is at the center of our conceptual understanding, and it is the realization that *the physical laws that apply to one part of our universe apply to all parts of the universe.*

The science of **cosmology** is the study of the universe itself, including its structure, history, origins, and fate. As we stressed in Chapter 1, the cosmological principle is not an article of faith. Rather, it is a testable scientific theory. An important prediction of the cosmological principle is that the conclusions we

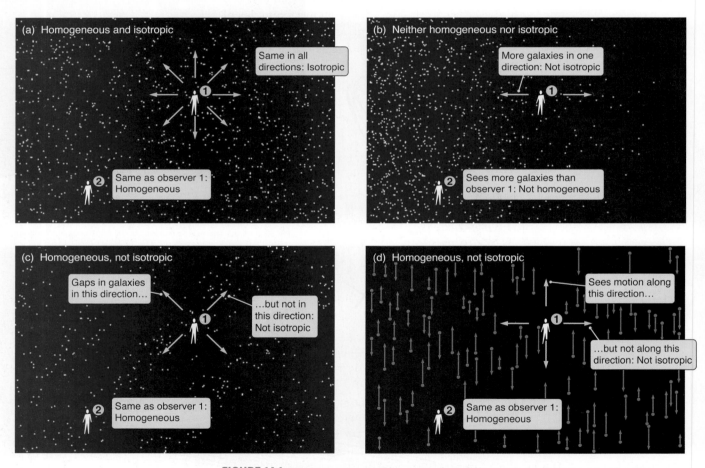

FIGURE 14.1 Homogeneity and isotropy in four different theoretical models of a universe. Blue arrows indicate the direction of view. (a) The distribution of galaxies is uniform, so this universe is both homogeneous and isotropic. (b) The density of galaxies is decreasing in one direction, so this universe is neither homogeneous nor isotropic. (c) The bands of galaxies lie along a unique axis, making this universe anisotropic. However, on large scales (the size of the box), the universe is the same everywhere, making it homogeneous. (d) The distribution of galaxies is uniform, but galaxies move along only one direction, so this universe also is not isotropic.

reach about our universe should be the same, whether we live in the Milky Way or in a galaxy billions of light-years away. In other words, if the cosmological principle is correct, then our universe will be **homogeneous**.

It is not easy to verify the prediction of homogeneity directly. We cannot travel to other galaxies to see whether conditions are the same. However, we can compare light arriving from closer and farther locations in the distant universe. For example, we can look at the way galaxies are distributed in distant space and ask whether that distribution is similar to the distribution nearby.

In addition to predicting that the universe is homogeneous, the cosmological principle requires that all observers (including us) have the same impression of the universe, regardless of the *direction* in which they are looking. If something is the same in all directions, then it is **isotropic**. This prediction of the cosmological principle is much easier to test directly than is homogeneity. For example, if galaxies were lined up in rows—a violation of the cosmological principle—we would get very different impressions, depending on the direction in which we looked. In most instances, isotropy goes hand in hand with homogeneity, and the cosmological principle requires them both. **Figure 14.1** shows examples of how the universe could have violated the cosmological principle by being either inhomogeneous or anisotropic, as well as examples of how the universe might satisfy the cosmological principle.

The isotropy and homogeneity of the distribution of galaxies in the universe are predictions of the cosmological principle that we can test directly. All of our observations show that the properties of the universe are basically the same, regardless of the direction in which we look. On very large scales, the universe appears homogeneous as well. The cosmological principle has withstood all our tests and forms the bedrock of modern cosmology.

14.2 We Live in an Expanding Universe

In 1925, using the newly finished 100-inch telescope on Mount Wilson, high above the then small city of Los Angeles, Edwin Hubble (**Figure 14.2**) was able to find some variable stars in the large neighboring galaxy of Andromeda. He recognized that these stars were very similar to the Cepheid variable stars in the Milky Way and the nearby Magellanic Cloud galaxies, though they appeared much fainter. Hubble used the period-luminosity relation for Cepheid variable stars (Chapter 13) to find distances to Andromeda and several other galaxies. These galaxies, similar in size to our own galaxy, were located at truly immense distances.

Vesto Slipher (1875–1969) used Lowell Observatory in Flagstaff, Arizona, to obtain spectra of these galaxies. Unsurprisingly, Slipher's galaxy spectra looked like the spectra of collections of stars with a bit of glowing interstellar gas mixed in. However, the emission and absorption lines in the spectra of these galaxies were seldom seen at the same wavelengths as in laboratory-generated spectra. Measurements of the Doppler velocities of these galaxies revealed that nearly all of the lines were shifted to longer, or redder, wavelengths, as shown in **Figure 14.3**. The galaxies were moving away from us, and the more distant they were, the faster they moved.

Recall from Chapter 5 that the Doppler shift causes the observed wavelengths of objects moving away from us to shift toward the red end of the spectrum. The wavelength in the laboratory is called the **rest wavelength** of the line, written λ_{rest}, and the **redshift** of a galaxy is written as *z*. Objects with higher values of

Observers everywhere should see the same universe.

The distribution of galaxies is isotropic and homogeneous.

FIGURE 14.2 Edwin Hubble, shown on the cover of the February 9, 1948, issue of *Time* magazine.

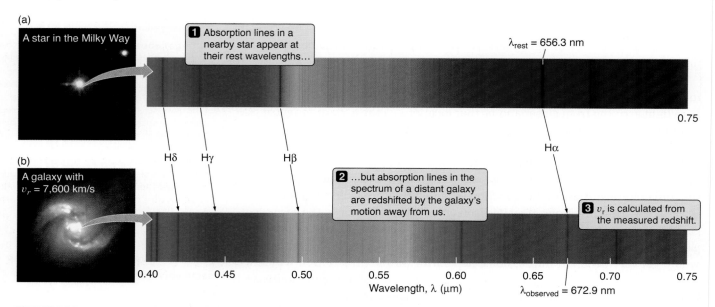

FIGURE 14.3 (a) A star in our galaxy, shown with its spectrum. (b) A distant galaxy, shown with its spectrum at the same scale as that of the star. Note that lines in the galaxy spectrum are redshifted to longer wavelengths.

Vocabulary Alert

Recession: In common language, this word is associated almost entirely with economics. Astronomers use it to refer to an object that is receding, or moving away. The recession velocity is therefore the velocity at which an object moves away.

▶‖ **AstroTour: Hubble's Law**

Light from distant galaxies is redshifted.

Hubble's law says that a galaxy's recession velocity is proportional to its distance.

z have greater redshifts and are therefore moving away from us more quickly. Hubble interpreted Slipher's redshifts as Doppler shifts, and he concluded that almost all of the galaxies in the universe are moving away from the Milky Way (see **Working It Out 14.1**). When he combined these measurements of galaxy recession velocities with his own estimates of the distances to these galaxies, he made one of the greatest discoveries in the history of astronomy. Hubble found that *the velocity at which a galaxy is moving away from us is proportional to the distance of that galaxy.* This simple relationship between distance and recession velocity has become known as **Hubble's law**. Hubble's law says that a galaxy 30 million parsecs (30 Mpc) away moves away from us twice as fast as a galaxy 15 Mpc away. (A parsec is 3.26 light-years, so a megaparsec is 3.26 million light-years.)

H_0 (which astronomers pronounce as "H naught") is the constant of proportionality between the distance and the speed, and it is called the **Hubble constant**. This constant is one of the most important numbers in cosmology, and many astronomical careers have been dedicated to trying to determine its value precisely and accurately.

All Observers See the Same Hubble Expansion

Hubble's law is a remarkable observation about the universe that has far-reaching implications. For one thing, Hubble's law helps us test the prediction that the universe is both isotropic and homogeneous. We can confirm its isotropy by observing that galaxies in one direction in the sky obey the same Hubble's law as galaxies in other directions in the sky. However, at first glance, you might think that Hubble's law suggests the universe is not homogeneous, because we seem to be sitting in a very special place: at the *center* of a tremendous expansion of space, with everything else in the universe streaming away from us. This initial

Working It Out 14.1 | Redshift: Calculating the Recession Velocity and Distance of Galaxies

The Doppler equation we learned for spectral lines showed that

$$v_r = \frac{\lambda_{observed} - \lambda_{rest}}{\lambda_{rest}} \times c$$

The fraction in front of the c is equal to z, the redshift. Substituting for the fraction, we get

$$v_r = z \times c$$

(Note: This correspondence works only for velocities much slower than the speed of light.)

Suppose we observe a hydrogen line with a rest wavelength of 122 nanometers (nm) in the spectrum of a distant galaxy. If the observed wavelength of the hydrogen line is 124 nm, then its redshift is

$$z = \frac{\lambda_{observed} - \lambda_{rest}}{\lambda_{rest}}$$

$$z = \frac{124 \text{ nm} - 122 \text{ nm}}{122 \text{ nm}}$$

$$z = 0.016$$

We can now calculate the recession velocity from this redshift:

$$v_r = z \times c = 0.016 \times 300{,}000 \text{ km/s} = 4{,}800 \text{ km/s}$$

How far away, though, is our distant galaxy? This is where Hubble's law and the Hubble constant ($H_0 = 72$ kilometers per second per megaparsec, or km/s/Mpc) come in. Hubble's law relates a galaxy's recession velocity to its distance, and can be expressed mathematically as $v_r = H_0 \times d_G$, where d_G is the distance to a galaxy measured in millions of parsecs. We can divide through by H_0 to get

$$d_G = \frac{v_r}{H_0}$$

$$d_G = \frac{4{,}800 \text{ km/s}}{72 \text{ km/s/Mpc}} = 67 \text{ Mpc}$$

From a simple measurement of the wavelength of a hydrogen line, we have learned that the distant galaxy is approximately 67 million parsecs away.

---

impression, however, is incorrect. *Hubble's law actually says that we are sitting in a uniformly expanding universe and that the expansion looks the same, regardless of our location.* To help you visualize this, we now turn to a useful model that you can build for yourself with materials you can probably find in your desk.

Figure 14.4 shows a long rubber band with paper clips attached along its length. If you stretch the rubber band, the paper clips, which represent galaxies in an expanding universe, get farther and farther apart. Imagine what this expansion would look like if you were an ant riding on paper clip A. As the rubber band is stretched, you notice that all of the paper clips are moving away from you. Clip B, the closest one, is moving away slowly. Clip C, located twice as far from you as B, is moving away twice as fast as B. Clip E, located four times as far away as B, is moving away four times as fast as B. From the perspective of an ant riding on clip A, all of the other paper clips on the rubber band are moving away with a velocity that is proportional to their distance. The paper clips located along the rubber band obey a Hubble-like law.

This demonstration of Hubble's law is a handy result, but the key insight comes from realizing there is nothing special about the perspective of paper clip A. If, instead, the ant were riding on clip E, clip D would be moving away slowly and clip A would be moving away four times as fast. Repeat this experiment for any paper clip along the rubber band, and you will arrive at the same result: the velocity at which other clips are moving away from the ant is proportional to their distance. The stretching rubber band, like the universe, is "homogeneous." The same Hubble-like law applies, regardless of where the ant is located.

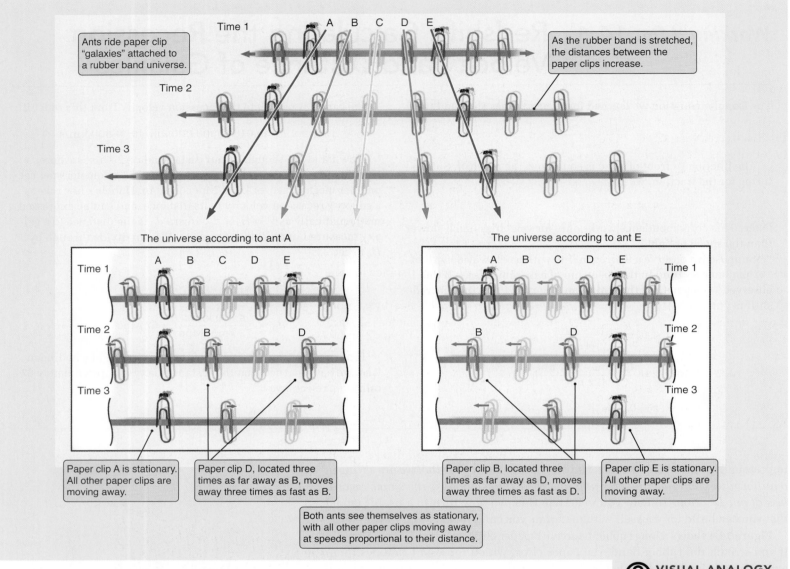

Ants ride paper clip "galaxies" attached to a rubber band universe.

As the rubber band is stretched, the distances between the paper clips increase.

Time 1
Time 2
Time 3

The universe according to ant A

A B C D E
Time 1
Time 2
B D
Time 3

Paper clip A is stationary. All other paper clips are moving away.

Paper clip D, located three times as far away as B, moves away three times as fast as B.

The universe according to ant E

A B C D E
Time 1
B D
Time 2
Time 3

Paper clip B, located three times as far away as D, moves away three times as fast as D.

Paper clip E is stationary. All other paper clips are moving away.

Both ants see themselves as stationary, with all other paper clips moving away at speeds proportional to their distance.

VISUAL ANALOGY

FIGURE 14.4 In this analogy of Hubble's law, a rubber band with paper clips evenly spaced along its length is stretched. As the rubber band stretches, an ant riding on clip A will observe clip C moving away twice as fast as clip B. Similarly, an ant riding on clip E will see clip C moving away twice as fast as clip D. Any ant will see itself as stationary, regardless of which paper clip it is riding, and it will see the other clips moving away with speed proportional to distance.

The universe is expanding uniformly.

The observation that nearby galaxies move away slowly and distant galaxies move away more rapidly does not say that we are at the center of anything. *Hubble's law means that the universe is expanding uniformly.* Any observer in any galaxy will see nearby galaxies moving away slowly and more distant galaxies moving away more rapidly. The same Hubble's law applies from their vantage point as applies from our vantage point on Earth.

Here is the only exception to this rule: in the case of galaxies that are gravitationally bound together, gravitational attraction dominates over the expansion of space. For example, the Andromeda Galaxy and the Milky Way are being pulled together by gravity. The Andromeda Galaxy is approaching at about 120 kilometers per second (km/s), so light from the Andromeda Galaxy is blueshifted. The fact that gravitational or electromagnetic forces can overwhelm the expansion of space also explains why the Solar System is not expanding, and neither are you.

Astronomers Build a Distance Ladder to Measure the Hubble Constant

Hubble's law tells us that our universe is expanding. But to know the present *rate* of the expansion, we need a good value for the Hubble constant, H_0. This requires knowing both the recession velocity and the distance of a large number of galaxies, including galaxies that are very far away.

Distances of remote objects are measured in a series of steps referred to as the **distance ladder**, which relates distances on a variety of scales, as illustrated in **Figure 14.5**. Within the Solar System, we can find distances using radar and signals from space probes. Once the distance to the Sun is known, we can use parallax (as discussed in Chapter 10) to measure distances to nearby stars, find their luminosities, and plot them on the H-R diagram. This equation calibrates the diagram, so that we need to know only a main-sequence star's temperature (relatively easy to measure from its spectrum) to find its luminosity. This information in turn enables us to estimate the star's distance by comparing its *apparent* brightness with its luminosity. This process for finding the distance to stars is known as **spectroscopic parallax**.

Moving farther out, we can measure the distance to relatively nearby galaxies using standard candles. More distant objects of the same type, such as O stars, globular clusters, planetary nebulae, novae, and variable stars, are fainter, or smaller. In the chapter-opening figure, a student is using this concept to find the relative distance to several spiral galaxies. Typically, astronomers use the brightness of objects rather than the size. Cepheid variables, an example of a standard candle, take us another step up the distance ladder and enable us to accurately measure distances to galaxies as far away as 30 Mpc. Even this is not far enough to determine a reliable value for the Hubble constant, but within that volume of space are many galaxies that we can scour for yet more powerful distance indicators. Among the best of these are Type Ia supernovae.

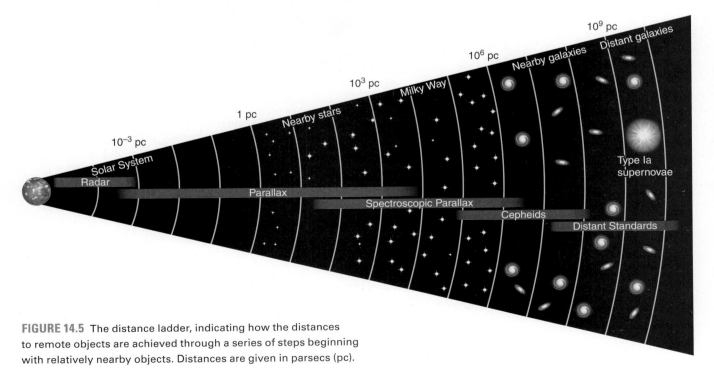

FIGURE 14.5 The distance ladder, indicating how the distances to remote objects are achieved through a series of steps beginning with relatively nearby objects. Distances are given in parsecs (pc).

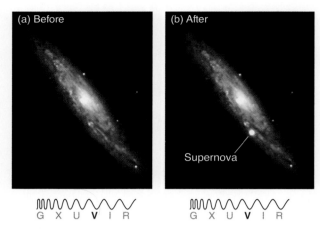

FIGURE 14.6 Galaxy NGC 3877 (a) before and (b) after the explosion of a Type Ia supernova. Type Ia supernovae are extremely luminous standard candles.

Recall from Chapter 12 that Type Ia supernovae occur when gas flows from an evolved star onto its white dwarf companion, pushing the white dwarf over the Chandrasekhar limit for the mass of an electron-degenerate object. When this happens, the overburdened white dwarf begins to collapse and then explodes. Because all Type Ia supernovae occur in white dwarfs of the same mass, we expect all such explosions to have about the same luminosity. This prediction is borne out by observations of Type Ia supernovae in galaxies with known distances. With a peak luminosity that outshines billions of Suns (**Figure 14.6**), Type Ia supernovae can be seen and measured with modern telescopes that can reach almost to the edge of the observable universe.

For the last few decades of the 20th century, astronomers working to obtain a value for H_0 were largely split into two camps. One group favored a Hubble constant of about 60 km/s/Mpc; a second camp viewed the data as supporting a value of about 110 km/s/Mpc. These two groups argued for years, each trying to obtain better and more convincing data, until Hubble Space Telescope observations in the mid-1990s began to converge on an intermediate value of about 72 km/s/Mpc. Most recently, other measures of distance have yielded results consistent with that value.

Figure 14.7 plots the measured recession velocities of galaxies against their measured distances. Because the velocity and the distance are proportional to

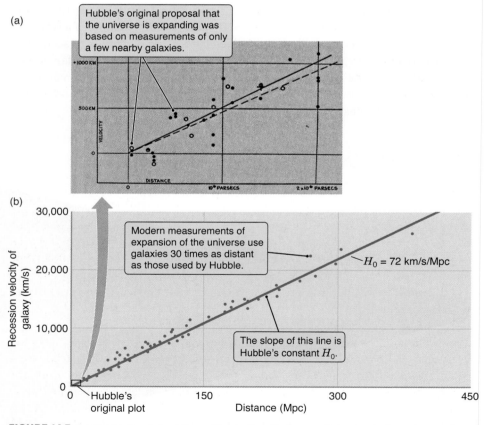

FIGURE 14.7 (a) Hubble's original figure illustrating that more distant galaxies are receding faster than less distant galaxies. (b) Modern data on galaxies up to 30 times farther away than those studied by Hubble show that recession velocity is proportional to distance.

each other, the points lie along a line on the graph with a slope equal to the proportionality constant H_0. Notice how well the data line up along the line. This tells us that the universe follows Hubble's law. Today, we have measured the Hubble constant to an accuracy of a few percent. The value is likely to be further refined in the years to come.

Current measurements put H_0 at 72 km/s/Mpc.

Hubble's Law Maps the Universe in Space and Time

Hubble's law gives us a practical tool for measuring distances to extremely remote objects. Once we know the value of H_0, we can use a straightforward measurement of the redshift of a galaxy to find its distance. In other words, once we know H_0, Hubble's law makes the once-difficult task of measuring distances in the universe relatively easy, allowing us to literally map the structure of the universe. This may seem like a logical impossibility, because we are using redshifts and distances to find H_0 and then using H_0 to find the distances. But we find H_0 from one set of galaxies and then we use that value to find the distance to a different, more distant set of galaxies.

Redshift tells us a galaxy's distance.

Hubble's law does more than place galaxies in space. It also places galaxies in time. Light travels at a huge but finite speed. Remember that when we look at the Sun, we see it as it existed 8 minutes ago. When we look at Alpha Centauri, the nearest stellar system beyond the Sun, we see it as it existed 4.3 years ago. When we look at the center of our galaxy, the picture we see is 27,000 years old. In general, when looking at a distant object, we speak of its **look-back time**—the time it has taken for the light from that object to reach our telescope. As we look into the distant universe, look-back times become very great indeed. The distance to a galaxy whose redshift $z = 0.1$ is 1.4 billion light-years (assuming $H_0 = 72$ km/s/Mpc), so the look-back time to that galaxy is 1.4 billion years. The look-back time to a galaxy where $z = 0.2$ is 2.7 billion years. As we look at objects with greater and greater redshifts, we see increasingly younger stages of our universe.

14.3 The Universe Began in the Big Bang

The most significant aspect of Hubble's law is what it tells us about the structure and origin of the universe itself. We know that all galaxies in the universe are moving apart, so if we could somehow run the "movie" of this separation backward in time, we would find the separation between the galaxies getting smaller as they become younger and younger. If we assume the speed of the expansion has been constant, then Hubble's law shows that about 6.8 billion years ago, when the universe was half its present age, all of the galaxies in the universe were separated from each other by half their present distances. Twelve billion years ago, all of the galaxies in the universe must have been separated from each other by about a tenth of their present distances. Assuming that galaxies have been moving apart at the same speed that we see today, then 13.7 billion years ago (a time equal to $1/H_0$), *there was no space between all the bits of matter and energy that make up today's universe* (see **Working It Out 14.2**). All that matter and energy, with no space to spread out in, was unimaginably dense and hot. The value of 1 divided by the Hubble constant is referred to as the **Hubble time**, which represents an estimate of the universe's age. Today's universe is hurtling outward from a tremendous expansion of space that took place 13.7 billion years

▶❚❚ AstroTour: **Hubble's Law**

Expansion and the Age of the Universe

We can use Hubble's law to estimate the age of the universe. Consider two galaxies located 30 Mpc ($d_G = 9.3 \times 10^{20}$ km) away from each other (**Figure 14.8**). If these two galaxies are moving apart from each other, then at some time in the past they must have been together in the same place at the same time. According to Hubble's law, and assuming that $H_0 = 72$ km/s/Mpc, the distance between these two galaxies is increasing at the rate

$$v_r = H_0 \times d_G$$
$$v_r = 72 \text{ km/s/Mpc} \times 30 \text{ Mpc}$$
$$v_r = 2{,}160 \text{ km/s}$$

Knowing the speed at which they are traveling, we can calculate the time it took for the two galaxies to become separated by 30 Mpc:

$$\text{Time} = \frac{\text{Distance}}{\text{Speed}} = \frac{9.3 \times 10^{20} \text{ km}}{2{,}160 \text{ km/s}} = 4.3 \times 10^{17} \text{ s}$$

Dividing by the number of seconds in a year, (about 3.16×10^7 s/yr) gives

$$\text{Time} = 1.4 \times 10^{10} \text{ yr}$$

In other words, *if* expansion of the universe has been constant, two galaxies that today are 30 Mpc apart started out at the same place about 14 billion years ago.

Now let's do the same calculation with two galaxies that are 60 Mpc apart (see Figure 14.8). These two galaxies are twice as far apart, but the distance between them is increasing twice as rapidly:

$$v_r = H_0 \times d_G = 72 \text{ km/s/Mpc} \times 60 \text{ Mpc} = 4{,}320 \text{ km/s}$$

Therefore,

$$\text{Time} = \frac{\text{Distance}}{\text{Speed}} = \frac{19 \times 10^{20} \text{ km}}{4{,}320 \text{ km/s}} = 4.4 \times 10^{17} \text{ s}$$

Dividing by the number of seconds in a year gives

$$\text{Time} = 1.4 \times 10^{10} \text{ yr}$$

Again, we calculate time as distance divided by speed (twice the distance divided by twice the speed) to find that these galaxies also took about 14 billion years to reach their current locations. We could do this calculation again and again for any pair of galaxies in the universe today. The farther apart the two galaxies are, the faster they are moving. But all galaxies took the same amount of time to get to where they are today.

If we work out the example using words instead of numbers, we can see why the answer is always the same. Because the velocity we are calculating comes from Hubble's law, velocity equals Hubble's constant multiplied by distance. Writing this out as an equation, we get

$$\text{Time} = \frac{\text{Distance}}{\text{Velocity}}$$

$$\text{Time} = \frac{\text{Distance}}{H_0 \times \text{Distance}}$$

Distance divides out to give

$$\text{Time} = \frac{1}{H_0}$$

We define $1/H_0$ as the Hubble time. This is one way of estimating the age of the universe.

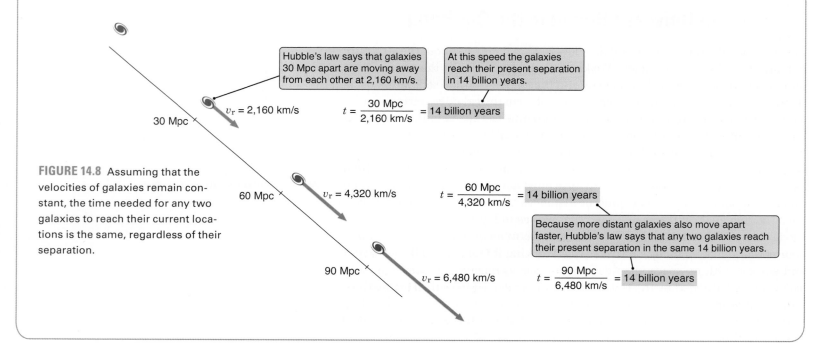

FIGURE 14.8 Assuming that the velocities of galaxies remain constant, the time needed for any two galaxies to reach their current locations is the same, regardless of their separation.

30 Mpc $v_r = 2{,}160$ km/s

Hubble's law says that galaxies 30 Mpc apart are moving away from each other at 2,160 km/s.

At this speed the galaxies reach their present separation in 14 billion years.

$$t = \frac{30 \text{ Mpc}}{2{,}160 \text{ km/s}} = 14 \text{ billion years}$$

60 Mpc $v_r = 4{,}320$ km/s

$$t = \frac{60 \text{ Mpc}}{4{,}320 \text{ km/s}} = 14 \text{ billion years}$$

Because more distant galaxies also move apart faster, Hubble's law says that any two galaxies reach their present separation in the same 14 billion years.

90 Mpc $v_r = 6{,}480$ km/s

$$t = \frac{90 \text{ Mpc}}{6{,}480 \text{ km/s}} = 14 \text{ billion years}$$

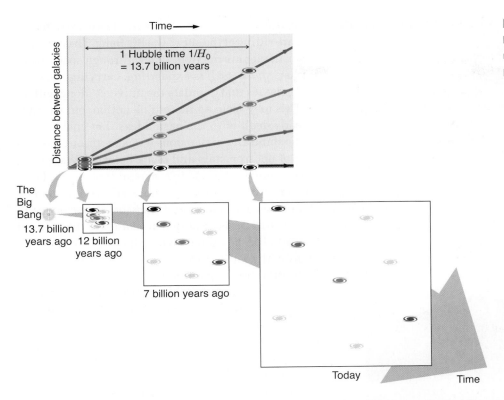

ago. This colossal event, which marked the beginning of our universe, is the **Big Bang (Figure 14.9).**

The idea of the Big Bang greatly troubled many astronomers in the early and middle years of the 20th century. Several different suggestions were put forward to explain the observed fact of Hubble expansion without resorting to the idea that the universe came into existence billions of years ago, and was extraordinarily hot and dense at that time. However, as more and more observations have come in and more discoveries about the structure of the universe have been made, the Big Bang theory has only grown stronger. Today only a very few serious workers in this field question whether the Big Bang took place. Virtually all the major predictions of the Big Bang theory (expansion of the universe being but one of them) have proven to be correct. As we will see, the Big Bang theory for the origin of our universe is now so well corroborated that it has crossed into the realm of scientific fact.

Expansion started with a Big Bang.

The implications of Hubble's law are striking. This single discovery forever changed our concept of the origin, history, and possible future of the universe in which we live. At the same time, Hubble's law has pointed to many new questions about the universe. To address them, we next need to consider exactly what we mean by the term *expanding universe*.

Galaxies Are *Not* Flying Apart through Space

At this point in our discussion, you may be picturing the expanding universe as a cloud of debris from an explosion flying outward through surrounding space. However, this is not an explosion in the usual sense of the word, and there is no surrounding space.

A common question about the Big Bang is, "Where did it take place?" The answer to this question, amazing as it seems, is that the Big Bang took place

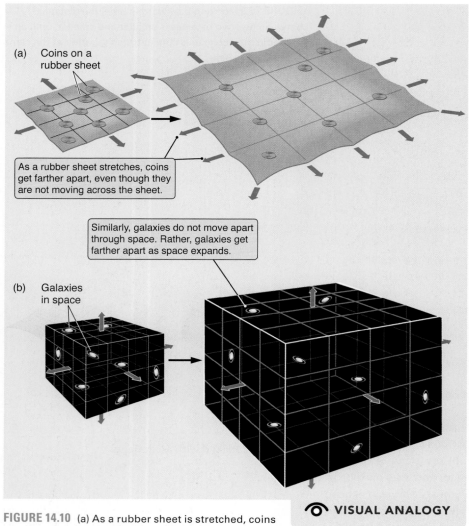

(a) Coins on a rubber sheet

As a rubber sheet stretches, coins get farther apart, even though they are not moving across the sheet.

Similarly, galaxies do not move apart through space. Rather, galaxies get farther apart as space expands.

(b) Galaxies in space

VISUAL ANALOGY

FIGURE 14.10 (a) As a rubber sheet is stretched, coins on its surface move farther apart, even though they are not moving with respect to the sheet itself. Any coin on the surface of the sheet will observe a Hubble-like law in every direction. (b) In analogous fashion, galaxies in an expanding universe are not flying apart through space. Rather, space itself is stretching.

The Big Bang happened everywhere.

everywhere. Wherever you are in the universe today, you are sitting at the site of the Big Bang. The reason is that galaxies are not flying apart through space at all. Rather, *space itself* is expanding, carrying the stars and galaxies that populate the universe along with it.

This may seem an incredible notion, but we have already dealt with the basic ideas that enable us to understand the expansion of space. In our discussion of neutron stars and black holes in Chapter 13, we encountered Einstein's general theory of relativity. General relativity says that space is distorted by the presence of mass, and that the consequence of this distortion is gravity. For example, the mass of the Sun, like any other object, distorts the geometry of spacetime around it; so Earth, coasting along in its inertial frame of reference, follows a curved path around the Sun. We illustrated this phenomenon in Figure 13.20 with the analogy of a ball placed on a stretched rubber sheet, showing how the ball distorted the surface of the sheet.

The surface of a rubber sheet can be distorted in other ways. Imagine a number of coins placed on a rubber sheet, as shown in **Figure 14.10**. Suppose we grab the edges of the sheet and begin pulling them outward. As the rubber sheet stretches, each coin remains at the same location on the surface of the sheet, but the distances between the coins increase. Two coins sitting close to each other move apart only slowly, while coins farther apart move away from each other more rapidly. The coins on the surface of a rubber sheet will obey a Hubble-like relationship as the sheet is stretched. Analogously, space in the universe is stretching, and galaxies are being carried along with the flow. This phenomenon is different from galaxies moving apart *through* space.

In the case of coins on a rubber sheet, there is a limit to how far we can stretch the sheet before it breaks. With space and the real universe, there is no such limit—the fabric of space can, in principle, go on expanding forever. Hubble's law is the observational consequence of the fact that the space between the galaxies is expanding.

Expansion Is Described with a Scale Factor

Suppose we place a ruler on the surface of the rubber sheet and draw a tick mark every centimeter, as in **Figure 14.11a**. If we want to know the distance between two points on the sheet, all we need to do is count the marks between the two points and multiply by 1 centimeter (cm) per tick mark.

As the sheet is stretched, however, the distance between the tick marks does not remain 1 cm. When the sheet is stretched to 150 percent of the size that it had when we drew our ruler, each tick mark is separated from its neighbors by 1½ times the original distance, or 1.5 cm. If we wanted to know the distance between two points, we could still count the marks; but we would have to increase the distance between tick marks by 1.5 to find the distance in centimeters. The **scale factor** of the sheet is now 1.5. If the sheet were twice the size that it was when we drew the ruler (**Figure 14.11b**), each mark would correspond to 2 cm of actual distance; the scale factor of the sheet would now be 2. The scale factor tells us the size of the sheet relative to its size at the time when we drew our ruler. The scale factor also tells us how much the distance between points on the sheet has changed.

We can apply this same idea to the universe. Suppose we choose today to lay out a "cosmic ruler" on the fabric of space, placing an imaginary tick mark every 10 Mpc. We define the scale factor of the universe at this time to be 1. In the past, when the universe was smaller, distances between the points in space marked by our cosmic ruler would have been less than 10 Mpc. The scale factor of that younger, smaller universe would have been less than 1 compared to the scale factor today. In the future, as the universe continues to expand, the distances between the tick marks on our cosmic ruler will grow to more than 10 Mpc, and the scale factor of the universe will be greater than 1. In this way we can use the scale factor, usually written as R_U, to keep track of the changing scale of the universe.

It is important to remember, when thinking about the expanding universe, that the laws of physics are themselves unchanged by the changing scale factor. For example, when we stretched out the rubber sheet, we did not change the properties of the coins on its surface. In like fashion, as space expands, the sizes and other physical properties of atoms, stars, and galaxies within it also remain unchanged.

Let's return now to the question of locating the center of expansion. Looking back in time, we see that the scale factor of the universe gets smaller and smaller, approaching zero as we get closer and closer to the Big Bang. The fabric of space that today spans billions of light-years spanned much smaller distances when the universe was young. When the universe was only a day old, all of the space that we see today amounted to a region only a few times the size of our Solar System. When the universe was a 50th of a second old, the vast expanse of space that makes up today's observable universe (and all the matter in it) occupied a volume only the size of today's Earth. As we continue to approach the Big Bang itself, going backward in time, the space that makes up today's observable universe becomes smaller and smaller—the size of a grapefruit, a marble, an atom, a proton. Every point in the fabric of space that makes up today's universe was right there at the beginning, a part of that unimaginably tiny, dense universe that emerged from the Big Bang.

These points bear repetition: *Where is the center of the Big Bang?* There is no center. The Big Bang did not occur at a specific point in space, because space itself came into existence with the Big Bang. *Where did the Big Bang happen?* It happened everywhere, including right where you are sitting. This is an important concept. If a particular point in today's universe marked the site of the Big Bang, that would be a very special point indeed. But there is no such point. The Big Bang happened everywhere. A Big Bang universe is homogeneous and isotropic, consistent with the cosmological principle.

Redshift Is Due to the Changing Scale Factor of the Universe

General relativity gives us a powerful tool for interpreting Hubble's great discovery of the expanding universe. It also forces us to rethink just what we mean when

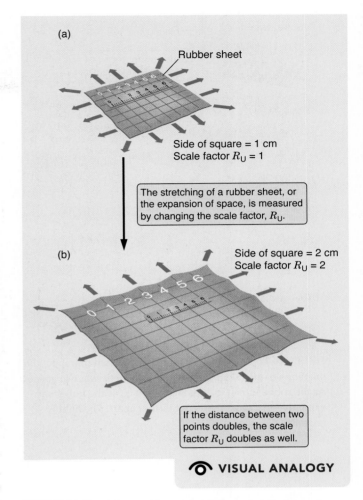

(a)

Rubber sheet

Side of square = 1 cm
Scale factor $R_U = 1$

The stretching of a rubber sheet, or the expansion of space, is measured by changing the scale factor, R_U.

(b)

Side of square = 2 cm
Scale factor $R_U = 2$

If the distance between two points doubles, the scale factor R_U doubles as well.

VISUAL ANALOGY

FIGURE 14.11 (a) On a rubber sheet, tick marks are drawn 1 cm apart. As the sheet is stretched, the tick marks move farther apart. (b) When the spacing between the tick marks is 2 cm, or twice the original value, we say that the scale factor of the sheet, R_U, has doubled. A similar scale factor, R_U, is used to describe the expansion of the universe.

Expansion does not affect the local physics of a galaxy, stars, atoms, or anything else.

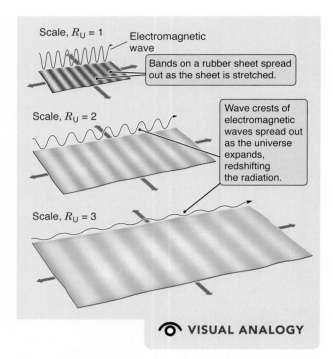

Scale, $R_U = 1$

Electromagnetic wave

Bands on a rubber sheet spread out as the sheet is stretched.

Scale, $R_U = 2$

Wave crests of electromagnetic waves spread out as the universe expands, redshifting the radiation.

Scale, $R_U = 3$

VISUAL ANALOGY

FIGURE 14.12 Bands drawn on a rubber sheet represent the positions of the crests of an electromagnetic wave in space. As the rubber sheet is stretched—that is, as the universe expands—the wave crests get farther apart. The light is redshifted.

we talk about the redshift of distant galaxies. Although it is true that the distance between galaxies is increasing as a result of the expansion of the universe, and that we can use the *equation* for Doppler shifts to measure the redshifts of galaxies, these redshifts are not due to Doppler shifts at all. As light comes toward us from distant galaxies, the scale factor of the space through which the light travels is constantly increasing; and as it does so, the distance between adjacent wave crests increases as well. The light is "stretched out" as the space it travels through expands.

Let's return to our rubber sheet analogy. If we draw a series of bands on the rubber sheet to represent the crests of an electromagnetic wave, as in **Figure 14.12**, we can watch what happens to the wave as the sheet is stretched out. By the time the sheet is stretched to twice its original size—that is, by the time the scale factor of the sheet is 2—the distance between wave crests has doubled. When the sheet has been stretched to three times its original size (a scale factor of 3), the wavelength of the wave will be three times what it was originally.

We apply this idea to light coming from a distant galaxy. When the light left the galaxy of its origin, the scale factor of the universe was smaller than it is today. The universe expanded while the light was in transit, and as it did so, the wavelength of the light grew longer in proportion to the increasing scale factor of the universe. The redshift of light from distant galaxies is therefore a direct measure of how much the universe has expanded since the time when the radiation left its source. Redshift measures how much the scale factor of the universe, R_U, has changed since the light was emitted.

14.4 Major Predictions of the Big Bang Theory Are Resoundingly Confirmed

Cosmological questions are some of the most fundamental questions we can ask. It is quite remarkable that we live in a time when we are finding real, testable answers to these questions by appealing to the empirical methods of science. It is essential, then, that the evidence supporting the theory of the Big Bang is of extraordinary quality. What evidence, apart from the observed expansion of the universe itself, requires us to accept that the Big Bang actually took place?

We See Radiation Left Over from the Big Bang

The story of the single most important confirmation of the Big Bang theory begins in the mid-1940s, when George Gamow (1904–1968) and his student Ralph Alpher (1921–2007) were thinking about the implications of Hubble expansion. They reasoned that since a compressed gas cools as it expands, the expanding universe should also be cooling. When the universe was very young and small, it must have consisted of an extraordinarily hot, dense gas. This hot, dense early universe would have been awash in blackbody radiation, which exhibits a blackbody spectrum.

Gamow and Alpher took this idea a step further. As the universe expanded, they reasoned, this radiation would have been redshifted to longer and longer wavelengths. Recall Wien's law from Chapter 6, which states that the temperature associated with blackbody radiation is inversely proportional to the peak wavelength. Shifting the wavelength of blackbody radiation to longer wavelengths is like shifting the temperature to lower values. As illustrated in **Figure 14.13**, dou-

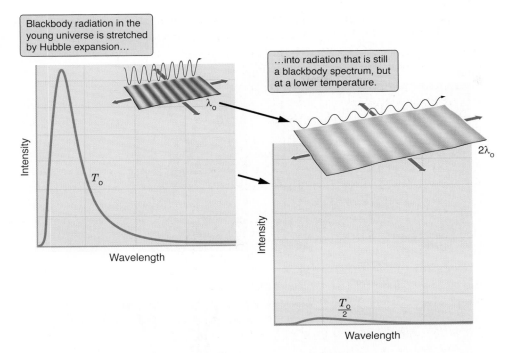

Blackbody radiation in the young universe is stretched by Hubble expansion…

…into radiation that is still a blackbody spectrum, but at a lower temperature.

λ_o

$2\lambda_o$

Intensity

Wavelength

T_o

Intensity

Wavelength

$\frac{T_o}{2}$

FIGURE 14.13 As the universe expanded, blackbody radiation left over from the hot young universe was redshifted to longer wavelengths. Redshifting a blackbody spectrum is equivalent to lowering its temperature.

bling the wavelength of the photons in a blackbody spectrum by doubling the scale factor of the universe is equivalent to cutting the temperature of the blackbody spectrum in half. In 1948, Alpher, Hans Bethe (1906–2005), and Gamow published a paper asserting that the radiation from the early universe should still be visible today and should have a blackbody spectrum with a temperature of about 5–10 kelvins (K). Early attempts to detect this radiation were unsuccessful.

In the early 1960s, two physicists at Bell Laboratories, Arno Penzias (b. 1933) and Robert Wilson (b. 1936), were trying to bounce radio signals off newly launched satellites. This hardly seems much of a feat today, when we routinely use cell phones and handheld GPS systems that communicate directly with satellites. At the time, however, the effort pushed technology to its limits. Penzias and Wilson needed a very sensitive microwave telescope for their work. Any spurious signals coming from the telescope itself might wash out the faint signals bounced off a satellite. Pictured in **Figure 14.14** along with their radio telescope, Penzias and Wilson worked tirelessly to eliminate all possible sources of interference originating from within their instrument. This work included such endless and menial tasks as keeping the telescope free of bird droppings and other extraneous material. Even so, Penzias and Wilson found that no matter how hard they tried to eliminate sources of extraneous noise, they could still detect a faint microwave signal when they pointed the telescope at the sky. Eventually, they came to accept that the signal they were detecting was real. The sky faintly glows in microwaves.

In the meantime, physicist Robert Dicke (1916–1997) and his colleagues at Princeton University also predicted a hot early universe, arriving independently at the same basic conclusions that Alpher and Gamow had reached two decades earlier. When Dicke and colleagues heard of the signal that Penzias and Wilson had found, they interpreted it as the radiation left behind by the hot early universe. The strength of the detected signal was consistent with the glow from a blackbody with a temperature of about 3 K, very close to the predicted value. Their results, published in 1965, reported the discovery of the glow left behind

FIGURE 14.14 Penzias and Wilson next to the Bell Labs radio telescope antenna with which they discovered the cosmic microwave background. This antenna is now a U.S. National Historic Landmark.

1 In the ionized early universe, light was trapped by free electrons. Radiation had a blackbody spectrum.

(a)

KEY
- Proton
- Electron
- Path of photon

(b)

2 When the universe recombined, it became transparent, and the blackbody radiation went its own way.

3 Recombination was like the fog suddenly clearing.

👁 **VISUAL ANALOGY**

FIGURE 14.15 The origin of cosmic microwave background radiation. (a) Before recombination, the universe was like a foggy day, except that the "fog" was a sea of hydrogen atoms. Radiation interacted strongly with free electrons and so could not travel far. The trapped radiation had a blackbody spectrum. (b) When the universe recombined, the fog cleared and this radiation was free to travel unimpeded.

by the Big Bang. Penzias and Wilson shared the 1979 Nobel Prize in Physics for their remarkable discovery.

This radiation left over from the early universe is called **cosmic microwave background (CMB) radiation**. When the universe was young, it was hot enough that all of the atoms in the universe were ions. Free electrons in such a plasma interact strongly with radiation, blocking its progress. At that time in the early universe, conditions within the universe were much like the conditions within a star: the universe was an opaque blackbody (**Figure 14.15a**).

As the universe expanded, the gas filling it cooled. By the time the universe was about a thousandth of its current size, the temperature had dropped to a few thousand kelvins, allowing protons and electrons to combine to form hydrogen

> The CMB is blackbody radiation that arose when the universe was hot and ionized.

atoms. This event, called the **recombination** of the universe, occurred when the universe was several hundred thousand years old (**Figure 14.15b**).

Hydrogen atoms are much less effective at blocking radiation than free electrons are, so when recombination occurred, the universe suddenly became transparent to radiation. Since that time, the radiation left behind from the Big Bang has been able to travel largely unimpeded throughout the universe. At the time of recombination, when the temperature of the universe was 3000 K, the radiation peaked at a wavelength of about 1 micrometer (μm). As the universe expands, this radiation is redshifted to longer wavelengths—and therefore, as in Figure 14.13, cooler temperatures. Today, the scale of the universe has increased a thousandfold since recombination, and the peak wavelength of the CMB has increased by a thousand times as well, to a value close to 1 millimeter (mm). The spectrum of the CMB still has the shape of a blackbody spectrum, but with a characteristic temperature of 2.73 K—only a thousandth what it was at the time of recombination.

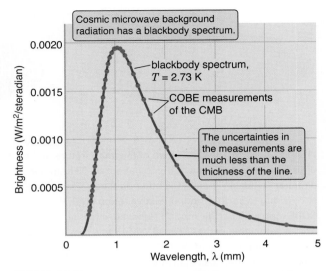

FIGURE 14.16 The spectrum of the CMB as measured by the Cosmic Background Explorer (COBE) satellite (red dots). The uncertainty in the measurement at each wavelength is much less than the size of a dot. The line running through the data is a blackbody spectrum with a temperature of 2.73 K.

Satellite Data Confirmed that the CMB Is Real

The presence of the cosmic microwave background radiation with a blackbody spectrum is a very strong prediction of the Big Bang theory. Penzias and Wilson had confirmed that a signal with the correct strength was there, but they could not say for certain whether the signal they saw had the spectral shape of a blackbody spectrum. It was not until the end of the 1980s that the predictions of Big Bang cosmology for the CMB were put to the ultimate test. A satellite called the Cosmic Background Explorer, or COBE, was launched in 1989. COBE made extremely precise measurements of the CMB at many wavelengths, from a few micrometers out to 1 cm. In January 1990, hundreds of astronomers gathered in a large conference room in Washington, D.C., at the winter meeting of the American Astronomical Society to hear the COBE team present its first results. Security surrounding the new findings had been tight, so the atmosphere in the room was electric. The tension did not last long; presentation of a single graph brought the room's occupants to their feet in a spontaneous ovation. John Mather (b. 1946) and George Smoot (b. 1945) were awarded a Nobel Prize for this work in 2006.

The data shown on that graph are reproduced in **Figure 14.16**. The small dots in the figure are the COBE measurements of the CMB at different frequencies. The uncertainty in each measurement is far less than the size of each dot. The line in the figure, which runs perfectly through the data points, is a blackbody spectrum with a temperature of 2.73 K. The agreement between theoretical prediction and observation is truly remarkable. The observed spectrum so perfectly matches the one predicted by Big Bang cosmology that there can be no real doubt we are seeing the residual radiation left behind from the hot, dense beginning of the early universe.

COBE unambiguously showed the CMB to have a blackbody spectrum at 2.73 K.

The CMB Measures Earth's Motion Relative to the Universe Itself

COBE provided us with much more than a measurement of the spectrum of the CMB. **Figure 14.17a** shows a map obtained by COBE of the CMB from the entire sky. The different colors in the map correspond to variations of less than 0.1 percent in the temperature of the CMB. Most of this range of temperature is present because one side of the sky looks slightly warmer than the opposite side of the

(a) (b) (c)

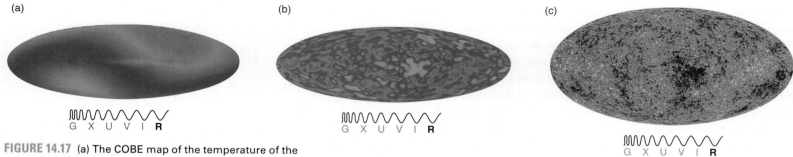

FIGURE 14.17 (a) The COBE map of the temperature of the CMB. The CMB is slightly hotter (by about 0.003 K) in one direction in the sky than in the other direction. This difference is due to Earth's motion relative to the CMB. (b) The COBE map with Earth's motion removed, showing tiny ripples remaining in the CMB. (c) WMAP has provided the highest resolution yet of the CMB. The radiation seen in this image was emitted less than 400,000 years after the Big Bang.

COBE found, and WMAP further revealed, tiny variations in the CMB resulting from the formation of structure during early times.

sky. This difference has nothing to do with the structure of the universe itself, but rather is the result of the motion of Earth with respect to the CMB.

The COBE map shows that one side of the sky is slightly hotter than the other because Earth and our Sun are moving at a velocity of 368 km/s in the direction of the constellation Crater. Radiation coming from the direction in which we are moving is slightly blueshifted (shifted to a higher characteristic temperature) by our motion, whereas radiation coming from the opposite direction is Doppler-shifted toward the red (or cooler temperatures). Our motion is due to a combination of factors, including the motion of our Sun around the center of the Milky Way and the motion of our galaxy relative to the CMB.

If we subtract this asymmetry from the COBE map, only slight variations in the CMB remain, as shown in **Figure 14.17b**. The brighter parts of this image are only about 1.00001 times brighter than the fainter parts. These slight variations are of crucial importance in the history of the universe. These tiny fluctuations in the CMB are the result of gravitational redshifts (see Chapter 13) caused by concentrations of mass that existed in the early universe. These concentrations later gave rise to galaxies and the rest of the structure that we see in the universe today.

Subsequent observations support the COBE findings. Beginning in 2001, the Wilkinson Microwave Anisotropy Probe (WMAP) satellite made more precise measurements of the variations of the CMB. **Figure 14.17c** shows the ripples measured by WMAP. These much higher-resolution maps have profound implications for our understanding of the origin of structure in the universe and enable us to determine several cosmological parameters. For example, the value of the Hubble constant we use in this book is the value inferred from the WMAP experiment.

The Big Bang Theory Correctly Predicts the Abundance of the Least Massive Elements

The next confrontation between the predictions of the Big Bang theory and observations of the universe came from a very different direction. When the universe was only a few minutes old, its temperature and density were high enough for nuclear reactions to take place. Collisions between protons in the early universe built up low-mass nuclei, including deuterium (heavy hydrogen) and isotopes of helium, lithium, beryllium, and boron. This process, called **Big Bang nucleosynthesis**, determined the final chemical composition of the matter that emerged from the hot phase of the Big Bang. No elements more massive than boron could have been formed in the Big Bang, because at that time the density in the universe was too low for reactions such as the triple-alpha process, which forms carbon in the interiors of stars (see Chapter 12), to occur. Therefore, all the more massive elements in the universe, including the atoms making up the bulk of our planet and ourselves, must have formed in subsequent generations of stars.

Figure 14.18 shows the calculated predictions of element abundances from Big Bang nucleosynthesis, plotted as a function of the present-day density of normal (luminous) matter in the universe. The first thing we see from this figure is that about 24 percent of the mass of the normal matter formed in the early universe should have ended up in the form of the very stable isotope ^{4}He, regardless of the density of matter in the universe. Indeed, when we look about us in the universe today, we find that about 24 percent of the mass of normal matter in the universe is in the form of ^{4}He, in complete agreement with the prediction of Big Bang nucleosynthesis.

Unlike helium, most isotope abundances depend sensitively on the density of normal matter in the universe. Comparing current abundances with models of isotope formation in the Big Bang helps pin down the density of the early universe. Beginning with the abundances of isotopes such as deuterium (^{2}H) and ^{3}He found in the universe today (shown as roughly horizontal pastel bands in Figure 14.18), and comparing them to different models' predictions about how abundant isotopes *should* be when formed at different densities (shown as saturated colors in Figure 14.18), we can find the density of normal matter (shown as the vertical yellow band in Figure 14.18). The best current measurements give a value of about 3.9×10^{-28} kilograms per cubic meter (kg/m^3) for the average density of normal matter in the universe today. This value lies well within the range predicted by the observations shown in Figure 14.18. Once again, the agreement is remarkable. And it holds for many different isotopes. Turning this around, we

▶❙❙ **AstroTour:** Big Bang Nucleosynthesis

Big Bang theory predicts a universe that is 24 percent helium by mass, which observations confirm.

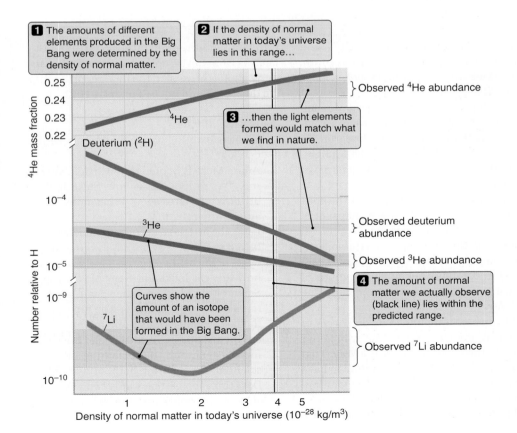

1 The amounts of different elements produced in the Big Bang were determined by the density of normal matter.

2 If the density of normal matter in today's universe lies in this range…

3 …then the light elements formed would match what we find in nature.

4 The amount of normal matter we actually observe (black line) lies within the predicted range.

Curves show the amount of an isotope that would have been formed in the Big Bang.

Observed ^{4}He abundance

Observed deuterium abundance

Observed ^{3}He abundance

Observed ^{7}Li abundance

^{4}He

Deuterium (^{2}H)

^{3}He

^{7}Li

^{4}He mass fraction: 0.25, 0.24, 0.23, 0.22

Number relative to H: 10^{-4}, 10^{-5}, 10^{-9}, 10^{-10}

Density of normal matter in today's universe (10^{-28} kg/m^3): 1, 2, 3, 4, 5

FIGURE 14.18 Observed and calculated abundances of the products of Big Bang nucleosynthesis, plotted against the density of normal matter in today's universe. Big Bang nucleosynthesis correctly predicts the amounts of these isotopes found in the universe today.

can begin with an observation of the amount of normal matter in and around galaxies and then use our understanding of the Big Bang to calculate what the chemical composition emerging from the Big Bang should have been. When we do this, the answer we get agrees remarkably well with the amounts of these elements we actually find in nature. This agreement also provides a powerful constraint on the nature of dark matter, which dominates the mass in the universe. As we see in the next chapter, dark matter cannot consist of normal matter made up of neutrons and protons; if it did, the density of neutrons and protons in the early universe would have been much higher, and the resulting abundances of light elements in the universe would have been much different from what we actually observe.

READING ASTRONOMY News

Cosmologists continue to study the CMB, using new, improved satellites with more sensitive detectors that see with greater resolution. These detectors are providing us with other exciting results along the way.

A Snapshot of the Sky from Billions of Years Ago

By **Amy Minsky,** *Montreal Gazette*

An international team of scientists, including some from Vancouver, unveiled the first wave of findings gleaned from Europe's Planck Space Telescope—a billion-dollar device designed to study the oldest source of light.

The ultimate goal is to reveal a picture of what the universe was like just after the Big Bang, and how it has evolved in the 14 billion years since.

Launched into orbit in 2009, the telescope has almost completed three of its four planned surveys of the entire sky. The telescope is looking through the "fog" of galaxies, stars, gas and dust standing in the way of that oldest light, which astronomers say is radiation that the Big Bang left over.

"I'm just so excited for all of this," said Adam Moss, a research associate in physics and astronomy at the University of British Columbia. "It's hard to describe how excited the team is and how much we're looking forward to the next steps."

The Canadian Space Agency funds one Canadian research team from UBC and another from the University of Toronto, both of which are part of the Planck science collaboration.

At a conference in Paris on Tuesday, members of the 15-nation team unveiled a catalogue of more than 15,000 of the "foreground" objects standing in the way of the oldest light. Many of these objects, Moss said, had never been seen before.

"We're looking at how these looked billions of years ago," he said. "It's like a snapshot of the sky from billions of years ago."

Moss said the catalogue will also help explain some objects that scientists know well but don't understand, such as the clumps of cold dust that string together and form the Milky Way.

The team also announced its discovery of a previously undetected gas that was found clinging to the edges of the Milky Way. That gas, they say, may eventually be found to have an impact on galaxy formation and evolution. They also witnessed cold dust clouds—among the coldest ever discovered—where stars were forming.

Planck continues to orbit 1.5 million kilometres from Earth, collecting data and snapping photos. Moss said the final report should be due at the end of 2012 or early 2013.

Evaluating the News

1. What is exciting about the objects in the catalog?
2. Adam Moss is quoted in the article as saying, "We're looking at how these looked billions of years ago." What does *these* apparently refer to?
3. Later, the article refers to "previously undetected gas that was found clinging to the edges of the Milky Way." Is Planck showing how this gas looked billions of years ago?
4. Scientists (and others) often complain that reporters misquote them. The answers to the last two questions point out an apparent inconsistency in the article. Does this inconsistency tell you something about the scientist, the reporter, both, or neither?
5. Does this inconsistency affect your impression of the quality of the science being done?

SUMMARY

14.1   Our observable universe contains hundreds of billions of galaxies, each with millions to hundreds of billions of stars. Observations suggest that our universe is homogeneous and isotropic, in agreement with the cosmological principle.

14.2   The expansion of the universe is governed by Hubble's law: a galaxy's recession velocity is proportional to its distance. The expansion of the universe produces the observed red-shifts, but this expansion does not affect the local physics or structure of objects.

14.3   The universe is expanding uniformly from a Big Bang, which occurred nearly 14 billion years ago. The Big Bang happened everywhere. It is not an explosion spreading out from a single point. Observed redshifts of distant galaxies are a result of the increasing scale factor of the universe.

14.4   We observe blackbody radiation at 2.73 K, which is the cooled remnant of the radiation present in the universe just after the Big Bang when it was much smaller and hotter. The CMB enables us to measure our velocity with respect to the background radiation, and it also shows evidence of the ripples that grew to become large-scale structures in the universe. Nuclear reactions of normal matter in the hot early universe produced all the helium we see today, as well as trace amounts of other light elements.

✧ SUMMARY SELF-TEST

1. In every direction that astronomers look, they see the same Hubble's law, with the same slope. This is an example of
   a. isotropy.
   b. homogeneity.
   c. hydrostatic equilibrium.
   d. energy balance.

2. No matter where you traveled in space, you would observe very similar distributions of galaxies. This is an example of
   a. isotropy.
   b. homogeneity.
   c. hydrostatic equilibrium.
   d. energy balance.

3. If the universe were not expanding, the relationship between the velocity of a galaxy and its distance (as in the Hubble's law plot) would
   a. be horizontal.
   b. follow a downward trend.
   c. first go up, then flatten out.
   d. start out flat, then fall.

4. Rank the following galaxies in order of distance from the Milky Way.
   a. a galaxy with a radial velocity of +200 km/s
   b. a galaxy with a radial velocity of +400 km/s
   c. a galaxy with a radial velocity of −50 km/s
   d. a galaxy with a radial velocity of +50 km/s

5. The cosmological redshift in Hubble's law is distinct from the redshift arising from the Doppler effect because
   a. the redshift happens as the photon moves through expanding space.
   b. the redshift happens as the galaxy moves through space.
   c. it's used for the light, not for sound.
   d. the two are not distinct; they are the same.

6. Which of the following adjectives describe the Big Bang? (Choose all that apply.)
   a. hot   b. dense
   c. loud   d. tiny
   e. vast   f. slow
   g. fast

7. Following are the three major observations related to the theory of the Big Bang that were discussed in this chapter. Which of these pieces led to the development of the theory, and which were predicted by the theory and subsequently observed?
   a. Hubble's law
   b. cosmic microwave background radiation
   c. the abundance of helium

8. Which of the following adjectives describe the early universe, as observed through the CMB? (Choose all that apply.)
   a. hot   b. cold
   c. dense   d. diffuse
   e. uniform on large scales f. uniform on small scales
   g. rapidly expanding   h. slowly expanding

QUESTIONS AND PROBLEMS

True/False and Multiple-Choice Questions

9. **T/F:** The universe is homogeneous and isotropic.

10. **T/F:** Cosmological redshift causes a change in the spectrum of a galaxy similar to the change that Doppler shift causes in the spectrum of a star.

11. **T/F:** No galaxies are observed to have a blueshift.

12. **T/F:** The fact that nearly all galaxies appear to be moving away from us means that we are at the center of the universe.

13. **T/F:** We don't know what the distant universe looks like now. We only know what it used to look like.

14. T/F: Hubble's law implies that the universe was once much denser than it is today.

15. T/F: The theory of the Big Bang predicted the existence of the cosmic microwave background.

16. Astronomers observe two galaxies, A and B. Galaxy A has a recession velocity of 2,500 km/s, while galaxy B has a recession velocity of 5,000 km/s. This means that
a. galaxy A is four times as far away as galaxy B.
b. galaxy A is twice as far away as galaxy B.
c. galaxy B is twice as far away as galaxy A.
d. galaxy B is four times as far away as galaxy A.

17. The distances to galaxies used to establish Hubble's law are found from
a. radar.
b. parallax.
c. main-sequence fitting.
d. standard candles, such as supernovae.

18. The Hubble constant is found from the
a. slope of the line fit to the data in Hubble's law.
b. y-intercept of the line fit to the data in Hubble's law.
c. dispersion in the data in Hubble's law.
d. inverse of the slope of the line fit to the data in Hubble's law.

19. The cosmic microwave background looks like a 2.73 K blackbody spectrum because
a. this was the temperature of the universe at the Big Bang.
b. the light has been redshifted since the Big Bang.
c. the light has been reddened by dust since the Big Bang.
d. Earth is receding slowly from the Big Bang.

20. The cosmic microwave background
a. comes from the moment the universe became transparent.
b. is a major confirmed prediction of the Big Bang.
c. reveals the conditions in the early universe.
d. all of the above

21. The helium abundance in the current universe
a. is due partly to the Big Bang and partly to stellar evolution.
b. is due only to the Big Bang.
c. is due only to stellar evolution.
d. is unknown at large scales.

Conceptual Questions

22. Look back at the chapter-opening figure. What other information would you need to obtain about these galaxies to determine the distance, once Hubble's law is known?

23. Describe the cosmological principle in your own words.

24. We see individual stars and galaxies. What, then, do astronomers mean when they say the universe is "homogeneous"?

25. The universe could be homogeneous without being isotropic. Explain what is meant by isotropic.

26. Imagine that you are standing in the middle of a dense fog.
a. Would you describe your environment as isotropic?
b. Would you describe it as homogeneous?
c. Explain your answers.

27. Early in the 20th century, astronomers discovered that most galaxies are moving away from the Milky Way (that is, they are redshifted). What was the significance of this discovery?

28. Edwin Hubble later made an even more important discovery: that the speed with which galaxies are receding is proportional to their distance. Why was this among the more important scientific discoveries of the 20th century?

29. If you lived in a remote galaxy, would you observe distant galaxies receding from your own galaxy, or would they appear to be approaching? Explain your answer.

30. Why is the Milky Way Galaxy not expanding along with the rest of the universe?

31. Why can we not use the measured radial velocities of nearby galaxies, such as the Andromeda Galaxy, to evaluate the Hubble constant (H_0)?

32. Discuss Figure 14.7. Would you have found the data in Figure 14.7a convincing when it first appeared? Compare Figure 14.7a to Figure 14.7b. Do the modern data show the same trend as Hubble's original data? Is this figure more convincing than Hubble's original data? Why or why not? Scientists often say that "more data are needed." Was this true in this case? Why or why not?

33. In Figure 14.7a, some of the data are below the x-axis (the line where velocity = 0). What does this tell you about those galaxies? Does this disprove Hubble's law? Explain.

34. Suppose that the universe was not expanding, but was instead contracting. Sketch what Figure 14.7b should look like in that case. Now, suppose that it was doing neither, but instead was static. Sketch Figure 14.7b for the case of a static universe.

35. Explain what astronomers mean by a *distance ladder*.

36. Examine Figure 14.5, which shows the cosmic distance ladder. Use the information in this figure to make a table that shows which distance method is used for which range of distances. In a third column, indicate which method covers the largest volume of space, and which covers the smallest.

37. As astronomers extend their distance ladder beyond 30 Mpc, they change their measuring standard from Cepheid variable stars to Type Ia supernovae. Why is this necessary?

38. What is meant by *Hubble time*?

39. Examine Figure 14.9. The black galaxy is at 0 distance at time $t = 0$ and then remains at 0 distance. Why?

40. As the universe expands from the Big Bang, we know that galaxies are not actually flying apart from one another. What is really happening?

41. The scale factor can be used as a "cosmic ruler" in describing the expansion of the universe. Explain the scale factor.

42. Knowing that you are studying astronomy, a curious friend asks where the center of the universe is located. You smile and answer, "Right here and everywhere." Explain in detail why you would give this answer.

43. The general relationship between radial velocity (v_r) and redshift (z) is $v_r = cz$. This simple relationship fails, however, for very distant galaxies with large redshifts. Explain why. (Hint: What happens to the speed of the galaxy in this simple relationship if z becomes greater than 1.0?)

44. Figure 14.10 shows that galaxies do not move through space, but instead are carried along by the expansion of space. Why is this distinction important?

45. What is the origin of the CMB?

46. Why is it significant that the CMB displays a blackbody spectrum?

47. As the sensitivity of our instrumentation increases, we are able to look ever farther into space and, therefore, ever further back in time. When we reach the era of recombination, however, we run into a wall and can see no further back in time. Explain why.

48. What is the significance of the tiny brightness variations that are observed in the CMB?

49. What important characteristics of the early universe are revealed by today's observed abundances of various isotopes, such as ^2H and ^3He?

Problems

50. Examine Figure 14.7. Is this a linear plot, or a logarithmic one? How do you know?

51. Using the data from Figure 14.3, find the shift in wavelength of the Hα line. Use this to confirm the recession velocity of the galaxy shown.

52. Hubble time ($1/H_0$) represents the age of a universe that has been expanding at a constant rate since the Big Bang. Assuming an H_0 value of 72 km/s/Mpc and a constant rate of expansion, calculate the age of the universe in years. (Note: 1 year $= 3.16 \times 10^7$ seconds, and 1 parsec $= 3.09 \times 10^{13}$ km.)

53. Throughout the latter half of the 20th century, estimates of H_0 ranged from 60 to 110 km/s/Mpc. Calculate the age of the universe in years for each of these estimated values of H_0.

54. Study Figure 14.8. At what distance would you find a galaxy with a speed of 8,640 km/s? What would the Hubble time be for this galaxy?

55. In Figure 14.11, a rubber sheet is shown as an analogy to help you think about the scale factor. Between the moments shown in parts (a) and (b), each square doubles in size on every edge. How does the area of a square change? Imagine that the sheet is now a block of rubber, expanding in three dimensions instead of two. How would the volume of a cube change between the moment shown in part (a) and the moment shown in part (b)?

56. The spectrum of a distant galaxy shows the Hα line of hydrogen ($\lambda_{rest} = 656.28$ nm) at a wavelength of 750 nm. Assume that $H_0 = 72$ km/s/Mpc.

a. What is the redshift (z) of this galaxy?
b. What is its recession velocity in kilometers per second?
c. What is the distance of the galaxy in mega-parsecs?

57. The spectrum of the CMB is shown as the red dots in Figure 14.16, along with a blackbody spectrum for a blackbody at temperature of 2.73 K. From the graph, determine the peak wavelength of the CMB spectrum. Use Wien's law to find the temperature of the CMB. How does this rough measurement that you just made compare to the accepted temperature of the CMB?

58. COBE observations show that our Solar System is moving in the direction of the constellation Crater at a speed of 368 km/s relative to the cosmic reference frame. What is the blueshift (negative value of z) associated with this motion?

59. Figure 14.18 includes a lot of information in one graph. Convert this figure into a table with the following columns: isotope, observed abundance (range), calculated abundance. Make one row for each isotope, and then fill in the table with the isotope data and calculations that lie along the black line that shows the observed density of normal matter.

60. If new observations suggested that the mass fraction of helium remains at 24 percent but the mass fraction of deuterium (^2H) is really 10^{-6}, how would our estimates of the density of normal matter in the universe be affected? Refer to Figure 14.18.

61. The average density of normal matter in the universe is 4×10^{-28} kg/m³. The mass of a hydrogen atom is 1.66×10^{-27} kilograms (kg). On average, how many hydrogen atoms are there in each cubic meter in the universe?

62. To get a feeling for the emptiness of the universe, compare its density (4×10^{-28} kg/m³) with that of Earth's atmosphere at sea level (1.2 kg/m³). How much denser is our atmosphere? Write this ratio using standard notation.

 SmartWork, Norton's online homework system, includes algorithmically generated versions of these questions, plus additional conceptual exercises. If your instructor assigns questions in SmartWork, log in at **smartwork.wwnorton.com**.

 StudySpace is a free and open website that provides a Study Plan for each chapter of **Understanding Our Universe**. Study Plans include animations, reading outlines, vocabulary flashcards, and multiple-choice quizzes, plus links to premium content in SmartWork and the ebook. Visit **wwnorton.com/studyspace**.

Exploration | Hubble's Law for Balloons

The expansion of the universe is extremely difficult to visualize, even for professional astronomers. In this exploration, you will use the surface of a balloon to get a feel for how an "expansion" changes distances between objects. Throughout this exploration, you must remember to think of the surface of the balloon as a two-dimensional object, much like the surface of Earth is a two-dimensional object for most people. The average person can move east or west, or north or south, but into Earth and out to space are not options. For this exploration, you will need a balloon, 11 small stickers, a piece of string, and a ruler. A partner is helpful as well. **Figure 14.19** shows some of the steps in this process.

Blow up the balloon partially, and *do not tie it shut*. Stick the 11 stickers on the balloon (these represent galaxies) and number them. Galaxy 1 is the reference galaxy.

Measure the distance between the reference galaxy and each of the numbered galaxies. The easiest way to do this is to use your piece of string. Lay it along the balloon between the two galaxies

and then measure the string. Record these data in the "Distance 1" column of a table like the one on the following page.

Simulate the expansion of your balloon universe by *slowly* blowing up the balloon the rest of the way. Have your partner count the number of seconds it takes you to do this, and record this number in the "Time elapsed" column of the table (each row has the same time elapsed, because the expansion occurred for the same amount of time for each galaxy). Tie it shut. Measure the distance between the reference galaxy and each numbered galaxy again. Record these data in the "Distance 2" column of the table.

Subtract the first measurement from the second. Record the difference in the table.

Divide this difference, which represents the distance traveled by the galaxy, by the time it took to blow up the balloon. A distance divided by a time gives you an average speed.

Make a graph of the velocity (on the *y*-axis) versus Distance 2 (on the *x*-axis) to get "Hubble's law for balloons." You may wish to roughly fit a line to these data to clarify the trend.

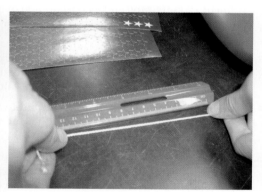
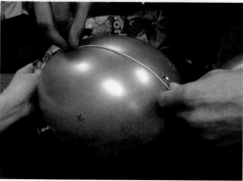

FIGURE 14.19 Students use balloons to develop their understanding of the expansion of space.

GALAXY NUMBER	DISTANCE 1	DISTANCE 2	DIFFERENCE	TIME ELAPSED	VELOCITY
1 (reference)	0	0	0		0
2					
3					
4					
5					
6					
7					
8					
9					
10					
11					

1. Describe your data. If you fit a line to them, would it be horizontal, trend upward, or trend downward?

2. Was there anything special about your reference galaxy? Was it different in any way from the others?

3. If you had picked a different reference galaxy, would the trend of your line be different? If you are not sure about the answer to this question, get another balloon and try it!

4. The expansion of the universe behaves similarly to the movement of the galaxies on the balloon. We don't want to carry the analogy too far, but there is one more thing to think about. In your balloon, you probably have some areas that expanded less than others because the material was thicker—there was more "balloon stuff" holding it together. How is this similar to some places in the actual universe?

15

The Realm of the Galaxies

It has been less than a century since we came to realize that the universe is filled with objects that closely resemble our own galaxy, the Milky Way. But just as stars vary based on their mass or on their stage of evolution, galaxies also come in many forms. After realizing that galaxies are huge collections of stars and dust, astronomers have found ways to classify these galaxies into various types and make sense of the differences among them. One of our goals is to understand the origin of each type of galaxy, just as we interpret the origins of the different types of stars.

✧ LEARNING GOALS

Stars are not spread uniformly through space. Instead, they are grouped into what the German philosopher Immanuel Kant called "island universes," which today we know to be galaxies. In the image at right, a student has collected images to follow the Hubble tuning fork, one way to organize these galaxies based on their shapes. By the end of this chapter, you should be able to reproduce such a figure, showing how S, E, and SB galaxies are different and what distinguishes an Sa galaxy from an Sc galaxy, for example. You should also be able to:

- Determine a galaxy's type, given information about the orbits of the stars it contains

- Explain why the arms of spiral galaxies, which form whenever the disk of a spiral galaxy is disturbed, are sites of star formation

- Describe the known properties of dark matter

- Explain the evidence that shows that most—perhaps all—large galaxies have supermassive black holes at their centers

- Describe the unified model of active galactic nuclei

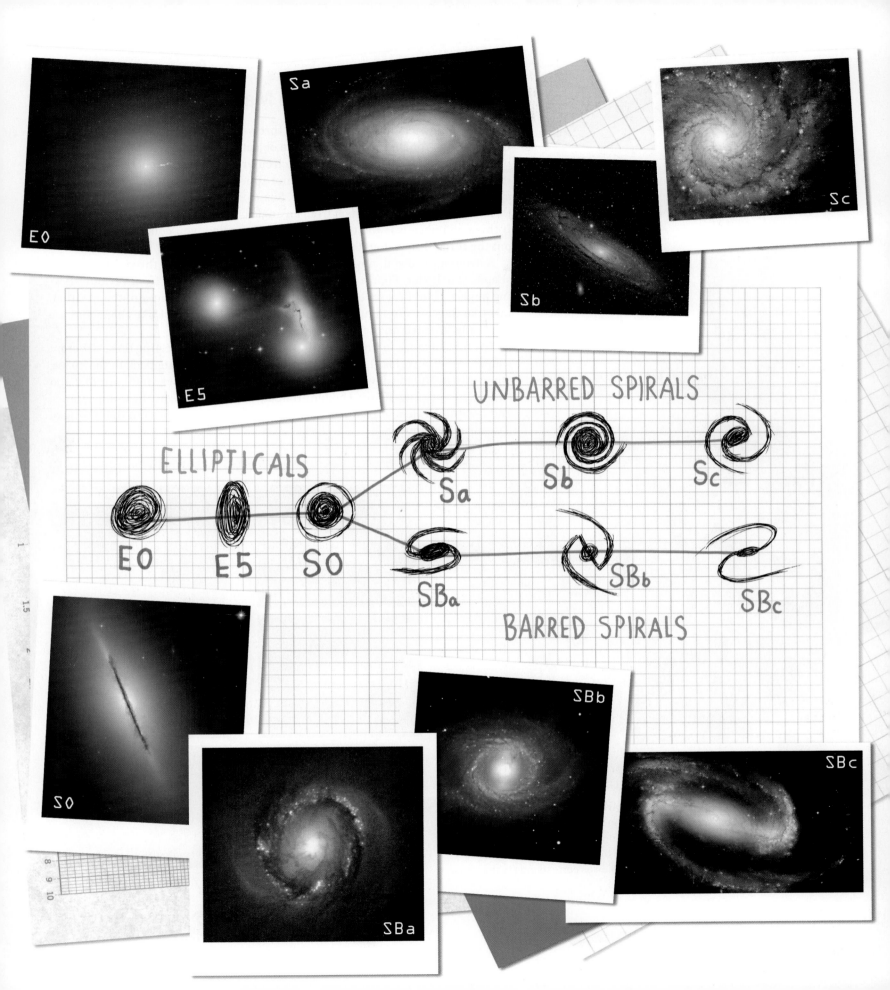

E0

Sa

Sc

E5

Sb

UNBARRED SPIRALS

ELLIPTICALS

Sa

Sb

Sc

E0 E5 S0

SBa

SBb

SBc

BARRED SPIRALS

S0

SBb

SBc

SBa

15.1 Galaxies Come in Many Types

Imagine taking a handful of coins and throwing them into the air, as shown in **Figure 15.1a**. You know that all of these objects are very much the same: dimes, pennies, nickels, quarters—all flat and circular. When you look at the objects falling through the air, however, they do not appear to be all the same. Some coins appear face on, and they look circular. Some coins appear edge on and look like thin lines. Most coins are seen from an angle between these two extremes. Even if this one image was the only information you had, you could use it to figure out the three-dimensional shape of a coin—flat and circular.

Astronomers use a similar method to discover the true three-dimensional shapes of galaxies. **Figure 15.1b** shows a set of galaxies seen from various viewing angles, from face on to edge on. We can infer from images of the sky that, just like the coins in Figure 15.1a, galaxies often have a disklike shape and are randomly oriented on the sky.

The first advance in understanding galaxies came from sorting these different shapes into categories. The classifications we use today date back to the 1930s, when Edwin Hubble devised a scheme much like that shown in **Figure 15.2** and the chapter-opening figure. Hubble grouped all galaxies according to appearance and positioned them on a diagram that resembles the fork used in the tuning of a musical instrument.

Galaxies are classified according to their appearance.

(a)

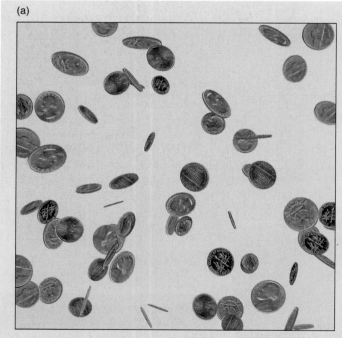

(b)

FIGURE 15.1 (a) A handful of coins thrown in the air provides a helpful analogy for the difficulties in identifying the shapes of certain types of galaxies. We see some face on, some edge on, and most somewhere in between. (b) Disk-shaped galaxies seen from various perspectives or angles. The variety of angles we see for galaxies corresponds to the range of perspectives for the coins in Figure 15.1a.

G X U V I R

VISUAL ANALOGY

On the bottom (or "handle") of this **tuning fork diagram** sit objects that are elliptical in three dimensions, something like an American football. These are called **elliptical galaxies** (labeled *E* on the diagram). They have numbered subtypes ranging from nearly spherical (E0) to quite elliptical (E7), and they show little evidence of the flat, round disk that we see in other types of galaxies.

On the two "tines" of the fork are the **spiral galaxies**, designated with an initial *S*. The defining feature of a spiral galaxy is a flattened, rotating disk. The spiral arms that give these galaxies their name lie in this disk. In addition to disks and arms, spiral galaxies have central **bulges**, which look like elliptical galaxies.

The distinction between spiral and elliptical galaxies is not always clear. Some galaxies are a combination of the two types, having stellar disks but no spiral arms. Hubble called these intermediate types **S0 galaxies** and placed them near the junction of his tuning fork. Further blurring the distinction, telescopic observations have revealed that many, if not most, elliptical galaxies contain small rotating disks at their centers. Elliptical and S0 galaxies share another similarity: neither produces many new stars.

Hubble noticed that the bulges of roughly half of all spiral galaxies are bar shaped. He called these galaxies **barred spirals** (SB) and placed them along the right-hand tine of the tuning fork, as shown in Figure 15.2. Spirals that lack a barlike bulge lie along the left-hand tine. Hubble positioned the spiral galaxies along the fork's tines according to the prominence of the central bulge and how tightly the spiral arms are wound. For example, Sa and SBa galaxies have the largest bulges and display tightly wound and smooth spiral arms. Sc and SBc galaxies have small central bulges and more loosely woven spiral arms, often very knotty in appearance.

Galaxies that fall into none of these classes are called **irregular galaxies** (Irr). As their name implies, irregular galaxies are often without symmetry in shape or structure, and they do not fit neatly on Hubble's tuning fork.

Originally, Hubble thought that his tuning fork diagram might do for galaxies what the H-R diagram had done for stars. This has not turned out to be the case—the tuning fork diagram does not describe the evolution of galaxies—but his classification scheme organized the study of these objects.

Stellar Motions Give Galaxies Their Shapes

A galaxy is not a solid object like a coin, but a collection of stars, gas, and dust. In an elliptical galaxy, stars are moving in all possible directions. Unlike planets in our Solar System, which move on nearly circular orbits about the Sun, stars in an elliptical galaxy follow orbits with a wide range of shapes, as shown in **Figure 15.3**. These orbits are more complex than the orbits of planets because the gravitational field within an elliptical galaxy does not come from a single central object. Taken together, all of these stellar orbits give an elliptical galaxy its shape.

Orbital speeds are also a factor. The faster the stars are moving, the more spread out the galaxy is. If the stars in an elliptical galaxy are moving in truly random directions, the galaxy will have a spherical shape. However, if stars tend to move faster in one direction than in others, the galaxy will be more spread out in that direction, giving it an elongated shape. These differences in stellar orbits cause some elliptical galaxies to be round, while others are elongated. As with the coins tossed in the air, the appearance of an elliptical galaxy in the sky does not necessarily tell us its true shape. For example, a galaxy might actually be shaped like an American football; but if we happen to see it end on, it will look round like a soccer ball.

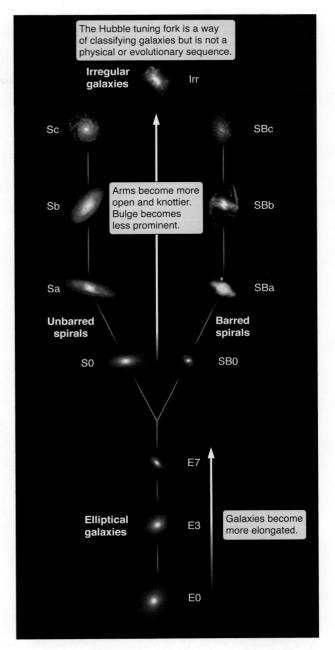

FIGURE 15.2 Tuning fork diagram showing Edwin Hubble's scheme for classifying galaxies according to their appearance. Elliptical galaxies form the "handle" of this tuning fork. Unbarred and barred spiral galaxies lie along the left and right tines of the fork, respectively. S0 galaxies lie along the bottom left and right tines. Irregular galaxies are not placed on the tuning fork.

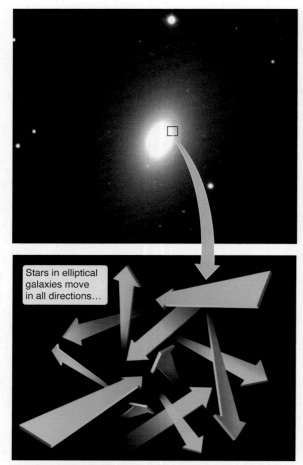

Stars in elliptical galaxies move in all directions…

…on complex, irregular orbits.

G X U V I R

FIGURE 15.3 Elliptical galaxies take their shape from the orbits of the stars they contain. The colored lines superimposed on the galaxy represent the complex, irregular orbits of its stars.

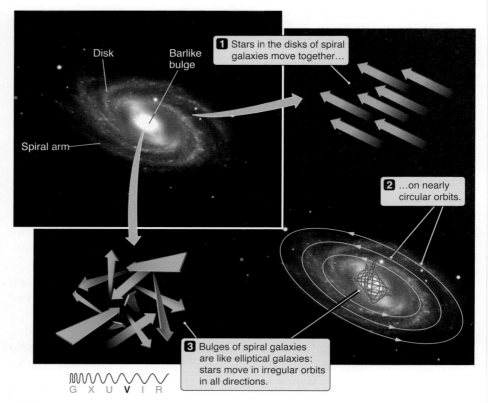

Disk   Barlike bulge

Spiral arm

1 Stars in the disks of spiral galaxies move together…

2 …on nearly circular orbits.

3 Bulges of spiral galaxies are like elliptical galaxies: stars move in irregular orbits in all directions.

G X U V I R

FIGURE 15.4 The components of a barred spiral galaxy. The orbits of stars in the rotating disk and the ellipse-like bulge are indicated.

The orbits of stars in the disks of spiral galaxies are quite different from those of stars in elliptical galaxies. The components of a spiral galaxy are shown in **Figure 15.4**. The defining feature of a spiral galaxy is that it has a flattened, rotating disk. Like the planets of our Solar System, most of the stars in the disk of a spiral galaxy follow nearly circular orbits and travel in the same direction about a concentration of mass at the center of the galaxy. But the stellar orbits in a spiral galaxy's central bulge are quite different from those in the galaxy's disk. As with elliptical galaxies, the gravitational field within the bulge does not come from a single object, and the stars therefore follow a wide variety of irregular orbits. The bulges of spiral galaxies are thus roughly spherical in shape.

Other Differences among Galaxies

In addition to the differences in their stellar orbits, there is another important distinction between spiral and elliptical galaxies. Most spiral galaxies contain large amounts of dust and cold, dense gas concentrated in the midplanes of their disks. Just as the dust in the disk of our own galaxy can be seen on a clear summer night as a dark band slicing the Milky Way in two (**Figure 15.5a**), the dust in an edge-on spiral galaxy appears as a dark, obscuring band running down the midplane of the disk (**Figure 15.5b**). The cold gas that accompanies the dust can also be seen in radio observations of spiral galaxies. In contrast, elliptical galaxies contain large amounts of very hot gas that we see primarily by observing the X-rays it emits.

The difference in shape between elliptical and spiral galaxies offers some insight into why the gas in ellipticals is hot, while in spirals it is cold. Just as gas settles into a disk around a forming star, cold gas settles into the disk of a spiral galaxy because of conservation of angular momentum. In contrast, elliptical galaxies do not have a net rotation, so the gas does not settle into a disk. In an elliptical galaxy, the only place that cold gas could collect is at the center. However, the density of stars in the center of elliptical galaxies is so high that evolving stars and Type Ia supernovae continually reheat the gas, preventing most of it from cooling off.

The colors of spiral and elliptical galaxies tell us a great deal about their star formation histories. Stars form from dense clouds of cold gas. Because the gas we see in elliptical galaxies is very hot, active star formation is not taking place in those galaxies today. The reddish colors of elliptical and S0 galaxies confirm that little or no star formation has occurred there for quite some time. The stars in these galaxies are an older population of lower-mass stars. The bluish colors of the disks of spiral galaxies, on the other hand, confirm that massive, young, hot stars are forming in the cold molecular clouds contained within the disk. Even though *most* of the stars in a spiral disk are old, the massive, young stars are so luminous that their blue light dominates what we see. When it comes to star formation, most irregular galaxies are like spiral galaxies. Some irregular galaxies are currently forming stars at prodigious rates, given their relatively small sizes.

The relationship between luminosity and size among the different types of galaxies is not straightforward. Galaxies range in luminosity from about a million up to a million million solar luminosities (10^6–10^{12} $L_\odot$) and in size from a few hundred to hundreds of thousands of parsecs. There is no distinct size difference between elliptical and spiral galaxies; about half of both types of galaxies fall within a similar range of sizes. Although it is true that the most luminous elliptical galaxies are more luminous than the most luminous spiral galaxies, there is considerable overlap in the range of luminosities among all Hubble types.

Mass is the single most important parameter in determining the properties and evolution of a star. In contrast, differences in mass and size do not lead to obvious differences among galaxies. Only subtle differences in color and concentration exist between large and small galaxies, making it difficult for us to distinguish which are large and which are small. Even when a smaller, nearby spiral galaxy is seen next to a larger, distant spiral (**Figure 15.6**), it can be hard to tell which is

FIGURE 15.5 (a) The dust in the plane of the Milky Way obscures our view toward the galactic center. (b) Similarly, the dust in the plane of the nearly edge-on spiral galaxy M104 is seen as a dark, obscuring band in the midplane of the galaxy.

Star formation is occurring in spiral galaxies but not in elliptical galaxies.

FIGURE 15.6 The mass or size of a spiral galaxy does not determine its appearance. Even though these galaxies appear to be similar in size and luminosity, galaxy (a) is much larger and 10 times more luminous. However, it is also 4 times more distant than the smaller galaxy (b).

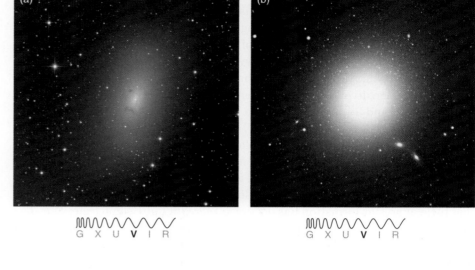

FIGURE 15.7 Dwarf elliptical galaxies (a) differ in appearance from giant elliptical galaxies (b). Giant elliptical galaxies are more centrally concentrated than dwarf elliptical galaxies.

which. Still, galaxies that have relatively low luminosity (less than 1 billion $L_\odot$) are called **dwarf galaxies** and more luminous galaxies are called **giant galaxies**. Only elliptical and irregular galaxies come in both types. Among spiral and S0 galaxies, we find only giants. It is relatively easy to tell the difference between a dwarf elliptical galaxy and a giant elliptical galaxy (as shown in **Figure 15.7**). Giant elliptical galaxies have a much higher density of stars, more centrally concentrated than in dwarf ellipticals.

15.2 Stars Form in the Spiral Arms of a Galaxy's Disk

From pictures of spiral galaxies outside of our own spiral Milky Way Galaxy, we might have guessed that most stars in a galaxy's disk are located in the spiral arms. This turns out not to be the case. **Figure 15.8** shows images of the Andromeda Galaxy taken in ultraviolet and visible light. Notice that whereas the spiral arms are relatively prominent in the UV image (Figure 15.8a), they are less prominent when viewed in visible light (Figure 15.8b). If we carefully count the actual numbers of stars, we find that although stars are slightly concentrated in spiral arms, this concentration is not strong enough to account for the prominence of the arms. In fact, the concentration of stars in the disks of spiral galaxies varies quite smoothly as it decreases outward from the center of the disk to the edge of the galaxy. However, molecular clouds, associations of O and B stars, and other structures associated with star formation are all concentrated in spiral arms. Spiral arms look so prominent when viewed in blue or UV light because they contain significant concentrations of young, massive, luminous stars.

Stars form when dense interstellar clouds become so massive and concentrated that they begin to collapse under the force of their own gravity. If stars form in spiral arms, then spiral arms must be places where clouds of interstellar gas pile up and are compressed. Such is indeed the case. There are many ways to trace the presence of gas in the spiral arms of galaxies. Pictures of face-on spiral galaxies, such as the one featured in **Figure 15.9a**, show dark lanes where clouds of dust block starlight. These lanes provide one of the best tracers of spiral arms. Spiral arms also show up in other tracers of concentrations of gas, such as radio emission from neutral hydrogen or from carbon monoxide (**Figure 15.9b**).

FIGURE 15.8 The Andromeda Galaxy in ultraviolet (a) and visible (b) light. Note that the spiral arms are most prominent in ultraviolet light, which is dominated by young hot stars, and in emission from interstellar clouds that are ionized by the radiation from young hot stars. The spiral arms are less prominent in visible light.

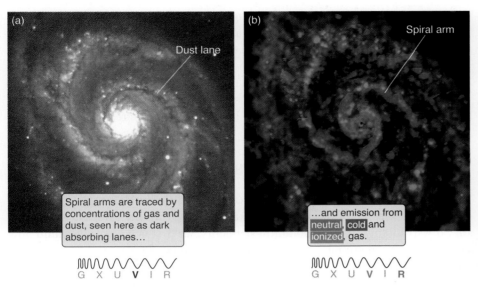

FIGURE 15.9 Two images of a face-on spiral galaxy showing the spiral arms. (a) This visible-light image also shows dust absorption. (b) This image of 21-cm emission shows the distribution of neutral interstellar hydrogen, CO emission from cold molecular clouds, and Hα emission from ionized gas.

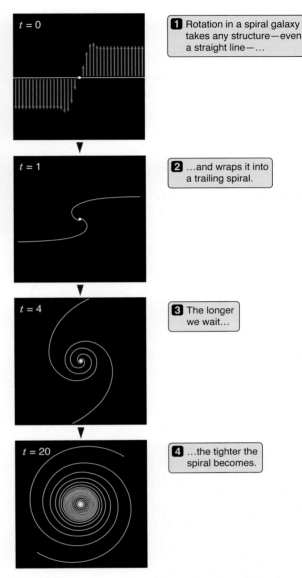

FIGURE 15.10 The differential rotation of a spiral galaxy will naturally take even an originally linear structure and wrap it into a progressively tighter spiral as time (*t*) goes by.

Any disturbance in the disk of a spiral galaxy will cause a spiral pattern because the disk rotates. Disks do not rotate like a solid body. Instead, material close to the center takes less time to travel around the galaxy than material farther out. **Figure 15.10** illustrates the point. As the model galaxy rotates, a straight line through the center becomes a spiral. In the time it takes for objects in the inner part of the galaxy to complete several rotations, objects in the outer parts of the galaxy may not have completed even a single revolution.

A spiral galaxy can be disturbed by, for example, gravitational interactions with other galaxies or a burst of star formation. However, a single disturbance will not produce a *stable* spiral-arm pattern. Spiral arms produced from one disturbance will wind themselves up completely in two or three rotations of the disk and then disappear. Other types of disturbances are repetitive, sustaining spiral structure indefinitely. For example, when the bulge in the center of a spiral galaxy is elongated (as seems to be the case for most spiral galaxies), then the bulge gravitationally disturbs the disk. As the disk rotates through this disturbance, repeated episodes of star formation occur and stable spiral arms form.

Spiral structure can also be created by star formation. Regions of star formation release energy into their surroundings through UV radiation, stellar winds, and supernova explosions. This energy compresses clouds of gas and triggers more star formation. Typically, many massive stars form in the same region at about the same time; their combined mass outflows and supernova explosions occur over only a few million years, creating large, expanding bubbles of hot gas. These bubbles concentrate the gas into dense clouds at their edges, causing more star formation. Rotation bends the resulting strings of star-forming regions into spiral structures.

Regular disturbances in the disks of spiral galaxies are called **spiral density waves**: regions of greater mass density and increased pressure in the galaxy's interstellar medium. These waves move around a disk in the pattern of a two-armed spiral. Because they are waves, it is the disturbance that moves, not the material.

Rotation in a disk galaxy naturally produces spiral structure.

As the disk orbits, material passes *through* the spiral density waves. The stars in the arm today are not the same stars that were in the arm 20 million years ago.

A spiral density wave has very little effect on stars, but it does compress the gas that flows through it. Stars form in the resulting compressed gas. Massive stars have such short lives (typically 10 million years or so) that they never drift far from the spiral arms. Less massive stars, on the other hand, have plenty of time to move away from their places of birth, filling in the rest of the disk.

Spiral density waves compress gas, triggering star formation.

15.3 Galaxies Are Mostly Dark Matter

Efforts to measure the masses of galaxies during the last decades of the 20th century led to some of the most remarkable and surprising findings in the history of astronomy. To understand these results, we first need to understand how astronomers go about measuring the mass of a galaxy. One method is to add up the mass of the stars, dust, and gas that we can see. A galaxy's spectrum is primarily composed of starlight, so we can find out what types of stars are in the galaxy. Stellar evolution then tells us how to turn the luminosity of the galaxy into an estimate of the total stellar mass. The physics of radiation from interstellar gas at X-ray, infrared, and radio wavelengths enables us to estimate the mass of these other components. Together, the stars, gas, and dust in a galaxy are called **luminous matter** (or simply **normal matter**) because this matter emits electromagnetic radiation.

However, this method does not allow us to determine a galaxy's total mass. Black holes, for example, would not be accounted for in this method, yet they still have mass. Fortunately, we have a method for determining mass that does not involve luminosity. Stars in disks follow orbits that are much like the Keplerian orbits of planets around their parent stars and binary stars around each other. To measure the mass of a spiral galaxy, we apply Kepler's laws, just as we do for those other systems (**Working It Out 15.1**).

Astronomers use Kepler's laws to measure galaxy mass.

Working It Out 15.1 | Finding the Mass of a Galaxy

The Sun is located 8,000 parsecs (1.6 billion astronomical units, AU) from the center of the Milky Way, and it takes 250 million years to orbit the center. From this information, we can find the mass of the Milky Way interior to the orbit of the Sun. In Working It Out 10.2, we used Newton's version of Kepler's law, which relates the total mass of an orbiting system ($m_1 + m_2$), in units of solar masses, to the orbital period (P) in years and the average radius of the orbit (A) in AU:

$$m_1 + m_2 = \frac{A^3_{\text{AU}}}{P^2_{\text{years}}}$$

In this case, the mass of the Sun is tiny compared to the mass of the galaxy interior to the Sun, so that $m_1 + m_2$ is essentially

equal to m_1, the mass of the Milky Way Galaxy interior to the Sun's orbit. Plugging in the values for the orbit of the Sun gives

$$m_1 = \frac{A^3_{\text{AU}}}{P^2_{\text{years}}}$$

$$m_1 = \frac{(1.6 \times 10^9 \text{ AU})^3}{(250 \times 10^6 \text{ years})^2}$$

$$m_1 = 6.6 \times 10^{10} \, M_\odot$$

The mass of the Milky Way, interior to the Sun, is 66 billion solar masses. The Sun is in the disk of the Milky Way, not at the edge, so the total mass of the Milky Way is larger than this. Still, this calculation indicates the vast amount of mass contained within the Milky Way.

We would also like to know how the mass is distributed in a galaxy. We might hypothesize that the mass and the light are distributed in the same way—assuming, that is, that the luminous mass in these galaxies is all the mass there is. Based on this hypothesis, we can make a prediction: The light of all galaxies, including spiral galaxies, is highly concentrated toward the center (**Figure 15.11a**). It would follow, then, that nearly all the mass of a spiral galaxy is contained in its center (**Figure 15.11b**). This situation is much like the Solar System, where nearly all the mass is in the Sun, in the middle. Therefore, we would predict fast orbital velocities near the center of the spiral galaxy and slower orbital velocities farther out (**Figure 15.11c**).

To test this prediction, we use the Doppler effect to measure orbital motions of stars, gas, or dust. Once we find the velocities, we can create a graph of the orbital velocity versus distance from the galaxy's center. This kind of graph is called a **rotation curve**.

Vera Rubin (b. 1928) pioneered work on galaxy rotation rates. She discovered that, contrary to our earlier prediction, the rotation velocities of spiral galaxies remain about the same out to the most distant measured parts of the galaxies (**Figure 15.11d**). Observations of 21-cm radiation from neutral hydrogen show that the rotation curves remain flat even well outside the extent of the visible disks. Our hypothesis that mass and light are distributed in the same way is wrong.

The logical next step is to turn the question around and ask, what mass distribution would cause this unexpected rotation curve? **Figure 15.12a** shows the result of such a calculation. In addition to the centrally concentrated luminous

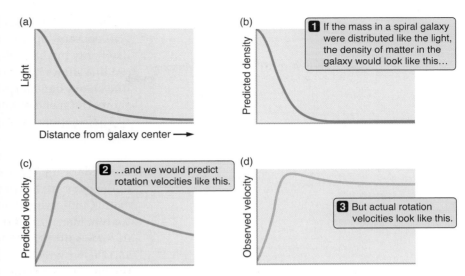

FIGURE 15.11 (a) The profile of visible light in a typical spiral galaxy. (b) The mass density of stars and gas located at a given distance from the galaxy's center. If stars and gas accounted for the entire mass of the galaxy, then the galaxy's rotation curve would be as shown in (c). However, galaxies have observed rotation curves more like the curve shown in (d).

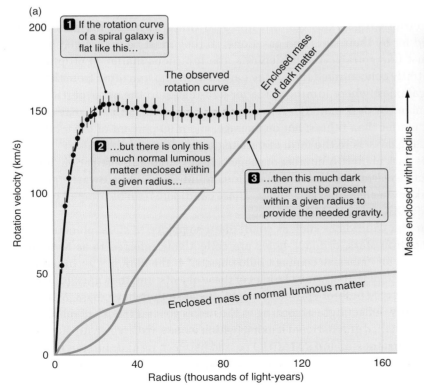

FIGURE 15.12 (a) We can use the flat rotation curve of the spiral galaxy NGC 3198 to determine the total mass within a given radius. Notice that the normal mass that can be accounted for by stars and gas is only part of the needed gravity. Extra dark matter is needed to explain the rotation curve. (b) In addition to the matter we can see, galaxies must be surrounded by halos containing a large amount of dark matter.

▶❚❚ **AstroTour: Dark Matter**

Most of the mass comprising galaxies is dark matter.

G X U V I R

FIGURE 15.13 Combined visible-light and X-ray images of elliptical galaxy NGC 1132. The false-color blue/purple halo is X-ray emission from hot gas surrounding the galaxy. The hot gas extends well beyond the visible light from stars.

matter, these galaxies must have a second component consisting of matter that does not show up in our census of stars, gas, and dust. This material, which reveals itself only by the influence of its gravity, is called **dark matter**. In the graph, the red line shows how much luminous mass is inside a particular radius. The blue line shows how much dark matter is inside a particular radius. The black line shows the speed of rotation *at* a particular radius.

The rotation curves of the inner parts of spiral galaxies match predictions based on their luminous matter, indicating that the inner parts of spiral galaxies are mostly luminous matter. Within the entire *visual* image of a galaxy, the mix of dark and luminous matter is about half and half. However, rotation curves in the outer part of the galaxy do not match our predictions based on luminous matter, indicating that the outer parts of spiral galaxies are mostly dark matter. Astronomers currently estimate that as much as 95 percent of the total mass in some spiral galaxies consists of a greatly extended **dark matter halo** (**Figure 15.12b**), far larger than the visible spiral portion of the galaxy located at its center. This is a startling statement. A spiral galaxy illuminates only the inner part of a much larger distribution of mass that is dominated by some type of matter we cannot see.

What about elliptical galaxies? Again, we want to compare the luminous mass, measured from the light we can see, with the gravitational mass, measured from the effects of gravity. Since elliptical galaxies do not rotate, we cannot use Kepler's laws to measure the gravitational mass. Instead, we notice that an elliptical galaxy's ability to hold onto its hot, X-ray-emitting gas depends on its mass: if the galaxy is not massive enough, the hot atoms and molecules will escape into intergalactic space. To find the mass of an elliptical galaxy, first we infer the total amount of gas from X-ray images, such as the blue and purple halo seen in **Figure 15.13**. Next we calculate the mass that is needed to hold onto the gas. We then compare that gravitational mass with the luminous mass. The amount of dark matter is the difference between what is needed to hold onto the gas and the observed luminous matter.

Some elliptical galaxies contain up to 20 times as much mass as can be accounted for by their stars and gas alone, so they must be dominated by dark matter, just like spirals. As with spirals, the luminous matter in ellipticals is more centrally concentrated than is the dark matter. The transition from the inner parts of galaxies (where luminous matter dominates) to the outer parts (which are dominated by dark matter) is remarkably smooth. Some galaxies may contain less dark matter than others; but on average, about 90 percent of the total mass in a typical galaxy is in the form of dark matter.

What is dark matter? A number of suggestions are under investigation: Jupiter-like objects, swarms of black holes, copious numbers of white dwarf stars, and exotic unknown elementary particles. These candidates can be lumped into two groups: MACHOs and WIMPs.

Dark matter candidates such as small main-sequence M stars, planets, white dwarfs, neutron stars, or black holes are collectively referred to as **MACHOs**, which stands for "massive compact halo objects." If the dark matter in our halo consists of MACHOs, there must be a lot of these objects, and they must each exert gravitational force but not emit much light. Because they have mass, MACHOs gravitationally deflect light according to Einstein's general theory of relativity, a phenomenon called gravitational lensing (which we saw in Chapter 13). If we were observing a distant star and a MACHO passed between us and the star, the star's light would be deflected and perhaps, if the geometry were just right, focused

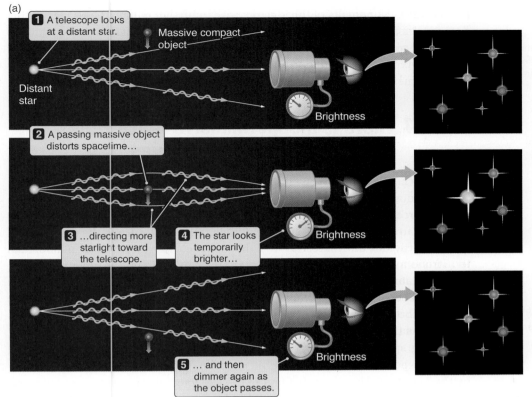

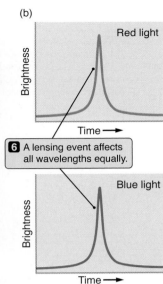

FIGURE 15.14 (a) The light from a distant star is affected by a compact object crossing our line of sight. (b) The observed light curves of a real star experiencing a lensing event.

by the intervening MACHO as it passed across our line of sight, as illustrated in **Figure 15.14a**. Because gravity affects all wavelengths equally, such lensing events should look the same in all colors, ruling out other causes of variability.

We would be remarkably lucky if such an event occurred just as we were observing a single distant star. However, astronomers monitored the stars in the Large and Small Magellanic Clouds (two of the small companion galaxies of the Milky Way Galaxy), observing tens of millions of stars for several years. They found a number of examples of events of the sort shown in **Figure 15.14b**, but not nearly enough to account for the amount of dark matter in the halo of our galaxy. Thus, it was concluded that the dark matter in our galaxy is probably *not* composed primarily of MACHOs.

This leaves the exotic unknown elementary particles commonly known as **WIMPs**, which stands for "weakly interacting massive particles." These particles are predicted to be something like neutrinos; they would barely interact with ordinary matter, yet would have some mass. WIMPs are currently the favored explanation because there are not enough MACHOs to account for the effects we observe. Experiments are under way at the Large Hadron Collider and on the International Space Station to prove the existence of such particles, and additional experiments are being done to detect such particles from our galactic halo as they pass through the Earth.

Some proposed explanations of the "missing mass" do not rely on dark matter. For example, Modified Newtonian Dynamics (MoND) calls for modifications to Newton's law of gravity that become apparent only on large scales. Observations at intermediate scales, for example in the Bullet Cluster where two galaxy clusters are colliding, show that the original formulation of MoND cannot explain the observation. Proponents of the idea propose to further modify MoND by adding hot neutrinos. As the science proceeds, MoND, as well as other proposed explanations, will continue to be tested as possible alternatives to dark matter. Eventually, all of the explanations but one will be ruled out, and the surviving explanation will become the theory that explains the problem of galaxy rotation curves.

15.4 There Is a Supermassive Black Hole at the Heart of Most Galaxies

Although they shine with the light of billions of stars, galaxies pale in comparison with the most brilliant beacons of all: **quasars**. *Quasar* is short for "quasi-stellar radio source," so named because astronomers first observed these mysterious objects as **unresolved** points at radio wavelengths. The story of the discovery of quasars provides an interesting insight into the discovery of new phenomena in astronomy and how conventional thinking can sometimes hinder progress.

In the late 1950s, radio surveys detected a number of bright, compact objects that at first seemed to have no optical counterparts. Improved radio positions revealed that these radio sources coincided with faint, very blue, starlike objects. Unaware of the true nature of these objects, astronomers called them "radio stars." Obtaining spectra of the first two radio stars was a laborious task, requiring 10-hour exposures with photographic plates. Astronomers were greatly puzzled by the results. Rather than displaying the expected absorption lines characteristic of blue stars, the spectra showed only a single pair of emission lines that were broad—indicating very rapid motions within these objects—and that did not seem to correspond to the lines of any known substances.

For several years astronomers believed they had discovered a new type of star, until one astronomer, Maarten Schmidt (b. 1929), realized that these broad spectral lines were the highly redshifted lines of ordinary hydrogen. The implications were astounding: these "stars" were not stars. They were extraordinarily luminous objects at enormous distances.

Quasars are phenomenally powerful, pouring forth the luminosity of a trillion to a thousand trillion (10^{12}–10^{15}) Suns. They are also very distant—the nearest quasar is about 300 million parsecs (Mpc) away. Literally billions of galaxies are closer to us than the nearest quasar. The distance to an object also tells us the amount of time that has passed since the light from that object left its source. Since we see quasars only at great distances, we know that they are quite rare in the universe at this time, but were once much more common. This discovery of the distribution of quasars was one of the first pieces of evidence to demonstrate that the universe has evolved over time.

There are several types of active galactic nuclei.

Today we know that quasars result from the most extreme form of activity that can occur in the hearts of large galaxies, as seen in **Figure 15.15**. Together, quasars and their less luminous but still active cousins are called **active galactic nuclei** (**AGNs**). Several distinct types of active nuclei have been identified in different types of galaxies.

Seyfert galaxies, named after Carl Seyfert (1911–1960), who discovered them in 1943, are spiral galaxies whose centers contain AGNs discernible in visible light as a distinct bright spot. The luminosity of a typical Seyfert nucleus can be 10 billion to 100 billion $L_\odot$, comparable to the luminosity of the rest of the galaxy as a whole. The luminosities of AGNs found in elliptical galaxies are similar to those of Seyfert nuclei (10 billion to 100 billion $L_\odot$). Unlike Seyfert nuclei, how-

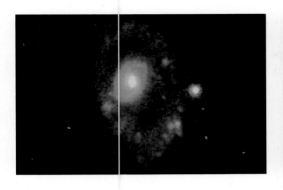
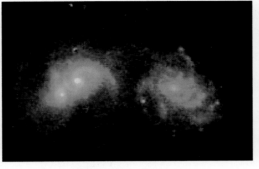

GXUVIR

FIGURE 15.15 The environments around quasars, which are found in the centers of galaxies. Those galaxies often show evidence of interactions with other galaxies.

ever, AGNs in elliptical galaxies are usually most prominent in the radio portion of the electromagnetic spectrum, earning them the name **radio galaxies**. Radio galaxies, and their distant, extremely luminous cousins the quasars, are often the sources of slender jets that extend outward millions of parsecs from the galaxy, powering twin lobes of radio emission such as those seen in **Figure 15.16**.

Much of the light from AGNs is synchrotron radiation, which comes from relativistic charged particles spiraling around in the direction of a magnetic field. This is the same type of radiation that we first encountered coming from Jupiter's magnetosphere and later saw again in such extreme environments as the Crab Nebula. The fact that AGNs accelerate large amounts of material to nearly the speed of light indicates that they are violent objects indeed. In addition to the continuous spectrum of synchrotron emission, the spectra of many quasars and Seyfert nuclei show emission lines that are smeared out by the Doppler effect across a wide range of wavelengths. This observation implies that gas in AGNs is swirling around the centers of these galaxies at speeds of thousands or even tens of thousands of kilometers per second.

AGNs Are the Size of the Solar System

The enormous radiated power and mechanical energy of active galactic nuclei are mind-boggling on their own, but they are made even more spectacular because all of this power emerges from a region that can be no larger than a light-day or so across—comparable in size to our own Solar System. How can we make such a claim? For one thing, quasars and other AGNs at the centers of galaxies appear only as unresolved points of light, even in our most powerful telescopes. For more evidence, we turn not to the sky but to the halftime show at a local football game.

Figure 15.17 illustrates a problem faced by every director of a marching band. When a band is all together in a tight formation at the center of the field, the notes you hear in the stand are clear and crisp; the band plays together beautifully. But as the band spreads out across the field, its sound begins to get mushy. This is not because the marchers are poor musicians. Rather, it is because sound travels at a finite speed. On a cold, dry December day, sound travels at a speed of about 330 meters per second (m/s). At this speed, it takes sound approximately one-third of a second to travel from one end of the football field to the other. Even if every musician on the field played a note at exactly the same instant in response to the

GXUVIR

FIGURE 15.16 Radio emission from a double-lobed radio galaxy (shown in red) is superimposed over an image of visible starlight from the galaxy (shown in blue). The lobes, powered by a relativistic jet streaming outward from the nucleus of the galaxy, extend to more than 1 Mpc from the galaxy's center.

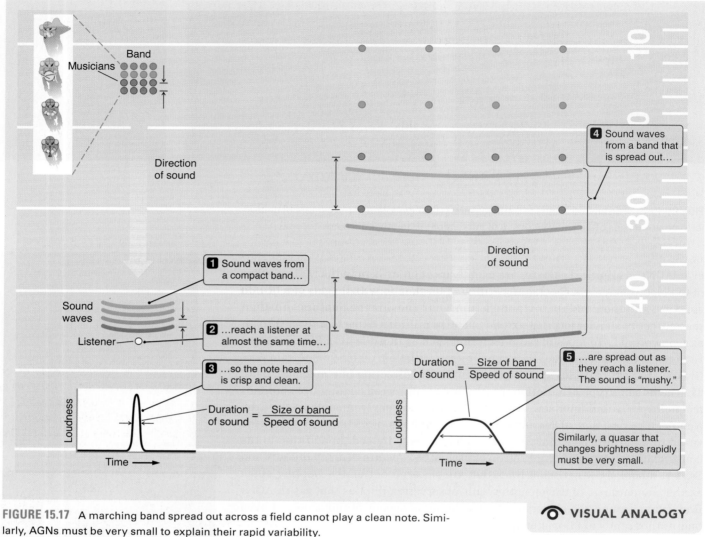

FIGURE 15.17 A marching band spread out across a field cannot play a clean note. Similarly, AGNs must be very small to explain their rapid variability.

VISUAL ANALOGY

▶‖ **AstroTour: Active Galactic Nuclei**

director's cue, in the stands you would hear the instruments close to you first but would have to wait longer for the sound from the far end of the field to arrive.

If the band is spread from one end of the field to the other, then the beginning of a note will be smeared out over about one-third of a second, or the difference in sound travel time from the near side of the field to the far side. If the band were spread out over two football fields, it would take about two-thirds of a second for the sound from the most distant musicians to arrive at your ear. If our marching band were spread out over a kilometer, then it would take roughly 3 seconds—the time it takes sound to travel a kilometer—for us to hear a crisply played note start and stop. Even with our eyes closed, it would be easy to tell whether the band was in a tight group or spread out across the field.

Exactly the same principle applies to the light we observe from active galactic nuclei. Quasars and other AGNs change their brightness dramatically over the course of only a day or two—and in some cases as briefly as a few hours. This rapid variability sets an upper limit on the size of the AGN, just as hearing clear music from the band tells us that the band musicians are close together. The AGN powerhouse must therefore be no more than a light-day or so across because if it were larger, the light we see could not possibly change in only a day or two. An AGN has the light of 10,000 galaxies pouring out of a region of space that would come close to fitting within the orbit of Neptune.

AGNs vary rapidly and so must be relatively small.

Supermassive Black Holes and Accretion Disks Run Amok

When astronomers first discovered AGNs, they put forward a variety of ideas to explain them. But as observations revealed the tiny sizes and incredible energy densities of AGNs, only one answer seemed to make sense: violent accretion disks surrounding **supermassive black holes**—black holes with masses from thousands to tens of billions of solar masses—power AGNs. We have run across accretion disks several times before. Accretion disks surround young stars, providing the raw material for planetary systems. Accretion disks around white dwarfs, fueled by material torn from their bloated evolving companions, lead to novae and Type Ia supernovae. Accretion disks around neutron stars and star-sized black holes a few kilometers across are seen as X-ray binary stars. Now take these examples and scale them up to a black hole with a mass of a billion solar masses and a radius comparable in size to the orbit of Neptune. Furthermore, imagine an accretion disk fed by substantial fractions of *entire galaxies* rather than the small amounts of material being siphoned off a star. *That* is an active galactic nucleus.

AGNs are powered by accretion onto supermassive black holes.

Astronomers have developed this basic picture of a supermassive black hole surrounded by an accretion disk into a more complete physical description called the **unified model of AGNs**. This model attempts to unify our understanding of all AGNs—quasars, Seyfert galaxies, and radio galaxies—within the same physical framework.

Figure 15.18a shows the various components of the unified model of AGNs. In this model, an accretion disk surrounds a supermassive black hole. Much farther out lies a large torus (doughnut) of gas and dust consisting of material that is feeding the central engine. Each different component of the unified model accounts for different observed properties of AGNs.

As material moves inward toward the supermassive black hole, conversion of gravitational energy heats the accretion disk to hundreds of thousands of kelvins, causing it to glow brightly in visible and ultraviolet light. Conversion of gravitational energy to thermal energy as material falls onto the accretion disk is also a source of X-rays, UV radiation, and other energetic emission. When we discussed the Sun, we marveled at the efficiency of fusion, which converts 0.7 percent of the mass of hydrogen into energy. In contrast, approximately 50 percent of the mass of infalling material around a supermassive black hole is converted to luminous energy. The rest of that mass is pulled into the black hole itself, causing it to grow even more massive.

The interaction of the accretion disk with the black hole gives rise to powerful radio jets. Throughout, twisted magnetic fields accelerate charged particles

FIGURE 15.18 (a) The unified model of active galactic nuclei. The appearance of the object changes when the disk and torus are viewed from the edge (b), at higher inclination (c), and close to face on (d). Astronomers believe that the viewing angle, the mass of the central black hole, and the rate at which it is being fed determine the properties of any AGN that we see.

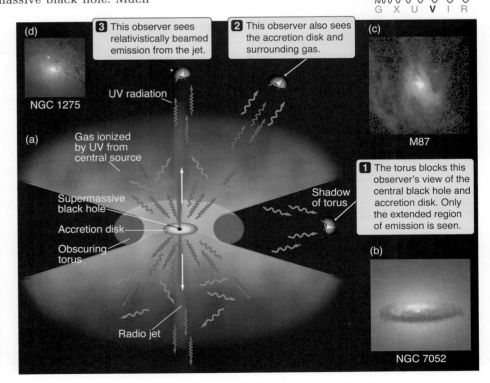

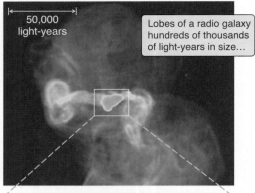

50,000 light-years

Lobes of a radio galaxy hundreds of thousands of light-years in size...

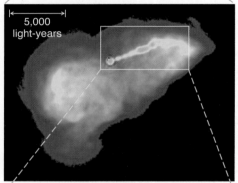

5,000 light-years

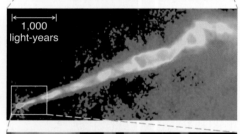

1,000 light-years

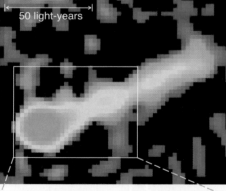

50 light-years

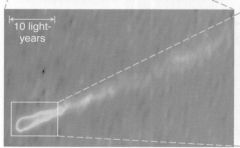

10 light-years

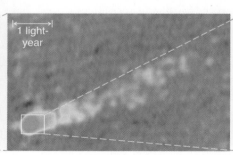

1 light-year

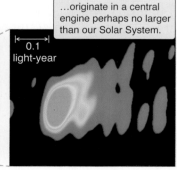

0.1 light-year

...originate in a central engine perhaps no larger than our Solar System.

G X U V I R

such as electrons and protons to relativistic speeds, accounting for the observed synchrotron emission. Gas in the accretion disk or in nearby clouds orbiting the central black hole at high speeds gives off emission lines that are smeared out by the Doppler effect into the broad lines seen in AGN spectra. This accretion disk surrounding a supermassive black hole is the "central engine" that leads to AGNs.

The outer torus plays a somewhat different role in the unified model. Located far from the inner turmoil of the accretion disk, and far larger than the central engine, some of the outer torus is ionized by UV light from the AGN. The most important conceptual role of the outer torus is that it allows a single model to explain many of the differences observed among various AGNs. The outer torus may obscure our view of the central engine in different ways, depending on the viewing angle. The unified model of AGNs can account for a wide range of AGN properties, including the average spectra of AGNs.

The AGN Features We See Depend on Our Perspective

In the unified model, when we view the AGN edge on, we see emission lines from the surrounding torus and other surrounding gas. We can also sometimes see the torus in absorption against the background of the galaxy. **Figure 15.18b** is a Hubble Space Telescope (HST) image of the inner part of the galaxy NGC 7052, showing what appears to be just such a shadow. From this nearly edge-on orientation, we cannot see the accretion disk itself, so we do not expect to see the Doppler-smeared lines that originate closer to the supermassive black hole. If jets are present in the AGN, however, we should be able to see these emerging from the center of the galaxy.

If we look at the inner accretion disk somewhat more face on (**Figure 15.18d**), we can see over the edge of the torus, getting a more direct look at the accretion disk and the location of the black hole. In this case, we see more of the synchrotron emission from the region around the black hole and the Doppler-broadened lines produced in and around the accretion disk. **Figure 15.18c** shows an HST image of one such object, called M87, at an intermediate inclination. M87 is a source of powerful jets that continue outward for about thirty thousand parsecs but originate in the tiny engine at the heart of the galaxy (**Figure 15.19**). Spectra of the disk at the center of this galaxy (**Figure 15.20**) show the rapid rotation of material around a central black hole with a mass of 3 billion solar masses ($3 \times 10^9 \, M_\odot$).

FIGURE 15.19 The visible jet from the galaxy M87 extends over 30,000 parsecs but originates in a tiny volume at the heart of the galaxy (far lower right).

The material in an AGN jet travels very close to the speed of light. As a result, relativistic effects are important. One of these is an extreme form of the Doppler effect called **relativistic beaming**: matter traveling at close to the speed of light concentrates any radiation it emits into a tight beam pointed in the direction in which it is moving. Because of relativistic beaming, an AGN jet coming toward us is much brighter than the one moving away from us. As a result, we often observe only one side of the jets from AGNs, even though the radio lobes of radio galaxies are always two-sided. The jet moving away is just too faint to observe.

In the rare instances when the accretion disk in a quasar or radio galaxy is viewed almost directly face on, relativistic beaming dominates our observations. Emission lines and other light coming from hot gas in the accretion disk are overwhelmed by the bright glare of jet emission beamed directly at us (as in Figure 15.18d).

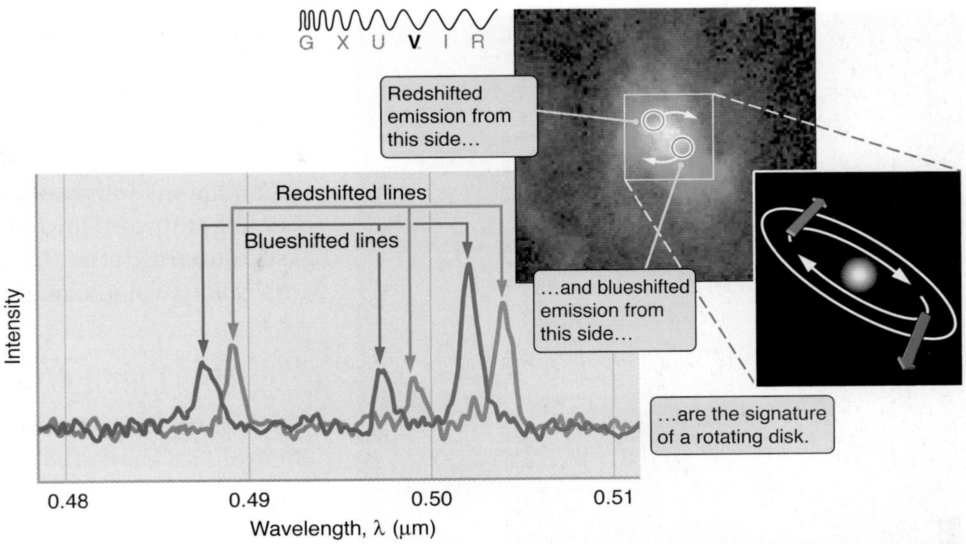

FIGURE 15.20   An HST image of the nearby radio galaxy M87. The high velocities observed provide direct evidence of rotation about a supermassive black hole at this galaxy's center.

Normal Galaxies and AGNs

The essential elements of an AGN are a central engine (an accretion disk surrounding a supermassive black hole) and a source of fuel (gas and stars flowing onto the accretion disk). Without a source of matter falling onto the black hole, an AGN would no longer be active. If we were to look at such an object, we would see a normal (not active) galaxy with a supermassive black hole sitting in its center.

Only a small percentage of present-day galaxies contain AGNs as luminous as the host galaxy. When the universe was younger, there were many more AGNs than exist today. If our understanding of AGNs is correct, then all the supermassive black holes that powered those dead AGNs should still be around. If we combine what we know of the number of AGNs in the past with ideas about how long a given galaxy remains in an active phase, we are led to predict that many—perhaps even *most*—normal galaxies today contain supermassive black holes. This is a somewhat startling prediction—galaxies like our own Milky Way once had active, supermassive black holes at their centers. Yet here is a prediction that can be tested.

A concentration of mass at the center of a galaxy should draw surrounding stars close to it. The central region of such a galaxy would be much brighter in the presence of such a mass than if stars alone were responsible for the gravitational field. Stars feeling the gravitational pull of a supermassive black hole in the center of a galaxy should also orbit at very high velocities and therefore show very large Doppler shifts. Astronomers have found evidence of this sort in every normal galaxy with a substantial bulge in which they have conducted a careful search. The masses inferred for these black holes range from 10,000 $M_\odot$ to 5 billion $M_\odot$. The mass of the supermassive black hole seems to be related to the mass of the elliptical-galaxy or spiral-galaxy bulge in which it is found. All large galaxies probably contain supermassive black holes. These observations confirm our prediction. They also tell us something remarkable about the structure and history of normal galaxies.

All large galaxies probably contain supermassive black holes.

Apparently the only difference between a normal galaxy and an active galaxy is whether the supermassive black hole at its center is being fed at the time we see that galaxy. The rarity of present-day galaxies with very luminous AGNs does not indicate which galaxies have the potential for AGN activity. Rather, it indicates which galaxy centers are being lit up at the moment. If we were to drop a large amount of gas and dust directly into the center of any large galaxy, this material would fall inward toward the central black hole, forming an accretion disk and a surrounding torus. The predicted result of this process is that the nucleus of this galaxy would change into an AGN.

Mergers and Interactions Make the Difference

Galaxies do not exist in isolation. Even our own Milky Way has several neighbors. **Figure 15.21** shows the havoc that interactions between galaxies can cause, pulling interacting galaxies into distorted shapes in which stars and gas are drawn out into sweeping arcs and tidal tails. Sometimes when galaxies interact, they pass by each other and go their separate ways. Such interactions can trigger the formation of spiral structure, and they can also slam clouds of interstellar gas together at high speeds, triggering additional star formation. Tidally distorted or interacting galaxies often contain regions of vigorous ongoing star formation. Sometimes when two galaxies interact, they merge to form a single, larger galaxy.

To account for the many large galaxies that we see today, interactions and mergers must have been much more prevalent when the universe was younger, which explains the larger number of AGNs that existed in the past. Computer models show that galaxy-galaxy interactions can cause gas located thousands of parsecs from the center of a galaxy to fall inward toward the center, where it can provide fuel for an AGN. During mergers, a significant fraction of a cannibalized galaxy might wind up in the accretion disk. HST images of quasars, like those in Figure 15.15, often show that quasar host galaxies are tidally distorted or are surrounded by other visible matter that is probably still falling into the galaxies. The most violent forms of AGN activity were likely most common in the early universe because that was when galaxies were forming, and large amounts of matter were constantly being drawn in by the gravity of newly formed galaxies. This process is still at work today. Galaxies that show evidence of recent interactions with other galaxies are more likely to house AGNs in their centers.

Our understanding of AGNs is far from complete. For example, the unified model does not account for why one quasar can be a powerful radio source while another, identical in all other respects, emits no radio waves that we can detect, even with the most sensitive radio telescopes. Also, we still cannot predict how long an outburst of AGN activity will last, or how often galaxies will undergo episodes of AGN activity. However, we *can* say confidently that any large galaxy, including our own, might be only a chance encounter away from becoming an AGN.

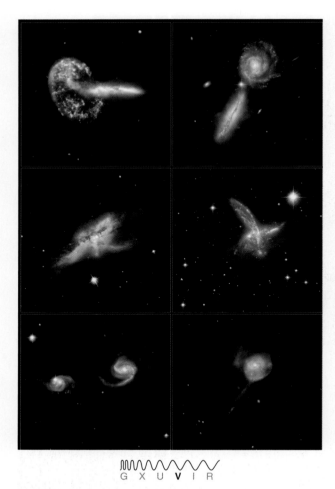

G X U V I R

FIGURE 15.21 These tidally interacting galaxies show severe distortions, including stars and gas drawn into long tidal tails.

Galaxy-galaxy interactions fuel AGN activity.

▶❙❙ **AstroTour: Galaxy Interactions and Mergers**

15.5 Galaxies Form Groups, Clusters, and Larger Structures

The vast majority of galaxies are parts of gravitationally bound collections of galaxies. The smallest and most common are called **galaxy groups**. A galaxy group contains up to several dozen galaxies, most of them dwarf galaxies. Our

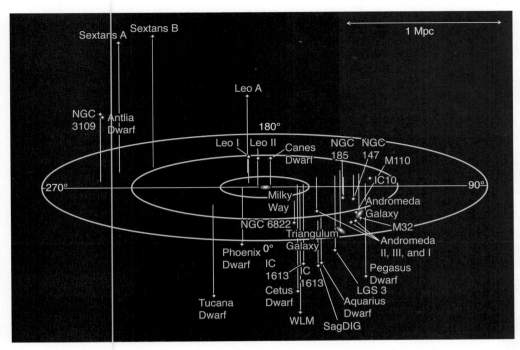

FIGURE 15.22 A graphical map of the galaxies in our own Local Group of galaxies. Most are dwarf galaxies. Spiral galaxies are shown in yellow.

Milky Way is a member of the **Local Group**, which consists of two giant spirals (the Milky Way and the Andromeda Galaxy), along with more than 30 smaller dwarf galaxies in a volume of space roughly 2 Mpc in diameter (**Figure 15.22**). Almost 98 percent of all the galaxy mass in the Local Group resides in the two giant galaxies.

Galaxy clusters consist of hundreds of galaxies, often with a more regular structure than is found in galaxy groups. Galaxy clusters are larger than groups, typically occupying a volume of space 3–5 Mpc across. In many ways, our own Local Group can be considered a small cluster of galaxies, and **Figure 15.23** shows its position relative to two well-known clusters (the Virgo Cluster and the Coma Cluster). As in groups, in a cluster the number of dwarf galaxies is larger than the number of giant galaxies. However, most of the *mass* in galaxy clusters resides in the giant galaxies. Although spiral galaxies are common in most systems, elliptical galaxies are prevalent in only about one-fourth of galaxy clusters. The Virgo Cluster, located 17 Mpc from the Local Group, is an example of a cluster containing mostly spiral galaxies. The more distant Coma Cluster is dominated by giant elliptical and S0 galaxies.

Understanding how clusters form and evolve requires observations beyond the visible. Clusters are bright in the X-ray region of the spectrum (**Figure 15.24a**), indicating that they are rich in hot gas. The amount of visible, luminous mass is not enough to keep this hot gas from escaping. Like stars in the disks of galaxies, cluster galaxies orbit the center of mass of the cluster much faster than can be accounted for by the luminous mass we observe. These observations indicate large concentrations of dark matter, which has other effects that can be observed directly. The total luminous and dark mass in clusters acts as a gravitational lens for light coming from galaxies in the background. These gravitational lenses

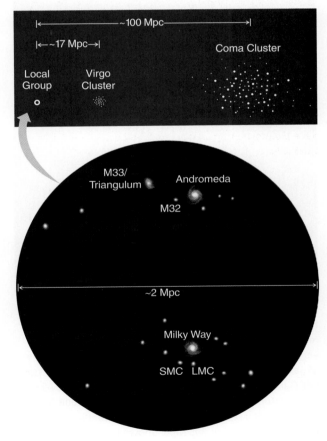

FIGURE 15.23 A close-up view of the Local Group (bottom) and its position relative to the Virgo Cluster, which is the nearest cluster, and to the more distant Coma Cluster.

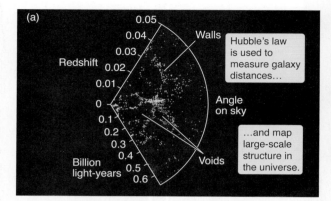

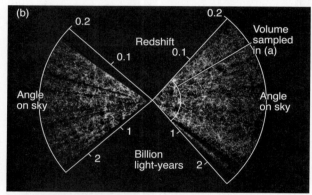

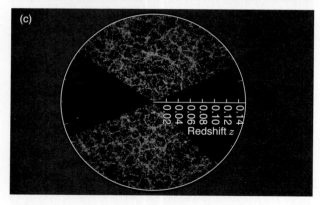

FIGURE 15.25 Redshift surveys use Hubble's law to map the universe. (a) In 1986, the Harvard-Smithsonian Center for Astrophysics redshift survey, called "A Slice of the Universe," was the first to show that clusters and superclusters of galaxies are part of even larger-scale structures. (b) The 2dF Galaxy Redshift Survey, completed in 2003, shows similar structures at even larger distances. (c) The 2008 Sloan Digital Sky Survey map of the universe extends outward to a distance of about 600 Mpc. Shown here is a sample of 67,000 galaxies colored according to the ages of their stars; the redder, more strongly clustered points show galaxies that are made of older stars.

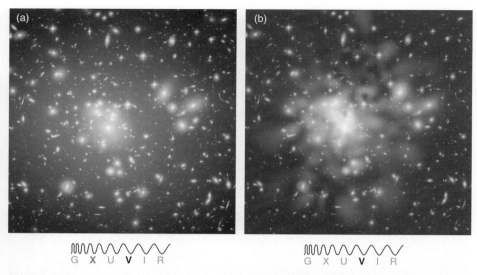

FIGURE 15.24 (a) Galaxy clusters like Abell 1689 are rich in hot gas, which emits X-rays (seen here in purple). (b) Gravitational lensing warps images of background galaxies into arcs. Astronomers use these arcs to trace the distribution of dark matter in the cluster. This is also Abell 1689, with the dark matter distribution shown in blue-white.

warp the images of background galaxies into distorted arcs. In **Figure 15.24b**, the blue-white glow shows the distribution of dark matter inferred from the arcs.

Clusters and groups of galaxies themselves bunch together to form enormous **superclusters**, which contain tens of thousands or even hundreds of thousands of galaxies and span regions of space typically more than 30 Mpc in size. Our Local Group is part of the Virgo Supercluster, which also includes the Virgo Cluster.

Hubble's law is a powerful tool for mapping the distribution of galaxies, groups, clusters, and superclusters in space. Using Hubble's law, all we need to determine the distance to a galaxy is a single spectrum from which we can measure the galaxy's redshift. We now know the redshifts to well over 1 million galaxies, and therefore we know the distances to well over 1 million galaxies. From this information, we can develop a map of the structure of the universe on the largest scales.

The Harvard-Smithsonian Center for Astrophysics conducted the first large redshift survey and, in 1986, presented the astronomical community with a "slice of the universe," as displayed in **Figure 15.25a**. The observations show that the clusters and superclusters, rather than being scattered randomly through space, are linked in an intricate network of "filaments" and "walls." The concentrations of galaxies, in turn, surround large **voids**, regions of space that are largely empty of galaxies. These voids are some of the largest "structures" seen in the universe. Though the voids may seem empty, we do not know that they are empty of *matter*—only that they are largely empty of observable galaxies. Clusters and superclusters are located within the walls and filaments. This structure, however, is not peculiar to the "nearby" universe. Subsequent surveys have looked at much larger volumes of space. **Figure 15.25b** shows the results of one such survey conducted with the Anglo-Australian Telescope at the Siding Spring Observatory in Australia. Results from a more recent survey, the Sloan Digital Sky Survey, are shown in **Figure 15.25c**. For as far out as our observations can currently measure, the universe has a porous structure reminiscent of a sponge. Together, galaxies and the larger groupings in which they are found are referred to as **large-scale structure**.

NASA's Great Observatories program covers a wide range of wavelengths from four different instruments. Sometimes, these instruments are used together to produce shockingly beautiful photographs of the cosmos. Often, these are images of galaxies, as in this posting from August 2010.

Colliding Galaxies Swirl in Dazzling New Photo

By **SPACE**.com

A spectacular new image of two colliding galaxies shows a cosmic region teeming with stellar activity.

The cosmic smash-up, which began more than 100 million years ago but is still occurring, has triggered the formation of millions of stars in the clouds of dust and gas within the galaxies. The composite image was created using data from several different space telescopes (**Figure 15.26**).

The most massive of these young stars speed through their evolution in only a few million years, dying a violent stellar death in supernova explosions.

The colliding Antennae galaxies are located about 62 million light-years away from Earth. In addition to the photo, NASA also released a video of the galaxy collision using the same data from the Chandra X-ray Observatory (blue), the Hubble Space Telescope (gold), and the Spitzer Space Telescope (red).

The X-ray image from Chandra shows vast clouds of hot, interstellar gas that are injected with rich deposits of elements, such as oxygen, iron, magnesium and silicon, created in supernova explosions. The enriched gas will be incorporated into new generations of young stars and planets.

The Antennae galaxies get their name from the long, wispy antenna-like "arms" that can be detected in wide-angle views of the system. These appendage features were produced by tidal forces that were generated from the cosmic collision.

The bright, point-like sources in the image are produced by material that is falling onto black holes and neutron stars, which are dead

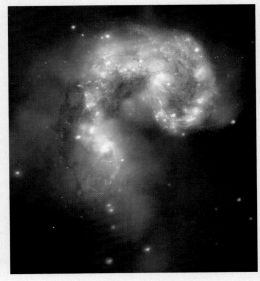

〰〰〰〰〰〰
G X U V I R

FIGURE 15.26 This beautiful new composite image of two colliding galaxies was released by NASA's Great Observatories. The collision between the Antennae galaxies, which are located about 62 million light-years (19 parsecs) from Earth, began more than 100 million years ago and is still occurring. Credit: NASA, ESA, SAO, CXC, JPL-Caltech, and STScI.

relics of massive stars. Some of the black holes in the Antennae galaxies may contain masses that are almost one hundred times that of the sun.

The data from the Spitzer telescope show infrared light from warm clouds of dust that have been heated by newly-formed stars, with the brightest clouds lying in the overlap region between the two colliding galaxies.

The Hubble data reveal old stars and star-forming regions in gold and white, while dusty filaments appear in brown. In the optical image, many of the fainter objects denote clusters containing thousands of stars.

The Chandra image was taken in Dec. 1999, the Spitzer image was taken in Dec. 2003, and the Hubble image was taken in July 2004 and Feb. 2005. Data from the three observatories were combined to make the new composite image.

Evaluating the News

1. Where would the Antennae galaxies lie on the Hubble tuning fork diagram?
2. In the article, the author states that the collision "has triggered the formation of millions of stars." How might we know that? In what part of the colliding galaxies is this star formation likely to be taking place?
3. These galaxies lie 62 million light-years (19 Mpc) away from Earth. What part of the distance ladder might have given this distance?
4. Explain why Chandra pictures show hot gas, Spitzer pictures show warm dust, and Hubble data show stars and some dust.
5. The author states that "The Antennae galaxies get their name from the long, wispy antenna-like 'arms' that can be detected in wide-angle views of the system. These appendage features were produced by tidal forces that were generated from the cosmic collision." What causes these tidal forces? Why do tidal forces produce long, wispy arms?
6. Is this image an important one for astronomers? For the public? For you? Explain the significance of this new image, created from old data.

SUMMARY

15.1 The shapes of galaxies and the types of orbits of their stars determine their Hubble classification. Galaxies are classified as elliptical (E), spiral (S), barred spiral (SB) or irregular (Irr). Currently, stars are forming in the disks of spiral galaxies but not in ellipticals or S0 galaxies.

15.2 Spiral arms are regions of intense star formation, and the arms are visible due to the concentration of bright young stars. Any significant disturbance in the rotating disk of a galaxy will lead to spiral structure. Regular disturbances called spiral density waves trigger star formation.

15.3 Most of the mass in galaxies does not reside in gas, dust, or stars; rather, about 90 percent of a galaxy's mass is in the form of dark matter, which does not emit or absorb light to any significant degree. The two main groups of candidates for the composition of dark matter are MACHOs (massive objects like planets, stars, and black holes) and WIMPs (exotic elementary particles).

15.4 Most—perhaps all—large galaxies have supermassive black holes at their centers. When gas accretes onto one of these supermassive black holes, the center of the galaxy becomes an active galactic nucleus, which can emit as much as a thousand times the light of the whole galaxy, all coming from a region the size of our Solar System. The unified model of active galactic nuclei proposes an outer torus of gas and dust that obscures our view of the AGN in different ways, depending on the angle at which we see it.

15.5 Galaxies reside in groups, clusters, and larger structures, which generally formed after galaxies did. Dark matter is present in large quantities in clusters and was discovered there. On very large scales, the structures in the universe reside mainly on filaments and walls surrounding large regions devoid of galaxies.

✧ SUMMARY SELF-TEST

1. A particular galaxy contains only stars in disordered orbits. This type of galaxy is most likely
   a. an elliptical.
   b. a spiral.
   c. an irregular.

2. In spiral galaxies, stars form predominantly in
   a. the arms of the disk.   **b.** the halo.
   c. the bulge.   **d.** the bar.

3. The rotation curves of galaxies tell us that galaxies are mostly
   a. stars.   **b.** dust and gas.
   c. dark matter.   **d.** black holes.

4. Which of the following are properties of dark matter?
   a. It absorbs all the light that falls on it.
   b. It reflects all the light that falls on it.
   c. It doesn't interact with light, as far as we can tell.
   d. It gravitationally attracts other matter.
   e. It gravitationally repels other matter.
   f. It doesn't interact gravitationally, as far as we can tell.

5. Supermassive black holes
   a. are extremely rare. There are only a handful in the universe.
   b. are completely hypothetical.
   c. occur in most, perhaps all, large galaxies.
   d. occur only in the space between galaxies.

6. Supermassive black holes
   a. gradually consume their host galaxies.
   b. are "fed" by disturbed gas when galaxies interact.
   c. are a recent development in the universe's history.
   d. are an as yet undetected hypothesis.

7. The light that comes to us from active galactic nuclei comes from a volume about the size of
   a. Earth.   **b.** the Solar System.
   c. a globular cluster.   **d.** the bulge of the Milky Way.

8. Place the following types of galaxy collections in order of increasing size:
   a. wall   **b.** cluster
   c. group   **d.** supercluster

QUESTIONS AND PROBLEMS

True/False and Multiple-Choice Questions

9. **T/F:** Galaxies are sometimes difficult to classify because their orientation affects their appearance.

10. **T/F:** The Hubble tuning fork diagram shows how galaxies evolve (change over time).

11. **T/F:** The orbits of stars in the bulge of a spiral galaxy are disordered.

12. **T/F:** Astronomers have evidence of dark matter in only a few galaxies so far.

13. **T/F:** Active galactic nuclei surround some supermassive black holes.

14. If a galaxy has some stars in ordered orbits, it is most likely a(n)
    a. spiral.   **b.** irregular.
    c. elliptical.   **d.** giant elliptical.

15. The orbits of stars in the bulges of spiral galaxies most closely resemble
    a. orbits of planets in the Solar System.
    b. orbits of stars in the disk of a spiral galaxy.
    c. orbits of stars in an elliptical galaxy.
    d. orbits of objects in the Kuiper Belt.

16. Most galaxies have sizes in the range of
    a. hundreds of parsecs.
    b. tens of thousands of parsecs.
    c. hundreds of thousands of parsecs.
    d. millions of parsecs.

17. The flat rotation curves of spiral galaxies imply that the distribution of mass resembles
    a. the Solar System; most mass is concentrated in the center.
    b. a wheel; the density remains the same as the radius increases.
    c. the light distribution of the galaxy; a large concentration occurs in the middle, but significant mass exists quite far out.
    d. an invisible sphere much larger than the visible galaxy.

18. Gravitational lensing allows astronomers to detect massive objects by
    a. temporarily increasing the brightness of the object.
    b. temporarily decreasing the brightness of the object.
    c. temporarily increasing the brightness of a background object.
    d. temporarily decreasing the brightness of a background object.
    e. changing the color of the object (shifting it to the red or blue).

19. Astronomers know active galactic nuclei are relatively small because
    a. variations in brightness occur quickly.
    b. variations in brightness occur slowly.
    c. variations in brightness are small.
    d. they have been directly imaged.

20. Emission lines from active galactic nuclei are both red- and blue-shifted because the nucleus
    a. is exploding.
    b. is imploding.
    c. is spinning.
    d. resembles an elliptical galaxy.

Conceptual Questions

21. Name and describe a common, everyday object that appears so dissimilar when viewed from different angles that you might think from these different views that it is actually a different object.

22. Name and describe three types of galaxies.

23. What are the principal morphological (structural) differences between spiral and elliptical galaxies?

24. How do E0 and E7 elliptical galaxies differ in shape?

25. What would be the shape of an elliptical galaxy whose stars were all traveling in random directions in their orbits? Explain your answer.

26. Sketch the galaxy shown in Figure 15.5b. Label the disk, the bulge, and the halo.

27. How does gas temperature differ between elliptical and spiral galaxies?

28. Explain why star formation in spiral galaxies takes place mostly in the spiral arms.

29. Some galaxies have regions that are relatively blue in color; other regions appear redder. Aside from color, what can you say about the differences between these regions?

30. Galaxies come in a large range of both luminosity and size. How might astronomers figure out which of these two properties varies more among galaxies?

31. Describe the spiral arms in a galaxy, and explain at least one of the mechanisms that create them.

32. Describe the difference between predicted and observed rotation curves of galaxies, as shown in Figure 15.11.

33. In describing galaxies, what do astronomers mean by *luminous* (or *normal*) matter?

34. With regard to its interaction with electromagnetic radiation, how does dark matter differ from normal matter?

35. What evidence do we have that most galaxies are composed largely of dark matter?

36. Contrast the rotation curve for a galaxy containing only normal matter with one containing dark matter.

37. How does a spiral galaxy's dark matter halo differ from its visible spiral component?

38. Name some of the candidates for the composition of dark matter.

39. How would you explain a quasar to a relative or friend?

40. Which is more luminous: a quasar or a galaxy with a hundred billion solar-type stars? Explain your answer.

41. The nearest quasar is about 300 million parsecs (Mpc) away. Why do we not see any that are closer?

42. What distinguishes a "normal" galaxy from one we call "active"—that is, one that contains an AGN?

43. Contrast the size of a typical AGN with the size of our own Solar System. How do we know how big an AGN is?

44. Describe what must be happening at the centers of galaxies that contain AGNs.

45. It is likely that most galaxies contain supermassive black holes, yet in many galaxies there is no obvious evidence for their existence. Why do some black holes reveal their presence while others do not?

46. Study Figure 15.25. Explain how these images show what astronomers mean when they say the universe is homogeneous and isotropic.

47. What are the main differences between galaxy groups and galaxy clusters?

48. Why is it reasonable to call our Local Group of galaxies a group rather than a cluster? Why is it reasonable to call our Local Group a cluster?

49. What is the difference between a galaxy cluster and a supercluster? Is our galaxy part of either? How do we know this?

50. How do astronomers use the following to measure the amount of dark matter contained in a cluster of galaxies?
    a. motions of individual members of the cluster
    b. extremely hot gas that fills the intergalactic space within the cluster
    c. gravitational lensing by the cluster

51. Can a galaxy be located inside a void? Explain.

52. Is it likely that voids are filled with dark matter? Why or why not?

Problems

53. Assume that there are 1 trillion (10^{12}) galaxies in the universe, that the average galaxy has a mass equivalent to 100 billion (10^{11}) average stars, and that an average star has a mass of 10^{30} kg.
    a. Ignoring dark matter, how much mass (in kilograms) does the universe contain?
    b. If the mass of an average particle of normal matter is 10^{-27} kg, how many particles are there in the entire universe?

54. Figure 15.10 shows a time sequence for a spiral galaxy's arms to wind themselves up. What is a reasonable unit for the time step in this diagram? Is it seconds? Years? Millions of years? Billions of years?

55. Suppose the number density of galaxies in the universe is, on average, 3×10^{-68} galaxies per cubic meter (m^3). If astronomers could observe all galaxies out to a distance of 10^{10} parsecs, how many galaxies would they find?

56. Figure 15.12 shows the distribution of dark and normal matter that must be present to create the observed rotation curve.
    a. Explain what is plotted in blue and in red. Is this the amount of mass present at that radius? If so, why does it continuously increase as the radius increases? If not, what is actually plotted here?
    b. At what radius is there the same enclosed mass of dark matter as normal matter?
    c. Imagine that the plot was many times wider than shown, and the radii extended billions of parsecs. What do you expect would happen to the blue line as you got very, very far from the galaxy?

57. The Keplerian speed of a distant star orbiting a galaxy is twice the speed that the galaxy's visible mass would suggest. What is the ratio of dark matter to normal matter within the star's orbit?

58. Use Figure 15.20 and the Doppler effect to estimate the rotation speed of the disk at the core of the galaxy. Note that the rest wavelength would be halfway between the redshifted and blueshifted lines.

59. The nearest known quasar is 3C 273. It is located in the constellation of Virgo and is bright enough to be seen in a medium-sized amateur telescope. With a redshift of 0.158, what is the distance of 3C 273 in parsecs?

60. The quasar 3C 273 has a luminosity of 10^{12} $L_\odot$. Assuming that the total luminosity of a large galaxy, such as the Andromeda Galaxy, is 10 billion times that of the Sun, compare the luminosity of 3C 273 with that of the entire Andromeda Galaxy.

61. Consider a hypothetical star orbiting 3C 273 at a distance of 100,000 AU with a period of 1,080 years. What is the mass of 3C 273 in solar masses?

62. A quasar has the same brightness as a foreground galaxy that happens to be 2 Mpc distant. If the quasar is 1 million times more luminous than the galaxy, what is the distance of the quasar?

63. You read in the newspaper that astronomers have discovered a "new" cosmological object that appears to be flickering with a period of 83 minutes. Having read this book, you are able to quickly estimate the maximum size of this object. How large can it be?

64. A solar-type star ($M_\odot = 2 \times 10^{30}$ kg), accompanied by its planets, approaches a supermassive black hole. As it crosses the event horizon, half of its mass falls into the black hole while the other half is completely converted to energy in the form of light. How much energy does the dying solar system send out to the rest of the universe?

65. A quasar has a luminosity of 10^{41} watts (W), or joules per second (J/s), and 10^8 $M_\odot$ to feed it. Assuming constant luminosity and that half of the mass is converted to light, estimate the quasar's lifetime.

66. A lobe in a visible jet from the galaxy M87 is observed at a distance of 2,000 parsecs from the galaxy's center and moving outward at a speed of 0.99 times the speed of light (0.99c). Assuming constant speed, how long ago was the lobe expelled from the supermassive black hole at the galaxy's center?

67. Figure 15.22 shows a map of the Local Group, centered on the Milky Way.
    a. What is the name of the galaxy closest to the Milky Way? How far away is it?
    b. What is the approximate diameter of the Local Group?
    c. The Milky Way has been placed at the center of this map because that is where we are. According to the map, is the Milky Way actually at the center of the local group? Explain.

SmartWork, Norton's online homework system, includes algorithmically generated versions of these questions, plus additional conceptual exercises. If your instructor assigns questions in SmartWork, log in at **smartwork.wwnorton.com**.

StudySpace is a free and open website that provides a Study Plan for each chapter of **Understanding Our Universe**. Study Plans include animations, reading outlines, vocabulary flashcards, and multiple-choice quizzes, plus links to premium content in SmartWork and the ebook. Visit **wwnorton.com/studyspace**.

Exploration | Galaxy Classification

Galaxy classification sounds simple, but it can become complicated when you actually attempt it. **Figure 15.27**, taken by the Hubble Space Telescope, shows a small portion of the Coma Cluster of galaxies. The Coma Cluster contains thousands of galaxies, each containing billions of stars. Some of the objects in this image (the ones with a bright cross) are foreground stars, in the Milky Way. Some of the galaxies in this image are far behind the Coma cluster. Working with a partner, you will classify the 20 or so brightest galaxies in this cluster.

First, make a map by laying a piece of paper over the image and numbering the 20 or so brightest (or largest) galaxies in the image. Copy this map so that you and your partner each have a list of the same galaxies.

Separately, classify the galaxies (label them galaxy 1, galaxy 2, etc.). If it is a spiral galaxy, what is its subtype: a, b, or c? If it is an elliptical, how elliptical is it? Make a table that contains the galaxy number, the type you have assigned, and any comments that will help you remember why you made that choice.

When you are done classifying, compare your list with your partner's. Now comes the fun part! Argue about the classifications until you agree—or until you agree to disagree.

If you find this activity interesting and rewarding, astronomers can use your help: visit www.galaxyzoo.org to get involved in a citizen science project to classify galaxies, some of which have never been viewed before by human eyes.

1. **Which galaxy type was easiest to classify?**

2. **Which galaxy type was hardest to classify?**

3. **What makes it hard to classify some of the galaxies?**

4. **Which galaxy type did you and your partner agree about most often?**

5. **Which galaxy type did you and your partner disagree about most often?**

6. **How might you improve your classification technique?**

FIGURE 15.27   A Hubble Space Telescope image of the Coma Cluster.

G X U **V** I R

16 Our Galaxy: The Milky Way

We live in a universe full of galaxies of many sizes and types, visible in our most powerful telescopes all the way to the edge of the observable universe. Yet when we gaze at the night sky unaided by telescopes, we do not see this universe of galaxies. Rather, the night sky is filled with a single galaxy—our home, the galaxy we call the Milky Way.

Are its spiral arms prominent? Is the bulge barlike, and is it large or small relative to the disk? These questions are easier to ask and answer for distant galaxies than for the Milky Way, because we are inside the Milky Way. Yet we have already learned far more about our galaxy than about any other. Stars, planets, and the interstellar medium—almost everything we know about them, we have learned within the context of our own galactic home. It is time to merge the perspective of our look outward at the universe of other galaxies with our knowledge of our own locality to better understand our Milky Way as a spiral galaxy.

✦ LEARNING GOALS

Of the hundreds of billions of galaxies in the universe, the one that means the most to us is our cosmic home, the Milky Way. It may be just another galaxy, but it is the only galaxy we can study at close range. The figure at right shows a photo of the Milky Way taken by a student. By the end of the chapter, you should be able to explain why this image implies that we live in a spiral galaxy. You should be able to identify the nature of the dark lanes in this photo. You should also be able to:

- Describe the components of the interstellar medium and how they emit, scatter, and absorb light

- Explain how we measure the size of the Milky Way using variable stars in globular clusters

- Sketch the galaxy's structure as revealed by Doppler-shifted radio emission

- Chart how the chemical composition of the Milky Way has evolved with time

- List differences in the age and chemical composition of groups of stars that tell us about the history of star formation in our galaxy

- Describe the properties of the black hole at the center of our galaxy

THE MILKY WAY

... as seen during last Saturday's star party.

(120sec exposure)

YOU ARE HERE

16.1 Space Is Not Empty

Vocabulary Alert

Extinction: In common language, extinction refers to the complete loss of a biological species—all its members are dead. Astonomers use the word to mean that light has been removed from the spectrum through a broad range of wavelengths.

In Chapter 5, we saw how the Sun formed from a protostar that condensed out of the interstellar medium. It is not too surprising, then, that the chemical composition of the interstellar medium in our region of the Milky Way is similar to the chemical composition of the Sun. In the interstellar medium about 90 percent of the atomic nuclei are hydrogen, and almost all of the remaining 10 percent are helium. More massive elements collectively account for only 0.1 percent of the atomic nuclei, or about 2 percent of the mass in the interstellar medium. Roughly 99 percent of this interstellar matter is gaseous, and this **interstellar gas** consists of individual atoms or molecules moving freely about, like the molecules in the air around us.

However, interstellar gas is far less dense than air. The air around you has an average density of about 2.5×10^{19} molecules per cubic centimeter (cm^3). The interstellar medium has an average density of less than 1 atom/cm^3. There is about as much material in a column of air between your eye and the floor beside you as there is interstellar gas in a column with the same area stretching from the Solar System to the center of the galaxy.

The Interstellar Medium Is Dusty

About 1 percent of the material in the interstellar medium is in the form of solid grains, referred to as **interstellar dust**. Ranging in size from little more than large molecules up to particles about 300 nm across, these solid grains more closely resemble the particles of soot from a candle flame than the dust that collects on a windowsill. (It would take several hundred "large" interstellar grains to span the thickness of a single human hair.) Interstellar dust begins to form when refractory materials (see Chapter 5) such as iron, silicon, and carbon stick together to form grains in dense, relatively cool environments like the outer atmospheres and "stellar winds" of cool, red giant stars—or in dense material thrown into space by stellar explosions. Once these grains are in the interstellar medium, other atoms and molecules may stick to them. This process is remarkably efficient: about half of all interstellar atoms more massive than helium (1 percent of the total mass of the interstellar medium) are found in interstellar grains.

If there is only as much material between us and the center of the galaxy as there is between your eye and the floor, then you might expect our view of distant objects to be as clear as your view of your own big toe. But interstellar dust is extremely effective at blocking light. If the air around you contained as much dust as a comparable mass of interstellar material, it would be so dirty that you would be hard-pressed to see your hand held up 10 cm in front of your face. If you go out on a dark July or August night and look closely at the Milky Way, visible as a faint band of diffuse light running through the constellation Sagittarius, you will see a dark "lane" running roughly down the middle of this bright band, splitting it in two. This dark band is the result of vast expanses of interstellar dust blocking our view of distant stars (see also the chapter-opening figure).

When interstellar dust gets in the way of radiation from distant objects, the effect is called **interstellar extinction**. Not all electromagnetic radiation suffers equally from interstellar extinction. **Figure 16.1** shows two images of the Milky Way: one taken in visible light, the other taken in the infrared (IR). The dark clouds that block the shorter-wavelength visible light seem to have vanished in

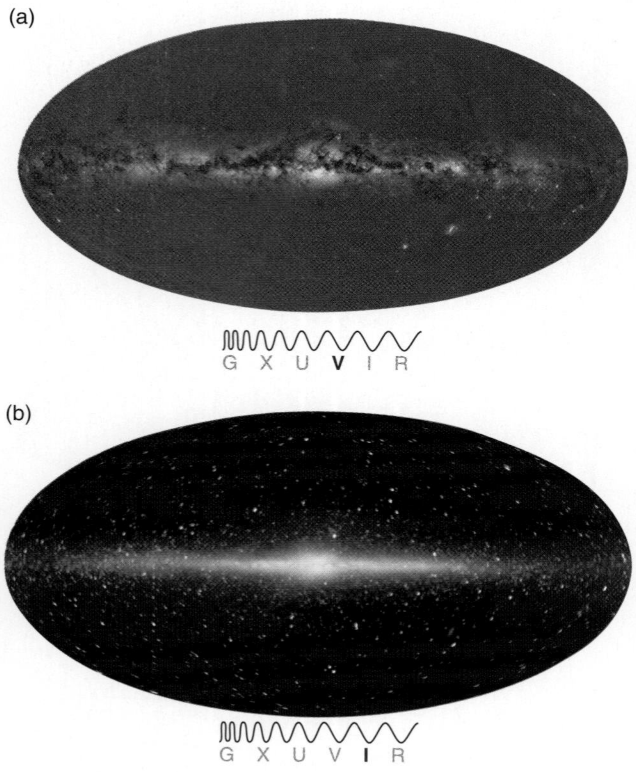
(a)
G X U V I R
(b)
G X U V I R

FIGURE 16.1 (a) An all-sky picture of the Milky Way, taken in visible light. The dark splotches blocking our view are dusty interstellar clouds. (b) The same field of view as seen in the near infrared. Infrared radiation penetrates interstellar dust, giving us a clearer view of the stars in our galaxy.

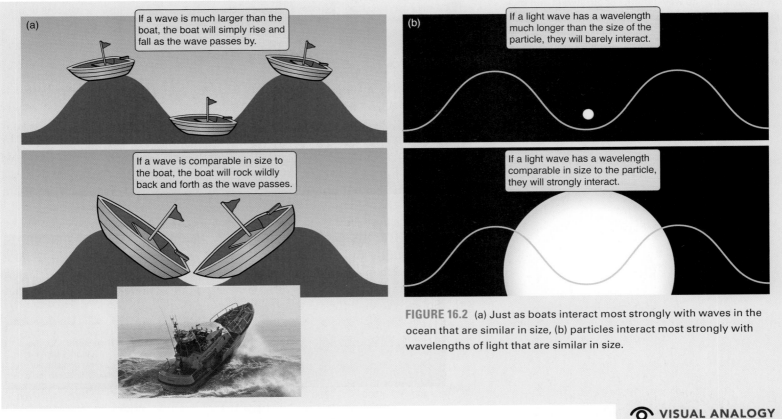

FIGURE 16.2 (a) Just as boats interact most strongly with waves in the ocean that are similar in size, (b) particles interact most strongly with wavelengths of light that are similar in size.

VISUAL ANALOGY

the longer-wavelength IR image in part (b), enabling us to see through the clouds to the center of our galaxy and beyond.

To understand why short-wavelength radiation is blocked by dust while long-wavelength radiation is not, think about a different kind of wave—waves on the surface of the ocean (**Figure 16.2a**). Imagine you are on the ocean in a boat. If the waves are much bigger than your boat, the swell causes you to bob gently up and down. But that is about all; there is no other interaction between the waves and the boat. The story is quite different if the waves are comparable in size to your boat. To picture this, imagine waves roughly half the size of your boat. Now the bow of the boat may be on a wave crest while the stern of the boat is in a trough, or vice versa. The boat tips wildly back and forth as the waves go by. If the size of the boat and the wavelength of the waves are a good match, even fairly modest waves will rock the boat. (You might have noticed this if you were ever in a canoe or rowboat when the wake from a speedboat came by.) Now imagine viewing these two situations from the perspective of the wave. The wave is hardly affected when it is much bigger than the boat, but it is strongly affected when it is small compared to the boat. The energy to drive the wild motions of the boat comes from the wave, so the motion of the wave is affected by the interaction.

The interaction of electromagnetic waves with matter is more involved than that of a boat rocking on the ocean, but the same basic idea often applies (**Figure 16.2b**). Tiny interstellar dust grains interact most strongly with ultraviolet and blue light, which have wavelengths comparable to the typical size of dust grains. For this reason, ultraviolet and blue light are effectively blocked by interstellar dust. Short wavelengths suffer heavily from interstellar extinction. Infrared and radio radiation, on the other hand, have wavelengths that are too long to interact

FIGURE 16.3 (a) The wavelengths of ultraviolet and blue light are close to the size of interstellar grains, so the grains effectively block this light. Grains are less effective at blocking longer-wavelength light. As a result, the spectrum of a star (b) when seen through an interstellar cloud (c) appears fainter and redder.

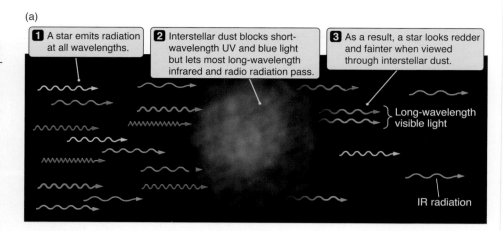

(a)

1 A star emits radiation at all wavelengths.

2 Interstellar dust blocks short-wavelength UV and blue light but lets most long-wavelength infrared and radio radiation pass.

3 As a result, a star looks redder and fainter when viewed through interstellar dust.

Long-wavelength visible light

IR radiation

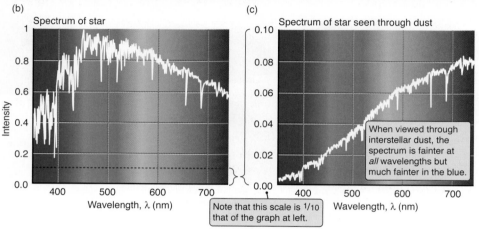

(b) Spectrum of star

(c) Spectrum of star seen through dust

When viewed through interstellar dust, the spectrum is fainter at *all* wavelengths but much fainter in the blue.

Note that this scale is ¹/₁₀ that of the graph at left.

Wavelength, λ (nm)

Wavelength, λ (nm)

Intensity

Long-wavelength radiation penetrates interstellar dust.

strongly with the tiny interstellar dust grains, so they travel largely unimpeded across great interstellar distances. To sum up, at visible and ultraviolet wavelengths, most of the galaxy is hidden from our view by dust. In the infrared and radio portions of the spectrum, however, we get a far more complete view.

In the visible part of the spectrum, blue light suffers more from extinction than does red light. As a result, as shown in **Figure 16.3**, an object viewed through dust looks redder (actually less blue) than it really is. This effect is called **reddening**.

Working It Out 16.1 | Finding the Temperature of Dust

Wien's law, introduced in Chapter 6, relates the temperature of an object to the peak wavelength of the emission. For warm dust at a temperature of 100 K,

$$\lambda_{\text{peak}} = \frac{2{,}900 \ \mu\text{m K}}{T}$$

$$\lambda_{\text{peak}} = \frac{2{,}900 \ \mu\text{m K}}{100 \ \text{K}}$$

$$\lambda_{\text{peak}} = 29 \ \mu\text{m}$$

But for cooler dust, at a temperature of 10 K,

$$\lambda_{\text{peak}} = \frac{2{,}900 \ \mu\text{m K}}{T}$$

$$\lambda_{\text{peak}} = \frac{2{,}900 \ \mu\text{m K}}{10 \ \text{K}}$$

$$\lambda_{\text{peak}} = 290 \ \mu\text{m}$$

The temperature and the peak wavelength are inversely proportional to each other, so that if the temperature drops, the peak wavelength gets longer. For the temperatures that are common for dust in the interstellar medium, the peak wavelength is in the micrometer (10^{-6} m) range. Micrometer wavelengths are 1,000 times longer than those of visible light, which are in the nanometer range.

Correcting for the fact that stars and other objects appear both fainter and redder than they would in the absence of dust can be one of the most difficult parts of interpreting astronomical observations, and it often adds to our uncertainty when measuring an object's properties.

Interstellar extinction may not be much of a concern at infrared wavelengths, but dust still plays an important role in infrared observations. Like any other solid object, grains of dust glow at wavelengths determined by their temperature. In interstellar space, dust is often heated by starlight and by the gas in which it is immersed to temperatures of tens to hundreds of kelvins (**Figure 16.4**). At a temperature of 100 K, Wien's law says that dust will glow most strongly at a wavelength of 29 micrometers (μm), whereas cooler dust—say, at a temperature of 10 K—glows most strongly at a wavelength of 290 μm (**Working It Out 16.1**). Both of these wavelengths are in the far-infrared part of the electromagnetic spectrum, so observations in the far-infrared show thermal radiation from dust (**Figure 16.5**).

Interstellar Gas Has Different Temperatures and Densities

The gas and dust that fill interstellar space within our galaxy are not spread out evenly. About half of all interstellar gas is concentrated in the dense regions called interstellar clouds (see Section 5.1), which fill only about 2 percent of the volume of interstellar space. The other half of the interstellar gas is spread out through the remaining 98 percent of the volume of interstellar space, and is called **intercloud gas**.

The properties of intercloud gas vary from place to place (**Table 16.1**). Some intercloud gas is extremely hot, with temperatures in the millions of kelvins, close to those found in the centers of stars. Even so, were you to find yourself adrift in an expanse of hot intercloud gas, your first concern would be freezing to death. The gas is hot, so the atoms that make it up are moving about very rapidly; if one of these atoms ran into you, it would hit you very hard. But there are so few atoms in a given volume of hot intercloud gas that they would very rarely collide with you, so this million-kelvin gas would do little to keep you warm. You would radiate energy away much faster than the gas around you could replace the lost energy.

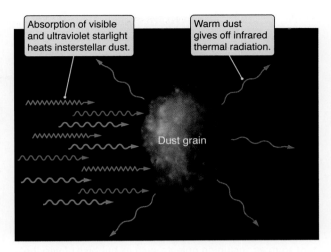

Absorption of visible and ultraviolet starlight heats interstellar dust.

Warm dust gives off infrared thermal radiation.

Dust grain

FIGURE 16.4 The temperature of interstellar grains is determined by equilibrium between absorbed and emitted radiation.

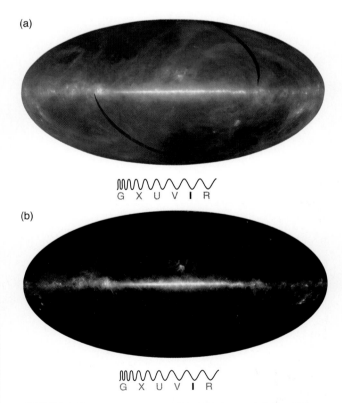

(a)

G X U V I R

(b)

G X U V I R

FIGURE 16.5 (a) Far-infrared (100 μm) images of the sky show clouds of warm, glowing interstellar dust around the Milky Way. (The black bands here and in Figure 16.6 are due to missing data.) (b) The galactic plane at shorter infrared wavelengths (9 μm).

TABLE | 16.1

Typical Properties of Components of the Interstellar Gas

COMPONENT OF OF INTERSTELLAR MEDIUM	TEMPERATURE	DENSITY	STATE OF HYDROGEN
Hot intercloud gas	~1,000,000 K	~0.005 atoms/cm^3	Ionized
Warm intercloud gas	~8000 K	0.01–1 atom/cm^3	Ionized or neutral
Cold intercloud gas	~100 K	1–100 atoms/cm^3	Neutral
Interstellar clouds	~10 K	100–1,000 atoms/cm^3	Molecular or neutral

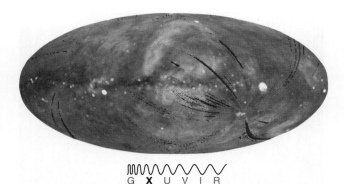

FIGURE 16.6 The faint X-ray glow that fills the sky is due largely to emission from the bubble of million-kelvin gas that surrounds our Solar System. Bright spots are more distant sources, including objects such as bubbles of very hot, high-pressure gas surrounding the sites of recent supernova explosions.

Extremely hot intercloud gas, heated by supernovae, occupies about half the volume of interstellar space. Our Solar System is passing through a bubble of hot intercloud gas that may be the remnant of a supernova that exploded 300,000 years ago. Hot intercloud gas glows faintly in the energetic X-ray portion of the electromagnetic spectrum. Orbiting X-ray telescopes observe the entire sky aglow with faint X-rays coming from our local bubble of million-kelvin hot gas (**Figure 16.6**).

Not all interstellar gas is as hot as our local bubble. Most of the remaining interstellar gas is warm. Ultraviolet starlight with wavelengths shorter than 91.2 nm has enough energy to ionize hydrogen, kicking the electron free of the nucleus. (Any "extra" energy that the photon has increases the kinetic energy of the electron.) If enough of these energetic photons are around, then any neutral hydrogen atom in the interstellar gas will soon be ionized by starlight. About half of the volume of warm intercloud gas is kept ionized by starlight. But if there is a large enough expanse of warm intercloud gas, then the ionizing photons in the starlight get "used up" by the outermost atoms, shielding the neutral gas nearer the center of the cloud from starlight, much as Earth's ozone layer shields the surface of Earth from harmful ultraviolet radiation from the Sun.

As starlight passes through gas, it produces absorption lines that tell us much about the gas's temperature, density, and chemical composition. Intercloud gas can also produce emission lines in regions of warm *ionized* gas, where protons and electrons constantly recombine into hydrogen atoms. Typically, the resulting hydrogen atom is left in an excited state. The atom then drops down to lower and lower energy states, emitting a photon at each step, so warm, ionized intercloud gas glows in emission lines of hydrogen (as well as other elements).

The faint emission in **Figure 16.7** comes mostly from warm, ionized intercloud gas. The bright spots are regions where intense ultraviolet radiation from massive, hot, luminous O and B stars ionizes even relatively dense interstellar clouds. These are called **H II** ("H two") **regions** because the hydrogen atoms are in their ionized, or second, state; that is, they have lost their electron. O stars live for only a few million years, so they usually do not move very far from where

FIGURE 16.7 Warm (about 8000 K) interstellar gas glows in an emission line of hydrogen. This image, showing the hydrogen emission from much of the northern sky, reveals how complex the structure of the interstellar medium is.

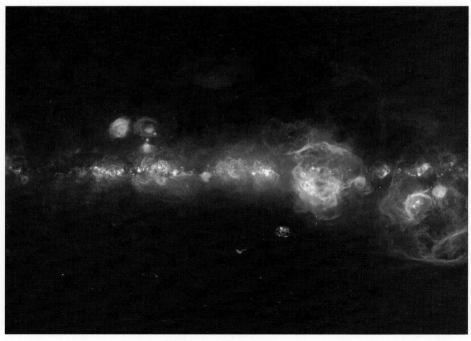

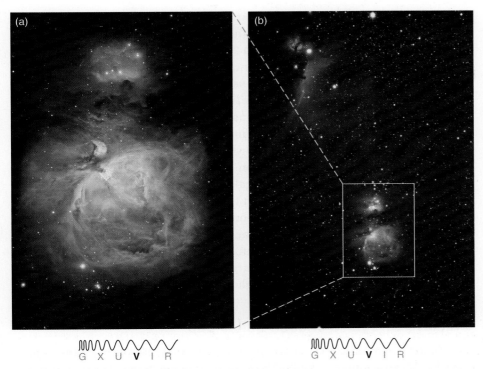

(a)

(b)

G X U **V** I R

G X U **V** I R

FIGURE 16.8 (a) The Orion Nebula is seen here as a glowing region of interstellar gas surrounding a cluster of young, hot stars. New stars are still forming in the dense clouds surrounding the nebula. (b) The Orion Nebula (at bottom) is only a small part of the larger Orion star-forming region. The dark Horsehead Nebula is seen at the upper left.

they formed. These H II regions are the very clouds from which the stars were born and are indicators of active star formation.

One of the closest H II regions to the Sun is the Orion Nebula, located 1,340 light years from the Sun in the constellation Orion (**Figure 16.8**). Almost all of the ultraviolet light that powers the nebula comes from a single hot star, and only a few hundred stars are forming in its immediate vicinity. In contrast, **Figure 16.9** shows a giant H II region called 30 Doradus, located in the Large Magellanic Cloud, a small companion galaxy to the Milky Way 160,000 light-years away. A dense star cluster containing thousands of hot, luminous stars powers this enormous cloud of ionized gas. If 30 Doradus were as close as the Orion Nebula, it would be bright enough in the nighttime sky to cast shadows!

Warm, *neutral* hydrogen gas gives off radiation in a different way than warm, ionized hydrogen. Many subatomic particles, including protons and electrons, have a property called *spin* that causes them to behave as though they are magnetized—as if each of these particles had a bar magnet, with a north and a south pole, built into it. A hydrogen atom can exist in only two different configurations: Either the magnetic "poles" of the proton and electron point in the *same* direction, or they point in *opposite* directions. When the poles point in opposite directions, the atom has slightly more energy than when they point in the same direction.

It takes energy to change a hydrogen atom from the lower-energy, aligned state to the higher-energy, unaligned state. In warm, neutral interstellar gas, interactions between atoms supply this energy. If left undisturbed long enough, a hydrogen atom in the higher-energy state will spontaneously jump to the lower-energy state, emitting a photon in the process. The energy difference between the two

G X U **V** I R

FIGURE 16.9 The 30 Doradus Nebula is located in a small companion galaxy to the Milky Way, 160,000 light-years from Earth. The ultraviolet radiation coming from the dense cluster of thousands of young, massive stars causes interstellar gas in this region of intense star formation to glow.

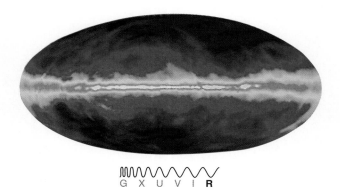

FIGURE 16.10 This radio image of the sky taken at a wavelength of 21 cm shows clouds of neutral hydrogen gas throughout our galaxy. Because radio waves penetrate interstellar dust, 21-cm observations are a crucial method for probing the structure of our galaxy.

Doppler velocities measured from 21-cm radiation show that we live inside a rotating galaxy.

magnetic states of a hydrogen atom is extremely small, so the radiation given off by these transitions has a wavelength in the radio region of the spectrum, at 21 cm.

In **Figure 16.10**, the sky is aglow with 21-cm radiation from neutral hydrogen. Because of its long wavelength, 21-cm radiation freely penetrates dust in the interstellar medium, enabling us to see neutral hydrogen throughout our galaxy. Measurements of the Doppler shift of the line indicate how fast the neutral hydrogen is moving toward us or away from us. These two attributes make the 21-cm line of neutral hydrogen important for understanding the structure of our galaxy.

The velocities of neutral interstellar hydrogen measured from 21-cm radiation are plotted in **Figure 16.11** as a function of the direction in which we are looking. Looking toward the center of the galaxy, we see that on one side hydrogen clouds are moving toward us, while on the other side clouds are moving away from us. This is the pattern of the rotation velocity of a disk. In other directions, the velocities we see are complicated by our moving vantage point within the disk and so are more difficult to interpret at a glance. Even so, observed velocities of neutral hydrogen enable us to measure our galaxy's rotation curve and even determine the structure present throughout the disk of our galaxy.

Clouds Are Regions of Cool, Dense Gas

As mentioned earlier, intercloud gas fills 98 percent of the volume of interstellar space but accounts for only half of the mass of interstellar gas. The remaining 50 percent of all interstellar gas is concentrated into much denser interstellar clouds

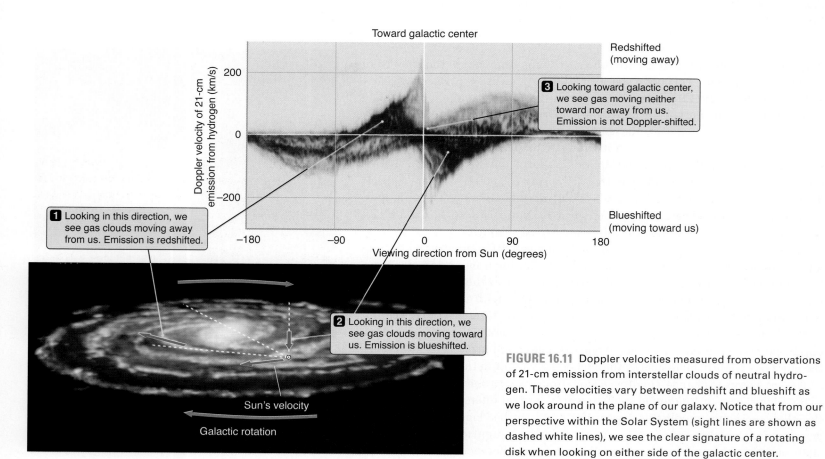

FIGURE 16.11 Doppler velocities measured from observations of 21-cm emission from interstellar clouds of neutral hydrogen. These velocities vary between redshift and blueshift as we look around in the plane of our galaxy. Notice that from our perspective within the Solar System (sight lines are shown as dashed white lines), we see the clear signature of a rotating disk when looking on either side of the galactic center.

that occupy only 2 percent of the volume of the galaxy. Most interstellar clouds are composed primarily of isolated neutral hydrogen atoms and are much cooler and denser than the warm intercloud gas.

On Earth it is uncommon to find atoms in isolation—most atoms are in molecules. However, in most of interstellar space, including most interstellar clouds, molecules do not survive for long. If interstellar gas is too hot, then any molecules that do exist will soon collide with other molecules or atoms that have enough energy to break the molecules apart. The temperature in a neutral hydrogen cloud may be low enough for some molecules to survive, but photons with enough energy to break molecules apart can penetrate neutral hydrogen clouds. Only in the hearts of the densest interstellar clouds, where dust effectively blocks even the relatively low-energy photons required to shatter molecules, does interstellar chemistry get its chance to build molecules out of atoms. These dark clouds, which we saw in Section 5.1, are for this reason called *molecular clouds*. Molecular clouds are often evident from their silhouettes against a background of stars (**Figure 16.12**).

Inside such dark, cold clouds, atoms combine to form a wide variety of molecules, most commonly molecular hydrogen (H_2). Molecules radiate primarily in the radio and infrared portions of the electromagnetic spectrum. Molecular emission lines are useful in the same way that atomic emission lines are useful: the wavelengths of the lines are an unmistakable fingerprint of the kind of molecule responsible for them.

In addition to molecular hydrogen, approximately 150 other molecules have been discovered in interstellar space. These molecules range from very simple structures such as carbon monoxide (CO) to carbon chains as long as $HC_{11}N$. Very large carbon molecules, made of hundreds of individual atoms, bridge the gap between large interstellar molecules and small interstellar grains. Visible light cannot escape from the molecular clouds where interstellar molecules are concentrated. However, the radio waves emitted by molecules escape easily from dark molecular clouds. Observations of molecular lines reveal the innermost workings of the densest and most opaque interstellar clouds.

FIGURE 16.12 In visible light, interstellar molecular clouds are seen in silhouette against a background of stars and glowing gas. (a) Light from background stars is blocked by dust and gas in nearby Barnard 68, a dense, dark molecular cloud. (b) Infrared wavelengths can penetrate much of this gas and dust, as seen in this zoomed, false-color image of Barnard 68.

G X U V I R

G X U V I R

Molecular clouds range from a few times the mass of our Sun to 10 million solar masses. The smallest molecular clouds may be less than half a light-year across, whereas the largest molecular clouds may be over a thousand light-years in size. **Giant molecular clouds** are typically a few hundred thousand times the mass of our Sun and are typically about 100–200 light-years across. Our galaxy contains several thousand giant molecular clouds and a much larger number of smaller ones. Still, molecular clouds fill only about 0.1 percent of interstellar space. These clouds may be rare, but they are extremely important because they are the cradles of star formation.

Magnetic Fields and Cosmic Rays Fill the Galaxy

The interstellar medium of the Milky Way is laced with measurable magnetic fields that are wound up and strengthened by the rotation of the galaxy's disk. This interstellar magnetic field, however, is a million times weaker than Earth's magnetic field. Charged particles spiral around magnetic fields, causing a net motion along the field rather than across it. Conversely, magnetic fields cannot freely escape from a cloud of gas containing charged particles. The dense clouds of interstellar gas in the midplane of the Milky Way anchor the galaxy's magnetic field to the disk, as shown in **Figure 16.13**. These magnetic fields trap **cosmic rays** within our galaxy. Despite their name, cosmic rays are not a form of electromagnetic radiation. (They were named before their true nature was known.) Rather, they are charged particles moving at nearly the speed of light.

Most cosmic-ray particles are protons, but some are nuclei of helium, carbon, and other elements. A few are high-energy electrons and other subatomic particles. Cosmic rays span an enormous range in particle energy. We can observe the lowest-energy ones, with energies as low as about 10^{-11} joules (J), by using interplanetary spacecraft. These energies correspond to the energy of a proton moving at a velocity of a few tenths of the speed of light. In contrast, the most energetic cosmic rays are 10 trillion (10^{13}) times as energetic as the lowest-energy cosmic rays. These high-energy cosmic rays are known from the showers of elementary particles that they cause when crashing through Earth's atmosphere.

Astronomers hypothesize that cosmic rays are accelerated to high energies by the shock waves produced in supernova explosions. The very highest-energy cosmic rays are as much as a hundred million times more energetic than any particle ever produced in a particle accelerator on Earth. These extremely high energies make them much more difficult to explain.

The disk of our galaxy glows from synchrotron radiation produced by cosmic rays (mostly electrons) spiraling around the galaxy's magnetic field. Such synchrotron radiation is seen in the disks of other spiral galaxies as well, telling us that they, too, have magnetic fields and populations of energetic cosmic rays. Even so, the very highest-energy cosmic rays are moving much too fast to be confined to our galaxy.

FIGURE 16.13 The mass of interstellar clouds anchors the magnetic field of the Milky Way to the disk of the galaxy. The magnetic field in turn traps the galaxy's cosmic rays, much as a planetary magnetosphere traps charged particles.

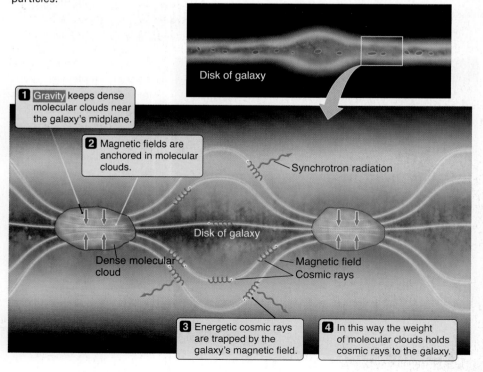

Disk of galaxy

1 Gravity keeps dense molecular clouds near the galaxy's midplane.

2 Magnetic fields are anchored in molecular clouds.

Synchrotron radiation

Disk of galaxy

Dense molecular cloud

Magnetic field
Cosmic rays

3 Energetic cosmic rays are trapped by the galaxy's magnetic field.

4 In this way the weight of molecular clouds holds cosmic rays to the galaxy.

Any such cosmic rays formed in the Milky Way soon stream away from the galaxy into intergalactic space. It is likely that some of the energetic cosmic rays reaching Earth originated in energetic events outside our galaxy.

The total energy of all of the cosmic rays in the galactic disk can be estimated from the energy of the cosmic rays reaching Earth. The strength of the interstellar magnetic field can be measured in a variety of ways, including the effect that it has on radio waves passing through the interstellar medium. These measurements indicate that in our galaxy, the magnetic-field energy and the cosmic-ray energy are about equal to each other. Both are comparable to the energy present in other energetic components of the galaxy, including the motions of interstellar gas and the total energy of electromagnetic radiation within the galaxy.

16.2 Measuring the Milky Way

As noted in earlier chapters, determining the distances to objects in the sky is a complex problem; but it is crucial for determining various properties of objects, such as luminosity or size. In Chapter 10 we learned how parallax is used to measure the distances to nearby stars within a few hundred light-years, and how spectroscopic parallax allows us to use the H-R diagram to find the distance to more distant stars. In Chapter 12 we learned about Type Ia supernovae, which always have the same luminosity, because the exploding star is always 1.4 solar masses. In Chapter 13, we learned about Cepheid variables, RR Lyrae stars, and period-luminosity relationships. Type Ia supernovae, Cepheid variables, and RR Lyrae stars are all examples of "standard candles": objects whose luminosity can be found without knowing their distance. We can then compare luminosity with apparent brightness to find out how far away they are.

Globular Clusters and the Size of the Milky Way Galaxy

A **globular cluster**, such as the one in **Figure 16.14**, is a large spheroidal group of stars held together by gravity. Many clusters can be seen through small telescopes. The motions of stars within a globular cluster are much like the motions of stars within an elliptical galaxy. However, globular clusters are much smaller and the stars are closer together.

Our galaxy has more than 150 cataloged globular clusters (and there are likely many more—dust in the disk of our galaxy may hide them from view). The known globular clusters have luminosities ranging from a low of about 1,000 solar luminosities (1,000 $L_\odot$) to a high of about 1 million $L_\odot$. A typical globular cluster consists of 500,000 stars packed into a volume of space with a radius of only 15 light-years—much more crowded than the average density within the Milky Way. Still, they contain only a small percentage (about 0.001 percent) of the Milky Way's stars.

About one-fourth of the globular clusters in the Milky Way reside in or near the disk. The rest occupy a large, spherical volume of space surrounding the disk and bulge, referred to as the halo of the Milky Way. These globular clusters lie within the dark matter halo we discussed in Chapter 15. Globular clusters are very luminous and lie outside of the dusty disk, so they can be easily seen at great distances. To actually find the distance, we need to look at the properties of the stars they contain.

G X U V I R

FIGURE 16.14 A Hubble Space Telescope image of the globular cluster M80.

(a)

(b)

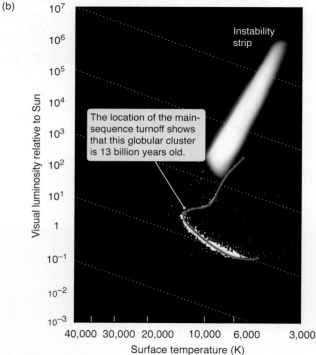

FIGURE 16.15 (a) The globular cluster M92 and (b) an H-R diagram of the stars it contains. The main-sequence turnoff indicates the age of the cluster.

Globular clusters contain good standard candles (mainly Cepheid variables) and are also good standard candles themselves. **Figure 16.15** shows globular cluster M92 and the H-R diagram for its stars. The diagram's main-sequence turnoff occurs for stars with masses of about 0.8 $M_\odot$, which corresponds to a main-sequence lifetime of close to 13 billion years. This age is typical for globular clusters, making them the oldest known objects in our galaxy. Globular clusters formed when the universe and our galaxy were very young. Compared to globular-cluster stars, our Sun, at about 5 billion years old, is a relatively young member of our galaxy.

In an H-R diagram of an old cluster, the horizontal branch crosses the instability strip. These RR Lyrae stars, which we saw in Chapter 13, are easy to spot in globular clusters because they are very luminous and have a distinctive light curve. American astronomer Henrietta Leavitt (1868–1921) determined that there is a relationship between the period and the luminosity for RR Lyrae stars. Harlow Shapley used this period-luminosity relationship to find the luminosities of RR Lyrae stars in globular clusters. He then used the inverse square law of radiation to combine these luminosities with measured brightnesses to determine the distances to globular clusters. Finally, Shapley cross-checked his results by noting that more distant clusters (as measured with his standard candle) also tended to appear smaller in the sky, as expected.

Shapley made a three-dimensional map of globular clusters from these distances and the locations of the clusters in the sky. This map showed that globular clusters occupy a roughly spherical region of space with a diameter of about 300,000 light-years. These globular clusters trace out the halo of the Milky Way Galaxy, as shown in **Figure 16.16**, which reflects the modern view of the globular-cluster distribution.

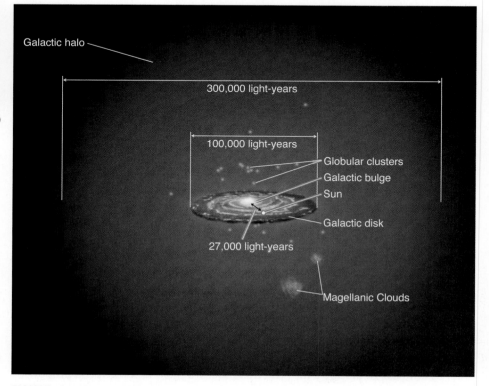

FIGURE 16.16 This diagram of the disk, bulge, and halo of the Milky Way Galaxy also shows the Magellanic Clouds and the location of the Sun within the Milky Way disk.

Globular clusters orbit the gravitational center of the galaxy, so the center of the distribution of globular clusters is the same as the gravitational center of the galaxy. The distance from the Sun to this central point is about 27,000 light years, placing us roughly halfway out toward the edge of the galactic disk (see Figure 16.16).

Infrared studies of the distributions and motions of stars toward the center of our galaxy show that the Milky Way has a substantial bar with a modest bulge at its center. Two major spiral arms—Scutum-Centaurus and Perseus—connect to the ends of the central bar and sweep through the galaxy's disk, just like the arms we saw in other spiral galaxies. Putting all the available information together, we conclude that the Milky Way is a middle-of-the-road giant barred spiral. From outside, our galaxy probably looks much like the one shown in **Figure 16.17**, and we would classify the Milky Way as an SBbc galaxy (about halfway between an SBb and an SBc).

We can use the rotation of our own galaxy to determine whether dark matter dominates the Milky Way as in most spiral galaxies. **Figure 16.18** shows the rotation curve of the Milky Way as inferred primarily from 21-cm observations. The orbital motion of the nearby Large Magellanic Cloud provides the outermost point in the rotation curve, at a distance of roughly 160,000 light-years from the center of the galaxy. Like other spiral galaxies, the Milky Way has a fairly flat rotation curve.

From the rotation curve of the Milky Way, we find that the galaxy's gravitational mass must be about $6–10 \times 10^{11} M_\odot$. However, the luminous mass, found from adding the masses of stars, dust and gas, is a much lower value (roughly one-tenth of the gravitational mass). Like other spiral galaxies, the Milky Way is mostly dark matter. Like other galaxies, visible matter dominates the inner part of our galaxy, and dark matter dominates its outer parts.

G X U **V** I R

FIGURE 16.17 From the outside, the Milky Way would look much like this barred spiral galaxy, NGC 6744.

The Milky Way is mostly dark matter.

16.3 Stars in the Milky Way

Our Sun is a middle-aged star located among other middle-aged stars that orbit around the galaxy within the disk. Yet near the Sun are other stars, usually much older, that are a part of the galactic halo and whose orbits are carrying them *through* the disk. Using the ages, chemical abundances, and motions of nearby stars, we can differentiate between disk and halo stars to learn more about the galaxy's structure.

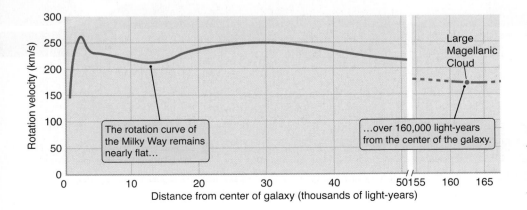

FIGURE 16.18 A plot showing rotation velocity versus distance from the center of the Milky Way. The most distant point comes from measurements of the orbit of the Large Magellanic Cloud. The nearly flat rotation curve indicates that dark matter dominates the outer parts of our galaxy.

We See Stars of Different Age and Different Chemical Composition

Most globular clusters are in the halo. With ages of up to 13 billion years, they are among the oldest objects known; in fact, no young globular clusters are seen. In contrast, **open clusters**, like the one shown in **Figure 16.19**, are much less tightly bound collections of a few dozen to a few thousand stars that are found in the disk of our galaxy. As with globular clusters, the stars in an open cluster all formed in the same region at about the same time. Open clusters have a wide range of ages. Some contain the very youngest stars known, while others contain stars somewhat older than the Sun. Because open clusters are loosely bound together, they are easily disrupted by the gravitational tug from nearby objects, so they do not survive long in the disk of our galaxy. The oldest open clusters are several billion years younger than the youngest globular clusters.

The difference in ages between globular and open clusters indicates that stars in the halo formed first, but this epoch of star formation did not last long. Star formation in the disk started later but has been continuing ever since. Star formation processes in the massive, compact globular clusters also must have been much different from the processes in the less massive, more scattered open clusters.

When the universe was very young, only the least massive elements existed. All elements more massive than boron must have formed by nucleosynthesis in stars. For this reason, the abundance of massive elements in the interstellar medium provides a record of all the star formation that has taken place up to the

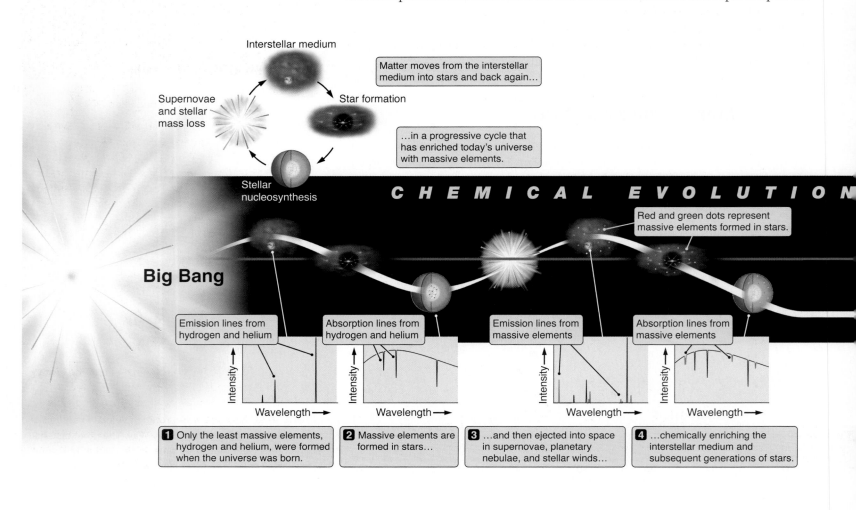

present time. Gas with high abundances of massive elements has gone through a great deal of stellar processing, whereas gas with low abundances of massive elements has not.

In turn, the abundance of massive elements in the atmosphere of a star provides a snapshot of the chemical composition of the interstellar medium *at the time the star formed*. (In main-sequence stars, material from the core does not mix with material in the atmosphere, so the abundances of chemical elements inferred from the spectra of a star are the same as the abundances in the interstellar gas from which the star formed.) As illustrated in **Figure 16.20**, the chemical composition of a star's atmosphere reflects the cumulative amount of star formation that has occurred up to the moment it formed.

If our ideas about the chemical evolution of the universe are correct, then stars in globular clusters, being among the earliest stars to form, should contain only very small abundances of massive elements. And that is exactly what we see. Some globular-cluster stars contain only 0.5 percent as much of these massive elements as our Sun has. This relationship between age and abundances of massive elements is evident throughout much of the galaxy. Lower abundances of heavy elements characterize not just globular-cluster stars, but all of the stars in our galaxy's halo. Within the disk, the chemical evolution of the Milky Way has continued as generations of stars have further enriched the interstellar medium with the products of their nucleosynthesis. Older disk stars typically have lower abundances of massive elements than younger disk stars. Similarly, older stars in the outer parts of our galaxy's bulge have lower abundances of massive elements than younger stars in the disk.

Star formation is generally more active in the inner part of the Milky Way than in the outer parts, because interstellar gas is denser in the inner part. If such activity has continued throughout the history of our galaxy, we might predict massive elements to be more abundant in the inner parts of our galaxy than in the outer parts. Observations of chemical abundances in the interstellar medium, based both on interstellar absorption lines in the spectra of stars

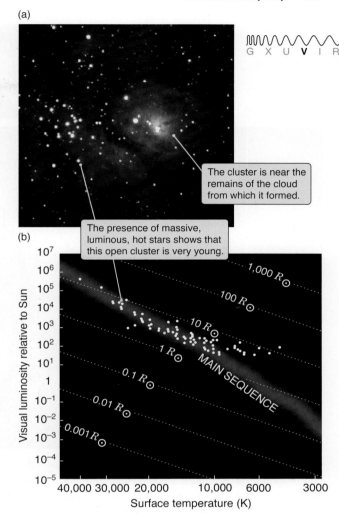

(a)

The cluster is near the remains of the cloud from which it formed.

(b)

The presence of massive, luminous, hot stars shows that this open cluster is very young.

FIGURE 16.19 (a) Open star cluster NGC 6530 is located 5,200 light-years away, in the disk of our galaxy. (b) The H-R diagram of stars in this cluster shows that it has an age of a few million years or less, because it shows no main-sequence turnoff.

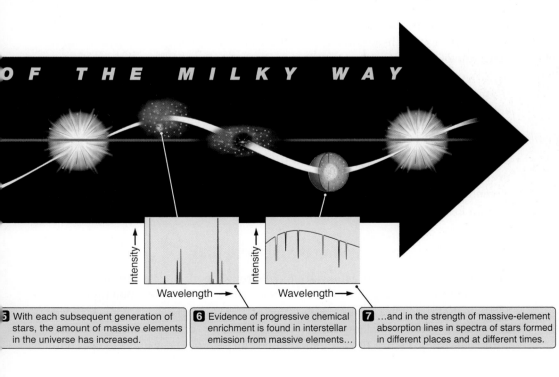

OF THE MILKY WAY

5 With each subsequent generation of stars, the amount of massive elements in the universe has increased.

6 Evidence of progressive chemical enrichment is found in interstellar emission from massive elements...

7 ...and in the strength of massive-element absorption lines in spectra of stars formed in different places and at different times.

FIGURE 16.20 As subsequent generations of stars form, live, and die, they enrich the interstellar medium with massive elements—the products of stellar nucleosynthesis. The chemical evolution of our galaxy and other galaxies can be traced in many ways, including by the strength of interstellar emission lines and stellar absorption lines.

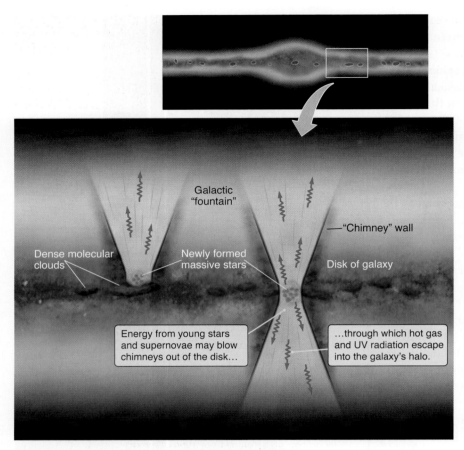

FIGURE 16.21 In the "galactic fountain" model of the disk of a spiral galaxy, gas is pushed away from the plane of the galaxy by energy released by young stars and supernovae and then falls back onto the disk.

Generations of stars must have formed even before globular clusters did.

and on emission lines in glowing H II regions, confirm this prediction. Astronomers have documented similar trends in other galaxies. These trends can be seen in the composition of stars as well.

Our basic idea that higher massive-element abundances should correspond to the more prodigious star formation in the inner galaxy seems correct, but the full picture is not this simple. The chemical composition of the interstellar medium at any location depends on a wealth of factors. New material falling into the galaxy might affect interstellar chemical abundances. Chemical elements produced in the inner disk might be blasted into the halo in great "fountains" powered by the energy of massive stars (as illustrated in **Figure 16.21**), only to fall back onto the disk elsewhere. Past interactions with other galaxies might have stirred the Milky Way's interstellar medium, mixing gas from those other galaxies with our own. The variations of chemical abundances within the Milky Way and other galaxies—and what these variations tell us about the history of star formation and nucleosynthesis—remain active topics of research.

Although the details are complex, several clear and important lessons can be learned from patterns in massive-element abundances in the galaxy. The first is that even the very oldest globular-cluster stars contain *some* amount of massive chemical elements. This implies that globular-cluster stars and other halo stars were not the first stars in our galaxy to form. At least one generation of massive, short-lived stars must have lived and died, ejecting newly synthesized massive elements into space, before even the oldest globular clusters formed. Further, every star less massive than about $0.8\ M_\odot$ that ever formed is still around today. Even so, we find *no* disk stars with exceptionally low massive-element abundances. We would have found these stars by now if they existed. The gas that wound up in the plane of the Milky Way must have seen a significant amount of star formation *before* it settled into the disk of the galaxy and made stars.

In this discussion, we have focused mainly on the variations in chemical abundances from place to place. These variations tell us a lot about the history of our galaxy and a lot about the origin of the material that we are made from. It is important to remember, however, that even a chemically "rich" star like the Sun, which is made of gas processed through approximately 9 billion years of previous generations of stars, is still composed of less than 2 percent massive elements. Luminous matter in the universe is still dominated by hydrogen and helium formed long before the first stars.

Looking at a Cross Section through the Disk

The youngest stars in our galaxy are most strongly concentrated in the galactic plane, defining a disk about 1,000 light-years thick (but over 100,000 light-years across—thin indeed). The older population of disk stars, distinguishable by lower abundances of massive elements, has a much "thicker" distribution, about 12,000 light-years thick. **Figure 16.22** illustrates how the population of stars changes with distance from the galactic plane. The youngest stars are concentrated closest to the plane of the galaxy because this is where the molecular clouds are. Older stars make up the thicker parts of the disk. There are two hypotheses for the origin of this thicker disk. One suggests that these stars formed in the midplane of the disk long ago but have since been kicked up out of the plane of the galaxy, primarily by gravitational interactions with massive molecular clouds (as shown in Figure 16.22). The other hypothesis suggests that these stars were acquired from the merging process that formed our galaxy.

A rotating cloud of gas naturally collapses into a thin disk when gas falling from one direction (above the disk) collides with gas falling from the other direction (below the disk). The same process applies to clouds of gas that are pulled by gravity toward the midplane of the disk of a spiral galaxy. Although stars are free to pass back and forth from one side of the disk to the other, cold, dense clouds of interstellar gas settle into the central plane of the disk. These clouds are seen as the concentrated dust lanes that lie in the middle of the disks of spiral galaxies (as seen in the chapter-opening figure), like tomato sauce lies between the cheese and the crust on a pizza. This thin lane slicing through the middle of the disk is the place both where new stars form and where new stars are found.

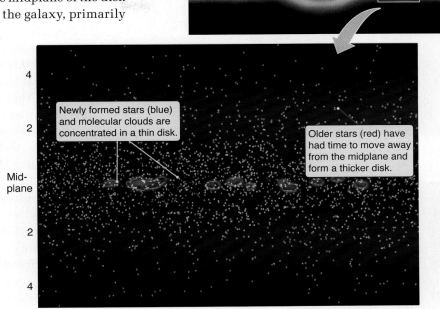

FIGURE 16.22 A vertical profile of the disk of the Milky Way Galaxy. Gas and young stars are concentrated in a thin layer in the center of the disk. Older populations of stars make up the thicker portions of the disk.

The interstellar medium is a dynamic place—energy from star-forming regions can shape it into impressively large structures. We mentioned earlier that energy from regions of star formation can impose interesting structure on the interstellar medium, clearing out large regions of gas in the disk of a galaxy. Many massive stars forming in the same region can blow "chimneys" out through the disk of the galaxy via a combination of supernova explosions and strong stellar winds. If enough massive stars are formed together, sufficient energy may be deposited to blast holes all the way through the plane of the galaxy. In the process, dense interstellar gas can be thrown high above this plane (see Figure 16.21). Maps of the 21-cm emission from neutral hydrogen in our galaxy and visible-light images of hydrogen emission from some edge-on external galaxies show numerous vertical structures in the interstellar medium of disk galaxies. These vertical structures are often interpreted as the "walls" of chimneys.

16.4 The Milky Way Hosts a Supermassive Black Hole

Dense clouds of dust and gas hide our visible-light view of the Milky Way's center. Yet it is here that we might expect to find a supermassive black hole, as is often found in other galaxies with central bulges. Fortunately, infrared, radio, and some X-ray radiation pass through dust (**Figure 16.23**). The X-ray view (Figure 16.23a) shows the location of a strong radio source called Sagittarius A* (abbreviated Sgr A*), which astronomers believe lies at the exact center of the Milky Way. The infrared image (Figure 16.23b) cuts through the dust to reveal the crowded, dense core of the galaxy containing hundreds of thousands of stars.

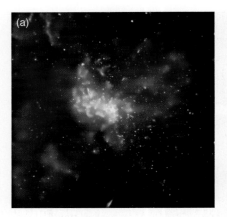

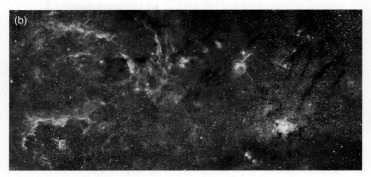

FIGURE 16.23 (a) An X-ray view of the Milky Way's central region showing the active source, Sgr A*, as the brightest spot at the middle of the image. Lobes of superheated gas (shown in red) are evidence of recent, violent explosions near Sgr A*. (b) This wide infrared view (890 light-years across) of the central core of the Milky Way shows hundreds of thousands of stars. The bright white spot at the far right marks the galaxy's center, home of a supermassive black hole. (c) Radio observations of the center of the Milky Way reveal wispy molecular clouds (purple) glowing from strong synchrotron emission. Cold dust (20–30 K) associated with molecular clouds is shown in orange. Diffuse infrared emission appears in blue-green. The galactic center (Sgr A*) lies within the bright area to the right of center.

Radio observations (Figure 16.23c) reveal synchrotron emission from wisps and loops of material distributed throughout the region. This is similar to the synchrotron emission seen from active galactic nuclei (AGNs), but at far lower levels.

The motions of stars closest to the Sgr A* source suggest a central mass very much greater than that of the few hundred stars orbiting there. Stars less than 0.1 light-year from the galaxy's center follow Kepler's laws. The closest stars studied are only about 0.01 light-years from the center, so close that their orbital periods are only about a dozen years. The positions of these stars change noticeably over time, and we can see them speed up as they whip around what can only be a supermassive black hole at the focus of their elliptical orbits (**Figure 16.24**). Using Kepler's third law, we estimate that the black hole at the center of our own galaxy is a relative lightweight, having a mass of "only" $3.7 \times 10^6 \, M_\odot$.

Clouds of interstellar gas at the galaxy's center are heated to millions of degrees by shock waves from supernova explosions and colliding stellar winds blown outward by young massive stars. Superheated gas produces X-rays, and the Chandra X-ray Observatory has detected more than 2,000 X-ray sources within the central region of the galaxy. These include frequent, short-lived X-ray flares near Sgr A* which provide direct evidence that matter falling toward the supermassive black hole fuels the energetic activity at the galaxy's center. Huge lobes of 20-million-K gas surrounding Sgr A* (shown in Figure 16.23a) indicate that several flares have occurred within the past 10,000 years. Nevertheless, this activity is not as intense as that seen in other galaxies with central black holes. Our supermassive black hole currently has no source of large amounts of material nearby, so it is relatively quiet. The inner Milky Way is a reminder that our galaxy was almost certainly "active" in the past and could become active once again.

16.5 The Milky Way Offers Clues about How Galaxies Form

A fundamental goal of stellar astronomy is to understand the life cycle of stars, including how stars form from clouds of interstellar gas. Galactic astronomy has the same basic goal. That is, astronomers would like very much to have a complete and well-tested theory of how our galaxy formed. Unfortunately, such a complete theory is not yet in hand. Even so, what researchers have seen so far offers many clues.

Globular clusters in the halo must have been among the first stars formed that still exist today. They are not concentrated in the disk or bulge of the galaxy, so they formed from clouds of gas well before those clouds had settled into the galaxy's disk. The presence of small amounts of massive elements in the atmospheres of halo stars indicates that at least one generation of stars must have lived and died *before* the formation of the halo stars we see today. We have yet to find any stars from that first generation in our galaxy today.

Our galaxy must have formed when the gas within a huge "clump" of dark matter collapsed into a large number of small protogalaxies. Some of these smaller clumps are still around today in the form of small, satellite dwarf galaxies near our own. The largest among them are the Large and Small Magellanic Clouds (**Figure 16.25**), which are easily seen by the naked eye and appear much like detached pieces of the Milky Way. Another companion, the Sagittarius Dwarf, is now plowing through the disk of the Milky Way on the other side of the bulge. It seems likely that at some point the Sagittarius Dwarf will become incorporated into the Milky Way—an indication that our galaxy is still growing. There are more than 20 of these satellite dwarf galaxies, although we can't be sure that all are gravitationally bound to the Milky Way. Many were unknown until very recently because of their low luminosity.

The remainder of these protogalaxies merged to form the barred spiral galaxy we call the Milky Way. In this process, stars were formed in the halo, in the bulge, and in the disk. The first stars to form ended up in the halo—some in globular clusters, but many not. Gas that settled into the disk of the Milky Way quickly formed several generations of stars. A dense, concentrated mass quickly grew into the supermassive black hole at the center of the galaxy. The details of this process are sketchy, but some calculations indicate that so much mass was concentrated in this small region that almost any sequence of events would have led to the formation of a massive black hole.

We need to be careful not to get ahead of ourselves. The Milky Way offers many clues about the way galaxies form, but much of what we know of the process comes from looking beyond our local system. Images of distant galaxies (which we see as they existed billions of years ago), as well as observations of the glow left behind by the formation of the universe itself, provide equally important pieces of the puzzle. We will have to leave the story of galaxy formation as we currently understand it unfinished, for now, as we turn our attention to the immensely larger structure of the universe as a whole.

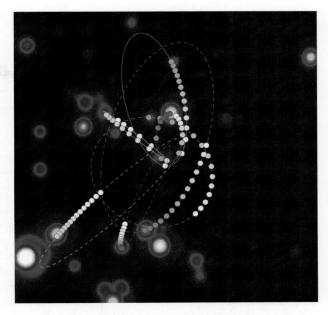

FIGURE 16.24 Orbits of several stars within 0.1 light-year (about 6,000 AU) of the Milky Way's center. The Keplerian motions of these stars reveal the presence of a 3.7-million-$M_\odot$ supermassive black hole at the galaxy's center. Colored dots show the measured positions of the stars over a 12-year interval.

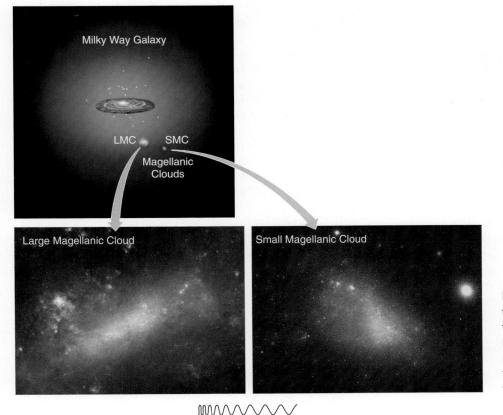

FIGURE 16.25 Our Milky Way is surrounded by more than 20 dwarf companion galaxies, the largest among them being the Large and Small Magellanic Clouds. The Magellanic Clouds were named for Ferdinand Magellan (c. 1480–1521), who headed an early European expedition that ventured far enough into the Southern Hemisphere to see them.

Hyperfast Star Kicked Out of Milky Way

By **Lisa Grossman**, *Wired Science*

New Hubble observations suggest a dramatic origin story for one of the fastest stars ever detected, involving a tragic encounter with a black hole, a lost companion and swift exile from the galaxy.

The star, HE 0437-5439, is one of just 16 so-called hypervelocity stars, all of which were thought to come from the center of the Milky Way. The Hubble observations allowed astronomers to definitively trace the star's origin to the heart of the galaxy for the first time.

Based on observations taken three and a half years apart, astronomers calculated that the star is zooming away from the Milky Way's center at a speed of 1.6 million miles per hour—three times faster than the sun [orbits the galaxy].

"The star is traveling at an absurd veloc-

FIGURE 16.26 HST image of the hyperfast star HE 0437-5439.

ity, twice as much as it needs to escape the galaxy's gravitational field," said hypervelocity star hunter Warren Brown of the Harvard-Smithsonian Center for Astrophysics, who found the first unbound star in 2005, in a press release. "There is no star that travels that quickly under normal circumstances—something exotic has to happen."

Earlier observations linked the star to a neighboring galaxy, the Large Magellanic Cloud. But Brown and his colleagues claim that the new Hubble observations settle the question of the star's origin squarely in favor of the Milky Way.

One reason the star's home was under debate is its bizarrely youthful appearance. Based on its speed, the star would have to be 100 million years old to have traveled from the Milky Way's center to its current location, 200,000 light-years away. But its mass—nine times that of our sun—and blue color mean it should have burned out after only 20 million years.

The new origin story reconciles the star's age and speed, and has all the makings of a melodrama. A hundred million years ago, astronomers suggest, the runaway star was a member of a triple-star system that veered disastrously close to the galaxy's central supermassive black hole. One member of the trio was captured, and its momentum was transferred to the remaining binary pair, which was hurled from the Milky Way at breakneck speed.

As time passed, the larger star evolved into a puffy red giant and devoured its partner. The two merged into the single, massive, blue star called a blue straggler that Hubble observed.

This bizarre scenario conveniently explains why the star looks so young. By merging into a blue straggler, the two origi-

nal stars managed to look like a star one-fifth its true age.

The findings were published online in a paper in *The Astrophysical Journal Letters*.

The team is hunting for the homes of four other unbound stars, all zooming around the fringes of the Milky Way.

"Studying these stars could provide more clues about the nature of some of the universe's unseen mass, and it could help astronomers better understand how galaxies form," said study coauthor Oleg Gnedin of the University of Michigan in a press release.

Evaluating the News

1. What's so special about the star described in the article?
2. The star is shown in **Figure 16.26**. From this image alone, would it be possible to determine that there was something special about this star?
3. In the article, Warren Brown states that the star is traveling at twice the speed needed to escape the Milky Way, and that something exotic must have happened to speed it up. Why would he make this argument? (Hint: What happens to objects moving faster than the escape velocity?)
4. What is unusual about the age of the star?
5. How might astronomers figure out the mass of the star? Why does that set a limit on how old the star should be?
6. What kind of observations could distinguish between an origin in the Large Magellanic Cloud and an origin in the Milky Way?
7. The supermassive black hole at the center of the Milky Way apparently hurled this star away from the center of the Milky Way. Is this likely to be a common occurrence? Explain your reasoning.

SUMMARY

16.1 The interstellar medium is a complex region, ranging from cold, relatively dense molecular clouds to hot, tenuous intercloud gas. Dust and gas in the interstellar medium blocks visible light but is transparent at longer, infrared wavelengths. Neutral hydrogen cannot be detected at visible and infrared wavelengths, but it is revealed by its 21-cm microwave emission. Our galaxy also contains energetic cosmic rays trapped by magnetic fields.

16.2 We live in the disk of the Milky Way, a barred spiral SBbc galaxy that is 100,000 light-years across. The Sun is about 27,000 light-years from the center of this galaxy. We use variable stars of known luminosity to find the distances to globular clusters, which enable us to measure the size of the Milky Way's extended halo.

16.3 The Doppler velocities of radio lines show that the rotation curve of the Milky Way is flat, like those of other galaxies, and that the mass of our galaxy is mostly in the form of dark matter. The chemical composition of the Milky Way has evolved with time, and there must have been a generation of stars before the oldest halo and globular-cluster stars we see today formed. Star formation is still actively occurring in the disk of our galaxy, leading to complex structures within the disk.

16.4–16.5 At the center of the galaxy is a massive black hole, which produces rapid orbital velocities of nearby stars. The Milky Way formed from a collection of smaller protogalaxies that collapsed out of a halo of dark matter.

✧ SUMMARY SELF-TEST

1. The cold intercloud gas is characterized by emission from
   a. ionized hydrogen.
   b. neutral hydrogen.
   c. ionized or neutral hydrogen.
   d. molecular hydrogen.

2. Interstellar dust
   a. blocks visible light, but allows infrared and radio light through.
   b. blocks radio or infrared light, but allows visible light through.
   c. blocks all types of light equally.
   d. permits all types of light to pass.

3. The hot interstellar medium is heated by
   a. the cosmic microwave background.   b. starlight.
   c. supernovae.   d. planetary nebulae.

4. Place in order the steps involved in finding the distance to the center of the Milky Way from variable stars in globular clusters.
   a. Use the luminosity and the brightness to find the distance to the globular clusters.
   b. Use the period-luminosity relation to find the luminosity of the variable stars.
   c. Find the distance to the center of the globular clusters, which is the distance to the center of the Milky Way.
   d. Find the direction to the center of the distribution of the globular clusters.
   e. Find the period of the variable stars.

5. Radio emission reveals that the Milky Way is
   a. an elliptical galaxy.
   b. an irregular galaxy resulting from a collision.
   c. a spiral with three arms.
   d. a barred spiral with three arms.

6. In general, the Milky Way
   a. has the same chemical composition as time passes.
   b. has more abundant hydrogen as time passes.
   c. has more abundant heavy elements as time passes.
   d. has less abundant heavy elements as time passes.

7. In general, older stars have lower ____ than younger stars.
   a. masses
   b. abundance of heavy elements
   c. luminosities
   d. rotation rates

8. In the disk of the Milky Way, stars are _____ and dust and gas are _____ than in the halo.
   a. younger; more diffuse   b. older; more diffuse
   c. older; denser   d. younger; denser

9. The supermassive black hole at the center of the Milky Way has a mass in the range of
   a. hundreds of solar masses.   b. thousands of solar masses.
   c. millions of solar masses.   d. billions of solar masses.

QUESTIONS AND PROBLEMS

True/False and Multiple-Choice Questions

10. **T/F:** Interstellar dust causes stars to appear bluer than they actually are.

11. **T/F:** Ages of globular clusters put a lower limit on the age of the Milky Way.

12. **T/F:** The Milky Way contains almost no dark matter.

13. **T/F:** An old globular cluster contains no massive, luminous hot stars.

14. **T/F:** The disk of the Milky Way is threaded with magnetic field lines.

15. Astronomers observe the interstellar medium itself primarily in
    a. radio and infrared light.   b. visible light.
    c. UV light.   d. gamma-ray light.

16. Interstellar gas produces _____ in the spectrum of a back-ground star.
    a. Doppler shifts   b. emission lines
    c. absorption lines   d. reddening

17. The most important thing about giant molecular clouds is that
    a. they have many types of molecules.
    b. they are regions of star formation.
    c. they are rare.
    d. they are opaque.

18. Looking toward the galactic center, we see no redshift or blue-shift. This tells us
    a. the center of the galaxy is motionless.
    b. the Sun and the center are not moving toward or away from each other.
    c. the Milky Way is stationary with respect to other galaxies.
    d. the Sun is stationary with respect to the center of the galaxy.

19. The Milky Way Galaxy
    a. has been exactly the same since it formed.
    b. has only changed its composition, but not its dynamics, since it formed.
    c. is still evolving, consuming nearby small galaxies.
    d. is finished evolving, having consumed all nearby galaxies.

Conceptual Questions

20. Look back at the chapter-opening figure. Explain the logic that leads us to determine that the Milky Way is a spiral galaxy, based on images like these.

21. Examine Figure 16.3a. On the left, photons of all colors are entering the dust cloud, carrying energy with it. On the right, only long-wavelength photons and infrared radiation are leaving the cloud. Explain what happens to the photons with shorter wavelengths.

22. Figure 16.1 shows the Milky Way in visible and near-infrared light. Figure 16.5 shows the Milky Way in infrared light. If you were to move farther out in the disk, even outside the galaxy, you would see much this same view. Figure 16.6, taken at X-ray wave-lengths, is different. If you moved to another location in the disk, you would likely see a very different view in the X-ray region of the spectrum. Explain what is different about the emission you are seeing in Figure 16.6.

23. The interstellar medium is approximately 99 percent gas and 1 percent dust. Why is it the dust and not the gas that blocks our visible-light view of the galactic center?

24. What is the primary source of heating of hot intercloud gas?

25. Interstellar gas typically cools by colliding with other gas atoms or grains of dust; during the collision, the atom loses energy and hence its temperature is lowered. How does this effect explain why tenuous gases are generally so hot, while dense gases tend to be so cold?

26. When a hydrogen atom is ionized, a single particle becomes two particles.

    a. Identify the two particles.
    b. If both particles have the same kinetic energy, which one is moving faster?

27. What causes a hydrogen atom to radiate a photon of 21-cm radio emission?

28. Why is it so difficult for astronomers to get an overall picture of the structure of our galaxy?

29. What do astronomers mean by *standard candle*?

30. Why are stars in globular clusters so useful as standard candles in determining the size of the galaxy?

31. Describe the distribution of globular clusters within the galaxy, and explain what that implies about the size of the galaxy and our distance from its center.

32. How do we know that the stars in globular clusters are among the oldest stars in our galaxy?

33. Compare globular and open clusters.
    a. What are the main differences in the gas out of which globu-lar and open clusters formed?
    b. Why do globular clusters have such high masses while open clusters have low masses?
    c. Are these issues related? Explain.

34. Old stars in the inner disk of our galaxy have higher abundances of massive elements than do young stars in the outer disk. Explain how this might happen. (Based on what you learned about the formation of elements, shouldn't old stars have a *lower* abundance of massive elements than young stars have?)

35. Do you think it is likely for astronomers to find a young star that has no massive elements in it? Why or why not? How about an old star?

36. How do 21-cm radio observations reveal the rotation of our galaxy?

37. What does the rotation curve of our galaxy say about the presence of dark matter in the galaxy?

38. Explain the observational evidence that shows we live in a spiral galaxy, not an elliptical galaxy.

39. In the Hubble scheme for classifying galaxies, how is the Milky Way classified?

40. What does the abundance of a star's massive elements tell us about the age of the star?

41. Where do we find the youngest stars in our galaxy?

42. Halo stars are found in the vicinity of the Sun. What observa-tional evidence distinguishes them from disk stars?

43. Are cosmic rays a form of electromagnetic radiation? Explain your answer.

44. Can a cosmic ray travel at the speed of light? Why or why not?

45. What is one source of synchrotron radiation in our galaxy, and where is it found?

46. Why must we use X-ray, infrared, and 21-cm radio observations to probe the center of our galaxy?

47. What is Sgr A* and how was it detected?

48. Explain the evidence for a supermassive black hole at the center of the Milky Way.

49. How does the mass of the supermassive black hole at the center of our galaxy compare with that found in most other spiral galaxies?

50. To observers in Earth's Southern Hemisphere, the Large and Small Magellanic Clouds look like detached pieces of the Milky Way. What are these "clouds," and why is it not surprising that they look so much like pieces of the Milky Way?

51. What is the origin of the Milky Way's satellite galaxies?

52. What has been the fate of most of the Milky Way's satellite galaxies?

53. Why are most of the Milky Way's satellite galaxies so difficult to detect?

54. Use your imagination to describe how our skies might appear if our Sun and Solar System were located
  a. near the center of the galaxy.
  b. near the center of a large globular cluster.
  c. near the center of a large, dense molecular cloud.

Problems

55. Calculate the peak wavelength of emission from dust at a temperature of 150 K.

56. The Sun completes one trip around the center of the galaxy in approximately 230 million years. How many times has our Solar System made the circuit since its formation 4.6 billion years ago?

57. Estimate the fraction of the volume of the Milky Way that is taken up by the disk, using the numbers and shapes shown in Figure 16.16.

58. Estimate the typical density of dust grains (grains per cubic centimeter) in the interstellar medium. A typical grain has a mass of 10^{-17} kilograms (kg). (Hint: You know the typical density of gas, and what fraction of the interstellar medium, by mass, is made of dust.)

59. A typical temperature of intercloud gas is 8000 K. Using Wien's law, find the peak wavelength of the thermal emission from this gas.

60. The Sun is located about 27,000 light-years from the center of the galaxy, and the galaxy's disk probably extends another 30,000 light-years farther out from the center. Assume that the Sun's orbit takes 230 million years to complete.
  a. With a truly flat rotation curve, how long would it take a globular cluster located near the edge of the disk to complete one trip around the center of the galaxy?
  b. How many times has that globular cluster made the circuit since its formation about 13 billion years ago?

61. Parallax measurements of the variable star RR Lyrae indicate that it is located 750 light-years from the Sun. A similar star observed in a globular cluster located far above the galactic plane appears 160,000 times fainter than RR Lyrae.
  a. How far from the Sun is this globular cluster?
  b. What does your answer to part (a) tell you about the size of the galaxy's halo compared to the size of its disk?

62. Although the flat rotation curve indicates that the total mass of our galaxy is approximately $8 \times 10^{11}\ M_\odot$, electromagnetic radiation associated with normal matter suggests a total mass of only $3 \times 10^{10}\ M_\odot$. Given this information, calculate the fraction of our galaxy's mass that is made up of dark matter.

63. Compare the H-R diagram for the young cluster NGC 6530 (see Figure 16.19) with H-R diagrams in previous chapters to determine how soon the most massive stars will explode, and how massive the protostars are that are still contracting onto the main sequence.

64. Given what you have learned about the distribution of massive elements in the Milky Way, and what you know about the terrestrial planets, where do you think such planets are most likely and least likely to form?

65. A cosmic-ray proton is traveling at nearly the speed of light: 3×10^8 meters per second (m/s).
  a. Using Einstein's familiar relationship between mass and energy ($E = mc^2$), show how much energy (in joules) the cosmic-ray proton would have if m were based only on the proton's rest mass (1.7×10^{-27} kg).
  b. The actual measured energy of the cosmic-ray proton is 100 J. What, then, is the relativistic mass of the cosmic-ray proton?
  c. How much greater is the relativistic mass of this cosmic-ray proton than the mass of a proton at rest?

66. One of the fastest cosmic rays ever observed had a speed of $[1.0 - (1.0 \times 10^{-24})] \times c$ (very, very close to c). Assume that the cosmic ray and a photon left a source at the same instant. To a stationary observer, how far behind the photon would the cosmic ray be after traveling for 100 million years?

67. A star in a circular orbit about the black hole at the center of our galaxy ($M_{BH} = 7.36 \times 10^{36}$ kg) has an orbital radius of 0.0131 light-year (1.24×10^{14} meters). What is the average speed of this star in its orbit?

68. How large is the black hole at the center of our galaxy (that is, where is its event horizon)?

69. A star is observed in a circular orbit about a black hole with an orbital radius of 1.5×10^{11} kilometers (km) and an average speed of 2,000 km/s. What is the mass of this black hole in solar masses?

 SmartWork, Norton's online homework system, includes algorithmically generated versions of these questions, plus additional conceptual exercises. If your instructor assigns questions in SmartWork, log in at **smartwork.wwnorton.com**.

 StudySpace is a free and open website that provides a Study Plan for each chapter of **Understanding Our Universe**. Study Plans include animations, reading outlines, vocabulary flashcards, and multiple-choice quizzes, plus links to premium content in SmartWork and the ebook. Visit **wwnorton.com/studyspace**.

Exploration | The Center of the Milky Way

Adapted from Learning Astronomy by Doing Astronomy, *by Ana Larson*

Astronomers once thought that the Sun was at the center of the Milky Way. A more accurate picture of the size and shape of our galaxy was constructed from observations of RR Lyrae stars in globular clusters by Harlow Shapley. In this Exploration, you will repeat this experiment and find the center of the Milky Way for yourself.

Table 16.2 shows the galactic longitude and the projected distance for globular clusters in the Milky Way. To arrive at these coordinates, astronomers imagine that the plane of the Milky Way is horizontal, like a pizza, and the Sun is at the origin in the middle of it. A line is drawn straight "down" from a globular cluster to the plane. The projected distance is the distance from the Sun to the point where the line hits the plane, and the galactic longitude tells which direction that point is in. The polar graph in **Figure 16.27** goes with this type of coordinate system. Projected distance is the distance out from the center in kiloparsecs (kpc), and the galactic longitude is the angle measured around the outside of the graph. The galactic longitudes of significant constellations are indicated on the edge of the graph.

Plot each data pair on the graph by finding the galactic longitude indicated outside the circle, then coming in toward the center to the projected distance and making a dot. (The two globular clusters in bold in the table have already been plotted for you as examples.) After plotting all of the globular clusters, estimate the center of their distribution and mark it with an X. This is the center of the Milky Way.

1. What is the approximate distance from the Sun to the center of the Milky Way?

2. What is the galactic longitude of the center of the Milky Way?

3. How do we know the Sun is not at the center of the Milky Way?

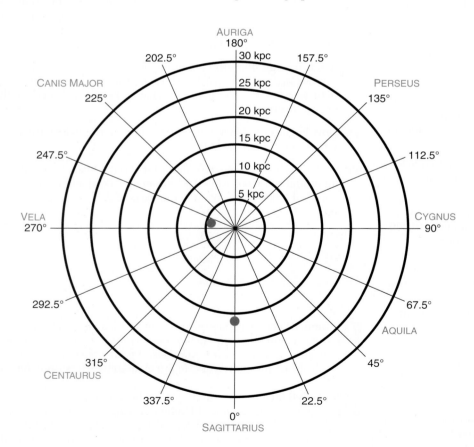

FIGURE 16.27 A polar graph used for plotting distance and direction.

TABLE 16.2

Globular Cluster Data

CLUSTER "NAME"	GALACTIC LONGITUDE (DEGREES)	PROJECTED DISTANCE (KPC)	CLUSTER "NAME"	GALACTIC LONGITUDE (DEGREES)	PROJECTED DISTANCE (KPC)
104	306	3.5	6273	357	7
362	302	6.6	**6287**	**0**	**16.6**
2808	283	8.9	6333	5	12.6
4147	**251**	**4.2**	6356	7	18.8
5024	333	3.4	6397	339	2.8
5139	309	5	6535	27	15.3
5634	342	17.6	6712	27	5.7
Pal 5	1	24.8	6723	0	7
5904	4	5.5	6760	36	8.4
6121	351	4.1	Pal 10	53	8.3
O 1276	22	25	Pal 11	32	27.2
6638	8	15.1	6864	20	31.5
6171	3	15.7	6981	35	17.7
6218	15	6.7	7089	54	9.9
6235	359	18.9	Pal 12	31	25.4
6266	353	11.6	288	147	0.3
6284	358	16.1	1904	228	14.4
6293	357	9.7	Pal 4	202	30.9
6341	68	6.5	4590	299	11.2
6366	18	16.7	5053	335	3.1
6402	21	14.1	5272	42	2.2
6656	9	3	5694	331	27.4
6717	13	14.4	5897	343	12.6
6752	337	4.8	6093	353	11.9
6779	62	10.4	6541	349	3.9
6809	9	5.5	6626	7	4.8
6838	56	2.6	6144	352	16.3
6934	52	17.3	6205	59	4.8
7078	65	9.4	6229	73	18.9
7099	27	9.1	6254	15	5.7

17 Modern Cosmology and the Origin of Structure

Cosmology is the study of the universe on the very grandest of scales, including its nature, origin, evolution, and ultimate destiny. We have already learned many things about the universe: that it is expanding, having originated in a Big Bang nearly 14 billion years ago; that it was once very hot, filled with thermal radiation that has now cooled to a temperature of 2.7 kelvins; that light elements in the universe were produced within the first few minutes after the Big Bang. In this chapter we take a closer look at the nature of the universe and how it has evolved over time, and we contemplate its ultimate fate.

✧ LEARNING GOALS

The universe that emerged from the Big Bang was incredibly uniform, wholly unlike today's universe of galaxies, stars, and planets. In this chapter we will find that complex structure is a natural, unavoidable consequence of the action of physical laws in our evolving universe. The figure at right shows the distribution of galaxies in the universe alongside a heap of soap bubbles, a commonly used analogy for the structure of the universe. By the end of the chapter, you should be able to explain in what ways the universe resembles soap bubbles. You should also be able to:

- Explain how the gravity due to the mass of the universe affects its history, shape, and fate

- Connect the accelerating expansion of our universe to a concept called dark energy

- Describe the reasoning that leads astronomers to hypothesize an early period of rapid expansion known as inflation

- List the order in which the fundamental forces of nature became distinct

- Describe how galaxies emerged in the early universe

- Describe the ultimate fate of our universe

- Explain the scientific status of ideas about parallel universes

PROF KEEPS
SAYING IT'S LIKE
SOAP
BUBBLES...

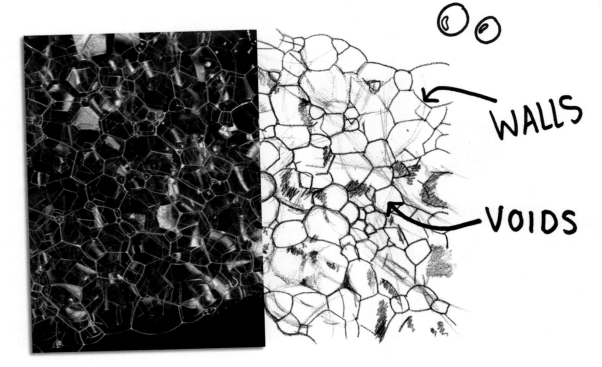

WALLS

VOIDS

17.1 The Universe Has a Destiny and a Shape

What is the fate of the universe? This is clearly one of the fundamental questions of modern cosmology. The answer depends in part on the distribution of mass on very large scales. The gravitational effect of this distributed mass is one of the factors that determine how the universe evolves.

To see how gravity affects the expansion of the universe, it will help to recall gravity's effects on the motion of projectiles. The fate of a projectile fired straight up from the surface of the Moon depends on its speed. If the speed is less than the Moon's escape velocity (2.4 kilometers per second, or km/s), the Moon's gravity will eventually stop the projectile and pull it back to the Moon's surface. But if the speed of the projectile is greater than the Moon's escape velocity, then the projectile will escape from the Moon entirely.

Gravity slows the expansion of the universe.

Just as the mass of the Moon gravitationally pulls on a projectile to slow its climb, the mass contained in the universe gravitationally slows the universe's expansion. If there is enough mass in the universe, then gravity will be strong enough to stop the expansion. The universe will slow, stop, and eventually collapse in on itself. But if there is not enough mass, then the expansion of the universe may slow, but it will never stop. The universe will expand forever.

A planet's mass and radius determine the escape velocity from its surface. The "escape velocity" of the universe is also determined by its mass and size—specifically, its average *density*. If the universe is denser on average than a particular value, called the **critical density**, then gravity is strong enough to eventually stop and reverse the expansion. If the universe is less dense than the critical density, gravity is too weak and the universe will expand forever.

The faster the universe is expanding, the more mass is needed to turn that expansion around. For this reason, the critical density depends on the value of the Hubble constant, H_0. Assuming that $H_0 = 72$ kilometers per second per megaparsec (km/s/Mpc) and that gravity is the only thing we have to worry about, the universe's critical density has a value of 8×10^{-27} kilograms per cubic meter (kg/m³), or fewer than 5 hydrogen atoms in every cubic meter. Rather than trying to keep track of such awkward numbers, we will instead talk about the *ratio* of the actual density of the universe to this critical density. We call this ratio Ω_{mass} (pronounced "omega sub mass"). Because it is a ratio of two densities, Ω_{mass} has no units.

The expansion of different possible mass-dominated universes is shown in **Figure 17.1**, which plots the scale factor R_U (see Chapter 14) versus time for different values of Ω_{mass}. If Ω_{mass} is greater than 1, the universe will eventually collapse. Conversely, if Ω_{mass} is less than 1, the universe will be slowed by gravity, but it will still expand forever. The dividing line, where Ω_{mass} equals 1, corresponds to a universe that expands more and more slowly but never quite stops.

Until the closing years of the 20th century, most astronomers thought that this straightforward application of gravity to the universe was all there was to the question of expansion and collapse. Researchers carefully measured the mass of galaxies and collections of

FIGURE 17.1 Three possible scenarios for the fate of the universe, based on the critical density of the universe but ignoring any cosmological constant.

$$\Omega_{mass} = \frac{\text{Actual density of a universe}}{\text{Critical density of the universe}}$$

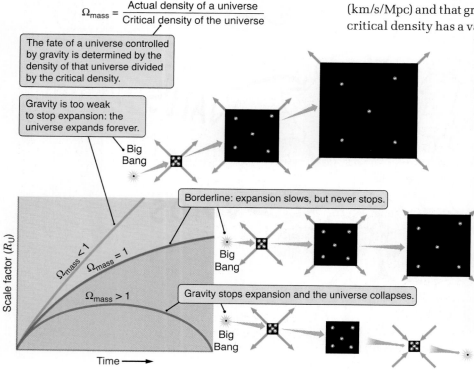

The fate of a universe controlled by gravity is determined by the density of that universe divided by the critical density.

Gravity is too weak to stop expansion: the universe expands forever.

Big Bang

Borderline: expansion slows, but never stops.

Big Bang

Gravity stops expansion and the universe collapses.

Big Bang

$\Omega_{mass} < 1$

$\Omega_{mass} = 1$

$\Omega_{mass} > 1$

Scale factor (R_U)

Time ⟶

galaxies in the hope that this would reveal the density and therefore the fate of the universe. The luminous matter gives a value for Ω_{mass} of about 0.02. Galaxies contain about 10 times as much dark matter as normal matter, so adding in the dark matter in galaxies pushes the value of Ω_{mass} up to about 0.2. When we include the mass of dark matter *between* galaxies, Ω_{mass} could increase to 0.3 or higher. Still, by this accounting there is only about one-third as much mass in the universe as is needed to stop the universe's expansion.

The Accelerating Universe Revives Einstein's "Biggest Blunder"

If the expansion of the universe is in fact slowing down, as these simple models using gravity predict, then when the universe was young it must have been expanding faster than it is today. Objects that are very far away (so that we see them as they were long ago) should therefore have larger velocities than our local Hubble's law would lead us to expect.

During the 1990s, astronomers tested this prediction. They measured the brightness of Type Ia supernovae in very distant galaxies and compared each of those brightnesses with the expected brightness based on the redshifts of those galaxies. The findings of these studies sent a wave of excitement through the astronomical community. Rather than showing that the expansion of the universe is slowing, the data indicated that it is *speeding up*. For this to be true, a force must be pushing the *entire universe* outward in opposition to gravity.

The idea of a repulsive force opposing the attractive force of gravity is not new. When Einstein used general relativity to calculate the structure of spacetime in the universe, he was greatly troubled. The theory clearly indicated that any universe containing mass could not be static, any more than a ball can hang motionless in the air. However, Einstein's formulation of spacetime came more than a decade before Hubble discovered the expansion of the universe, and the conventional wisdom at the time was that the universe was indeed static—that it neither expands nor collapses.

To force his new general theory of relativity to allow for a static universe, Einstein inserted a "fudge factor" called the **cosmological constant** into his equations. The cosmological constant acts as a repulsive force in the equations, opposing gravity and allowing galaxies to remain stationary despite their mutual gravitational attraction.

When Hubble announced his discovery that the universe is expanding, Einstein realized his mistake. General relativity demands that the structure of the universe be dynamic. Instead of inventing the cosmological constant, Einstein could have *predicted* that the universe must either be expanding or contracting with time. What a coup it would have been for his new theory to successfully predict such an amazing and previously unsuspected result. He called the introduction of his fudge factor, the cosmological constant, the "biggest blunder" of his career as a scientist. It is ironic that with our measurements of the brightness of Type Ia supernovae, Einstein's biggest blunder has returned to center stage. The repulsive force represented by the infamous cosmological constant is just what is needed to describe a universe that is expanding at an ever-accelerating rate. Today, we write this constant as Ω_Λ (pronounced "omega sub lambda").

If Ω_Λ is not zero, something is effectively pushing outward from within the universe, adding to its expansion. Gravity will have a harder time turning the expansion around. **Figure 17.2** shows plots of the scale factor R_U versus time that

The expansion of the universe is accelerating.

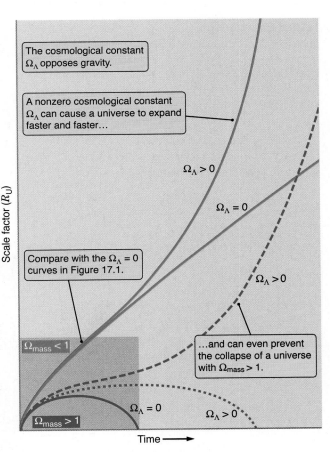

FIGURE 17.2 Plots of scale factor R_U versus time for cosmologies with and without a cosmological constant, Ω_Λ. If there is enough mass in a universe, gravity could still overcome the cosmological constant and cause that universe to collapse. Any universe without enough mass to eventually collapse will instead end up expanding at an ever-increasing rate.

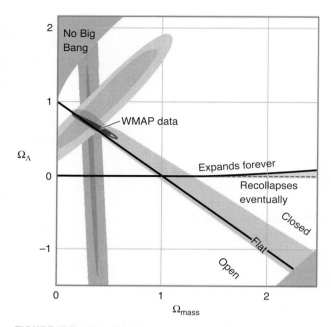

FIGURE 17.3 Current observations from different sources— Type Ia supernovae (yellow), measurements of mass in galaxies and clusters (orange), and detailed observations of the structure of the cosmic microwave background (pink and bright red)—suggest that the best current estimate for Ω_Λ is about 0.7 and about 0.3 for Ω_{mass}, which means that the expansion of the universe is accelerating.

The universe appears to be accelerating and will expand forever.

are similar to those shown in Figure 17.1, but now we have included the effects of a nonzero cosmological constant. If Ω_{mass} is sufficiently large, a universe will collapse back on itself, regardless of whether Ω_Λ is zero. In contrast, if a universe expands forever, its evolution will depend on whether or not Ω_Λ is zero.

While a universe is young and compact, gravity dominates the effect of the cosmological constant. As a universe expands, gravity gets weaker because the mass spreads out. Since the cosmological constant remains, well, constant, it becomes increasingly important. Unless gravity is able to turn the expansion around, the cosmological constant wins in the end, causing the expansion to continue accelerating forever. Even if Ω_{mass} is greater than one, a large enough cosmological constant could overwhelm gravity and make the universe expand forever.

When Einstein added the cosmological constant to his equations of general relativity, he considered it a new fundamental constant, similar to Newton's universal gravitational constant *G*. Today we realize that the vacuum can have distinct physical properties of its own. For example, the vacuum can have a nonzero energy even in the total absence of matter. We call this energy **dark energy**. Dark energy produces exactly the kind of repulsive force that Einstein's cosmological constant calls for.

Figure 17.3 shows the range of values for Ω_{mass} and Ω_Λ that are allowed by current observations. Each colored region represents the data from a different experiment. Values of Ω_{mass} and Ω_Λ outside of these regions are ruled out by these experiments, so the allowed values of Ω_{mass} and Ω_Λ must lie within the area on the graph where all of these regions overlap. The allowed values for Ω_{mass} and Ω_Λ are about 0.3 and 0.7, respectively. These values are most tightly constrained by the data from WMAP, in bright red. The expansion of our universe is dominated by the effect of dark energy.

The dark diagonal line in Figure 17.3 shows where $\Omega_{mass} + \Omega_\Lambda = 1$. The experiments tightly constrain the value of $\Omega_{mass} + \Omega_\Lambda$ to lie along this line. This means that on very large scales, the universe we live in is "flat"—it is only locally warped by the mass inside, like the rubber sheet we learned about in Chapter 13. This, in turn, means that ordinary Euclidean geometry describes circles, triangles and parallel lines in the universe as a whole.

Other geometries exist. For example, the universe could have been positively curved on the largest scales, like the surface of a ball, so that the circumference of a circle is less than 2π times the radius. Parallel lines would converge (think of lines of longitude on Earth), and the sum of a triangle's angles would be greater than 180°. Or the universe could have been negatively curved, like the middle of a Pringles potato chip. In this case, the circumference of a circle is more than 2π times the radius. Parallel lines would diverge, and the sum of a triangle's angles would be less than 180°.

17.2 Inflation

A century ago, astronomers were struggling just to get a handle on the size of the universe. Today we have a comprehensive theory that ties together many diverse facts about nature: the constancy of the speed of light, the properties of gravity, the motions of galaxies, and even the origins of the very atoms we are made of. The case for the Big Bang is compelling. Even so, as our knowledge of the expansion of the universe has grown and our observations of the cosmic microwave background have improved, a number of puzzles have arisen. The solution to these puzzles has forced us to consider some remarkable ideas about how our universe expanded when it was very young.

The Universe Is Much Too Flat

The first puzzle that we run into when observing our universe is that it is too flat. In fact, the universe is much too close to being exactly flat for this to have happened by chance. Any deviation from flatness would grow over time, so if the universe originally had a value of $\Omega_{mass} + \Omega_\Lambda$ even slightly different from 1, the value would by now be drastically different, and easily detectable. For the present-day value of $\Omega_{mass} + \Omega_\Lambda$ to be as close to 1 as it is, $\Omega_{mass} + \Omega_\Lambda$ could not have differed from 1 by more than one part in 100,000 when the universe was 2,000 years old. When the universe was 1 second old, it had to be flat by at least one part in 10 billion. At even earlier times, it had to be much flatter still. This is too special a situation to be the result of chance—a fact referred to in cosmology as the **flatness problem**. *Something* about the early universe must have *forced* $\Omega_{mass} + \Omega_\Lambda$ to have a value incredibly close to 1.

The Cosmic Microwave Background Is Much Too Smooth

The second problem faced by our cosmological models is that the cosmic microwave background is surprisingly smooth. Following the discovery of the cosmic microwave background (CMB) in the 1960s, many observers turned their attention to mapping this background glow. At first they were reassured as result after result showed that the temperature of the CMB is remarkably constant (variations of less than one part in 3,000), regardless of where one looks in the sky. Yet over time, this strong confirmation of Big Bang cosmology came to challenge our view of the early universe. Once we remove our motion relative to the CMB from the picture, the CMB is not just smooth—it is *too* smooth.

The early universe was subject to the quantum mechanical **uncertainty principle**. The uncertainty principle says that as we look at a system at smaller and smaller scales, the properties of that system become less and less well determined. As we look at smaller and smaller scales in the universe, conditions *must* fluctuate in unpredictable ways. In particular, quantum mechanics says that the younger the universe, the more dramatic those fluctuations become. When the universe was young, it could *not* have been smooth. There *must* have been dramatic variations, or "ripples," in the density and temperature of the universe from place to place. When we look at the universe today, we should see the fingerprint of those early ripples imprinted on the cosmic microwave background—but we do not. The fact that the CMB is so smooth is referred to as the **horizon problem** in cosmology: different parts of the universe are too much like other parts of the universe—distant parts that should have been "over the horizon" and beyond the reach of any signals that might have smoothed out the early quantum fluctuations.

The CMB is smoother than the early universe should have been.

Inflation Solves the Problems

In the early 1980s, Alan Guth (b. 1947) offered a solution to cosmology's flatness and horizon problems. Guth suggested that for a brief time, the young universe expanded at a rate *far* in excess of the speed of light. This rapid expansion is

Inflation was a period of rapid expansion of the universe.

called **inflation**. Between the ages of about 10^{-35} and 10^{-33} second, the scale factor R_U of the universe increased by a factor of at least 10^{30} and perhaps very much more. In that incomprehensibly brief instant, the observable universe grew from 10-trillionths the size of the nucleus of an atom to a region about 3 meters across. That is like a grain of very fine sand growing to the size of today's *entire universe*—all in a billionth of the time it takes light to cross the nucleus of an atom. During inflation, space *itself* expanded so rapidly that the distances *between* points in space increased faster than the speed of light. Inflation does not violate the rule that nothing can travel *through* space at greater than the speed of light, because the space itself was expanding.

To understand how inflation solves the flatness and horizon problems, imagine that you are an ant living in the *two-dimensional* universe defined by the surface of a golf ball, as shown in **Figure 17.4**. Your universe would have two very apparent characteristics: First, it would be obviously positively curved. If you were to walk around the circumference of a circle in your two-dimensional universe and then measure the radius of the circle, you would find the circumference to be less than 2π times the radius. If you were to draw a triangle in your universe, the sum of its angles would be greater than 180°. The second obvious

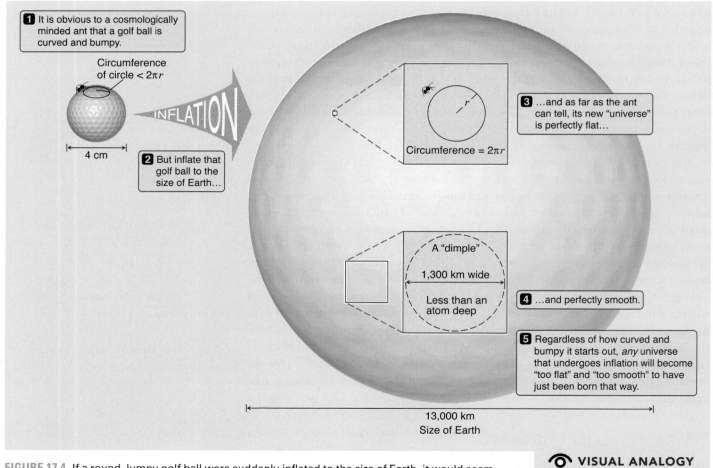

1 It is obvious to a cosmologically minded ant that a golf ball is curved and bumpy.

Circumference of circle < $2\pi r$

4 cm

INFLATION

2 But inflate that golf ball to the size of Earth...

Circumference = $2\pi r$

3 ...and as far as the ant can tell, its new "universe" is perfectly flat...

A "dimple"

1,300 km wide

Less than an atom deep

4 ...and perfectly smooth.

5 Regardless of how curved and bumpy it starts out, *any* universe that undergoes inflation will become "too flat" and "too smooth" to have just been born that way.

13,000 km
Size of Earth

VISUAL ANALOGY

FIGURE 17.4 If a round, lumpy golf ball were suddenly inflated to the size of Earth, it would seem extraordinarily flat and smooth to an ant on its surface. Similarly, after undergoing inflation, any universe would seem both extremely flat and extremely smooth, regardless of the exact geometry and irregularities it started with.

characteristic would be the dimples, approximately a half-millimeter deep, on the surface of the golf ball.

Now imagine that your golf-ball universe suddenly grew to the size of Earth. First, the curvature of your universe would no longer be apparent. An ant (or person) walking around on the surface of Earth would be hard-pressed to tell that Earth is not flat. The circumference of a circle would be 2π times its radius, and there would be 180° in a triangle. In the case of inflationary cosmology, the universe after inflation would be extraordinarily flat (that is, with $\Omega_{mass} + \Omega_\Lambda$ extraordinarily close to 1) *regardless* of what the geometry of the universe was before inflation. Because the universe was inflated by a factor of at least 10^{30}, $\Omega_{mass} + \Omega_\Lambda$ immediately after inflation must have been equal to 1 to within one part in 10^{60}, which is flat enough for $\Omega_{mass} + \Omega_\Lambda$ to remain close to 1 today. If inflation occurred, then today's universe is not flat by chance. It is flat because *any* universe that underwent inflation would become flat.

Huge expansion also smooths out inhomogeneities.

What about the horizon problem? When our golf ball universe inflated to the size of Earth, the dimples that covered the surface of the golf ball were stretched out as well. Instead of being a half-millimeter or so deep and a few millimeters across, these dimples now are only an atom deep but are hundreds of kilometers across. Again, our ant would be hard-pressed to detect any dimples at all. In the case of the real universe, inflation took the large fluctuations caused by quantum uncertainty and stretched them out so much that they are not measurable in today's (local) universe. The slight irregularities in the CMB are the faint ghosts of quantum fluctuations that occurred as the universe inflated.

An early era of inflation in the history of the universe offers a handy way of solving the horizon and flatness problems, but it seems quite remarkable that the universe should have undergone a period during which it expanded at such an "astronomical" rate. The cause of inflation lies in the fundamental physics that governed the behavior of matter and energy at the earliest moments of the universe.

17.3 The Earliest Moments

There are four fundamental forces in nature, and everything in the universe is a result of their action. Chemistry and light are products of the **electromagnetic force** acting between protons and electrons in atoms and molecules. The energy produced in fusion reactions in the heart of the Sun comes from the **strong nuclear force** that binds together the protons and neutrons in the nuclei of atoms. The **weak nuclear force** governs *beta decay* of nuclei, in which a neutron decays into a proton, an electron, and an antineutrino. Finally, there is **gravity**, which plays such a major role in astronomy. We must explore backward in time, toward the Big Bang itself, to understand how these forces came into being. This exploration will allow us to put together a timeline of the history of the universe from earliest times to the present.

There are four fundamental forces in nature.

In Chapter 4 we spoke of electromagnetism in terms of electric and magnetic fields, but we also spoke of the quantum mechanical description of light as a stream of particles called photons. Because there is only one reality, both of these descriptions of electromagnetism have to coexist. The branch of physics that deals with this reconciliation is called **quantum electrodynamics (QED)**.

QED treats charged particles almost as if they were baseball players engaged in an endless game of catch. As baseball players throw and catch baseballs, they experience forces. Similarly, in QED, charged particles "throw" and "catch" an endless stream of "virtual photons." The QED description of the electromagnetic interaction between two charged particles is an average of all the possible ways that the particles could throw photons back and forth. The resulting force acts, over large scales, like the classical electric and magnetic fields described by Maxwell's theory. Physicists speak of the electromagnetic force being "mediated by the exchange of photons." As is always the case with quantum mechanics, the world described by QED is hard to picture. Even so, QED is one of the most accurate, well tested, and precise branches of physics. As of this writing, not even the tiniest measurable difference between the predictions of the theory and the outcome of an actual experiment has been found.

The central idea of QED—forces mediated by the exchange of carrier particles—provides a template for understanding two of the other three fundamental forces in nature. The electromagnetic and weak nuclear forces have been combined into a single theory called **electroweak theory**. This theory predicts the existence of three particles—labeled W^+, W^-, and Z^0—that mediate the weak nuclear force. In the 1980s, physicists identified these particles in laboratory experiments, thus confirming the essential predictions of electroweak theory.

The strong nuclear force is described by a third theory, called **quantum chromodynamics (QCD)**. In this theory, particles such as protons and neutrons are composed of more fundamental building blocks, called **quarks**, that are bound together by the exchange of another type of carrier particle, dubbed **gluons**. Together, electroweak theory and QCD are referred to as the **standard model** of particle physics. Excluding gravity, the standard model explains all the observed interactions of matter and has made many predictions that were subsequently confirmed by laboratory experiments. However, the standard model leaves many questions unanswered, such as why strong interactions are so much stronger than weak interactions.

A Universe of Particles and Antiparticles

For modern theories of particle physics to make any sense, every type of particle in nature must also have an alter ego—an **antiparticle**. (We saw the electron's antiparticle, the positron, in nuclear reactions within the Sun in Chapter 11. A positron is identical to an electron except that it has a positive charge instead of a negative charge.) One fascinating property of particle-antiparticle pairs is that if you bring such a pair together, the two particles will annihilate each other.

When a particle-antiparticle pair annihilates, the mass of the two particles is converted into energy in accord with Einstein's special theory of relativity ($E = mc^2$), and the energy is carried away by a pair of photons (as calculated in **Working It Out 17.1**). The reverse process is also possible: two photons may collide to produce a particle and its antiparticle—by a process called **pair production**.

Now we apply this idea to a hot universe awash in a bath of blackbody radiation. When the universe was less than about 100 seconds old and had a temperature greater than a billion kelvins, it was filled with energetic photons that were constantly colliding, creating electron-positron pairs; and these electron-positron pairs were constantly annihilating each other, creating pairs of gamma-ray photons. The whole process reached an equilibrium, determined strictly by temperature, in which pair creation and pair annihilation exactly balanced each other. Rather than being filled only with a swarm of photons, at this time the universe was filled with a swarm of photons, electrons, and positrons, as illustrated in **Figure 17.5a**. Earlier, when the universe was even hotter, photons would have produced a swarm of protons and antiprotons.

As the universe cooled, there was no longer enough energy to support the cre-

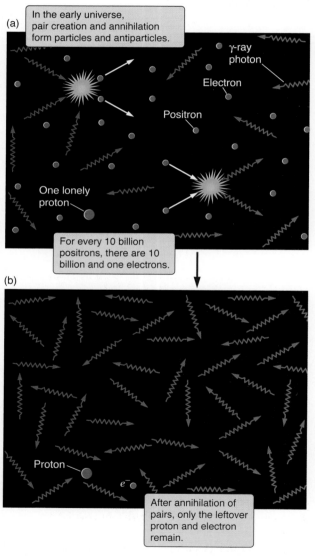

FIGURE 17.5 (a) A swarm of electrons, positrons, and photons in the very early universe. For every 10 billion positrons, there are 10 billion and one electrons. Before this time, an era of proton-antiproton annihilation left this small piece of universe with only one proton and no antiprotons. (b) After the electrons and positrons annihilate, only the one electron is left, along with the remaining proton and many photons.

ation of particle pairs, so the swarm of particles and antiparticles that filled the early universe annihilated each other and were not replaced. When this cooling happened, every proton should have been annihilated by an antiproton; and then, at still cooler temperatures, every electron should have been annihilated by a positron. This was almost the case, but not quite. For every electron in the universe today, there were 10 billion and one electrons in the early universe, but only 10 billion positrons. This one-part-in-10-billion excess of electrons over positrons meant that when electron-positron pairs finished annihilating each other, some electrons were left over—enough to account for all the electrons in all the atoms in the universe today (**Figure 17.5b**). Similarly, there was an excess of protons over antiprotons in the early universe, and the protons we see today are all that is left from the annihilation of proton-antiproton pairs.

> For every 10 billion positrons there were 10 billion and one electrons.

If the standard model of particle physics were a complete description of nature, then the imbalance of one part in 10 billion between matter and antimatter would not have been present in the early universe. The symmetry between matter and antimatter would have been complete. No matter at all would have survived into today's universe, and we would not exist. The fact that you are reading this page demonstrates that something more needs to be added to the model.

The symmetry between matter and antimatter may be broken in a theory that joins the electroweak and strong nuclear forces together, in much the same way that electroweak theory unified our understanding of electromagnetism and the weak nuclear force. Such a theory, which combines three of the fundamental forces into a single unified force, is referred to as a **grand unified theory** (**GUT**). Grand unified theories break the symmetry and explain why the universe is composed of matter rather than antimatter. There are several competing GUTs, and only the very simplest of GUTs have been ruled out.

Working It Out 17.1 | What Type of Photons Result from Pair Annihilation?

Using what we've learned so far, we can determine the type of photons that result when an electron annihilates with a positron. Imagine that the two particles are barely moving when they collide. This is the simplest case, since only the masses of the particles need to be considered. The mass of an electron is 9.11×10^{-31} kg. The mass of the positron is identical, so the total mass involved is

$$m = 2 \times m_e$$
$$m = 2 \times (9.11 \times 10^{-31} \text{ kg})$$
$$m = 1.82 \times 10^{-30} \text{ kg}$$

We can find the equivalent energy by using $E = mc^2$:

$$E = m \times c^2$$
$$E = (1.82 \times 10^{-30} \text{ kg}) (3.00 \times 10^8 \text{ m/s})^2$$
$$E = 1.64 \times 10^{-13} \text{ J}$$

In this type of collision, two photons typically are created, and each of them has half of this total energy: $E = 8.20 \times 10^{-14}$ J.

From $E = hf$ and $c = f\lambda$, we can find that the energy of a photon is related to its wavelength:

$$\lambda = hc/E$$
$$\lambda = (6.63 \times 10^{-34} \text{ J·s}) \times (3.00 \times 10^8 \text{ m/s}) / (8.20 \times 10^{-14} \text{ J})$$
$$\lambda = 2.43 \times 10^{-12} \text{ m}$$

This is in the gamma-ray region of the spectrum, so the annihilation of electrons and positrons results in the emission of gamma-ray photons.

What about our assumption that the electron and positron were nearly stationary before they collided? If the particles were instead moving before they collided, they would have had additional energy in the form of kinetic energy. The photons resulting from the collision would then be even more energetic, making their wavelength even smaller. Since all photons shorter than 10^{-10} m are gamma rays, these photons would still be gamma rays.

The conditions described by all versions of GUT held sway when the universe was *very* young (younger than about 10^{-35} second) and *very* hot (hotter than about 10^{27} K). Enough energy was available for particles to be freely created. During this time, the distinction among the electromagnetic, weak nuclear, and strong nuclear forces had not yet come into being. There was only the one unified force.

As we move backward in time, we have yet to cross one threshold. How does gravity fit into this scheme? General relativity provides a beautifully successful description of gravity that correctly predicts the orbits of planets, describes the ultimate collapse of stars, and even enables us to calculate the structure of the universe. Yet general relativity's description of gravity "looks" very different from our theories of the other three forces. Rather than talking about the exchange of photons or gluons or other carrier particles, general relativity talks about the smooth, continuous canvas of spacetime that events are painted upon.

Gravity does not fit into the GUT picture.

We might be tempted to say, "Oh well. Gravity works one way and the other forces work another way, and that is how the universe happens to be." In practice, that is exactly what we do when we call on quantum mechanics to tell us about the properties of atoms and then use relativity to describe the passage of time or the expansion of the universe. Even the era of GUTs is described perfectly if we treat gravity as a separate force. As we push back even closer to the moment of the Big Bang, however, this coexistence between relativity and quantum mechanics turns instead into a conflict.

Toward a Theory of Everything

When the universe was younger than about 10^{-42} second, its density was incomprehensibly high. The observable universe was so small that 10^{60} universes would have fit into the volume of a single proton. Under these extreme conditions, quantum physics is required to describe not just particles, but spacetime itself. The failure of general relativity to describe this early universe is much like the failure of Newtonian mechanics to describe the structure of atoms. An electron in an atom must be thought of in terms of probabilities rather than certainties. Similarly, there is no unique history for the earliest moments after the Big Bang. This era in the history of the universe is referred to as the **Planck era**, signifying that we can understand the structure of the universe during this period only by using the ideas of quantum mechanics.

The conflict between general relativity and quantum mechanics brings us to the current limits of human knowledge. Known physics can take us back to a time when the universe was a millionth of a trillionth of a trillionth of a trillionth of a second old; but to push back any further, we need something new. We need a theory that combines general relativity and quantum mechanics into a single theoretical framework unifying all four of the fundamental forces. To understand the earliest moments of the universe, we need a **theory of everything (TOE)**. We don't have one.

String theory is a possible theory of everything.

A successful theory of everything would do more than unify general relativity with quantum mechanics. It would tell us which of the possible GUTs is correct and would tell us the nature of dark matter and dark energy. A successful theory of everything would also explain the how, when and why of inflation. Physicists are currently grappling with what a TOE might look like. One contender for the title is **string theory**. Here, elementary particles are viewed not as points but as tiny loops called "strings." A guitar string vibrates in one way to play an F, another way to play a G, and yet another way to play an A. According to string theory, different types of elementary particles are like different "notes" played by vibrating loops of string.

In principle, string theory provides a way to reconcile general relativity and quantum mechanics, but there is a price to pay for this potential advantage. To make string theory work, we have to imagine that these tiny loops of string are vibrating in a universe with nine spatial dimensions. (Adding time to the list would make our universe 10-dimensional.) How can that be, when we clearly experience only three spatial dimensions? Whereas the three spatial dimensions that we know spread out across the vastness of our universe, the other six spatial dimensions predicted by string theory wrap tightly around themselves (**Figure 17.6**), extending no further today than they did a brief instant after the Big Bang.

To better understand this bizarre notion, imagine what it would be like to live in a three-dimensional universe (like the one we experience) in which one of those dimensions extended for only a tiny distance. Living in such a universe would be like living within a thin sheet of paper that extended billions of light-years in two directions but was far smaller than an atom in the third. In such a universe, we would easily be aware of length and width—we could move in those directions at will. In contrast, we would have no freedom to move in the third dimension at all, and we might not even recognize that the third dimension existed. Perhaps our only inkling of the true nature of space would come from the fact that in order to explain the results of particle physics experiments, we would have to assume that particles extended into a third, unseen dimension. If string theory is correct, we see three spatial dimensions extending possibly forever, but we are unaware of the fact that each point in our three-dimensional space also has a tiny extent in six other dimensions.

String theory is only a pale shadow of the sort of well-tested theories that we have made use of throughout this book. In some respects, string theory is no more than a promising idea providing direction to theorists searching for a TOE. We will probably never be able to build particle accelerators that enable us to directly search for the most fundamental particles predicted by a TOE. The energies required are simply too high. Fortunately, nature has provided us with the ultimate particle accelerator: the Big Bang itself.

Order "Froze Out" of the Cooling Universe

To understand the very earliest moments in the history of the universe, we have been looking backward to higher and higher energies and correspondingly to earlier and earlier times. We have seen how the four forces become unified in stages—from four separate forces to electroweak theory to GUTs and then to a TOE. We find that our universe started out with one (as yet unknown) theory of everything; and as the universe expanded and cooled, the various forces came into being. **Figure 17.7** illustrates how the four fundamental forces emerged in the evolving universe. In the first 10^{-43} second after the Big Bang, as described by the TOE, the physics of elementary particles and the physics of spacetime were one and the same. As the universe expanded and cooled, gravity parted ways with the forces described by the GUT. Spacetime took on the properties described by general relativity. Inflation may also have been taking place at this time.

As the universe continued to expand and its temperature fell further, less and less energy was available for the creation of particle-antiparticle pairs. When the particles responsible for GUT interactions could no longer form, the strong force split off from the electroweak force. Somewhere along the line, as the unity of the

FIGURE 17.6 It is virtually impossible to visualize how six spatial dimensions wrapped up into structures far smaller than the nucleus of an atom would look. Here such geometries are projected onto the two-dimensional plane of the paper.

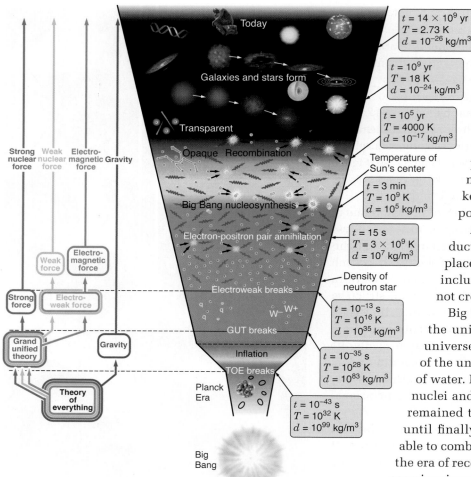

FIGURE 17.7 Eras in the evolution of the universe. As the universe expanded and cooled following the Big Bang, it went through a series of phases determined by what types of particles could be created freely at that temperature. Later, the structure of the universe was set by the gravitational collapse of material to form galaxies and stars and by the chemistry made possible by elements formed in stars.

Ripples in the early universe were the seeds of galaxies and large-scale structure.

original TOE was lost, the symmetry between matter and antimatter was broken. As a result, the universe ended up with more matter than antimatter.

The next big change took place when the particles responsible for unifying the electromagnetic and weak nuclear forces split out, leaving these two forces independent of each other. All four fundamental forces of nature that govern today's universe were now separate. At one 10-trillionth of a second, the temperature of the universe had fallen to 10^{16} K. It was a full minute or two before the universe cooled to the billion-kelvin mark, below which not even pairs of electrons and positrons could form.

Although the universe was now too cool for pair production, it was still hot enough for nuclear reactions to take place. These reactions formed the least massive elements, including helium, lithium, beryllium, and boron, but could not create more massive elements.

Big Bang nucleosynthesis had come to an end by the time the universe was 5 minutes old, and the temperature of the universe had dropped below about 800 million K. The density of the universe at this point had fallen to only about a tenth that of water. Normal matter in the universe now consisted of atomic nuclei and electrons, awash in a bath of radiation. The universe remained this way for the next several hundred thousand years, until finally the temperature dropped so far that electrons were able to combine with atomic nuclei to form neutral atoms. This was the era of recombination, which we see directly when we look at the cosmic microwave background.

17.4 Gravity Forms Large-Scale Structure

The large-scale structure of walls and voids in this chapter's opening figure, which we first encountered at the end of Chapter 15, cries out for explanation. Cosmologists have proposed a number of ideas. Early on, it was suggested that voids were the result of huge expanding blast waves from tremendous explosions that might have occurred in the early universe. The correct answer has turned out to be less fanciful, but far more satisfying: large-scale structure is the fingerprint of gravity.

In our discussion of star formation in Chapter 5, we learned about gravitational instabilities. Star formation can begin with a molecular cloud with clumps inside it. Gravity causes those clumps to collapse faster than their surroundings: gravity can turn density variations of clouds into stars. The same gravitational instability can turn density variations of the universe into galaxies.

The slight ripples in the cosmic microwave background almost certainly result from quantum mechanical variations that imprinted structure on the early universe at the time of inflation. These variations provided the "clumps" or "seeds" from which galaxies and collections of galaxies grew. In many ways, it is truly amazing that quantum mechanics, the physics that governs atomic nuclei, atoms, and molecules but is almost undetectable in our daily lives, is responsible for seeding the very largest structures that we can see in our universe.

It is one thing to say that galaxies and larger structures formed from gravitational instabilities that began with slight irregularities in the early universe. It is quite another to turn this statement into a real scientific theory with testable predictions. To accomplish that, we combine the ideas we want to test with the laws of physics, construct a model, and then compare the predictions of that model with observations of the universe.

To build a model of the formation of large-scale structure, we have to begin with three key pieces of information. First we have to decide what universe we are going to model. That is, what values of Ω_{mass} and Ω_A are we going to assume? These values determine how rapidly the universe expands. The more rapidly a universe expands, or the less mass it contains, the more difficult it will be for gravity to pull material together into galaxies and larger-scale structures.

Second, we need to know how large and how concentrated the early clumps were. There are several ways to approach this question. Satellite data provide a good picture of structure in the CMB, from which we can infer what the early clumps in the universe must have looked like. Alternatively, models of inflation make predictions of the structure that will emerge after the rapid expansion. These predictions are especially important to test because they tie together the large-scale structure of today's universe with our most basic ideas about what the universe was like in the briefest instant after the Big Bang. Currently, we know enough to say that the early universe was "clumpier" on smaller (galaxy-sized) scales than it was on the scales of the clusters, superclusters, filaments, and voids. This means that smaller structures formed first, whereas larger structures took more time to form. The idea of small structures forming first and larger structures forming later is referred to as **hierarchical clustering**. Hierarchical clustering has become one of the most important themes in our growing understanding of how structure in the universe formed.

Third, we need a complete list of the types and amounts of ingredients that existed in the early universe. We need to know the balance between radiation, normal matter, and dark matter. We also need to make some choices about the nature of the dark matter that we use in our model.

Once we have these three pieces of information—the values of Ω_{mass} and Ω_A, the way matter fluctuations developed, and the nature and mixture of the different forms of matter we are using—the rest is physics and calculations. Despite some real difficulties in carrying out this strategy, we are aiming to answer this key question: What choices lead to a model universe that most resembles the real universe in which we live?

Dark Matter Is Essential to Galaxy Formation

When we observe the cosmic microwave background using spacecraft like COBE and WMAP, we find variations in the background radiation of about one part in 100,000. Theoretical models show that these variations are far too small to explain the structure we see in today's universe. Gravity is not strong enough to grow galaxies and clusters of galaxies from such poor "seeds." These models indicate that for ripples in the density of the universe to have formed today's galaxies, the density of those ripples must have been at least 0.2 percent greater than the average density of the universe at the time of recombination. If that is true, the variations in the CMB today should be at least 30 times larger than what is observed. At first glance this might seem to be irreconcilable with our understanding of the origin of structure in the universe, but it is instead a crucial result that ties together several pieces of the puzzle.

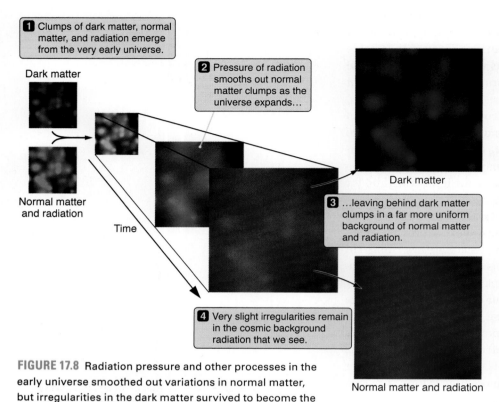

1 Clumps of dark matter, normal matter, and radiation emerge from the very early universe.

Dark matter

Normal matter and radiation

Time

2 Pressure of radiation smooths out normal matter clumps as the universe expands…

Dark matter

3 …leaving behind dark matter clumps in a far more uniform background of normal matter and radiation.

4 Very slight irregularities remain in the cosmic background radiation that we see.

Normal matter and radiation

FIGURE 17.8 Radiation pressure and other processes in the early universe smoothed out variations in normal matter, but irregularities in the dark matter survived to become the seeds of galaxy formation.

Dark matter provides the foundation for observed structure.

Dark matter turns out to be an essential ingredient enabling us to understand the origin of structure. It first appeared back in Chapter 15 as an ad hoc construction—an annoyance, really—used by astronomers to explain the oddly flat rotation curves of spiral galaxies. Dark matter dominates normal matter on the scale of clusters of galaxies as well. Now we find that irregularities in normal matter in the early universe were too slight to form galaxies. So where will we turn for an answer to the question of how structure in the universe *did* form? You guessed it: dark matter.

In Chapters 15 and 16, we learned that there is much more dark matter than normal matter in the universe. Yet our discussion of Big Bang nucleosynthesis in Chapter 14 revealed that the amount of normal matter we see in the universe predicts just the right abundances of light elements. Therefore, dark matter cannot be made of normal matter consisting of neutrons, protons, and electrons. If it were, it would have affected the formation of chemical elements in the early universe. The abundances of several isotopes of the least massive elements would be quite different from what we find in nature. Dark matter must be something else—something that has no electric charge (so it does not interact with electromagnetic radiation) and that interacts only feebly with normal matter. Clumps of such dark matter in the early universe would not have interacted with radiation or normal matter, so we would not see them in the CMB, and they would not have been smoothed out by pressure waves and radiation (**Figure 17.8**). Dark matter solves the problems of modeling the formation of galaxies and clusters of galaxies.

There Are Two Classes of Dark Matter

Here is the story of galaxy formation in a nutshell. Dark matter in the early universe was much more strongly clumped than normal matter. Within a few million years after recombination, these dark matter clumps pulled in the surrounding normal matter. Later, gravitational instabilities caused these clumps to collapse. The normal matter in the clumps went on to form visible galaxies. This story seems plausible enough, but the details of how this happened depend greatly on the properties of dark matter itself. Even though we do not yet know exactly what dark matter in the universe is made of, we can talk about two broad classes of dark matter that are based on how it behaves.

One possibility is **cold dark matter**, which consists of feebly interacting particles that are moving about relatively slowly, like the atoms and molecules in a cold gas. There are several candidates for cold dark matter. Cold dark matter might consist of tiny black holes produced in the early universe. Few physicists and cosmologists favor this idea, however. Most think instead that cold dark matter consists of an unknown elementary particle. One candidate is the **axion**, an as yet undetected particle first proposed to explain some observed properties of neutrons. Axions should have very low mass, and they would have been produced in great abundance in the Big Bang. Another candidate is the **photino**, an elementary particle related to the photon. Some theories of particle physics predict that the photino exists and has a mass about 10,000 times that of the proton. Our state of knowledge about the particles that make up cold dark matter could soon change; photinos might be detected in current particle accelerators, and experiments are under way to search for axions and photinos that are trapped in the dark matter halo of our galaxy.

Hot dark matter consists of particles that are moving very rapidly. Neutrinos are one example of hot dark matter. We have seen that neutrinos interact with matter so feebly that they are able to flow freely outward from the center of the Sun. There is no question that the universe is filled with neutrinos, which might account for as much as 5 percent of the mass of the universe. Although this percentage is not high enough to account for all of the dark matter in the universe, it may still have had a noticeable effect on the formation of structure.

Slow-moving particles are more easily held by gravity than are fast-moving particles, so particles of cold dark matter clump together more easily into galaxy-sized structures than do particles of hot dark matter. On the largest scales of massive superclusters, both hot dark matter and cold dark matter can form the kinds of structures we see; but on much smaller scales, only cold dark matter can clump enough to produce structures like the galaxies we see filling the universe. To account for the formation of today's galaxies, we need cold dark matter.

Cold dark matter consists of relatively massive, slowly moving particles.

Hot dark matter consists of less massive, rapidly moving particles.

A Galaxy Forms within a Collapsing Clump of Dark Matter

We can best see how models of galaxy formation work by following the events predicted by the models step-by-step. Consider a universe made up of 90 percent cold dark matter and 10 percent normal matter, clumped together in a manner consistent with observations of the cosmic microwave background. On the scale of an individual galaxy, the effect of the cosmological constant is so small that it can be ignored.

Figure 17.9a shows one clump of dark matter at the time of recombination. The dark matter is less uniformly distributed than normal matter; but overall, the distribution of matter is still remarkably uniform. By a few million years after recombination (**Figure 17.9b**), the universe of our model calculation has expanded severalfold. Spacetime is expanding, so the clump of dark matter is also expanding. However, the clump of dark matter is not expanding as rapidly as its surroundings are, because its self-gravity has slowed down its expansion. The clump now stands out more with respect to its surroundings. The gravity of the dark matter clump has begun to pull in normal matter as well. By the stage shown in **Figure 17.9c**, normal matter is clumped in much the same way as dark matter.

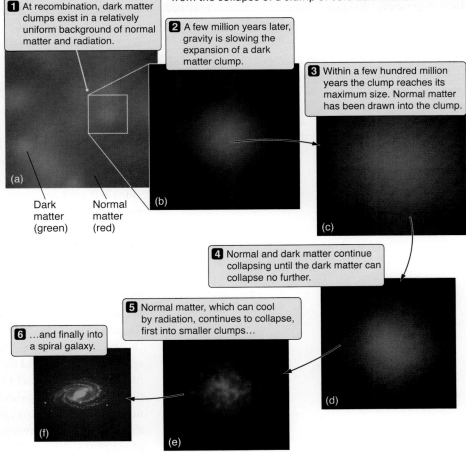

FIGURE 17.9 Stages in the formation of a spiral galaxy from the collapse of a clump of cold dark matter.

1 At recombination, dark matter clumps exist in a relatively uniform background of normal matter and radiation.

2 A few million years later, gravity is slowing the expansion of a dark matter clump.

3 Within a few hundred million years the clump reaches its maximum size. Normal matter has been drawn into the clump.

4 Normal and dark matter continue collapsing until the dark matter can collapse no further.

5 Normal matter, which can cool by radiation, continues to collapse, first into smaller clumps...

6 ...and finally into a spiral galaxy.

Dark matter (green)   Normal matter (red)

G **X** U V I R

FIGURE 17.10 The Bullet Cluster of galaxies, at a redshift of $z = 0.3$, represents the collision of two giant clusters of galaxies in the process of merging. The galaxies in this system constitute less mass than the X-ray-emitting gas (shown in red), which is itself much less massive than the dark matter (shown in blue) measured from the gravitational lensing by this cluster.

A ball thrown in the air will slow, stop, and then fall back to Earth. In like fashion, the clump of dark matter will stop expanding when its own self-gravity slows and then stops its initial expansion. By the time the universe is about a billion years old, the clump of dark matter has reached its maximum size and is beginning to collapse (see Figure 17.9c). The collapse of the dark matter clump stops when the clump is about half its maximum size, however, because the particles making up the cold dark matter are moving too rapidly to be pulled in any closer (**Figure 17.9d**). The clump of cold dark matter is now given its shape by the orbits of its particles, in the same way that an elliptical galaxy is given its shape by the orbits of the stars it contains.

Unlike dark matter (which cannot emit radiation), the normal matter in the clump is able to radiate away energy, cool, and collapse. Small concentrations of normal matter within the dark matter collapse under their own gravity to form clumps of normal matter that range from the size of globular clusters to the size of dwarf galaxies. These clumps of normal matter then fall inward toward the center of the dark matter clump, as shown in **Figure 17.9e**. According to models, gas in our universe can cool quickly enough to fall in toward the center of the dark matter clump only if the clump has a mass of $10^8–10^{12}$ solar masses ($M_\odot$). This is just the range of masses of observed galaxies. This agreement between theory and observation is an important success of the theory that galaxies form from cold dark matter.

Gravitational interaction between irregular clumps tugs on each protogalactic clump, so that it has a little bit of rotation when it begins its collapse. As normal matter falls inward toward the center of the dark matter clump, this rotation forces much of the gas to settle into a rotating disk (**Figure 17.9f**), just as the collapsing cloud around a protostar settles into an accretion disk. The disk formed by the collapse of each protogalactic clump becomes the disk of a spiral galaxy.

17.5 Large-Scale Structure Forms through Mergers

As in star formation, galaxies do not always form in isolation. It is likely that a clump often produced more than one galaxy and that these galaxies later interacted or even merged. The tidal interactions between the galaxies and the collisions between gas clouds in the galaxies probably triggered many regions of star formation throughout the combined system. The collision and merging of two giant clusters is nicely illustrated in **Figure 17.10**, which shows the Bullet Cluster of galaxies at a redshift of $z = 0.3$. The stars contained within the galaxies of this system have less total mass than the X-ray-emitting gas, which is itself much less massive than the dark matter, whose mass we deduce from the gravitational lensing produced by the cluster. The cluster's galaxies, hot gas, and dark matter are widely displaced from one another due to the violence of the collision that formed this cluster.

The merging of galaxies also answers another puzzle. We have seen how spiral galaxies could form in the early universe, but what about the formation of elliptical galaxies? Ellipticals are now thought to result from the merger of two or more spiral galaxies. **Figure 17.11** shows a computer simulation of such a merger. If the merging galaxies are not originally spinning in the same direction, then the resulting merged galaxy loses its disk-like character. The dark matter halos of the galaxies merge, and the stars eventually settle down into the blob-

Ellipticals form from the mergers of spirals.

like shape of an elliptical galaxy. As this picture suggests, elliptical galaxies are known to be more common in dense clusters where mergers are likely to have been more frequent.

The same mergers and interactions that formed the giant galaxies we see today also provided the fuel to power the quasars and other active galactic nuclei (AGNs) that were common in the early universe. With galaxies crashing together, supermassive black holes forming, quasars flooding the young universe with intense radiation and powerful jets, and star formation running amok, galaxy formation in the young universe was a violent, messy business. This conclusion has clear consequences for what we should expect to see when looking back to a time in the history of the universe when galaxies were actively forming. Rather than the well-formed spirals and ellipticals that dominate today's universe, the early universe should have contained many clumpy, irregular objects—which are indeed what we see in images of the young universe (**Figure 17.12**).

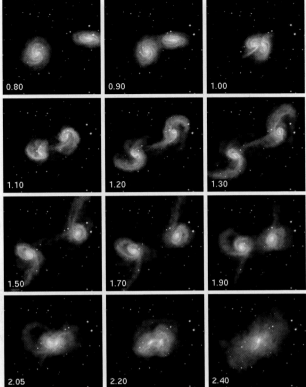

FIGURE 17.11 In this computer simulation, two spiral galaxies merge to form an elliptical galaxy. Numbers indicate fractional time: 0.00 is when the interaction began, and 1.00 is the time it took for the galaxies to fall together.

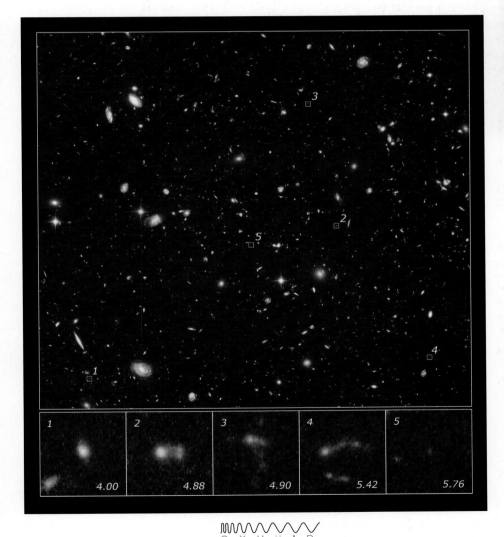

FIGURE 17.12 The Hubble Ultra Deep Field provides a glimpse into the universe's distant past. As examples of the youngest galaxies in the process of forming and merging, insets 1–5 show high-redshift objects ranging from $z = 4.00$ to $z = 5.76$.

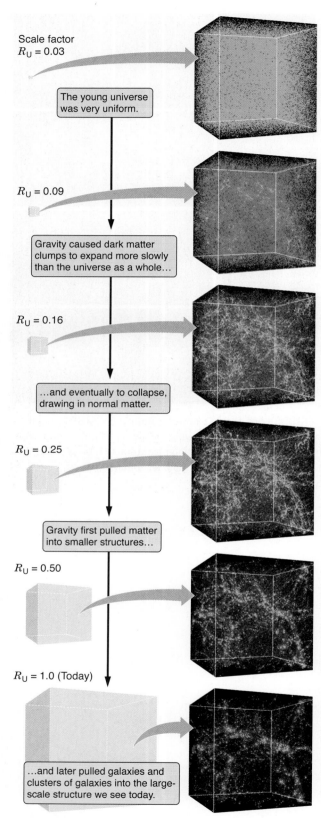

Scale factor
$R_U = 0.03$

The young universe was very uniform.

$R_U = 0.09$

Gravity caused dark matter clumps to expand more slowly than the universe as a whole…

$R_U = 0.16$

…and eventually to collapse, drawing in normal matter.

$R_U = 0.25$

Gravity first pulled matter into smaller structures…

$R_U = 0.50$

$R_U = 1.0$ (Today)

…and later pulled galaxies and clusters of galaxies into the large-scale structure we see today.

FIGURE 17.13 A computer simulation of the formation of very large-scale structure in a universe filled with cold dark matter.

Star Formation Leaves a Gap in the Models

We know that most of the normal matter in a galaxy winds up as stars; but when and where do those stars form, and how long does it take? Although we can say a great deal about how individual stars form in our galaxy now, we do not yet understand the differences between star formation in the early universe and the star formation going on around us today. In a universe devoid of massive elements, there were no dense, dusty molecular clouds and no spiral disks in which such clouds might congregate. Instead, the first stars must have formed from the collapse of clouds of hydrogen and helium gas within the overall clump of matter they were a part of.

The fact that the atmospheres of even the oldest halo stars in the Milky Way contain some amount of massive elements tells us that a significant amount of star formation must have taken place *very* early during the collapse of the Milky Way. If we could closely inspect a collapsing protogalactic clump in the stages depicted in Figure 17.9d and e, we would expect to see stars forming and supernovae exploding in the collapsing halo. The halo stars that we see today must have formed while the Milky Way was still collapsing, before the gas from which they formed coalesced into the disk. Disk stars, all of which have relatively high abundances of massive elements, formed later.

Uncertainties in our understanding of star formation complicate comparisons between models of galaxy formation and observations of the early universe. When we talk about structure on scales much larger than galaxies, star formation is less of an issue. To understand the formation of clusters and larger structures, we need worry only about the way that galaxy-sized clumps of matter fall together under the force of gravity.

Figure 17.13 shows the results of one computer simulation that follows the motions of millions of clumps of dark matter as they fall through space under their mutual gravitational attraction. The scale of this simulation is so large that we cannot follow the outcome of individual galaxies but instead are looking at the overall distribution of dark matter. The simulation shows that the smaller-scale structures develop first. During the first few billion years, dark matter falls together into structures comparable in size to today's clusters of galaxies. Only later do the spongelike filaments, walls, and voids become well defined. Look back at the chapter-opening figure to see how this distribution of dark matter, which we cannot see directly, has influenced the distribution of luminous matter, which we can see.

The similarities between the results of the models and observations of large-scale structure are quite remarkable. Only model universes with certain combinations of shape, mass, nature of ripples, type of dark matter, and values for the cosmological constant produce structure similar to what we actually see. This is an important result. Our models contain assumptions consistent with our knowledge of the early universe, and they predict the formation of large-scale structure similar to what we actually see in today's universe. **Figure 17.14** summarizes the galaxy formation process, in which smaller objects form first and merge into ever-larger structure, leading ultimately to the Hubble Ultra Deep Field (HUDF) picture shown in Figure 17.12.

Drawing on this and previous chapters, take a moment to trace the history of the universe, from its earliest beginnings to the formation of Earth. It's a remarkable story, and it's a triumph of humanity that we understand it even in the broadest terms. The level of detail at which we are beginning to understand it in modern

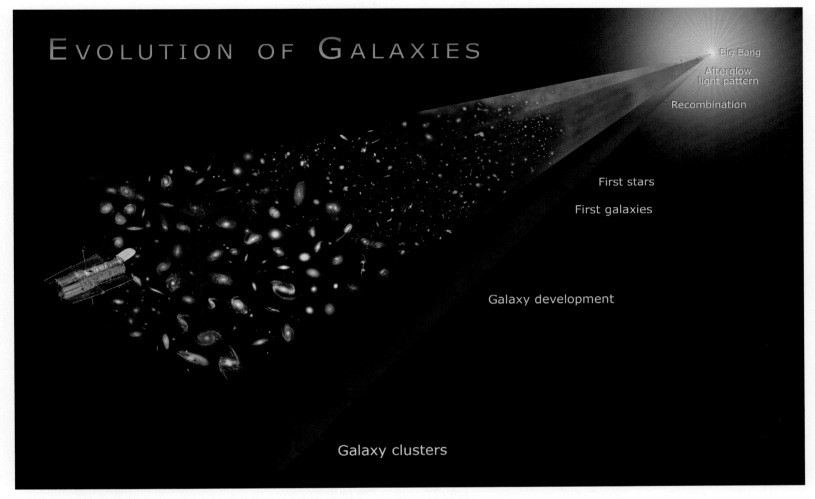

FIGURE 17.14 A schematic view of how structure formed in the universe, from smaller systems to larger ones, leading ultimately to the kind of Hubble Ultra Deep Field picture shown in Figure 17.12.

times is even more remarkable. As Carl Sagan (1934–1996) said, "We are a way for the cosmos to know itself." You now know what he meant, and you are a part of that knowledge.

17.6 Thinking More Broadly about Our Universe

It is natural to wonder about our place and time in the universe. How will our universe evolve into the future? What existed before the Big Bang? Do other universes coexist with our own? We can address some questions using conventional physics; others require more speculation.

The Deep Future

Using well-established physics, we can calculate how the existing structures in the universe will evolve over a very long time. One such calculation is illustrated

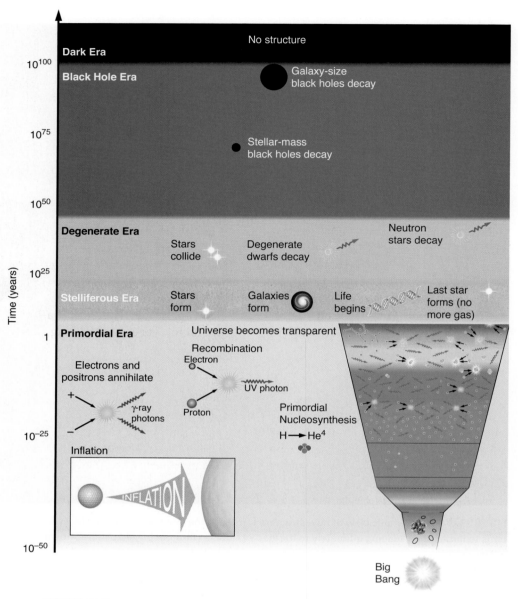

Time (years)

Dark Era — No structure

10^{100} — Black Hole Era — Galaxy-size black holes decay

10^{75} — Stellar-mass black holes decay

10^{50}

Degenerate Era — Stars collide — Degenerate dwarfs decay — Neutron stars decay

10^{25} — Stelliferous Era — Stars form — Galaxies form — Life begins — Last star forms (no more gas)

1 — Primordial Era — Universe becomes transparent — Recombination — Electron — Proton — UV photon — Electrons and positrons annihilate — γ-ray photons — Primordial Nucleosynthesis — H → He^4

10^{-25} — Inflation

10^{-50} — Big Bang

FIGURE 17.15 How structure in the universe evolves—from the Primordial Era to the present and then to the future Dark Era.

in **Figure 17.15**. During the first era of the universe, the Primordial Era—the first several hundred thousand years after the Big Bang and before recombination—the universe was a swarm of radiation and elementary particles. Today we live during the second era, the Stelliferous Era ("Era of Stars"); but this era, too, will end. Some 100 trillion (10^{14}) years from now the last molecular cloud will collapse to form stars, and a mere 10 trillion years later the least massive of these stars will evolve to form white dwarfs.

Following the Stelliferous Era, most of the normal matter in the universe will be locked up in brown dwarfs and degenerate stellar objects: white dwarfs and neutron stars. During this Degenerate Era, the occasional star will still flare up as ancient substellar brown dwarfs collide, merging to form low-mass stars that burn out in a trillion years or so. However, the main source of energy during this era will come from the decay of protons and neutrons and the annihilation of particles of dark matter. Even these processes will eventually run out of fuel. In 10^{39} years, white dwarfs will have been destroyed by proton decay, and neutron stars will have been destroyed by the beta decay of neutrons.

As the Degenerate Era comes to an end, the only significant concentrations of mass left will be black holes. These will range from small ones with the masses of single stars to greedy monsters that grew during the Degenerate Era to have masses as large as those of galaxy clusters. During the period that follows—the Black Hole Era—these black holes will slowly evaporate into elementary particles through the emission of Hawking radiation. A black hole with a mass of a few solar masses will evaporate into elementary particles in 10^{65} years, and galaxy-sized black holes will evaporate in about 10^{98} years. By the time the universe reaches an age of 10^{100} years, even the largest of the black holes will be gone. A universe vastly larger than ours will contain little but photons with colossal wavelengths, neutrinos, electrons, positrons, and other waste products of black hole evaporation. The Dark Era will have arrived as the universe continues to expand forever—into the long, cold, dark night of eternity.

Multiverses

Is our universe the only one? Since we've defined the universe as "everything," what does it mean to say "multiple universes"? Are there parallel universes, either separated in space or even occupying exactly the same space as ours? Many cosmologists think seriously about the idea of **multiverses**, or collections of parallel universes.

Let's begin with the simplest example of such parallel universes, illustrated in **Figure 17.16**. The age of the universe—that is, the amount of time that has passed since the Big Bang—is 13.7 billion years. Light reaching us today can have traveled a maximum of 13.7 billion light-years (Gly). Therefore, our observable universe—everything that we can possibly observe today—must be within a sphere having a radius of 13.7 Gly. We cannot possibly see anything farther away than that. The observational evidence suggests that the geometry of space is flat. If this is true, then the universe is truly infinite in size and must therefore contain an infinite number of similar spheres. As dark energy causes the universe to expand faster and faster, the separate observable universes move farther apart and will never overlap. These parallel universes are simply too far away for us ever to be able to observe them.

What are these other parallel universes like? We can say several things based on what we have learned about the observable universe. First, if the cosmological principle holds, then on large scales each of these observable universes should look similar to our own, although details may be very different. Still, in a truly infinite universe there must be an infinite number of observable universes exactly like ours, with an exact copy of you reading an identical version of *Understanding Our Universe*. We know this because if our own observable universe is cooler everywhere than about 10^8 K, then there can be no more than 10^{118} particles in the observable universe, and there are only so many ways that a finite number of particles can be distributed. If you then ask how far you must go before you

Ultimately, all structure in the universe will decay.

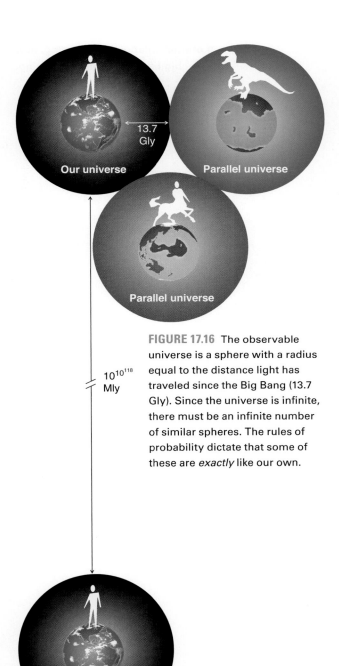

FIGURE 17.16 The observable universe is a sphere with a radius equal to the distance light has traveled since the Big Bang (13.7 Gly). Since the universe is infinite, there must be an infinite number of similar spheres. The rules of probability dictate that some of these are *exactly* like our own.

are sure to find an observable universe just like—and we mean *identical to*—our own, the answer is about $10^{10^{118}}$ Mly. Yes, that's 10 raised to the power 10^{118}. So, in an infinite universe—as enormous as it might be—we still know how far away an identical copy of you must be.

The inflationary universe model forms the basis of a second type of multiverse. Imagine a universe that undergoes **eternal inflation**, with no beginning or end to the inflation. If such a universe exists, then quantum fluctuations may cause some regions to expand more slowly than the rest of the universe. As a result, such a region may form a bubble whose inflating phase will soon end. In this scenario, we would be living inside such a region, and our Big Bang would just be the condensation of our bubble within the eternally inflating universe.

One nice feature of eternal inflation is that it neatly answers the question of what there was before the Big Bang. Since the universe has been inflating and will continue to inflate forever, there is no beginning or end. Our own bubble, or parallel universe, separated from the rest of the universe at a time we call the Big Bang, but other bubbles are constantly separating and becoming their own parallel-universe big bangs.

It is fair to ask whether the idea of parallel universes, or multiverses, is really science. Throughout this book we have emphasized that any legitimate scientific theory must be testable and ultimately falsifiable. Are there tests capable of proving these multiverse ideas to be wrong? Possibly. For example, we do test the first multiverse, involving distinct observable universes, when we measure the isotropy of the CMB or the large-scale distribution of galaxies. Since this multiverse idea follows inevitably from an infinite, expanding universe, it makes predictions that can be tested. The eternal inflation model is harder to test, because we will never directly observe its parallel universes. But if we obtain a theory of everything that predicts eternal inflation, and if that theory of everything is itself falsifiable, then there will be a connection between eternal inflation and observation. There is considerable debate within the scientific community as to whether models such as eternal inflation can be truly falsified.

> If eternal inflation is happening, new big bangs are forever forming.

Dark energy is an as yet poorly understood phenomenon, so of course scientists are working hard either to detect it more directly or to disprove its existence altogether. They use every tool at their disposal to try to do this, including the mighty Green Bank Telescope, the largest steerable telescope in the world.

New Galaxy Maps to Help Find Dark Energy Proof?

By **Ker Than,** *National Geographic News*

A newly developed technique could one day help astronomers use giant sound waves to test theories of dark energy, the mysterious force thought to be causing the universe to fly apart faster over time.

Called intensity mapping, the technique looks for unique radio emissions of hydrogen gas in galaxies and galaxy clusters to map the large-scale structure of the universe. Hydrogen is the lightest and most abundant element in the universe, and it tends to cluster around galaxies because of their strong gravitational pull.

Until now, astronomers have been mapping large cosmic structures by identifying the galaxies and clusters themselves—a process akin to mapping forests on Earth by counting individual trees. The new method is more like mapping forests by looking for large patches of green.

By speeding up large-scale mapping efforts, the method should reveal how structures in the universe have evolved since the big bang.

In particular, the new maps might help astronomers detect variations in matter distribution caused by enormous sound waves called baryon acoustic oscillations, or BAOs, which were created shortly after the big bang.

Using the cosmic sound waves as rulers to measure the size of the universe over time can help scientists understand how dark energy has affected the cosmos.

Sound Waves as Cosmic Rulers

In the new proof-of-concept study, Tzu-Ching Chang of the Canadian Institute for Theoretical Astrophysics in Toronto and colleagues used the Robert C. Byrd Green Bank Telescope in West Virginia to pick up hydrogen radio emissions in or near thousands of galaxies at once.

The team was also able to measure signals emitted when the universe was only about seven billion years old.

"These observations detected more hydrogen gas than all the previously detected hydrogen in the universe, and at distances ten times farther than any radio wave-emitting hydrogen seen before," study co-author Ue-Li Pen, of the University of Toronto, said in a statement.

The team hopes the method, with refinement, will detect hydrogen emissions from even further back in time, perhaps before the formation of the very first galaxies.

And by scanning large enough regions of the sky, the method may uncover the imprints of BAOs propagating through the universe, the scientists say.

As BAOs traveled through the primordial gases of the early universe, they affected the matter they passed through, causing it to clump up in some places more than in others. This pattern should show up in the distribution of today's galaxies and galaxy clusters.

Because scientists know BAOs repeat every 450 million light-years, "one can measure the BAO signature at different [ages of the universe] and use it as a standard ruler to determine the size of the universe at that time," study leader Chang said.

"And because at later times the expansion of the universe is mostly driven by dark energy, BAO measurements at one or several [time points] can thus help study the properties of dark energy."

Current theories of dark energy, for example, state that the expansion rate is dependent on the density of matter in the universe. Since the early universe was smaller and more tightly packed, scientists hope the BAO data will show that the early universe's expansion rate was slower than it is now.

Evaluating the News

1. What's new about the intensity mapping method?
2. What kind of "unique radio emissions of hydrogen gas" are they probably mapping? (Hint: You may need to review the information about the interstellar medium.)
3. The author is reporting on a proof-of-concept study. Was this study successful?
4. How will this study help scientists understand dark energy?
5. In the last sentence, the author states that scientists "hope" the data will show that the expansion rate was slower in the distant past. Considering what you learned about the scientific method in Chapter 1, comment on whether "hopeful" is an appropriate scientific attitude.

SUMMARY

17.1   Both gravity and the cosmological constant (or dark energy) determine the fate of the universe. Observations suggest that rather than slowing down, the expansion of the universe is accelerating.

17.2   The very early universe may have gone through a brief but dramatic period of exceptionally rapid expansion, called inflation. If this is true, inflation would explain both the flatness and the homogeneity of the universe we see today.

17.3   Understanding the very earliest moments in the universe requires that we also understand how the four fundamental forces of nature were all unified into one basic phenomenon in early times.

17.4   The first galaxies formed through the gravitational collapse of clumps of cold dark matter that arose from the early universe.

17.5   Observed galaxies come from complex mergers. The visible gas in galaxies cools and falls inward to form the visible stars, which are surrounded by a dark matter halo. Over time, the mutual gravitational attraction of clumps of dark matter produced the walls, filaments, and voids that characterize the large-scale structure observed today.

17.6   Structure will continue to form in our universe for at least another 10^{15} years, but our current understanding of quantum effects suggests that they will ultimately lead to the decay of all structure, from planets to black holes. Cosmologists also speculate about the possible existence of multiverses, collections of parallel universes.

✦ SUMMARY SELF-TEST

1. If there is enough mass in the universe that the density is higher than the critical density, the universe will
   a. expand forever.
   b. expand, but gradually slow down.
   c. eventually collapse.
   d. neither expand nor contract.

2. When astronomers discovered that the universe was _____, they had to revive Einstein's cosmological constant.
   a. accelerating
   b. expanding
   c. contracting
   d. decelerating

3. Which of the following are properties of the young universe as revealed by the cosmic microwave background? Select all that apply.
   a. It was hot.
   b. It was cold.
   c. It was dense.
   d. It was diffuse.
   e. It was uniform on large scales.
   f. It was "clumpy" on large scales.

4. Which of the following problems led astronomers to hypothesize inflation? Select all that apply.
   a. the solar neutrino problem
   b. the horizon problem
   c. the expansion problem
   d. the flatness problem
   e. the acceleration problem

5. Place the following forces in order of their "freeze out" in the first moments after the Big Bang:
   a. gravity
   b. strong nuclear force
   c. weak nuclear force
   d. electromagnetic force

6. Which of the following are candidates for the composition of cold dark matter? Select all that apply.
   a. neutrinos
   b. electrons
   c. axions
   d. neutrons
   e. photinos
   f. tiny black holes
   g. protons

7. Our universe will
   a. expand forever.
   b. expand for a long time and then collapse.
   c. expand, but gradually slow down.
   d. neither expand nor contract.

QUESTIONS AND PROBLEMS

True/False and Multiple-Choice Questions

8. **T/F:**   The cosmological constant makes the universe accelerate.

9. **T/F:**   In our universe, $\Omega_{\mathrm{mass}} + \Omega_{\Lambda}$ is as close to zero as we can measure.

10. **T/F:**   Inflation is the theory that the universe is expanding today.

11. **T/F:**   The early universe was filled with almost equal numbers of matter and antimatter particles.

12. **T/F:**   The universe became transparent when it cooled below the temperature of the Sun.

13. If Ω_{mass} is 0.7 and Ω_Λ is 0.5 today, then the universe will
    a. expand forever.
    b. expand and then contract.
    c. expand, but gradually slow down.
    d. remain static.

14. If the universe is dominated by dark energy, it will
    a. expand forever.
    b. expand and then contract.
    c. expand, but gradually slow down.
    d. remain static.

15. If the universe is dominated by matter, it will
    a. expand forever.
    b. expand and then contract.
    c. expand, but gradually slow down.
    d. remain static.

16. A positron is related to an electron in that
    a. it has all the same properties except opposite mass.
    b. it has all the same properties except opposite charge.
    c. it has all the same properties except opposite spin.
    d. it has all the same properties except how it interacts with light.

17. The horizon problem states that
    a. the ratio of matter to antimatter is too large.
    b. the cosmological constant is too close to one.
    c. the cosmic microwave background is too bumpy.
    d. the cosmic microwave background is too uniform.

18. On the largest scales, galaxies in the universe are distributed
    a. uniformly.
    b. along filaments and walls.
    c. in disconnected clumps.
    d. along lines extending radially outward from the center.

19. Galaxies in the young universe were _____ galaxies in the universe today.
    a. just like
    b. smaller and more irregular than
    c. far more numerous than
    d. larger and more prototypical than

Conceptual Questions

20. As applied to the universe, what is the meaning of *critical density*?

21. What set of circumstances would cause an expanding universe to reverse its expansion and collapse?

22. What is the principal difference between normal matter and dark matter?

23. Describe the observational evidence suggesting that Einstein's cosmological constant (a repulsive force) may be needed to explain the historical expansion of the universe.

24. Describe what astronomers mean by dark energy.

25. If the universe is being forced apart by dark energy, why isn't our galaxy, Solar System, or planet being torn apart?

26. What is the flatness problem, and why has it been a problem for cosmologists?

27. During the period of inflation, the universe may have briefly expanded at 10^{30} (a million trillion trillion) or more times the speed of light. Why did this ultrarapid expansion not violate Einstein's special theory of relativity, which says that neither matter nor communication can travel faster than the speed of light?

28. Why is high-energy physics important to our understanding of the early universe?

29. Name the four fundamental forces in nature.

30. The fundamental forces of the universe are generally assumed not to change. How would the fate of the universe be affected if Newton's gravitational constant changed with time?

31. Of the four fundamental forces in nature, which one depends on electric charge?

32. The standard model cannot explain why neutrinos have mass, or why electron-positron asymmetry existed in the early universe. Do these failings make it an incomplete theory? Should all of its predictions be ignored until the theory can solve these remaining issues?

33. Explain how quantum electrodynamics describes the electromagnetic interaction between two charged particles.

34. What are quarks?

35. Explain what happens when you bring a particle and an antiparticle together.

36. Why are there so few antiparticles in the universe?

37. Explain the process of pair creation.

38. Describe the Planck era.

39. What are the basic differences between a grand unified theory (GUT) and a theory of everything (TOE)?

40. Consider the term *string theory* in light of the discussion in Chapter 1. Many scientists object to using the word *theory* to describe string theory. Why?

41. As the sensitivity of our instrumentation increases, we are able to look ever farther into space and, therefore, ever further back in time. However, we can see no further back in time than the era of recombination. Explain why.

42. Suppose you could view the early universe when galaxies were first forming. How would it be different from the universe we see today?

43. As clumps containing cold dark matter and normal matter collapse, they heat up. When a clump collapses to about half its maximum size, the increased thermal motion of particles tends to inhibit further collapse. Whereas normal matter can overcome this effect and continue to collapse, dark matter cannot. Explain the reason for this difference.

44. Imagine that there are galaxies in the universe composed mostly of dark matter with relatively few stars or other luminous normal matter. If this were true, how might we learn of the existence of such galaxies?

45. How are the processes of star formation and galaxy formation similar? How do they differ?

46. What is the origin of large-scale structure?

47. Why is dark matter so essential to the galaxy formation process?

48. Which do we believe is the correct evolutionary sequence: (a) small star clusters formed first, which were bound together into galaxies, which were later bound together in clusters and superclusters; or (b) supercluster-sized regions collapsed to form clusters, which then later collapsed to form galaxies, which formed small clusters of stars? Justify your answer.

49. Why does the current model of large-scale structure require that we include the effects of dark matter?

50. Why do we think that some hot dark matter exists?

51. Describe the stages of galaxy formation for both the dark matter and gas components.

52. How does a roughly spherical cloud of gas collapse to form a disk-like, rotating spiral galaxy?

53. How do astronomers believe elliptical galaxies formed? Does this formation process explain the observed differences between spirals and ellipticals?

54. Based on our understanding of galaxy formation, describe how galaxies should appear as you look further back in time. Are the features you described observed?

55. How do we know that the dark matter in the universe must be composed mostly of cold—rather than hot—dark matter?

56. Using broad strokes, describe the process of structure formation in the universe, starting at recombination (half a billion years after the Big Bang) and ending today.

57. How can we be certain that gravity, and not the other fundamental forces, is responsible for large-scale structure?

58. We have never observed a star with zero heavy metals in its atmosphere. What does this fact imply about the history of star formation in the early universe?

59. Previous chapters painted a fairly comprehensive picture of how and why stars form. Why, then, is it difficult to model the star formation history of a young galaxy? Is this difficulty a failure of our theories?

Problems

60. Figure 17.2 contains a lot of information in one graph.
    a. What do the orange lines tell you?
    b. What do the blue lines tell you?
    c. The uppermost blue line shoots off toward the top of the graph. What does this tell you about the universe described by the line?
    d. The bottom blue line rises and then drops back to zero. What does this tell you about the universe described by the line?
    e. Of all these lines, which one most likely describes our actual universe today?

61. Figure 17.3 is an important graph in cosmology. It places the data from three independent types of experiments on the same graph.

a. All of the data overlap near the line labeled "Flat." What conclusions can we draw from this fact?
    b. Suppose that the latest data, represented by bright red ellipses, were located at $\Omega_\Lambda = 2$ instead. What would the response of astronomers be? What would this mean about our models of the universe based on prior data?
    c. The upper left of the diagram is the region in which a universe would not have required a Big Bang. Do the data allow us to entertain this possibility for our own universe?
    d. According to the data represented on this graph, will our universe recollapse eventually?

62. On a globe of Earth, we can easily determine whether the surface of Earth is positively curved, negatively curved, or flat. Measure the angles between two lines of longitude and the equator. Now follow those two lines of longitude north to the North Pole, and measure the angle between them. Add the three angles in your triangle together. Is the sum more or less than 180°? What type of curvature does the surface of Earth have?

63. How many hydrogen atoms need to be in 1 cubic meter (m^3) of space to equal the critical density of the universe?

64. The proton and antiproton each have the same mass, $m_p = 1.67 \times 10^{-27}$ kg. What is the energy (in joules) of each of the two gamma rays created in a proton-antiproton annihilation?

65. There are about 500 million CMB photons in the universe for every hydrogen atom. What is the equivalent mass of these photons? Is it large enough to factor into the overall density of the universe?

66. Suppose you brought together a gram of ordinary-matter hydrogen atoms (each composed of a proton and an electron) and a gram of antimatter hydrogen atoms (each composed of an antiproton and a positron). Keep in mind that 2 grams is less than the mass of a dime.
    a. Calculate how much energy (in joules) would be released as the ordinary matter and antimatter hydrogen atoms annihilated one another.
    b. Compare this amount of energy with the energy released by a 1-megaton hydrogen bomb (1.6×10^{14} J).

67. One GUT predicts that on average a proton will decay in about 10^{31} years, which means if you have 10^{31} protons, you should see one decay per year. The Super-Kamiokande observatory in Japan holds about 20 million kg of water in its main detector, and it did not see any decays in 5 years of continual operation. What limit does this observation place on proton decay and on the GUT described here?

68. If 300 million neutrinos fill each cubic meter of space, and if neutrinos account for only 5 percent of the mass density (including dark energy) of the universe, estimate the mass of a neutrino.

69. What is the approximate mass of
    a. an average group of galaxies?
    b. an average cluster?
    c. an average supercluster?

70. The lifetime of a black hole varies in direct proportion to the cube of the black hole's mass. How much longer does it take a supermassive black hole of 3 million solar masses to decay compared to a stellar black hole of 3 solar masses?

71. Currently, the Milky Way and Andromeda galaxies (M31) are separated by about 0.7 million parsecs and moving toward each other at about 120 kilometers per second (km/s). Estimate how long it may take for the two to collide. Why do you think this may or may not be a good estimate of how long it will take these galaxies to fully merge?

 SmartWork, Norton's online homework system, includes algorithmically generated versions of these questions, plus additional conceptual exercises. If your instructor assigns questions in SmartWork, log in at **smartwork.wwnorton.com**.

 StudySpace is a free and open website that provides a Study Plan for each chapter of **Understanding Our Universe**. Study Plans include animations, reading outlines, vocabulary flashcards, and multiple-choice quizzes, plus links to premium content in SmartWork and the ebook. Visit **wwnorton.com/studyspace**.

Exploration | The Story of a Proton

Now that you have surveyed the current astronomical understanding of the universe, you are prepared to put the pieces together to make a story of how you came to be sitting in your chair, holding this book, reading these pages. It is valuable to take a moment to work your way backward through the book, from the Big Bang through all the intervening steps that had to occur, to the beginning of the book, which started with looking at the sky.

1. **In the Big Bang, how is a proton formed?**

2. **How might that proton become part of one of the first stars?**

3. **Suppose that proton later becomes part of a carbon atom in a 4-$M_\odot$ star. Through what type of nebula does it pass before returning to the interstellar medium?**

4. **Suppose that carbon atom then becomes part of the molecular-cloud core forming the Sun and the Solar System. What two physical processes dominate the core's collapse as the Solar System forms and that carbon atom becomes part of a planet?**

5. **Beginning with the Big Bang, create a timeline that traces the full history of a proton that becomes a part of the nucleus of a carbon atom on Earth.**

This is in essence the astronomical story of how Earth formed. We will address what comes next—how you have come to be on Earth at this time, studying the sky—in Chapter 18.

18

Life in the Universe

We have followed the origin of structure in the universe from the earliest moments after the Big Bang, when the fundamental forces of nature came to be, to the formation of galaxies and other large-scale structure visible today. We saw how stars (including our Sun) formed from clouds of gas and dust within these galaxies, and how planets (including our Earth) formed around those stars. We learned of the geological processes that shaped early Earth into the planet we know. In short, we have traced the origin of structure from the instant the universe came into existence up until the modern day. But there is one piece of the story that we have left out. Although the focus of our study is *astronomy*, no discussion of how structure evolved in the universe would be complete without some consideration of the origin of the particular type of structure we call **life**.

✧ LEARNING GOALS

Throughout this book, we have developed an understanding of the universe and all it contains. Still, one very important and very personal component remains: you, the reader of this book. What combination of events—some probable, others much less likely— has led to your existence on a small, rocky planet orbiting a typical middle-aged star? Are you and your fellow humans unique, or are there others like you inhabiting planets scattered throughout the vast universe? At right, a student has traveled to Australia and snapped a picture of stromatolites, one of the earliest forms of life on Earth. After reading this chapter, you should be able to put this picture in the context of our search for life on other worlds. You should also be able to:

- List the requirements for life as we know it to exist

- Explain why terrestrial life likely originated in Earth's oceans

- List the characteristics of the habitable zone of a solar system and the habitable zone within a galaxy

- Describe some of the methods used to search for extraterrestrial life, and explain the significance of our current null results

- Explain why all life on Earth must eventually come to an end

STROMATOLITES

Shark Bay, Australia

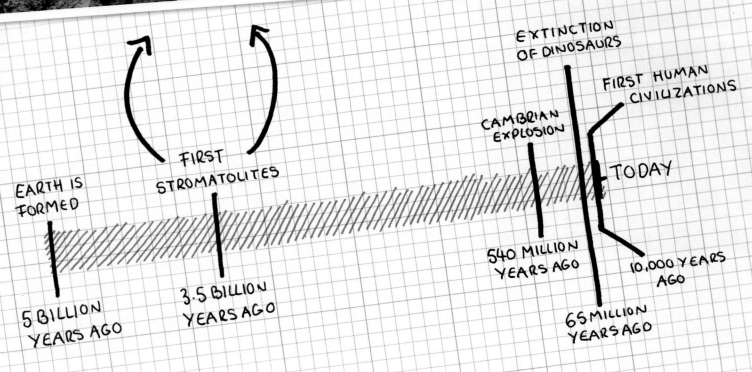

EXTINCTION
OF DINOSAURS

FIRST HUMAN
CIVILIZATIONS

CAMBRIAN
EXPLOSION

FIRST
STROMATOLITES

EARTH IS
FORMED

TODAY

5 BILLION
YEARS AGO

3.5 BILLION
YEARS AGO

540 MILLION
YEARS AGO

65 MILLION
YEARS AGO

10,000 YEARS
AGO

FIGURE 18.1 Chemist Stanley Miller, with his experiment simulating an early-Earth atmosphere.

FIGURE 18.2 (a) Life on Earth may have arisen near oceanic geothermal vents like this one. Similar environments might exist elsewhere in the Solar System. (b) Living organisms around such vents, like the giant tube worms shown here, rely on geothermal rather than solar energy for their survival.

18.1 Life on Earth

Before delving into the questions of how life first appeared on Earth and how it has since evolved, we should first ask, How do we define *life*? The answer might at first seem obvious, but it is not immediately self-evident. Many scientists will tell us that there is no universal definition of life. Although we may have a feeling for what we mean when speaking about terrestrial life, a universal definition must encompass very different forms of life that may occur elsewhere in the universe. To arrive at a complete definition, we must consider not only life-forms that we are completely unaware of but also those that exceed the limits of our imagination. We will speculate about alien life later in the chapter.

A Definition of Life

Life is a complex biochemical process that draws energy from the environment in order to survive and reproduce. With the assistance of special molecules such as ribonucleic acid (RNA) and deoxyribonucleic acid (DNA), organisms are able to evolve. All terrestrial life involves carbon-based chemistry and employs liquid water as its biochemical "solvent."

With at least a basic idea of what we mean by *life*, we turn to the question of its origins. As we learned in earlier chapters, Earth's secondary atmosphere was formed in part by carbon dioxide and water vapor that poured forth as a product of volcanism; a heavy bombardment of comets likely added large quantities of water, methane and ammonia to the mix. These are all simple molecules, incapable of carrying out life's complex chemistry. However, early Earth had abundant sources of energy, such as lightning and ultraviolet solar radiation, that could tear these relatively simple molecules apart, creating fragments that could reassemble into molecules of greater mass and complexity. As rain carried them out of the atmosphere, these heavier organic molecules ended up in Earth's oceans, forming a "primordial soup."

In 1952, American chemists Harold Urey (1893–1981) and Stanley Miller (1930–2007) attempted to create conditions similar to what they thought existed on early Earth. To a laboratory jar containing liquid water as an "ocean," they added methane, ammonia, and hydrogen as a primitive atmosphere; electric sparks simulated lightning as a source of energy (**Figure 18.1**). Within a week, the Urey-Miller experiment yielded 11 of the 20 basic amino acids that link together to form proteins, the structural molecules of life. Other organic molecules that are components of nucleic acids, the precursors of RNA and DNA, also appeared in the mix. We now know that Earth's early secondary atmosphere contained carbon dioxide and nitrogen rather than hydrogen, but recent experiments with more realistic atmospheric compositions have produced results similar to those of Urey and Miller. From laboratory experiments such as these, scientists have developed various models in which life got its start in Earth's oceans, rich in prebiotic organic molecules.

The details of where and how these prebiotic molecules evolved into the molecules of life are not so clear. Some biologists think life began in the ocean depths, where volcanic vents provided the hydrothermal energy needed to create the highly organized molecules responsible for biochemistry (**Figure 18.2**). Others believe that life originated in tide pools, where lightning and ultraviolet radiation supplied the energy (**Figure 18.3**). In either case, short strands of self-

FIGURE 18.3 Life may have begun in tide pools.

replicating molecules may have formed first, later evolving into RNA and finally into DNA, the huge molecule that serves as the biological "blueprint" for self-replicating organisms.

A few scientists think that perhaps life on Earth was "seeded" from space in the form of microbes brought here by meteoroids or comets. Although this hypothesis might tell us how life came to Earth, it does not explain its origin elsewhere in the Solar System or beyond. There is no scientific evidence at this time to support the seeding hypothesis.

The First Life

If life did indeed get its start in Earth's oceans, when did it happen? In Chapter 5 we learned that following its formation roughly 4.6 billion years ago, young Earth suffered severe bombardment by Solar System debris for several hundred million years. These were hardly the conditions under which life could form and gain a foothold. However, once the bombardment had abated and Earth's oceans had appeared, there were opportunities for living organisms to evolve. The earliest *indirect* evidence for terrestrial life is carbonized material found in Greenland rocks dating back to 3.85 billion years ago. Stronger and more direct evidence for early life appears in the form of fossilized masses of simple microbes called stromatolites, which date back to about 3.5 billion years ago. Fossilized stromatolites have been found in western Australia and southern Africa, and living examples still exist today (see the chapter-opening figure). We may never know the precise date when life first appeared on Earth, but current evidence suggests that the earliest life-forms appeared within a billion years after the formation of the Solar System, and shortly after the end of young Earth's catastrophic bombardment by leftover planetesimals.

The earliest organisms were extremophiles, life-forms that not only survive, but thrive, under extreme environmental conditions. Extremophiles include organisms living in subfreezing environments or in water temperatures as high

Life on Earth may have begun more than 3.5 billion years ago.

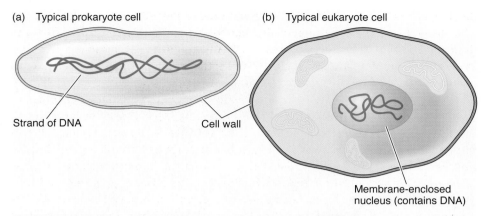

(a) Typical prokaryote cell   (b) Typical eukaryote cell

Strand of DNA   Cell wall

Membrane-enclosed
nucleus (contains DNA)

FIGURE 18.4 (a) A simple prokaryotic cell contains little more than the cell's genetic material. (b) A eukaryotic cell contains several membrane-enclosed structures, including a nucleus that houses the cell's genetic material.

> Oxygen in Earth's atmosphere is a result of photosynthesis by living organisms.

as 120°C, which occur in the vicinity of deep-ocean hydrothermal vents. Other extremophiles are found under the severe conditions of extraordinary salinity, pressure, dryness, acidity, or alkalinity. Among these early life-forms was an ancestral form of cyanobacteria, single-celled organisms otherwise known as blue-green algae. Cyanobacteria photosynthesized carbon dioxide and released oxygen as a waste product. Oxygen, however, is a highly reactive gas, and the newly released oxygen was quickly removed from Earth's atmosphere by the oxidation of surface minerals. Only when the exposed minerals could no longer absorb more oxygen did atmospheric levels of oxygen begin to rise. Oxygenation of Earth's atmosphere and oceans began about 2 billion years ago, and the current level was reached only about 250 million years ago (see Chapter 7). Without cyanobacteria and other photosynthesizing organisms, Earth's atmosphere would be as oxygen-free as the atmospheres of Venus and Mars.

Biologists comparing genetic DNA sequences find that terrestrial life is divided into two types: prokaryotes and eukaryotes. Prokaryotes, such as bacteria, are simple organisms that consist of free-floating DNA inside a cell wall; they lack both cell structure and a nucleus (**Figure 18.4a**). Eukaryotes, which include animals, plants and fungi, have a more complex form of DNA contained within the cell's membrane-enclosed nucleus (**Figure 18.4b**). The first eukaryote fossils date from about 2 billion years ago, coincident with the rise of free oxygen in the oceans and atmosphere, although the first *multicellular* eukaryotes did not appear until a billion years later.

Life Becomes More Complex

All life on Earth, whether prokaryotic or eukaryotic, shares a similar genetic code that originated from a common ancestor. DNA sequencing enables biologists to trace backward to the time when different types of life first appeared on Earth and to identify the species these life-forms evolved from. Scientists have used DNA sequencing to establish the evolutionary tree of life, which describes the interconnectivity of all species. The tree has revealed some interesting relationships; for example, it places animals (including us) as most similar to fungi, which branched off the evolutionary tree after slime molds and plants. The earliest primates branched off from other mammals about 70 million years ago, and the great apes (gorillas, chimpanzees, bonobos, and orangutans) split off from the lesser apes about 20 million years ago. DNA tests show that humans and chimpanzees share about 98 percent of their DNA; the two groups are believed to have evolved from a common ancestor about 6 million years ago.

Living creatures in Earth's oceans remained much the same—a mixture of single-celled and relatively primitive multicellular organisms—for more than 3 billion years after the first appearance of terrestrial life. Then, between 540 million and 500 million years ago, the number and diversity of biological species increased spectacularly. Biologists call this event the Cambrian explosion. The trigger of this sudden surge in biodiversity remains unknown, but possibilities include rising oxygen levels, an increase in genetic complexity, major climate change, or a combination of these. The "Snowball Earth" hypothesis suggests that before the Cambrian explosion, Earth was in a period of extreme cold between about 750 million and 550 million years ago and was covered almost entirely by ice. Astrophysicists speculate that the Sun's output at that time was 5–10 percent weaker than it is at present. During this period, many animals may have died out, thus making it easier for new species to adapt and thrive. Another possibility is that a marked increase in atmospheric oxygen (see Figure 7.5) would have been accompanied by a corresponding increase in stratospheric ozone that, as we learned in Chapter 7, shields us from deadly solar ultraviolet radiation. With a protective ozone layer in place, life was free to leave the oceans and move to land (**Figure 18.5**).

The first plants appeared on land about 475 million years ago, and large forests and insects go back 360 million years. The age of dinosaurs began 230 million years ago and ended abruptly 65 million years ago, when a small asteroid or comet collided with Earth and exterminated more than 70 percent of all existing plant and animal species. Mammals were the big winners in the aftermath. Our earliest human ancestors appeared a few million years ago, and the first civilizations occurred a mere 10,000 years ago. Our industrial society, barely more than two centuries old, is but a footnote in the history of life on our planet.

The Cambrian explosion marks an abrupt increase in the diversity of living creatures.

FIGURE 18.5 *Tiktaalik Roseae*, a fish with limblike fins and ribs, was an animal in a mid-evolutionary step of leaving the water for dry land.

Humans are here today because of a series of events that have occurred throughout the history of the universe. Some of these events seem inevitable, such as the creation of heavy elements by earlier generations of stars and the formation of life-supporting planets, including Earth. Other events may have been less likely, such as the creation of self-replicating molecules that led to Earth's earliest life. A few events stand out as random, such as the life-destroying impact by a piece of space debris 65 million years ago. This event—fortunate for us, but not for the dinosaurs—made possible the evolution of advanced forms of mammalian life and, ultimately, our existence as human beings.

You exist because of a series of events—some inevitable and others more random.

Evolution Is a Means of Change

Imagine that just once during the first few hundred million years after the formation of Earth, a single molecule formed by chance somewhere in Earth's oceans. That molecule had a very special property: Chemical reactions between that molecule and other molecules in the surrounding water resulted in that molecule's making a copy of itself. Now there were two such molecules. Chemical reactions would produce copies of each of these molecules as well, making four. Four became 8, 8 became 16, 16 became 32, and so on. By the time the original molecule had copied itself just 100 times, over a *million trillion trillion* (10^{30}) of these molecules would have existed. That is 100 million times more of these molecules than there are stars in the observable universe. (Such unconstrained replication would in fact be highly unlikely, primarily due to the limited availability of raw materials needed for reproduction and competition from similar molecules.)

To create life, only one self-replicating molecule needed to form by chance.

Chemical reactions are not perfect. Sometimes when a molecule replicates, the new molecule is not an exact duplicate of the old one. The likelihood that a copying error will occur while a molecule is replicating increases significantly with the number of copies being made. The imperfection in the attempted copy is called a **mutation**. Most of the time such an error is devastating, leading to a molecule that can no longer reproduce at all. But occasionally a mutation is actually helpful, leading to a molecule that is *better* at duplicating itself than the original was. Even if imperfections in the copying process crop up only once every 100,000 times that a molecule reproduces itself, and even if only one out of 100,000 of these errors turns out to be beneficial, after only 100 generations there will still be a hundred million trillion (10^{20}) errors that, by blind luck, might improve on the original molecule. Copies of each improved molecule will inherit the change. These molecules have a form of **heredity**—the ability of one generation of structure to pass on its characteristics to future generations.

As molecules of the early Earth continued to interact with their surroundings and make copies of themselves, they split into many different varieties. Eventually, the descendants of our original molecule became so numerous that the building blocks they needed in order to reproduce became scarce. In the face of this scarcity of resources, varieties of molecules that were more successful than others in reproducing themselves became more numerous. Varieties that could break down other varieties of self-replicating molecules and use them as raw material were especially successful in this world of limited resources. Competition, predation, and cooperation entered the picture. After a few generations, certain molecules dominated the mix while less successful varieties became less and less common. This process, in which better-adapted molecules thrive and less well-adapted molecules die out, is referred to as **natural selection**.

Success breeds success.

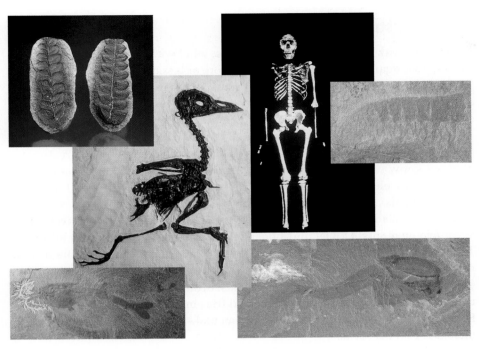

FIGURE 18.6 Fossils record the history of the evolution of life on Earth.

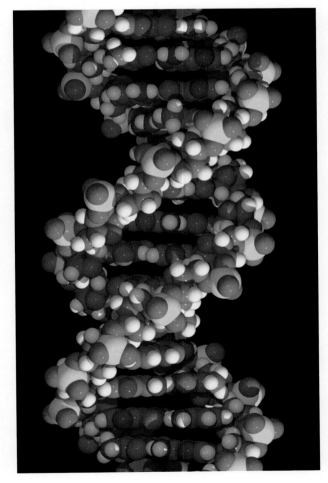

FIGURE 18.7 DNA is the blueprint for life.

Four billion years is a long time—enough time for the combined effects of heredity and natural selection to shape the descendants of that early self-copying molecule into a huge variety of complex, competitive, successful structures. Geological processes on Earth, such as sedimentation, have preserved a fossil record of the history of these structures (**Figure 18.6**). Among these descendants are "structures" capable of thinking about their own existence and unraveling the mysteries of the stars.

The molecules of DNA (**Figure 18.7**) that make up the chromosomes in the nuclei of the cells throughout your body are direct descendants of those early self-duplicating molecules that flourished in the oceans of a young Earth. Although the game played by the molecules of DNA in your body is far more elaborate than the game played by those early molecules in Earth's oceans, the fundamental rules remain the same. We now realize that this process is inevitable: any system that combines the elements of mutation, heredity, and natural selection *must* and *will* evolve.

18.2 Life beyond Earth

The story of the formation and evolution of life cannot be separated from the narrative of astronomy. We know that we live in a universe full of stars, and that systems of planets orbit many—probably most—of those stars. The evolution of life on Earth is but one of many examples that we have encountered of the emergence of structure in an evolving universe. This point leads naturally to one of the more profound questions that we can ask about the universe: Has life arisen elsewhere? To explore this question, we need to take a closer look at the chemistry of life on Earth.

The Chemistry of Life

When we speak about the chemistry of life, what we really mean is life itself. All known organisms are composed of a more or less common suite of complex chemicals, and very complex ones at that. We begin by looking at ourselves. Approximately two-thirds of the atoms in our bodies are hydrogen (H), about one-fourth are oxygen (O), a tenth are carbon (C), and a few hundredths are nitrogen (N). The remaining elements, and there are several dozen of them, make up only 0.2 percent of our total inventory. It turns out that all living creatures—that we know of—are an assemblage of molecules composed almost entirely of these four elements (sometimes called CHON), along with small amounts of phosphorus and sulfur. Some of these molecules are enormous. Consider DNA, which is responsible for our genetic code. DNA is made entirely from only *five* elements: CHON and phosphorus. But the DNA in each cell of our bodies is composed of combinations of *tens of billions* of atoms of these same five elements. Then there are proteins, the huge molecules responsible for the structure and function of living organisms. Proteins are long chains of smaller molecules called amino acids. Terrestrial life employs 20 specific amino acids, which also contain no more than five elements—in this case CHON plus sulfur instead of phosphorus.

Can this be all there is to our body chemistry? No; the chemistry of life is far too complex to get by with a mere half-dozen atomic species. Many others are present in smaller amounts but are essential to the complicated chemical processes that make living organisms tick. They include sodium, chlorine, potassium, calcium, magnesium, iron, manganese, and iodine. Finally, there are the so-called trace elements, such as copper, zinc, selenium, and cobalt. Trace elements also play a crucial role in life chemistry but are needed in only tiny amounts.

We know that the infant universe was composed basically of hydrogen and helium and very little else. But fast-forward to a later time—some 9 billion years later. By this time, all the heavier chemical elements essential to life were present and available in the molecular cloud that gave birth to our Solar System. As we learned in Chapters 12 and 13, those heavy elements—up to and including iron—were forged in the nuclear furnaces of earlier generations of low- and high-mass stars and were then dispersed into space. At times, this dispersal was passive. For example, low-mass stars, such as dying red giants, lose their gravitational grip on their overly extended atmospheres. Along with hydrogen and helium, the newly created heavy elements are blown off into space, eventually finding their way into molecular clouds. Other dispersals were more violent. Most of the trace elements essential to biology are more massive than iron, so they are not produced in the interiors of main-sequence stars. They are instead created within a matter of minutes during the violent supernova explosions that mark the death of high-mass stars. These elements, too, are thrown into the chemical mix found in molecular clouds.

Heavy elements produced by generations of stars provide the raw materials for life—but what kind of life? The only form that we have discussed so far is terrestrial, because it is the only one we know of. As we have already pointed out, terrestrial life is organic, or carbon based. What, then, is so special about carbon? Carbon is the lightest among the tetravalent atoms, which can bond with as many as four other atoms or molecules. (*Tetra* means "four," and *valence* refers to an atom's ability to attach to other atoms or molecules.) If the attached molecules also contain carbon, the result can be an enormous variety of long-chain

> All known life is composed primarily of only six elements.

molecules. This great versatility enables carbon to form the complex molecules that provide the basis for terrestrial life's chemistry.

There could be carbon-based forms of extraterrestrial life that have chemistries quite different from our own. For example, there are countless varieties of amino acids beyond the 20 employed by terrestrial life. Furthermore, molecules other than RNA and DNA may be capable of self-replication.

Science fiction writers often speculate about silicon-based life-forms because silicon, like carbon, is tetravalent. As a potential life-enabling atom, silicon has both advantages and disadvantages when compared to carbon. An important advantage is that silicon-based molecules remain stable at much higher temperatures than carbon-based molecules, possibly enabling silicon-based life to thrive in high-temperature environments, such as on planets that orbit close to their parent star. But silicon has a serious disadvantage that makes silicon-based life less likely. Silicon is a larger and more massive atom than carbon, and it cannot form long chains of atoms as complex as those based on carbon. Any type of silicon-based life would be much simpler than the carbon-based life we have here on Earth. Even so, we cannot rule out the possibility of silicon-based life existing in high-temperature niches somewhere within the universe.

Finally, although carbon's unique properties make it readily adaptable to the chemistry of life on Earth, we don't know what other chemistries life might adopt. Life is highly adaptable and tenacious; when it comes to the form that extraterrestrial life might take, nothing should surprise us.

> **Life-forms other than carbon based are possible.**

Life within Our Solar System

In our quest to find evidence of extraterrestrial life, the logical place to start would be right here in our own Solar System. As humans we have long been curious about the possibility of life beyond Earth. Some early conjectures seem ridiculous, considering what we now know about the Solar System. Two centuries ago, the eminent astronomer Sir William Herschel, discoverer of Uranus, proclaimed, "We need not hesitate to admit that the Sun is richly stored with inhabitants." In 1877, Italian astronomer Giovanni Schiaparelli (1835–1910) observed what appeared to be linear features on Mars and dubbed them *canali*, meaning "channels" in Italian. In one of astronomy's great ironies, the famous American observer of Mars, Percival Lowell (1855–1916), misinterpreted Schiaparelli's *canali* as "canals," suggesting that they were constructed by intelligent beings. Initial public fascination with Martians turned to hysteria in 1938 when Orson Welles aired H. G. Wells's fictitious *War of the Worlds* as "live" radio news coverage of militant Martians invading Earth. Panic ensued when many listeners believed that the "invasion" was actually happening.

The next several decades saw only modest progress in the search for life in the Solar System. During the mid-20th century, ground-based telescopes discovered that Mars possesses an atmosphere and water, both considered essential for any terrestrial-type life to get its start and evolve. During the 1960s, the United States and the Soviet Union sent reconnaissance spacecraft to the Moon, Venus, and Mars, but the instrumentation carried aboard these spacecraft was more suited to learning about the physical and geological properties of these bodies than to searching for life. Serious efforts to look for signs of life—past or present—would have to await advanced spacecraft with specialized bio-instrumentation.

In the meantime, astronomers and biologists alike were discussing where to look and what to look for; and thus was born the science of **astrobiology**, the study of the origin, evolution, distribution, and future of life in the universe. Astrobiologists knew that Mercury and the Moon lacked atmospheres, thus ruling them out as possibilities. The giant planets and their moons were thought to be too remote and too cold to sustain life. Venus, they knew, was far too hot. But Mars seemed just right. In the mid-1970s, two American *Viking* spacecraft were sent to Mars with detachable landers containing a suite of instruments designed to find evidence of a terrestrial type of life. (Some scientists were critical of the specific sites chosen, claiming that higher-latitude locations, where water ice might exist, would have been preferable.) When the *Viking* landers failed to find evidence of life on Mars, hopes faded for finding life on any other body orbiting our Sun.

Since that time, however, optimism has been renewed. A better understanding of the history of Mars indicates that at one time the planet was wetter and warmer, leading many scientists to believe that fossil life or even living microbes might yet be buried under the planet's surface. The first decade of the 21st century saw a return to Mars and a continuation in the search for evidence of current or preexisting life. In 2008, NASA's *Phoenix* spacecraft landed at a far northern latitude, inside the planet's arctic circle, where specialized instruments dug into and analyzed the martian water-ice permafrost. *Phoenix* found that the martian arctic soil has a chemistry similar to the Antarctic dry valleys on Earth, where life exists deep below the surface at the ice-soil boundary. Aqueous minerals, such as calcium carbonate, reveal that ancient oceans were once present on the planet. However, *Phoenix* did not find direct evidence of life, so the long-standing question of whether there is life on Mars remains unanswered.

> **Mars may harbor extraterrestrial life.**

The 1980s brought NASA's instrumented robots to the outer Solar System, and what they found surprised many astrobiologists. Although the outer planets themselves did not appear to be habitats for life, some of their moons became objects of special interest. Jupiter's moon Europa is covered with a layer of water ice that appears to overlie a great ocean of liquid water. Impacts by comet nuclei may have added a mix of organic material, another essential ingredient for life. Once thought to be a frozen, inhospitable world, Europa is now a candidate for biological exploration. Saturn's moon Titan appears to be rich in organic chemicals, many of which are thought to be precursor molecules of a type that existed on prebiotic Earth. The message from Titan was clear: the chemistry necessary for life likely exists elsewhere in the Solar System.

> **Life might exist on icy moons in the outer Solar System.**

Two decades later, the *Cassini* mission found additional evidence for a variety of prebiotic molecules in Titan's atmosphere, and it identified another potential site for life: Saturn's tiny moon Enceladus. The spacecraft detected water-ice crystals spouting from cryovolcanoes near Enceladus's south pole. Liquid water must lie beneath its icy surface, and Enceladus therefore joins Europa as a possible habitat of extremophile life, perhaps life similar to that found near geothermal vents deep within Earth's oceans.

Our efforts to find evidence of life on other bodies within our own Solar System have so far been unsuccessful, but the quest continues. The discovery of life on even one Solar System body beyond Earth would be exciting: if life arose independently *twice* in the same planetary system, then life might exist throughout the universe.

Habitable Zones around Parent Stars

Searching for life within our own Milky Way Galaxy is a daunting task. Nevertheless, astronomers are narrowing the search by searching for planets with environments conducive to the formation and evolution of life (at least, life as we understand it), while eliminating planets that are clearly unsuitable for supporting life. Most of the extrasolar planets discovered so far have environments that are much too harsh to provide a haven for any known life-forms.

Astronomers search for planetary systems that are stable. Planets in stable systems remain in nearly circular orbits that preserve relatively uniform climatological and oceanic environments. Planets in very elliptical orbits can experience wild temperature swings that could be detrimental to the survival of life. A stable temperature that maintains the existence of water in a liquid state might be important. We know that liquid water was essential for the formation and evolution of life on Earth. Of course, we don't know if liquid water is an absolute requirement for life elsewhere, but it's a good starting point when we think about where to look. The region around a star that provides a range of temperatures in which liquid water can exist is called the star's **habitable zone**. On planets that are too close to their parent star, water would exist only as a vapor—if at all. On planets that are too far from their star, water would be permanently frozen as ice. Still another consideration is planet size. Large planets such as Jupiter retain most of their light gases—hydrogen and helium—during formation and so become gas giants without a surface. Planets that are very small may have insufficient surface gravity to retain their atmospheric gases and so end up like our Moon.

In our own Solar System, we can see that Venus, which orbits at 0.7 times Earth's distance from the Sun, has become an inferno because of its runaway greenhouse effect (see Chapter 7). Any liquid water that might once have existed on Venus has long since evaporated and been lost to space. Mars orbits about 1.5 times farther from the Sun than Earth, and all of the water that we see on Mars today is frozen. But the orbit of Mars is more elliptical and variable than Earth's, giving the planet a greater variety of climate, including long-term cycles that might occasionally permit liquid water to exist. Most astrobiologists put the habitable zone of our Solar System at about 0.9–1.4 astronomical units (AU), which includes Earth but just misses Venus and Mars. Yet this range may be too narrow because sources of energy other than sunlight, such as tidal heating, might provide the power necessary for sustaining life. For example, extremophiles could be thriving in liquid water beneath the surfaces of some icy moons of Jupiter and Saturn.

Astronomers must also think about the type of star they are observing in their search for life-supporting planets (**Figure 18.8**). Stars that are less massive than the Sun and thus cooler will have narrower habitable zones, lessening the chance that a habitable planet will just happen to form within that slender zone. Stars that are more massive than the Sun are hotter and will have a larger habitable zone. However, a star's main-sequence lifetime depends on its mass. For example, a star of 2 $M_\odot$ would enjoy relative stability on the main sequence for only about a billion years before the helium flash incinerated everything around it. On Earth, a billion years was long enough for bacterial life to form and cover the planet, but insufficient for anything more advanced to evolve. Of course, we don't know if evolution might happen at a different pace elsewhere; we have only our one terrestrial case as an example. Still, main-sequence lifetime is a sufficiently strong consideration that most astronomers prefer to focus their efforts on stars with longer lifetimes—specifically, spectral types F, G, and K.

Finally, some astronomers think about a "galactic habitable zone," referring to a star's location within the Milky Way Galaxy. Stars that are situated too far from the galactic center may have protoplanetary disks with insufficient quantities of heavy elements—such as oxygen and silicon (silicates), iron, and nickel—

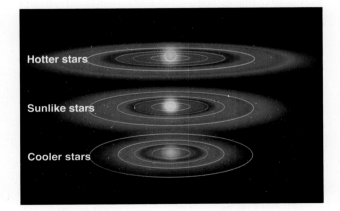

FIGURE 18.8 The distance and extent of a habitable zone (green) surrounding a star depends on the star's temperature—in other words, its mass. Regions too close to the star are too hot (red), and those too far away are too cold (blue).

that make up terrestrial-type planets like Earth. Stars that are too close to the galactic center also face problems. These regions experience less star formation and therefore less recycling of heavy elements. Perhaps more serious is the high-energy radiation environment near the galactic center (X-ray and gamma ray), which is damaging to RNA and DNA. Even so, for many stars the galactic habitable zone may not remain as a permanent home if they tend to migrate within the galaxy and change their distance from the galactic center over time. In short, astronomers try to narrow down the vast numbers of stars as they conduct their search, but they acknowledge that these types of arguments—based on what worked well for planet Earth—might not be applicable when we are looking at other planetary systems.

The search for Earth-like planets from space is already under way. In 2006, the European Space Agency (ESA) launched CoRoT, the first space telescope dedicated to extrasolar planet detection. CoRoT monitors nearby stars for changes in brightness that would indicate the presence of a transiting extrasolar planet (see Chapter 5). The ESA spacecraft has already made several discoveries, including one planet that is only 1.7 times the size of Earth. In 2009, NASA launched the Kepler spacecraft into a solar orbit that trails Earth in its own orbit about the Sun. Staying well behind Earth keeps our planet from getting in Kepler's field of view, thereby enabling uninterrupted monitoring of Kepler's target stars. Over the next several years, the spacecraft's photometer will simultaneously and continually monitor the brightness of more than 100,000 stars, looking for transiting extrasolar planets. Kepler can detect planets as small as 0.8 times the size of Earth. By early 2011, Kepler had discovered more than 1,000 suspected extrasolar planetary systems, 20 percent of which included multiple planets. Based on these data, it appears that about 1 in 15 extrasolar planets may be Earth-sized.

18.3 The Search for Signs of Intelligent Life

Given the right conditions—an Earth-type planet orbiting within a habitable zone around a solar-type star—life of some kind might invariably arise. It took less than a billion years for life to form in Earth's oceans. This earliest life was primitive, and it took another 3.5 billion years for it to climb the evolutionary ladder and arrive at a species with the intelligence and technological capability to contemplate the universe that gave it birth and begin a search through the vastness of space for others like itself. As humans, we have reached that stage at a time when our star is only halfway through its lifetime. Such would not be the case for life arising on planets orbiting stars much more massive than the Sun. A 3 $M_\odot$ star has a main-sequence lifetime of only 600 million years. Our terrestrial experience suggests that just as life was getting its start on one of the star's habitable planets, it would quickly be doomed as the star reached the end of its brief period of stability. A star 1½ times as massive as the Sun has a main-sequence lifetime of a couple of billion years, likely insufficient for any life to reach the equivalent of Earth's Cambrian explosion. If we are to search for signs of intelligent life, we should look to planets surrounding stars of about a solar mass or smaller. And now that we have some ideas about where to search for intelligent life, how do we establish communication?

During the 1970s, humans made preliminary efforts. The *Pioneer 10* and *Pioneer 11* spacecraft, which will probably spend eternity drifting through interstellar space, each carry the plaque shown in **Figure 18.9**. It describes

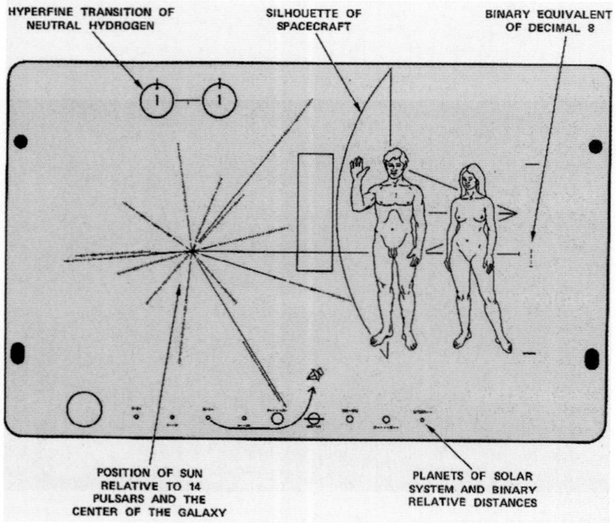

FIGURE 18.9 The plaque included with the *Pioneer 10* and *Pioneer 11* probes, which were launched in the early 1970s and will eventually leave the Solar System to travel through the millennia in interstellar space.

ourselves and our location to any future interstellar traveler who might happen to find it. Another message to the cosmos accompanied the two *Voyager* spacecraft on identical phonograph records that contain greetings from planet Earth in 60 languages, samples of music, animal sounds, and a message from then-President Jimmy Carter. Interestingly, these messages created concern among some politicians and nonscientists who felt that scientists were dangerously advertising our location in the galaxy, even though radio signals had been broadcast into space for nearly 80 years at the time. The messages were also criticized by some philosophers, who claimed that we were making ridiculous anthropomorphic assumptions about the aliens being sufficiently like us to decode these messages. (Ironically, most Earthlings could not play those phonograph records today.)

Sending messages on spacecraft may not be the most efficient way to make contact with the universe, but it was a significant gesture nonetheless. A somewhat more practical effort was made in 1974, when astronomers used the 300-meter-wide dish of the Arecibo radio telescope to beam a message (**Figure 18.10**) toward the star cluster M13. (If someone from M13 answers, we will know in about 50,000 years.)

The Drake Equation

The first serious effort to search for intelligent extraterrestrial life was made by astronomer Frank Drake (b. 1930) in 1960. Drake used what was then astronomy's most powerful radio telescope to listen for signals from two nearby stars. Although his search revealed nothing unusual, it prompted him to develop an equation that bears his name. This equation is an excellent way to organize our thoughts about whether intelligent life might exist elsewhere, because it includes the things we need to know to even make a guess about how many intelligent civilizations there might be. The **Drake equation** estimates the number (N) of intelligent, communicating civilizations that may exist within the Milky Way Galaxy:

$$N = R^* \times f_p \times n_e \times f_l \times f_i \times f_c \times L$$

The seven factors on the right side of the equation are the conditions that Drake thought must be met for a civilization to exist:

1. R^* is the number of stars suitable for the development of intelligent life that form in our galaxy each year. This is roughly 7 per year and is the least contentious of all these terms.
2. f_p is the fraction of those stars that form planetary systems. Questions remain, but we do know that planets form as a natural by-product of star formation and that many—perhaps most—stars have planets. We will assume that f_p is between 0.5 and 1.
3. n_e is the number of worlds, per stellar system, with an environment suitable for life. If we look at the Solar System, we might decide this number is about 2. Earth definitely did develop life, and Mars might have. Of course, the Solar System is just one example, and we don't really know what this term should be.
4. f_l is the fraction of suitable worlds on which life actually arises. Remember that just a single self-replicating molecule may be enough to get the ball rolling. Many biochemists now believe that if the right chemical and environ-

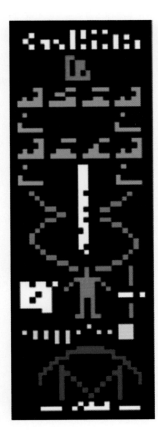

FIGURE 18.10 The message we beamed toward the star cluster M13 in 1974. This binary-encoded message contains the numbers 1–10, hydrogen and carbon atoms, some interesting molecules, DNA, a human with description, basics of our Solar System, and basics of the Arecibo telescope. A reply may be forthcoming in about 50,000 years.

mental conditions are present, then life *will* develop. If they are correct, then f_l is close to 1. For f_l we will use a range of 0.1 to 1.

5. f_i is the fraction of those planets harboring life that eventually develop intelligent life. Intelligence is certainly the kind of survival trait that might often be strongly favored by natural selection. On the other hand, on Earth it took about 4 billion years—roughly half the expected lifetime of our star—to evolve tool-building intelligence. The correct value for f_i might be close to 0.01, or it might be closer to 1. The truth is, we just don't know.

6. f_c is the fraction of intelligent life-forms that develop technologically advanced civilizations—that is, civilizations that send communications into space. With only one example of a technological civilization to work with, f_c is hard to estimate. We will consider estimates for f_c between 0.1 and 1.

7. L is the number of years that such civilizations exist. This factor is certainly the most difficult of all to estimate because it depends on the long-term stability of advanced civilizations. We have had a technological civilization on Earth for about 100 years, and during that time we have developed, deployed, and used weapons with the potential to eradicate our civilization and render Earth hostile to life for many years to come. We have also so degraded our planet's ecosystem that many respectable biologists and climatologists wonder if we are nearing the brink. Do all technological civilizations destroy themselves within a thousand years? Conversely, if most technological civilizations learn to use their technology for survival rather than self-destruction, might they instead survive for a million years? For L in our calculation, we will use a range between 1,000 years and 1 million years.

As illustrated in **Figure 18.11**, the conclusions we draw using the Drake equation depend a great deal on the assumptions we make. For the most pessimistic of our estimates, the Drake equation sets the number of technological civilizations in our galaxy at about 1. If this is correct, then we are the *only* technological

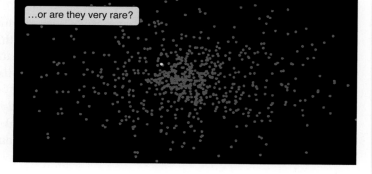

FIGURE 18.11 Two estimates of the existence of intelligent civilizations in our galaxy based on the Drake equation. Notice how widely these estimates vary, given optimistic and pessimistic assumptions about the seven factors (see the text) affecting the prevalence of intelligent extraterrestrial life. White dots show stars with possible civilizations.

civilization in the Milky Way. (Because we are here, we know that *N* must be at least one.) Such a universe would still be full of intelligent life. With a hundred billion galaxies in the observable universe, even these pessimistic assumptions would mean that a billion technological civilizations exist out there somewhere. On the other hand, we would have to go a *very* long way (10 million parsecs or so) to find our nearest neighbors.

At the other extreme, what if we take the most optimistic view, assuming that intelligent life arises and survives everywhere it gets the chance? The Drake equation then says that there should be roughly 15 million technological civilizations in our galaxy alone. In this case, our nearest neighbors may be "only" 40 or 50 light-years away. If scientists in that civilization are listening to the universe with their own radio telescopes, hoping to answer the question of other life in the universe for themselves, then as you read this page they may be puzzling over an episode of the original *Star Trek* television series that began in 1966.

If we did run across another technologically advanced civilization, what would it be like? Looking back at the Drake equation, we can see it is highly unlikely that we have neighbors nearby, unless civilizations typically live for many thousands or even millions of years. In this case, any civilization that we encountered would almost certainly have been around for much longer than we have. Having survived that long, would its members have learned the value of peace, or would they have developed strategies for controlling their pesky neighbors? Science-fiction books and movies are filled with amusing and thoughtful stories that explore what life in the universe might be like (**Figure 18.12**). For the moment, though, let's set aside such speculation to look at the question as scientists. How might we go about finding a real answer?

FIGURE 18.12 The classic 1951 film *The Day the Earth Stood Still* portrayed intelligent extraterrestrials who threatened to destroy Earth if we carried our violent ways into space.

Technologically Advanced Civilizations

During lunch with colleagues, the famous physicist Enrico Fermi (1901–1954), a firm believer in extraterrestrial beings, is reported to have asked, "If the universe is teeming with aliens . . . where is everybody?" Fermi's question—first posed in 1950 and sometimes called the "Fermi paradox"—remains unanswered. Consider also the following closely related question: If intelligent life-forms are common but interstellar travel is difficult or impossible, why don't we detect their radio transmissions? We have failed so far to detect any alien radio signals—but it is not for lack of trying.

Drake's original project of listening for radio signals from intelligent life around two nearby stars has grown over the years into a much more elaborate program that is referred to as the Search for Extraterrestrial Intelligence, or **SETI**. Scientists from around the world have thought carefully about what strategies might be useful for finding life in the universe. Most of these have focused on the idea of using radio telescopes to listen for signals from space that bear an unambiguous signature of an intelligent source. Some have listened intently at certain key frequencies, such as the frequency of the interstellar 21-centimeter line from hydrogen gas. The assumption behind this approach is that if a civilization wanted to be heard, its denizens would tune their broadcasts to a channel that astronomers throughout the galaxy should be listening to. More recent searches have made use of advances in technology to record as broad a range of radio signals from space as possible. Analysts then use computers to search these databases for patterns that might suggest the signals are intelligent in origin.

Unlike much astronomical research, SETI receives its funding from private rather than government sources, and SETI researchers have found ingenious

SETI scientists search for intelligent, communicating civilizations in the Milky Way.

FIGURE 18.13 When complete, the Allen Telescope Array will listen for evidence of intelligent life from as many as a million stellar systems.

ways to continue the search for extraterrestrial civilizations with limited resources. One project, known as SETI@home, involves the use of hundreds of thousands of personal computers around the world to analyze the institute's data. SETI screen saver programs installed on personal computers worldwide download radio observations from the SETI Institute over the Internet, analyze these data while the computers' owners are off living their lives, and then report the results of their searches back to the Institute. Perhaps the first sign of intelligent life in the universe might be found by a computer sitting on your desk.

SETI's Allen Telescope Array (ATA), named for Microsoft cofounder Paul Allen, who provided much of the initial financing for the project, uses a similar low-cost approach. A joint venture between the SETI Institute and the University of California, the ATA consists of a "farm" of small, inexpensive radio dishes like those used to capture signals from orbiting communication satellites (**Figure 18.13**). The ATA can search the sky 24 hours a day, 7 days a week, for signs of intelligent life. Unfortunately, funding for ATA ran out in April 2011, but SETI scientists are seeking other funding to continue the project. If they are successful in their efforts, eventually there will be 350 dishes, each with a diameter of 6.1 meters and collectively yielding a total signal-receiving area greater than that of a 100-meter radio telescope. Just as your brain can sort out sounds coming from different directions, this array of radio telescopes will be able to determine the direction a signal is coming from, allowing it to listen to many stars at the same time. Over several years' time, astronomers using the ATA are expected to survey as many as a million stars, hoping to find a civilization that has sent a signal in our direction. If reality is anything like the more optimistic of the assumptions we used in evaluating the Drake equation, this project will stand a good chance of success.

Finding even *one* nearby civilization in our galaxy—that is, a *second* technological civilization in our own small corner of the universe—could mean that the universe as a whole is literally teeming with intelligent life. SETI may not be in the mainstream of astronomy, and the likelihood of its success is difficult to predict, but its potential payoff is enormous. Few discoveries would do more to change our understanding of ourselves than certain knowledge that we are not alone.

Science fiction is filled with tales of humans who leave Earth to "seek out new life and new civilizations." Unfortunately, these scenarios are not realistic. The distances to the stars and their planets are enormous; to explore a significant sample of stars would require extending the physical search over tens or hundreds of light-years. As we learned in Chapter 13, special relativity limits how fast we can travel. The speed of light is the limit, and even at that rate it would take over 4 years to get to the *nearest* star. Time dilation would favor the astronauts themselves, and they would return to Earth younger than if they had stayed at home. But suppose they visited a star 15 light-years distant. Even if they traveled at speeds close to that of light, by the time they returned to Earth, 30 years would have passed. Some science fiction enthusiasts get around this problem by invoking "warp speed" or "hyperdrive" to travel faster than the speed of light, or by using wormholes as shortcuts across the galaxy—but there is absolutely no evidence that any of these shortcuts are possible.

Life as We Do Not Know It

So far, we have confined our discussion to the search for life as we know it, as it may exist now. We do this for sound scientific reasons, primarily having to do with falsifiability. Very different forms of life, or life in the distant past or future, are purely speculative and nonscientific. This does not mean these ideas are uninteresting. They are just not falsifiable, and therefore not science.

For example, imagine for a moment that intelligent life evolved amid the swarm of exotic particles that filled the universe shortly after the Big Bang. Such creatures would have been far smaller than today's atoms and perhaps would have lived out their lives in 10^{-40} second or so. To those creatures the universe—all 3 meters of it—would have seemed incomprehensibly vast. These creatures and their entire civilization would have had to evolve, live, and die out in a tiny fraction of the time that it takes for a single synapse in our brains to fire. If such creatures ever calculated the conditions in *today's* universe, they would have imagined a frozen time when the temperature of the universe was only a thousandth of a trillionth of a trillionth of what they knew—a time when most matter had ceased to exist altogether and the tiny fraction that remained was spread out over a volume of space 10^{75} times greater than that of their universe. In short, such creatures would have looked forward and seen *our* universe as the frozen, desolate future. It seems doubtful they could have foreseen the existence of stars, galaxies, planets, and intelligent creatures for whom a single thought took longer than the entire history of the universe they knew.

Now turn the tables, and think *forward* to a time when the universe is 10^{50} times older than it is today. Who can say that there will not be life then? Perhaps the life-forms of this distant future will be spread across countless trillions of light-years of space, with lives that unfold over untold eons of time. For such organisms, if they ever exist, it will be *we* who are the impossible creatures, alive for the briefest of instants, still immersed in the momentary fireball of the Big Bang.

Ideas like these are interesting and fun to contemplate. They bring perspective, and they sometimes inspire questions that lead to scientific inquiry. But they are not, of themselves, scientific, no matter how scientific they may sound. Recall Chapter 1, where we defined scientific ideas as being first and foremost falsifiable. There are no tests we can apply to prove that such creatures did not (or will not) exist; therefore, the idea of them is not a scientific hypothesis.

Now consider more common examples. Some people claim that aliens have already visited us: tabloid newspapers, books, and websites are filled with tales of UFO sightings, government conspiracies and cover-ups, alleged alien abductions, and UFO religious cults. However, none of these reports meet the basic standards of science. They are not falsifiable—they lack verifiable evidence and repeatability—and we must conclude that there is no scientific evidence for any alien visitations.

When you encounter discussions of past, future, or very different alien life-forms, it is useful to ask whether the ideas are falsifiable, and whether the evidence is verifiable and repeatable. If not, the discussions are speculation, not science.

| *18.4* The Fate of Life on Earth

In this book, we have used our understanding of physics and cosmology to look back through time and watch as structure formed throughout the universe. We have peered into the future and seen the ultimate fate of the universe, from its enormous clusters of galaxies down to its tiniest components. Now we examine our own destiny and contemplate the fate that awaits Earth, humanity, and the star on which all life depends.

Eventually, the Sun Must Die

About 5 billion years from now, the Sun will end its long period of relative stability. Shedding its identity as the passive star that has nurtured life on Earth, the Sun will expand to become a red giant and later an asymptotic giant branch (AGB) star, swelling to hundreds of times its present size. The giant planets, orbiting outside the extended red giant atmosphere, may survive in some form. Even so, they will suffer the blistering radiation from a Sun grown thousands of times more luminous than it is today.

The terrestrial planets will not fare as well. Some—perhaps all—of the worlds of the inner Solar System will be engulfed by the expanding Sun. Just as an artificial satellite is slowed by drag in Earth's tenuous outer atmosphere and eventually falls to the ground in a dazzling streak of white-hot light, so, too, will a terrestrial planet caught in the Sun's atmosphere be consumed by the burgeoning star. If this is Earth's fate, our home world will leave no trace other than a slight increase in the amount of massive elements in the Sun's atmosphere. As the Sun loses more and more of its atmosphere in an AGB wind, our atoms may be expelled back into the reaches of interstellar space from which they came, perhaps to become incorporated into new generations of stars, planets, and even life itself.

Another planetary fate is possible, however. In this scenario, as the red giant Sun loses mass in a powerful wind, its gravitational grasp on the planets will weaken, and the orbits of both the inner and outer planets will spiral outward. If Earth moves out far enough, it may survive as a seared cinder, orbiting the white dwarf that the Sun will become. Barely larger than Earth and with its nuclear fuel exhausted, the white dwarf Sun will slowly cool, eventually becoming a cold, inert sphere of degenerate carbon, orbited by what remains of its retinue of planets. Thus the ultimate outcome for our Earth—consumed in the heart of the Sun or left behind as a frigid, burned rock orbiting a long-dead white dwarf—is not yet certain.

> Far-future Earth will be consumed by the Sun or left as an icy cinder.

The Future of Life on Earth

Life on our planet has even less of a future, for it will not survive long enough to witness the Sun's departure from the main sequence. Well before that cataclysmic event takes place, the Sun's luminosity will begin to rise. As solar luminosity increases, so will temperatures on all the planets, including our own. Eventually Earth's temperatures will climb so high that all animal and plant life will perish. Even the extremophiles that inhabit the oceanic depths will die as the oceans boil away. Models of the Sun's evolution are still too imprecise to predict with certainty when that fatal event will occur, but the end of all terrestrial life may be only 1 billion or 2 billion years away. That is, of course, a comfortably distant time from now, but it is well short of the Sun's departure from the main sequence.

It is far from certain, however, that the descendants of today's humanity will even be around a billion years from now. Some of the threats that await us come from beyond Earth. For the remainder of the Sun's life, the terrestrial planets, including Earth, will continue to be bombarded by asteroids and comets. Perhaps a hundred or more of these impacts will involve kilometer-sized objects, capable of causing the kind of devastation that eradicated the dinosaurs and most other species 65 million years ago. Although these events may create new surface scars, they will have little effect on the integrity of Earth itself. Earth's geological record is filled with such events, and each time they happen, life manages to recover and reorganize.

It seems likely, then, that some form of life will survive to see the Sun begin its march toward instability. However, individual species do not necessarily fare so well when faced with cosmic cataclysm. If the descendants of humankind survive, it will be because we chose to become players in the game by changing the odds of such planetwide biological upheavals. We are rapidly developing technology that could enable us to detect most threatening asteroids and modify their orbits well before they could strike Earth. Comets are more difficult to guard against because long-period comets appear from the outer Solar System with little warning. To offer protection, various means of defending ourselves would have to be in place, ready to be used on very short notice. So far, we have been slow to take such threats seriously. Although impacts from kilometer-sized objects are infrequent, objects only a few dozen meters in size, carrying the punch of a several-megaton bomb, strike Earth about once every 100 years.

We might protect ourselves from the fate of the dinosaurs, but in the long run the descendants of humanity will either leave this world or die out. Planetary systems surround other stars, and all that we know tells us that many other Earth-like planets should exist throughout our galaxy. Colonizing other planets is currently the stuff of science fiction, but if our descendants are ultimately to survive the death of our home planet, off-Earth colonization must become science fact at some point in the future.

Although humankind may soon be capable of protecting Earth from life-threatening comet and asteroid impacts, in other ways we are our own worst enemy. We are poisoning the atmosphere, the water, and the land that form the habitat for all terrestrial life. As our population grows unchecked, we are occupying more and more of Earth's land and consuming more and more of its resources while sending thousands of species of plants and animals to their extinction each year. At the same time, human activities are dramatically affecting the balances of atmospheric gases. The climate and ecosystem of Earth constitute a finely balanced, complex system capable of exhibiting chaotic behavior. The fossil record shows that Earth has undergone sudden and dramatic climatic changes in response to even minor perturbations. Such drastic changes in the overall balance of nature would certainly have consequences for our own survival. For the first time in human history, we possess the means to unleash nuclear or biological disasters that could threaten the very survival of our species. In the end, the fate of humanity will depend more than anything on whether we accept stewardship of ourselves and of our planet.

Figure 18.14 is the very famous "pale blue dot" picture, taken by the *Voyager 1* spacecraft from beyond the orbit of Neptune at a distance of more than 40 AU. The beams in the picture are sunlight scattered off the spacecraft. The arrow points to a dot, which is Earth—the only place in the entire universe where life is confirmed to exist. Compare the size of that dot to the size of the universe. Compare the history of life on Earth to the history of the universe. Compare Earth's future with the fate of the universe. Astronomy is humbling. We are so incredibly tiny. And yet we are so incredibly unique, as far as we know. Think for a moment about what that means to you. This may be the most important lesson the universe has to teach us.

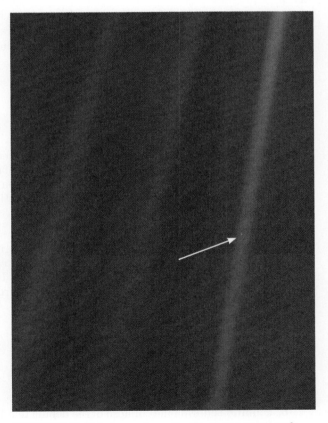

FIGURE 18.14 This image from the *Voyager 1* spacecraft shows the Earth from a distance of 3.8 billion miles, well past the orbit of Neptune. The streaks in this image are scattered sunlight. The "pale blue dot" in the rightmost streak is Earth.

READING ASTRONOMY News

Scientist: SETI Should Switch "Channels"

By **UPI.com**

The SETI Institute, listening to the cosmos for signs of signals from alien civilizations, may be monitoring the wrong "channels," a U.S. astrophysicist says.

Gregory Benford of the University of California, Irvine, says such a civilization wanting to announce its presence would transmit "cost-optimized" narrowly focused signals, not the continuous omni-directional signals the SETI program has been scanning for, a university release said Wednesday.

"This approach is more like Twitter and less like War and Peace," James Benford,

Gregory Benford's twin and fellow physicist, says.

Such short, targeted blips, dubbed Benford beacons, should be the targets of SETI efforts, a growing number of scientists say.

And the Benfords also suggest concentrating the search on our own Milky Way galaxy, especially its center where 90 percent of its stars are located.

"The stars there are a billion years older than our sun," Gregory Benford says, "which suggests a greater possibility of contact with an advanced civilization than does pointing SETI receivers outward to the newer and less crowded edge of our galaxy."

Evaluating the News

1. What is Dr. Gregory Benford's primary critique of current SETI efforts?
2. According to the article, is this critique backed up by any scientific studies? How might you find out if there was actually science behind this critique?
3. Does the article describe an "idea" or a "theory"? Explain.
4. Do you agree with the concluding statement that SETI should search toward the center of the galaxy? Why might this be a good idea? Why might it be a bad one?

SUMMARY

18.1 Life is basically a form of complex carbon-based chemistry, made possible by special molecules capable of reproducing themselves. Life likely formed in Earth's oceans, evolving chemically from an organic "soup" of prebiotic molecules into self-replicating organisms.

18.2 All terrestrial life is composed mostly of only six elements: carbon, hydrogen, oxygen, nitrogen, sulfur, and phosphorus. Life-forms that are very different from ours, including those based on silicon chemistry, cannot be ruled out. Space-based instruments have begun the search for Earth-

like planets. Astronomers prefer to focus their search for extraterrestrial life on Earth-like planets orbiting in habitable zones surrounding solar-type stars.

18.3 The Drake equation helps us organize our estimates and assumptions determining the likelihood of intelligent life in the Milky Way Galaxy. Although our galaxy may be teeming with intelligent life, none has yet been detected.

18.4 Long before the Sun ends its period of stability on the main sequence, all terrestrial life will have perished after an increasingly luminous Sun makes Earth uninhabitable.

SUMMARY SELF-TEST

1. The term *primordial soup* refers to
   a. conditions just after the Big Bang.
   b. conditions during the collapse of the nebula that formed the Sun.
   c. conditions just after the formation of Earth.
   d. conditions in early Earth's oceans.

2. The Urey-Miller experiment produced _____ in a laboratory jar.
   a. life
   b. RNA and DNA
   c. amino acids
   d. proteins

3. Scientists think that terrestrial life probably originated in Earth's oceans because
   a. all the chemical pieces were there.
   b. energy was available there.
   c. earliest evidence for life on Earth comes in ocean-dwelling forms.
   d. all of the above.

4. *Natural selection* means that
   a. poorly adapted forms of life die.
   b. well-adapted forms of life are selected and encouraged.
   c. forms of life are designed for their environment.
   d. environments are designed for the life-forms that inhabit them.

5. Any system with heredity, mutation, and natural selection
   a. will change over time.
   b. will improve over time.
   c. will decay over time.
   d. will develop intelligence.

6. The fact that we have not detected alien civilizations yet tells us that
   a. they are not there.
   b. they are rare.
   c. we are in a "blackout," and they are not talking to us.
   d. we don't know enough yet to draw any conclusions.

7. All life on Earth *must* eventually come to an end because
   a. humans will make the planet uninhabitable.
   b. asteroids will make the planet uninhabitable.
   c. new species will arise that we won't recognize as life.
   d. the sun will make the planet uninhabitable.

QUESTIONS AND PROBLEMS

True/False and Multiple-Choice Questions

8. **T/F:** All life on Earth branches from the same evolutionary tree.

9. **T/F:** All that is required for life to begin is a solitary, self-replicating molecule.

10. **T/F:** Astronomers do not consider the possibility of non-carbon-based life.

11. **T/F:** There are several places other than Earth where life might exist in our Solar System.

12. **T/F:** It is likely that we will discover many civilizations that are not advanced.

13. **T/F:** Today, we can use the Drake equation to find the actual number of intelligent communicating civilizations in the galaxy.

14. **T/F:** It is impossible that humans will exist on Earth 7 billion years from now.

15. The difference between prokaryotes and eukaryotes is that
    a. prokaryotes have no DNA.
    b. prokaryotes have no cell wall.
    c. prokaryotes have no nucleus.
    d. prokaryotes do not exist today.

16. In the phrase "theory of evolution," the word *theory* means that evolution
    a. is an idea that can't be tested scientifically.
    b. is an educated guess to explain natural phenomena.
    c. probably doesn't happen anymore.
    d. is a well-tested, well-corroborated scientific explanation of natural phenomena.

17. Mutations are
    a. changes to a life-form's DNA.
    b. always deadly for the life-form.
    c. always deadly to the species.
    d. always beneficial to the life-form.
    e. always beneficial to the species.

18. The habitable zone is the place around a star where
    a. life has been found.
    b. atmospheres can contain oxygen.
    c. liquid water exists.
    d. liquid water can exist on the surface of a planet.

19. The length of time an intelligent, communicating civilization lasts affects _____ in the Drake equation.
    a. the value of R^*
    b. the value of f_i
    c. the value of f_c
    d. the value of L

20. The most critical threat to the survival of humanity is
    a. comets.
    b. asteroids.
    c. increasing solar luminosity.
    d. humanity itself.

Conceptual Questions

21. Why do we generally talk about molecules such as DNA or RNA when discussing life on Earth?

22. Provide a definition for *life* that does not rely on RNA or DNA. Is this definition sufficient to account for new forms that may be discovered in the future, either on Earth or elsewhere?

23. How do we suspect that the building blocks of DNA first formed on Earth?

24. When do we believe life first appeared on Earth? What evidence supports this belief?

25. Today, most known life enjoys moderate climates and temperatures. Compare this environment to some of the conditions in which early life developed.

26. Discuss some extremophiles living on Earth today.

27. How was Earth's carbon dioxide atmosphere changed into today's oxygen-rich atmosphere? How long did that transformation take?

28. What are the similarities and differences between prokaryotes and eukaryotes?

29. Tracing back on the evolutionary tree, what kinds of life do we find humans are similar to, and when did the evolutionary steps leading from those kinds of life to humans happen?

30. What was the Cambrian explosion, and what might have caused it?

31. Why do you suppose plants and forests appeared in large numbers before large animals did?

32. Why was evolution inevitable on Earth?

33. What were the general conditions needed on our planet for life to arise?

34. Is biological evolution under way on Earth today? If so, how might humans continue to evolve?

35. Where did all the atoms in your body come from?

36. The *Viking* spacecraft did not find evidence of life on Mars when it visited that planet in the late 1970s, nor did the *Phoenix* lander when it examined the martian soil in 2009. Does this imply that life never existed on the planet? Why or why not?

37. What is a habitable zone? What defines its boundaries?

38. In searching for intelligent life elsewhere, why is listening with radio telescopes currently our favored method?

39. Why is it likely that life on Earth as we know it will end long before the Sun runs out of nuclear fuel?

40. Most life on Earth today relies directly or indirectly on the Sun. For example, the food chain of life near the ocean's surface relies on algae, which use photosynthesis. Organisms living on ocean floors generally feed from surface material that sank, so these bottom dwellers rely indirectly on the Sun. Are there any forms of life today that do not rely in any way on solar energy for life? If so, what does this imply about the prospects of life in remote places in our Solar System?

41. Suppose the early Earth had oceans of ammonia (NH_3) rather than water (H_2O). Could life have evolved in these conditions? Why or why not? What if Earth had been totally dry—that is, with no liquid of any kind? Explain how this may have affected the formation of terrestrial life.

42. Knowing the history and composition of Mars, speculate on whether life evolved there before the planet became geologically dead. If so, how complex did the life become?

43. Describe the properties of a planet, and its star, that would make it more probable for life to have evolved there.

44. Using your textbook (see Appendix 3) or the Internet, suggest a few nearby stars that could be potential candidates for life-bearing planets. Explain why you selected each one.

45. Why do you think we sent a coded radio signal to the globular cluster M13 in 1974—rather than, say, to a nearby star?

Problems

46. If the chance that a given molecule will make a copying error is one in 100,000, how many generations are needed before, on average, at least one mutation has occurred?

47. Consider an organism Beta that, because of a genetic mutation, has a 5 percent greater probability of survival than its nonmutated form, Alpha. Alpha has only a 95 percent probability (p_r) of reproducing itself compared to Beta. After n generations, Alpha's population within the species would be $S_p = (p_r)^n$ compared to Beta's. Calculate Alpha's relative population after 100 generations. (Note: You may need a scientific calculator or help from your instructor to evaluate the quantity 0.95^{100}.)

48. To fully appreciate the power of heredity, mutation, and natural selection, consider Alpha's relative population (from Problem 47) after 5,000 generations if Beta has a mere 0.1 percent survivability advantage over Alpha.

49. Use the Drake equation to calculate your own "most likely" number of intelligent civilizations. To do this, choose numbers for each parameter that you consider most likely.

50. Use the Drake equation to calculate your own "least likely" number of intelligent civilizations. To do this, choose numbers for each parameter that you consider least likely. If you calculate a number less than one, you know that you have been too pessimistic. Why?

 SmartWork, Norton's online homework system, includes algorithmically generated versions of these questions, plus additional conceptual exercises. If your instructor assigns questions in SmartWork, log in at **smartwork.wwnorton.com**.

 StudySpace is a free and open website that provides a Study Plan for each chapter of **Understanding Our Universe**. Study Plans include animations, reading outlines, vocabulary flashcards, and multiple-choice quizzes, plus links to premium content in SmartWork and the ebook. Visit **wwnorton.com/ studyspace**.

Exploration | Fermi Problems and the Drake Equation

The Drake equation is a way of organizing our thoughts about whether there might be other intelligent, communicating civilizations in the galaxy. This type of thinking is very useful for getting estimates of a value, particularly when analyzing systems for which counting is not actually possible. The types of problems that can be solved in this way are often called Fermi Problems after Enrico Fermi, who was mentioned in this chapter. For example, we might ask, "What is the circumference of Earth?"

You could Google this, or you could already "know" it, or you might look it up in this textbook. Alternatively, you could very carefully measure the shadow of a stick in two locations at the same time on the same day. Or you could drive around the planet. Or you could reason this way:

How many time zones are between New York and Los Angeles?

3 time zones. You know this from traveling, or from television.

How many miles is it from New York to Los Angeles?

3,000 miles. You know this from traveling, or living in the world.

So, how many miles per time zone?

3,000/3 = 1,000

How many time zones in the world?

24, because there are 24 hours in a day, and each time zone marks an hour.

So what is the circumference of the Earth?

24,000 miles, because there are 24 time zones, each 1,000 miles wide.

The measured circumference is 24,900 miles, which agrees to within 4 percent.

Following are several Fermi Problems. Time yourself for an hour, and work as many of them as possible. (You don't have to do them in order!)

1. **How much has the mass of the human population on Earth increased in the last year?**

2. **How much energy does a horse consume in its lifetime?**

3. **How many pounds of potatoes are consumed in the United States annually?**

4. **How many cells are there in the human body?**

5. **If your life earnings were given to you by the hour, how much is your time worth per hour?**

6. **What is the weight of solid garbage thrown out by American families each year?**

7. **How fast does human hair grow (in feet per hour)?**

8. **If all the people on Earth were crowded together, how much area would we cover?**

9. **How many people could fit on Earth if every person occupied 1 square meter of land?**

10. **How much carbon dioxide (CO_2) does an automobile emit each year?**

11. **What is the mass of Earth?**

12. **What is the average annual cost of an automobile including overhead (maintenance, looking for parking, cleaning, etc.)?**

13. **How much ink was used printing all the newspapers in the United States today?**

PERIODIC TABLE OF THE ELEMENTS

Key:
- 1 — Atomic number
- H — Symbol
- Hydrogen — Name
- 1.00794 — Average atomic mass
- Metals
- Metalloids
- Nonmetals

1 / 1A	2 / 2A	3 / 3B	4 / 4B	5 / 5B	6 / 6B	7 / 7B	8 (8B)	9 (8B)	10 (8B)	11 / 1B	12 / 2B	13 / 3A	14 / 4A	15 / 5A	16 / 6A	17 / 7A	18 / 8A
1 H Hydrogen 1.00794																	2 He Helium 4.002602
3 Li Lithium 6.941	4 Be Beryllium 9.012182											5 B Boron 10.811	6 C Carbon 12.0107	7 N Nitrogen 14.0067	8 O Oxygen 15.9994	9 F Fluorine 18.9984032	10 Ne Neon 20.1797
11 Na Sodium 22.98976928	12 Mg Magnesium 24.3050											13 Al Aluminum 26.9815386	14 Si Silicon 28.0855	15 P Phosphorus 30.973762	16 S Sulfur 32.065	17 Cl Chlorine 35.453	18 Ar Argon 39.948
19 K Potassium 39.0983	20 Ca Calcium 40.078	21 Sc Scandium 44.955912	22 Ti Titanium 47.867	23 V Vanadium 50.9415	24 Cr Chromium 51.9961	25 Mn Manganese 54.938045	26 Fe Iron 55.845	27 Co Cobalt 58.933195	28 Ni Nickel 58.6934	29 Cu Copper 63.546	30 Zn Zinc 65.38	31 Ga Gallium 69.723	32 Ge Germanium 72.64	33 As Arsenic 74.92160	34 Se Selenium 78.96	35 Br Bromine 79.904	36 Kr Krypton 83.798
37 Rb Rubidium 85.4678	38 Sr Strontium 87.62	39 Y Yttrium 88.90585	40 Zr Zirconium 91.224	41 Nb Niobium 92.90638	42 Mo Molybdenum 95.96	43 Tc Technetium [98]	44 Ru Ruthenium 101.07	45 Rh Rhodium 102.90550	46 Pd Palladium 106.42	47 Ag Silver 107.8682	48 Cd Cadmium 112.411	49 In Indium 114.818	50 Sn Tin 118.710	51 Sb Antimony 121.760	52 Te Tellurium 127.60	53 I Iodine 126.90447	54 Xe Xenon 131.293
55 Cs Cesium 132.9054519	56 Ba Barium 137.327	57 La Lanthanum 138.90547	72 Hf Hafnium 178.49	73 Ta Tantalum 180.94788	74 W Tungsten 183.84	75 Re Rhenium 186.207	76 Os Osmium 190.23	77 Ir Iridium 192.217	78 Pt Platinum 195.084	79 Au Gold 196.966569	80 Hg Mercury 200.59	81 Tl Thallium 204.3833	82 Pb Lead 207.2	83 Bi Bismuth 208.98040	84 Po Polonium [209]	85 At Astatine [210]	86 Rn Radon [222]
87 Fr Francium [223]	88 Ra Radium [226]	89 Ac Actinium [227]	104 Rf Rutherfordium [261]	105 Db Dubnium [262]	106 Sg Seaborgium [266]	107 Bh Bohrium [264]	108 Hs Hassium [277]	109 Mt Meitnerium [268]	110 Ds Darmstadtium [271]	111 Rg Roentgenium [272]	112 Cn Copernicium [285]	114 Uuq Ununquadium [289]		116 Uuh Ununhexium [292]			

6 Lanthanides

58 Ce Cerium 140.116	59 Pr Praseodymium 140.90765	60 Nd Neodymium 144.242	61 Pm Promethium [145]	62 Sm Samarium 150.36	63 Eu Europium 151.964	64 Gd Gadolinium 157.25	65 Tb Terbium 158.92535	66 Dy Dysprosium 162.500	67 Ho Holmium 164.93032	68 Er Erbium 167.259	69 Tm Thulium 168.93421	70 Yb Ytterbium 173.05	71 Lu Lutetium 174.967

7 Actinides

90 Th Thorium 232.03806	91 Pa Protactinium 231.03588	92 U Uranium 238.02891	93 Np Neptunium [237]	94 Pu Plutonium [244]	95 Am Americium [243]	96 Cm Curium [247]	97 Bk Berkelium [247]	98 Cf Californium [251]	99 Es Einsteinium [252]	100 Fm Fermium [257]	101 Md Mendelevium [258]	102 No Nobelium [259]	103 Lr Lawrencium [262]

We have used the U.S. system as well as the system recommended by the International Union of Pure and Applied Chemistry (IUPAC) to label the groups in this periodic table. The system used in the United States includes a letter and a number (1A, 2A, 3B, 4B, etc.), which is close to the system developed by Mendeleev. The IUPAC system uses numbers 1–18 and has been recommended by the American Chemical Society (ACS). While we show both numbering systems here, we use the IUPAC system exclusively in the book. Elements with atomic numbers higher than 112 have been reported but not yet fully authenticated.

PROPERTIES OF PLANETS, DWARF PLANETS, AND MOONS

Physical Data for Planets and Dwarf Planets

PLANET	EQUATORIAL RADIUS (km)	EQUATORIAL RADIUS ($R/R_\oplus$)	MASS (kg)	MASS ($M/M_\oplus$)	AVERAGE DENSITY (RELATIVE TO WATER*)	ROTATION PERIOD (DAYS)	TILT OF ROTATION AXIS (DEGREES, RELATIVE TO ORBIT)	EQUATORIAL SURFACE GRAVITY (RELATIVE TO EARTH[†])	ESCAPE VELOCITY (km/s)	AVERAGE SURFACE TEMPERATURE (K)
Mercury	2,440	0.383	3.30×10^{23}	0.055	5.427	58.65	0.01	0.378	4.3	340 (100, 725)[§]
Venus	6,052	0.949	4.87×10^{24}	0.815	5.243	243.02[‡]	177.36	0.907	10.36	737
Earth	6,378	1.000	5.97×10^{24}	1.000	5.513	1.000	23.44	1.000	11.19	288 (183, 331)[§]
Mars	3,397	0.533	6.42×10^{23}	0.107	3.934	1.0260	25.19	0.377	5.03	210 (133, 293)[§]
Ceres	475	0.075	9.47×10^{20}	0.0002	2.100	0.378	3.0	0.029	0.51	200
Jupiter	71,492	11.209	1.90×10^{27}	317.83	1.326	0.4135	3.13	2.364	59.5	165
Saturn	60,268	9.449	5.68×10^{26}	95.16	0.687	0.4440	26.73	0.916	35.5	134
Uranus	25,559	4.007	8.68×10^{25}	14.536	1.270	0.7183[‡]	97.77	0.889	21.3	76
Neptune	24,764	3.883	1.02×10^{26}	17.148	1.638	0.6713	28.32	1.12	23.5	58
Pluto	1,160	0.182	1.30×10^{22}	0.0021	2.030	6.387[‡]	122.53	0.083	1.23	40
Haumea	~700	0.11	4.0×10^{21}	0.0007	~3	0.163	?	0.045	0.84	<50
Makemake	750	0.12	4.18×10^{21}	0.0007	~2	0.32	?	0.048	0.8	~30
Eris	1,200	0.188	1.5×10^{22} (est.)	0.0025 (est.)	~2	>0.3?	?	0.082	~1.3	30

*The density of water is 1,000 kg/m³.

[†]The surface gravity of Earth is 9.81 m/s².

[‡]Venus, Uranus, and Pluto rotate opposite to the directions of their orbits. Their north poles are south of their orbital planes.

[§]Where given, values in parentheses give extremes of recorded temperatures.

Orbital Data for Planets and Dwarf Planets

PLANET	MEAN DISTANCE FROM SUN (A^*)		ORBITAL PERIOD (P) (SIDEREAL YEARS)	ECCENTRICITY	INCLINATION (DEGREES, RELATIVE TO ECLIPTIC)	AVERAGE SPEED (km/s)
	(10^6 km)	(AU)				
Mercury	57.9	0.387	0.2408	0.2056	7.005	47.36
Venus	108.2	0.723	0.6152	0.0068	3.395	35.02
Earth	149.6	1.000	1.0000	0.0167	0.000	29.78
Mars	227.9	1.524	1.8808	0.0934	1.850	24.08
Ceres	413.7	2.765	4.6027	0.079	10.587	17.88
Jupiter	778.3	5.203	11.8626	0.0484	1.304	13.06
Saturn	1,426.7	9.537	29.4475	0.0539	2.485	9.64
Uranus	2,870.7	19.189	84.0168	0.0473	0.772	6.80
Neptune	4,498.0	30.070	164.7913	0.0086	1.769	5.43
Pluto	5,906.4	39.48	247.9207	0.2488	17.14	4.67
Haumea	6,428.1	43.0	281.9	0.198	28.22	4.50
Makemake	6,789.7	45.3	305.3	0.164	29.00	4.39
Eris	10,183	68.05	561.6	0.4339	43.82	3.43

$^*A$ is the semimajor axis of the planet's elliptical orbit.

Properties of Selected Moons*

PLANET	MOON	ORBITAL PROPERTIES		PHYSICAL PROPERTIES		
		P (DAYS)	A (10^3 km)	R (km)	M (10^{20} kg)	RELATIVE DENSITY† (g/cm³) (WATER = 1.00)
Earth (1 moon)	Moon	27.32	384.4	1,737.5	735	3.34
Mars (2 moons)	Phobos	0.32	9.38	$13.4 \times 11.2 \times 9.2$	0.0001	1.9
	Deimos	1.26	23.46	$7.5 \times 6.1 \times 5.2$	0.00002	1.5
Jupiter (64 known moons)	Metis	0.30	127.97	21.5	0.00012	3
	Amalthea	0.50	181.40	$131 \times 73 \times 67$	0.0207	0.8
	Io	1.77	421.80	1,822	893	3.53
	Europa	3.55	671.10	1,561	480	3.01
	Ganymede	7.15	1,070	2,631	1,482	1.94
	Callisto	16.69	1,883	2,410	1,080	1.83
	Himalia	250.56	11,461	85	0.067	2.6
	Pasiphae	744‡	23,624	30	0.0030	2.6
	Callirrhoe	759‡	24,102	4.3	0.00001	2.6

(*continued*)

Properties of Selected Moons* (continued)

PLANET	MOON	ORBITAL PROPERTIES		PHYSICAL PROPERTIES		
		P (DAYS)	*A* (10³ km)	*R* (km)	*M* (10²⁰ kg)	RELATIVE DENSITY† (g/cm³) (WATER = 1.00)
Saturn (62 known moons)	Pan	0.58	133.58	14	0.00005	0.42
	Prometheus	0.61	139.38	74.0 × 50.0 × 34.0	0.0016	0.5
	Pandora	0.63	141.70	55 × 44 × 31	0.0014	0.5
	Mimas	0.94	185.54	198	0.38	1.15
	Enceladus	1.37	238.04	252	1.08	1.6
	Tethys	1.89	294.67	533	6.17	0.97
	Dione	2.74	377.42	562	11.0	1.48
	Rhea	4.52	527.07	764	23.1	1.23
	Titan	15.95	1,222	2,575	1,346	1.88
	Hyperion	21.28	1,501	205 × 130 × 110	0.0559	0.54
	Iapetus	79.33	3,561	736	18.1	1.08
	Phoebe	550.3‡	12,948	107	0.08	1.6
	Paaliaq	687.5	15,024	11	0.0001	2.3
Uranus (27 known moons)	Cordelia	0.34	49.80	20	0.0004	1.3
	Miranda	1.41	129.90	236	0.66	1.21
	Ariel	2.52	190.90	579	12.9	1.59
	Umbriel	4.14	264.96	585	12.2	1.46
	Titania	8.71	436.30	789	34.2	1.66
	Oberon	13.46	583.50	761	28.8	1.56
	Setebos	2,225‡	17,420	24	0.0009	1.5
Neptune (13 known moons)	Naiad	0.29	48.0	48 × 30 × 26	0.002	1.3
	Larissa	0.55	73.5	108 × 102 × 84	0.05	1.3
	Proteus	1.12	117.6	218 × 208 × 201	0.5	1.3
	Triton	5.88‡	354.8	1,353	214	2.06
	Nereid	360.13	5,513.82	170	0.3	1.5
Pluto (4 moons)	Charon	6.39	19.60	604	15.2	1.65
Haumea (2 moons)	Namaka	18	25.66	85	0.018	~1
	Hi'iaka	49	49.88	170	0.179	~1
Eris	Dysnomia	15.8	37.4	50–125?	?	?

*Innermost, outermost, largest, and/or a few other moons for each planet.

†The density of water is 1,000 kg/m³.

‡Irregular moon (has retrograde orbit).

NEAREST AND BRIGHTEST STARS

Stars within 12 Light-Years of Earth

NAME*	DISTANCE (ly)	SPECTRAL TYPE†	RELATIVE VISUAL LUMINOSITY‡ (SUN = 1.000)	APPARENT MAGNITUDE	ABSOLUTE MAGNITUDE
Sun	1.58×10^{-5}	G2V	1.000	−26.74	4.83
Alpha Centauri C (Proxima Centauri)	4.24	M5.5V	0.000052	11.09	15.53
Alpha Centauri A	4.36	G2V	1.5	0.01	4.38
Alpha Centauri B	4.36	K0V	0.44	1.34	5.71
Barnard's star	5.96	M4Ve	0.00044	9.53	13.22
CN Leonis	7.78	M5.5	0.000020	13.44	16.55
BD +36-2147	8.29	M2.0V	0.0056	7.47	10.44
Sirius A	8.58	A1V	22	−1.43	1.47
Sirius B	8.58	DA2	0.0025	8.44	11.34
BL Ceti	8.73	M5.5V	0.000059	12.54	15.40
UV Ceti	8.73	M6.0	0.000039	12.99	15.85
V1216 Sagittarii	9.68	M3.5V	0.00050	10.43	13.07
HH Andromedae	10.32	M5.5V	0.00010	12.29	14.79
Epsilon Eridani	10.52	K2V	0.28	3.73	6.19
Lacaille 9352	10.74	M1.5V	0.011	7.34	9.75
FI Virginis	10.92	M4.0V	0.00033	11.13	13.51
EZ Aquarii A	11.26	M5.0V	0.000047	13.33	15.64
EZ Aquarii B	11.26	M5e	0.000050	13.27	15.58
EZ Aquarii C	11.26	—	0.000025	14.03	16.34
Procyon A	11.40	F5IV-V	7.31	0.38	2.66
Procyon B	11.40	DA	0.00054	10.70	12.98
61 Cygni A	11.40	K5.0V	0.086	5.21	7.49
61 Cygni B	11.40	K7.0V	0.040	6.03	8.31

(*continued*)

Stars within 12 Light-Years of Earth (*continued*)

NAME*	DISTANCE (ly)	SPECTRAL TYPE†	RELATIVE VISUAL LUMINOSITY‡ (SUN = 1.000)	APPARENT MAGNITUDE	ABSOLUTE MAGNITUDE
Gliese 725 A	11.52	M3.0V	0.0029	8.90	11.16
Gliese 725 B	11.52	M3.5V	0.0014	9.69	11.95
Groombridge 34 A	11.62	M1.5V	0.0063	8.08	10.32
Groombridge 34 B	11.62	M3.5V	0.00041	11.06	13.30
Epsilon Indi A	11.82	K5Ve	0.15	4.69	6.89
Epsilon Indi B (brown dwarf)	11.82	T1.0	—	—	—
Epsilon Indi C (brown dwarf)	11.82	T6.0	—	—	—
DX Cancri	11.82	M6.5V	0.000014	14.78	16.98
Tau Ceti	11.88	G8V	0.45	3.49	5.68
Gliese 1061	11.99	M5.5V	0.000067	13.09	15.26

*Stars may carry many names, including common names (such as Sirius), names based on their prominence within a constellation (such as Alpha Canis Majoris, another name for Sirius), or names based on their inclusion in a catalog (such as BD +36-2147). Addition of letters A, B, and so on, or superscripts indicates membership in a multiple-star system.

†Spectral types such as M3 are discussed in Chapter 10. Other letters or numbers provide additional information. For example, *V* after the spectral type indicates a main-sequence star, and III indicates a giant star. Stars of spectral type T are brown dwarfs.

‡*Luminosity* in this table refers only to radiation in "visual" light.

The 25 Brightest Stars in the Sky

NAME	COMMON NAME	DISTANCE (ly)	SPECTRAL TYPE	RELATIVE VISUAL LUMINOSITY* (SUN = 1.000)	APPARENT VISUAL MAGNITUDE	ABSOLUTE VISUAL MAGNITUDE
Sun	Sun	1.58×10^{-5}	G2V	1.000	−26.8	4.82
Alpha Canis Majoris	Sirius	8.60	A1V	22.9	−1.47	1.42
Alpha Carinae	Canopus	313	F0II	13,800	−0.72	−5.53
Alpha Bootis	Arcturus	36.7	K1.5IIIFe-0.5	111	−0.04	−0.29
Alpha¹ Centauri	Rigel Kentaurus	4.39	G2V	1.50	−0.01	4.38
Alpha Lyrae	Vega	25.3	A0Va	49.7	0.03	0.58
Alpha Aurigae	Capella	42.2	G5IIIe+G0III	132	0.08	−0.48
Beta Orionis	Rigel	770	B8Ia	40,200	0.12	−6.69
Alpha Canis Minoris	Procyon	11.4	F5IV-V	7.38	0.34	2.65
Alpha Eridani	Achernar	144	B3Vpe	1,090	0.50	−2.77
Alpha Orionis	Betelgeuse	427	M1-2Ia-Iab	9,600	0.58	−5.14
Beta Centauri	Hadar	350	B1III	8,700	0.60	−5.03
Alpha Aquilae	Altair	16.8	A7V	11.1	0.77	2.21
Alpha Tauri	Aldebaran	65.1	K5+III	151	0.85	0.63
Alpha Virginis	Spica	262	B1III-IV+B2V	2,230	1.04	−3.55
Alpha Scorpii	Antares	604	M1.5Iab-Ib+B4Ve	11,000	1.09	−5.28
Beta Geminorum	Pollux	33.7	K0IIIb	31.0	1.15	1.09
Alpha Piscis	Fomalhaut	25.1	A3V	17.2	1.16	1.73
Alpha Cygni	Deneb	3,200	A2Ia	260,000	1.25	−8.73
Beta Crucis	Becrux	353	B0.5III	3,130	1.30	−3.92
Alpha² Centauri	Alpha Centauri B	4.39	K1V	0.44	1.33	5.71
Alpha Leonis	Regulus	77.5	B7V	137	1.35	−0.52
Alpha Crucis	Acrux	321	B0.5IV	4,000	1.40	−4.19
Epsilon Canis Majoris	Adara	431	B2II	3,700	1.51	−4.11
Gamma Crucis	Gacrux	88.0	M3.5III	142	1.59	−0.56

Sources: Data from *The Hipparcos and Tycho Catalogues*, 1997, European Space Agency SP-1200; SIMBAD Astronomical Database (http://simbad.u-strasbg.fr/simbad); and Research Consortium on Nearby Stars (www.chara.gsu.edu/RECONS).
Luminosity in this table refers only to radiation in "visual" light.

| STAR MAPS

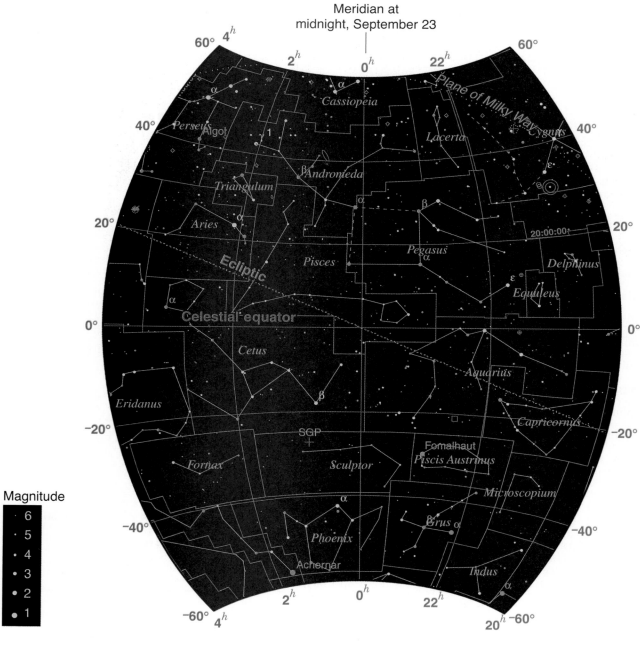

Figure A4.1 The sky from right ascension 20^h to 04^h and declination −60° to +60°.

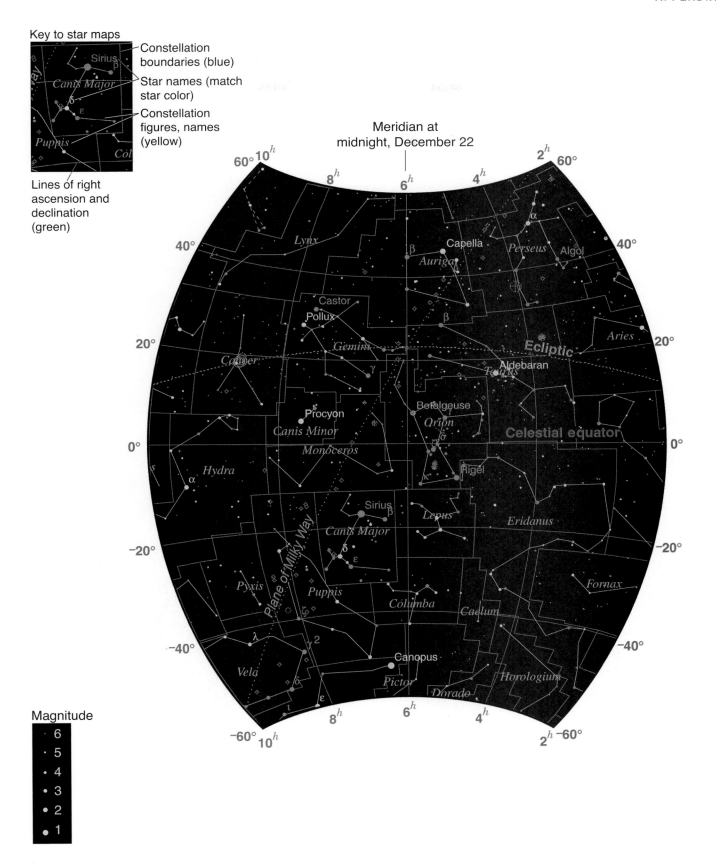

Figure A4.2 The sky from right ascension 02ʰ to 10ʰ and declination −60° to +60°.

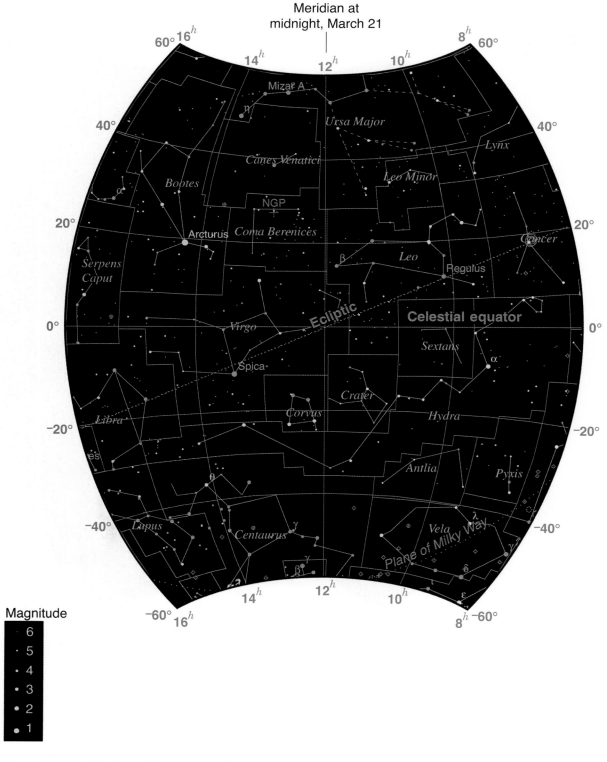

Figure A4.3 The sky from right ascension 08ʰ to 16ʰ and declination −60° to +60°.

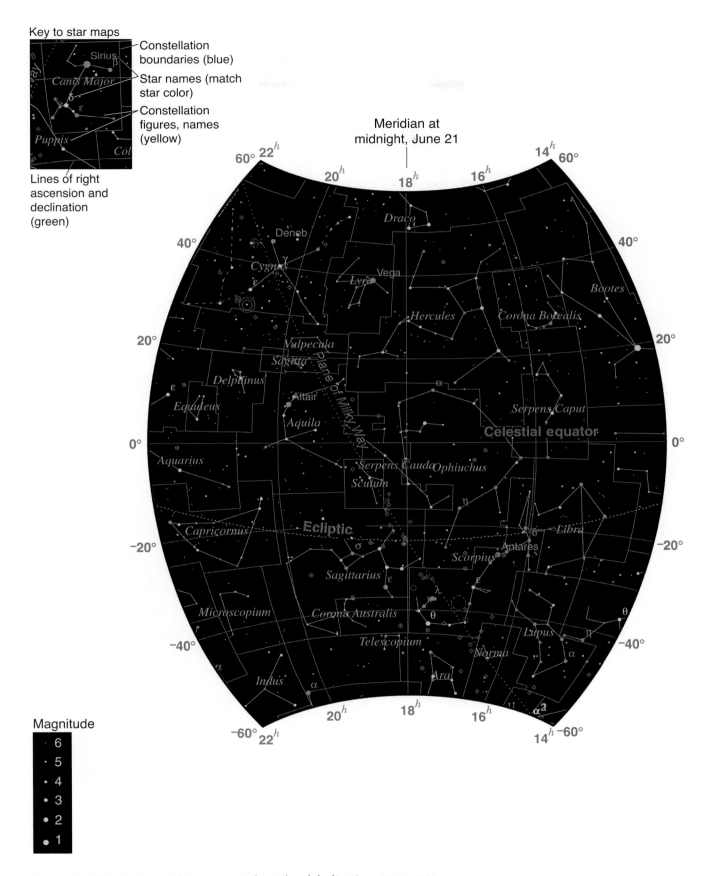

Key to star maps

Constellation boundaries (blue)

Star names (match star color)

Constellation figures, names (yellow)

Lines of right ascension and declination (green)

Meridian at midnight, June 21

Magnitude
- 6
- 5
- 4
- 3
- 2
- 1

Figure A4.4 The sky from right ascension 14^h to 22^h and declination −60° to +60°.

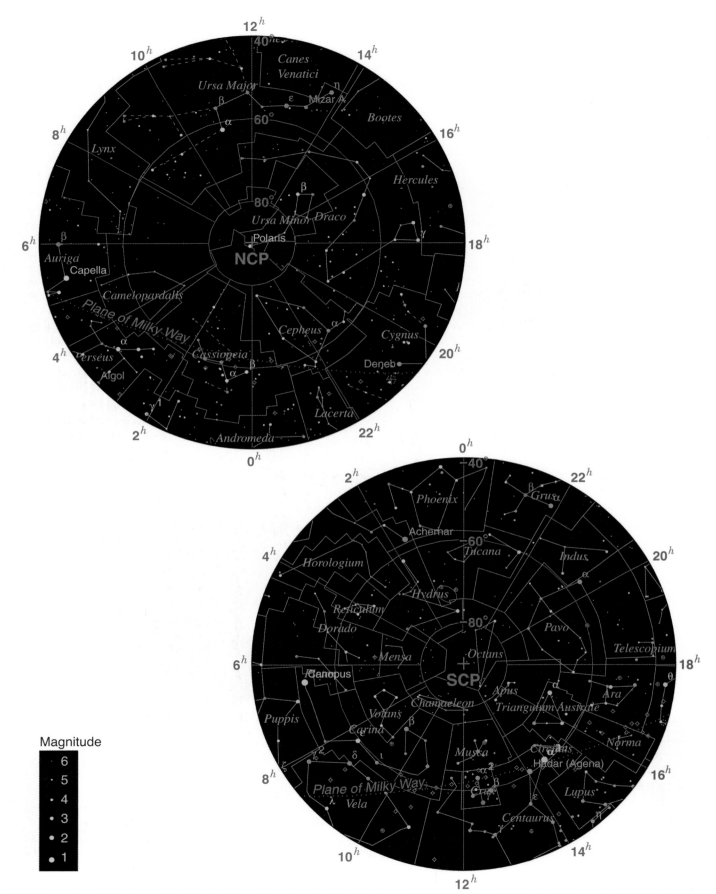

Figure A4.5 The regions of the sky north of declination +40° and south of declination −40°. NCP, north celestial pole; SCP, south celestial pole.

A

aberration of starlight The apparent displacement in the position of a star that is due to the finite speed of light and Earth's orbital motion around the Sun.

absolute magnitude A measure of the intrinsic brightness of a celestial object, generally a star. Specifically, the apparent brightness of an object, such as a star, if it were located at a standard distance of 10 parsecs (pc). Compare *apparent magnitude*.

absolute zero The temperature at which thermal motions cease. The lowest possible temperature. Zero on the Kelvin temperature scale.

absorption The capture of electromagnetic radiation by matter. Compare *emission*.

absorption line An intensity minimum in a spectrum that is due to the absorption of electromagnetic radiation at a specific wavelength determined by the energy levels of an atom or molecule. Compare *emission line*.

acceleration The rate at which the speed and/or direction of an object's motion is changing.

accretion disk A flat, rotating disk of gas and dust surrounding an object, such as a young stellar object, a forming planet, a collapsed star in a binary system, or a black hole.

achondrite A stony meteorite that does not contain chondrules. Compare *chondrite*.

active comet A comet nucleus that approaches close enough to the Sun to show signs of activity, such as the production of a coma and tail.

active galactic nucleus (AGN) A highly luminous, compact galactic nucleus whose luminosity may exceed that of the rest of the galaxy.

active region An area of the Sun's chromosphere anchoring bursts of intense magnetic activity.

adaptive optics Electro-optical systems that largely compensate for image distortion caused by Earth's atmosphere.

AGB See *asymptotic giant branch*.

AGN See *active galactic nucleus*.

albedo The fraction of electromagnetic radiation incident on a surface that is reflected by the surface.

algebra A branch of mathematics in which numeric variables are represented by letters.

alpha particle A ^{4}He nucleus, consisting of two protons and two neutrons. Alpha particles get their name from the fact that they are given off in the type of radioactive decay referred to as "alpha decay."

altitude The location of an object above the horizon, measured by the angle formed between an imaginary line from an observer to the object and a second line from the observer to the point on the horizon directly below the object.

Amors A group of asteroids whose orbits cross the orbit of Mars but not the orbit of Earth. Compare *Apollos* and *Atens*.

amplitude In a wave, the maximum deviation from its undisturbed or relaxed position. For example, in a water wave the amplitude is the vertical distance from crests to the undisturbed water level.

angular momentum A conserved property of a rotating or revolving system whose value depends on the velocity and distribution of the system's mass.

annular solar eclipse The type of solar eclipse that occurs when the apparent diameter of the Moon is less than that of the Sun, leaving a visible ring of light ("annulus") surrounding the dark disk of the Moon. Compare *partial solar eclipse* and *total solar eclipse*.

Antarctic Circle The circle on Earth with latitude 66.5° south, marking the northern limit where at least 1 day per year is in 24-hour daylight. Compare *Arctic Circle*.

anticyclonic motion The rotation of a weather system resulting from the Coriolis effect as air moves outward from a region of high atmospheric pressure. Compare *cyclonic motion*.

antimatter Matter made from antiparticles.

antiparticle An elementary particle of antimatter identical in mass but opposite in charge and all other properties to its corresponding ordinary matter particle.

aperture The clear diameter of a telescope's objective lens or primary mirror.

aphelion (pl. **aphelia**) The point in a solar orbit that is farthest from the Sun. Compare *perihelion*.

Apollos A group of asteroids whose orbits cross the orbits of both Earth and Mars. Compare *Amors* and *Atens*.

apparent magnitude A measure of the apparent brightness of a celestial object, generally a star. Compare *absolute magnitude*.

apparent retrograde motion A motion of the planets with respect to the "fixed stars," in which the planets appear to move westward for a period of time before resuming their normal eastward motion. The heliocentric model explains this effect much more simply than the geocentric model.

arcminute (arcmin) A minute of arc ('), a unit used for measuring angles. An arcminute is 1/60 of a degree of arc.

arcsecond (arcsec) A second of arc ("), a unit used for measuring very small angles. An arcsecond is 1/60 of an arcminute, or 1/3,600 of a degree of arc.

Arctic Circle The circle on Earth with latitude 66.5° north, marking the southern limit where at least 1 day per year is in 24-hour daylight. Compare *Antarctic Circle*.

ash The products of fusion that collect in the core of a star.

asteroid Also called *minor planet*. A primitive rocky or metallic body (planetesimal) that has survived planetary accretion. Asteroids are parent bodies of meteoroids.

asteroid belt Also called *main asteroid belt*. The region between the orbits of Mars and Jupiter that contains most of the asteroids in our Solar System.

astrobiology An interdisciplinary science combining astronomy, biology, chemistry, geology, and physics to study life in the cosmos.

astrology The belief that the positions and aspects of stars and planets influence human affairs and characteristics, as well as terrestrial events.

astronomical seeing A measurement of the degree to which Earth's atmosphere degrades the resolution of a telescope's view of astronomical objects.

astronomical unit (AU) The average distance from the Sun to Earth: approximately 150 million kilometers (km).

astronomy The scientific study of planets, stars, galaxies, and the universe as a whole.

astrophysics The application of physical laws to the understanding of planets, stars, galaxies, and the universe as a whole.

asymptotic giant branch (AGB) The path on the H-R diagram that goes from the horizontal branch toward higher luminosities and lower temperatures, asymptotically approaching and then rising above the red giant branch.

Atens A group of asteroids whose orbits cross the orbit of Earth but not the orbit of Mars. Compare *Amors* and *Apollos*.

atmosphere The gravitationally bound, outer gaseous envelope surrounding a planet, moon, or star.

atmospheric greenhouse effect A warming of planetary surfaces produced by atmospheric gases that transmit optical solar radiation but partially trap infrared radiation.

atmospheric probe An instrumented package designed to provide in situ measurements of the chemical and/or physical properties of a planetary atmosphere.

atmospheric window A region of the electromagnetic spectrum in which radiation is able to penetrate a planet's atmosphere.

atom The smallest piece of a chemical element that retains the properties of that element. Each atom is composed of a nucleus (neutrons and protons) surrounded by a cloud of electrons.

AU See *astronomical unit*.

aurora Emission in the upper atmosphere of a planet from atoms that have been excited by collisions with energetic particles from the planet's magnetosphere.

autumnal equinox 1. One of two points where the Sun crosses the celestial equator. 2. The day on which the Sun appears at this location, marking the first day of autumn (about September 23 in the Northern Hemisphere and March 21 in the Southern Hemisphere). Compare *vernal equinox*.

axion A hypothetical elementary particle first proposed to explain certain properties of the neutron and now considered a candidate for cold dark matter.

B

backlighting Illumination from behind a subject as seen by an observer. Fine material such as human hair and dust in planetary rings stands out best when viewed under backlighting conditions.

bar A unit of pressure. One bar is equivalent to 10^5 newtons per square meter—approximately equal to Earth's atmospheric pressure at sea level.

barred spiral A spiral galaxy with a bulge having an elongated, barlike shape.

basalt Gray to black volcanic rock, rich in iron and magnesium.

beta decay 1. The decay of a neutron into a proton by emission of an electron (beta ray) and an antineutrino. 2. The decay of a proton into a neutron by emission of a positron and a neutrino.

Big Bang The event that occurred 13.7 billion years ago that marks the beginning of time and the universe.

Big Bang nucleosynthesis The formation of low-mass nuclei (H, He, Li, Be, B) during the first few minutes following the Big Bang.

Big Rip A hypothetical cosmic event in which all matter in the universe, from stars to subatomic particles, is progressively torn apart by expansion of the universe.

binary star A system in which two stars are in gravitationally bound orbits about their common center of mass.

binding energy The minimum energy required to separate an atomic nucleus into its component protons and neutrons.

biosphere The global sum of all living organisms on Earth (or any planet or moon). Compare *hydrosphere* and *lithosphere*.

bipolar outflow Material streaming away in opposite directions from either side of the accretion disk of a young star.

black hole An object so dense that its escape velocity exceeds the speed of light; a singularity in spacetime.

blackbody An object that absorbs and can reemit all electromagnetic energy it receives.

blackbody spectrum Also called *Planck spectrum*. The spectrum of electromagnetic energy emitted by a blackbody per unit area per second, which is determined only by the temperature of the object.

blueshift The Doppler shift toward shorter wavelengths of light from an approaching object. Compare *redshift*.

Bohr model A model of the atom, proposed by Niels Bohr in 1913, in which a small positively charged nucleus is surrounded by orbiting electrons, similar to a miniature solar system.

bolide A very bright, exploding meteor.

bound orbit A closed orbit in which the velocity is less than the escape velocity. Compare *unbound orbit*.

bow shock 1. The boundary at which the speed of the solar wind abruptly drops from supersonic to subsonic in its approach to a planet's magnetosphere; the boundary between the region dominated by the solar wind and the region dominated by a planet's magnetosphere. 2. The interface between strong collimated gas and dust outflow from a star and the interstellar medium.

brightness The apparent intensity of light from a luminous object. Brightness depends on both the *luminosity* of a source and its distance. Units at the detector: watts per square meter (W/m^2).

brown dwarf A "failed" star that is not massive enough to cause hydrogen fusion in its core. An object whose mass is intermediate between that of the least massive stars and that of supermassive planets.

bulge The central region of a spiral galaxy that is similar in appearance to a small elliptical galaxy.

C

C See *Celsius*.

C-type asteroid An asteroid made of material that has largely been unmodified since the formation of the Solar System; the most primitive type of asteroid. Compare *M-type asteroid* and *S-type asteroid*.

caldera The summit crater of a volcano.

carbon-nitrogen-oxygen (CNO) cycle One of the ways in which hydrogen is converted to helium (hydrogen burning) in the interiors of main-sequence stars. See also *proton-proton chain* and *triple-alpha process*.

carbon star A cool red giant or asymptotic giant branch star that has an excess of carbon in its atmosphere.

carbonaceous chondrite A primitive stony meteorite that contains chondrules and is rich in carbon and volatile materials.

Cassini Division The largest gap in Saturn's rings, discovered by Jean-Dominique Cassini in 1675.

catalyst An atomic and molecular structure that permits or encourages chemical and nuclear reactions but does not change its own chemical or nuclear properties.

CCD See *charge-coupled device*.

celestial equator The imaginary great circle that is the projection of Earth's equator onto the celestial sphere.

celestial sphere An imaginary sphere with celestial objects on its inner surface and Earth at its center. The celestial sphere has no physical existence but is a convenient tool for picturing the directions in which celestial objects are seen from the surface of Earth.

Celsius (C) Also called *centigrade scale*. The arbitrary temperature scale—defined by Anders Celsius (1701–1744)—that defines 0°C as the freezing point of water and 100°C as the boiling point of water at sea level. Unit: °C. Compare *Fahrenheit* and *Kelvin scale*.

center of mass The location associated with an object system at which we may regard the entire mass of the system as being concentrated. The point in any isolated system that moves according to Newton's first law of motion.

centigrade scale See *Celsius*.

centripetal force A force directed toward the center of curvature of an object's curved path.

Cepheid variable An evolved high-mass star with an atmosphere that is pulsating, leading to variability in the star's luminosity and color.

Chandrasekhar limit The upper limit on the mass of an object supported by electron degeneracy pressure; approximately 1.4 solar mass ($M_\odot$).

chaos Behavior in complex, interrelated systems in which tiny differences in the initial configuration of a system result in dramatic differences in the system's later evolution.

charge-coupled device (CCD) A common type of solid-state detector of electromagnetic radiation that transforms the intensity of light directly into electric signals.

chemistry The study of the composition, structure, and properties of substances.

chondrite A stony meteorite containing chondrules. Compare *achondrite*.

chondrule A small, crystallized, spherical inclusion of rapidly cooled molten droplets found inside some meteorites.

chromatic aberration A detrimental property of a lens in which rays of different wavelengths are brought to different focal distances from the lens.

chromosphere The region in the Sun's atmosphere located between the photosphere and the corona.

circular velocity The orbital velocity needed to keep an object moving in a circular orbit.

circumpolar Referring to the part of the sky, near either celestial pole, that can always be seen above the horizon from a specific location on Earth.

circumstellar disk See *protoplanetary disk*.

classical mechanics The science of applying Newton's laws to the motion of objects.

classical planets The eight major planets of the Solar System: Mercury, Venus, Earth, Mars, Jupiter, Saturn, Uranus, and Neptune.

climate The state of an atmosphere averaged over an extended time. Compare *weather*.

closed universe A finite universe with a curved spatial structure such that the sum of the angles of a triangle always exceeds 180°. Compare *flat universe* and *open universe*.

CMB See *cosmic microwave background radiation*.

CNO cycle See *carbon-nitrogen-oxygen cycle*.

cold dark matter Particles of dark matter that move slowly enough to be gravitationally bound even in the smallest galaxies. Compare *hot dark matter*.

color index The color of a celestial object, generally a star, based on the ratio of its brightness in blue light (b_B) to its brightness in "visual" (or yellow-green) light (b_V). The difference between an object's blue (B) magnitude and visual (V) magnitude, B-V.

coma (pl. comae) The nearly spherical cloud of gas and dust surrounding the nucleus of an active comet.

comet A complex object consisting of a small, solid, icy nucleus; an atmospheric halo; and a tail of gas and dust.

comet nucleus A primitive planetesimal, composed of ices and refractory materials, that has survived planetary accretion. The "heart" of a comet, containing nearly the entire mass of the comet. A "dirty snowball."

comparative planetology The study of planets through comparison of their chemical and physical properties.

complex system An interrelated system capable of exhibiting chaotic behavior. See also *chaos*.

composite volcano A large, cone-shaped volcano formed by viscous, pasty lava flows alternating with pyroclastic (explosively generated) rock deposits. Compare *shield volcano*.

compound lens A lens made up of two or more elements of differing refractive index, the purpose of which is to minimize chromatic aberration.

conservation law A physical law stating that the amount of a particular physical quantity (such as energy or angular momentum) of an isolated system does not change over time.

conservation of angular momentum The physical law stating that the amount of angular momentum of an isolated system does not change over time.

conservation of energy The physical law stating that the amount of energy of an isolated, closed system does not change over time.

constant of proportionality The multiplicative factor by which one quantity is related to another.

constellation An imaginary image formed by patterns of stars; any of 88 defined areas on the celestial sphere used by astronomers to locate celestial objects.

constructive interference A state in which the amplitudes of two intersecting waves reinforce one another. Compare *destructive interference*.

continental drift The slow motion (centimeters per year) of Earth's continents relative to each other and to Earth's mantle. See also *plate tectonics*.

continuous radiation Electromagnetic radiation with intensity that varies smoothly over a wide range of wavelengths.

convection The transport of thermal energy from the lower (hotter) to the higher (cooler) layers of a fluid by motions within the fluid driven by variations in buoyancy.

convective zone A region within a star in which energy is transported outward by convection.

conventional greenhouse effect The solar heating of air in an enclosed space, such as a closed build-ing or car, resulting primarily from the inability of the hot air to escape.

core 1. The innermost region of a planetary interior. 2. The innermost part of a star.

core accretion A process for forming giant planets, whereby large quantities of surrounding hydrogen and helium are gravitationally captured onto a massive rocky core.

Coriolis effect The apparent displacement of objects in a direction perpendicular to their true motion as viewed from a rotating frame of reference. On a rotating planet, different latitudes rotating at different speeds cause this effect.

corona The hot, outermost part of the Sun's atmosphere.

coronal hole A low-density region in the solar corona containing "open" magnetic field lines along which coronal material is free to stream into interplanetary space.

coronal mass ejection An eruption on the Sun that ejects hot gas and energetic particles at much higher speeds than are typical in the solar wind.

cosmic microwave background (CMB) radiation Isotropic microwave radiation from every direction in the sky having a 2.73-kelvin (K) blackbody spectrum. The CMB is residual radiation from the Big Bang.

cosmic ray A very fast-moving particle (usually an atomic nucleus); cosmic rays fill the disk of our galaxy.

cosmological constant A constant, introduced into general relativity by Einstein, that characterizes an extra, repulsive force in the universe due to the vacuum of space itself.

cosmological principle The (testable) assumption that the same physical laws that apply here and now also apply everywhere and at all times, and that there are no special locations or directions in the universe.

cosmological redshift (z) The redshift that results from the expansion of the universe rather than from the motions of galaxies or gravity (see *gravitational redshift*).

cosmology The study of the large-scale structure and evolution of the universe as a whole.

Crab Nebula The remnant of the Type II supernova explosion witnessed by Chinese astronomers in A.D. 1054.

crescent Any phase of the Moon, Mercury, or Venus in which the object appears less than half illuminated by the Sun. Compare *gibbous*.

Cretaceous-Tertiary (K-T) boundary The boundary between the Cretaceous and Tertiary periods in Earth's history. This boundary corresponds to the time of the impact of an asteroid or comet and the extinction of the dinosaurs.

critical density The value of the mass density of the universe that, ignoring any cosmological constant, is just barely capable of halting expansion of the universe.

crust The relatively thin, outermost, hard layer of a planet, which is chemically distinct from the interior.

cryovolcanism Low-temperature volcanism in which the magmas are composed of molten ices rather than rocky material.

cyclone See *hurricane*.

cyclonic motion The rotation of a weather system resulting from the Coriolis effect as air moves toward a region of low atmospheric pressure. Compare *anticyclonic motion*.

Cygnus X-1 A binary X-ray source and probable black hole.

D

dark energy A form of energy that permeates all of space (including the vacuum) producing a repulsive force that accelerates the expansion of the universe.

dark matter Matter in galaxies that does not emit or absorb electromagnetic radiation. Dark matter is thought to comprise most of the mass in the universe. Compare *luminous matter*.

dark matter halo The centrally condensed, greatly extended dark matter component of a galaxy that contains up to 95 percent of the galaxy's mass.

daughter product An element resulting from radioactive decay of a more massive *parent element*.

decay 1. The process of a radioactive nucleus changing into its daughter product. 2. The process of an atom or molecule dropping from a higher-energy state to a lower-energy state. 3. The process of a satellite's orbit losing energy.

density The measure of an object's mass per unit of volume. Units: kilograms per cubic meter (kg/m^3).

destructive interference A state in which the amplitudes of two intersecting waves cancel one another. Compare *constructive interference*.

differential rotation Rotation of different parts of a system at different rates.

differentiation The process by which materials of higher density sink toward the center of a molten or fluid planetary interior.

diffraction The spreading of a wave after it passes through an opening or past the edge of an object.

diffraction limit The limit of a telescope's angular resolution caused by diffraction.

diffuse ring A sparsely populated planetary ring spread out both horizontally and vertically.

direct imaging A technique for detecting extrasolar planets by observing them directly with telescopes.

dispersion The separation of rays of light into their component wavelengths.

distance ladder A sequence of techniques for measuring cosmic distances; each method is calibrated using the results from other methods that have been applied to closer objects.

Doppler effect The change in wavelength of sound or light that is due to the relative motion of the source toward or away from the observer.

Doppler redshift See *redshift*.

Doppler shift The amount by which the wavelength of light is shifted by the Doppler effect.

Drake equation A prescription for estimating the number of intelligent civilizations existing elsewhere.

dust devil A small tornado-like column of air containing dust or sand.

dust tail A type of comet tail consisting of dust particles that are pushed away from the comet's head by radiation pressure from the Sun. Compare *ion tail*.

dwarf galaxy A small galaxy with a luminosity ranging from 1 million to 1 billion solar luminosities ($L_\odot$). Compare *giant galaxy*.

dwarf planet A body with characteristics similar to those of a classical planet except that it has not cleared smaller bodies from the neighboring regions around its orbit. Compare *planet* (definition 2).

dynamic equilibrium A state in which a system is constantly changing but its configuration remains the same because one source of change is exactly balanced by another source of change. Compare *static equilibrium*.

dynamo A device that converts mechanical energy into electric energy in the form of electric currents and magnetic fields. The "dynamo effect" is thought to create magnetic fields in planets and stars by electrically charged currents of material flowing within their cores.

E

eccentricity (e) A measure of the departure of an ellipse from circularity; the ratio of the distance between the two foci of an ellipse to its major axis.

eclipse 1. The total or partial obscuration of one celestial body by another. 2. The total or partial obscuration of light from one celestial body as it passes through the shadow of another celestial body.

eclipse season Any time during the year when the Moon's line of nodes is sufficiently close to the Sun for eclipses to occur.

eclipsing binary A binary system in which the orbital plane is oriented such that the two stars appear to pass in front of one another as seen from Earth. Compare *spectroscopic binary* and *visual binary*.

ecliptic 1. The apparent annual path of the Sun against the background of stars. 2. The projection of Earth's orbital plane onto the celestial sphere.

ecliptic plane The plane of Earth's orbit around the Sun. The ecliptic is the projection of this plane on the celestial sphere.

effective temperature The temperature at which a black body, such as a star, appears to radiate.

ejecta 1. Material thrown outward by the impact of an asteroid or comet on a planetary surface, leaving a crater behind. 2. Material thrown outward by a stellar explosion.

electric field A field that is able to exert a force on a charged object, whether at rest or moving. Compare *magnetic field*.

electric force The force exerted on a charged particle by an electric field. Compare *magnetic force*.

electromagnetic force The force, including both electric and magnetic forces, that acts on electrically charged particles. One of four fundamental forces of nature. The force mediated by photons.

electromagnetic radiation A traveling disturbance in the electric and magnetic fields caused by accelerating electric charges. In quantum mechanics, a stream of photons. Light.

electromagnetic spectrum The spectrum made up of all possible frequencies or wavelengths of electromagnetic radiation, ranging from gamma rays through radio waves and including the portion our eyes can use.

electromagnetic wave A wave consisting of oscillations in the electric-field strength and the magnetic-field strength.

electron (e^-) A subatomic particle having a negative charge of 1.6×10^{-19} coulomb (C), a rest mass of 9.1×10^{-31} kilograms (kg), and mass-equivalent energy of 8×10^{-14} joules (J). The antiparticle of the *positron*. Compare *proton* and *neutron*.

electron-degenerate Describing the state of material compressed to the point at which electron density reaches the limit imposed by the rules of quantum mechanics.

electroweak theory The quantum theory that combines descriptions of both the electromagnetic force and the weak nuclear force.

element One of 92 naturally occurring substances (such as hydrogen, oxygen, and uranium) and more than 20 human-made ones (such as plutonium). Each element is chemically defined by the specific number of protons in the nuclei of its atoms.

elementary particle One of the basic building blocks of nature that is not known to have sub-structure, such as the *electron* and the *quark*.

ellipse A conic section produced by the inter-section of a plane with a cone when the plane is passed through the cone at an angle to the axis other than 0° or 90°.

elliptical galaxy A galaxy of Hubble type "E" class, with a circular to elliptical outline on the sky containing almost no disk and a population of old stars. Compare *irregular galaxy, S0 galaxy*, and *spiral galaxy*.

emission The release of electromagnetic energy when an atom, molecule, or particle drops from a higher-energy state to a lower-energy state. Compare *absorption*.

emission line An intensity peak in a spectrum that is due to sharply defined emission of electro-magnetic radiation in a narrow range of wave-lengths. Compare *absorption line*.

empirical science Scientific investigation that is based primarily on observations and experimental data. It is descriptive rather than based on theo-retical inference.

energy The conserved quantity that gives objects and systems the ability to do work. Units: joules (J).

energy transport The transfer of energy from one location to another. In stars, energy transport is carried out mainly by radiation or convection.

entropy A measure of the disorder of a system related to the number of ways a system can be rearranged without its appearance being affected.

equator The imaginary great circle on the surface of a body midway between its poles that divides the body into northern and southern hemispheres. The equatorial plane passes through the center of the body and is perpendicular to its rotation axis. Compare *meridian*.

equilibrium The state of an object in which physical processes balance each other so that its properties or conditions remain constant.

equinox Literally, "equal night." 1. One of two positions on the ecliptic where it intersects the celestial equator. 2. Either of the two times of year (the *autumnal equinox* and *vernal equinox*) when the Sun is at one of these two positions. At this time, night and day are of the same length every-where on Earth. Compare *solstice*.

equivalence principle The principle stating that there is no difference between a frame of reference that is freely floating through space and one that is freely falling within a gravitational field.

erosion The degradation of a planet's surface topography by the mechanical action of wind and/or water.

escape velocity The minimum velocity needed for an object to achieve a parabolic trajectory and thus permanently leave the gravitational grasp of another mass.

eternal inflation The idea that a universe might inflate forever. In such a universe, quantum effects could randomly cause regions to slow their expan-sion, eventually stop inflating, and experience an explosion resembling our Big Bang.

event A particular location in spacetime.

event horizon The effective "surface" of a black hole. Nothing inside this surface—not even light—can escape from a black hole.

evolutionary track The path that a star follows across the H-R diagram as it evolves through its lifetime.

excited state An energy level of a particular atom, molecule, or particle that is higher than its ground state. Compare *ground state*.

exoplanet See *extrasolar planet*.

extrasolar planet Also called *exoplanet*. A planet orbiting a star other than the Sun.

F

F See *Fahrenheit*.

Fahrenheit (F) The arbitrary temperature scale—defined by Daniel Gabriel Fahrenheit (1686–1736)—that defines 32°F as the melting point of water and 212°F as the boiling point of water at sea level. Unit: °F. Compare *Celsius* and *Kelvin scale*.

fault A fracture in the crust of a planet or moon along which blocks of material can slide.

filter An instrument element that transmits a limited wavelength range of electromagnetic radiation. For the optical range, such elements are typically made of different kinds of glass and take on the hue of the light they transmit.

first quarter Moon The phase of the Moon in which only the western half of the Moon, as viewed from Earth, is illuminated by the Sun. It occurs about a week after a new Moon. Compare *third quarter Moon*.

fissure A fracture in the planetary lithosphere from which magma emerges.

flat rotation curve A rotation curve of a spiral galaxy in which rotation rates do not decline in the outer part of the galaxy, but remain relatively constant to the outermost points.

flat universe An infinite universe whose spatial structure obeys Euclidean geometry, such that the sum of the angles of a triangle always equals 180°. Compare *closed universe* and *open universe*.

flatness problem The surprising result that the sum of Ω_{mass} plus Ω_Λ is extremely close to unity in the present-day universe; equivalent to saying that it is surprising the universe is so close to being exactly flat.

flux The total amount of energy passing through each square meter of a surface each second. Units: watts per square meter (W/m²).

flux tube A strong magnetic field contained within a tubelike structure. Flux tubes are found in the solar atmosphere and connecting the space between Jupiter and its moon Io.

flyby A spacecraft that first approaches and then continues flying past a planet or moon. Flybys can visit multiple objects, but they remain in the vicinity of their targets only briefly. Compare *orbiter*.

focal length The optical distance between a telescope's objective lens or primary mirror and the plane (called the focal plane) on which the light from a distant object is focused.

focal plane The plane, perpendicular to the optical axis of a lens or mirror, in which an image is formed.

focus (pl. foci) 1. One of two points that define an ellipse. 2. A point in the focal plane of a telescope.

force A push or a pull on an object.

frame of reference A coordinate system within which an observer measures positions and motions.

free fall The motion of an object when the only force acting on it is gravity.

frequency The number of times per second that a periodic process occurs. Unit: hertz (Hz), 1/s.

full Moon The phase of the Moon in which the near side of the Moon, as viewed from Earth, is fully illuminated by the Sun. It occurs about two weeks after a *new Moon*.

G

galaxy A gravitationally bound system that consists of stars and star clusters, gas, dust, and dark matter; typically greater than 1,000 light-years across and recognizable as a discrete, single object.

galaxy cluster A large, gravitationally bound collection of galaxies containing hundreds to thousands of members; typically 10–15 mega-light-years (Mly) across. Compare *galaxy group* and *supercluster*.

galaxy group A small, gravitationally bound collection of galaxies containing from several to a hundred members; typically 4–6 mega-light-years (Mly) across. Compare *galaxy cluster* and *supercluster*.

gamma ray Electromagnetic radiation with higher frequency, higher photon energy, and shorter wavelength than all other types of electromag-netic radiation.

gas giant A giant planet formed mostly of hydrogen and helium. In our Solar System, Jupiter and Saturn are the gas giants. Compare *ice giant*.

gauss A basic unit of magnetic flux density.

general relativistic time dilation The verified prediction that time passes more slowly in a gravitational field than in the absence of a gravitational field. Compare *time dilation*.

general relativity See *general theory of relativity*.

general theory of relativity Sometimes referred to as simply *general relativity*. Einstein's theory explaining gravity as the distortion of spacetime by massive objects, such that particles travel on the shortest path between two events in spacetime. This theory deals with all types of motion. Compare *special theory of relativity*.

geocentric A coordinate system having the center of Earth as its center. Compare *heliocentric*.

geodesic The path an object will follow through spacetime in the absence of external forces.

geometry A branch of mathematics that deals with points, lines, angles, and shapes.

giant galaxy A galaxy with luminosity greater than about 1 billion solar luminosities ($L_\odot$). Compare *dwarf galaxy*.

giant molecular cloud An interstellar cloud composed primarily of molecular gas and dust, having hundreds of thousands of solar masses.

giant planet One of the largest planets in the Solar System (Saturn, Jupiter, Uranus, or Neptune), typically 10 times the size and many times the mass of any *terrestrial planet* and lacking a solid surface.

gibbous Any phase of the Moon, Mercury, or Venus in which the object appears more than half illuminated by the Sun. Compare *crescent*.

global circulation The overall, planet-wide circulation pattern of a planet's atmosphere.

globular cluster A spherically symmetric, highly condensed group of stars, containing tens of thousands to a million members. Compare *open cluster*.

gluon The particle that carries (or, equivalently, mediates) interactions due to the strong nuclear force.

gossamer ring An extremely tenuous planetary ring found beyond Jupiter's main ring.

grand unified theory (GUT) A unified quantum theory that combines the strong nuclear, weak nuclear, and electromagnetic forces but does not include gravity.

granite Rock that is cooled from magma and is relatively rich in silicon and oxygen.

grating An optical surface containing many narrow, closely and equally spaced parallel grooves or slits that spectrally disperse reflected or transmitted light.

gravitational lens A massive object that gravitationally focuses the light of a more distant object to produce multiple brighter, magnified, possibly distorted images.

gravitational lensing The bending of light by gravity. Can be used to detect extrasolar planets.

gravitational potential energy The stored energy in an object that is due solely to its position within a gravitational field.

gravitational redshift The shifting to longer wavelengths of radiation from an object deep within a gravitational well.

gravitational wave A wave in the fabric of spacetime emitted by accelerating masses.

gravity 1. The mutually attractive force between massive objects. 2. An effect arising from the bending of spacetime by massive objects. 3. One of four fundamental forces of nature.

great circle Any circle on a sphere that has as its center the center of the sphere. The celestial equator, the meridian, and the ecliptic are all great circles on the sphere of the sky, as is any circle drawn through the zenith.

Great Red Spot The giant, oval, brick-red anticyclone seen in Jupiter's southern hemisphere.

greenhouse effect See *atmospheric greenhouse effect* and *conventional greenhouse effect*.

greenhouse gas One of a group of atmospheric gases such as carbon dioxide that are transparent to visible radiation but absorb infrared radiation.

Gregorian calendar The modern calendar. A modification of the Julian calendar decreed by Pope Gregory XIII in 1582. By this time the less accurate Julian calendar had developed an error of 10 days over the 13 centuries since its inception.

ground state The lowest possible energy state for a system or part of a system, such as an atom, molecule, or particle. Compare *excited state*.

GUT See *grand unified theory*.

H

H II region A region of interstellar gas that has been ionized by UV radiation from nearby hot massive stars.

H-R diagram The Hertzsprung-Russell diagram, which is a plot of the luminosities versus the surface temperatures of stars. The evolving properties of stars are plotted as tracks across the H-R diagram.

habitable zone The distance from its star at which a planet must be located in order to have a temperature suitable for life; often assumed to be temperatures at which water exists in a liquid state.

Hadley circulation A simplified, and therefore uncommon, atmospheric global circulation that carries thermal energy directly from the equator to the polar regions of a planet.

half-life The time it takes half a sample of a particular radioactive parent element to decay to a daughter product.

halo The spherically symmetric, low-density distribution of stars and dark matter that defines the outermost regions of a galaxy.

harmonic law See *Kepler's third law*.

Hawking radiation Radiation from a black hole.

Hayashi track The path that a protostar follows on the H-R diagram as it contracts toward the main sequence.

head The part of a comet that includes both the nucleus and the inner part of the coma.

heat death The possible eventual fate of an open universe, in which entropy has triumphed and all energy- and structure-producing processes have come to an end.

heavy element Also called *massive element*. 1. In astronomy, any element more massive than helium. 2. In other sciences (and sometimes also in astronomy), any of the most massive elements in the periodic table, such as uranium and plutonium.

Heisenberg uncertainty principle The physical limitation that the product of the position and the momentum of a particle cannot be smaller than a well-defined value, Planck's constant (h).

heliocentric A coordinate system having the center of the Sun as its center. Compare *geocentric*.

helioseismology The use of solar oscillations to study the interior of the Sun.

helium flash The runaway explosive burning of helium in the degenerate helium core of a red giant star.

Herbig-Haro (HH) object A glowing, rapidly moving knot of gas and dust that is excited by bipolar outflows in very young stars.

heredity The process by which one generation passes on its characteristics to future generations.

hertz (Hz) A unit of frequency equivalent to cycles per second.

Hertzsprung-Russell diagram See *H-R diagram*.

HH object See *Herbig-Haro object*.

hierarchical clustering The "bottom-up" process of forming large-scale structure. Small-scale structure first produces groups of galaxies, which in turn form clusters, which then form superclusters.

high-mass star A star with a main-sequence mass greater than about 8 solar masses ($M_\odot$). Compare *low-mass star*.

high-velocity star A star belonging to the halo found near the Sun, distinguished from disk stars by moving far faster and often in the direction opposite to the rotation of the disk and its stars.

homogeneous In cosmology, describing a universe in which observers in any location would observe the same properties.

horizon The boundary that separates the sky from the ground.

horizon problem The puzzling observation that the cosmic background radiation is so uniform in all directions, despite the fact that widely separated regions should have been "over the horizon" from each other in the early universe.

horizontal branch A region on the H-R diagram defined by stars burning helium to carbon in a stable core.

hot dark matter Particles of dark matter that move so fast that gravity cannot confine them to the volume occupied by a galaxy's normal luminous matter. Compare *cold dark matter.*

hot Jupiter A large, Jovian-type extrasolar planet located very close to its parent star.

hot spot A place where hot plumes of mantle material rise near the surface of a planet.

Hubble constant (H_0) The constant of proportionality relating the recession velocities of galaxies to their distances. See also *Hubble time.*

Hubble time An estimate of the age of the universe from the inverse of the Hubble constant, $1/H_0$.

Hubble's law The law stating that the speed at which a galaxy is moving away from us is proportional to the distance of that galaxy.

hurricane Also called *cyclone* or *typhoon.* A large tropical cyclonic system circulating counterclockwise in the Northern Hemisphere and clockwise in the Southern Hemisphere. Hurricanes can extend outward from their center to more than 600 kilometers (km) and generate winds in excess of 300 kilometers per hour (km/h).

hydrogen burning The release of energy from the nuclear fusion of four hydrogen atoms into a single helium atom.

hydrogen shell burning The fusion of hydrogen in a shell surrounding a stellar core that may be either degenerate or fusing more massive elements.

hydrosphere The portion of Earth that is largely liquid water. Compare *biosphere* and *lithosphere.*

hydrostatic equilibrium The condition in which the weight bearing down at a particular point within an object is balanced by the pressure within the object.

hypothesis A well-thought-out idea, based on scientific principles and knowledge, that leads to testable predictions. Compare *theory.*

Hz See *hertz.*

I

ice The solid form of a volatile material; sometimes the *volatile material* itself, regardless of its physical form.

ice giant A giant planet formed mostly of the liquid form of volatile substances (ices). In our Solar System, Uranus and Neptune are the ice giants. Compare *gas giant.*

ideal gas law The relationship of pressure (P) to number density of particles (n) and temperature (T) expressed as $P = nkT$, where k is Boltzmann's constant.

igneous activity The formation and action of molten rock (magma).

impact crater The scar of the impact left on a solid planetary or moon surface by collision with another object. Compare *secondary crater.*

impact cratering A process involving collisions between solid planetary objects.

index of refraction (n) The ratio of the speed of light in a vacuum (c) to the speed of light in an optical medium (v).

inert gas A gaseous element that combines with other elements only under conditions of extreme temperature and pressure. Examples include helium, neon, and argon.

inertia The tendency for objects to retain their state of motion.

inertial frame of reference 1. A frame of reference that is not accelerating. 2. In general relativity, a frame of reference that is falling freely in a gravitational field.

inflation An extremely brief phase of ultra-rapid expansion of the very early universe. Following inflation, the standard Big Bang models of expansion apply.

infrared (IR) radiation Electromagnetic radiation with frequencies and photon energies occurring in the spectral region between those of visible light and microwaves. Compare *ultraviolet radiation.*

instability strip A region of the H-R diagram containing stars that pulsate with a periodic variation in luminosity.

integration time The time interval during which photons are collected and added up in a detecting device.

intensity (of light) The amount of radiant energy emitted per second per unit area. Units for electromagnetic radiation: watts per square meter (W/m²).

intercloud gas A low-density region of the interstellar medium that fills the space between interstellar clouds.

interference The interaction of two sets of waves producing high and low intensity, depending on whether their amplitudes reinforce (*constructive interference*) or cancel (*destructive interference*).

interferometer Also called *interferometric array.* A group or array of separate but linked optical or radio telescopes whose overall separation determines the angular resolution of the system.

interferometric array See *interferometer.*

interstellar cloud A discrete, high-density region of the interstellar medium made up mostly of atomic or molecular hydrogen and dust.

interstellar dust Small particles or grains (0.01–10 micrometers [μm]) of matter, primarily carbon and silicates, distributed throughout interstellar space.

interstellar extinction The dimming of visible and ultraviolet light by interstellar dust.

interstellar gas The tenuous gas, far less dense than air, comprising 99 percent of the matter in the interstellar medium.

interstellar medium The gas and dust that fill the space between the stars within a galaxy.

inverse square law The rule that a quantity or effect diminishes with the square of the distance from the source.

ion An atom or molecule that has lost or gained one or more electrons.

ion tail A type of comet tail consisting of ionized gas. Particles in the ion tail are pushed directly away from the comet's head in the antisolar direction at high speeds by the solar wind. Compare *dust tail.*

ionize The process by which electrons are stripped free from an atom or molecule, resulting in free electrons and a positively charged atom or molecule.

ionosphere A layer high in Earth's atmosphere in which most of the atoms are ionized by solar radiation.

IR Infrared.

iron meteorite A metallic meteorite composed mostly of iron-nickel alloys. Compare *stony-iron meteorite* and *stony meteorite.*

irregular galaxy A galaxy without regular or symmetric appearance. Compare *elliptical galaxy, S0 galaxy,* and *spiral galaxy.*

irregular moon A moon that has been captured by a planet. Some irregular moons revolve in the opposite direction from the rotation of the planet, and many are in distant, unstable orbits. Compare *regular moon.*

isotopes Forms of the same element with differing numbers of neutrons.

isotropic In cosmology, describing a universe whose properties observers find to be the same in all directions.

J

J See *joule*.

jansky (Jy) The basic unit of flux density. Units: watts per square meter per hertz ($W/m^2/Hz$).

jet 1. A stream of gas and dust ejected from a comet nucleus by solar heating. 2. A collimated linear feature of bright emission extending from a protostar or active galactic nucleus.

joule (J) A unit of energy or work. $1\ J = 1$ newton meter.

Jy See *jansky*.

K

K See *kelvin*.

K-T boundary See *Cretaceous-Tertiary boundary*.

KBO See *Kuiper Belt object*.

kelvin (K) The basic unit of the Kelvin scale of temperature.

Kelvin scale The temperature scale—defined by William Thomson, better known as Lord Kelvin (1824–1907)—that uses Celsius-sized degrees, but defines 0 K as absolute zero instead of as the melting point of water. Compare *Celsius* and *Fahrenheit*.

Kepler's first law A rule of planetary motion, inferred by Johannes Kepler, stating that planets move in orbits of elliptical shapes with the Sun at one focus.

Kepler's laws The three rules of planetary motion inferred by Johannes Kepler from the data acquired by Tycho Brahe.

Kepler's second law Also called *law of equal areas*. A rule of planetary motion, inferred by Johannes Kepler, stating that a line drawn from the Sun to a planet sweeps out equal areas in equal times as the planet orbits the Sun.

Kepler's third law Also called *harmonic law*. A rule of planetary motion inferred by Johannes Kepler that describes the relationship between the period of a planet's orbit and its distance from the Sun. The law states that the square of the period of a planet's orbit, measured in years, is equal to the cube of the semimajor axis of the planet's orbit, measured in astronomical units: $(P_{years})^2 = (A_{AU})^3$.

kinetic energy (E_K) The energy of an object due to its motions. $E_K = \frac{1}{2}\ mv^2$. Units: joules (J).

Kirkwood gap A gap in the main asteroid belt related to orbital resonances with Jupiter.

Kuiper Belt A disk-shaped population of comet nuclei extending from Neptune's orbit to perhaps several thousand astronomical units (AU) from the Sun. The highly populated innermost part of the Kuiper Belt has an outer edge approximately 50 AU from the Sun.

Kuiper Belt object (KBO) Also called *trans-Neptunian object*. An icy planetesimal (comet nucleus) that orbits within the Kuiper Belt beyond the orbit of Neptune.

L

Lagrangian equilibrium point One of five points of equilibrium in a system consisting of two massive objects in nearly circular orbit around a common center of mass. Only two Lagrangian points (L_4 and L_5) represent stable equilibrium. A third smaller body located at one of the five points will move in lockstep with the center of mass of the larger bodies.

lander An instrumented spacecraft designed to land on a planet or moon. Compare *rover*.

large-scale structure Observable aggregates on the largest scales in the universe, including galaxy groups, clusters, and superclusters.

latitude The angular distance north (+) or south (−) from the equatorial plane of a nearly spherical body.

law of equal areas See *Kepler's second law*.

law of gravitation See *universal law of gravitation*.

leap year A year that contains 366 days. Leap years occur every 4 years when the year is divisible by 4, correcting for the accumulated excess time in a normal year, which is approximately 365¼ days long.

length contraction The relativistic compression of moving objects in the direction of their motion.

Leonids A November meteor shower associated with the dust debris left by comet Tempel-Tuttle. Compare *Perseids*.

libration The apparent wobble of an orbiting body that is tidally locked to its companion (such as Earth's Moon) resulting from the fact that its orbit is elliptical rather than circular.

life A biochemical process in which living organisms can reproduce, evolve, and sustain themselves by drawing energy from their environment. All terrestrial life involves carbon-based chemistry, assisted by the self-replicating molecules ribonucleic acid (RNA) and deoxyribonucleic acid (DNA).

light All electromagnetic radiation, which comprises the entire electromagnetic spectrum.

light-year (ly) The distance that light travels in 1 year—about 9 trillion kilometers (km).

limb The outer edge of the visible disk of a planet, moon, or the Sun.

limb darkening The darker appearance caused by increased atmospheric absorption near the limb of a planet or star.

line of nodes 1. A line defined by the intersection of two orbital planes. 2. The line defined by the intersection of Earth's equatorial plane and the plane of the ecliptic.

lithosphere The solid, brittle part of Earth (or any planet or moon), including the crust and the upper part of the mantle. Compare *biosphere* and *hydrosphere*.

lithospheric plate A separate piece of Earth's lithosphere capable of moving independently. See also *continental drift* and *plate tectonics*.

Local Group The small group of galaxies of which the Milky Way and the Andromeda galaxies are members.

long-period comet A comet with an orbital period of greater than 200 years. Compare *short-period comet*.

longitudinal wave A wave that oscillates parallel to the direction of the wave's propagation. Compare *transverse wave*.

look-back time The time that it has taken the light from an astronomical object to reach Earth.

low-mass star A star with a main-sequence mass of less than about 8 solar masses ($M_\odot$). Compare *high-mass star*.

luminosity The total flux emitted by an object. Unit: watts (W). See also *brightness*.

luminosity class A spectral classification based on stellar size, from the largest supergiants to the smallest white dwarfs.

luminosity-temperature-radius relationship A relationship among these three properties of stars indicating that if any two are known, the third can be calculated.

luminous matter Also called *normal matter*. Matter in galaxies—including stars, gas, and dust—that emits electromagnetic radiation. Compare *dark matter*.

lunar eclipse An eclipse that occurs when the Moon is partially or entirely in Earth's shadow. Compare *solar eclipse*.

lunar tide A tide on Earth that is due to the differential gravitational pull of the Moon. Compare *solar tide*. See also *tide* (definition 2).

ly See *light-year*.

M

M-type asteroid An asteroid that was once part of the metallic core of a larger, differentiated body that has since been broken into pieces; made mostly of iron and nickel. Compare *C-type asteroid* and *S-type asteroid*.

MACHO Literally, "massive compact halo object." MACHO's include brown dwarfs, white dwarfs, and black holes, which are candidates for being considered dark matter. Compare *WIMP*.

magma Molten rock, often containing dissolved gases and solid minerals.

magnetic field A field that is able to exert a force on a moving electric charge. Compare *electric field*.

magnetic force A force associated with, or caused by, the relative motion of charges. Compare *electric force*. See also *electromagnetic force*.

magnetosphere The region surrounding a planet that is filled with relatively intense magnetic fields and plasmas.

magnitude A system used by astronomers to describe the brightness or luminosity of stars. The brighter the star, the smaller its magnitude.

main asteroid belt See *asteroid belt*.

main sequence The strip on the H-R diagram where most stars are found. Main-sequence stars are fusing hydrogen to helium in their cores.

main-sequence lifetime The amount of time a star spends on the main sequence, fusing hydrogen into helium in its core.

main-sequence turnoff The location on the H-R diagram of a single-aged stellar population (such as a star cluster) where stars have just evolved off the main sequence. The position of the main-sequence turnoff is determined by the age of the stellar population.

mantle The solid portion of a rocky planet that lies between the crust and the core.

mare (pl. **maria**) A dark region on the Moon, composed of basaltic lava flows.

mass 1. Inertial mass: the property of matter that resists changes in motion. 2. Gravitational mass: the property of matter defined by its attractive force on other objects. According to general relativity, the two are equivalent.

mass-luminosity relationship An empirical relationship between the luminosity (L) and mass (M) of main-sequence stars expressed as a power law—for example, $L \propto M^{3.5}$.

mass transfer The transfer of mass from one member of a binary star system to its companion. Mass transfer occurs when one of the stars evolves to the point that it overfills its Roche lobe, so that its outer layers are pulled toward its binary companion.

massive element See *heavy element*.

matter 1. Objects made of particles that have mass, such as protons, neutrons, and electrons. 2. Anything that occupies space and has mass.

Maunder Minimum The period from 1645 to 1715, when very few sunspots were observed.

megabar A unit of pressure equal to 1 million bars.

mega-light-year (Mly) A unit of distance equal to 1 million light-years.

meridian The imaginary arc in the sky running from the horizon at due north through the zenith to the horizon at due south. The meridian divides the observer's sky into eastern and western halves. Compare *equator*.

mesosphere The layer of Earth's atmosphere immediately above the stratosphere, extending from an altitude of 50 kilometers (km) to about 90 km.

meteor The incandescent trail produced by a small piece of interplanetary debris as it travels through the atmosphere at very high speeds. Compare *meteorite* and *meteoroid*.

meteor shower A larger-than-normal display of meteors, occurring when Earth passes through the orbit of a disintegrating comet, sweeping up its debris.

meteorite A *meteoroid* that survives to reach a planet's surface. Compare *meteor* and *meteoroid*.

meteoroid A small cometary or asteroidal fragment ranging in size from 100 micrometers (μm) to 100 meters. When entering a planetary atmosphere, the meteoroid creates a *meteor*, which is an atmospheric phenomenon. Compare *meteor* and *meteorite*; also *planetesimal* and *zodiacal dust*.

micrometer (μm) Also called *micron*. 10^{-6} meter; a unit of length used for the wavelength of electromagnetic radiation. Compare *nanometer*.

micron See *micrometer*.

microwave radiation Electromagnetic radiation with frequencies and photon energies occurring in the spectral region between those of infrared radiation and radio waves.

Milky Way Galaxy The galaxy in which our Sun and Solar System reside.

minor planet See *asteroid*.

minute of arc See *arcminute*.

Mly See *mega-light-year*.

μm See *micrometer*.

modern physics Usually, the physical principles, including relativity and quantum mechanics, that have been developed since James Maxwell's equations were published.

molecular cloud An interstellar cloud composed primarily of molecular hydrogen.

molecular-cloud core A dense clump within a molecular cloud that forms as the cloud collapses and fragments. Protostars form from molecular-cloud cores.

molecule Generally, the smallest particle of a substance that retains its chemical properties and is composed of two or more atoms. A very few types of molecules, such as helium, are composed of single atoms.

momentum The product of the mass and velocity of a particle. Units: kilograms times meters per second (kg m/s).

moon A less massive satellite orbiting a more massive object. Moons are found around planets, dwarf planets, asteroids, and Kuiper Belt objects.

multiverse A collection of parallel universes that together comprise all that is.

mutation In biology, an imperfect reproduction of self-replicating material.

N

N See *newton*.

nadir The point on the celestial sphere located directly below an observer, opposite the *zenith*.

nanometer (nm) One billionth (10^{-9}) of a meter; a unit of length used for the wavelength of light. Compare *micrometer*.

natural selection The process by which forms of structure, ranging from molecules to whole organisms, that are best adapted to their environment become more common than less well-adapted forms.

NCP See *north celestial pole*.

neap tide An especially weak tide that occurs around the time of the first or third quarter Moon when the gravitational forces of the Moon and the Sun on Earth are at right angles to each other, thus producing the least pronounced tides. Compare *spring tide*. See also *tide* (definition 2).

near-Earth asteroid An asteroid whose orbit brings it close to the orbit of Earth. See also *near-Earth object*.

near-Earth object (NEO) An asteroid, comet, or large meteoroid whose orbit intersects Earth's orbit.

nebula (pl. **nebulae**) A cloud of interstellar gas and dust, either illuminated by stars (bright nebula) or seen in silhouette against a brighter background (dark nebula).

nebular hypothesis The first plausible theory of the formation of the Solar System, proposed by Immanuel Kant in 1734. Kant hypothesized that the Solar System formed from the collapse of an interstellar cloud of rotating gas.

NEO See *near-Earth object*.

neutrino A very low-mass, electrically neutral particle emitted during beta decay. Neutrinos interact with matter only very feebly and so can penetrate through great quantities of matter.

neutrino cooling The process in which thermal energy is carried out of the center of a star by neutrinos rather than by electromagnetic radiation or convection.

neutron A subatomic particle having no net electric charge, and a rest mass and rest energy nearly equal to that of the proton. Compare *electron* and *proton*.

neutron star The neutron-degenerate remnant left behind by a Type II supernova.

new Moon The phase of the Moon in which the Moon is between Earth and the Sun, and from Earth we see only the side of the Moon not being illuminated by the Sun. Compare *full Moon*.

newton (N) The force required to accelerate a 1-kilogram (kg) mass at a rate of 1 meter per second per second (m/s²). Units: kilograms times meters per second squared (kg m/s²).

Newton's first law of motion The law, formulated by Isaac Newton, stating that an object will remain at rest or will continue moving along a straight line at a constant speed until an unbalanced force acts on it.

Newton's laws See *Newton's first law of motion*, *Newton's second law of motion*, and *Newton's third law of motion*.

Newton's second law of motion The law, formulated by Isaac Newton, stating that if an unbalanced force acts on a body, the body will have an acceleration proportional to the unbalanced force and inversely proportional to the object's mass: $a = F/m$. The acceleration will be in the direction of the unbalanced force.

Newton's third law of motion The law, formulated by Isaac Newton, stating that for every force there is an equal and opposite force.

nm See *nanometer*.

normal matter See *luminous matter*.

north celestial pole (NCP) The northward projection of Earth's rotation axis onto the celestial sphere. Compare *south celestial pole*.

North Pole The location in the Northern Hemisphere where Earth's rotation axis intersects the surface of Earth. Compare *South Pole*.

nova (pl. novae) A stellar explosion that results from runaway nuclear fusion in a layer of material on the surface of a white dwarf in a binary system.

nuclear burning Release of energy by fusion of low-mass elements.

nuclear fusion The combination of two less massive atomic nuclei into a single more massive atomic nucleus.

nucleosynthesis The formation of more massive atomic nuclei from less massive nuclei, either in the Big Bang (Big Bang nucleosynthesis) or in the interiors of stars (stellar nucleosynthesis).

nucleus (pl. nuclei) 1. The dense, central part of an atom. 2. The central core of a galaxy, comet, or other diffuse object.

objective lens The primary optical element in a telescope or camera that produces an image of an object.

oblateness The flattening of an otherwise spherical planet or star caused by its rapid rotation.

obliquity The inclination of a celestial body's equator to its orbital plane.

observational uncertainty The fact that real measurements are never perfect; all observations are uncertain by some amount.

Occam's razor The principle that the simplest hypothesis is the most likely; named after William of Occam (circa 1285–1349), the medieval English cleric to whom the idea is attributed.

Oort Cloud A spherical distribution of comet nuclei stretching from beyond the Kuiper Belt to more than 50,000 astronomical units (AU) from the Sun.

opacity A measure of how effectively a material blocks the radiation going through it.

open cluster A loosely bound group of a few dozen to a few thousand stars that formed together in the disk of a spiral galaxy. Compare *globular cluster*.

open universe An infinite universe with a negatively curved spatial structure (much like the surface of a saddle) such that the sum of the angles of a triangle is always less than 180°. Compare *closed universe* and *flat universe*.

orbit The path taken by one object moving around another object under the influence of their mutual gravitational or electric attraction.

orbital resonance A situation in which the orbital periods of two objects are related by a ratio of small integers.

orbiter A spacecraft that is placed in orbit around a planet or moon. Compare *flyby*.

organic Describing a substance, not necessarily of biological origin, that contains the element carbon.

P

P wave See *primary wave*.

pair production The creation of a particle-antiparticle pair from a source of electromagnetic energy.

paleomagnetism The record of Earth's magnetic field as preserved in rocks.

parallax 1. The apparent shift in the position of one object relative to another object, caused by the changing perspective of the observer. 2. In astronomy, the displacement in the apparent position of a nearby star caused by the changing location of Earth in its orbit.

parent element A radioactive element that decays to form more stable *daughter products*.

parsec (pc) The distance to a star with a parallax of 1 arcsecond using a base of 1 astronomical unit (AU). One parsec is approximately 3.26 light-years.

partial lunar eclipse A lunar eclipse in which the Moon passes through the penumbra of Earth's shadow. Compare *total lunar eclipse*.

partial solar eclipse The type of eclipse that occurs when Earth passes through the penumbra of the Moon's shadow, so that the Moon blocks only a portion of the Sun's disk. Compare *annular solar eclipse* and *total solar eclipse*.

pc See *parsec*.

peculiar velocity The motion of a galaxy relative to the overall expansion of the universe.

penumbra (pl. penumbrae) 1. The outer part of a shadow, where the source of light is only partially blocked. 2. The region surrounding the *umbra* of a sunspot. The penumbra is cooler and darker than the surrounding surface of the Sun but not as cool or dark as the umbra of the sunspot.

perihelion (pl. perihelia) The point in a solar orbit that is closest to the Sun. Compare *aphelion*.

period The time it takes for a regularly repetitive process to complete one cycle.

period-luminosity relationship The relationship between the period of variability of a pulsating variable star, such as a Cepheid or RR Lyrae variable, and the luminosity of the star. Longer-period Cepheid or RR Lyrae variables are more luminous than their shorter-period cousins.

Perseids A prominent August meteor shower associated with the dust debris left by comet Swift-Tuttle. Compare *Leonids*.

phase One of the various appearances of the sunlit surface of the Moon or a planet caused by the change in viewing location of Earth relative to both the Sun and the object. Examples include crescent phase and gibbous phase.

photino An elementary particle related to the photon. One of the leading candidates for cold dark matter.

photochemical reaction A chemical reaction driven by the absorption of electromagnetic radiation.

photodissociation The breaking apart of molecules into smaller fragments or individual atoms by the action of photons.

photoelectric effect An effect whereby electrons are emitted from a substance illuminated by photons above a certain critical frequency.

photometry The process of measuring the brightness of a source of light, generally over a specific range of wavelength.

photon Also called *quantum of light*. A discrete unit or particle of electromagnetic radiation. The energy of a photon is equal to Planck's constant (h) multiplied by the frequency (f) of its electromagnetic radiation: $E_{photon} = h \times f$. The photon is the particle that mediates the electromagnetic force.

photosphere The apparent surface of the Sun as seen in visible light.

physical law A broad statement that predicts a particular aspect of how the physical universe behaves and that is supported by many empirical tests. See also *theory*.

pixel The smallest picture element in a digital image array.

Planck era The early time, just after the Big Bang, when the universe as a whole must be described with quantum mechanics.

Planck spectrum See *blackbody spectrum*.

Planck's constant (h) The constant of proportionality between the energy of a photon and the frequency of the photon. This constant defines how much energy a single photon of a given frequency or wavelength has. Value: $h = 6.63 \times 10^{-34}$ joule-second.

planet 1. A large body that orbits the Sun or other star that shines only by light reflected from the Sun or star. 2. In the Solar System, a body that orbits the Sun, has sufficient mass for self-gravity to overcome rigid body forces so that it assumes a spherical shape, and has cleared smaller bodies from the neighborhood around its orbit. Compare *dwarf planet*.

planet migration The theory that a planet can move from its formation distance around its parent star to a different distance though gravitational interactions with other bodies or loss of orbital energy from interaction with gas in the protoplanetary disk.

planetary nebula The expanding shell of material ejected by a dying asymptotic giant branch star. A planetary nebula glows from fluorescence caused by intense ultraviolet light coming from the hot, stellar remnant at its center.

planetary system A system of planets and other smaller objects in orbit around a star.

planetesimal A primitive body of rock and ice, 100 meters or more in diameter, that combines with others to form a planet. Compare *meteoroid* and *zodiacal dust*.

plasma A gas that is composed largely of charged particles but also may include some neutral atoms.

plate tectonics The geological theory concerning the motions of lithospheric plates, which in turn provides the theoretical basis for continental drift.

positron A positively charged subatomic particle; the antiparticle of the *electron*.

power The rate at which work is done or at which energy is delivered. Unit: watts (W) or joules per second (J/s).

precession of the equinoxes The slow change in orientation between the ecliptic plane and the celestial equator caused by the wobbling of Earth's axis.

pressure Force per unit area. Units: newtons per square meter (N/m^2) or bars.

primary atmosphere An atmosphere, composed mostly of hydrogen and helium, that forms at the same time as its host planet. Compare *secondary atmosphere*.

primary mirror The principal optical mirror in a reflecting telescope. The primary mirror determines the telescope's light-gathering power and resolution. Compare *secondary mirror*.

primary wave Also called *P wave*. A longitudinal seismic wave, in which the oscillations involve compression and decompression parallel to the direction of travel (that is, a pressure wave). Compare *secondary wave*.

principle A general idea or sense about how the universe is that guides us in constructing new scientific theories. Principles can be testable theories.

prograde motion 1. Rotational or orbital motion of a moon that is in the same sense as the planet it orbits. 2. The counterclockwise orbital motion of Solar System objects as seen from above Earth's orbital plane. Compare *retrograde motion*.

prominence An archlike projection above the solar photosphere often associated with a sunspot.

proportional Describing two things whose ratio is a constant.

proton (p or p⁺) A subatomic particle having a positive electric charge of 1.6×10^{-19} coulomb (C), a mass of 1.67×10^{-27} kilograms (kg), and a rest energy of 1.5×10^{-10} joules (J). Compare *electron* and *neutron*.

proton-proton chain One of the ways in which hydrogen burning can take place. This is the most important path for hydrogen burning in low-mass stars such as the Sun. See also *carbon-nitrogen-oxygen cycle* and *triple-alpha process*.

protoplanetary disk Also called *circumstellar disk*. The remains of the accretion disk around a young star from which a planetary system may form.

protostar A young stellar object that derives its luminosity from the conversion of gravitational energy to thermal energy, rather than from nuclear reactions in its core.

pulsar A rapidly rotating neutron star that beams radiation into space in two searchlight-like beams. To a distant observer, the star appears to flash on and off, earning its name.

pulsating variable star A variable star that undergoes periodic radial pulsations.

Q

QCD See *quantum chromodynamics*.

QED See *quantum electrodynamics*.

quantized Describing a quantity that exists as discrete, irreducible units.

quantum chromodynamics (QCD) The quantum mechanical theory describing the strong nuclear force and its mediation by gluons. Compare *quantum electrodynamics*.

quantum efficiency The fraction of photons falling on a detector that actually produces a response in the detector.

quantum electrodynamics (QED) The quantum theory describing the electromagnetic force and its mediation by photons. Compare *quantum chromodynamics*.

quantum mechanics The branch of physics that deals with the quantized and probabilistic behavior of atoms and subatomic particles.

quantum of light See *photon*.

quark The building block of protons and neutrons.

quasar Short for *quasi-stellar radio* source. The most luminous of the active galactic nuclei, seen only at great distances from our galaxy.

R

radial velocity The component of velocity that is directed toward or away from the observer.

radian The angle at the center of a circle subtended by an arc equal to the length of the circle's radius. Therefore, 2π radians equals 360°, and 1 radian equals approximately 57.3°.

radiant The direction in the sky from which the meteors in a meteor shower seem to come.

radiation belt A toroidal ring of high-energy particles surrounding a planet.

radiative transfer The transport of energy from one location to another by electromagnetic radiation.

radiative zone A region in the interior of a star through which energy is transported outward by radiation.

radio galaxy A type of elliptical galaxy with an active galactic nucleus at its center and having very strong emission (10^{35}–10^{38} watts [W]) in the radio part of the electromagnetic spectrum. Compare *Seyfert galaxy*.

radio telescope An instrument for detecting and measuring radio frequency emissions from celestial sources.

radio wave Electromagnetic radiation in the extreme long-wavelength region of the spectrum, beyond the region of microwaves.

radioisotope A radioactive element.

radiometric dating Use of radioactive decay to measure the ages of materials such as minerals.

ratio The relationship in quantity or size between two or more things.

ray 1. A beam of electromagnetic radiation. 2. A bright streak emanating from a young impact crater.

recombination 1. The combining of ions and electrons to form neutral atoms. 2. An event early in the evolution of the universe in which hydrogen and helium nuclei combined with electrons to form neutral atoms. The removal of electrons caused the universe to become transparent to electromagnetic radiation.

red giant A low-mass star that has evolved beyond the main sequence and is now fusing hydrogen in a shell surrounding a degenerate helium core.

red giant branch A region on the H-R diagram defined by low-mass stars evolving from the main sequence toward the horizontal branch.

reddening The effect by which stars and other objects, when viewed through interstellar dust, appear redder than they actually are. Reddening is caused by the fact that blue light is more strongly absorbed and scattered than red light.

redshift Also called *Doppler redshift*. The shift toward longer wavelengths of light by any of several effects, including Doppler shifts, gravitational redshift, or cosmological redshift. Compare *blueshift*.

reflecting telescope A telescope that uses mirrors for collecting and focusing incoming electromagnetic radiation to form an image in their focal planes. The size of a reflecting telescope is defined by the diameter of the primary mirror. Compare *refracting telescope*.

reflection The redirection of a beam of light that is incident on, but does not cross, the surface between two media having different refractive indices. If the surface is flat and smooth, the angle of incidence equals the angle of reflection. Compare *refraction*.

refracting telescope A telescope that uses objective lenses to collect and focus light. Compare *reflecting telescope*.

refraction The redirection or bending of a beam of light when it crosses the boundary between two media having different refractive indices. Compare *reflection*.

refractory material Material that remains solid at high temperatures. Compare *volatile material*.

regular moon A moon that formed together with the planet it orbits. Compare *irregular moon*.

relative humidity The amount of water vapor held by a volume of air at a given temperature compared (stated as a percentage) to the total amount of water that could be held by the same volume of air at the same temperature.

relative motion The difference in motion between two individual frames of reference.

relativistic Describing physical processes that take place in systems traveling at nearly the speed of light or located in the vicinity of very strong gravitational fields.

relativistic beaming The effect created when material moving at nearly the speed of light beams the radiation it emits in the direction of its motion.

remote sensing The use of images, spectra, radar, or other techniques to measure the properties of an object from a distance.

resolution The ability of a telescope to separate two point sources of light. Resolution is determined by the telescope's aperture and the wavelength of light it receives.

rest wavelength The wavelength of light we see coming from an object at rest with respect to the observer.

retrograde motion 1. Rotation or orbital motion of a moon that is in the opposite sense to the rotation of the planet it orbits. 2. The clockwise orbital motion of Solar System objects as seen from above Earth's orbital plane. Compare *prograde motion*.

ring An aggregation of small particles orbiting a planet or star. The rings of the four giant planets of the Solar System are composed variously of silicates, organic materials, and ices.

ring arc A discontinuous, higher-density region within an otherwise continuous, narrow ring.

ringlet A narrowly confined concentration of ring particles.

Roche limit The distance at which a planet's tidal forces exceed the self-gravity of a smaller object, such as a moon, asteroid, or comet, causing the object to break apart.

Roche lobe The hourglass- or figure eight-shaped volume of space surrounding two stars, which constrains material that is gravitationally bound by one or the other.

rotation curve A plot showing how the orbital velocity of stars and gas in a galaxy changes with radial distance from the galaxy's center.

rover A remotely controlled instrumented vehicle designed to traverse and explore the surface of a terrestrial planet or moon. Compare *lander*.

RR Lyrae variable A variable giant star whose regularly timed pulsations are good predictors of its luminosity. RR Lyrae stars are used for distance measurements to globular clusters.

S

S-type asteroid An asteroid made of material that has been modified from its original state, likely as the outer part of a larger, differentiated body that has since broken into pieces. Compare *C-type asteroid* and *M-type asteroid*.

S wave See *secondary wave*.

S0 galaxy A galaxy with a bulge and a disk-like spiral, but smooth in appearance like ellipticals. Compare *elliptical galaxy*, *irregular galaxy*, and *spiral galaxy*.

satellite 1. An object in orbit about a more massive body. 2. A moon.

scale factor (R_U) A dimensionless number proportional to the distance between two points in space. The scale factor increases as the universe expands.

scattering The random change in the direction of travel of photons, caused by their interactions with molecules or dust particles.

Schwarzschild radius The distance from the center of a nonrotating, spherical black hole at which the escape velocity equals the speed of light.

scientific method The formal procedure—including hypothesis, prediction, and experiment or observation—used to test (attempt to falsify) the validity of scientific hypotheses and theories.

scientific notation The standard expression of numbers with one digit (which can be zero) to the left of the decimal point and multiplied by 10 to the exponent required to give the number its correct value. Example: $2.99 \times 10^8 = 299,000,000$.

SCP See *south celestial pole*.

second law of thermodynamics The law stating that the entropy or disorder of an isolated system always increases as the system evolves.

second of arc See *arcsecond*.

secondary atmosphere An atmosphere that formed—as a result of volcanism, comet impacts, or another process—sometime after its host planet formed. Compare *primary atmosphere*.

secondary crater A crater formed from ejecta thrown from an *impact crater*.

secondary mirror A small mirror placed on the optical axis of a reflecting telescope that returns the beam back through a small hole in the *primary mirror*, thereby shortening the mechanical length of the telescope.

secondary wave Also called *S wave*. A transverse seismic wave, which involves the sideways motion of material. Compare *primary wave*.

seismic wave A vibration due to an earthquake, a large explosion, or an impact on the surface that travels through a planet's interior.

seismometer An instrument that measures the amplitude and frequency of seismic waves.

self-gravity The gravitational attraction among all the parts of the same object.

semimajor axis Half of the longer axis of an ellipse.

SETI The Search for Extraterrestrial Intelligence project, which uses advanced technology combined with radio telescopes to search for evidence of intelligent life elsewhere in the universe.

Seyfert galaxy A type of spiral galaxy with an active galactic nucleus at its center; first discovered in 1943 by Carl Seyfert. Compare *radio galaxy*.

shepherd moon A moon that orbits close to rings and gravitationally confines the orbits of the ring particles.

shield volcano A volcano formed by very fluid lava flowing from a single source and spreading out from that source. Compare *composite volcano*.

short-period comet A comet with an orbital period of less than 200 years. Compare *long-period comet*.

sidereal day The length of time the Earth takes to rotate to the same position relative to the stars: 23^h56^m. Compare *solar day*.

sidereal period An object's orbital or rotational period measured with respect to the stars. Compare *synodic period*.

silicate One of the family of minerals composed of silicon and oxygen in combination with other elements.

singularity The point where a mathematical expression or equation becomes meaningless, such as the denominator of a fraction approaching zero. See also *black hole*.

solar abundance The relative amount of an element detected in the atmosphere of the Sun, expressed as the ratio of the number of atoms of that element to the number of hydrogen atoms.

solar day The 24-hour period of Earth's axial rotation that brings the Sun back to the same local meridian where the rotation started. Compare *sidereal day*.

solar eclipse An eclipse that occurs when the Sun is partially or entirely blocked by the Moon. Compare *lunar eclipse*.

solar flare Explosions on the Sun's surface associated with complex sunspot groups and strong magnetic fields.

solar maximum (pl. maxima) The time, occurring about every 11 years, when the Sun is at its peak activity, meaning that sunspot activity and related phenomena (such as prominences, flares, and coronal mass ejections) are at their peak.

solar neutrino problem The historical observation that only about a third as many neutrinos as predicted by theory seemed to be coming from the Sun.

Solar System The gravitationally bound system made up of the Sun, planets, dwarf planets, moons, asteroids, comets, and Kuiper Belt objects, along with their associated gas and dust.

solar tide A tide on Earth that is due to the differential gravitational pull of the Sun. Compare *lunar tide*. See also *tide* (definition 2).

solar wind The stream of charged particles emitted by the Sun that flows at high speeds through interplanetary space.

solstice Literally, "sun standing still." 1. One of the two most northerly and southerly points on the ecliptic. 2. Either of the two times of year (the *summer solstice* and *winter solstice*) when the Sun is at one of these two positions. Compare *equinox*.

south celestial pole (SCP) The southward projection of Earth's rotation axis onto the celestial sphere. Compare *north celestial pole*.

South Pole The location in the Southern Hemisphere where Earth's rotation axis intersects the surface of Earth. Compare *North Pole*.

spacetime The four-dimensional continuum in which we live, and which we experience as three spatial dimensions plus time.

special relativity See *special theory of relativity*.

special theory of relativity Sometimes referred to as simply *special relativity*. Einstein's theory explaining how the fact that the speed of light is a constant affects nonaccelerating frames of reference. Compare *general theory of relativity*.

spectral type A classification system for stars based on the presence and relative strength of absorption lines in their spectra. Spectral type is related to the surface temperature of a star.

spectrograph A device that spreads out the light from an object into its component wavelengths. See also *spectrometer*.

spectrometer A *spectrograph* in which the spectrum is generally recorded digitally by electronic means.

spectroscopic binary A binary star pair whose existence and properties are revealed only by the Doppler shift of its spectral lines. Most spectroscopic binaries are close pairs. Compare *eclipsing binary* and *visual binary*.

spectroscopic parallax Use of the spectroscopically determined luminosity and the observed brightness of a star to determine the star's distance.

spectroscopic radial velocity method A technique for detecting extrasolar planets by examining Doppler shifts in the light from stars.

spectroscopy The study of electromagnetic radiation from an object in terms of its component wavelengths.

spectrum (pl. spectra) 1. The intensity of electromagnetic radiation as a function of wavelength. 2. Waves sorted by wavelength.

speed The rate of change of an object's position with time, without regard to the direction of movement. Units: meters per second (m/s) or kilometers per hour (km/h). Compare *velocity*.

spherically symmetric Describing an object whose properties depend only on distance from the object's center, so that the object has the same form viewed from any direction.

spin-orbit resonance A relationship between the orbital and rotation periods of an object such that the ratio of their periods can be expressed by simple integers.

spiral density wave A stable, spiral-shaped change in the local gravity of a galactic disk that can be produced by periodic gravitational kicks from neighboring galaxies or from nonspherical bulges and bars in spiral galaxies.

spiral galaxy A galaxy of Hubble type "S" class, with a discernible disk in which large spiral patterns exist. Compare *elliptical galaxy*, *irregular galaxy*, and *S0 galaxy*.

spoke One of several narrow radial features seen occasionally in Saturn's B ring. Spokes appear dark in backscattered light and bright in forward, scattering light, indicating that they are composed of tiny particles. Their origin is not well understood.

sporadic meteor A meteor that is not associated with a specific meteor shower.

spreading center A zone from which two tectonic plates diverge.

spring tide An especially strong tide that occurs near the time of a new or full Moon, when lunar tides and solar tides reinforce each other. Compare *neap tide*. See also *tide* (definition 2).

stable equilibrium An equilibrium state in which the system returns to its former condition after a small disturbance. Compare *unstable equilibrium*.

standard candle An object whose luminosity either is known or can be predicted in a distance-independent way, so its brightness can be used to determine its distance via the inverse square law of radiation.

standard model The theory of particle physics that combines electroweak theory with quantum chromodynamics to describe the structure of known forms of matter.

star A luminous ball of gas that is held together by gravity. A normal star is powered by nuclear reactions in its interior.

star cluster A group of stars that all formed at the same time and in the same general location.

static equilibrium A state in which the forces within a system are all in balance so that the system does not change. Compare *dynamic equilibrium*.

Stefan-Boltzmann constant (σ) The proportionality constant that relates the flux emitted by an object to the fourth power of its absolute temperature. Value: $5.67 \times 10^{-8}\,W/(m^2\,K^4)$ (W = watts, m = meters, K = kelvin).

Stefan-Boltzmann law The law stating that the amount of electromagnetic energy emitted from the surface of a body, summed over the energies of all photons of all wavelengths emitted, is proportional to the fourth power of the temperature of the body.

stellar mass loss The loss of mass from the outermost parts of a star's atmosphere during the course of its evolution.

stellar occultation An event in which a planet or other Solar System body moves between the observer and a star, eclipsing the light emitted by that star.

stellar population A group of stars with similar ages, chemical compositions, and dynamic properties.

stereoscopic vision The way an animal's brain combines the different information from its two eyes to perceive the distances to objects around it.

stony-iron meteorite A meteorite consisting of a mixture of silicate minerals and iron-nickel alloys. Compare *iron meteorite* and *stony meteorite*.

stony meteorite A meteorite composed primarily of silicate minerals, similar to those found on Earth. Compare *iron meteorite* and *stony-iron meteorite*.

stratosphere The atmospheric layer immediately above the *troposphere*. On Earth it extends upward to an altitude of 50 kilometers (km).

string theory The theory that conceives of particles as strings in 10 dimensions of space and time; the current contender for a theory of everything.

strong nuclear force The attractive short-range force between protons and neutrons that holds atomic nuclei together; one of the four fundamental forces of nature, mediated by the exchange of gluons. Compare *weak nuclear force*.

subduction zone A region where two tectonic plates converge, with one plate sliding under the other and being drawn downward into the interior.

subgiant A giant star smaller and lower in luminosity than normal giant stars of the same spectral type. Subgiants evolve to become giants.

subgiant branch A region of the H-R diagram defined by stars that have left the main sequence but have not yet reached the red giant branch.

sublimation The process in which a solid becomes a gas without first becoming a liquid.

subsonic Moving within a medium at a speed slower than the speed of sound in that medium. Compare *supersonic*.

summer solstice 1. One of two points where the Sun is at its greatest distance from the celestial equator. 2. The day on which the Sun appears at this location, marking the first day of summer (about June 21 in the Northern Hemisphere and December 22 in the Southern Hemisphere). Compare *winter solstice*.

sungrazer A comet whose perihelion is within a few solar diameters of the surface of the Sun.

sunspot A cooler, transitory region on the solar surface produced when loops of magnetic flux break through the surface of the Sun.

sunspot cycle The approximate 11-year cycle during which sunspot activity increases and then decreases. This is one-half of a full 22-year cycle, in which the magnetic polarity of the Sun first reverses and then returns to its original configuration.

supercluster A large conglomeration of galaxy clusters and galaxy groups; typically, more than 100 million light-years (Mly) in size and containing tens of thousands to hundreds of thousands of galaxies. Compare *galaxy cluster* and *galaxy group*.

superluminal motion The appearance (though not the reality) that a jet is moving faster than the speed of light.

supermassive black hole A black hole of 1,000 solar masses ($M_\odot$) or more that resides in the center of a galaxy, and whose gravity powers active galactic nuclei.

supernova (pl. **supernovae**) A stellar explosion resulting in the release of tremendous amounts of energy, including the high-speed ejection of matter into the interstellar medium. See also *Type Ia supernova* and *Type II supernova*.

supersonic Moving within a medium at a speed faster than the speed of sound in that medium. Compare *subsonic*.

superstring theory See *string theory*.

surface brightness The amount of electromagnetic radiation emitted or reflected per unit area.

surface wave A seismic wave that travels on the surface of a planet or moon.

symmetry 1. The property that an object has if the object is unchanged by rotation or reflection about a particular point, line, or plane. 2. In theoretical physics, the correspondence of different aspects of physical laws or systems, such as the symmetry between matter and antimatter.

synchronous rotation The case in which the period of rotation of a body on its axis equals the period of revolution in its orbit around another body. A special type of spin-orbit resonance.

synchrotron radiation Radiation from electrons moving at close to the speed of light as they spiral in a strong magnetic field; named because this kind of radiation was first identified on Earth in particle accelerators called "synchrotrons."

synodic period An object's orbital or rotational period measured with respect to the Sun. Compare *sidereal period*.

T

T Tauri star A young stellar object that has dispersed enough of the material surrounding it to be seen in visible light.

tail A stream of gas and dust swept away from the coma of a comet by the solar wind and by radiation pressure from the Sun.

tectonism Deformation of the lithosphere of a planet.

telescope The basic tool of astronomers. Working over the entire range from gamma rays to radio, astronomical telescopes collect and concentrate electromagnetic radiation from celestial objects.

temperature A measure of the average kinetic energy of the atoms or molecules in a gas, solid, or liquid.

terrestrial planet An Earth-like planet, made of rock and metal and having a solid surface. In our Solar System, the terrestrial planets are Mercury, Venus, Earth, and Mars. Compare *giant planet*.

theoretical model A detailed description of the properties of a particular object or system in terms of known physical laws or theories. Often, a computer calculation of predicted properties based on such a description.

theory A well-developed idea or group of ideas that are tied solidly to known physical laws and make testable predictions about the world. A very well-tested theory may be called a *physical law*, or simply a fact. Compare *hypothesis*.

theory of everything (TOE) A theory that unifies all four fundamental forces of nature: strong nuclear, weak nuclear, electromagnetic, and gravitational forces.

thermal conduction The transfer of energy in which the thermal energy of particles is transferred to adjacent particles by collisions or other interactions. Conduction is the most important way that thermal energy is transported in solid matter.

thermal energy The energy that resides in the random motion of atoms, molecules, and particles, by which we measure their temperature.

thermal equilibrium The state in which the rate of thermal-energy emission by an object is equal to the rate of thermal-energy absorption.

thermal motion The random motion of atoms, molecules, and particles that gives rise to thermal radiation.

thermal radiation Electromagnetic radiation resulting from the random motion of the charged particles in every substance.

thermosphere The layer of Earth's atmosphere at altitudes greater than 90 kilometers (km), above the mesosphere. Near its top, at an altitude of 600 km, the temperature can reach 1000 K.

third quarter Moon The phase of the Moon in which only the eastern half of the Moon, as viewed from Earth, is illuminated by the Sun. It occurs about one week after the full Moon. Compare *first quarter Moon*.

tidal bulge A distortion of a body resulting from tidal stresses.

tidal locking Synchronous rotation of an object caused by internal friction as the object rotates through its tidal bulge.

tidal stress Stress due to differences in the gravitational force of one mass on different parts of another mass.

tide 1. The deformation of a mass due to differential gravitational effects of one mass on another because of the extended size of the masses. 2. On Earth, the rise and fall of the oceans as Earth rotates through a tidal bulge caused by the Moon and Sun.

time dilation The relativistic "stretching" of time. Compare *general relativistic time dilation*.

TNO See *trans-Neptunian object*.

TOE See *theory of everything*.

topographic relief The differences in elevation from point to point on a planetary surface.

tornado A violent rotating column of air, typically 75 meters across with 200- kilometer-per-hour (km/h) winds. Some tornadoes can be more than 3 km across, and winds up to 500 km/h have been observed.

torus A three-dimensional, doughnut-shaped ring.

total lunar eclipse A lunar eclipse in which the Moon passes through the umbra of Earth's shadow. Compare *partial lunar eclipse*.

total solar eclipse The type of eclipse that occurs when Earth passes through the umbra of the Moon's shadow, so that the Moon completely blocks the disk of the Sun. Compare *annular solar eclipse* and *partial solar eclipse*.

transform fault The actively slipping segment of a fracture zone between lithospheric plates.

transit method A technique for detecting extrasolar planets by observing a star's decrease in brightness when a planet passes in front of it.

trans-Neptunian object (TNO) See *Kuiper Belt object*.

transverse wave A wave that oscillates perpendicular to the direction of the wave's propagation. Compare *longitudinal wave*.

triple-alpha process The nuclear fusion reaction that combines three helium nuclei (alpha particles) together into a single nucleus of carbon. See also *carbon-nitrogen-oxygen cycle* and *proton-proton chain*.

Trojan asteroid One of a group of asteroids orbiting in the L_4 and L_5 Lagrangian points of Jupiter's orbit.

tropical year The time between one crossing of the vernal equinox and the next. Because of the precession of the equinoxes, a tropical year is slightly shorter than the time that it takes for Earth to orbit once about the Sun.

Tropics The region on Earth between latitudes 23.5° south and 23.5° north, and in which the Sun appears directly overhead twice during the year.

tropopause The top of a planet's troposphere.

troposphere The convection-dominated layer of a planet's atmosphere. On Earth, the atmospheric region closest to the ground within which most weather phenomena take place. Compare *stratosphere*.

tuning fork diagram The two-pronged diagram showing Hubble's classification of galaxies into ellipticals, S0s, spirals, barred spirals, and irregular galaxies.

turbulence The random motion of blobs of gas within a larger cloud of gas.

Type Ia supernova A supernova explosion in which no trace of hydrogen is seen in the ejected material. Most Type Ia supernovae are thought to be the result of runaway carbon burning in a white dwarf star onto which material is being deposited by a binary companion.

Type II supernova A supernova explosion in which the degenerate core of an evolved massive star suddenly collapses and rebounds.

typhoon See *hurricane*.

U

ultraviolet (UV) radiation Electromagnetic radiation with frequencies and photon energies greater than those of visible light but less than those of X-rays, and wavelengths shorter than those of visible light but longer than those of X-rays. Compare *infrared radiation*.

umbra (pl. umbrae) 1. The darkest part of a shadow, where the source of light is completely blocked. 2. The darkest, innermost part of a sunspot. Compare *penumbra*.

unbalanced force The nonzero net force acting on a body.

unbound orbit An orbit in which the velocity is greater than the escape velocity. Compare *bound orbit*.

uncertainty principle See *Heisenberg uncertainty principle*.

unified model of AGNs A model in which many different types of activity in the nuclei of galaxies are all explained by accretion of matter around a supermassive black hole.

uniform circular motion Motion in a circular path at a constant speed.

unit A fundamental quantity of measurement—for example, metric units or English units.

universal gravitational constant (G) The constant of proportionality in the universal law of gravity. Value: $G = 6.673 \times 10^{-11}$ newtons times meters squared per kilogram squared (N m²/kg²).

universal law of gravitation The law stating that the gravitational force between any two objects is proportional to the product of their masses and inversely proportional to the square of the distance between them: $F \propto (m_1 m_2/r^2)$.

universe All of space and everything contained therein.

unstable equilibrium An equilibrium state in which a small disturbance will cause a system to move away from equilibrium. Compare *stable equilibrium*.

UV Ultraviolet.

V

vacuum A region of space that contains very little matter. In quantum mechanics and general relativity, however, even a perfect vacuum has physical properties.

variable star A star with varying luminosity. Many periodic variables are found within the instability strip on the H-R diagram.

velocity The rate and direction of change of an object's position with time. Units: meters per second (m/s) or kilometers per hour (km/h). Compare *speed*.

vernal equinox 1. One of two points where the Sun crosses the celestial equator. 2. The day on which the Sun appears at this location, marking the first day of spring (about March 21 in the Northern Hemisphere and September 23 in the Southern Hemisphere). Compare *autumnal equinox*.

virtual particle A particle that, according to quantum mechanics, comes into existence only momentarily. According to theory, fundamental forces are mediated by the exchange of virtual particles.

visual binary A binary system in which the two stars can be seen individually from Earth. Compare *eclipsing binary* and *spectroscopic binary*.

void A region in space containing little or no matter. Examples include regions in cosmological space that are largely empty of galaxies.

volatile material Sometimes called *ice*. Material that remains gaseous at moderate temperature. Compare *refractory material*.

volcanism The occurrence of volcanic activity on a planet or moon.

vortex (pl. vortices) Any circulating fluid system. Specifically, 1. an atmospheric anticyclone or cyclone; 2. a whirlpool or eddy.

W

W See *watt*.

waning The changing phases of the Moon as it becomes less fully illuminated between full Moon and new Moon as seen from Earth. Compare *waxing*.

watt (W) A measure of *power*. Units: joules per second (J/s).

wave A disturbance moving along a surface or passing through a space or a medium.

wavefront The imaginary surface of an electromagnetic wave, either plane or spherical, oriented perpendicular to the direction of travel.

wavelength The distance on a wave between two adjacent points having identical characteristics. The distance a wave travels in one period. Unit: meter.

waxing The changing phases of the Moon as it becomes more fully illuminated between new Moon and full Moon as seen from Earth. Compare *waning*.

weak nuclear force The force underlying some forms of radioactivity and certain interactions between subatomic particles. It is responsible for radioactive beta decay and for the initial proton-proton interactions that lead to nuclear fusion in the Sun and other stars. One of the four fundamental forces of nature, mediated by the exchange of *W* and *Z* particles. Compare *strong nuclear force*.

weather The state of an atmosphere at any given time and place. Compare *climate*.

weight 1. The force equal to the mass of an object multiplied by the local acceleration due to gravity. 2. In general relativity, the force equal to the mass of an object multiplied by the acceleration of the frame of reference in which the object is observed.

white dwarf The stellar remnant left at the end of the evolution of a low-mass star. A typical white dwarf has a mass of 0.6 solar mass ($M_\odot$) and a size about equal to that of Earth; it is made of nonburning, electron-degenerate carbon.

Wien's law A relationship describing how the peak wavelength, and therefore the color, of electromagnetic radiation from a glowing blackbody changes with temperature.

WIMP Literally, "weakly interacting massive particle." A hypothetical massive particle that interacts through gravity but not with electromagnetic radiation and is a candidate for dark matter. Compare *MACHO*.

winter solstice 1. One of two points where the Sun is at its greatest distance from the celestial equator. 2. The day on which the Sun appears at this location, marking the first day of winter (about December 22 in the Northern Hemisphere and June 21 in the Southern Hemisphere). Compare *summer solstice*.

X

X-ray Electromagnetic radiation having frequencies and photon energies greater than those of ultraviolet light but less than those of gamma rays, and wavelengths shorter than those of UV light but longer than those of gamma rays.

X-ray binary A binary system in which mass from an evolving star spills over onto a collapsed companion, such as a neutron star or black hole. The material falling in is heated to such high temperatures that it glows brightly in X-rays.

Y

year The time it takes Earth to make one revolution around the Sun. A solar year is measured from equinox to equinox. A sidereal year, Earth's true orbital period, is measured relative to the stars.

Z

zenith The point on the celestial sphere located directly overhead from an observer. Compare *nadir*.

zero-age main sequence The strip on the H-R diagram plotting where stars of all masses in a cluster begin their lives.

zodiac The constellations lying along the plane of the ecliptic.

zodiacal dust Particles of cometary and asteroidal debris less than 100 micrometers (µm) in size that orbit the inner Solar System close to the plane of the ecliptic. Compare *meteoroid* and *planetesimal*.

zodiacal light A band of light in the night sky caused by sunlight reflected by zodiacal dust.

zonal wind The planet-wide circulation of air that moves in directions parallel to the planet's equator.

Chapter 1

Summary Self-Test

1. a night's sleep
2. b-d-a-c-e
3. falsify
4. b
5. c
6. mathematics

Conceptual Questions

19. For example, the distance from the Sun to Neptune is about the time needed to fly from New York City to London.
21. 2.5 million years. This is the amount of time it takes for light that leaves Andromeda to reach us.
25. A "theory" is generally understood to mean an idea a person has, whether or not there is any proof, evidence, or way to test it. A "scientific theory" is an explanation for an occurrence in nature, must be based on observations and data, and must make testable predictions.
27. This suggests that one of the fields has an incorrect theory, basis, or understanding, and the two fields need to be carefully considered to reconcile this discrepancy.
33. The cosmological principle essentially states that the universe will look the same to every observer inside it.

Problems

47. 5.9×10^{17} mi.
49. 5.25×10^6 mm. The nearest star is 15.75 km and Andromeda is 1.3×10^7 km.
51. 5,333 km/hr or 6.7 times faster.
55. $86,400 \rightarrow 8.64 \times 10^4$ and $0.0123 \rightarrow 1.23 \times 10^{-2}$.
57. 0.24 sec

Chapter 2

Summary Self-Test

1. d
2. d
3. tilt of the Earth's axis
4. a
5. a
6. b
7. positions; motion
8. a-d-b-c
9. closest; farthest

Conceptual Questions

25. north celestial pole, south celestial pole, celestial equator
33. Seasons would be imperceptible.
45. Eclipses occur only when the Earth-Sun and Earth-Moon planes line up.
49. Moonlight is reflected sunlight.
53. Zero.

Problems

57. 50 min later
63. 15 min. 13.8 min. 1.2 min.
71. (a) north of 66.5° latitude. (b) close to the summer solstice.
75. (a) yes. (b) yes. (c) no.
77. 316 yr.

Chapter 3

Summary Self-Test

1. b-a-c
2. All choices are the same.
3. d
4. e
5. a
6. a

Conceptual Questions

17. Acceleration is how quickly the speed or direction of an object is changing.
19. The planet experiences constant acceleration.
23. The Moon's gravity is 1/7 that of Earth.
25. A *bound* orbit will repeat itself; an *unbound* orbit never returns.
27. The object came from outside the Solar System.

Problems

29. (a) 6.94 m/s². (b) 8,328 N. (c) The road pushes against the tires.
33. (a) 200 kph. (b) 200 kph.
39. 85 kg, 314 N
41. 0.04 times as strong
43. (a) 0.71 yr. (b) 41.7 km/s.

Chapter 4

Summary Self-Test

1. c
2. d
3. a, c, e

4. e-c-b-d-a
5. (a) 3. (b) 4. (c) 2. (d) 6. (e) 7. (f) 1. (g) 5.

Conceptual Questions

17. distance
21. gamma rays; radio waves
23. chromatic aberration
27. Refraction happens when a wave travels at an angle through media with different speeds of propagation.
31. Light has to pass through our turbulent atmosphere.

Problems

39. 380 m; 3.05 m
41. 2 million times greater
45. 480 m
47. 5.5×10^{11} bits per sec
51. 27 cm

Chapter 5

Summary Self-Test

1. b-c-a-d-e-f
2. b
3. b
4. temperature; pressure
5. a

Conceptual Questions

27. a small parcel of gas in hydrostatic equilibrium
37. gravitational potential, kinetic energy, thermal energy
41. containing carbon
45. spectroscopic radial velocities; transits; microlensing; direct imaging
49. Stars are very bright and far away, while planets appear very close to their stars. Thus it is difficult to mask out the light of the star and still see close enough to it to see the reflected light of a planet.

Problems

55. The Sun's spin is 5.6 percent of Jupiter's orbital angular momentum.
57. $\dfrac{L_{\text{Jupiter}}}{L_{\text{Earth}}} = 727$
63. 1 percent
65. (a) 4.9 times the volume. (b) 4.9 times the mass.

67. (a) $D_{\text{Osiris}} = 1.82 \times 10^5$ km. (b) Osiris is 34 percent larger than Jupiter.

Chapter 6

Summary Self-Test

1. b
2. b
3. d
4. e
5. c
6. c
7. c-b-e-d-a

Conceptual Questions

31. Igneous rocks can be radiometrically dated, but sedimentary and metamorphic rocks can't.
35. radioactive decay and friction
41. transverse waves can't penetrate liquids
47. bluer is always hotter
55. Mars is too cold and its atmospheric pressure is too low for liquid water to exist.

Problems

57. (a) Volume $= 1.08 \times 10^{21}$ m³. (b) Area $= 5.09 \times 10^{14}$ m². (c) Volume would become 8 times larger, and the surface area 4 times larger.
61. 110 W/m²
65. 5800 K
67. (a) 9.35 μm. (b) 131 W.
71. 120 million years ago

Chapter 7

Summary Self-Test

1. d-b-c-f-a-e
2. d
3. c
4. d
5. b

Conceptual Questions

29. remnants of the original atmospheres, or gas/dust kicked up by recent impacts
31. impacts from icy comets
35. methane, water, ammonia
43. Charged particles from the solar wind are trapped in the magnetosphere and then dumped into the upper atmosphere.
49. extreme heat

Problems

57. (a) 64 pennies. (b) 1.84×10^{19}
59. The two graphs track each other fairly reasonably. A trend like this is suggestive of cause-and-effect but not proof.
61. (a) 271.7 K. (b) 156.9 K.
63. (a) -21 C. (b) We did not account for the greenhouse effect.

Chapter 8

Summary Self-Test

1. gas; ice
2. false
3. temperature; pressure
4. a-c-f-e-b-d
5. b
6. b
7. All are correct.
8. a

Conceptual Questions

23. Terrestrials are small, solid, close to the Sun, and dense; they have thin secondary atmospheres, weak magnetic fields, few moons, and no rings. Giants are large, gaseous, far from the Sun, and low density; they have thick primary atmospheres, strong magnetic fields, and numerous rings and moons.
27. moons and artificial satellites
37. Saturn and Neptune will be like Earth; Uranus will have extreme seasons.
47. The rings are inside the Roche limit and will never make one ring.
53. The E-ring is fed by Enceladus.

Problems

55. 632 AU. This is 21 times further than Neptune's orbit.
57. (a) 1,320 Earths.

$$\text{(b)} \ \frac{\rho_J}{\rho_\oplus} = \frac{M_J/V_J}{M_\oplus/V_\oplus} = \frac{M_J}{M_\oplus}\frac{V_\oplus}{V_J} = \frac{318 M_\oplus}{M_\oplus}\frac{V_\oplus}{1{,}320 V_\oplus} = 0.24.$$

59. Pressure increases more rapidly on Neptune than Uranus. Saturn has the slowest rate of change, Neptune has the fastest.
61. westerly wind at 499 km/h
63. 60 K

Chapter 9

Summary Self-Test

1. Dwarf planets, moons, asteroids, comets
2. active
3. massive
4. b
5. d
6. a

Conceptual Questions

27. Europa's surface is covered with water ice and shows indirect evidence for liquid water underneath.
31. Gravitational "stirring" from nearby Jupiter keeps the objects in the asteroid belt well mixed.
35. Kuiper Belt objects are too far from the Sun for their volatile gases to sublimate into gas.
39. Asteroids are loose agglomerations of refractory rock and metal held together only by self-gravity. Comet nuclei are "dirty snowballs" that are composed primarily of ices and organic materials held together by a loose rocky matrix.
43. We pass through the dust tails of comets; the dust flecks slamming into the atmosphere produce the shower of lights we also call "shooting stars."

Problems

51. (a) 2.5 km/s. (b) 2.5 times faster than the speed of material in Io's vents.
53. (a) surface area of 4.14×10^{13} m² and a volume of 2.51×10^{19} m³. (b) 1.24×10^{11} m³. (c) 2×10^8 yr. (d) at least 20 times.
55. 2.7 times larger. The ring was probably formed by tidal disruption of a medium-sized moon rather than by slow escape of particles from nearby moons.
59. (a) 2.2 AU (Encke), 17.9 AU (Halley), 178.3 AU (Hale-Bopp). (b) 35.8 AU (Halley), 356.6 AU (Hale-Bopp). (c) Hale-Bopp
63. 25 million H-bombs equals one comet nucleus impact.

Chapter 10

Summary Self-Test

1. b
2. a
3. b
4. d
5. velocities; period; distance
6. b
7. d
8. c
9. c

Conceptual Questions

25. From Mars and Jupiter, we could measure the distance to stars farther away. From Venus, we could measure only the distance to closer ones.
31. If two stars have the same luminosity but one is larger, then the larger one must be cooler. If they have the same size, then the brighter one must be hotter.
35. Red photons are shown with a long wavelength while ultraviolet photons have short wavelengths. In a real cloud of gas, photons will travel randomly in all directions and not just leave in one direction.
41. If the inclination is small, then we can ignore it. If it is moderate, then we will overestimate the masses of the binary stars.
45. (a) They are too low-mass to be able to sustain nuclear fusion in their cores. (b) They produce so much thermonuclear energy in their interiors that they would quickly blow off their outermost layers.

Problems

51. Star C would have twice the parallax of star A and 4 times the parallax of star B.

55. 4.22 yrs

57. Rigel is twice as distant as Betelgeuse so to appear the same brightness, Rigel must be about 4 times more luminous. Betelgeuse is red and so much cooler than Rigel, so it must be much larger.

61. For every Sun-like star, there are 10^{-4} stars with 10,000 times the luminosity. If there are 10^6 Sun-like stars, there are 100 stars with 10,000 solar luminosities.

65. halfway between the two masses; 2/3 of the way toward the heavier mass

Chapter 11

Summary Self-Test

1. b
2. e = a = g; f; h = b; d = c
3. c
4. a
5. c
6. c
7. a
8. a

Conceptual Questions

35. The oceans are not hot enough.

41. There are very few fissile nuclei within the Sun, or any other star.

47. The neutrino will always reach Earth first, since neutrinos travel straight through a star or a planet without interacting with any other matter.

55. a constant, low-density flow of charged particles ejected from the Sun's atmosphere

59. The Earth orbits the Sun, and so it moves through space as the Sun rotates.

Problems

65. (a) $1,650 \text{ kg/m}^3$. (b) 65 percent denser than water.

69. (a) 8.3 min after being created. (b) about 100,000 yrs later.

71. (a) height above the base of the photosphere. (b) In the upper atmosphere of the Sun, temperature and density do not track each other, meaning that the emission of energy is not a blackbody process.

75. (a) 1.1 g. (b) 900 bombs' worth in one textbook.

77. a few hundred thousand years. This is so much shorter than the age of the Earth that it is not a viable mechanism.

Chapter 12

Summary Self-Test

1. a
2. a
3. b

4. d
5. c-f-h-d-i-a-e-g-b
6. d-c-f-a-b-e-g

Conceptual Questions

23. Low-mass, low-luminosity stars are more common because they are longer lived.

27. Faster hydrogen burning means the star will grow larger and more luminous to allow the extra energy created to escape; but, since the temperature of the core remains constant, the star's surface temperature will drop.

33. Higher luminosity implies faster rates of nuclear burning, which means the star has much less time to live.

37. The star will go on to burn heavier elements up to iron and then will become a Type-II supernova. On the H-R diagram the star will make small loops just between a horizontal-branch area and a giant-branch area as it burns oxygen and higher elements.

41. Unstable equilibrium. If you move even slightly to one side of the equilibrium point, gravity will pull you further to that side and further away from equilibrium.

Problems

47. (a) 56 billion yrs. (b) 110 million yrs. (c) 360,000 yrs. These ages are on the same order as the actual stars in Table 12.1.

49. (a) 83 km/s. (b) 37 km/s. (c) Mass loss increases with increasing size, since the surface gravity drops.

53. yes; distance, by spectroscopic parallax

57. 224 km

59. (a) $1.8 \times 10^{13} L_\odot$. (b) $1.8 \times 10^5 L_\odot$. (c) $1.8 \times 10^{-3} L_\odot$. (d) $1.8 \times 10^{-11} L_\odot$.

Chapter 13

Summary Self-Test

1. c
2. e-a-f-b-c-d
3. d
4. d
5. b
6. a, b, d, e, f
7. b

Conceptual Questions

37. In each successive burning cycle, there are fewer nuclei to be fused and the increasing core temperature burns through the fuel more rapidly.

41. Both. First, the core implodes; then the resulting release of energy drives a shock wave through the star that blows it up.

47. Binding energy is the amount of energy needed to break a nucleus into its components. The difference in binding energies

between the beginning and end products of a nuclear reaction is the energy given off in that reaction.

51. Time dilation at high speeds. The faster the muon travels, the longer it travels before its internal clock says it is time to decay.

57. The effects of relativity show themselves only in extreme circumstances, like traveling close to the speed of light or being right next to a black hole or neutron star.

Problems

61. (a) 3.8×10^{23} kg/min. (b) 5.25 times the mass of the moon per minute.

63. (a) 30,300 pc. (b) We cannot use parallax to measure the distance to the farther star because it is over 1,000 pc away.

65. 7×10^{13} J

67. 1,560 km. This is much larger than the pulsar's radius of about 10 km.

69. Earth would have a radius around 113 m, which is about the size of a football field, as claimed.

Chapter 14

Summary Self-Test

1. a
2. b
3. a
4. c-d-a-b
5. a
6. a, b, d
7. (a) led to the development; (b) and (c) were predicted and observed.
8. a, c, e

Conceptual Questions

25. Isotropic means that if we look out into space on very large scales, the distribution of matter that we see in one direction is the same as what we would see in the other direction.

29. The observer would see all galaxies expanding away, just as we do.

33. They have negative velocity, and are probably bound (gravitationally) to the Milky Way. These few galaxies do not disprove Hubble's law, since they are not part of the Hubble flow.

37. As one moves to farther distances, brighter objects are required so that they can still be seen.

45. The CMB is the redshifted, leftover energy from the early universe.

Problems

51. 7,588 km/s

53. 1.63×10^{10} yr if $H_0 = 60$, 8.9×10^9 yr if $H_0 = 110$

57. 1.1 mm; 2.63 K

61. 0.24 atoms/m³

Chapter 15

Summary Self-Test

1. a
2. a
3. c
4. c, d
5. c
6. b
7. b
8. c-b-d-a

Conceptual Questions

25. Random motions in three-dimensions translate into a spherical shape.
33. Luminous matter is matter that interacts with light.
37. The halo is a spheroidal mass distribution that is *much* larger than the galactic disk itself. The spiral component, on the other hand, is a flat disk.
43. About the size of our Solar System, based on AGN time variability, since an object can't be any larger than the timescales of variability, otherwise those signals would be washed out.
49. A supercluster is composed of many clusters. Clusters are small enough that the galaxies are orbiting each other uniformly, while a supercluster has not had time to "relax" into this state. We are part of the local "group," which is a little too poor in large galaxies to be called a cluster. We are part of the "local supercluster" that also contains the Virgo cluster, as evidenced by our relative motion with respect to Virgo caused by the gravity from that cluster.

Problems

53. (a) 10^{53} kg. (b) 10^{80} particles.
55. 10^{11} galaxies
57. 4 times more dark matter than luminous matter
63. 83 light-minutes or about 10 AU
65. 2.8 million years

Chapter 16

Summary Self-Test

1. b
2. a
3. c
4. e-b-a-d-c
5. d
6. c

7. b
8. d
9. c

Conceptual Questions

27. Hyperfine structure (or the "spin flip" of an electron in a hydrogen atom)
37. Our rotation curve is flat, implying that up to 90 percent of the Milky Way is dark matter.
39. SBb or SBc
47. The massive object at the very center of the galaxy; it was first detected as a strong radio source, and later in x-rays.
51. Remnants of the small protogalaxies that first formed in our dark-matter halo

Problems

55. 19.3 μm
57. About 0.02 percent
61. (a) 300,000 ly. (b) The halo is about 3 times larger than the disk.
65. (a) 1.53×10^{-10} J. (b) 1.11×10^{-15} kg. (c) 644 billion times.
67. 2094 km/s

Chapter 17

Summary Self-Test

1. c
2. a
3. a, c, e
4. b, d
5. a-b-c-d
6. c, e, f
7. a

Conceptual Questions

25. Just as our planet, galaxy, or Solar System are not being torn apart by the expansion of space, so too are they stable against the repulsive force of dark energy: gravity holds them together. Also, a large collection of matter is not a vacuum, so the amount of dark energy present in say, the Sun, is very low.
29. gravity, strong nuclear, weak nuclear, electromagnetic
37. If a photon contains more energy than the combined mass-energy of a particle and its antiparticle, then the photon can become a particle-antiparticle pair that flies off in different directions, conserving mass-energy and momentum.
49. Inhomogeneities in the density of normal matter were never strong enough to cause the structure seen today.

57. The nuclear forces are too short ranged to work over cosmic distances, and electromagnetic forces work only for charged particles; most matter in the universe is neutral.

Problems

63. 5 atoms per cubic meter
65. 10^{-30} kg, or about 1,000 times less than an H atom. We can ignore it.
67. 2.7×10^{34} yrs
69. (a) 10^{13} solar masses. (b) $10^{14} - 10^{15}$ solar masses. (c) $10^{16} - 10^{18}$ solar masses.
71. 5.4 billion years. Since the galaxies accelerate during this time, and a merger requires a number of passes of each galaxy through the other, this is only a rough estimate.

Chapter 18

Summary Self-Test

1. d
2. c
3. d
4. a
5. b
6. d
7. d

Conceptual Questions

27. Cyanobacteria slowly broke CO_2 molecules apart to form O_2. This process required about 4 billion years to reach today's oxygen levels.
29. We are similar to chimpanzees (6 million years ago), which were preceded by primates (70 million years ago), which were preceded by the arrival of animals and fungi (1 billion years ago).
33. a reasonably stable environment, abundant building blocks, a "solvent" in which the chemistry could work, and a lot of time
35. H and He are mostly from the Big Bang. Everything else is from giant-branch winds and supernovae.
37. The region of space around a star where we think life could form in Earth-like conditions. In broad terms, the boundaries are the distances at which water becomes a solid (too cold) or a gas (too hot) but the details also have to do with the energy output of the star, and the properties of the planet such as albedo and the greenhouse strength.

Problems

47. Alphas will be 0.59 percent as numerous as Betas.

Text Credits

Rebecca Boyle: "New Telescope Relies on Never-Before-Manufactured Material; No Problem, Says NASA." From Popular Science, September 29, 2010. Used by permission of Wright's Media.

Denise Chow: "Giant Propellers Discovered in Saturn's Rings." From Space.com, July 8, 2010. Used by permission of Tech Media Network.

European Southern Observatory: "Unraveling the Mystery of Massive Star Birth." From ESO.org, July 14, 2010. Used by permission of the European Southern Observatory.

Lisa Grossman: "Hyperfast Star Kicked Out of Milky Way." From Wired.com, July 22, 2010. Copyright © 2010 Condé Nast Publications. All rights reserved. Originally published in Wired.com. Reprinted by permission.

Amina Khan: "'Moon Is Wetter, Chemically More Complex than Thought' NASA Says." From The Los Angeles Times, October 22, 2010. Used by permission of the Los Angeles Times.

Marianne LeVine: "Professors Weigh in on Controversial Paper." From The Stanford Daily, January 12, 2011. Used by permission of The Stanford Daily.

Zoe Macintosh: "Sun Eruption That May Have Spawned Zombie Satellite Identified." From Space.com. Used by permission of Tech Media Network.

Thomas H. Maugh II: "Orbiting Telescope Spots Possible Planets." From The Los Angeles Times, June 16, 2010. Used by permission of the Los Angeles Times.

Alex Morales: "Chilean Quake Likely Shifted Earth's Axis, NASA Scientist Says." From Bloomberg Business Week, March 1, 2010. Used by permission of The YGS Group.

Amy Minsky: "First Findings Revealed as Scientists Peer into Universe's Oldest Light." From The Vancouver Sun, January 11, 2011. Material reprinted with the express permission of Postmedia News, a division of Postmedia Network, Inc.

Shahrzad Noorbaloochi: "Stanford Study Measures Expertise in Climate Change Debate." From The Epoch Times, June 29, 2010. Used by permission of The Epoch Times.

Dennis Overbye: "Vote Makes It Official: Pluto Isn't What It Used to Be." From The New York Times, August 25, 2006. Copyright © 2006 The New York Times. All rights reserved. Used by permission and protected by the copyright laws of the United States.

Ian Sample: "What a Scorcher—Hotter, Heavier and Millions of Times Brighter than the Sun." From The Guardian, July 22, 2010. Copyright Guardian News & Media Ltd. 2010.

Jason Schreiber: "Reminder: Clear Your Car of Snow." From Union Leader, January 14, 2011. Used by permission of the Union Leader Library.

Space.com Staff: "Scientists May Be Missing Many Star Explosions." From Space.com, Space.com, December 27, 2010, and "Colliding Galaxies Swirl in Dazzling New Photo." From August 5, 2010. Used by permission of Tech Media Network.

Ker Than: "New Galaxy Maps to Help Find Dark Energy Proof?" From National Geographic News, July 21, 2010. Copyright © 2010 National Geographic Stock.

Figure Credits

Chapter 1
opener Ron Proctor; (Topographic Map) USGS/Utah Geological Survey; **1.2a** Neil Ryder Hoos; **1.2b** Neil Ryder Hoos; **1.2c** Owen Franken/Corbis; **1.2d** Monkey Business Images/Dreamstime.com; **1.2e** PhotoDisc; **1.2f** American Museum of Natural History; **1.2g** American Museum of Natural History; **1.2h** NASA/JPL/Caltech; **1.3** Ron Watts/Corbis; **1.4** Michael J. Tuttle (NASM/NASA); **1.5** NRAO/AUI, James J. Condon, John J. Broderick, and George A. Seielstad; **1.6** Fermilab; **1.8** Bettmann/Corbis; **1.9** Matheisl/Getty Images; **1.10** Craig Lovell/Corbis

Chapter 2
opener Photo Researchers, Inc.; **2.9** (left) Pekka Parviainen/Photo Researchers, Inc., (right) David Nunuk/Photo Researchers, Inc.; **2.18** Roger Ressmeyer/Corbis; **2.19** Nick Quinn; **2.20** Dennis L. Mammana; **2.22** Anthony Ayiomamitis (TWAN); **2.24** Nicolaus Copernicus (1473–1543) (oil on canvas), Pomeranian School, (16th century)/Nicolaus Copernicus Museum, Frombork, Poland/Lauros/Giraudon/The Bridgeman Art Library; **2.25** Tunç Tezel

Chapter 3
opener NASA; **3.1** The Granger Collection, New York; **3.2** SSPL/Jamie Cooper/The Image Works

Chapter 4
opener © Roger Ressmeyer/CORBIS; **4.9** ASU Physics Instructional Resource Team; **4.13a** Science Photo Library; **4.13b** Hulton-Deutsch Collection/Corbis; **4.14b** Jean-Charles Cuillandre (CFHT); **4.14c** Junenoire Photography; **4.15** Eyewire Collection/Getty Images; **4.17** ASU Physics Instructional Resource Team; **4.19a, b** (bottom) Gemini Observatory, US National Science Foundation, and University of Hawaii Institute for Astronomy, (top) Keck Observatory; **4.21a** RogerRessmeyer/Corbis; **4.21b** David Parker/Photo Researchers, Inc.; **4.22** Photo by Dave Finley, courtesy National Radio Astronomy Observatory and Associated Universities, Inc.; **4.23** ESO

Chapter 5
opener NASA, ESA, N. Smith (University of California, Berkeley), and the Hubble Heritage Team (STScI/AURA); NASA, ESA, M. Robberto (STScI), and the HST Orion Nebula Team; **5.5** Jeff Hester and Paul Scowen (Arizona State University), Bradford Smith (University of Hawaii), Roger Thompson (University of Arizona), and NASA; **5.8a** STScI photo: Karl Stapelfeldt (JPL); **5.8b** D. Padgett (IPAC/Caltech), W. Brandner (IPAC), K. Stapelfeldt (JPL) and NASA; **5.8c** D. Padgett (IPAC/Caltech), W. Brandner (IPAC), K. Stapelfeldt (JPL) and NASA; **5.8d** D. Padgett (IPAC/Caltech), W. Brandner (IPAC), K. Stapelfeldt (JPL) and NASA; **5.9** Photograph by Pelisson, SaharaMet; **5.11** Reuters/Corbis; **5.14a** Karl Stapelfeldt/JPL/NASA; **5.14b** NASA/JPL-Caltech/UIUC; **5.18** NASA/Johns Hopkins University Applied Physics Laboratory/Carnegie Institution of Washington; **5.19** NASA, ESA, J.R. Graham and P. Kalas (University of California, Berke-ley), and B. Matthews (Hertzberg Institute of Astrophysics); **5.24** C. Marois, National Research Council, Canada; **5.25** NASA, ESA, P. Kalas, J. Graham, E. Chiang, E. Kite (University of California, Berkeley), M. Clampin (NASA Goddard Space Flight Center), M. Fitzgerald (Lawrence Livermore National Laboratory), and K. Stapelfedt and J. Krist (NASA Jet Propulsion Laboratory)

Chapter 6
opener ESO; **6.1** NASA; **6.2** Stockli, Nelson, Hasler, Goddard Space Flight Center/NASA; **6.3** (top left) Jim Wark/Visuals Unlimited/Corbis, (top right) Frank Lukasseck/Corbis, (bottom left) Stuart Wilson/Photo Researchers, Inc.; **Table 6.1** (left to right) NASA/Johns Hopkins University Applied Physics Laboratory/Carnegie Institution of Washington; NASA-JPL; **6.1c** M-Sat Ltd/Photo Researchers, Inc.; NASA/USGS; Science Source/Photo Researchers Inc; **6.4b** NASA/JSC; **6.4c** NASA/Johns Hopkins University Applied Physics Laboratory/Carnegie Institution of Washington; **6.5a-f** Montes De Oca & Associates; **6.6** Photograph by D. J. Roddy and K. A. Zeller, USGS, Flagstaff, AZ.; **6.7** NASA/JSC; **6.8** NASA/JPL; **6.9** NASA/JPL/Caltech; **6.13** Joe Tucciarone/Science Photo Library/Photo Researchers, Inc.; **6.18** Grant Heilman Photography; **6.19** Donald Duckson/Visuals Unlimited/Corbis; **6.24** NASA/JSC; **6.25** NASA/Johns Hopkins University Applied Physics Laboratory/Smithsonian Institution/Carnegie Institution of Washington; **6.26a** ESA/DLR/FU Berlin (G. Neukum); **6.26b** NSSDC/NASA; **6.27** NASA/Magellan Image/JPL; **6.29** NASA/JSC; **6.30** K. C. Pau; **6.31** NASA/Johns Hopkins University Applied Physics Laboratory/Carnegie Institution of Washington.; **6.32** ESA/DLR/FU Berlin (G. Neukum); **6.33a** NASA/JPL/Malin Space Science Systems; **6.33b** NASA/MOLA Science Team; **6.34** NASA/JPL-Caltech; **6.35** NASA/JPL/University of Arizona; **6.36a** NASA/JPL/Malin Space Science Systems; **6.36b** NASA/JPL/Cornell; **6.37** ESA/DLR/FU Berlin (G. Neukum); **6.38** NASA/JPL-Caltech/University of Arizona/Texas A&M University

Chapter 7
opener Senior Airman Joshua Strang (USAF)/Wikimedia Commons; **7.3** NASA/NSSDC/GSFC; **7.5** (left to right) Bob Blaylock/Wikipedia Commons; Wenche Eikrem and Jahn Throndsen, University of Oslo/Wikimedia Commons; PLoS journals/Wikipedia Commons; xvazquez/Wikipedia Commons; De Agostini Picture Library/De Agostini/Getty Images; Shizhao/Wikimedia commons; Wikimedia Commons; **7.8a** NASA; **7.8b** Raymond Gehman/Corbis; **7.11** NASA/NSSDC; **7.12** NASA/JPL/Caltech; **7.13** NASA/JPL/Caltech; **7.14** NASA/STScI; **7.15** "Graph: Global Temperature Is Increasing in Response to Added Greenhouse Gas over Time." Based on Climate Change 2001: The Scientific Basis. Contribution of Working Group I to the Third Assessment Report of the Intergovernmental Panel on Climate Change, Figure 2.22. Cambridge University Press; **7.16** Robert A. Rohde/Global Warming Art

Chapter 8
opener NASA/JPL (Cassini); **Table 8.1** (left to right) NASA/JPL; NASA/JPL; NASA/JPL; NASA/JPL; NASA, ESA, and the Hubble Heritage Team (STScI/AURA); NASA/JPL/

University of Arizona; **8.5** NASA/JPL/Caltech and NASA/JPL/Caltech; **8.6** CICLOPS/NASA/JPL/University of Arizona; **8.7** NASA/JPL/Space Science Institute; **8.8a** NASA/JPL/Space Science Institute; **8.8b** Courtesy of Cassini Imaging Central Laboratory for Operations; **8.9a** NASA, L. Sromovsky, and P. Fry (University of Wisconsin–Madison); **8.9b** NASA, L. Sromovsky, and P. Fry (University of Wisconsin–Madison); **8.10** NASA/JPL; **8.13a** NASA/JPL/Caltech; **8.14** NASA, ESA, L. Sromovsky and P. Fry (University of Wisconsin), H. Hammel (Space Science Institute), and K. Rages (SETI Institute); **8.19** N. M. Scheider, J. T. Trauger, Catalina Observatory; **8.20a** J. Clarke (University of Michigan), NASA; **8.20b** Courtesy of John Clark, Boston University and NASA/STScI; **8.23** NASA/JPL/Space Science Institute; **8.24** NASA/JPL/Caltech; **8.25** NASA/JPL/Caltech; **8.26** NASA/JPL/Space Science Institute; **8.27** NASA/JPL

Chapter 9
opener Original artwork by Ron Proctor; **9.1a** Dr. R. Albrecht, ESA/ESO Space Telescope European Coordinating Facility/NASA; **9.1b** Dr. R. Albrecht, ESA/ESO Space Telescope European Coordinating Facility/NASA; **9.3a** M. Brown (Caltech), C. Trujillo (Gemini), D. Rabinowitz (Yale), NSF, NASA; **9.4** NASA, ESA, J. Parker (Southwest Research Instittue), P. Thomas (Cornell University), and L. McFadden (University of Maryland, College Park); **9.5** NASA/JPL/Caltech; **9.6** NASA/JPL/Caltech; **9.7a** NASA/JPL/Space Science Institute; **9.7b** NASA/JPL/Space Science Institute; **9.8** NASA/JPL/Caltech; **9.9** NASA/JPL; **9.10** NASA/JPL/Space Science Institute; **9.11** NASA/JPL; **9.12a** NASA/JPL/ESA/University of Arizona; **9.12b** NASA/JPL; **9.13** NASA/JPL/ESA/ University of Arizona; **9.14** NASA; **9.16** NASA/JPL/Caltech; **9.17** NEAR Project, NLR, JHUAPL, Goddard SVS, NASA; **9.18a** NASA/JPL/ University of Arizona; **9.18b** ESA/DLR/FU Berlin (G. Neukum); **9.21** Courtesy of Terry Acomb; **9.23a** Dr. Robert McNaught; **9.23b** Don Goldman; **9.24** NASA/NSSDC/GSFC; **9.25** NASA/JPL/Caltech; **9.26a** NASA/JPL/Caltech/UMD; **9.26b** NASA/JPL-Caltech/UMD; **9.27** NASA/JPL-Caltech/UMD; **9.28** NASA/JPL/MPIA; **9.29a** NASA/JPL/MPIA; **9.29b** NASA/STScI; **9.29c** R. Evans, J. T. Trauger, H. Hammel and the HST Comet Science Team, and NASA; **9.30** Courtesy of the Wolbach Library, Harvard-Smithsonian Center for Astrophysics, Cambridge, MA; **9.31b** Barrie Rokeach/Getty Images; **9.32a** Courtesy of Ron Greeley; **9.32b** Courtesy of Ron Greeley; **9.32c** Courtesy of Ron Greeley; **9.32d** Courtesy of Ron Greeley; **9.33** NASA/JPL/Cornell; **9.35** Larry Marschall and Christy Zuidema, Gettysburg College**Chapter 10: opener** Roberto Mura/Wikimedia Commons; Penumbra Design, Inc.; **10.18** David H McKinnon PhD/www.csu.edu.au/telescope

Chapter 11
opener SOHO (NASA, ESA); **11.9b** NOAO/AURA/NSF; **11.10a** Mila Zinkova/Wikipedia; **11.12** Nigel Sharp, NOAO/NSO/Kitt Peak FTS/AURA/NSF; **11.14a** Stefan Seip, www.astromeeting.de; **11.14b** © Laura Kay; **11.14c** © Laura Kay; **11.15** NASA/Photo Researchers, Inc.; **11.16** SOHO/ESA/NASA; **11.18a** NASA/SDO/Solar Dynamics Observatory; **11.18b** NASA/SDO/Solar Dynamics Observatory; **11.21a** NASA; **11.21b** SOHO/ESA/NASA; **11.22a** SOHO/ESA/NASA; **11.22b** SOHO-EIT Consortium, ESA, NASA; **11.23a** NASA/SDO/Solar Dynamics Observatory; **11.23b** NASA/SDO/Solar Dynamics Observatory; **11.24** SOHO/ESA and NASA

Chapter 12
opener NOAO; **12.9a** Minnesota Astronomical Society; **12.9b** NASA and The Hubble Heritage Team (STScI/AURA); **12.11** (top left) Dr. Raghvendra Sahai (JPL) and Dr. Arsen

R. Hajian (USNO), NASA and The Hubble Heritage Team (STScI/AURA)

Chapter 13
opener (left) Brad Farrington, (right) Louisiana Delta Community College; **13.5** N. Smith, J. A. Morse (U. Colorado) et al., NASA; **13.6** NASA/Goddard Space Flight Center; **13.9a** Anglo-Australian Observatory, photographs by David Malin; **13.9b** Anglo-Australian Observatory, photographs by David Malin; **13.10a** Joachim Trumper, Max-Planck-Institut für extraterrestrische Physik; **13.10b** Courtesy Nancy Levenson and colleagues, American Astronomical Society; **13.10c** Jeff Hester, Arizona State University, and NASA; **13.14a** European Southern Observatory, ESO PR Photo 40f199; **13.14b** Jeff Hester, Arizona State University, and NASA; **13.14c** Jeff Hester, Arizona State University, and NASA; **13.25** ESA/Hubble

Chapter 14
opener The Hubble Heritage Team (AURA/STScI/NASA); **14.2** TIMEPIX/Getty Images; **14.3** Courtesy of Jeff Hester; **14.6** WIYN/NOAO/NSF, WIYN Consortium, Inc.; **14.14** Lucent Technologies' Bell Labs; **14.15** Ann and Rob Simpson; **14.17a** NASA COBE Science Team; **14.17b** NASA COBE Science Team; **14.17c** NASA COBE Science Team; **14.19** (all) Courtesy of The Author

Chapter 15
opener NASA, ESA, and the Hubble Heritage Team (STScI/AURA); NASA, ESA, and R. Sharples (University of Durham); Adam Evans/Wikimedia Commons; NASA, ESA, and the Hubble Heritage (STScI/AURA)-ESA/Hubble Collaboration; G. Fritz Benedict, Andrew Howell, Inger Jorgensen, David Chapell (University of Texas), Jeffery Kenney (Yale University), Beverly J. Smith (CASA, University of Colorado),NASA; Joseph D. Schulman/Wikimedia Commons; **15.1a** (top left) NASA; (center left) NASA; (bottom left) NASA; (top right) Canada-France-Hawaii Telescope/J.-C. Cuillandre/Coelum; (center right) NASA; (bottom right) NASA; **15.2** (top left) NASA, ESA, and the Hubble Heritage Team (STScI/AURA); (top right) NASA, ESA, A.Aloisi (STScI/ESA), and the Hubble Heritage (STScI/AURA)-ESA/Hubble Collaboration; (bottom left) NASA, ESA, and the Hubble Heritage (STScI/AURA)-ESA/Hubble Collaboration; (bottom right) NASA, ESA, and the Hubble Heritage Team (STScI/AURA)-ESA/Hubble Collaboration; **15.3** NOAO/AURA/NSF; **15.4** David Malin; **15.5a** Jeff Barton/Wikimedia Commons; 15.5b NASA and the Hubble Heritage Team (STScI/AURA); **15.6a** European Southern Observatory (ESO); **15.6b** European Southern Observatory (ESO); **15.7a** Johannes Schedler/Panther Observatory; **15.7b** Johannes Schedler/Panther Observatory; **15.8a** NASA/Swift/Stefan Immler (GSFC) and Erin Grand (UMCP); **15.8b** Bill Schoening, Vanessa Harvey/REU program/NOAO/AURA/NSF; **15.9a** Todd Boroson/NOAO/AURA/NSF; **15.9b** Richard Rand, University of New Mexico; **15.11** David Malin; **15.13** X-ray: NASA/CXC/Penn State/G. Garmire; Optical: NASA/ESA/STScI/M.West; **15.15** Courtesy of Dr. Michael Brotherton; **15.16** NRAO/Alan Bridle et al.; **15.18b** Courtesy Bill Kell, University of Alabam, and NASA; **15.18c** Holland Ford, STScI/Johns Hopkins University and NASA; **15.19** F.Owen, NRAO, with J.Biretta, STScI, and J.Eilek, NMIMT; **15.20** Holland Ford, STScI/Johns Hopkins University; Richard Harms, Applied Research Corp.; Zlatan Tsvetanov, Arthur Davidsen, and Gerard Kriss at Johns Hopkins University; Ralph Bohlin and George Hartig at STScI; Linda Dressel and Ajay K. Kochhar at Applied Research Corp. in Landover, MD; and Bruce Margon from the University of Washing-

ton, Seattle, NASA; **15.21** NASA, ESA, the Hubble Heritage (STScI/AURA)-ESA/Hubble Collaboration, and A.Evans (University of Virginia, Charlottesville/ NRAO/Stony Brook University); **15.23 (top)** Anglo-Australian Observatory; **15.23 (bottom)** Omar Lopez-Cruz and Ian Shelton/NOAO/AURA/NSF; **15.24a** X-ray: NASA/CXC/MIT/E.-H Peng. Optical: NASA/STScI ; **15.24b** NASA, ESA, D. Coe (NASA JPL/California Institute of Technology, and Space Telescope Science Institute), N. Benitez (Institute of Astrophysics of Andalusia, Spain), T. Broadhurst (University of the Basque Country, Spain), and H. Ford; **15.25a** Max Tegmark/SDSS Collaboration; **15.25b** Max Tegmark/SDSS Collaboration; **15.25c** Jordan Raddick/SDSS; **15.26** NASA, ESA, SAO, CXC, JPL-Caltech, and STScI; **15.27** NASA, ESA, and the Hubble Heritage Team (STScI/AURA)

Chapter 16
opener F. Ringwald/ Sierra Remote Observatories **16.1a** Courtesy of Dr. A. Mellinger; **16.1b** NASA/NSSDC/GSFC Diffuse Infrared Background Experiment (DIRBE) instrument on the Cosmic Background Explorer; **16.2** (inset) Thomas Colla/AP PHOTO; **16.5a** NSSDC/GSFC Infrared Astronomical Satellie (IRAS); **16.5b** JAXA; **16.6** NSSDC/KKGSFC Roentgen Satellite (ROSAT); **16.7** Dr. Douglas Finkbeiner; **16.8a** Courtesy of Stefan Seip; **16.8b** Courtesy of Emmanuel Mallart www.astrophoto-mallart.com & www.axisinstruments.com; **16.9** ESO Telescope Systems Division; **16.10** C. Jones, Smithsonian Astrophysical Observator; **16.12a** ESO; **16.12b** ESO; **16.14** Hubble Heritage Team (AURA/STScI/NASA); **16.15a** NOAO/AURA/NSF; **16.16** Courtesy of Jeff Hester; **16.17** Anglo-Australian Observatory; **16.19a** The Electronic Universe Project;**16.23a** NASA/CXC/MIT/F.K.Baganoff; **16.23b** Hubble, NASA, ESA, & D.Q.Wang (U.Mass, Amherst); Spuitzer: NASA, JPL, & S.Stolovy (SSC/Caltech); **16.23c** NRAO/AUI; **16.24** This image was created by Prof. Andrea Ghez and her research team at UCLA and is from data sets obtained with the W. M. Keck Telescopes. Image creators include Andrea Ghez, Angelle Tanner, Seth Hornstein, and Jessica Lu; **16.25** Anglo-Australian Observatory, David Malin; **16.26** NASA, ESA, and Z. Levay (STScI)

Chapter 17
opener Thomas Jarrett (IPAC/Caltech); Andre Karwath/Creative Commons; **17.6** From *The Elegant Universe* by Brian Greene ©1999 by Brian R. Greene. Used by permission of W.W. Norton & Company, Inc.; **17.10** NASA/CXC/CfA/M.Markevitch et al. and ASA/STScI;Magellan/U.Arizona/ D.Clowe et al.; Lnesing Map: NASA/STScI; ESO WFI; Magellan/U.Arizona/D.Clowe et al.; **17.11** Images produced by Donna Cox Smithsonian Institution and Motorola Corporation; **17.12** NASA, ESA, and N.Pirzkal (STScI/ESA)

Chapter 18
opener Paul Harrison/Creative Commons; **18.1** David McNew/REUTERS; **18.2a** Michael Perfit, University of Florida, Robert Embley/NOAA; **18.2b** Woods Hole Oceanographic Institution, Deep Submergence Operations Group, Dan Fornari; **18.3** Gaertner/Alamy; **18.5** Courtesy of National Science Foundation; **18.6** (top left) Mark A.Schneider/Visuals Unlimited; (center middle left) K.Sandved/Visuals Unlimited; (bottom left) Ken Lucas/Visuals Unlimited; (top right) Science VU/National Museum of Kenya; (middle right) Ken Lucas/Visuals Unlimited; (bottom right) Ken Lucas/Visuals Unlimited; **18.7** Kenneth E.Ward/Biografx/Photo Researchers, Inc.; **18.10** NASA; **18.11** F.Drake (UCSC) et al., Arecibo Observatory (Cornell, NAIC); **18.12** Twentieth Century Fox Film Corporation/Photofest; **18.13** Seti/NASA; **18.14** NASA/JPL